Lecture Notes in Physics

Springer-Verlag
Berlin Heidelberg GmbH

The Editorial Policy for Proceedings

The series Lecture Notes in Physics reports new developments in physical research and teaching – quickly, informally, and at a high level. The proceedings to be considered for publication in this series should be limited to only a few areas of research, and these should be closely related to each other. The contributions should be of a high standard and should avoid lengthy redraftings of papers already published or about to be published elsewhere. As a whole, the proceedings should aim for a balanced presentation of the theme of the conference including a description of the techniques used and enough motivation for a broad readership. It should not be assumed that the published proceedings must reflect the conference in its entirety. (A listing or abstracts of papers presented at the meeting but not included in the proceedings could be added as an appendix.)

When applying for publication in the series Lecture Notes in Physics the volume's editor(s) should submit sufficient material to enable the series editors and their referees to make a fairly accurate evaluation (e.g. a complete list of speakers and titles of papers to be presented and abstracts). If, based on this information, the proceedings are (tentatively) accepted, the volume's editor(s), whose name(s) will appear on the title pages, should select the papers suitable for publication and have them refereed (as for a journal) when appropriate. As a rule discussions will not be accepted. The series editors and Springer-Verlag will normally not interfere with the detailed editing except in fairly obvious cases or on technical matters.

Final acceptance is expressed by the series editor in charge, in consultation with Springer-Verlag only after receiving the complete manuscript. It might help to send a copy of the authors' manuscripts in advance to the editor in charge to discuss possible revisions with him. As a general rule, the series editor will confirm his tentative acceptance if the final manuscript corresponds to the original concept discussed, if the quality of the contribution meets the requirements of the series, and if the final size of the manuscript does not greatly exceed the number of pages originally agreed upon. The manuscript should be forwarded to Springer-Verlag shortly after the meeting. In cases of extreme delay (more than six months after the conference) the series editors will check once more the timeliness of the papers. Therefore, the volume's editor(s) should establish strict deadlines, or collect the articles during the conference and have them revised on the spot. If a delay is unavoidable, one should encourage the authors to update their contributions if appropriate. The editors of proceedings are strongly advised to inform contributors about these points at an early stage.

The final manuscript should contain a table of contents and an informative introduction accessible also to readers not particularly familiar with the topic of the conference. The contributions should be in English. The volume's editor(s) should check the contributions for the correct use of language. At Springer-Verlag only the prefaces will be checked by a copy-editor for language and style. Grave linguistic or technical shortcomings may lead to the rejection of contributions by the series editors. A conference report should not exceed a total of 500 pages. Keeping the size within this bound should be achieved by a stricter selection of articles and not by imposing an upper limit to the length of the individual papers. Editors receive jointly 30 complimentary copies of their book. They are entitled to purchase further copies of their book at a reduced rate. As a rule no reprints of individual contributions can be supplied. No royalty is paid on Lecture Notes in Physics volumes. Commitment to publish is made by letter of interest rather than by signing a formal contract. Springer-Verlag secures the copyright for each volume.

The Production Process

The books are hardbound, and the publisher will select quality paper appropriate to the needs of the author(s). Publication time is about ten weeks. More than twenty years of experience guarantee authors the best possible service. To reach the goal of rapid publication at a low price the technique of photographic reproduction from a camera-ready manuscript was chosen. This process shifts the main responsibility for the technical quality considerably from the publisher to the authors. We therefore urge all authors and editors of proceedings to observe very carefully the essentials for the preparation of camera-ready manuscripts, which we will supply on request. This applies especially to the quality of figures and halftones submitted for publication. In addition, it might be useful to look at some of the volumes already published. As a special service, we offer free of charge LaTeX and TeX macro packages to format the text according to Springer-Verlag's quality requirements. We strongly recommend that you make use of this offer, since the result will be a book of considerably improved technical quality. To avoid mistakes and time-consuming correspondence during the production period the conference editors should request special instructions from the publisher well before the beginning of the conference. Manuscripts not meeting the technical standard of the series will have to be returned for improvement.

For further information please contact Springer-Verlag, Physics Editorial Department II, Tiergartenstrasse 17, D-69121 Heidelberg, Germany

Piotr Garbaczewski Marek Wolf
Aleksander Weron (Eds.)

Chaos – The Interplay Between Stochastic and Deterministic Behaviour

Proceedings of the XXXIst Winter School
of Theoretical Physics
Held in Karpacz, Poland
13–24 February 1995

Springer

Editors

Piotr Garbaczewski
Marek Wolf
Institute of Theoretical Physics
University of Wrocław
Pl. M. Borna 9
PL-50-205 Wrocław, Poland

Aleksander Weron
H. Steinhaus Center Poland for Stochastic Methods
Technical University
Wybrzeże Wyspiańskiego 27
PL-50-370 Wrocław, Poland

Cataloging-in-Publication Data applied for.

Die Deutsche Bibliothek - CIP-Einheitsaufnahme

Chaos : the interplay between stochastical and deterministic behaviour ; proceedings of the XXXIst Winter School of Theoretical Physics, held in Karpacz Poland, 13 - 24 February 1995 / Piotr Garbaczewski ... (ed.).
(Lecture notes in physics ; Vol. 457)
ISBN 978-3-662-14030-7 ISBN 978-3-540-44722-1 (eBook)
DOI 10.1007/978-3-540-44722-1

NE: Garbaczewski, Piotr [Hrsg.]; Zimowa Szkoła Fizyki Teoretycznej <31, 1995, Karpacz>; GT

ISBN 978-3-662-14030-7

Originally published by Springer-Verlag Berlin Heidelberg New York in 1995
Softcover reprint of the hardcover 1st edition 1995

Typesetting: Camera-ready by the authors
SPIN: 10508220 55/3142-543210 - Printed on acid-free paper

PREFACE

Theoretical science up to the end of the nineteenth century can be viewed as the modeling of natural phenomena by deterministic solutions of differential equations. At that time it was at that time commonly thought that if all initial data could only be collected, one would be able to predict the future with certainty. The theory of chaos is fascinating if for no other reason than its blurring of the long-held distinction between random and deterministic phenomena. It is potentially capable of explaining very complex processes with simple, parsimonious models. However, there are many important questions about the new theory that remain currently unresolved, and it is likely that some of these issues will never find solution. The question of choosing between deterministic versus stochastic modeling of a process under study is such an issue, but the stochastic modeling presently has a clear advantage because of the rich variety of model-fitting tools provided by mathematical statistics and modern probability theory.

The discovery of chaos in mechanical systems and the subsequent development of mathematical and physical theories of nonequilibrium phenomena, of the relationship between a physical and algorithmic complexity, supported by the mathematical experimentation on powerful computers, may yet prove to be events of great importance. Undoubtedly, the quantitative description of chaos is one of the triumphs of nonlinear science. On the other side, there does not appear to be a universally accepted, mathematical nor physical definition of chaos. There are different ways to quantify what one might mean by complex or unpredictable behaviour. The primary concept appears to be the notion of sensitivity to initial conditions, albeit the detailed reason of this sensitivity in any particular problem, be it deterministic or stochastic, must be separately settled and justified by phenomenological data. Seemingly a trivial task on the theoretical level, it has proved to be a real problem, while fitting mathematical models to experimental situations, since the distinction between deterministic and essentially stochastic manifestations of chaos is not sharp.

Chaotic and random phenomena provide the area where certain frontiers of physics and mathematics can fully overlap on problems that are fundamental and yet have practical implications. The XXXI Karpacz Winter School of Theoretical Physics: "CHAOS: the Interplay Between Stochastics, Classics and Quanta" was organized by the Institute of Theoretical Physics of the Wrocław University during the period: February 13 – February 24, 1995 in Karpacz, Poland. The aim of this school was to provide a forum for the presentation and discussion of recent results on chaotic and essentially random phenomena with emphasis on their interaction (interplay) and application in a number of physical problems. A presentation of related mathematical lectures was an integral part of the School.

This volume contains the written and often extended versions of lectures, and contributed papers (seminar talks) presented in the course of Karpacz'95. A great majority of papers present new results in the field and the remaining are expository in nature. The collection demonstrates the broad range of interests of the school participants but, following our intention, focuses on the unifying *interplay* idea, expressed in the School title, to the extent that makes any traditional chapter-style organization of the material unprofitable. Therefore a major division into Lectures and Seminars is followed by the alphabetical ordering of contributions.

However, for the benefit of a potential readership some general hints towards the main subject-matter streamlines of the School, modulo their mingling and braiding, might be helpful. Therefore we shall give a brief survey of the content of the volume.

Its title is slightly different from that originally coined for Karpacz '95, and emphasizes the interplay between deterministic and stochastic manifestations of chaos. Neither of these terms should be assigned a narrow meaning. We commented before on the unsharpness of traditional intuitions about chaos, and in addition one should realise that what is meant by deterministic or stochastic usually remains ambiguous unless more detailed physical specifications are given to quantities involved in the theoretical (mathematical) framework.

We extract what can be interpreted as the (non-disjoint) subject-matter indicatives for the invited lectures presented in the course of Karpacz'95:

- Random walks, Brownian motion, diffusions, jump and compound jump processes, Lévy processes and their generalisations (Beck, Belavkin, Blanchard, Cetto, Garbaczewski, Grigolini, Haba, Hu, Klafter, Lasota, Reichl, Weron, Zambrini)
- Deterministic versus stochastic dynamics, classical and quantum (Beck, Cetto, Garbaczewski, Haba, Grigolini, Zambrini)
- Classical versus quantum stochastics (Belavkin, Blanchard, Graham, Grigolini)
- Semiclassical chaos (Graham, Grigolini, Jauslin, Lai, Reichl, Tomsovic)
- Turbulence, dissipation, order (Beck, Hu, Manneville)
- Nonlinear oscillations and strange attractors (Kapitaniak, Manneville, Szemplińska-Stupnicka, Vavriv)
- Mathematics of fractals and chaos (Hu, Keane, Lasota, Mrozek)
- Quantum qualitative measures of chaos–entropies and characteristic exponents (Vilela Mendes, Roepstorff)
- Computer experiments and theory (Mrozek, Tomsovic, Weron)

Seminars appear to complement these topics in a natural way.

We thank all the invited lecturers, contributing speakers and other participants for the lively meeting they created. It is this interaction between physicists, chemists, mathematicians, and computer scientists (and engineers) which enriches and propels this field.

The School was made possible through the financial support from a number of Institutions which are acknowledged separately. We have another obligation to express our gratitude to at least a few colleagues, members of the Organizing Committee, Dr. A. Borowiec, Dr. M. Mozrzymas and Mr. P. Siemion, while extending our thanks to many other, unnamed persons. Last but not least, we would like to thank Ms Anna Jadczyk for her invaluable editorial assistance.

The Editors
Wrocław, May 1995

Acknowledgment

The Organizers of the XXXI-st Winter School of Theoretical Physics in Karpacz, wish to express their gratitude to the following organizations for their financial support:

Dr. Wilhelm Heinrich Heraeus und Else Heraeus–Stiftung,
Institute of Theoretical Physics, University of Wrocław,
H. Steinhaus Center for Stochastic Methods, Technical University of Wrocław,
Ministry of National Education of Poland,
University of Wrocław,
Polish Academy of Sciences,
Institute of Automation of Power Systems (IASE),
Foundation for the University of Wrocław,
Foundation for the Karpacz Winter School of Theoretical Physics.

CONTENTS

Part I: LECTURES[1]

[1] The names of the speakers are underlined where there are more than one author

[1] The names of the speakers are underlined where there are more than one author

LECTURES

Stochastic Processes from Deterministic Dynamics

Christian Beck

School of Mathematical Sciences,
Queen Mary and Westfield College
University of London,
Mile End Road, London E1 4NS, England

Abstract: We deal with classes of smooth deterministic mappings that generate Brownian motion and Langevin processes in an appropriate scaling limit. We investigate the non–Gaussian corrections that occur if the scaling limit is not completely performed. We deal with higher–order correlation functions, and describe how to calculate probability densities in a perturbative way. As a physical application we describe a chaotic cascade model for fully developed turbulent flows. The slightly asymmetric non–Gaussian corrections of our model, produced by the underlying deterministic chaotic dynamics, come out in the same way as in various turbulence experiments. We also deal with possible applications of our approach in Euclidean quantum mechanics. The effects of a chaotic dynamics generating the Wiener process of the Feynman–Kac formula on a very small time scale are discussed.

1 Introduction

Regarding the initial value as a random variable, any deterministic chaotic dynamical system generates ***by definition*** a stochastic process. Remember that a stochastic process is just defined to be a deterministic function of the time and of a random variable [1]. The initial condition of the dynamical system, when distributed according to some probability measure, can be regarded as this random variable. Just the difference to the 'classical' theory of stochastic processes is that chaotic dynamical systems usually do not generate simple stochastic processes. Rather, the processes are generically non–Markovian, non–Gaussian, and the invariant probability densities (=stationary distributions) may have a fractal structure with infinitely many singularities.

In these proceedings I describe some low–dimensional smooth deterministic mappings for which one can rigorously prove that they generate Brownian motion and Langevin processes, i.e. simple stochastic processes, in an appropriate scaling limit. The scaling limit roughly means that we look at the trajectories 'from far away'. An example is shown in Fig 1. Here we see trajectories $(y_n^{(1)}, y_n^{(2)})$

generated by the 4–dimensional smooth deterministic mapping[1]

$$x_{n+1}^{(i)} = 1 - 2x_n^{(i)2}$$
$$y_{n+1}^{(i)} = y_n^{(i)} + \nu x_n^{(i)} \tag{1}$$

for two different initial values $(x_0^{(1)}, x_0^{(2)}) \in [-1,1]^2$. ν is a small parameter. The question is: Do these deterministic trajectories represent true Brownian motion? Or are they just approximating Brownian motion? The answer is that actually both is true: In the scaling limit one can rigorously prove that the above system generates Brownian motion. On the other hand, if we look at the trajectory on a microscopic scale, complicated higher–order correlations and non–Gaussian corrections will show up. How to deal with these, is described in the following.

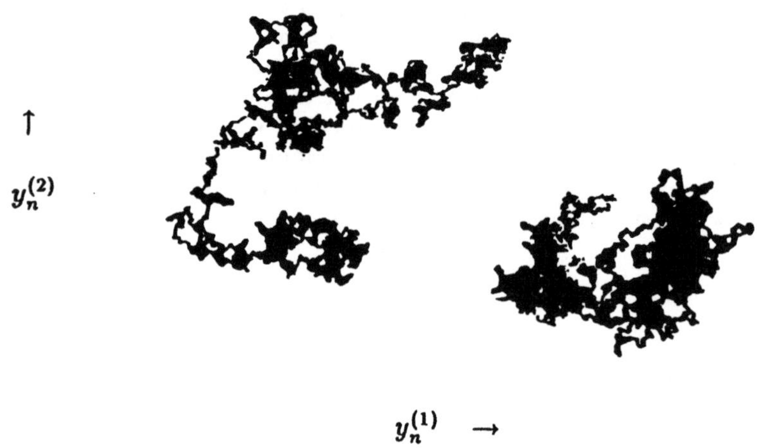

Fig. 1. Diffusion trajectories generated by the deterministic dynamical system (1).

In section 2 I will introduce the relevant class of dynamical systems, and it is shown that they generate Langevin processes in the scaling limit. Three different examples will be treated, namely deterministic mappings that asymptotically generate the Ornstein–Uhlenbeck process, the Wiener process, and a Langevin process with a double–well potential, respectively. In section 3 I briefly look at the structure of higher–order correlations of a system conjugated to a Bernoulli shift. A complex tree structure arises. In section 4 I proceed to probability densities and describe how to obtain the Fokker–Planck equation from the Perron–Frobenius equation, as well as the next–order equation describing

[1] Certainly there are also other approaches to generate 'deterministic diffusion', but the approach presented here appears to be the easiest and most straightforward one.

the deterministic chaotic corrections in the vicinity of the Fokker–Planck limit case. In section 5 I extend the model to a selfsimilar spatio–temporal version. This provides a tool to explain the slight asymmetry of velocity probability distributions measured in fully developed turbulent flows. The model, based on a hierarchical momentum balance equation, for the first time explains velocity distributions of fully developed 3–dimensional turbulent flows in a *quantitatively* correct way. Finally, in section 6 I deal with quantum mechanical applications.

2 Dynamical Systems of Langevin Type

In the Langevin approach, one studies the stochastic differential equation

$$\dot{Y} = A(Y) + L(t) \ . \tag{2}$$

Here Y takes values in $\mathbf{R}^k$, A is a vector–valued function, and $L(t)$ is Gaussian white noise. Let us here study a deterministic chaotic analogue of the Langevin equation. We consider the equation

$$\dot{Y} = A(Y) + L_\tau(t) \ , \tag{3}$$

where the rapidly fluctuating force $L_\tau(t)$ is given in a deterministic chaotic way as follows:

$$L_\tau(t) = \nu \sum_{n=1}^{\infty} f(x_{n-1})\delta(t - n\tau) \tag{4}$$

$$x_{n+1} = T(x_n) \ . \tag{5}$$

δ denotes Dirac's delta function, $\tau > 0$ is the time difference between subsequent impulses, and the x_n evolve in a deterministic way according to some chaotic mapping $T : X \to X$. The function f projects from the phase space X onto the $\mathbf{R}^k$.

If the initial value x_0 is fixed, then $L_\tau(t)$ is completely deterministic. On the other hand, if x_0 is regarded as a random variable, distributed according to some probability distribution, then —by definition—$L_\tau(t)$ is a stochastic process. Hence in this case eq. (3) is a stochastic differential equation. But certainly it is not the Langevin equation, but a more complicated equation, since $L_\tau(t)$ is not Gaussian white noise, but something more complicated.

We now write eq. (3) in a more convenient way. Integrating from $n\tau + 0$ to $(n+1)\tau + 0$ we see that eq. (3) can equivalently be written as the mapping

$$\begin{aligned} x_{n+1} &= T(x_n) \\ y_{n+1} &= g(\tau, y_n) + \nu f(x_n) \ , \end{aligned} \tag{6}$$

where $g(t, y_0)$ is the solution of the initial value problem

$$\begin{aligned} \dot{g} &= A(g) \\ g(0, y_0) &= y_0 \end{aligned} \tag{7}$$

and

$$y_n := Y(n\tau + 0) \ . \tag{8}$$

gives a stroboscopic picture of the process $Y(t)$. For continuous time t,

$$Y(t) = g(t - n\tau, y_n) \qquad n = \lfloor t/\tau \rfloor \tag{9}$$

($\lfloor \ \rfloor$: integer part). The mapping given by eq. (6) is called a *dynamical system of Langevin type.*

As a simple example, let us consider the 1–dimensional problem of a particle of mass 1 moving in a damping medium under deterministic chaotic kicks. $Y(t)$ can be interpreted as the velocity of the particle. We have a linear damping force

$$A(Y) = -\gamma Y \qquad (\gamma > 0) \tag{10}$$

and the solution of the initial value problem (7) is given by $g(t, y_0) = e^{-\gamma t} y_0$. Hence the corresponding 2–dimensional mapping is

$$\begin{aligned} x_{n+1} &= T(x_n) \\ y_{n+1} &= \lambda y_n + \nu f(x_n) \ , \end{aligned} \tag{11}$$

where $\lambda = e^{-\gamma\tau}$ is a parameter. The mapping (11) is called a *dynamical system of linear Langevin type.* We recognize the general structure of a map of Kaplan–Yorke type [2]. For example, we may choose $T(x) = 1 - 2x^2$, $f(x) = x$, $\nu = 1$. The attractor obtained for $\lambda = 0.4$ is shown in Fig. 2.

As it is obvious from the figure, this chaotic dynamics is much more complicated than that of the linear Langevin equation $\dot{Y} = -\gamma Y + L(t)$, which just generates the Ornstein–Uhlenbeck process, a Gaussian Markov process. But there is one important limit case where the complicated dynamics generated by the mapping (11) actually reduces to a simple Langevin dynamics. This is described by the following theorem [4]:

Theorem 1 *If T has the φ-mixing property (see below), and if f is sufficiently smooth (see below), then, setting $\nu = \tau^{1/2}$, we have $\lim_{\tau \to 0} L_\tau(t) = L(t)$ (Gaussian white noise) ($t = n\tau$ finite). Thus in this limit the y–variable of the map (6) generates the Langevin process (2).*

The proof of this theorem is based on a functional central limit theorem for weakly dependent variables, which can be found in Billingsley's book [3]. The important progress made in [4] was to apply this limit theorem to mappings of type (6), obtaining Langevin processes in the scaling limit.

The φ–mixing property is a slightly stronger property than the ordinary strong mixing property. For technical details, see [3, 4]. Certainly any system conjugated to a Bernoulli shift is φ–mixing, but conjugacy to a Bernoulli shift is not a necessary condition. The function f should be continuous, but also this can be weakened [4]. The initial distribution of the x_0 should be absolutely continuous with respect to the natural invariant measure of T, otherwise it is

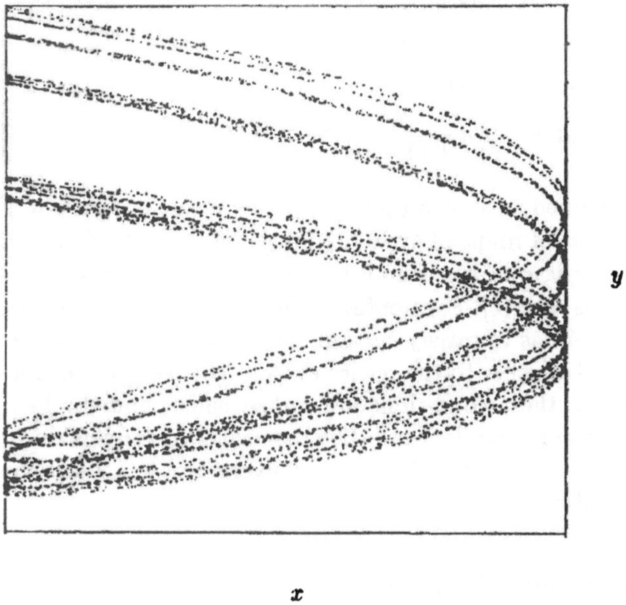

Fig. 2. Strange attractor of the mapping (11) with $T(x) = 1 - 2x^2$, $f(x) = x$, $\nu = 1$, $\lambda = 0.4$.

arbitrary (due to the strong mixing property of T, any smooth initial distribution converges to the natural invariant density).

From the theorem it follows immediately that the y-variable of the mapping (11) with $T(x) = 1 - 2x^2$, $f(x) = x$ and $\nu = \tau^{1/2}$ generates the Ornstein–Uhlenbeck process in the limit $\tau \to 0$ (or $\lambda = e^{-\gamma\tau} \to 1$). Clearly, in Fig. 2, where $\lambda = 0.4$, we are still far away from this limit case. However, decreasing τ, the attractor starts looking more 'stochastic', and the marginal probability density of y exhibits a transition scenario from a complicated distribution to a smooth Gaussian distribution (for pictures, see [5]).

Can we also generate the Wiener process by a simple deterministic dynamics? Certainly, we can. In this case we just consider the case $A(y) = 0$. The solution of the initial value problem (7) with $A = 0$ is $g(t, y_0) = y_0$, thus in this case the map (6) of Langevin type is

$$\begin{aligned} x_{n+1} &= T(x_n) \\ y_{n+1} &= y_n + \nu f(x_n) \ . \end{aligned} \tag{12}$$

For T φ–mixing, f smooth, and $\nu = \tau^{1/2}$, it generates the Wiener process in the limit $\tau \to 0$ (in the y-variable). A 2-dimensional version, with $T(x) = 1 - 2x^2$ and $f(x) = x$, was actually plotted in Fig. 1.

As a third example, we may look onto the chaotically driven motion of a particle in a double well potential $V(y) = -\frac{1}{2}y^2 + \frac{1}{4}y^4$. By this we mean that the force A is given by $A(y) = -\frac{\partial}{\partial y}V(y) = y - y^3$. The solution of the initial value problem (7) is $g(t, y_0) = y_0 \left\{y_0^2 + (1 - y_0^2)e^{-2t}\right\}^{-1/2}$. Thus the corresponding map of Langevin type is

$$\begin{aligned} x_{n+1} &= T(x_n) \\ y_{n+1} &= \frac{y_n}{\sqrt{y_n^2 + (1 - y_n^2)e^{-2\tau}}} + \nu f(x_n) \; . \end{aligned} \tag{13}$$

This map has been studied as an example in [7]. The main emphasis of [7] was to study mixing properties of maps of type (6) for general forces A. For the linear case, i.e. if $A(y) = -\gamma y$, one can easily prove that if T is mixing, the entire system (6) is mixing as well [6]. For nonlinear functions A this is not so obvious any more, and indeed A has to satisfy certain conditions that guarantee that the mixing property of T transfers to the entire system (6) (see [7] for details). For a double well potential, these conditions are *not* satisfied, and the behaviour of the system (13) can be non–ergodic (and thus non–mixing) for certain parameter values. Nevertheless, for $\nu = \tau^{1/2} \to 0$ there is convergence to a Langevin process in a double well potential.

3 Higher–Order Correlations

The example of eq. (12) is simple enough to learn more about the typical structure of a chaotic stochastic process. For simplicity, let us choose $f(x) = x$. Iterating the mapping for the initial value $y_0 = 0$, we can write the y–variable as

$$y_n = \nu \sum_{j=0}^{n-1} x_j \tag{14}$$

From $t = n\tau$ we have $\nu = \tau^{1/2} \sim 1/\sqrt{n}$, and thus we just obtain a properly rescaled sum of iterates x_j. If the x_j were independent random variables, the convergence to the Wiener process would be a trivial consequence of the central limit theorem for independent random variables. But even for strongly chaotic mappings such as $T(x) = 1 - 2x^2$, the x_j are *not* independent! Theorem 1 just states that the dependence is weak enough to still guarantee the convergence to Brownian motion in the limit $\tau \to 0$, $t = n\tau$ finite.

The stochastic process generated by a chaotic mapping $T : X \to X$ with invariant measure μ is completely characterized by the hierarchy of all higher order correlation functions

$$\langle x_{n_1} x_{n_2} \cdots x_{n_r} \rangle = \int_X d\mu(x_0) T^{n_1}(x_0) T^{n_2}(x_0) \cdots T^{n_r}(x_0) \; , \tag{15}$$

or, equivalently, by the characteristic functional. Even for the simplest examples (e.g., $T(x) = 2x \bmod 1$ or $T(x) = 1 - 2x^2$) the explicit evaluation of the integral on the right–hand side for arbitrary r turns out to be very complicated. This may serve as a vague explanation for the fact that so far almost all investigations on correlations of dynamical systems have been restricted to two–point functions, although chaotic processes are generically non–Gaussian, and hence all higher moments and correlation functions are necessary for a complete understanding. The first complete treatment of *all* higher–order correlations of a mapping was given in [8], where the following was proved:

Theorem 2 *Let $x_{n+1} = 1 - 2x_n^2$ and let L_r denote the union of all unlabelled double binary forests with r leaves. Then we have the following equivalence:*

$$\langle x_{n_1} x_{n_2} \cdots x_{n_r} \rangle \neq 0 \Longleftrightarrow (n_1, \ldots, n_r) \in L_r$$

Hence the system $T(x) = 1 - 2x^2$ only possesses correlations on the subspace of tuples that can be represented as double binary forests. The first sets L_r are shown in Fig. 3. Still I have to explain how to 'translate' the double binary forests to tuples. This can be best understood by examples (the precise 'Feynman rules' are given in [8]). As an example, the first tree in the set L_5, plotted in Fig. 3, represents the tuple $(n, n-1, n-2, n-3, n-3)$ and all possible permutations of it, since the leaves (represented by small black spheres) have distances 0,1,2,3,3 from the root. Similarly, the second tree in L_5 represents the tuple $(n, n-2, n-2, n-2, n-2)$ and all possible permutations. The theorem says that the 5–point functions $\langle x_n x_{n-1} x_{n-2} x_{n-3}^2 \rangle$ and $\langle x_n x_{n-2}^4 \rangle$ do not vanish. This is remarkable, because the average $\langle x_n \rangle$ vanishes. Notice that for independent random variables all r–point functions would factorize, and thus the above 5–point functions would certainly vanish for independent random variables with average 0. Chaotic systems generate much more complicated processes with nontrivial higher–order correlations!

4 How to Calculate Probability Densities

Remember that the time evolution of densities ρ_n under a mapping $F : \Omega \to \Omega$ is governed by the Perron–Frobenius operator L:

$$\rho_{n+1} = L\rho_n \tag{16}$$

For 2–dimensional mappings F, this equation reads

$$\rho_{n+1}(x', y') = \sum_{(x,y) \in F^{-1}(x',y')} \frac{\rho_n(x, y)}{|\det DF|} , \tag{17}$$

where DF is the Jacobian. For maps of linear Langevin type, with $\nu = \tau^{1/2}$, this equation can be written as follows:

$$\lambda \rho_{n+1}(x', y) = \sum_{x \in T^{-1}(x')} \frac{1}{|T'(x)|} \rho_n(x, \lambda^{-1}(y - \tau^{1/2} x)) . \tag{18}$$

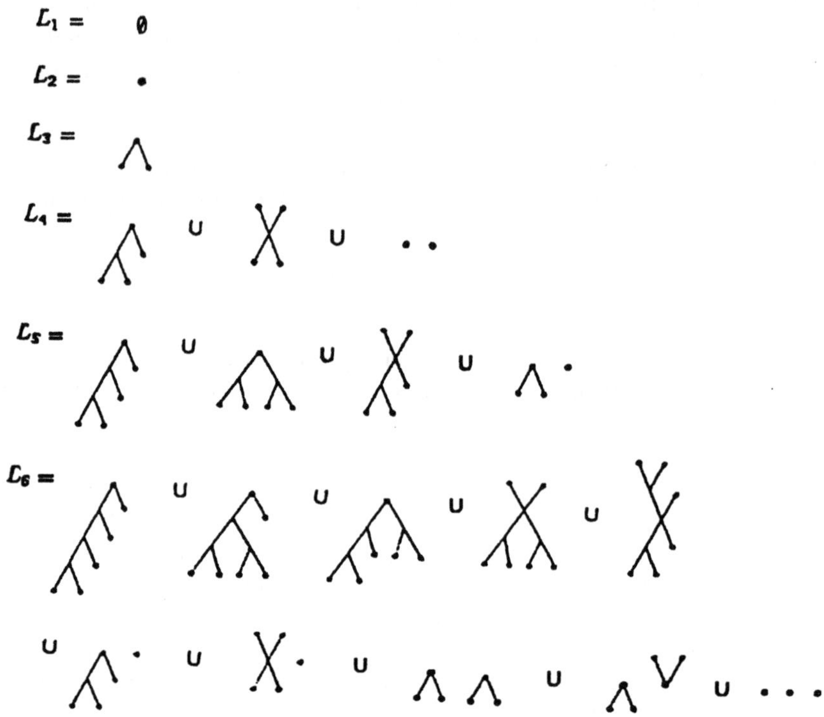

Fig. 3. The sets L_r of double binary forests with r leaves for $r = 1, 2, \ldots, 6$, representing non–vanishing r–point functions of the mapping $T(x) = 1 - 2x^2$.

$\rho_n(x, y)$ is a density on the 2–dimensional phase spce Ω. Let us introduce a continuous–time suspension $\rho(x, y, t)$ defined by

$$\rho_n(x, y) = \rho(x, y, t) \qquad (t = n\tau) \ . \tag{19}$$

For small enough τ, all density functions are expected to become smooth analytical functions, since the Gaussian function is smooth and analytic, and so is the SRB measure of the map T. Hence we make the ansatz

$$\rho(x, y, t) = \varphi(x, y, t) + \tau^{1/2}a(x, y, t) + \tau b(x, y, t) + \tau^{3/2}c(x, y, t) + O(\tau^2) \tag{20}$$

Here φ, a, b, c are τ–independent functions. With $\lambda = e^{-\gamma\tau}$, and choosing for simplicity $\gamma = 1$, we expand the Perron–Frobenius equation (18) in $\tau^{1/2}$, compare the various orders of $\tau^{1/2}$, and integrate over x:

$$p_0(y, t) := \int dx \, \varphi(x, y, t) \tag{21}$$

$$\alpha(y,t) := \int dx\, a(x,y,t) \tag{22}$$

$$\beta(y,t) := \int dx\, b(x,y,t) \tag{23}$$

$$\gamma(y,t) := \int dx\, c(x,y,t) \tag{24}$$

This means, we just look at the marginal distribution of the y–variable

$$\varrho(y,t) = p_0(y,t) + \tau^{1/2}\alpha(y,t) + \tau\beta(y,t) + \tau^{3/2}\gamma(y,t) + O(\tau^2) \tag{25}$$

After a lengthy, but straightforward calculation [20] one obtains for symmetric maps T with a symmetric invariant density h the following result: The functions p_0, α, and b obey the equations

$$0 = \frac{\partial}{\partial y}(y p_0(y,t)) + \frac{1}{2}\langle x^2\rangle \frac{\partial^2}{\partial y^2} p_0(y,t) - \frac{\partial}{\partial t} p_0(y,t) \tag{26}$$

$$\frac{\partial}{\partial y}\int dx\, x b(x,y,t) = \frac{\partial}{\partial y}(y\alpha(y,t)) + \frac{1}{2}\langle x^2\rangle \frac{\partial^2}{\partial y^2}\alpha(y,t) - \frac{\partial}{\partial t}\alpha(y,t) \tag{27}$$

$$b(x',y,t) = \sum_{x\in T^{-1}(x')} \frac{1}{|T'(x)|}\left\{ b(x,y,t) + h(x)\frac{1}{2}(x^2 - \langle x^2\rangle)\frac{\partial^2}{\partial y^2} p_0(y,t)\right\} \ . \tag{28}$$

These equations are an *exact* consequence of the ansatz (20) and the Perron–Frobenius equation (18). Notice that eq. (26) is the Fokker–Planck equation. That was to be expected, since Theorem 1 guarantees the convergence to a Langevin process in the limit $\tau \to 0$. The new thing is that we have found a general equation for the first–order correction term $\alpha(y,t)$. It satisfies an 'inhomogeneous' Fokker–Planck equation with a source term. The source term is determined by the functional equation (28). These are the general equations governing the Fokker–Planck limit case and its vicinity!

Surprisingly enough, for special mappings one can find explicit solutions of the equations. We restrict ourselves to the stationary case $t \to \infty$, where the densities become time–independent. For $T(x) = 1 - 2x^2$, one finds the solution of eq. (28) as [20]

$$b(x,y) = x h(x)\sqrt{\frac{2}{\pi}}(1-4y^2)e^{-2y^2} \ , \tag{29}$$

where $h(x) = 1/(\pi\sqrt{1-x^2})$ is the invariant density of T. From this we can evaluate the left–hand side of the inhomogeneous Fokker–Planck equation (27), getting finally

$$\alpha(y) = \sqrt{\frac{2}{\pi}}e^{-2y^2}\left(2y - \frac{8}{3}y^3\right) \ . \tag{30}$$

Notice that $\alpha(y)$ is an odd function, it is an *asymmetric* first–order correction to the Gaussian limit case $p_0(y) = \sqrt{\frac{2}{\pi}}e^{-2y^2}$. One can do the same analysis for the second-order correction $\beta(y,t)$. The final perturbative result for the stationary probability density is

$$\varrho(y) = \sqrt{\frac{2}{\pi}} \left\{ 1 + \tau^{1/2} \left(-\frac{8}{3} y^3 + 2y \right) + \tau \left(\frac{32}{9} y^6 - \frac{31}{3} y^4 + \frac{15}{2} y^2 - \frac{37}{48} \right) + O(\tau^{3/2}) \right\} e^{-2y^2} . \quad (31)$$

The coincidence with a numerically obtained histogram of iterates of the y-variable is excellent (Fig. 4). In particular, the slight asymmetry of the distribution is reproduced correctly.

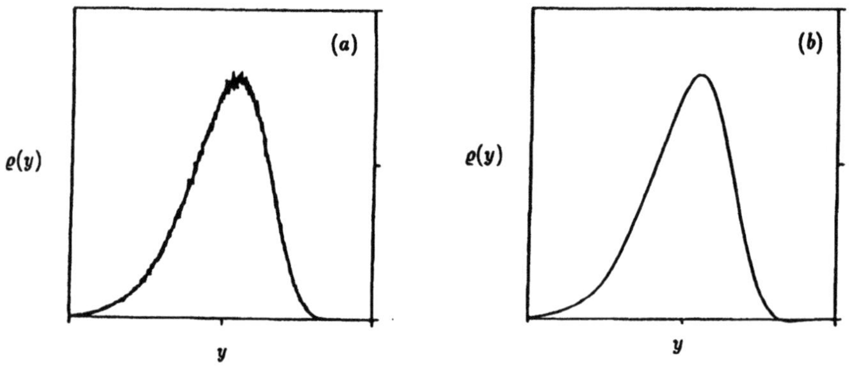

Fig. 4. Stationary probability density of the y-variable of a dynamical system of linear Langevin type for $T(x) = 1 - 2x^2$, $f(x) = x$, $\gamma = 1$ and $\tau^{1/2} = 0.324$. (a) Histogram of the iterates. (b) Analytical result obtained in second-order perturbation theory.

5 Fully Developed Turbulent Flows

In fully developed turbulent flows, probability densities of velocity differences are also observed to have slightly asymmetric deviations from a Gaussian shape. These experimentally measured probability densities cannot be explained by numerical simulations of the Navier–Stokes equations, because the best computers, integrating the Navier–Stokes equations by brute force, reach Reynolds numbers $Re_\lambda \approx 100$ (with CPU times of a month or so [10]), whereas the true physical experiments reach $Re_\lambda \approx 3000$. Nature is still much better than a computer! Hence we need appropriate models to explain the experimentally measured velocity distributions. The model described in the following is based on a dynamical system of Langevin type.

To model turbulence, the approach of the previous sections has to be extended to infinite dimensional (spatially extended) systems. Moreover, the self-similarity of the turbulent energy cascade as well as the random fluctuations of the energy transfer must be taken into account. In a turbulent flow, the velocity field $\mathbf{v}(\mathbf{x},t)$ is strongly fluctuating both in space and time. Let us look at two points in the liquid that are separated by a vector $\mathbf{r}$. The component of the velocity field in $\mathbf{r}$ direction is denoted by v_r. The velocity difference in $\mathbf{r}$–direction beween the two points $\mathbf{x}$ and $\mathbf{x}+\mathbf{r}$ is denoted as

$$u(r,t) = v_r(\mathbf{x},t) - v_r(\mathbf{x}+\mathbf{r},t) \ . \tag{32}$$

It is a rapidly fluctuating scalar field. Neglecting boundary conditions, its stochastic properties are invariant under translation. Also, assuming isotropy, the stochastic properties just depend on the length r of the vector $\mathbf{r}$, rather than its direction. This is why we just write $u(r,t)$. Keeping r constant, we may abbreviate the notation and write $u(t)$ instead of $u(r,t)$. The dynamics of $u(t)$ is basically determined by two competing effects:

1. In the absence of external or internal driving forces velocity differences tend to relax to the laminar state $u=0$. Thus we have

$$\dot{u} = -\gamma u, \qquad \gamma > 0 \ . \tag{33}$$

2. There are time dependent chaotic forces in the turbulent liquid that drive the velocity differences and prevent the relaxation to the laminar state. Let us denote the resulting force in $\mathbf{r}$–direction (i.e., the difference of the forces at $\mathbf{x}$ and $\mathbf{x}+\mathbf{r}$) by $F(t)$. We obtain

$$\dot{u} = -\gamma u + F(t) \ . \tag{34}$$

Eq. (34) looks similar to a linear Langevin equation, but certainly there are major differences:

1. The force $F(t)$ is not Gaussian white noise, but a chaotic force changing on a typical time scale τ.
2. The relaxation time scale γ^{-1} and the time scale τ of the force have comparable orders of magnitude.
3. In the Langevin theory u is the velocity. In our model u is the *velocity difference.*
4. We are on a macroscopic scale, rather than on a mesoscopic scale.
5. $F(t)$ fluctuates due to the nonlinearity of the Navier–Stokes equation, rather than due to collisions with molecules.

From a numerical point of view, it is much easier to deal with chaotic mappings than with chaotic flows. For this reason, let us discretize the time and assume that the force $F(t)$ is a deterministic chaotic kick force:

$$F(t) = \nu \sum_{n=1}^{\infty} f(x_{n-1})\delta(t-n\tau) \tag{35}$$

$$x_{n+1} = T(x_n) \tag{36}$$

Just as in section 2 we get

$$\begin{aligned} x_{n+1} &= T(x_n) \\ u_{n+1} &= \lambda u_n + \nu f(x_n) \ , \end{aligned} \tag{37}$$

where $u_n = u(n\tau + 0)$ and $\lambda = e^{-\gamma\tau} < 1$. We will typically choose for T the fully developed logistic map $T(x) = 1 - 2x^2$ and $\nu f(x) = x$.

As usual for fully developed 3–dimensional flows, let us consider a cascade model [11, 12]. This means, we assume that fully developed turbulence is characterized by the existence of a selfsimilar set of eddies on various scales. The various levels of this cascade are labelled by an index k. The larger k, the smaller is the size of the corresponding eddy. The eddy at level k is called 'mother eddy' and the corresponding smaller eddies at level $k + 1$ are called 'daughter eddies'.

Let us denote by $u_n^{(k)}$ the velocity difference of two points in the liquid that are separated by a distance r_k, where r_k scales (in the inertial range) as

$$r_k = 2^{-k} \ . \tag{38}$$

Our basic model assumption is that the momentum loss at level k serves as a chaotic driving force at level $k + 1$. Moreover, a particular one of the daughter eddies at level $k + 1$ gets only a random fraction $\xi_n^{(k)}$ of the momentum loss of the mother eddy at level k, the other part is transfered to other daughter eddies at level $k + 1$. This can be expressed by the following coupled map lattice [13]:

$$\begin{aligned} x_{n+1} &= T(x_n) \\ u_{n+1}^{(1)} &= \lambda_1 u_n^{(1)} + x_n \\ u_{n+1}^{(k)} &= \lambda_k u_n^{(k)} + c\xi_n^{(k-1)}(1 - \lambda_{k-1})u_n^{(k-1)} \qquad (k = 2, \ldots, K) \end{aligned} \tag{39}$$

Here $\xi_n^{(k)}$ is a random variable on $[0, 1]$. It is multiplied by a coupling constant c. To be as general as possible, we allow the damping λ to depend on the scale, expressed by the index k. The simplest choice is to take for the $\xi_n^{(k)}$ uniformly distributed independent random variables. But they can also be generated by the binary shift map or other strongly chaotic maps on the unit interval— the results do not depend on details of this choice. Notice that eq. (39) has the general structure of a coupled map lattice with one–sided random couplings.

The physical meaning of the coupled map lattice (39) is that of a hierarchical momentum balance equation: During one time unit, the momentum loss of the mother eddy at level $k - 1$ is $(1 - \lambda_{k-1})u_n^{(k-1)}$ in appropriate units. A random fraction of this momentum (determined by the factor $c\xi_n^{(k-1)}$) serves as a driving force for a particular daughter eddy at level k, the other part is transfered to the other daughter eddies.

Still we have to specify how the parameters λ_k depend on the scale index k. A reasonable choice is to model the well known increase of dissipation on smaller scales by assuming that $\gamma = \gamma(k)$ changes linearly with the scale index k, or, in other words, that λ_k decreases exponentially:

$$\lambda_k = e^{-\gamma_0 \tau k} \tag{40}$$

It is plausible to assume that $\gamma_0\tau$ is a measure for the inverse Reynolds number Re_λ^{-1}, since

$$\gamma_0\tau \sim \frac{\nu}{L^2}\tau \sim \frac{\nu}{vL} = Re_\lambda^{-1} \tag{41}$$

Here ν denotes the kinematic viscosity, v is a typical velocity and L is a typical length scale. Actually, $\gamma_0\tau$ could also be a more complicated function of the inverse Reynolds number, the precise relation is unknown, but this doesn't matter for our purposes.

Fig. 5 shows various probability distributions of velocity differences measured by Castaing, Gagne and Hopfinger in a jet experiment at Reynolds number $Re_\lambda = 852$ [9].The distance r/η (η: Kolmogorov scale) varies from $r/\eta = 100$ (a) to $r/\eta = 23.6$ (b) and $r/\eta = 3.3$ (c). The thin lines in the figures show histograms obtained by iterating the coupled map lattice (39) for $\gamma_0\tau = 0.022$, $k = 7$ (a), $k = 10$ (b), and $k = 17$ (c). As usual, we have chosen for the map T the fully developed logistic map. All distributions are normalized and have variance 1 (by this the constant c is eliminated). The quantitative agreement with the experimental data is remarkably good. Notice that only two parameters can be varied for the coupled map lattice ($\gamma_0\tau$ and k), and similarly just two relevant parameters are changed in the experiments (Re_λ and r). In our simulation we keep the parameter $\gamma_0\tau$ constant and vary the index k, since the Reynolds number is kept constant in the experiment. Increasing k, we observe the typical transition scenario from distributions similar to a slightly asymmetric Gaussian to distributions with stretched exponential tails. In contrast to many other intermittency models, the coupled map lattice yields the transition scenario in a *quantitatively* correct way, i.e. the simulated distributions coincide precisely with the experimentally measured distributions. Relation (38) appears to be well satisfied in the inertial range, whereas there is a more complicated (unknown) relation between r and k in the dissipative range.

In [13] further systematic comparisons with other experiments at different Reynolds numbers were carried out. All experimental data could be reproduced by the coupled map lattice in a quantitatively correct way. One can also look at the scaling exponents ζ_m of the moments, defined by

$$\langle u(r)^m \rangle \sim r^{\zeta_m} \quad , \tag{42}$$

in the inertial range. The coupled map lattice does not only exactly reproduce the shape of the observed probability distributions, but also the scaling exponents ζ_m come out correctly.

Notice that if we had chosen for the x_n statistically independent random variables with average 0 rather than the iterates of a chaotic map, all odd moments of u would vanish, and no asymmetry of the invariant density of u would arise at all. Thus, the agreement with the experimental data would be much worse in this case. The slight asymmetry can be understood in terms of non–trivial higher–order correlations due to the underlying chaotic dynamics [13]. Since τ (i.e. Re_λ^{-1}) is finite, these deterministic corrections are relevant.

We may ask: Why does this simple model yield such good results? Actually, the Navier–Stokes equation becomes singular for $Re_\lambda \to \infty$. So nobody knows

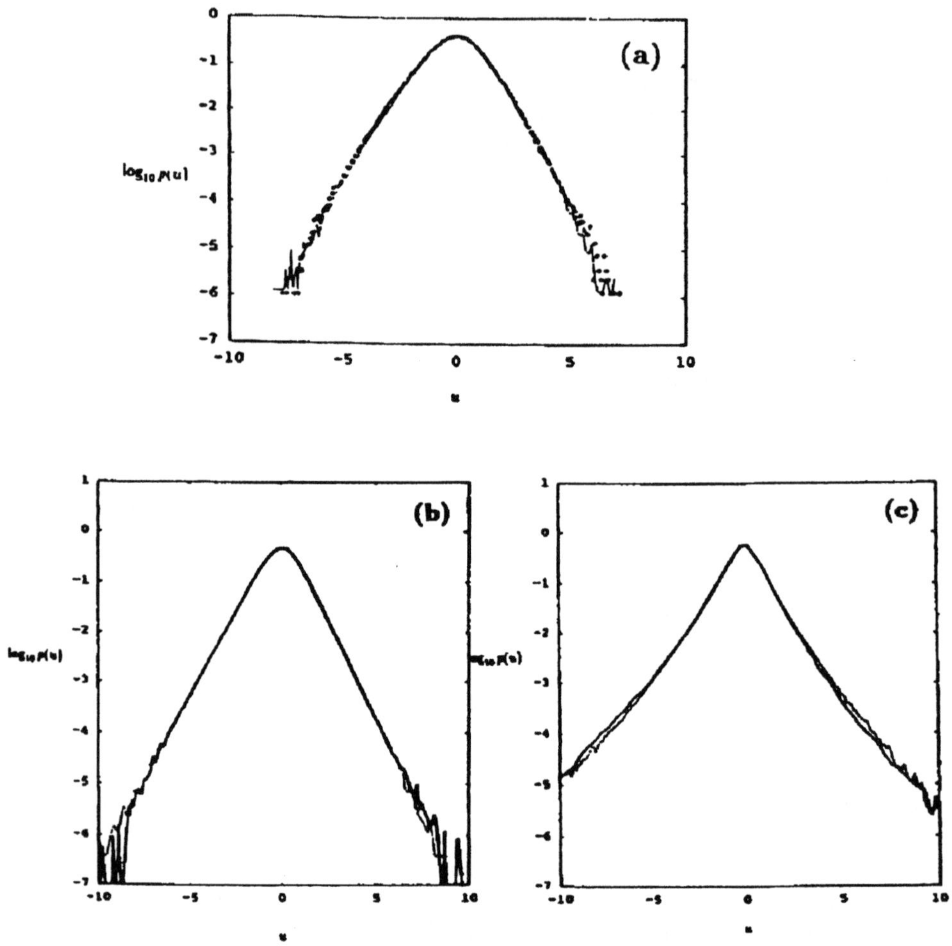

Fig. 5. Logarithm of the probability density of velocity differences measured in a fully developed turbulent flow at distances r/η =100 (a), 23.6 (b), 3.3 (c). The dotted line and the thick solid lines represent the physical experiment. The thin solid lines are densities obtained by iterating the coupled map lattice (39) for $\gamma_0 \tau = 0.022$ and $k = 7$ (a), 10 (b), 17 (c).

what happens. The Navier–Stokes equation might even not be valid in this limit, and ought to be replaced by a suitable dynamical system.

I have two conjectures, why the model works so well in the limit of very large Reynolds numbers.

Conjecture A: The local driving forces in a turbulent flow, although completely deterministic, might try to reach a state of *strongest possible random properties* in the limit $Re_\lambda \to \infty$. Thus a mapping such as $T(x) = 1 - 2x^2$, conjugated to a Bernoulli shift, is a good approximation of the true dynamics.

Conjecture B: We know that the Gaussian limit distribution is a *universal* limit distribution, i.e. it is independent of details of the map T (for example, the φ–mixing property is sufficient, other details of the mapping are irrelevant). Maybe the behaviour in the *vicinity* of this Gaussian fixed point is universal, too. This would mean that large classes of dynamical systems (perhaps even including the Navier–Stokes dynamics) approach the Gaussian limit dynamics in the same (slightly asymmetric) way.

6 Chaotic Dynamics and Euclidean Quantum Mechanics

We saw that deterministic dynamical systems can be a perfect source of Brownian motion and Langevin processes, provided we perform the scaling limit. We may thus think about the possibility that the Brownian motion underlying the various formulations of Euclidean quantum mechanics might ultimately be produced by chaotic dynamical systems on a very small time scale. I will work this out for the Feynman–Kac approach [14, 15, 16], but certainly one can also do it for more sophisticated approaches such as the one presented by Zambrini [17].

To simulate an N–dimensional quantum system with potential V, let me consider the following $2N+1$–dimensional mapping [18]

$$Q: \quad \begin{aligned} x_{n+1} &= T(x_n) \\ y_{n+1} &= y_n + \sigma^{-1}\sqrt{\tfrac{\hbar\tau}{m}}\, x_n \\ z_{n+1} &= z_n \exp\left\{-\tfrac{\tau}{\hbar}V(y_n)\right\} \end{aligned} \tag{43}$$

with the initial value $z_0 = 1$. Q is called *Euclidean quantum map*. By the notation of eq. (43) we actually mean that T consists of N φ–mixing 1–dimensional mappings. The variance σ^2 is given by $\sigma^2 = \langle x_0^2 \rangle + 2\sum_{n=1}^{\infty}\langle x_0 x_n\rangle$. Knowing that the y–variable of the mapping generates the N–dimensional Wiener process in the scaling limit, it is easy to see that the expectation of the z–variable of this system converges to the integrated Euclidean propagator $G_t(y, y_0)$ of a quantum system with potential V:

$$\int d^N\mu(x_0) z_n(x_0, y_0) = \langle z_n\rangle \to \int d^N y G_t(y, y_0) \quad (\tau \to 0) \tag{44}$$

Indeed, this is just a translation of the Feynman–Kac formula [14, 15, 16] into a dynamical systems context. Hence, to calculate the integrated propagator of a quantum system, one can simply iterate the quantum map Q several times for an ensemble of initial values x_0, averaging the variable z_n over the various trajectories.

Euclidean quantum maps are appropriate tools to calculate expectations of observables, e.g. ground state energies. For simplicity, let us assume that there is a discrete energy spectrum. The propagator can be expanded in terms of eigenfunctions Ψ_j of the Hamilton operator as follows:

$$G_t(y, y_0) = \sum_{j=0}^{\infty} \Psi_j(y)\Psi_j^*(y_0)e^{-\frac{E_j t}{\hbar}} \tag{45}$$

($E_0 < E_1 < E_2 < \cdots$: discrete energy eigenvalues). For large $t = n\tau$

$$\int d^N y\, G_t(y, y_0) = \Psi_0^*(y_0) \int d^N y\, \Psi_0(y)e^{-\frac{E_0 t}{\hbar}} + O(e^{-\frac{E_1 t}{\hbar}}) \tag{46}$$

provided the wave functions are integrable. Thus in the scaling limit

$$\log\langle z_n\rangle = -E_0 \frac{\tau}{\hbar} \cdot n + C + O(e^{-(E_1 - E_0)\frac{\tau}{\hbar}\cdot n}) \tag{47}$$
$$(\tau \text{ small, } t = n\tau \text{ large}),$$

or

$$\lim_{t\to\infty} \lim_{\tau=\frac{t}{n}\to 0} -\frac{\hbar}{n\tau} \log\langle z_n\rangle = E_0 \ . \tag{48}$$

Knowing E_0, $E_1 - E_0$ can be obtained via

$$\log\left|\log\langle z_n\rangle - C + E_0\frac{\tau}{\hbar}n\right| = -(E_1 - E_0)\frac{\tau}{\hbar}\cdot n + const + O(e^{-(E_2 - E_1)\frac{\tau}{\hbar}\cdot n}) \ . \tag{49}$$

In principle, one can continue obtaining the whole energy spectrum.

This algorithm, based on the map (43), was introduced in 1991 [18]. I tested it for both independent random variables x_n and chaotic iterates x_n, and presented some numerical results for ground state energies of the harmonic and anharmonic oscillator, as well as for the hydrogen atom. In 1992, Korzeniowski et al. [19] applied and extended the algorithm to many-electron atoms. They were able to determine the ground state energy of Helium with a precision of six digits, the first excited state with a precision of five digits, and the ground state energies of Li, Be and B with a precision of 4 digits. These are quite good results which can compete with the most powerful algorithms known.

Are Euclidean quantum maps just a powerful algorithm to calculate ground state energies, or is there anything deeper behind? Suppose τ is an extremely small time scale (e.g., the Planck scale or less) on which we might have a chaotic dynamics generating the Wiener process responsible for a Feynman–Kac dynamics. Certainly, we would not notice this in 'daily–life' quantum mechanics, since the time scale is so small that the Wiener process is perfectly approximated. Nevertheless, one can ask what would be the tiny corrections to ordinary quantum mechanics for finite τ. For this, one has to do a perturbative expansion quite similar to that in section 4. Instead of starting from the Perron–Frobenius operator, one has to start from a generalization of the Perron–Frobenius operator (Ruelle's transfer operator), which depends on the potential under consideration, and expand it in a perturbative way in the small time scale parameter $\tau^{1/2}$. This has been done in [21]. The result is quite similar as for the Fokker–Planck case: In zeroth–order one obtains the Euclidean Schrödinger equation, the next order correction obeys an 'inhomogeneous' Schrödinger equation with an external source term. The source term obeys a functional equation, which is the same as eq. (28), just that $p_0(y,t)$ is replaced by the zeroth-order Schrödinger wave function $\Psi(y,t)$. The equations can be solved for special mappings such as $T(x) = 1 - 2x^2$.

Of course, all this is just a formal perturbative approach, but perhaps there may be some physical applications in the future. In any case, it should be clear that this is a totally different approach to 'quantum chaos', which has nothing to do with the conventional semiclassical approach. The approach presented here can also be generalized to quantum field theories [22].

References

1. N.G. van Kampen, *Stochastic Processes in Physics and Chemistry*, North-Holland, Amsterdam (1981).
2. J.L. Kaplan and J.A. Yorke, in *Functional Differential Equations and the Approximation of Fixed Points*, Lecture Notes in Mathematics, vol. 730, p. 228 (1979).
3. P. Billingsley, *Convergence of Probability Measures*, Wiley, New York (1968).
4. C. Beck and G. Roepstorff, Physica **145A**, 1 (1987).
5. C. Beck and F. Schlögl, *Thermodynamics of Chaotic Systems*, Cambridge University Press, Cambridge (1993).
6. C. Beck, Commun. Math. Phys. **130**, 51 (1990).
7. C. Beck, G. Roepstorff and C. Schroer, Physica **72D**, 211 (1994).
8. C. Beck, Nonlinearity **4**, 1131 (1991).
9. B. Castaing, Y. Gagne and E.J. Hopfinger, Physica **46D**, 177 (1990).
10. A. Vincent and M. Meneguzzi, J. Fluid Mech. **225**, 1 (1991).
11. U. Frisch, P.L. Sulem and M. Nelkin, J. Fluid Mech. **87**, 719 (1978).
12. R. Benzi, G. Paladin, G. Parisi and A. Vulpiani, J. Phys. **A17**, 3521 (1984).
13. C. Beck, Phys. Rev. **49E**, 3641 (1994).
14. R.P. Feynman and A.R. Hibbs, *Quantum Mechanics and Path Integrals*, McGraw-Hill, New York (1965).
15. L.S. Schulman, *Techniques and Applications of Path Integration*, Wiley, New York (1981).
16. G. Roepstorff, *Path Integral Approach to Quantum Physics*, Springer, Berlin (1994).
17. J. Zambrini, these proceedings (1995).
18. C. Beck, Physica **50D**, 1 (1991).
19. A. Korzeniowski, J.L. Fry, D.E. Orr, and N.G. Fazleev, Phys. Rev. Lett. **69**, 893 (1992).
20. C. Beck, *From the Perron-Frobenius Equation to the Fokker-Planck Equation*, to appear in J. Stat. Phys. (1995).
21. C. Beck, *From Ruelle's Transfer Operator to the Schrödinger Operator*, to appear in Physica D (1995).
22. C. Beck, *Chaotic Quantization of Field Theories*, to appear in Nonlinearity (1995).

The Interplay of Classical and Quantum Stochastics: Diffusion, Measurement and Filtering

V.P. Belavkin

Mathematics Department,
University of Nottingham, UK

Abstract: A brief presentation of the basic concepts in quantum probability theory is given in comparison to the classical one. The notion of quantum white noise, its explicit representation in Fock space, and necessary results of noncommutative stochastic analysis and integration are outlined.

Algebraic differential equations that unify the quantum non Markovian diffusion with continuous non demolition observation are derived. A stochastic equation of quantum diffusion filtering generalising the classical Markov filtering equation to the quantum flows over arbitrary *–algebra is obtained.
Key words and phrases: quantum probability and noise, noncommutative stochastic calculus, quantum diffusion and flows, continuous nondemolition processes, a posteriori state diffusion and spontaneous localisation.

Introduction

Beginning in the mid seventies, modern probability theory has, (along with traditional subjects, such as dynamical systems with random perturbations), been also concerned with fundamentally new stochastic objects — quantum dynamic systems with an inherently probabilistic nature. Mathematically, the concept of quantum probability arises not because of the lack of information for a complete description of the object, the instability of chaotic motion or the inaccuracy of measurement but is due to the noncommutativity of the algebra of random variables which are represented by the operators in the Hilbert space. As quantum probability theory is an intrinsically stochastic theory, it cannot be stated within the framework of the Kolmogorov axioms [1], which assume the fundamentally deterministic description of the classical systems under the given point states $\omega \in \Omega$. It is based on different axioms [2], [3] namely, the Neumann axioms, whose greater generality can be demonstrated even in the case of a finite number of alternative elementary events $\omega = 1, \ldots, n$.

Let us illustrate for this simple case how the classical probability space $(\Omega, \mathcal{F}, \mathbf{P})$ can be represented as a special case of the quantum space that is defined by the triple $(\mathrm{H}, \mathcal{A}, \mathbf{E})$. Here H is a (finite–dimensional) complex space of column–vectors $h = [\eta^i]$, $\eta^i \in \mathbf{C}$ with scalar product $(g|h) = \sum \bar{\zeta}^i p_i \eta^i \equiv g^* h$ for $g = [\zeta^i] \in \mathrm{H}$ defined by the weights (probabilities) $p_i > 0$, $\mathcal{A}$ is an associative, but not necessarily commutative matrix algebra $X = [\xi^i_k]$ closed under the involution $X \mapsto X^*$ defined by the Hermitian conjugation

$$(X^* g|h) = (g|Xh), [\xi^i_k]^* = [p_i^{-1} \bar{\xi}^k_i p_k] \ ,$$

and with matrix $I = [\delta^i_k]$ as the identity $I \in \mathcal{A}$, and $\mathbf{E}[X] = (e|Xe)$ is a positive normalized functional ($\mathbf{E}[X^* X] \geq 0$, $\mathbf{E}[I] = 1$) of expectation of noncommuting variables X defined by a fixed unit vector $e \in \mathrm{H}$, where $\|e\|^2 := (e|e) = 1$.

Classical random variables $x : \Omega \to \mathbf{C}$ can also be described by the multiplication operators $X = \widehat{x}$, $(\widehat{x}h)(\omega) = x(\omega)\, h(\omega)$ in the complex Hilbert space $\mathrm{H} = L^2(\Omega, \mathcal{F}, \mathbf{P})$ of $\mathcal{F}$ –measurable $\mathbf{P}$–square–integrable functions $h : \Omega \to \mathbf{C}$,

$$\|h\|^2 = \int |h(\omega)|^2 \mathbf{P}(d\omega) = (h|h) < \infty \ .$$

Their expectations $\mathbf{E}[\widehat{x}] = \int x(\omega)\, \mathbf{P}(d\,\omega)$ are defined as $(e|\widehat{x}e)$ by the unit function $e(\omega) \equiv 1$ which is normalized with respect to any probability measure $\mathbf{P}$. Thus, however, only commutative operator algebras $\mathcal{A}$ are obtained, whose elements are given in the finite case of $\Omega = \{1, \ldots, n\}$ by all the diagonal matrices $\widehat{x} = [\xi(i)\delta^i_k]$ with the commutative product, corresponding to the pointwise multiplication of the functions $x(\omega) = \xi(i)$, where $\omega = i$.

Conversely, any quantum probability space $(\mathrm{H}, \mathcal{A}, \mathbf{E})$ can be reduced to the classical one $(\Omega, \mathcal{F}, \mathbf{P})$ only in the case of the commutativity of the algebra $\mathcal{A}$; in the finite–dimensional case this is realised by simultaneous reduction of the commuting matrices $X \in \mathcal{A}$ to diagonal form $[\xi(i)\delta^i_k]$. The probabilities p_i of the elementary events $\omega = i$ in the diagonal representation are defined by the restriction $p_i = \mathbf{E}[P_i]$ of the functional $\mathbf{E}[X] = \sum \xi(i) p_i$ on projective matrices $P_j = [\xi_j(i)\, \delta^k_i]$, $\xi_j(i) = \delta^i_j$.

In this article a quantum analog of diffusion and the problem of its continuous measurement and stochastic filtering, that gives the solution of the Zeno paradox [4] (as a result of establishing an a posteriori stationary state), are considered within the framework of the noncommutative algebraic approach. A derivation of the stochastic equation is given for a nonnormalized a posteriori quantum state, which is obtained in [5] by renormalising the basic equation of nonlinear quantum filtering [6]. The solution of the equation has been found for the case of linear quantum diffusion of canonical commutation relations, obtained previously for the quantum Gaussian case by means of linear Markov filtering methods in [7], [8].

In presenting the second (basic) section, we deliberately avoided the questions concerning the sufficient conditions for the dense definition of the unbounded infinitesimal generators that guarantee the uniqueness of solutions of quantum stochastic and operator equations; this is beyond the scope of this article. We only point out that in the first and second sections these questions are not relevant (see [9]) for the Markovian case with complete pre–Hilbert domain D in the initial Hilbert space H, corresponding to boundedness of the operators

L and H in $\mathrm{D} = \mathrm{H}$. Moreover, a solution exists for an unbounded algebra $\mathcal{A}$ of canonical commutation relations, which is considered in the third section, in the framework of a quantum calculus of kernels for the operators L, $H \in \mathcal{A}$ in the initial Fock scale $\{\mathrm{F}_\xi | \xi > 1\}$[10] if their inductive limit $\cup \mathrm{F}_\xi$ is chosen as D. Besides, the explicitly solvable model of this section with linear unbounded generators L and H, does not require the estimates obtained in these scales.

For completeness the notation and explicite methods of quantum stochastic integration and the proof of their estimates in Fock scale [5] are briefly presented in the Appendix. The comprehensive statement of the author's general approach, outlined above, and the estimates for the integrals can be found in [10], [11]. The earlier results on quantum stochastic calculus in the framework of Hudson and Parthasarathy approach [12], are reviewed in [13].

The approach presented generalizes the results obtained in [14] for purely quantum diffusion to the case of an arbitrary initial algebra $\mathcal{A}$. This enables a unified description of quantum and classical diffusion, their observation and filtering as special algebraic cases. However for simplicity the paper is restricted to one–dimensional variant of the infinite–dimensional quantum diffusion filtering [14]. The results explain the spontaneous localisation of a free quantum Brownian particle during the continuous observation of its coordinate; this was analysed earlier in [16] by the method of solving the a posteriori Schrödinger equation [15].

1 Quantum Diffusion and Nondemolition Measurement

1.1 Basic Notation

Let H be a complex Hilbert space and $\mathrm{D} \subseteq \mathrm{H}$ be a dense subspace defined as an inductive limit (see appendix 1) of some scale $\{\mathrm{H}_\xi | \xi > 1\}$ in the space H. Let the initial algebra $\mathcal{A}$ of noncommutative random variables describing a 'quantum object' at the initial moment $t = 0$ be represented by an involutive subalgebra $\mathcal{A} \subseteq \mathcal{B}(\mathrm{D})$ of linear operators $X : \mathrm{D} \to \mathrm{D}$, $X^* \in \mathcal{A}$, having (an inductively) continuous conjugate $X^* : \mathrm{D} \to \mathrm{D}$, $(X^*\chi|\psi) = (\chi|X\psi)$ with respect to the scalar product in H, with an identity operator $I \in \mathcal{A}$.

Let us denote by $\mathcal{H} = \mathrm{H} \otimes \mathcal{F}$ the tensor product H and the Fock space $\mathcal{F} = \Gamma(\mathcal{K})$ over the Hilbert space $\mathcal{K} = L^2(\mathbf{R}_+)$ of a 'quantum noise' $\widehat{w}_t(g)$, $g \in \mathcal{K}$, and let the pre–Hilbert space $\mathcal{D}$ be an inductive limit of the Hilbert scale $\mathcal{H}_\xi = \mathrm{H}_\xi \otimes \mathcal{F}_\xi$, $\xi > 1$, where $\{\mathcal{F}_\xi\}$ is the natural Fock scale (see appendix 2) over $\mathcal{K}$. We shall consider the quantum noise as a set $\{\widehat{w}_t(g) | g \in L^2(\mathbf{R}_+)\}$ of Brownian motions $t \mapsto \widehat{w}_t(g)$, represented in $\mathcal{F}$ by self–adjoint operators

$$\widehat{w}_t(g) = \int_0^t \left(g(r)\,\mathrm{d}\widehat{a}_r^* + \bar{g}(r)\,\mathrm{d}\widehat{a}_r\right) \equiv \widehat{a}_t^*(g) + \widehat{a}_t(g^*) \ ,$$

with a Gaussian state on the algebra generated by them, which is induced by the vacuum function $\delta_\emptyset \in \mathcal{F}$. Here $\{\widehat{a}_r, \widehat{a}_r^* | r \in \mathbf{R}_+\}$ are canonical operators of creation $\widehat{a}_r^*$ and annihilation $\widehat{a}_r$ in $\mathcal{F}$ (see Appendix 3) called quantum stochastic integrators, and $g^*(t) = \bar{g}(t)$. Note that each operator function $t \mapsto \widehat{w}_t(g)$ that has commutative values $[\widehat{w}_s(g), \widehat{w}_t(g)] = 0$ is equivalent to a classical Brownian motion with intensity $|g(t)|^2$, with respect to the vacuum vector $e = \delta_\emptyset$. This follows from the formula

$$e^{i\widehat{w}(g)} = e^{i\widehat{a}^*(g)} e^{-\frac{1}{2}\|g\|^2} e^{i\widehat{a}(g^*)}$$

and $e^{\widehat{a}(f)}\delta_\emptyset = \delta_\emptyset$ for any $f \in \mathcal{K}$, due to which the quantum characteristic function

$$\mathbf{E}[e^{i\widehat{w}_t(g)}] = (\delta_\emptyset | e^{i\widehat{w}_t(g)} \delta_\emptyset)$$

coincides with the classical Gaussian characteristic function

$$\int \exp\{i \int_0^t g(r) \mathrm{d}w_r\} \mathbf{P}(\mathrm{d}\omega) = \exp\left\{ -\frac{1}{2} \int_0^t |g(r)|^2 \mathrm{d}\, r \right\}$$

of the standard Wiener process w_t. However, the different Brownian motions $\widehat{w}_t(f)$ and $\widehat{w}_t(g)$ with $f^*g \neq g^*f$ do not have any classical representation on a single probability space $(\Omega, \mathcal{F}, \mathbf{P})$ because of noncommutativity (see A3 in Appendix 3):

$$[\widehat{w}_t(f), \widehat{w}_t(g)] = \int_0^t \Big(\bar{f}(r) g(r) - \bar{g}(r) f(r) \Big) \mathrm{d}\, r \ .$$

Definition 1. Let $\{\mathcal{A}_t | t \in \mathbf{R}_+\}$ be an increasing set of involutive subalgebras $\mathcal{A}_t \subseteq \mathcal{A}_s$, $t \leq s$ of the operators $X_t \in \mathcal{B}(\mathcal{D})$ generated by operators $X \in \mathcal{A}$, $\widehat{w}_t(g)$, $g \in \mathcal{K}$, such that

$$X_t \in \mathcal{A}_t \Leftrightarrow [X_t, Y] = 0 \ , \qquad \forall Y \in \mathcal{B}(\mathcal{D}) : [X_0, Y] = 0 = \big[W_t(g), Y \big] \ ,$$

where the operators $X_0 \in \mathcal{A}_0$, $W_t(g)$, $g \in L^2(0,t]$ are assumed to act in $\mathcal{H} = \mathrm{H} \times \mathcal{F}$ as $X_0 \otimes \hat{1}$ and $I \otimes \widehat{w}_t(g)$. A measurable operator function $F(t) : \mathcal{D} \to \mathcal{D}$ is called adapted if $F(t) \in \mathcal{A}_t$ for almost all $t \in \mathbf{R}_+$.

We shall consider here only quantum stochastic integrals of the form

$$\imath_0^t(F, D) = \int_0^t \left(F(r)\, \mathrm{d}A_r + D(r)\, \mathrm{d}A_r^* \right) \ , \tag{1.1}$$

where $A_r = I \otimes \widehat{a}_r$, $A_r^* = I \otimes \widehat{a}_r^*$ and F, D are locally square–integrable (see Appendix 4) together with adjoint F^*, D^* adapted operator–functions $\mathbf{R}_+ \to \mathcal{A}_t$. Note that on the exponential vectors, described by the product–functions $h(\tau) = k^{\otimes}(\tau)\, \psi$, where $\psi \in \mathrm{D}$ and $k \in L^2(\mathbf{R}_+)$, the itegrals (1.1) are weakly defined as the usual operator integrals

$$(h \mid \imath_0^t(F, D)\, h) = \int_0^t \Big(h \mid [F(r)\, k(r) + D(r)\, \bar{k}(r)]\, h \Big)\, \mathrm{d}\, r \ .$$

This gives in particular, $\mathbf{E}[\imath_0^t(F, D)] = 0$ for $\mathbf{E}[X] = (e|Xe)$, where $e(\tau) = \delta_\emptyset(\tau)\, \psi$ is the exponential vector, corresponding to $k = 0$. Such operator integrals on the exponential domain were constructed by Hudson and Parthasarathy [12] for the case of bounded $F(t), D(t)$.

For the adapted integrals (1.1), the quantum Ito formula [10]–[12] can be obtained as in the case of (A.5) (see Appendix 5), corresponding to $X_t(t) = 0$ for almost all t. This formula defines (see Appendix 5) the pointwise multiplication

$$X(t)\,Y(t) = X(0)\,Y(0) + \int_0^t \mathrm{d}(XY)\,(r)$$

of the adapted operator–functions

$$X(t) = X(0) + \int_0^t \mathrm{d}X\,(r)\ , \qquad Y(t) = Y(0) + \int_0^t \mathrm{d}Y\,(r)$$

in terms of the product $\mathrm{d}X(t)\,\mathrm{d}Y(t) = D(t)^* F(t)\,\mathrm{d}\,t$ of their stochastic differentials $\mathrm{d}X = D^*\mathrm{d}A + D\,\mathrm{d}A^*$, $\mathrm{d}Y = F^*\mathrm{d}A + F\,\mathrm{d}A^*$:

$$\begin{aligned}\mathrm{d}(XY) = \mathrm{d}XY + X\mathrm{d}Y + \mathrm{d}X\mathrm{d}Y = D^*F\,\mathrm{d}\,t + (D^*Y + XF^*)\,\mathrm{d}A \\ +(DY + XF)\,\mathrm{d}A^*\ . \end{aligned} \tag{1.2}$$

The classical Ito formula for the stochastic integrals $I_0^t(f,\omega) = \int_0^t f(r,\omega)\,\mathrm{d}w_r$ with respect to the standard Wiener process $w = \{w_t | t \in \mathbf{R}_+\}$ can be obtained from (1.2) by the Segal one–to–one transformation $\omega : \widehat{w}_t \mapsto w_t$, where $\widehat{w}_t = \widehat{a}_t + \widehat{a}_t^*$. The latter represents the adapted operator integrals $\imath_0^t(\widehat{f},\widehat{f}) = \int_0^t \widehat{f}(r)\,\mathrm{d}\,\widehat{w}_r$ for the non–anticipated functionals $\widehat{f}(t) = f(t,\widehat{w})$ of commuting selfadjoint operators $\widehat{w} = \{\widehat{w}_t\}$ with $\|\widehat{f}\delta_\emptyset\|_t^2 := \int_0^t \|\widehat{f}(r)\,\delta_\emptyset\|^2 < \infty$ in the form of the Ito integrals $\omega[\imath_0^t(\widehat{f},\widehat{f})] = I_0^t(f,\omega)$, so that

$$\left\|\imath_0^t(\widehat{f},\widehat{f})\,\delta_\emptyset\right\|^2 = \|\widehat{f}\,\delta_\emptyset\|_t^2 = \int \left\|I_0^t(f,\omega)\right\|^2 \mathbf{P}(\mathrm{d}\,\omega)\ ,$$

where $\mathbf{P}$ is the standard Wiener probability measure.

1.2 Quantum Diffusion

Quantum stochastic evolution in the open system $\{\mathcal{A}_t\}$ is described by an adapted family $\{\iota(t) | t \in \mathbf{R}_+\}$ of $*$–representations $\iota(t) : X \mapsto X\,(t)$ of the initial algebra $\mathcal{A} \subseteq \mathcal{B}(\mathrm{D})$ into $\mathcal{B}(\mathcal{D})$, i.e. of linear maps $\mathcal{A} \to \mathcal{A}_t$, with the properties:

$$\iota(t, X^*X) = \iota(t,X)^*\iota(t,X)\ , \qquad \iota(t,I) = I_0 := I \otimes \hat{1}\ .$$

It is called diffusion motion if the operator–valued functions $t \mapsto X(t)$ have the quantum stochastic differentials of the form

$$\mathrm{d}X(t) + C(t)\,\mathrm{d}\,t = D^*(t)\,\mathrm{d}A_t + D(t)\,\mathrm{d}A_t^*\ . \tag{1.3}$$

Here $C(t) = \gamma(t,X)$ is an adapted operator–valued function, locally integrable ($p = 1$) for every $X \in \mathcal{A}$ defined by the linear maps $\gamma(t) : \mathcal{A} \to \mathcal{A}_t$, $t \in \mathbf{R}_+$. $D^*(t) = \delta^*(t,X)$, $D(t) = \delta(t,X)$ are adapted operator–valued functions, locally square–integrable for each $X \in \mathcal{A}$, defined by the linear maps $\delta^*(t)$, $\delta(t) : \mathcal{A} \to \mathcal{A}_t$.

Define the output process $Y = \{Y(t) | t \in \mathbf{R}_+\}$, which is subject to measurement and described by a commutative family of (essentially) selfadjoint operators $Y(t) = Y(t)^*$ on $\mathcal{D}$ with the initial condition $Y(0) = 0$ and stochastic differentials

$$\mathrm{d}Y(t) = G(t)\,\mathrm{d}t + F^*(t)\,\mathrm{d}A_t + F(t)\,\mathrm{d}A_t^* \ . \tag{1.4}$$

Here $G(t) = G(t)^* \in \mathcal{A}_t$ is essentially self–adjoint and locally–integrable ($p = 1$). $F(t) \in \mathcal{A}_t$ is locally square–integrable together with its conjugate: $F^*(t) = F(t)^*$; $G(t)$ and $F(t)$ are adapted operator–valued functions of $t \in \mathbf{R}_+$.

Unlike the classical case, not every involutive subalgebra $\mathcal{B}$ of $\mathcal{A}$, but only a central one $\mathcal{B} \subset \mathcal{A} \cap \mathcal{A}'$, defines the conditional expectations $\mathbf{E}[X|\mathcal{B}]$ for any state vector $e \in \mathrm{H}$ as the positive projections $\mathcal{A} \to \mathcal{B}$ which are compatible with $\mathbf{E}[X] = (e|Xe)$ such that $\mathbf{E}[\mathbf{E}[X|\mathcal{B}]] = \mathbf{E}[X]$ for all $X \in \mathcal{A}$. Hence, not every stochastic process described by the equation (1.4), can be considered as an output process for quantum diffusion, defined by equation (1.3), but only that for which the posterior expectations of $X(t)$ with respect to the observation $Y(s)$, $s \leq t$, exist.

Definition 2. A process $Y(t)$ is called *causal, or nondemolition* with respect to the process $X(t)$ if

$$[X(t), Y(s)] := X(t)Y(s) - Y(s)X(t) = 0 \tag{1.5}$$

for all $t \geq s$, $s \in \mathbf{R}_+$.

The nondemolition condition together with the self–nondemolition of Y, i.e. with the commutativity $[Y(t), Y(s)] = 0$, $\forall t, s$, is necessary and sufficient [11] for the existence of the conditional expectations $\widehat{\pi}(t, X) = \mathbf{E}[X(t)|\mathcal{B}_t]$ for the operators $X(t) = \iota(t, X)$, with respect to the $*$–algebras

$$\mathcal{B}_t = \{Y \in \mathcal{B}(\mathcal{D}) \mid [X, Y] = 0 \ , \quad \forall X \in \mathcal{B}(\mathcal{D}) : [X, Y(s)] = 0, \ \forall s \leq t\}$$

generated by the family $\{Y(s) | s \leq t\}$ and for every initial vector–function $e \in \mathcal{D}$, $\|e\| = 1$.

Proposition 1. The integrals

$$X(t) = X - \int_0^t (C(r)\,\mathrm{d}r - D^*(r)\,\mathrm{d}A_r - D(r)\,\mathrm{d}A_r^*) \ , \qquad X \in \mathcal{A}$$

define the $*$–representations $\iota(t) : X \mapsto X(t)$ iff the linear maps

$$\gamma(t) : X \longmapsto C(t) \ , \qquad \delta^*(t) : X \longmapsto D^*(t) \ , \qquad \delta(t) : X \longmapsto D(t)$$

satisfy the following differential conditions

$$\begin{aligned}
&\text{(i)} \ \ \gamma(t, X^*) = \gamma(t, X)^* \ , \qquad \delta(t, X^*) = \delta^*(t, X)^* \ , \qquad \forall X \in \mathcal{A} \ , \\
&\text{(ii)} \ \gamma(t, X^*X) = \iota(t, X)^*\gamma(t, X) + \gamma(t, X)^*\iota(t, X) - \delta(t, X)^*\delta(t, X) \ , \\
&\qquad \delta(t, X^*X) = \iota(t, X)^*\delta(t, X) + \delta(t, X^*)\,\iota(t, X) = \delta^*(t, X^*X)^* \ , \\
&\text{(iii)} \ \gamma(t, I) = 0 \ , \qquad \delta(t, I) = 0 = \delta^*(t, I) \ , \quad \forall t \in \mathbf{R}_+ \ .
\end{aligned}$$

The process $X(t)$ satisfies the condition (1.5) iff the stochastic derivations $C(t)$, $D^*(t)$, $D(t)$ also satisfy the condition (1.5) as $X(t)$ with respect to the nondemolition process $Y(t)$ for all $X \in \mathcal{A}$, and the derivatives G, F^*, F in (1.4) satisfy the differential nondemolition conditions

$$[X(t), F^*(t)] = 0 = [F(t), X(t)] \ , \qquad \forall t \in \mathbf{R}_+ \ ,$$
$$D^*(t)\,F(t) - F^*(t)\,D(t) = [G(t), X(t)] \ . \tag{1.6}$$

PROOF. The stochastic differentials $\mathrm{d}X(t) = X(t+\mathrm{d}t) - X(t)$ of the linear $*$–maps $\iota(t) : X \mapsto X(t)$ are defined by the linear $*$–maps $\gamma(t)$, $\delta^*(t)$, $\delta(t)$ by virtue of linear independence of the fundamental differentials $\mathrm{d}t$, $\mathrm{d}A_t$ and $\mathrm{d}A_t^*$. The conditions (ii) are found by applying the Ito formula (1.2) to $X(t)^* X(t)$:

$$\begin{aligned}\mathrm{d}\left(X(t)^* X(t)\right) &= \mathrm{d}X(t)^* \mathrm{d}X(t) + \mathrm{d}X(t)^* X(t) + X(t)^* \mathrm{d}X(t)\\ &= \left[\delta(t,X)^*\delta(t,X) - \gamma(t,X)^*\iota(t,X) - \iota(t,X)^*\gamma(t,X)\right]\mathrm{d}t\\ &\quad + \mathrm{d}\,\imath_0^t\left(\delta(X)^*\iota(X) + \iota(X)^*\delta^*(X),\ \delta(X^*)\,\iota(X) + \iota(X)^*\delta(X)\right).\end{aligned}$$

By equating the stochastic derivatives of this differential and

$$\mathrm{d}\,\iota(t, X^*X) = \mathrm{d}\,\imath_0^t\left(\delta^*(X^*X), \delta(X^*X)\right) - \gamma(X^*X)\,\mathrm{d}t \ ,$$

we obtain that δ^* and δ are the derivations of the algebra $\mathcal{A}$, and $-\gamma$ has the positive–definite dissipator

$$\iota(X)^*\gamma(X) + \gamma(X)^*\iota(X) - \gamma(X^*X) = \delta(X)^*\delta(X) \ .$$

The condition (iii) follows from $\mathrm{d}\,\iota(t, I) = 0$ because of the independence of γ, δ^*, δ.

If $Y(t)$ is a nondemolition process for $X(t)$, then

$$[\mathrm{d}X(t), Y(s)] = [X(t+\mathrm{d}t), Y(s)] - [X(t), Y(s)] = 0$$

with $t \geq s$; hence the nondemolition for C, D^*, D:

$$[C(t), Y(s)] = 0 \ , \qquad [D^*(t), Y(s)] = 0 \ , \qquad [D(t), Y(s)] = 0 \ , \quad \forall t \geq s \ ,$$

follows by commutativity of $Y(s)$ with the independent differentials $\mathrm{d}t$, $\mathrm{d}A_t$, $\mathrm{d}A_t^*$. Applying equation (1.2) to the differential of the commutator $[X(t),\ Y(t)] = 0$ we obtain (taking into account the equality $[\mathrm{d}X(t),\ Y(t)] = 0$):

$$\begin{aligned}\mathrm{d}\,[X(t), Y(t)] &= [\mathrm{d}X(t), \mathrm{d}Y(t)] + [\mathrm{d}X(t), \mathrm{d}Y(t)] + [X(t), \mathrm{d}Y(t)]\\ &= \Big(D^*(t)\,F(t) - F^*(t)\,D(t) + [X(t), G(t)]\Big)\,\mathrm{d}t\\ &\quad + \mathrm{d}\,\imath_0^t([X, F^*], [X, F]) = 0 \ ,\end{aligned}$$

which yields the differential self–nondemolition conditions (1.6).

1.3 The Markov Case

The quantum diffusion (1.3) with coefficients

$$C(t) = \iota(t, C_t) \ , \qquad D^*(t) = \iota(t, D_t^*) \ , \qquad D(t) = \iota(t, D_t) \ ,$$

corresponds to the Markov stochastic evolution (in strong sense). Here $C_t, D_t^*, D_t \in \mathcal{A}$ are defined by the structural maps

$$\gamma_t : X \mapsto C_t \ , \qquad \delta_t^* : X \mapsto D_t^* \ , \qquad \delta_t : X \mapsto D_t \ ,$$

for which the conditions (i)–(iii) indicated above were obtained by Hudson and Evans in [14]. The self–nondemolition conditions (1.6) give the restrictions for the coefficients G and F^*, F in this case. We shall restrict ourselves to consideration of the standard case $F(t) = I_0 = F^*(t)$ of the indirect measurement

$$Y(t) = \int_0^t \iota(t, G_r)\, \mathrm{d}\, r + I \otimes \widehat{w}_t \tag{1.7}$$

of the diffusion of a square–integrable initial process defined locally by $G_t \in \mathcal{A}$ over the standard Wiener process w_t represented in $\mathcal{H}$ by the operators $I \otimes \widehat{w}_t = A_t + A_t^*$. It is not hard to prove [5] that $Y(r) = V_t(I \otimes \widehat{w}_r) V_t^*$, $\forall t > r \in \mathbf{R}_+$ by the uniqueness of the stochastic operator equation

$$\mathrm{d}V_t + \frac{1}{8} G(t)\, V_t \mathrm{d}\, t = i\, G(t)\, V_t \mathrm{d}\, \widehat{u}_t, \qquad V_0 = I \ ,$$

where

$$G(t) = \iota(t, G_t) \ , \qquad \widehat{u}_t = \frac{i}{2}\left(\widehat{a}_t^* - \widehat{a}_t\right) = \frac{1}{2}\, \widehat{w}_t(i) \ .$$

The above implies the local unitary equivalence of the processes $A_t + A_t^*$ and $Y(t)$, which is always the case for the locally norm–square–integrable operator–functions $G_t : \mathrm{H} \to \mathrm{H}$ [9].

Corollary 1. In the case under consideration the condition (1.6) completely defines the structure of the inner derivation $X \mapsto [X, G_t]$ for $\delta_t - \delta_t^*$:

$$\delta_t(X) = \frac{1}{2}\, [X, G_t] - \alpha_t(X) \ , \tag{1.8}$$

where $\alpha_t(X^*) = \alpha_t(X)^*$, $\forall X \in \mathcal{A}$ is a $*$–derivation of the algebra $\mathcal{A}$. The conditions (i)–(iii) here also define the structure of the maps $\gamma_t : \mathcal{A} \to \mathcal{A}$ in the form

$$\begin{aligned} \gamma_t(X) &= \frac{1}{2} \lambda_t(X) - \beta_t(X) \ , \\ \lambda_t(X) &= \frac{1}{4}\left[G_t, [G_t, X]\right] + G_t \alpha_t(X) + \alpha_t(X)\, G_t - \alpha_t^2(X) \ , \end{aligned} \tag{1.9}$$

where $\alpha_t^2(X) = \alpha_t(\alpha_t(X))$, $\beta_t : \mathcal{A} \to \mathcal{A}$ is some $*$-derivation

$$\beta_t(X^* X) = X^* \beta_t(X) + \beta_t(X^*)\, X \ , \qquad \forall X \in \mathcal{A} \ .$$

PROOF. By taking into account the differentiation property

$$[G_t, X^*X] = X^*[G_t, X] + [G_t, X^*]X \ ,$$

and similarly for $\alpha_t(X^*X)$ we obtain

$$\begin{aligned}\lambda_t(X^*X) &= \frac{1}{4}[G_t, X^*[G_t, X] + [G_t, X^*]X] \\ &\quad + G_t\left(X^*\alpha_t(X) + \alpha_t(X^*)X\right) + \left(X^*\alpha_t(X) + \alpha_t(X^*)X\right)G_t \\ &\quad - \left(X^*\alpha_t^2(X) + 2\alpha_t(X)^*\alpha_t(X) + \alpha_t^2(X)^*X\right) \\ &= X^*\lambda_t(X) + \lambda_t(X)^*X - \frac{1}{2}[X, G_t]^*[X, G_t] \\ &\quad + [X, G_t]^*\alpha_t(X) + \alpha_t(X)^*[X, G_t] - 2\alpha_t(X)^*\alpha_t(X) \\ &= X^*\lambda_t(X) + \lambda_t(X)^*X - 2\delta_t(X)^*\delta_t(X) \ .\end{aligned}$$

Hence $\gamma_t^0 = \frac{1}{2}\lambda_t$ possesses the property (ii) of the map γ_t:

$$\gamma_t^0(X^*X) = X^*\gamma_t^0(X) + \gamma_t^0(X)^*X - \delta_t(X)^*\delta_t(X) \ ,$$

and the $*$-property $\gamma_t^0(X^*) = \gamma_t^0(X)^*$, as does γ_t in (iii). From here the result that $\beta_t = \gamma_t^0 - \gamma_t$ is a $*$-derivation follows.

The maps γ_t are called generators for the Lindblad equation

$$\mathrm{d}\mu_0^t/\mathrm{d}t + \mu_0^t \circ \gamma_t = 0 \ ,$$

where $\mu\circ\gamma(X) = \mu(\gamma(X))$, which is an algebraic analog of the Kolmogorov equation. This is satisfied by the operators $X_0^t = \mu_0^t(X)$, $X \in \mathcal{A}$ of the conditional expectation $\mu_0^t(X) = (\delta_\emptyset | X(t)\,\delta_\emptyset)$, with respect to the vacuum function $\delta_\emptyset \in \mathcal{F}$ defined by

$$X_0^t\psi = \left[\iota(t, X)\,h\right](\emptyset) \ , \qquad \forall\, h = \psi\otimes\delta_\emptyset, \quad \psi \in \mathrm{D} \ .$$

The differential conditions obtained for Markov diffusion are necessary for the existence of a unique solution $X(t)$ of equation (1.3) for all $X(0) = X \in \mathcal{A}$ and so, for the Lindblad equation. They are sufficient in the case [14] of constant α_t, β_t and G_t and the boundedness of the algebra $\mathcal{A}$ (for instance, with $\mathrm{D} = \mathrm{H}$). Moreover, as was proved in [10], the maps $\iota(t) : X \mapsto X(t)$ are representations of $\mathcal{A}$ in $\mathcal{A}_t$, satisfying the condition of nondemolition (1.5) for all $X \in \mathcal{A}$. This is also true under significantly more general conditions of local p–integrability over the norm of the operators $G_t : \mathrm{H} \to \mathrm{H}$ (with $p = 2$) and the maps α_t ($p = 2$) and β_t ($p = 1$) from $\mathcal{A} \subseteq \mathcal{B}(\mathrm{H})$ to $\mathcal{A}$.

In the case of inner derivations

$$\alpha_t(X) = i[S_t, X] \ , \qquad \beta_t(X) = i\Big[H_t + \frac{1}{4}(S_tG_t + G_tS_t), X\Big] \ ,$$

(as is the case for the von–Neumann algebra $\mathcal{A}$), a quantum Markov diffusion is defined by structure maps of the type

$$\delta_t(X) = [X, L_t] \ , \quad \delta_t^*(X) = [L_t^*, X] \ ,$$

$$\gamma_t(X) = \frac{1}{2}\left(L_t^*[L_t, X] + [X, L_t^*]\,L_t\right) + i[X, H_t] \ , \tag{1.10}$$

where

$$L_t = \frac{1}{2}G_t + iS_t \ , \qquad L_t^* = \frac{1}{2}G_t - iS_t \ , \qquad H_t = H_t^* \in \mathcal{A}_t \ .$$

2 Quantum Diffusion and Filtering

2.1 The a Posteriori Dynamics

The quantum diffusion (1.3) under nondemolition measurement (1.4), is described by the classical random variables $x_t(\omega) = \langle X(t)\rangle_t(\omega)$ of the conditional expectations $\langle X(t)\rangle_t = \mathbf{E}[X(t)|\mathcal{B}_t]$, for the operators $X(t) = \iota(t, X)$, on the trajectories $\omega \in \Omega$ of the process $Y(t)$, with respect to initial vector-valued function $e_0 = \psi_0 \otimes \delta_\emptyset$, $\psi_0 \in \mathrm{D}$. As was established for the first time in [5], [6], the random process $x_t : \Omega \to \mathbf{C}$, considered as a stochastic map $X(t) \mapsto x_t(\omega)$, satisfies the Ito filtering equation for quantum diffusion

$$\mathrm{d}\left\langle\iota(t,X)\right\rangle_t + \left\langle\gamma(t,X)\right\rangle_t \mathrm{d}t = \left\langle\kappa(t,X) - \left\langle G(t)\right\rangle_t \iota(t,X)\right\rangle_t \mathrm{d}\widetilde{Y} \qquad (2.1)$$

with respect to the stochastic map $\langle\cdot\rangle_t : X(t) \mapsto x_t$. In this equation $\widetilde{Y}(t) = Y(t) - \int_0^t \langle G(r)\rangle_r \mathrm{d}r$ is an innovation martingale for the observed process (1.4), and $\kappa(t) : \mathcal{A} \to \mathcal{B}(\mathrm{D})$ is a linear $*$-map, which in the case of $F(t) = I_0 = F^*(t)$ is of the particularly simple form:

$$\kappa(t,X) = \frac{1}{2}\left(G(t)\,X(t) + X(t)\,G(t)\right) - \alpha(t,X) \ , \qquad (2.2)$$

where $\alpha(t,X)$ is defined by the $*$-derivation $\delta(t)+\delta^*(t) = -2\alpha(t)$. Equation (2.1), derived in [5] by means of the martingale methods of quantum nonlinear filtering, extends the basic equation (8.10), [18] of the optimal diffusion filtering to the case of the noncommutative operator algebras $\mathcal{A}$. By complete analogy with the classical case the quantum Ito formula was used with the innovation process and the representation theorem [11], which requires the conditional expectations $x_t(\omega)$ to exist. This requirement is met by the self-nondemolition condition (1.5) for all $s \leq t$ which is trivially satisfied in the commutative case $s > t$.

In the quantum Markov case of the indirect measurement (1.7) the conditional expectation $x_t(\omega) = \pi_t(X,\omega)$ of the operators $X(t)$ can be found as in the classical case by solving an autonomous stochastic equation for the a posteriori state $\widehat{\pi}_t(X) = \langle\iota(t,X)\rangle_t$. The latter is defined on the trajectories ω as a linear stochastic positive normalized map $\omega(\widehat{\pi}_t(X)) = x_t(\omega)$ of the algebra $\mathcal{A}$ into $\mathbf{C}$ satisfying the condition

$$\int x_t(\omega)\,y(\omega)\,\mathbf{P}_0^t(\mathrm{d}\omega) = \left(e_0 \mid X(t)\,Y e_0\right) \ , \qquad \forall X \in \mathcal{A} \ .$$

In this equation, $Y \in \mathcal{B}_t$ is any bounded operator in the algebra of the observed $\mathcal{B}_t$, $y(\omega) = \omega(\widehat{y})$ is the Segal transformation of the operator $\widehat{y}$ in $\mathcal{F}$ that corresponds to the unitary-equivalent operator $I \otimes \widehat{y} = V_t^* Y V_t$, and $\mathbf{P}_0^t$ is an induced (by the unitary transformation) probability measure on the trajectories

$\omega|[0,t) := \{w_r | r \in [0,t)\}$, restricted to the interval $[0,t)$ with respect to the initial vector-state

$$\int y(\omega)\, \mathbf{P}_0^t(\mathrm{d}\,\omega) = \left(V_t^*(\psi_0 \otimes \delta_\emptyset)\right) \mid (I \otimes \widehat{y})\, V_t^*(\psi_0 \otimes \delta_\emptyset)) \ .$$

2.2 The Filtering Equation

Let us sketch the essentials in the derivation of a stochastic Markov quantum filtering equation, obtained for the general output process in [9]. First, we shall prove that the vacuum conditional expectation $\mu_g^t(X) = (\delta_\emptyset | \pi_g(t,X)\, \delta_\emptyset)$ of the product $\pi_g(t,X) = \iota(t,X) e_g(t)$, where

$$e_g(t) = \exp\left\{ \int_0^t g(r)\, \mathrm{d}Y(r) - \frac{1}{2} g(r)^2 \mathrm{d}\, r \right\} , \tag{2.3}$$

satisfies the linear evolution equation

$$\frac{\mathrm{d}}{\mathrm{d}\, t} \mu_g^t(X) + \mu_g^t \circ \gamma_t(X) = \mu_g^t\left(G_t \cdot X - \alpha_t(X)\right) g(t) \ , \tag{2.4}$$

where $\mu_g^0(X) = X$, and $G \cdot X = (GX + XG)/2$. Let us assume the uniqueness of this solution, which is always true for locally p-integrable bounded (over the norm) maps G_t, α_t $(p=2)$ and β_t $(p=1)$. We shall prove that $\mu_g^t(X)$ is the mathematical expectation of the product $\widehat{\mu}_g^t(X) = \widehat{\mu}^t(X)\, \hat{e}_g^t$, of the stochastic operators $\mu^t(X,\omega) = \omega[\widehat{\mu}^t(X)]$ (which satisfy a quantum filtering equation) with the exponentials

$$e_g^t(\omega) = \exp \int_0^t \left[g(r)\, \mathrm{d}w_r - \frac{1}{2} g(r)^2 \right] \mathrm{d}\, r = \omega(\hat{e}_g^t) \ ,$$

(which are defined with respect to the trajectories $\omega : t \mapsto w_t$ of the standard Wiener process w_t, $t \in \mathbf{R}_+$). The above means that the output process $Y(r)$, restricted by any $t \in \mathbf{R}_+$, is absolutely continuous with respect to the standard restricted process $w^t = \{w_r | r < t\}$. This follows from the unitary equivalence $Y(r) = V_t Y_r V_t^*, \forall r < t$ and $Y_r = I \otimes \widehat{w}_r$, $\forall t \in \mathbf{R}_+$, representing the output up to a time t with respect to the initial vector-function $e_0 = \psi_0 \otimes \delta_\emptyset$ and $e_0^t = V_t^* e_0$ correspondingly. The probability density $\rho_0^t(\omega) = \mathbf{P}_0^t(\mathrm{d}\,\omega)/\mathbf{P}(\mathrm{d}\,\omega)$, for the measurement of the trajectory $\{w_r | r < t\}$, of the process Y on the interval $[0,t)$, is defined with respect to the standard Wiener probability measure $\mathbf{P}(\mathrm{d}\,\omega)$ by the formula $\rho_0^t(\omega) = \varphi_0^t(I,\omega)$. Here $\varphi_0^t(\omega) = \varphi_0 \circ \mu^t(\omega)$ is the stochastic functional $\varphi_0^t(X,\omega) = (\psi_0 | \mu^t(X,\omega)\, \psi_0)$, which corresponds to the initial state $\varphi_0(X) = (\psi_0 | X \psi_0)$ on the algebra $\mathcal{A}$. Finally, we deduce a nonlinear equation for the a posteriori state $\pi_t(\omega)$ using the ordinary Ito formula and the normalization of the stochastic functional $\varphi_0^t(\omega)$.

Theorem 1. *Let the equation (2.4)has the unique solution $X_g^t = \mu_g^t(X)$, corresponding to the initial condition $\mu_g^0(X) = X$ for each $X \in \mathcal{A}$ and $g \in \mathcal{K}$. Then it coincides with the vacuum expectation $X_g^t = (\delta_\emptyset|\pi_g(t,X)\,\delta_\emptyset)$, where $\pi_g(t,X) = \iota(t,X)e_g(t)$ and is defined by the Wiener average*

$$X_g^t = \int_0^t X^t(\omega)\exp\left\{\int_0^t g(r)\,\mathrm{d}w_r - \frac{1}{2}g(r)^2\mathrm{d}\,r\right\}\mathbf{P}(\mathrm{d}\,\omega) \qquad (2.5)$$

over the continuous trajectories $\omega \in \Omega$. Here $X^t(\omega) = \omega[\widehat{\mu}^t]$ is the Segal transformation of the solution $\widehat{X}^t = \widehat{\mu}^t(X)$ to the operator filtering equation

$$\mathrm{d}\,\widehat{\mu}^t(X) + \widehat{\mu}^t \circ \gamma_t(X)\,\mathrm{d}\,t = \widehat{\mu}^t\left(G_t \cdot X - \alpha_t(X)\right)\mathrm{d}\,\widehat{w}_t \qquad (2.6)$$

with initial condition $\widehat{\mu}^0(X) = X$. In this case the linear stochastic equation (2.6) also has a unique solution in the Ito sense, which defines almost everywhere ($\rho_0^t(\omega) \neq 0$) for each $\psi_0 \in \mathrm{H}$ the a posteriori state

$$\pi_t(X,\omega) = \frac{(\psi_0|X^t(\omega)\,\psi_0)}{\rho_0^t(\omega)} \ ,$$

where the probability density $\rho_0^t(\omega) = (\psi_0|I^t(\omega)\,\psi_0)$ is given by the positive operator $I^t(\omega) = \mu^t(I,\omega)$, satisfying the martingale property $\int I^t(\omega)\mathbf{P}(\mathrm{d}\omega|w^r) = I^r(\omega), \forall r < t$.

PROOF. First we find a quantum stochastic equation for $X_g(t) = \pi_g(t,X)$ using the Ito formula

$$\mathrm{d}e_g(t) = g(t)\,e_g(t)\,\mathrm{d}Y(t)\ , \qquad e_g(0) = 1\ ,$$

where

$$\mathrm{d}Y(t) = G(t)\,\mathrm{d}\,t + F(t)^*\mathrm{d}A_t + \mathrm{d}A_t^*F(t)\ .$$

We obtain according to (1.2)

$$\begin{aligned}
\mathrm{d}\left(X(t)\,e_g(t)\right) &= \mathrm{d}X(t)\,\mathrm{d}\ e_g(t) + \mathrm{d}X(t)\,e_g(t) + X(t)\,\mathrm{d}\ e_g(t) \\
&= (g\,D^*F - C + gXG)(t)\,e_g(t)\,\mathrm{d}\,t \\
&\quad + \mathrm{d}\,\imath_0^t(D^* + X\,F^*gD + X\,Fg)\,e_g(t) \\
&= \pi_g\Big(t, (G_tX + X\,G_t)\,\frac{g(t)}{2} - \alpha_t(X)\,g(t) - \gamma_t(X)\Big)\,\mathrm{d}\,t \\
&\quad + \pi_g\left(t, \delta_t^*(X) + g(t)\,X\right)\mathrm{d}A_t + \pi_g\left(t, \delta_t(X) + g(t)\,X\right)\mathrm{d}A_t^*,
\end{aligned}$$

where the explicit form (1.8) has been used for

$$D^* = \delta^*(X)\ , \qquad D = \delta(X)\ , \qquad C = \gamma(X)\ , \qquad GX + D = G\cdot X - \alpha(X)$$

with $F = I_0 = F^*$.

Taking into account the martingale property of the quantum stochastic integral (1.1) with respect to the vacuum-vector $\delta_\emptyset \in \mathcal{F}$, we find the equation (2.4) for the operator

$$\mu_g^t(X) : \psi \longmapsto \mu_g^t(X)\,\psi = [\pi_g(t,X)h](\emptyset)$$

in H is implied by the action of $\pi_g(t,X) = X(t)\, e_g(t)$ on $h = \psi \otimes \delta_\emptyset$:

$$\mathrm{d}\,\mu_g^t(X) = \mu_g^t\big((G_t{\cdot}X - \alpha_t(X)\, g(t) - \gamma_t(X)\big)\mathrm{d}t \ , \qquad \forall X \in \mathcal{A} \ ,$$

where $\gamma_t : \mathcal{A} \to \mathcal{A}$ is defined in the form (1.10).

Now, if $X^t(\omega) = \mu^t(X,\omega)$ satisfies the stochastic equation

$$\mathrm{d}X^t(\omega) + C^t(\omega)\,\mathrm{d}\,t = D^t(\omega)\,\mathrm{d}\,\widehat{w}_t \ , \qquad X^0(\omega) = X \ ,$$

we can derive a differential for

$$\widehat{\mu}_g^t(X) = \widehat{\mu}^t(X)\,\widehat{e}_g^t \ , \qquad \widehat{e}_g^t = \exp\left\{\int_0^t \left(g(r)\,\mathrm{d}\,\widehat{w}_r - \frac{1}{2}\,g(r)^2\mathrm{d}\,r\right)\right\} \ ,$$

by means of the Ito formula of classical stochastic calculus.

$$\mathrm{d}X^t(\omega) + C^t(\omega)\,\mathrm{d}\,t = D^t(\omega)\,\mathrm{d}\,\widehat{w}_t \ , \qquad X^0(\omega) = X \ .$$

Using $\mathrm{d}\,\widehat{e}_g^t = g(t)\widehat{e}_g^t\,\mathrm{d}\,\widehat{w}_t$ we have

$$\begin{aligned}\mathrm{d}(\widehat{X}^t\widehat{e}_g^t) &= \mathrm{d}\,\widehat{X}^t\mathrm{d}\,\widehat{e}_g^t + \mathrm{d}\,\widehat{X}^t\widehat{e}_g^t + \widehat{X}^t\mathrm{d}\,\widehat{e}_g^t \\ &= \big(g(t)\,\widehat{D}^t - \widehat{C}^t\big)\,\widehat{e}_g^t\mathrm{d}\,t + \big(\widehat{D}^t + \widehat{X}^t g(t)\big)\,\widehat{e}_g^t\mathrm{d}\,\widehat{w}_t \ ,\end{aligned}$$

what can be written in the form of the stochastic equation

$$\mathrm{d}X_g^t\,(\omega) + (C_g^t\,(\omega) - D_g^t\,(\omega)\,g\,(t))\mathrm{d}t = (D_g^t\,(\omega) + X_g^t\,(\omega)\,g\,(t))\mathrm{d}w_t$$

for $X_g^t\,(\omega) = \omega\left(\widehat{X}^t\widehat{e}_g^t\right)$. Hence, the mathematical expectation (2.5) of $X_g^t(\omega) = X^t(\omega)e_g^t(\omega)$ with respect to the Gaussian measure $\mathbf{P}$ of the standard Wiener process w satisfies the equation

$$\mathrm{d}\,\mu_g^t(X) = \big(D_g^t g(t) - C_g^t\big)\,\mathrm{d}\,t \ , \qquad \mu_g^0(X) = X \ .$$

Comparison of this equation with equation (2.4) gives the coefficients

$$C_g^t = \int C^t(\omega)\,e_g^t(\omega)\,\mathbf{P}(\mathrm{d}\,\omega) \ , \qquad D_g^t = \int D^t(\omega)\,e_g^t(\omega)\,\mathbf{P}(\mathrm{d}\,\omega)$$

in the form:

$$C_g^t = \mu_g^t\big(\gamma_t(X)\big) \ , \qquad D_g^t = \mu_g^t\big(G_t{\cdot}X - \alpha_t(X)\big) \ .$$

Consequently $C^t(\omega) = \omega(\widehat{C}^t)$, $D^t(\omega) = \omega(\widehat{D}^t)$ are the coefficients

$$\widehat{C}^t = \widehat{\mu}^t\big(\gamma_t(X)\big) \ , \qquad \widehat{D}^t = \widehat{\mu}^t\big(G_t{\cdot}X - \alpha_t(X)\big) \ ,$$

that define an equation for $X^t(\omega) = \mu^t(X,\omega)$ in the form of (2.6). The solution $I^t(\omega) = \mu^t(I,\omega)$ to this equation for the initial condition $X = I$ defines a positive operator-valued diffusive process $I^t(\omega) = \omega(\widehat{\mu}^t(I))$ which satisfies to the martingale equation $\mathrm{d}I^t\,(\omega) = G^t\,(\omega)\,\mathrm{d}w_t$ with the initial condition $I^0\,(\omega) = I$, where $G^t\,(\omega) = \mu^t\,(G,\omega)$ and the properties $\alpha_t\,(I) = 0 = \gamma_t\,(I)$ are substituted into (2.6). Thus the Theorem 1 is proved.

2.3 The Classical Case

The remark that follows provides an explanation why equation (2.6) is a noncommutative analog of the Zakai filtering equation.

Remark 1. Let $\mathcal{A}$ be a commutative algebra equivalent to the space $C^\infty(\mathbf{R}^d)$ of infinitely-differentiable functions $x : \mathbf{R}^d \to \mathbf{C}$ with the pointwise product, and the involution $x^*(z) = \bar{x}(z)$. Then equation (2.6) is an operator representation of the Zakai equation

$$\mathrm{d}\,\mu^t_{z_0} + \frac{1}{2}\Delta_t \mu^t_{z_0}\mathrm{d}t = (g_t + \nabla_t)\,\mu^t_{z_0}\mathrm{d}w_t \ , \qquad \mu^0_{z_0} = \delta_{z_0} \ , \tag{2.7}$$

for the nonnormalized a posteriori distribution $\mu^t_{z_0}(\mathrm{d}z,\omega)$ of the Markov diffusion process $z(t)$ described by the stochastic equation

$$\mathrm{d}z + c_t(z)\,\mathrm{d}t = a_t(z)\,\mathrm{d}v_t, \qquad z(0) = z_0 \ , \tag{2.8}$$

with the indirect measurement

$$\mathrm{d}y(t) = g_t(z(t))\mathrm{d}t + \mathrm{d}w_t$$

defined by the standard Wiener process $w_t = -v_t$. Here

$$\int x(z)\,\nabla_t\mu(\mathrm{d}z) = -\int \sum_{k=1}^{d} a^k(z)\,x'_k(z)\,\mu(\mathrm{d}z) \ ,$$

$$\int x(z)\,\Delta_t\mu(\mathrm{d}z) = +2\int \sum_{k=1}^{d} c^k_t(z)\,x'_k(z)\,\mu(\mathrm{d}z) - \int \sum_{k,l=1}^{d} a^k_t(z)\,a^l_t(z)\,x''_{kl}(z)\,\mu(\mathrm{d}z) \ ,$$

with $x'_k = \partial_k x$, $x''_{kl} = \partial_k\partial_l x$. The integral $\int \mu^t_{z_0}(\mathrm{d}z,\omega) = \rho^t_{z_0}(\omega)$ defines the probability density of the output process $y(r,\omega)$ on the interval $0 \le r < t$ with respect to the Wiener distribution $\mathbf{P}(\mathrm{d}\,\omega)$ with given initial state $z_0 \in \mathbf{R}^d$, $\delta_{z_0}(\mathrm{d}z) = 1$ with $z_0 \in \mathrm{d}z$, $\delta_{z_0}(\mathrm{d}z) = 0$, and $z_0 \notin \mathrm{d}z$.

Indeed, in the case of the commutative algebra $\mathcal{A} \simeq C^\infty(\mathbf{R}^d)$, G_t is the multiplication operator by the given function $g_t(z)$ of the state $z \in \mathbf{R}^d$ of the Markov process $z(t)$. This process has the generator $\gamma_t(X)(z) = [\Gamma_t x](z)$, defined by the diffusion operator Γ_t on the measurable functions $x : z \mapsto x(z)$, $x \in C^\infty(\mathbf{R}^d)$. The indirect measurement of $g_t(z)$ is given by the output process

$$y(t) = \int_0^t g_r(z(r))\,\mathrm{d}r + w_t \ .$$

Since $[G_t, X] = 0$, $\forall X \in \mathcal{A}$, $\delta_t = -\alpha_t$ is a real derivation

$$\alpha_t(X)\,(z) = a_t(z)\,\partial x(z) := a^k_t(z)\,\partial_k x(z) \ , \qquad \partial_k = \frac{\partial}{\partial z^k} \ ,$$

and $\gamma_t = G_t \cdot \alpha_t - \beta_t - \frac{1}{2}\alpha_t^2$, where $G_t \cdot \alpha_t(X)(z) = g_t(z)\alpha_t(X)(z)$, the operation

$$\Gamma_t = (a_t g - b_t)\,\partial - \frac{1}{2}\,(a_t\partial)^2 = c_t^k \partial_k - \frac{1}{2}\,a_t^k a_t^l \partial_k \partial_l \tag{2.9}$$

is a standard generator of the diffusion process $z(t)$, with $c_t^k = a_t^l(g_t\delta_l^k - \frac{1}{2}\partial_l a_t^k) - b_t^k$.

Note that the noise $v_t = -w_t$ in the classical system appeared essentially the same as in the observation channel because it was represented in the Fock space of the Wiener process w_t. In order to represent a classical stochastic system in the same way with the noise $v_t \neq 0$ which is independent of w_t, it is necessary to start from the Fock space $\mathcal{F} = \Gamma(\mathcal{K})$ over $\mathcal{K} = \mathbf{C}^m \otimes L^2(\mathbf{R}_+)$ with multiplicity $m \geq 2$, as is the case in [14].

2.4 The a Posteriori Equation

In the general case the filtering equation (2.6) defines the nonnormalized a posteriori state $\widehat{\varphi}_0^t = \varphi_0 \circ \widehat{\mu}_0^t$, which is the vector-state $\widehat{\varphi}_0^t(X) = (\widehat{\psi}_0^t | X \widehat{\psi}_0^t)$ for all $\varphi_0(X) = (\psi_0 | X \psi_0)$ in the case of inner derivations (1.10). The normalized a posteriori state $\widehat{\pi}_t(X) = \widehat{\varphi}_0^t(X)/\widehat{\varphi}_0^t(I)$ satisfies the (nonlinear) a posteriori equation

$$\mathrm{d}\,\widehat{\pi}_t(X) + \widehat{\pi}_t \circ \gamma_t(X)\,\mathrm{d}t = \widehat{\pi}_t\big(\kappa_t(X) - \widehat{\pi}_t(G_t)\,X\big)\,\mathrm{d}\,\widetilde{w}_t \ , \tag{2.10}$$

with initial condition $\widehat{\pi}_0(X) = \varphi_0(X)$, where

$$\kappa_t(X) = G_t \cdot X - \alpha_t\,(X)\,, \mathrm{d}\,\widetilde{w}_t = \mathrm{d}w_t - \widehat{\pi}_t(G_t)\,\mathrm{d}t \ .$$

The nonlinear stochastic equation (2.10) is the Markov case of the general a posteriori diffusive equation (2.1) with innovating martingale $\mathrm{d}\widetilde{Y}$ represented by $\mathrm{d}\widetilde{w}$. It can be deduced from the linear one (2.6) for the nonnormalized state $\widehat{\varphi}_t(X) = \varphi_0(\widehat{\mu}_0^t(X))$ by applying the classical Ito formula to the product $\widehat{\varphi}_t(X) = \widehat{\rho}_0^t \widehat{\pi}_t(X)$ and noting that the positive martingale $\widehat{\rho}_0^t = \widehat{\varphi}_0^t(I)$ has the stochastic differential $\mathrm{d}\,\widehat{\rho}_0^t = \widehat{\varphi}_0^t(G_t)\mathrm{d}\,\widehat{w}_t$. Indeed,

$$\begin{aligned}
\mathrm{d}\,\widehat{\varphi}_0^t(X) &= \widehat{\rho}_0^t \mathrm{d}\,\widehat{\pi}_t(X) + \mathrm{d}\,\widehat{\rho}_0^t \widehat{\pi}_t(X) + \widehat{\rho}_0^t \mathrm{d}\,\widehat{\pi}_t(X) \\
&= \widehat{\varphi}_0^t(G_t)\,\widehat{\pi}_t \circ \widetilde{\kappa}_t(X)\,\mathrm{d}t + \big(\widehat{\varphi}_0^t(G_t)\,\widehat{\pi}_t(X) + \widehat{\rho}_0^t \widehat{\pi}_t \circ \widetilde{\kappa}_t(X)\big)\,\mathrm{d}\,\widehat{w}_t \\
&\quad - \widehat{\rho}_0^t\big(\widehat{\pi}_t \circ \gamma_t(X) + \widehat{\pi}_t \circ \widetilde{\kappa}_t(X)\,\widehat{\pi}_t(G_t)\big)\,\mathrm{d}t \\
&= \varphi_0^t\big(G_t \cdot X - \alpha_t(X)\big)\,\mathrm{d}\,\widehat{w}_t - \widehat{\varphi}_0^t \circ \gamma_t(X)\,\mathrm{d}t \ ,
\end{aligned}$$

where $\widetilde{\kappa}_t(X) = \kappa_t(X) - \widehat{\pi}_t(G_t)X$. Note that in the deduction of the equation the following relation was used:

$$\widehat{\rho}_0^t \widehat{\pi}_t \circ \widetilde{\kappa}_t(X)\,\widehat{\pi}_t(G_t) = \widehat{\pi}_t \circ \widetilde{\kappa}_t(X)\,\widehat{\varphi}_0^t(G_t) \ .$$

3 Appendix

1. Let $\{H_\xi | \xi > 1\}$ be a continuous family of Hilbert subspaces $H_\xi \subseteq H$ with nondecreasing norms: $\eta \leq \xi \Leftrightarrow \|\psi\|_\eta \leq \|\psi\|_\xi, \forall \psi \in H$. It will be called a *scale of the Hilbert space* H with a scalar product $(\psi|\psi) = \lim_{\xi\downarrow 1} \|\psi\|_\xi^2$. An *inductive limit* $\lim_{\xi\downarrow 1} H_\xi$ of the Hilbert scale $\{H_\xi\}$ is defined as a pre-Hilbert space $D = \cup_{\xi>1} H_\xi$, provided with the inductive convergence:

$$\psi_n \to 0 \Leftrightarrow \exists \xi > 1 : \|\psi_n\|_\xi \to 0 \ .$$

The operator $X : D \to D$ is called (*inductively*) *continuous* if $X\psi_n \to 0$ for any convergent sequence $\{\psi_n\}$ that tends to zero. This means, that the restriction of X to any subspace $H_\zeta \in \{H_\xi\}$ is a continuous map into some $H_\xi \subseteq D$, i.e. for any $\zeta > 1$ there exists $\xi > 1$, such that

$$\|X\|_\zeta^\xi := \sup_{\psi \in \mathcal{H}_\zeta} \left\{ \frac{\|X\psi\|_\xi}{\|\psi\|_\zeta} \right\} < \infty \ . \tag{A.1}$$

The set $\mathcal{B}(D)$ of all continuous operators X, having the Hermitian-conjugate operators X^* on the pre-Hilbert space D form an associative algebra with the identity $I \in \mathcal{B}(D)$ and the involution $X^{**} = X$. This algebra is a C^*-algebra of the bounded operators only if $D = H$.

2. Let Γ_+ be the set of all the chains $\tau_n = (t_1, \ldots, t_n)$, $t_i \in \mathbf{R}_+$, $t_1 < \cdots < t_n$ of length $n < \infty$, identified with finite subsets $\tau \subset \mathbf{R}_+$, $\tau = \{t_1, \ldots, t_n\}$ of the cardinality $|\tau| = n \in \{0, 1, \ldots\}$. We denote by $d\tau = \prod_{t\in\tau} dt$ the positive σ-finite measure on $\Gamma_+ = \sum_{n=0}^\infty \Gamma_n$, which is defined as a sum $\sum_{n=0}^\infty d\tau_n$ of the measures $d\tau_n = dt_1 \cdots dt_n$ on $\Gamma_n = \{|\tau| = n\}$ with the only atom $d\tau_0 = 1$ on the empty chain $\tau_0 = \emptyset$ corresponding to $|\tau| = 0$. The Hilbert space $H = L^2(\Gamma_+)$ of the square-integrable functions $f : \Gamma_+ \to \mathbf{C}$, $(f|f) < \infty$, where $(f|f) = \int |f(\tau)|^2 d\tau$,

$$\int g(\tau)\, d\tau := \sum_{n=0}^\infty \int\cdots\int_{0<t_1<\cdots<t_n<\infty} g(t_1, \ldots, t_n)\, dt_1 \cdots dt_n \ , \tag{A.2}$$

is naturally identified with the *Fock space* $\mathcal{F} = \Gamma(\mathcal{K})$, of the sequences $f = \{\varphi_n | n = 0, 1, \ldots\}$ of symmetric continuations $\varphi_n : \mathbf{R}_+^n \to \mathbf{C}$ of the functions $f(\tau_n)$, with the scalar product

$$(f \mid f) = \sum_{n=0}^\infty \frac{1}{n!} \int_0^\infty \cdots \int_0^\infty |\varphi(t_1, \ldots, t_n)|^2 dt_1 \cdots dt_n \ .$$

The Hilbert scale $\{\mathcal{F}_\xi | \xi > 1\}$ of the dense subspaces

$$\mathcal{F}_\xi = \left\{ f \in \mathcal{F} \mid \|f\|_\xi^2 = \int \xi^{|\tau|} |f(\tau)|^2 d\tau < \infty \right\}$$

is called Fock [9]–[11] over the Hilbert space $\mathcal{K} = L^2(\mathbf{R}_+)$. The function $f(\tau) = \delta_\emptyset(\tau)$ (where $\delta_\emptyset(\tau) = 1$ for $\tau = \emptyset$, $\delta_\emptyset(\tau) = 0$ for $\tau = \emptyset$), normalised ($\|f\|_\xi^2 = \int \xi^{|\tau|} \delta_\emptyset(\tau)\, d\tau = 1$) with respect to any ξ, is called the *vacuum* function.

3. The most important examples of the continuous operators, and their conjugates $\mathcal{F}_+ \to \mathcal{F}_+$ on the inductive limit $\mathcal{F}_+ = \cup_{\xi>1}\mathcal{F}_\xi$, are the 'quantum annihilation' operators $\widehat{a}_t$ on the interval $[0,t)$, which act as the integrals

$$[\widehat{a}_t f](\tau) = \int_0^t f(\tau \sqcup r)\,\mathrm{d}r \ , \qquad \forall f \in \mathcal{F} \ .$$

Here $\tau \sqcup r$ is a chain, defined almost everywhere ($r \notin \tau$) as a union $\{\tau, r\}$ of the single-point chain $r \in \mathbf{R}_+$ with some chain $\tau = \{t_1, \dots, t_n\}$. Although the operators $\widehat{a}_t$ possess a complete system of exponential eigenfunctions

$$f(\tau) = k^{\otimes}(\tau) := \prod_{t\in\tau} k(t) \ , \qquad k \in l^2(\mathbf{R}_+) \ ,$$

they are not normal, i.e. they do not commute with their adjoint operators $\widehat{a}_t^*$, defined by the finite sums

$$[\widehat{a}_t^* f](\tau) = \sum_{r\in\tau}^{r<t} f(\tau\backslash r) \ , \qquad \forall f \in \mathcal{F}_+ \ ,$$

where $\tau\backslash r = \{t \in \tau | t \neq r\}$. Polynomials of the operators $\{\widehat{a}_t^*, \widehat{a}_t | t \in \mathbf{R}_+\}$ form the *Weyl algebra* over the simple functions $g \in L^2(\mathbf{R}_+)$, generating the whole algebra $\mathcal{B}(\mathcal{F}_+)$ through the triviality of the commutant:

$$\{Y \in \mathcal{B}(\mathcal{F}_+) \mid X \in \{\widehat{a}_t^*, \widehat{a}_t\} \Leftrightarrow [X,Y] = 0\} = \mathbf{C}\hat{1} \ .$$

Linear elements of the algebra $\mathcal{B}(\mathcal{F}_+)$, defined as the quantum Wiener integrals

$$\widehat{r}(f,g) = \int_0^\infty (f(t)\,\mathrm{d}\widehat{a}_t + g(t)\,\mathrm{d}\widehat{a}_t^*) = \widehat{a}(f) + \widehat{a}^*(g)$$

of the square-integrable functions $f, g : \mathbf{R}_+ \to \mathbf{C}$, together with the identity operator $\hat{1}$, form the *-representation $\widehat{r}(f,g)^* = \widehat{r}(g^*, f^*)$ of the canonical commutation relations

$$[\widehat{r}(f,g), \widehat{r}(f,g)^*] = (\|f\|^2 - \|g\|^2)\,\hat{1} \qquad \text{(A.3)}$$

with respect to the involution $(f,g)^\sharp = (g^*, f^*)$. The vacuum function $\delta_\emptyset \in \mathcal{F}$ induces a Gaussian state on $\mathcal{B}(\mathcal{F}_+)$, defined by the characteristic functional

$$(\delta_\emptyset \mid e^{i\widehat{r}(f,g)}\delta_\emptyset) = \exp\left\{\frac{1}{2}\,(\|f\|^2 - \|g\|^2)\right\}$$

on the Hilbert space of pairs $\eta = (f,g)$ with the norm² $|\eta|^2 = \frac{1}{2}(\|f\|^2 + \|g\|^2)$ and symplectic bilinear form $s(\eta', \eta)$ which equals $s(\eta^\sharp, \eta) = \frac{1}{i}(\|f\|^2 - \|g\|^2)$ for $\eta' = (g^*, f^*)$.

4. The measurable function $F : t \mapsto F(t)$ whose values are the operators $\mathcal{D} \to \mathcal{D}$ of the inductive limit $\mathcal{D} = \cup_{\xi>1}\mathcal{H}_\xi$ is called locally *p-integrable in the inductive scale* $\{\mathcal{H}_\xi = \mathrm{H}_\xi \otimes \mathcal{F}_\xi\}$, if, for any $\zeta > 1$, there exists a $\xi > 1$, such that

$$\|F\|_{\zeta,t}^{\xi,p} := \left(\int_0^t (\|F(r)\|_\zeta^\xi)^p \mathrm{d}r\right)^{1/p} < \infty \ , \qquad \forall t \in \mathbf{R}_+ \ .$$

In particular, this condition means that the operators $F(t) : \mathcal{D} \to \mathcal{D}$ are continuous for almost all $t \in \mathbf{R}_+$. For such square-integrable functions the quantum stochastic integrals $\widehat{a}_t^*(F)$, $\widehat{a}_t(F)$ on $\{\widehat{a}_r^* | r < t\}$ and $\{\widehat{a}_r | r < t\}$ are defined to be the operators $\mathcal{D} \to \mathcal{D}$:

$$[\widehat{a}_t^*(F)\,h](\tau) = \sum_{r\in\tau}^{r<t} [F(r)\,h]\,(\tau) \ , \qquad \forall\, h \in \mathcal{D} \ ,$$

$$[\widehat{a}_t(F)\,h](\tau) = \int_0^t [F(r)\,\dot{h}(r)]\,(\tau) \ . \tag{A.4}$$

In this equation $h \mapsto \dot{h}(r)$ is a Maliven derivative, defined in the Fock representation $h : \Gamma_+ \to \mathrm{D}$ of the elements $h \in \mathcal{D}$ almost everywhere ($r \notin \tau$) by the vector-function $\dot{h}(r) \in \mathcal{D}$ as $\dot{h}(r, \tau) = h(r \sqcup \tau)$. The continuity of the operators $\widehat{a}_t(F)$, $\widehat{a}_t^*(F)$ in $\mathcal{D}$ follows directly from the estimates

$$\|\widehat{a}_t(F)\|_{\zeta+\varepsilon}^{\xi} \le \sqrt{1/\varepsilon}\,\|F\|_{\zeta,t}^{\xi,2} \ , \qquad \forall\, \varepsilon > 0 \ ,$$

$$\|\widehat{a}_t^*(F)\|_{\zeta}^{\xi-\varepsilon} \le \sqrt{\xi/\varepsilon}\,\|F\|_{\zeta,t}^{\xi,2} \ , \qquad \forall\, \varepsilon < \xi \ .$$

obtained in [5], [9]–[11]. From these estimates, if there exists $\xi > 1$ for any $\zeta > 1$ such that $\|F\|_{\zeta,t}^{\xi,2}$, $\|D\|_{\zeta,t}^{\xi,2} < \infty$ then, for any $\zeta^+ > 1$, there exists a $\xi_- > 1$, for which the operator

$$\imath_0^t(F, D) = \widehat{a}_t(F) + \widehat{a}_t^*(D)$$

is bounded from $\mathcal{H}_{\zeta^+}$ to $\mathcal{H}_{\xi_-}$. To be precise, by choosing for every $\zeta^+ > 1$ a $\zeta > 1$, such that $\zeta < \zeta^+$, and an $\varepsilon > \xi - 1$, such that $\varepsilon < \zeta^+ - \zeta$, we obtain

$$\|\imath_0^t(F, D)\|_{\zeta^+}^{\xi_-} \le \sqrt{\xi/\varepsilon}\,\left(\|F\|_{\zeta,t}^{\xi,2} + \|D\|_{\zeta,t}^{\xi,2}\right)$$

for any ξ_- on the nonempty interval $(1, \xi - \varepsilon]$.

5. If $D(t) \in \mathcal{B}(\mathcal{D})$ for almost all $t \in \mathbf{R}_+$ and the adjoint function $D^*(t) = D(t)^*$ is also locally integrable on $\mathcal{D}$, then $\widehat{a}_t^*(D) \in \mathcal{B}(\mathcal{D})$ and $\widehat{a}_t^*(D)^* = \widehat{a}_t(D^*)$. This means that the integral $\imath_0^t(D^*, D)$ is (formally) self-adjoint. Moreover, the set of integrals

$$X(t) = \imath_0^t(D^*, D) + \int_0^t G(r)\,\mathrm{d}r \ ,$$

where $G : \mathbf{R}_+ \to \mathcal{B}(\mathcal{D})$ is locally integrable ($p = 1$), together with the G^*, function, forms a $*$-algebra with respect to the pointwise operator product $(X^*X)(t) = X(t)^*X(t)$. This product is defined by the quantum nonadapted Ito formula [10]

$$\begin{aligned} X(t)^*X(t) &= \imath_0^t(D^*X + X^*D, F^*D + X^*D) \\ &+ \int_0^t [G^*X + X_r^*D + D^*D + D^*X_r + X^*G](r)\,\mathrm{d}r \ , \end{aligned} \tag{A.5}$$

which corresponds to the case of the locally square-integrable operator function $t \mapsto X_t(t)$, $X_t^*(t) = X(t)_t^*$. Here $X \mapsto X_t$ means a derivation, which is defined by

$$[X_t h](\tau) = [Xh](\tau \sqcup t) - [X\dot{h}(t)](\tau)$$

for almost all $\tau \in \Gamma$, $t \notin \tau$. In this instance, the case $X_t(t) = 0 = X_t^*(t)$ for all $t \in \mathbf{R}_+$ corresponds to the adaptive property of the operator functions $F(t), D(t)$ and $G(t)$.

Comprehensive information on explicit stochastic integration in the Fock scale is given in [20].

References

1. A.N. Kolmogorov, *The Basic Concepts of Probability Theory*, 2nd edition, M. Nauka, 1974, 120 pages.
2. J. von Neumann, *Mathematical foundations of Quantum Mechanics*, M. Nauka, 1964, 367 pages.
3. A.S. Holevo, *Probability and Stochastic aspects of Quantum Theory*, M. Nauka, 1980, 320 pages.
4. B. Misra and E.C.G. Sudarshan, *Zeno's paradox in quantum theory*, J. Math. Phys., **18**, (4) 756–763 (1977).
5. V.P. Belavkin, *Stochastic Calculus of Quantum Input-Output Processes and Quantum Nondemolition Filtering*, Itogi nauki i tekhniki Modern problems of mathematics series. Newest achievements, **36**, 29–68 (1989).
6. V.P. Belavkin, *Nondemolition measurements, nonlinear filtering and dynamic programming of quantum stochastic processes*, Lect. Notes Control and Inform. Sci., **121**, 245–261 (1989).
7. V.P. Belavkin, *Optimal Quantum Filtering of Markov Signals*, Problemy upravleniya i teoriya inform., **7**, (5) 345–360 (1978).
8. V.P. Belavkin, *Nondemolition measurement and control in quantum dynamical systems*, In: Information Complexity and Control in Quantum Physics. (Udine, 1985). Ed. by A. Blaquiere. Vienna: Springer, 311–329. (CISM Courses and Lectures. Vol. 294), (1987).
9. V.P. Belavkin, *Stochastic posterior equations of quantum nonlinear filtering*, in: Probability Theory and Mathematical Statistics. (Proceedings of the Fifth Vilnius Conference.)/ Ed. by B. Grigelionis et al. Utrecht/Vilnius: VSP/Mokslas, p. 91–109.
10. V.P. Belavkin, *A nonadapted stochastic calculus and nonstationary quantum evolution*, in: Proceedings of the Sixth Quantum Probability Conference. (Trento, 1989.) Singapore: World Scientific Publisher, 1991.
11. V.P. Belavkin, *Quantum stochastic calculus and quantum nonlinear filtering*, J. Multivar. Anal., **42**, (2) 171–201 (1992).
12. R.L. Hudson and K.R. Parthasarathy, *Quantum Ito's formula and stochastic evolutions*, Commun. Math. Phys., **93**, (3) 301–323 (1984).
13. K.R. Parthasarathy, *An Introduction to Quantum Stochastic Calculus*, Birkhäuser, 1992.
14. V.P. Belavkin, *Quantum Continual Measurements and a Posteriori Collapse on CCR*, Commun. Math. Phys., **146**, (3) 611–635 (1992).

15. V.P. Belavkin, *A posterior Schrödinger equation for continuous nondemolition measurement*, J. Math. Phys., **31**, (12) 2930–2934 (1990).
16. V.P. Belavkin and P. Staszewski, *Nondemolition observation of a free quantum particle*, Phys. Rev., **A45**, 1347–1356 (1992).
17. M.P. Evans and R.L. Hudson, *Multidimensional quantum diffusions*, Lect. Notes Math., **1303**, 69–88 (1988).
18. R.S. Liptser and A.N. Shiriaev, *Statistics of Random Processes*, M. Nauka, 1974, page 696.
19. R.L. Stratonovitch, *Conditional Markov Processes and their Application to Optimal Control Theory*, M. Moscow State University, 1966.
20. V.P. Belavkin, *Chaotic States and Stochastic Integration in Quantum Systems*, Uspekhi matem. nauk, **47**, (1) 47–106 (1992).

Event Enhanced and Piecewise Deterministic Quantum Theory or the Right Jump at the Right Place

Ph. Blanchard[1] *and A. Jadczyk*[2]

[1] Fac. of Physics and BiBoS
University of Bielefeld
D–33615 Bielefeld, Germany

[2] Institute of Theoretical Physics
University of Wrocław
PL–50–204 Wrocław, Poland

Abstract: The purpose of Event Enhanced Quantum Theory (EEQT) is to allow for a precise meaning to the concepts of "event", "experiment" and "measurement". Within EEQT one obtains not only Liouville equations describing the continuous dynamics of statistical ensembles but also a unique minimal piecewise deterministic random Markov process (PDP) than can be used for computer simulations of real time series for experiments on individual quantum systems. EEQT is therefore particularly relevant to today's experimental Quantum Physics since new technology needs new laws and its range of applications is rather wide. As an example a cloud chamber model will be discussed. In a particular, homogeneous, case this model contains GRW spontaneous localization model. All probabilistic interpretation of Standard Quantum Theory can be derived from the formalism of EEQT. Moreover EEQT has no need for observers or minds. EEQT is a precise and predictive theory not only giving enhanced answers but also inviting asking new questions for example on the grand vision of a Quantum Theory of history à la Gell Mann–Hartle or on Connes' version of the Standard Model. In conclusion EEQT is a minimal extension of the Standard Quantum Theory that accounts for events and satisfies the needs of human experience and modern technology.

1 Greatness and Troubles with Orthodox Quantum Theory

To start at the beginning we have first to say that Orthodox Quantum Theory (OQT) has proved to be incredibly powerful, practical and successful in the description of the properties of atoms, molecules and elementary particles. There seems to be no limit to the versatility of the Schrödinger equation. OQT was and is outstandingly successful in *computing* not only those phenomena for which it was invented but also numerous others in Physics, Chemistry and Biology making therefore wonderful advance in technology possible. On the other side the interpretation of OQT is still today the most controversial problem in the foundations of physics. Its successes resemble those of the Ptolemeian System

– which was also no doubt successful. It was John Bell's point of view that no formulation of OQT was free of fatal flaws [1, 2]. N. Bohr was persuaded from the consistency of OQT. For A. Einstein OQT does not allow a complete description of Nature and was only the first step of the final theory. "Red black magic calculus" is how he described OQT in a letter in 1925. "The fact that an adequate representation of Quantum Mechanics has been so long delayed is no doubt caused by the fact that Niels Bohr brainwashed a whole generation of theorists into thinking that the job was done fifty years ago" wrote Murray Gell Mann 1979. OQT considers two types of incompatible time evolution U and R, U denoting the unitary evolution resulting from Schrödinger's equation and R the reduction of the quantum state. U is linear, deterministic, local, continuous and time reversal invariant, while R is probabilistic, non–linear, discontinuous and acausal. For a fundamental physical theory this situation is not very satisfactory but J. Bell claimed that OQT works for all practical purposes (FAPP). Two options are possible for completing OQT. According to John Bell [2] "Either the wave function is not everything or it is not right ...". Gisin and Percival formulated the thesis that "the Schrödinger equation is no longer the best for all practical purposes" [3].

As emphasized already by E. Schrödinger [4] completely missing in OQT is an explanation for experimental facts i.e. a description of the actual individual times series of events. What is needed is a consistent framework which can be used to retrieve the existence of events despite the probabilistic character of quantum physics. A theory allowing the description of single systems is today necessary since advances in technology make fundamental experiments on individual quantum systems possible. Events can be recorded and have definitely no place in OQT. The importance of the events have been stressed by several authors. H.P. Stapp emphasized the role of events in the world process [5–7]. R. Haag [8] drawed attention to the fact that "an event in quantum physics is discrete and irreversible" and that "we must assume that the arrow of time is encoded in the fundamental laws ...". He also suggested [9] that "transformation of possibilities into facts must be an essential ingredient which must be included in the fundamental formulation of the theory". Each event must have in any case three characteristic properties:

- it is classical
- it is discrete
- it is irreversible .

For us the adjective "classical" has a well defined meaning: To each particular experimental situation corresponds a minimal set of events revealing us the Heisenberg transition from the possible to the actual and these events which can be recorded obey the rules of classical logic of Aristotle and Boole. Indeed an event must obey to the classical "yes–no" logic; to be an event it must never be in a superposition of being happened and being unhappened. Since an event must happen wholly it is necessary discrete. Finally each event is irreversible because it must have left a trace. Even if this trace can be erased, the very act of erasing will change the future, not the past. We believe that the events, and nothing but events are pushing forward the arrow of time. Irreversible laws are fundamental and reversibility is only an approximation. This fact has been recently once more emphasized by H. Rauch [10] referring to a work of H.A. Lorentz [11]. A typical event is for instance a track in a bubble chamber making elementary particles visible, or the click of a detector. Once the three characteristics of an event are accepted – and they are completely evident for any experimenter – it becomes clear what is necessary if we want to enhance OQT so as to include events into it.

First we must allow the formalism to include from the beginning classical quantities. Indeed we believe that is better when this assumption is done openly rather than indirectly as it is almost ever the case. Second we must take into account an irreversible coupling between classical and quantum degrees of freedom. Let us stress that in our opinion this minimal irreversibility is not the manifestation of noise, chaos or environment but expresses simply the universal fact that information must be paid with dissipation. Finally we need an important third step, namely to learn how to describe finite time series of events from which expectation values can be computed. Moreover, as human beings, we want not only to be able to compute statistical properties of ensembles but also to be in position to simulate finite time series for individual systems since we cannot enter twice the same place in the same stream of time.

In the following section we will describe in a condensed form what seems to us to be the minimal enhancement of quantum theory completing the needs of human experience and modern technology. In [12–18] we proposed mathematical and physical rules to describe

- the two kinds of time evolution of quantum systems namely continuous and stochastic
- the flow of information from the quantum system to the classical event-space
- the control of quantum states and processes by classical parameters.

2 All You Have in Mind to Know About EEQT

It is one of the aims of this section to express in a condensed form and partially informal way the philosophical backbone that can be extracted from the several models discussed in [12–20]. In EEQT the quantum system Q is coupled to a "classical" space C, where events do happen. We consider the total system $\Sigma = Q \times C$. Let us denote by X_c the classical event–manifold and by $\mathcal{H}_q$ the Hilbert space associated to the quantum system Q. A classical pure state is nothing else as a point in X_c and the coordinates of this point correspond exactly to the properties of C.

From the structural and from the mathematical point of view, the three most essential features of EEQT are

- tensoring of the non commutative quantum algebra of observables $\mathcal{A}_q = \mathcal{L}(\mathcal{H}_q)$ with a commutative algebra of continuous functions $\mathcal{A}_C = C(X_C)$.

We consider therefore the behaviour associated to the total algebra of observables $\mathcal{A}_t = \mathcal{L}(\mathcal{H}_q) \otimes C(X_C)$.

- replacing Schrödinger's unitary dynamics of pure states with a suitable completely positive semigroup $\alpha_t = e^{tL}$ describing the time evolution of ensembles. Time evolution of ensembles of coupled systems, prepared by the same algorithm, is described by a Liouville equation in the Hilbert space $\mathcal{H}_t$ of the total system with $\mathcal{H}_t = \mathcal{H}_q \otimes L^2(X_C)$. The main characteristic of the dynamics is that it does not map pure states into pure states but it is well defined on the level of the density matrices where it preserves convexity, positivity and trace.
- interpreting the continuous time evolution of the statistical states of the total system Σ in terms of a PDP, Markov process taking values in the set of pure states $S_1(\mathcal{H}_q) \otimes X_C$ of the total system with $S_1(\mathcal{H}_q) = \{\psi \in \mathcal{H}_q \mid \| \psi \| = 1\}$. The process consists of pairs (quantum jump, classical event) interrupting random periods of Schrödinger–type continuous in general non–linear dynamics. Time evolution of the PDP is derived from the Liouville equation. At random times distributed according to a specific inhomogeneous Poisson process jumps occur.

There are jumps of the quantum state vectors and also at the same time jumps of the states of C. But we do not observe Hilbert space vectors. Are quantum jumps "real"? They are "real" but belong to the "implicate order". Events are real and belong to "explicate order". Indeed the classical jumps we can see (to measure a quantity we must look at it) and these classical events can be recorded if necessary. Knowing this PDP one can answer many (perhaps even all) kinds of questions about time correlations of events as well as simulate numerically the possible histories of individual systems. Within EEQT there need not be cat paradoxes anymore – cats are allowed to behave as cats; we cannot predict individual events as they are random, but we can simulate the observations of individual systems. The pure quantum states can be viewed as a hidden variable and EEQT as a purely classical theory accounting for quantum phenomena. In EEQT the word measurement instead of being banned as suggested by J. Bell can be given now a precise and acceptable meaning. An *experiment* is a completely positive (CP) coupling between a quantum system and a classical event space. One observes then the classical system C and attempts to learn from it about characteristic of state and of dynamics of the quantum system Q. A *measurement* is an experiment that is used for a particular purpose namely for determining values, or statistical distribution of values of given physical quantities. The aim of any well performed measurement is therefore to get a maximum of information on Q and to pay for it with a minimum of dissipation.

By partial tracing each state ρ of the total system projects onto an effective quantum state $\hat{\rho} \equiv \Pi_q(\rho)$ and an effective classical $\mu \equiv \Pi_C(\rho)$.

Let us consider dynamics. The time evolution of the total system is given by a semigroup $\phi_t = e^{tL}$ of CP maps of $\mathcal{A}_{tot}$ where L is a Lindblad generator (cf. [18]). There is a simple method of constructing appropriate couplings. Suppose that we consider quantum properties F_α , $\alpha = 1, 2, \ldots$ that we want to measure. For each α we try to find a transformation of X_C with the following interpretation: If the quantum system Q has property F_α while the classical–event system is in a state x then C switches from x to a new state $\alpha(x)$. Denoting $\rho(t) = \phi_t(\rho_0)$ the time evolution of the states is given by the Liouville equation

$$\dot{\rho}_x(t) = -i[H_x, \rho_x(t)] + \sum_\alpha V_\alpha \rho_{\alpha(x)}(t) V_\alpha^*$$
$$-\frac{1}{2}\{\Lambda, \rho_x\}$$

where we have denoted

$$\Lambda = \sum V_\alpha^* V_\alpha \ .$$

The Liouville equation can be therefore written as a sum

$$\dot{\rho}(t) = -i[H \ , \rho(t)] + (\dot{\rho}(t))_{irr} \ .$$

The first term $-i[H_x, \rho_x(t)]$, diagonal in x, contains all deterministic parts of the time evolution, while the second accounts for irreversible processes associated to the coupling. It is important to notice that if the quantum Hamiltonian does not depend on the state of the classical system i.e. $H_x \equiv H$ for each $x \in X_C$, and if the maps α of X_C are one to one and onto, then the Liouville equation can be summed up over x which implies that the time evolution for the effective quantum states $\hat{\rho}$ obtained by partial tracing separates and we have

$$\dot{\hat{\rho}}(t) = -i[H, \hat{\rho}(t)] + \sum_\alpha V_\alpha \hat{\rho}(t) V_\alpha^* - \frac{1}{2}\{\Lambda, \hat{\rho}\}.$$

It should be stressed that this separating property of the Liouville equation describing the dynamics of the total system $\Sigma = Q \times C$ need in general not to hold. That is what happens in the SQUID–tank model [14].

<u>**Remark**</u> Assuming, for simplicity, that C has only finite number of states (which may be viewed as "pointer positions") $x = 1, \ldots, m$, an event is a change of state of C. Thus there are $m^2 - m$ possible events. An experiment is then described by a family H of quantum Hamiltonians and a family of $m^2 - m$ quantum operators g_{xy} with $g_{xx} = 0$.

A complete general theory of dissipative couplings of quantum systems to classical ones, does not yet exist. The best we can do is to study a lot of examples. In the following section we will sketch a characteristic situation describing a model for a cloud chamber. For every example we have considered, a PDP has been constructed that takes place on the space of pure states of the total system and which reproduces the Liouville equation by averaging. The theory of PDP is described in a recent book by M.H. Davis [21]. A PDP is determined by its local characteristics namely

i) A vector field which determines a flow on the state space

ii) A jump rate function λ
iii) A transition probability matrix Q.

Each observable A of the total system defines now a function $f_A(\psi, x)$ on the space $S = \{(\psi, x)\}$ of pure states of the total system. It can be shown that the time evolution $\frac{d}{dt} f_A(\psi, X)$ for observable can be written in a Davis form.

Let us now describe the PDP on pure states of the total system $\Sigma = Q \times C$ that leads to the Liouville equation after averaging over paths. For a derivation we refer to [20]. Suppose at initial time $t = 0$ the quantum system is in the pure state $\psi_0 \in \mathcal{H}_q$ and the classical system is in the pure state $\delta_{x_0} \approx X_0$. Then the time evolution of the quantum state is given by

$$\psi(t) = \frac{e^{-iH_{x_0}t - \frac{\Lambda}{2}t}}{\left\| e^{-iH_{x_0}t - \frac{\Lambda}{2}t} \right\|}$$

while the classical system remains at x_0 until a jump occurs at some random time t_1. The random jump time t_1 is governed by a nonhomogeneous Poisson process for which the probability $p(t, t+\Delta t)$ for the jump to occur in the time interval $(t, t+\Delta t)$, provided it did not occurred yet, is given by the formula

$$p(t, t+\Delta t) = 1 - e^{-\int_t^{t+\Delta t} \lambda(\psi(s))ds}$$

where

$$\lambda(\psi) \equiv <\psi, \Lambda\psi> \ .$$

When the jump occurs at $t = t_1$ then the classical system C jumps from x_0 to $\alpha_1(x_0)$ while quantum state vector jumps at the same time from its actual value $\psi(t_1)$ to $\psi_1 = V\alpha_1\psi(t_1)/ \parallel V\alpha_1\psi(t_1) \parallel$ and the process starts. The probability p_α of choosing a particular value of α is given by

$$p_\alpha = \frac{\parallel V_\alpha \psi(t_1) \parallel^2}{\lambda(\psi(t_1))} \ .$$

A fully satisfactory mathematical justification of the uniqueness of the minimal PDP starting from the Liouville equation can be given. The uniqueness of the PDP follows from the special form of the Liouville equation in EEQT. It describes transfer of information from Q to C without introducing unnecessary dissipation reflected by the fact that there should be zeros on the diagonal of the coupling V–matrix. Starting with a pure state (ψ, x) of the total system after time dt we obtain a mixed state; there will be mixing along classical, which is uniquely decomposable and mixing along quantum which is non uniquely decomposable. But it can be shown that mixing along classical is of the order dt and on the other hand mixing along quantum is only of the order $(dt)^2$ which implies infinitesimally unicity [22]. For a mathematically rigorous global proof see [23]. In other words our dissipation does not result from quantum noise but is nothing else as the necessary minimal price to pay for any bit of information received from the quantum system.

3 Cloud Chamber Model

As an illustration of EEQT we will now sketch the description of a non-relativistic cloud chamber model. For more details we refer to [19,20]. The points $a \in E_d \subset \mathbf{R}^3$ will parametrize detectors and finite sets of points will play the role of $x - s$ of the previous section. For each $a \in E_d$, in other words for each detector, let there be given a function $f_a(x)$ on E_d. Physically $f_a(x)$ describes the sensitivity of the detector located at the point a. For instance we could take typically for $f_a(x)$ a Gaussian function

$$f_a(x) = \lambda^{1/2}\left(\frac{\alpha}{\pi}\right)^{3/2} e^{-\alpha(x-a)^2} .$$

The height would then be approximately inversely proportional to the square root of the response time of the local counter while the width to its spatial extension. A point limit will then correspond to the limit $f_a^2(x) \to \lambda\delta(x-a)$, where δ denotes the Dirac measure in a. Our classical system is a continuous medium of 2–states detectors. At each point a the medium is one of two states: "on" represented by $\binom{1}{0}$ or "off" represented by $\binom{0}{1}$. The space of classical events can be identified with the space of finite subsets of E_d from which it follows that the total system is described by families $\{\rho_\Gamma\}_{\Gamma\subset E}, \Gamma$ finite subset of E_d, such that $\sum_\Gamma Tr\rho_\Gamma = 1$. What remains to be specified to define our model is the transformation $\alpha : \Gamma \to \alpha(\Gamma)$ but there is a natural choice namely each α flips the detector state at $x = a$. We can write it also as $\alpha(\Gamma) = \{a\}\Delta\Gamma$, where Δ denotes the symmetric difference.

The quantum mechanical Hilbert space is $\mathcal{H}_q = L^2(E_d, dx)$. The Liouville equation governing the time evolution of the total system is given by

$$\dot\rho_\Gamma = -i[H_\Gamma, \rho_\Gamma] + (\dot\rho_\Gamma)_{irr}$$

where $(\dot\rho_\Gamma)_{irr}$ is given by a Lindblad generator

$$\begin{aligned}(\dot\rho_\Gamma)_{irr} &= L_{int}(\rho)\\ &= \int_{E_d} da f_a \rho_{\Gamma\Delta\{a\}} f_a - \frac{1}{2}\{\Lambda, \rho_\Gamma\}\end{aligned}$$

with

$$\Lambda(x) = \int_{Ed} f_a^2(x)dx .$$

We can also construct the associated minimal PDP. We obtain a Davis generator corresponding to rate function $\lambda(\psi) =< \psi, \Lambda\psi >$ and probability kernel with non–zero elements given by

$$Q(\psi, \Gamma; d\psi', \alpha(\Gamma)) = \frac{\| f_a\psi \|^2}{\lambda(\psi)}\delta\left(\psi' - \frac{f_a\psi}{\| f_a\psi \|}\right) d\psi' .$$

Time evolution between jumps is given by

$$\psi_t = \frac{e^{-iHt - \frac{\Lambda t}{2}}\psi_0}{\left\| e^{-iHt - \frac{\Delta t}{2}}\psi_0 \right\|} .$$

The jump consists of a pair (classical event, quantum jump). The classical medium jump at a with probability density $p(a, \psi_{t_1}) = \| f_a \psi_{t_1} \|^2 / \lambda(\psi_{t_1})$ (flip of the detector) while the quantum part of the jump is jump of the Hilbert space vector ψ_{t_1} to $f_a \psi_{t_1} / \| f_a \psi_{t_1} \|$ and the process starts again. We recognize the von Neumann–Lüders projection postulate. The random jump time t_1 is governed by the inhomogeneous Poisson process with intensity $\lambda(\psi_{t_1})$. the probability for the jump to take place at place a is given by

$$p_a = \frac{\| f_a \psi_t \|^2}{\int \| f_a \psi_t \|^2 \, da} .$$

For $f_a^2(x) \to \delta(x - a)$ we recover $p_a = | \psi_t(a) |^2$ i.e. the Born interpretation of the wave function.

In [19,20] we discuss the case of a passive homogeneous medium for which H does not depend on the actual state of the medium. This implies that the reduced quantum state separates. For constant jump rate we recover the spontaneous localization model à la GRW [24].

4 Summary and Conclusions

Our models of coupling quantum systems to classical event–spaces can be surely criticized as being too phenomenological. Nevertheless EEQT is a minimal extension of OQT satisfying the need of human experience and modern technology and supplying the interface between the quantum world and the events. The cloud chamber model and EEQT do not involve observers and minds. What EEQT needs is computing power and an effective random number generator. Let us in this context formulate an open fundamental question: Can random numbers generators be avoided and replaced by deterministic algorithms of simple and clear meaning?

The fact that our models avoiding the concepts of observers and minds are totally objective does not imply that we do not appreciate the importance of the mind–body problem. In our opinion understanding the problem of mind needs also Quantum Theory and perhaps even more. Our feeling is that models in the spirit of EEQT can have applications in biology. Living organisms are coherent open systems with a program dependent of molecular recording processes. Variations take place there on the quantum level and are translated and amplified to generate macroscopic variations ....

The second important advantage of EEQT is that it provides as sketched in Section 2 its own interpretation. Moreover EEQT allows the description of single systems via PDP and provides in this way very effective methods for numerical simulation of experimental events.

After suitable transformation we believe that EEQT can be acceptable even by quantum purists who claim the universality of quantum theory and do not recognize the fact that there are classical events. They may consider our C as a "preferred basis" and then notice that the completely positive semi-group describing the dynamics always respect this basis. They will also appreciate the fact that only in special situations the effective evolution of the reduced quantum state $\hat{\rho} \equiv \Pi_1(\rho)$ separates but that this separation is not at all necessary in EEQT. In [12, 13, 15–20] we discuss following experiments and topics in the EEQT framework: Measurement–like processes, Stern–Gerlach experiment, Quantum Zeno Effect, EPR, SQUID–tank interaction, efficiency versus accuracy by measurement, simultaneous measurement of non–commuting observables, meaning of the wave function. In EEQT we need only to postulate that events can be observed. All the rest can be derived from this assumptions. All quantum mechanical probabilistic interpretation can be derived from the formalism of EEQT. EEQT invites also to ask new questions since we are tempted to consider the PDP as a "world process" to the entire universe including all kinds of "observers". The questions to be asked now are: what is time? What is classical? What is V describing the binamics? Of course we cannot provide answers but we can provide hints [22].

To draw a conclusion let us say that provided EEQT correctly accounts for experimental results it offers some new ways of seeing things and new mathematics providing an additional perspective on the duality between the potential and the actual, statistical ensembles and individual systems, waves and particles and the deterministic and the random. In his 1933 Spencer lecture A. Einstein mentioned the success not only of classical mechanics but also of the statistical interpretation of quantum theory. He added "I still believe in the possibility of giving a model of reality which shall represent events themselves and not merely the probability of their occurrence". EEQT can be considered as a step in this direction.

References

1. J. Bell, 'Towards an exact quantum mechanics' , in *Themes in Contemporary Physics II. Essays in honor of Julian Schwinger's 70th birthday*, Deser, S., and Finkelstein, R.J. Ed. , World Scientific, Singapore 1989.
2. J. Bell, 'Against measurement' , in *Sixty–Two Years of Uncertainty. Historical, Philosophical and Physical Inquiries into the Foundations of Quantum Mechanics*, Proceedings of a NATO Advanced Study Institute, August 5–15, Erice, Ed. Arthur I. Miller, NATO ASI Series B vol. 226 , Plenum Press, New York 1990.
3. N. Gisin and I.C. Percival, 'Stochastic Wave–Equations Versus Parallel World Components' , Phys. Lett. **A 175**, 144–145 (1993).
4. E. Schrödinger, 'The Philosophy of Experiment' , Il Nuovo Cimento **1**, 5–15 (1995).
5. H.P. Stapp, 'Bell's Theorem and World Process', Nuovo Cimento **29**, 270–276 (1975).

6. H.P. Stapp, 'Theory of Reality', Found. Phys. **7**, 313–323 (1977).
7. H.P. Stapp, 'Mind, Matter and Quantum Mechanics', Springer Verlag, Berlin 1993.
8. R. Haag, 'Events, histories, irreversibility', in *Quantum Control and Measurement*, Proc. ISQM Satellite Workshop, ARL Hitachi, August 28–29, 1992, Eds. H. Ezawa and Y. Murayama, North Holland, Amsterdam 1985.
9. R. Haag, 'Fundamental Irreversibility and the Concept of Events', Commun. Math. Phys. **132**, 245–251 (1990).
10. H. Rauch, 'Neutroneninterferometric = ein Labor der Quantenmechanik' , Phys. Bl. **50**, Nr. 5, 439–443 (1994).
11. H.A. Lorentz, *Theorie der Strahlung*, Akad. Verlag Leipzig 1927, page 48.
12. Ph. Blanchard and A. Jadczyk, 'On the interaction between classical and quantum systems', Phys. Lett. **A 175**, 157–164 (1993).
13. Ph. Blanchard and A. Jadczyk, 'Strongly coupled quantum and classical systems and Zeno's effect', Phys. Lett. **A 183**, 272–276 (1993).
14. Ph. Blanchard and A. Jadczyk, 'How and When Quantum Phenomena Become Real', in Proc. Third Max Born Symp. *Stochasticity and Quantum Chaos*, Sobotka 1993, pp. 13–31, Eds. Z. Haba et all., Kluwer Publ. 1994.
15. Ph. Blanchard and A. Jadczyk, 'Classical and quantum intertwine', in *Proceedings of the Symposium on Foundations of Modern Physics*, Cologne, June 1993, Eds. Busch, P., Lahti, P., and Mittelstaedt, P., World Scientific, Singapore 1993.
16. Ph. Blanchard and A. Jadczyk, 'From quantum probabilities to classical facts', in *Advances in Dynamical Systems and Quantum Physics*, Capri, May 1993, Ed. R. Figari, World Scientific (1994).
17. Ph. Blanchard and A. Jadczyk, 'Event–Enhanced Formalism of Quantum Theory or Columbus Solution to the Quantum Measurement Problem', in *Quantum Communications and Measurement*, Proc. Int. Workshop, Notttingham, July 1994, Eds. Belavkin, V. and Hudson, R.L. , Plenum 1994.
18. A. Jadczyk, 'Topics in Quantum Dynamics', Preprint CPT–Marseille 94/P. 3022, BiBoS 635/5/94, hep–th 9406204.
19. A. Jadczyk, 'Particle Tracks, Events and Quantum Theory', preprint RIMS 989, hep-th 9407157.
20. A. Jadczyk, 'On Quantum Jumps, Events and Spontaneous Localization Models', to appear in Found. Phys. , May 1995, also Preprint ESI–Wien 119, hep–th 9408020.
21. M.H.A. Davis, *Markov models and optimization*, Monographs on Statistics and Applied Probability, Chapman and Hall, London (1993).
22. Ph. Blanchard and A. Jadczyk.: 'Event–Enhanced–Quantum Theory and Piecewise Deterministic Dynamics', Preprint BiBoS 660/10/84, hep–th 9409189, submitted to Ann. der Physik.
23. A. Jadczyk, G. Kondrat and R. Olkiewicz, 'On uniqueness of the jump process in quantum measurement theory', to appear.
24. G.C. Ghirardi, A. Rimini and T. Weber, 'Unified dynamics for microscopic and macroscopic systems', Phys. Rev. **D 34**, 470–491 (1986).

Wave Mechanics: The Interplay Between Stochastics and Quanta

A.M. Cetto and L. de la Peña

Instituto de Física, UNAM
Apartado postal 20–364, 01000 Mexico D.F.

Abstract: The authors make an attempt to understand the wave aspects of quantum mechanics within the framework of stochastic electrodynamics. The electron, considered originally as a point particle, is seen to perform a fine vibration of wavenumber $k_c = mc/\hbar$ sustained by the zeropoint radiation field; its interaction with the field waves of frequencies close to $\omega_c = ck_c$ gives rise to a modulation wave with de Broglie's wavelength. A heuristic analysis leads to the construction of the wave equation for the modulation amplitude. A few simple examples are given to illustrate the appearance of interference phenomena and quantization as the result of demanding that a standing modulation characterizes any stationary dynamical situation.

1 Introduction

Within the framework of stochastic electrodynamics (SED), the random zero-point electromagnetic radiation field of energy $\frac{1}{2}\hbar\omega$ per mode is assumed to constitute an essential ingredient for an understanding of the behaviour of matter at the quantum level. And indeed, in the course of time a good number of results have been obtained in SED on this hypothesis, which are in quantitative agreement with the quantum description and provide an intuitive physical understanding of characteristic quantum phenomena (see [1] for a review; for a selection of more recent work see [2]).

More recently, a novel approach to SED has been developed, in the form of a nonperturbative treatment for the particle embedded in the zeropoint field [3], in an attempt to solve the problems faced by the theory when dealing with nonlinear forces (notable the Coulomb force). The basic idea is that the background field close to the particle is so fundamentally modified in the long run by radiation reaction, that a treatment based on classical trajectories perturbed by the free vacuum field becomes meaningless. With this approach it has been possible to establish contact with the usual formalism of quantum mechanics, under certain conditions and approximations. Specifically, and as a practical procedure, one looks for *stationary* solutions to the stochastic dynamical (Abraham–Lorentz type) equation for the particle, which can be expressed in terms of a set of characteristic frequencies with *nonrandom* values. This requisite leads to a set of algebraic equations which, in the radiationless approximation, coincide fully with Heisenberg's equations of motion when expressed in matrix form. The solutions turn out to satisfy the demand of detailed energy balance. The mechanical system is seen to respond *linearly* to the zeropoint field, without frequency mixing; it behaves as if composed of a set of linear oscillators, with the characteristic frequencies and oscillator strengths given by the well–known equations of the quantum formalism. The mode amplitudes of the effective field engaged in maintaining these stationary quantum states are of course no longer the amplitudes of the original, free zeropoint field; in particular, correlations appear between different field modes.

Up to this point the theory has thus been able to address two most conspicuous properties of the quantum domain, namely, the random properties of matter that underlie the statistical nature of the usual description, and the quantization phenomenon, efficiently and succintly expressed by the Heisenberg equations of motion. However, there is a further most characteristic quantum feature that has received little attention, namely the undulatory behaviour of matter. Needless to say, the undulatory nature of the zeropoint field itself, which is a basic constituent of the SED system, is expected to play an important role in this connection. Having learned that some of the most basic quantum properties of matter can be understood as acquired by its interaction with the vacuum, it would be even surprising if this field did not in some form impress also its *wave* properties on the particles. This simple consideration, when properly handled, should allow us to understand some of the most mysterious aspects of quantum systems, as we shall discuss in the present paper; initial attempts along the same direction within the framework of SED can be found in [4] and [5].

2 Vibrations Induced by the Zeropoint Field

We recall that in de Broglie's theory of matter waves [6], a wave of Compton's frequency $\omega_C = mc^2/\hbar$ is associated with the point electron at rest, and by a Lorentz transformation the phase of this wave for a moving electron is shown to have just de Broglie's wavelength. This wave is considered by de Broglie to be a physical entity, although he does not further specify the nature of it. The point of view that will be presented here is somewhat different: we start by recognizing that the electron is a radiating particle embedded in the zeropoint field; this provides in principle the possibility to understand de Broglie's wave as a phenomenon closely linked to the radiation field in permanent interaction with the electron. Before discussing the dynamics of the problem, let us briefly introduce the description of the field that will be used in the following.

2.1 Description of the Radiation Field

Since, as will be seen below, one is basically interested in the stationary solution of the dynamical equations, the most suitable procedure is a time Fourier analysis, which implies a spectral decomposition of the electromagnetic field. In free space it is normal to consider a finite volume of rectangular shape and perform a discrete development of the field components in terms of plane waves. However, it is convenient to write first a more general development that allows for other possible geometries of interest, and specialize afterwards, when convenient, to the plane–wave case.

Assume the space of interest to be delimited by boundaries of arbitrary geometry, on which the fields satisfy appropriate boundary conditions (typically, those of perfectly conducting walls); the vector potential for the field inside this volume can be written in the form

$$\mathbf{A}(\mathbf{x},t)=\sum_{\alpha}\left(\tilde{\mathbf{A}}_{\alpha}(\mathbf{x},\omega_{\alpha})e^{-i\omega_{\alpha}t}+\tilde{\mathbf{A}}_{\alpha}^{*}(\mathbf{x},\omega_{\alpha})e^{i\omega_{\alpha}t}\right) . \tag{1}$$

The subindex α represents the set of required indices for the specific development, including the polarization. Substituting in the Maxwell equations, with

$$\mathbf{B}=\nabla\times\mathbf{A}, \quad \mathbf{E}=-\nabla\Phi-\frac{1}{\mathbf{c}}\frac{\partial\mathbf{A}}{\partial\mathbf{t}} \tag{2}$$

in the Coulomb gauge, and introducing the transversality condition, one gets

$$\nabla^{2}\tilde{\mathbf{A}}_{\alpha}+k_{\alpha}^{2}\tilde{\mathbf{A}}_{\alpha}=0, \quad c^{2}k_{\alpha}^{2}=\omega_{\alpha}^{2}, \quad \nabla\cdot\tilde{\mathbf{A}}_{\alpha}=0 . \tag{3}$$

These equations along with the boundary conditions define a Hermitian eigenvalue problem; hence the modes corresponding to different eigenvalues are orthogonal. This allows to express the solutions in terms of a family of orthonormal functions, which is achieved writing

$$\tilde{\mathbf{A}}_{\alpha}(\mathbf{x},\omega_{\alpha})=b_{\alpha}\mathbf{G}_{\alpha}(\mathbf{x},\omega_{\alpha}) , \tag{4}$$

so that

$$\mathbf{A}(\mathbf{x},t)=\sum_{\alpha}\left(\mathbf{G}_{\alpha}(\mathbf{x},\omega_{\alpha})b_{\alpha}e^{-i\omega_{\alpha}t}+\mathbf{G}_{\alpha}^{*}(\mathbf{x},\omega_{\alpha})b_{\alpha}^{*}e^{i\omega_{\alpha}t}\right) , \tag{5}$$

and selecting the functions $\mathbf{G}_{\alpha}$ to satisfy the set of equations

$$\nabla^{2}\mathbf{G}_{\alpha}+k_{\alpha}^{2}\mathbf{G}_{\alpha}=0 , \tag{6}$$

$$\nabla\cdot\mathbf{G}_{\alpha}=0, \quad \int d^{3}x\,\mathbf{G}_{\alpha}^{*}\cdot\mathbf{G}_{\beta}=\delta_{\alpha\beta} . \tag{7}$$

In addition, products of any two $\mathbf{G}_{\alpha}$'s or any two $\mathbf{G}_{\alpha}^{*}$'s integrate to zero. The values of the amplitude coefficients b_{α} are fixed by the specific application. From equations (2) one obtains for the electric and magnetic fields:

$$\mathbf{E}(\mathbf{x},t)=i\sum_{\alpha}\frac{\omega_{\alpha}}{c}\left(\mathbf{G}_{\alpha}(\mathbf{x},\omega_{\alpha})b_{\alpha}e^{-i\omega_{\alpha}t}-\mathbf{G}_{\alpha}^{*}(\mathbf{x},\omega_{\alpha})b_{\alpha}^{*}e^{i\omega_{\alpha}t}\right) \tag{8}$$

$$\mathbf{B}(\mathbf{x},t)=\sum_{\alpha}\left(\nabla\times\mathbf{G}_{\alpha}(\mathbf{x},\omega_{\alpha})b_{\alpha}e^{-i\omega_{\alpha}t}+\nabla\times\mathbf{G}_{\alpha}^{*}(\mathbf{x},\omega_{\alpha})b_{\alpha}^{*}e^{i\omega_{\alpha}t}\right) . \tag{9}$$

Specifically, the vacuum field is a stochastic radiation field which averages to zero and possesses an energy $\frac{1}{2}\hbar\omega$ per normal mode. In the absence of material objects that alter its structure, it can be represented in terms of plane waves of the form

$$\mathbf{E} = i\sum_{n,\sigma}\sqrt{\frac{\pi\hbar\omega_n}{V}}\hat{\epsilon}_{n\sigma}a^0_{n\sigma}\exp i\left(\mathbf{k}_n\cdot\mathbf{x}-\omega_n t\right)+\text{c.c.}, \tag{10}$$

where V is the normalization volume, $\hat{\epsilon}_{n\sigma}$ is the polarization vector, n, σ stand for the wavenumber and polarization indices, and the energy per normal mode is obtained by calculating the average of the Hamiltonian:

$$\langle H\rangle = \frac{1}{8\pi}\int d^3x\left\langle \mathbf{E}^2+\mathbf{B}^2\right\rangle .$$

The amplitudes $a^0_{n\sigma}$ are random variables assumed to have the following statistical properties:

$$\begin{aligned} &\langle a^0_{n\sigma}\rangle = 0; && \langle a^{0*}_{n\sigma}\rangle = 0\ ;\\ &\langle a^0_{n\sigma}a^0_{n'\sigma'}\rangle = 0\ ; && \langle a^{0*}_{n\sigma}a^{0*}_{n'\sigma'}\rangle = 0\ ;\\ &\langle a^0_{n\sigma}a^{0*}_{n'\sigma'}\rangle = \delta_{nn'}\delta_{\sigma\sigma'}\ , \end{aligned} \tag{11}$$

where the average is taken over all realizations of the field. The last three equations express the statistical independence of the different modes. (When convenient, one uses a single greek letter instead of the double subindex used here to characterize a mode; thus, $a^0_{n\sigma}\to a^0_{\alpha}$.) The field described by Eqs. (10) and (11) is homogeneous, isotropic, and has a spectral energy density

$$\rho(\omega) = \frac{\hbar\omega^3}{2\pi^2c^3}\ . \tag{12}$$

In the presence of boundaries the spatial density of modes is modified, the field ceases to be homogeneous and the spectral energy density is accordingly affected, although the average energy per normal mode remains unchanged. This is the origin of the Casimir force between conducting objects, for example (see, e.g., [7]).

2.2 Approximate Abraham–Lorentz Equation and Frequency Cutoff

It has been customary in SED to work with the stochastic Abraham–Lorentz equation

$$m\ddot{\mathbf{x}} = m\tau\,\dddot{\mathbf{x}} + \mathbf{F}(\mathbf{x}) + \mathbf{F}_L \tag{13}$$

where $\mathbf{F}(\mathbf{x})$ is the external applied force, $\mathbf{F}_L$ represents the Lorentz force exerted on the particle by the random vacuum field and $m\tau\,\dddot{\mathbf{x}}$ is the radiation reaction force, with $\tau = 2e^2/3mc^3$ ($\approx 10^{-23}$s for the electron). The problem of solving Eq. (13) is normally simplified by considering the electric component only of the Lorentz force and writing it in the long–wavelength approximation, under the assumption that the relevant components of this field have a wavelength very long compared with the dimensions of the system; this procedure serves to study the low-frequency motions of the charged particle.

However, we would like to recall at this point that the Abraham–Lorentz equation is only approximate and that its problems with causality and runaway solutions can be considered to be a consequence of the approximations introduced to derive it. Indeed, a detailed study of the dynamics of the composite system shows that the interaction of the point particle with the electromagnetic field (including its own one) dresses the particle not only by changing its mass, but also by endowing it with an *effective structure*, that arises from the fine oscillations generated by the *very high-frequency* modes of the self-field [8]. Inversely, it can be shown that any charged particle with structure performs an oscillatory movement due to radiation reaction, with frequency of the order of Compton's frequency $\omega_C = mc^2/\hbar$[9]. So instead of ignoring these modes as is normally done, we propose to study their specific dynamical effect on the particle. But before doing this, let us quickly look at the general relation between the electric structure (real or effective) of a particle and the appearance of a cutoff in the frequency of the field modes acting on the particle.

To establish this relation in a quantitative way, assume that the particle has a structure represented by a charge density $\rho_s(\mathbf{r})$ (normalized to unity) and write this density in terms of its Fourier transform

$$\rho_s(\mathbf{r}) = \int d^3k\, \tilde{\rho}_s(\mathbf{k}) e^{i\mathbf{k}\cdot\mathbf{r}}, \quad \int d^3r\, \rho_s(\mathbf{r}) = 1 \,. \tag{14}$$

Similarly, write the local electric field as

$$\mathbf{E}(\mathbf{r}) = \int d^3k\, \tilde{\mathbf{E}}(\mathbf{k}) e^{i\mathbf{k}\cdot\mathbf{r}} \,. \tag{15}$$

The effective electric field on the particle is then given by this field averaged over the structure:

$$\mathbf{E}_{\text{eff}}(\mathbf{r}) = \int d^3r' \rho_s(\mathbf{r}')\mathbf{E}(\mathbf{r}+\mathbf{r}') = 8\pi^3 \int d^3k\, \tilde{\rho}_s(\mathbf{k})\tilde{\mathbf{E}}(\mathbf{k}) e^{i\mathbf{k}\cdot\mathbf{r}} \,, \tag{16}$$

where we have used the convolution theorem for Fourier transforms

$$\int d^3r' F(\mathbf{r}+\mathbf{r}')G^*(\mathbf{r}') = (2\pi)^3 \int d^3k\, \tilde{F}(\mathbf{k})\tilde{G}^*(\mathbf{k}) e^{i\mathbf{k}\cdot\mathbf{r}} \,. \tag{17}$$

A comparison of equations (15) and (16) shows that the consideration of a structure represented by $\rho_s(\mathbf{r})$ is equivalent to the introduction of a cutoff factor

$$K(\mathbf{k}) = 8\pi^3 \tilde{\rho}_s(\mathbf{k}) \,. \tag{18}$$

Assume for simplicity a structure with spherical symmetry, so that this factor depends only on the magnitude of $\mathbf{k}$, $k = \omega/c$. Then, for example for the charge distribution $\rho_s(r) = \left(\pi a^2\right)^{3/2} e^{-r^2/a^2}$ one gets $K(\omega) = e^{-a^2\omega^2/4c^2} = e^{-(\omega/\omega_c)^2}$ with $\omega_c = 2c/a$. This formula establishes an inverse relationship between the size of the charge structure and the value of the cutoff frequency for the specific model; of course the numerical coefficient depends on the charge distribution, but up to a factor of order one the disclosed relationship is general, $a\omega_c \sim c$.

The appearance of the cutoff means that the field components of very high frequencies $\omega \gtrsim c/a$, where a represents the extent of the structure, uncouple from the particle and become ineffective. These frequencies correspond to time intervals shorter than a/c, so that if $a > \tau c \approx r_c$, that is, if the particle extends over a radius larger than its classical radius, then all noncausal effects associated with the Abraham–Lorentz equation disappear. According to this discussion, even a cutoff frequency as high as $c/r_c \sim mc^2/\alpha\hbar = 137\omega_C$ ($\omega_C = mc^2/\hbar$) would solve the problem of causality; however, there are various arguments that favour a lower value for the cutoff. In particular, the nonrelativistic description fails at frequencies ω such that the energy $\hbar\omega$ becomes of order mc^2; at the velocities associated with such wavelengths also the magnetic effects previously neglected become appreciable. Considerations of this kind and other more general ones [8] suggest a cutoff frequency in the region of $\omega_C = mc^2/\hbar$; in fact, below this selection will prove to be the appropriate one.

In addition, from the formula for the mass correction obtained along the usual derivation of the Abraham–Lorentz equation for the point particle

$$\frac{\delta m}{m} = \frac{2\tau}{\pi} \int d\omega \tag{19}$$

it is clear that such a frequency cutoff gives a finite and reasonable value for the mass correction coefficient, of order α. Hence for a particle with a finite extension also the mass correction is finite.

It should be clear that in the present context the cutoff is introduced not because it is possible or practical, as is often the case in specific calculations, but because it is the only way to recover the validity and consistency of the description. Of course, the cutoff can have its origin in a combination of several factors, where the effective structure of the radiating particle is one of them but others may be generated by dynamical effects (for an example see [10]). In the case of the electron, it is known from high–energy scattering experiments probing very small distances that its extension is not bigger than $\sim 10^{-15}$ cm, which is two orders of magnitude smaller than the classical radius. To closely pack the hypothetical constituents of the electron into such a small structure would require fantastic kinetic energies, exceeding $10^4 mc^2$ —as follows from the Heisenberg inequalities—, so the picture of a *jiggling* point electron seems indeed more natural; this is, actually, the point of view adopted usually in QED (see, e.g., [11], p. 403).

We shall now show with a very simple example how this effective cutoff provides a mechanism leading to sustained high–frequency oscillations of the particle when it is embedded in the zeropoint radiation field. Even if the treatment is nonrelativistic, the example has the heuristic value of exhibiting the oscillations without pretending to capture the details of the effective structure of the electron.

2.3 Particle Within Conducting Walls

We shall study the dynamics of an electron (or any charged point particle) that is embedded in a field subject to boundary conditions. The motivation for this study is that the electron itself imposes boundary conditions on the surrounding field, thus endowing it with a spatial structure; however, it is more convenient for calculational purposes to look for simpler boundary geometries than that of an assumedly spherical electron. The simplest boundary conditions one can use for the electromagnetic field are those imposed by a planar surface made of a perfectly conducting material, as has been done repeatedly in connection with the theory of dispersion and van der Waals forces. Let us therefore assume here that our particle is close to a perfectly conducting wall but is otherwise free, which means that no other interaction between it and the wall is being considered; the only force acting on it is the Lorentz force due to the vacuum and to its own radiation field. For a nonrelativistic study of the motion of the particle we use the equation

$$m\ddot{\mathbf{x}} = e\mathbf{E}_{\mathrm{rr}} + e\mathbf{E} \ , \tag{20}$$

where $\mathbf{E}$ is the electric component of the vacuum field and $\mathbf{E}_{\mathrm{rr}}$ is the field radiated by the particle. In the long–wavelength approximation and assuming no boundary conditions, the latter term would have the familiar form $m\tau\dddot{\mathbf{x}}$, as in Eq. (13), but in the present case it assumes a more complicated structure.

The fields can be written in terms of normal modes, as above; to work with sums rather than integrals we temporarily assume the existence of a second wall at a distance L which in due time we take as infinite. Then the zeropoint field is given by equation (8); for the radiated field one gets by a Hamiltonian procedure applied to the composite (particle+field) system [8]

$$\mathbf{E}_{\mathrm{rr}}(t) = -2\pi e \sum_{\mathbf{n},\sigma} \mathbf{G}_{\mathbf{n}\sigma}(\mathbf{x})) \int_0^\infty ds\, \mathbf{G}^*_{\mathbf{n}\sigma}(\mathbf{x}(t-s)) \cdot \dot{\mathbf{x}}(t-s) e^{i\omega_n s} + \text{c.c.} \tag{21}$$

For particles that move slowly enough (i.e. nonrelativistically) along their trajectory, one can approximate

$$\left|\dot{\mathbf{G}}_{\mathbf{n}\sigma}\right| \sim \omega_n \left|\frac{\dot{x}}{c}\right| |\mathbf{G}_{\mathbf{n}\sigma}| \ll \omega_n |\mathbf{G}_{\mathbf{n}\sigma}| \ ; \tag{22}$$

and thus take $\mathbf{G}^*_{\mathbf{n}\sigma}(\mathbf{x}(t-s))$ out of the integral in (21).

Neglecting an initial transient term, the stationary solution of equation (20) has then the form

$$\dot{\mathbf{x}} = \mathbf{v} + \sum_{\mathbf{n},\sigma} (f_n \mathbf{G}_{\mathbf{n}\sigma} a_{\mathbf{n}\sigma} + \text{c.c.}) \ , \tag{23}$$

with $\mathbf{v}$ a systematic (constant) velocity,

$$f_n = \frac{e\,(\pi\hbar\omega_n)^{1/2}}{m\omega_n\,(1+i\delta_n)} \ , \tag{24}$$

and

$$\delta_n = \frac{4\pi e^2}{3m\omega_n} \sum_{\mathbf{m},\lambda} |\mathbf{G}_{\mathbf{m}\lambda}|^2 \int_0^\infty ds\, e^{i\omega_n s} \cos\omega_m s \ . \tag{25}$$

The average kinetic energy associated with the stochastic motion is, by equations (11)–(25),

$$\langle T_s\rangle = \frac{1}{2}m\left\langle (\dot{\mathbf{x}} - \mathbf{v})^2\right\rangle = \frac{\pi\hbar e^2}{m} \sum_{\mathbf{n},\sigma} \frac{|\mathbf{G}_{\mathbf{n}\sigma}|^2}{\omega_n\,|1 + i\delta_n|^2} \ . \tag{26}$$

Let us assume that the walls are perfectly conducting, up to a certain (very high) frequency ω_c at which they become transparent. Different electric or magnetic properties of the material would lead to a modification of the $\mathbf{G}_{\mathbf{n}\sigma}$, and the calculations would become more involved. However, the end effect would be substantially the same, since as will become clear below, what is important is that the walls impress a spatial structure on the modes of the field, up to the frequency ω_c. For a field confined between two vertical, parallel conducting plates of area S, separated a distance L one has

$$\mathbf{G}_{n1} = \sqrt{\frac{2}{LS}}\, e^{i\mathbf{k}_\rho\cdot\rho}\left(\hat{\mathbf{k}}_\rho \times \hat{\mathbf{x}}\right)\sin\xi,$$

$$\mathbf{G}_{n2} = \sqrt{\frac{2}{LS}}\, e^{i\mathbf{k}_\rho\cdot\rho}\left(\frac{k_x}{k_n}\hat{\mathbf{k}}_\rho \sin\xi + i\frac{k_\rho}{k_n}\hat{\mathbf{x}}\cos\xi\right) \ , \tag{27}$$

with ρ lying in the present case on the yz–plane of the plates and $\hat{\mathbf{x}}$ perpendicular to it; x is the distance of the particle to the left plane, and index $\sigma = 1, 2$ has been chosen to represent waves of linear polarization. The wave vector is $\mathbf{k}_n = \mathbf{k}_\rho + \mathbf{k}_x$, with $k_x = \pi n_x/L$, n_x an integer, and $\xi = k_x x$. From equations (27) one gets

$$\sum_\sigma |\mathbf{G}_{\mathbf{n}\sigma}|^2 = \frac{2}{LS}\left(1 - \frac{k_x^2}{k_n^2}\cos 2\xi\right) \ . \tag{28}$$

As the area of the walls is made infinite, the triple sum over the indices of the normal modes (n_x, n_y, n_z) transforms into:

$$\frac{1}{S}\sum_{\mathbf{n}} \underset{S\to\infty}{\longrightarrow} \frac{1}{4\pi}\sum_{n_x}{}' \int_{k_x^2}^{k_c^2} dk_n^2 \ , \tag{29}$$

where $\sum'$ is a sum over $n_x \geq 0$ with the first term multiplied by $\frac{1}{2}$, and $k_c = \omega_c/c$ is the upper limit for the wave number. The expression for the kinetic energy becomes then, by equations (26) and (28) and with $\omega = ck$:

$$\langle T_s\rangle = \frac{e^2\hbar}{2mc^2L}\sum_{n_x}{}' \int_{\omega_x}^{\omega_c} d\omega_n^2 \frac{1 - \frac{\omega_x^2}{\omega_n^2}\cos 2\xi}{\omega_n\,|1 + i\delta_n|^2} \ , \tag{30}$$

with δ_n obtained from equations (25) and (28), giving Re$\delta_n = \tau_x\omega_n$, where τ_x depends on x but is always of the order of τ (in the absence of boundary conditions, $\tau_x = \tau = 2e^2/3mc^3$), and Im δ_n of the order of $\tau\omega_c$. Since this term represents a radiative correction to the mass of the particle, it may be included it in the renormalized mass, and one can thus write

$$\langle T_s \rangle = \frac{e^2\hbar}{mc^2 L} \sum_{n_x}{}' \int_{\omega_x}^{\omega_c} d\omega \frac{1 - \frac{\omega_x^2}{\omega_n^2}\cos 2\xi}{1 + \tau_x^2 \omega_n^2} \ . \tag{31}$$

After integrating one obtains

$$\langle T_s \rangle = \frac{e^2\hbar}{mc^2 \tau_x L} \sum_{n}^{N}{}' F(n) \ , \tag{32}$$

where to lowest order in τ,

$$F(n) = \eta_c \left(1 - \frac{n}{N} - \frac{n}{N}\cos\zeta n + \frac{n^2}{N^2}\cos\zeta n \right) \ , \tag{33}$$

with

$$\eta_c = \tau_x \omega_c, \quad N = \frac{L\omega_c}{\pi c}, \quad \zeta = \frac{2\pi x}{L} \ . \tag{34}$$

N is the number of smallest wavelengths that fit between the two walls; since ω_c is very large, N has a huge value, whereas the parameter η_c is much smaller than one. For example, for $\omega_c \sim \omega_C, \eta_c = \tau_x\omega_c \sim e^2/\hbar c = \alpha$; observe, incidentally, that the mass correction referred to above becomes also of order α. Performing the summation one gets

$$\sum_{n=0}^{N}{}' F(n) = \frac{\eta_c N}{2} \times$$

$$\times \left(1 + \frac{N + N\cos\zeta N + \cos(N+1)\zeta + \cos\zeta N}{2N^3 \sin^2\frac{\zeta}{2}} - \frac{\sin(N+1)\zeta\cos\frac{\zeta}{2}}{2N^3\sin^3\frac{\zeta}{2}} \right) \ . \tag{35}$$

Very close to any of the walls this expression goes to

$$\sum_{n=0}^{N}{}' F(n) \underset{\zeta \to 0, 2\pi}{\longrightarrow} \frac{1}{3} N \eta_c \ , \tag{36}$$

whereas in the space between them,

$$\sum_{n=0}^{N}{}' F(n) = \eta_c \frac{N}{2} \left(1 + \frac{\cos^2\frac{N\zeta}{2}}{N^2\sin^2\frac{\zeta}{2}} \right), \quad \zeta \neq 0, 2\pi \ , \tag{37}$$

and the average kinetic energy becomes, by equation (32),

$$\langle T_s \rangle = \frac{\alpha}{2\pi} \frac{\hbar^2 \omega_c^2}{mc^2} \left(1 + \frac{\cos^2 k_c x}{N^2 \sin^2(\pi x/L)} \right) \equiv \langle T_1 \rangle + \langle T_2 \rangle \ , \quad x \neq 0, L \ . \tag{38}$$

The first term $\langle T_1 \rangle$ is a mass correction that is independent of the wall separation and of the position of the particle; it exists even in the absence of any body. On the other hand, the second term $\langle T_2 \rangle$ oscillates with the high wave number k_c, even for large separations L, so that in the presence of a single wall it attains the value:

$$\langle T_2 \rangle \underset{L\to\infty}{\longrightarrow} \frac{\alpha}{2\pi} \frac{\hbar^2}{mx^2} \cos^2 k_c x \; . \tag{39}$$

This result shows that the particle is set to vibrate with wave number k_c.

2.4 Vibrations of Neutral Particles

Similar calculations can be performed for neutral particles having a charge distribution. Take, for instance, a small electric vibrating dipole of natural oscillation frequency $\omega_0 \ll \omega_c$, moving freely in the vicinity of a conducting wall. To study the motion of the dipole moment $\mathbf{q} = e\mathbf{x}$, one must introduce into equation (20) an additional harmonic force term:

$$m\ddot{\mathbf{q}} = -m\omega_0^2 \mathbf{q} + e^2 \mathbf{E_{rr}} + e^2 \mathbf{E} \tag{40}$$

and in equation (21) $e\dot{\mathbf{x}}$ must be replaced by $\dot{\mathbf{q}}$. Writing the stochastic part of the solution of equation (40) as previously,

$$\mathbf{q} = \sum_{\mathbf{n},\sigma} \left(g_n \mathbf{G}_{\mathbf{n}\sigma} a_{\mathbf{n}\sigma} + \text{c.c.} \right) \; , \tag{41}$$

one obtains

$$g_n = \frac{-ie^2 \left(\pi\hbar\omega_n \right)^{1/2}}{m\Delta_n} \tag{42}$$

with $\Delta_n = \omega_n^2 - \omega_0^2 - i\tau_x\omega_n^3$; the function τ_x is as before. The average force of the random background field on the dipole is given by

$$\langle \mathbf{F} \rangle = \left\langle \left(\mathbf{q} \cdot \nabla \right) \mathbf{E} + \frac{1}{c} \dot{\mathbf{q}} \times \mathbf{B} \right\rangle$$

$$= -\frac{\pi e^2 \hbar}{m} \sum_{\mathbf{n},\sigma} \frac{\omega_n}{\Delta_n} \left(\left(\mathbf{G}_{\mathbf{n}\sigma} \cdot \nabla \right) \mathbf{G}^*_{\mathbf{n}\sigma} + \mathbf{G}_{\mathbf{n}\sigma} \times \left(\nabla \times \mathbf{G}^*_{\mathbf{n}\sigma} \right) \right) + \text{c.c.} \tag{43}$$

With $\mathbf{G}_{\mathbf{n}\sigma}$ given by equations (27), this leads after some simplifications to

$$\langle \mathbf{F} \rangle = -\nabla \left\langle U(\mathbf{x}) \right\rangle \; , \tag{44}$$

with

$$\langle U \rangle = -\frac{e^2 \hbar}{mc^2 L} \sum_{\mathbf{n},\sigma}{}' \int_{\omega_x}^{\omega_c} d\omega \frac{\omega_x^2 (\omega^2 - \omega_0^2) \cos 2\xi}{(\omega^2 - \omega_0^2)^2 + \tau_x^2 \omega^6} \; , \tag{45}$$

where $\xi = \omega_x x/c$, $\omega_x = \pi n_x c/L$, and the sum extends to $N = L\omega_c/\pi c$. In the limit $\omega_0 = 0$, equation (45) coincides with the expression (31) for $\langle T_s \rangle$, and hence we obtain the same end result, equation (38). For usual values of ω_0, which are much smaller than ω_c, the additional (ω_0–dependent) contribution to U is negligible. Hence the electric dipole exhibits essentially the same oscillation phenomenon as the electric charge, so that the whole treatment applies not only to charged particles, but also to neutral particles having electromagnetic interactions; this is an important result for SED.

Incidentally, note that instead of following the above procedure one could have chosen to calculate the total average energy of the dipole: $\langle T_s + V_s \rangle = \frac{m}{2e^2} < \dot{\mathbf{q}}^2 + \omega_0^2 \mathbf{q}^2 >$, with $\mathbf{q}$ given by equation (41), in the same approximation $\omega_0 \ll \omega_c$. By this method one arrives at the result obtained above, plus the constant ground–state energy of the oscillator $\frac{3}{2}\hbar\omega_0$, due to the contribution of the integrand at the pole, $\omega = \omega_0$.

2.5 The Vibrating Electron

The calculations leading to equation (39) show that it is mainly the field modes of wave numbers close to k_c that are effective in producing the fine vibrations of the particle, the action of the low–frequency field being irrelevant in this respect. This has two important consequences. In the first place, it means that the same phenomenon can be induced on the particle by any appropriate object (of whatever size and shape) capable of affecting the spatial structure of these higher–frequency modes in the close vicinity of the particle. This object may be the particle itself, if it is assumed that it imposes some sort of boundary conditions on the field. Under this assumption the presence of a second body is not necessary to induce the fine vibrations of the particle.

In the second place, observe from (26) that $\langle T_s \rangle$ does not depend on the state of motion, represented by the first term on the right–hand side of (23); it depends only on intrinsic properties of the particle and the zeropoint field. Indeed, the only condition for the validity of the result (39) is that (22) holds; hence it is valid in general for a particle moving with a nonrelativistic systematic velocity $\dot{x}$, which may itself be a function of time as a result of an external force acting on the particle. This conclusion is confirmed by a simple observation: for an order–of–magnitude calculation one can model an external binding force with a term of the form $-m\omega_{\text{eff}}^2 x$, with ω_{eff} appropriately defined, but always much smaller than ω_c. Now for the jiggling motion discussed here, with frequency ω_c, $|\ddot{x}| \simeq |\omega_c^2 x| \gg |\omega_{\text{eff}}^2 x|$; hence the free particle is an appropriate model for the high–frequency motion, and the details of the specific dynamics are irrelevant.

It is therefore possible to formally isolate the two motions, previously identified as the low– and the high–frequency components; however, below we will find that there is a profound relationship between them, though a very indirect one. Note that the ratio of the 'low' to the 'high' frequencies is of order (typical atomic energy$/mc^2$) $\sim \alpha^2$; this large gap separates the two scales of wavelengths that together conform the quantum phenomenon. They are 'complementary' in the sense that the low frequencies refer to mechanical and spectral properties characteristic of atoms and other bound systems, whereas the high frequencies refer basically to the oscillations of the atomic constituents themselves, whether bound or not.

This simple model for the motion of the free particle as a superposition of a constant (systematic) velocity and the jittering movement, is in agreement with what is observed in experimental situations; consider, for instance, the high precision attained in modern electron beam technology, where velocity deviations of less than one part in 10^6 are normal. In conclusion, the model electron performs a very fine jiggle, superposed to an approximately smooth motion involving frequencies that are smaller by several orders of magnitude.

In a strict sense, we have not obtained the result we were looking for, since instead of an oscillation in time with frequency ω_c we got a spatial vibration with wave number $k_c = \omega_c/c$. This seems to be as far as one can go with a crude nonrelativistic model; note, however, that for a relativistic vibration with velocity of order c both results would be equivalent. This is not as absurd as it may look at a first glance; jerky movements of the electron with velocity c have been considered not only in relation with zitterbewegung, but also in relativistic models of the spinning electron [12] and other instances, such as relativistic treatments of the electron as a stochastic particle. Two examples of the latter kind of theories, widely different in their conceptual framework, but both involving motions with speeds arbitrarily close to c, are given in Ref. [13]. We should mention that other mechanisms to generate zitterbewegung–like oscillations have been explored in the SED literature; see e.g. [14]. For many years, Cavalleri has insisted that zitterbewegung should be considered a fundamental property lying at the basis of quantum phenomena [15]; this point has been taken up in [4].

3 Genesis of de Broglie's Wavelength

Our purpose now is to understand why the vibrations with frequency ω_C play such a very special role in the behavior of the electron, as was proposed by de Broglie. The kinematic analysis is parallel to that of de Broglie's; it should be borne in mind, though, that there is a difference between the two theories, inasmuch as the present one deals with waves of the electromagnetic radiation field instead of a 'clock' wave of unspecified nature.

Let us consider the simplest possible example, that of a free electron moving with velocity v; as already discussed above, one can actually consider this velocity to be a slowly varying function of time (or of the coordinates, depending on the description). Also, there is no problem in considering a neutral point particle instead of a charged one. Now since the particle is performing fine oscillations with frequency ω_C, it radiates at this frequency. In a *stationary* state of motion, such radiation must be compensated by the interaction with the vacuum field. This means that the particle interacts intensely with the modes of frequency ω_C as measured in its proper frame. The specific mechanism of this interaction is irrelevant for the kinematics, and it will certainly be different for a neutral particle and for an electron; what is important is that the particle interacts selectively with just these modes. Now suppose that the electron moves with velocity v with respect to the laboratory and along the $+x$ direction, and consider for the moment the two waves of the zeropoint field which propagate along the $+x$ and $-x$ directions; their frequencies in the laboratory frame are respectively:

$$\omega_+ = \gamma\omega_C(1+\beta), \quad \omega_- = \gamma\omega_C(1-\beta) \ , \tag{46}$$

with $\gamma = (1-\beta^2)^{-1/2}$ and $\beta = v/c$. In a more convenient notation one may write

$$\omega_\pm = \omega_z \pm \omega_B; \quad \omega_z = \gamma\omega_C; \quad \omega_B = \gamma\beta\omega_C \ , \tag{47}$$

where $\gamma\omega_C$ is a shifted frequency associated to the Lorentz contraction of time and ω_B is a Doppler shift. The wave number transforms accordingly into

$$k_\pm = \gamma k_C \pm k_B, \quad k_B = \gamma\beta k_C \ . \tag{48}$$

As seen from the reference frame of the electron, the two waves form a standing wave of wavelength $\lambda_C = 2\pi/\omega_C$; seen from the laboratory, they still propagate in opposite directions but with frequencies $\omega_\pm$ and corresponding wave numbers $k_\pm$. To describe the ensuing travelling wave, we write the random mode amplitudes a_n^0 of the vacuum field in the form $a_n^0 = e^{i\theta_n}$, with uniformly distributed random phases θ_n such that $\langle e^{i(\theta_n-\theta_{n'})}\rangle = \delta_{nn'}$; the magnitude of the amplitude is taken as constant only for simplicity —as is frequently done in the SED literature—, though this is of little importance in the present context. The superposition of the waves φ_+, φ_- travelling in opposite directions gives

$$\varphi(v) = \varphi_+ + \varphi_- = 4\cos(\omega_z t - k_B x + \theta_1)\cos(\omega_B t - \gamma k_C x + \theta_2) \ , \tag{49}$$

with $\theta_{1,2} = \frac{1}{2}(\theta_+ \pm \theta_-)$ two random phases, statistically independent. For $v = 0$ the above expression reduces to

$$\varphi(0) = 4\cos(\omega_C t + \theta_1)\cos(k_C x - \theta_2) \ , \tag{50}$$

representing the original standing wave; Eq. (49), instead, represents a travelling wave that is temporally and spatially modulated. Keeping x fixed one sees that the carrier amplitude is modulated by a factor that oscillates in time with the Doppler frequency ω_B. On the other hand, keeping t fixed one sees that the fine oscillations (of wave number γk_C) are spatially modulated, k_B being the corresponding wave number; hence there is a modulation of wavelength

$$\lambda_B = \frac{2\pi}{k_B} = \frac{\lambda_C}{\gamma\beta} = \frac{m_0\lambda_C c}{p} \ , \tag{51}$$

where $p = m_0\gamma v = m_0 c\gamma\beta$ is the linear momentum of the particle. (An index 0 is added here to the rest mass, to distinguish it from $m = \gamma m_0$). Notice that λ_B originates in the Doppler shift of the frequency; for $v = 0$, there is no Doppler shift and λ_B becomes infinite. It is not difficult to verify that all these results remain valid when the velocity v is a slowly changing function of time or position; the only requisite is that the fractional changes of β with t or x be very small with respect to $2\pi/\omega_C$ or λ_C, respectively, conditions that are well satisfied in nonrelativistic motions. Although only waves along the direction of the particle's motion were considered in equation (49), it is an easy matter to show that waves propagating orthogonally (or in any other direction) produce the same phenomenon.

Now we observe that the first factor in Eq. (49) represents a wave with frequency ω_z given by the dispersion relation

$$\omega_z^2 = \omega_C^2 + c^2 k_B^2 \,, \tag{52}$$

and its group velocity is therefore equal to the velocity of the particle:

$$v_g = \frac{\partial \omega_z}{\partial k_B} = \frac{c^2 k_B}{\omega_z} = v \tag{53}$$

while its phase velocity u is given, as follows from equations (47) and (48), by

$$u = \frac{\omega_z}{k_B} = \frac{c^2}{v} \,. \tag{54}$$

Further, the phase of the time modulation factor travels with the velocity of the particle, as follows from (47):

$$u_m = \frac{\omega_B}{\gamma k_C} = v \,. \tag{55}$$

Thus the velocity of the particle determines the basic features of the carrier wave, namely, the group velocity v and the phase velocity u, and also the wavelength of the modulation, according to (51). Note that of all these properties only λ_B depends on the cutoff frequency. With $\omega_C = mc^2/\hbar$, as suggested in the above discussion, the modulation wavelength is given by the familiar de Broglie formula:

$$\lambda_B = \frac{2\pi\hbar}{p} = \frac{h}{mv} \tag{56}$$

which constitutes the basic principle of wave mechanics. This is, in essence, the mechanism by which the high–frequency phenomena become linked to the low–frequency domain of quantum mechanics. Although the selection of ω_C has at this stage a degree of arbiotrariness, the demand of consistency between the predictions derived from this part of the theory and the rest of SED leaves no other choice.

The foregoing discussion assigns a very specific physical meaning to the de Broglie's wave: it is the *modulation* of the wave formed by the Lorentz–transformed, Doppler–shifted superposition of the whole set of random stationary electromagnetic waves of frequency ω_C with which the electron interacts selectively; the properties of this modulation depend of course on the relative velocity v of the electron in the laboratory. Although the jiggling frequency ω_C and the frequency shift $\gamma\omega_C$ are of relativistic nature, the modulation arises essentially as a consequence of Doppler shifts on the component waves, and its motion can be studied safely in the nonrelativistic approximation, as long as $v \ll c$. The modulated wave is of course as real as the component waves that give rise to it, which are part of the zeropoint field; hence, it should exhibit all phenomena typical of waves.

3.1 The Wave Equation

Along the preceding discussion the dynamics of particle and wave have become intertwined: the motion of a particle with velocity v through the zeropoint field gives rise to a modulated wave in the laboratory frame; in its turn, this wave acts on the particle that is selectively coupled to it. Hence any influence of the surroundings on this wave can in principle have a dynamical effect on the particle. For instance, the building up of a stationary modulated wave is essential for the existence of a stationary state of motion of a bound particle; this and other more complicated wavelike effects on the particle will be commented upon further below. It is clear that a more detailed description of the wave associated with the particle is convenient at this point.

We start by observing that from the two factors comprising equation (49), which are of a very different scale, it is the first one, representing the spatial envelope of the modulated wave, which should interest us; at the scale of quantum mechanics the fine spatial vibrations introduce at most a finite effective radius and a finite upper response frequency for the particle, as discussed above. Now this first factor depends on both particle and wave properties; in particular it has a group velocity equal to v and a phase velocity $u = c^2/v$. This information can be used to construct a wave equation for the amplitude ϕ associated with this factor. Note that since we are isolating one of the factors that constitute the physical wave $\varphi(v)$, the amplitude ϕ will *not* in itself represent an electromagnetic wave; its geometrical meaning is clear from the fact that the phase velocity is $u > c$. The corresponding (three–dimensional) wave equation is

$$\nabla^2\phi - \frac{1}{u^2}\frac{\partial^2\phi}{\partial t^2} = 0\ . \tag{57}$$

From Eq. (49) it follows that $\partial^2\phi/\partial t^2 = -\omega_z^2\phi$, since it has been assumed that the time variation of v is slow, which means that $\left|\dot{k}_B/k_B\right| \ll \omega_z$; hence one gets with ω_z given by (47):

$$\nabla^2\phi + \frac{\omega_C^2}{c^2}\gamma^2 v^2\phi = 0\ , \tag{58}$$

or, introducing $E = p^2/2m + V$ with $p = m\gamma v$,

$$\nabla^2\phi + \frac{2m_0}{\hbar^2}(E - V)\phi = 0\ . \tag{59}$$

Note that the stationary equation just obtained still contains the energy of the *particle*, although it has been derived for a *wave amplitude*. The particle maintains its (almost) pointlike character and moves in space following a certain trajectory. The associated wave, on the other hand, can have a considerable spatial extension; indeed, for a free electron of well–defined velocity the wave train is in principle infinitely long, in the absence of barriers, according to the above equations. Even in a more realistic picture describing an ensemble of electrons with nearly the same velocity $\mathbf{v}$, both the length and the width of the wave train extend for a good number of wavelengths, as was already inferred by Thomson in 1930 from the results of his early diffraction experiments. With the present-day techniques used in high-resolution electron microscopy for the production of monochromatic electron beams, it is possible to obtain a lateral coherence typically of 10^3 wavelengths or more, and wave packets much longer than this [16]; but it is important to note that these practical limits are due to the angular spread and the energy dispersion of the actual electron beam (represented by the corresponding wave packet). In other words, at the present degree of precision the wave associated with a single (free) electron has indeed an infinite length, in agreement with the description provided by Eq. (59).

On the other hand, it is well known that the Schrödinger equation

$$E\psi = -\frac{\hbar^2}{2m}\nabla^2\psi + V\psi \,; \tag{60}$$

with ψ the probability amplitude, gives a correct statistical description of the particle aspects of a quantum system. Since the wave amplitude ϕ of Eq. (59) and the probability amplitude ψ of Eq. (60) must satisfy the same general mathematical requisites of continuity and square integrability, we see that in the stationary case $\phi \propto \psi$, and thus the probability density $\rho = \psi^*\psi$ becomes proportional to the wave intensity $\phi^*\phi$. This conclusion is strongly reminiscent of de Broglie's (1956) principle of the double solution.

The Schrödinger equation is therefore seen to describe simultaneously the amplitude of the *modulation wave* associated with a particle in terms of a wave amplitude ϕ, and the configuration space distribution for an *ensemble of particles* in terms of the probability amplitude ψ. However, this is not to say that the quantum corpuscle is (or behaves as) either a particle or a wave, in correspondence with the specificities of the setup. Quite on the contrary, both wave and particle have their place in nature and in theory, without excluding each other; they are clearly distinct and coexisting entities, each one with its own life and fate, though closely connected to each other through the electromagnetic interactions. We have thus two complementary descriptions, which taken together seem to embrace the complex behaviour of the electron.

The quantum transition frequencies are normally several orders of magnitude smaller than the Compton frequency. Nonrelativistic quantum mechanics affords thus a description over distances much larger than Compton's wavelength and times much longer than the inverse of Compton's frequency. Such description contrasts with the starting point used here, where the field modes of extremely high frequencies are essential to account for the wavelike behaviour of matter. At the level of quantum mechanics the high-frequency waves remain hidden, but their large-scale effect becomes apparent through the modulated wave. These two alternative and complementary views of the quantum system recall the almost simultaneously born original theories of Schrödinger and Heisenberg, and though each of these two views puts its accent on different features, they share the fundamental equation, namely the Schrödinger equation, which becomes thus the unifying element.

4 The Wave Properties of Particles

4.1 Wave Stationarity and Phase Quantization

Although it has been well known for many decades, it continues to be amazing that the wave equation derived to describe the modulation amplitude fully coincides with the Schrödinger equation that describes the behaviour of (an ensemble of) particles. Here we shall dig a little into the heuristics of this marvelous 'coincidence' and establish contact with some aspects of usual quantum mechanics, with the help of a few simple problems; we shall try to see, in particular, how the quantization rules emerge, with the aim of reaching a more intuitive understanding of this most basic phenomenon. Of course, one could say that formally such contact with quantum mechanics has already been established by arriving at the Schrödinger equation; however, we may refresh our physical intuition through the analysis of these simple examples.

Consider first a particle and its associated wave, in an infinite potential well in one dimension of width L. The particle moves with a systematic velocity v; since the walls are perfectly rigid, it bounces back and forth between them at constant speed. In a stationary situation the associated zeropoint modulation will form a stationary pattern, which according to Eq. (49) is of the form $\sim \cos(k_B x)$. Now since the particle is interacting with the stationary wave, it will remain in its stationary state of motion only if it encounters this wave in the same condition after each complete cycle, or after a displacement $\Delta x = 2L$. With the phase of the stationary modulation given by $k_B x$, this condition of stationarity means

$$k_B \Delta x = 2Lk_B = 2\pi n \; , \tag{61}$$

with n an integer, or (we exclude $n = 0$ because it means $v = 0$)

$$n\lambda_B = 2L, \qquad n = 1, 2, 3, \ldots \tag{62}$$

With λ_B given by Eq. (56) this result reads (in the nonrelativistic approximation)

$$p = mv = \frac{nh}{2L}, \quad \text{or} \quad E = \frac{p^2}{2m} = \frac{h^2 n^2}{8mL^2} \; , \tag{63}$$

which means that only for certain (discrete) values of v does the condition of stationarity hold; the corresponding allowed values of E coincide with the result obtained in quantum mechanics for the energy of a particle in an infinite potential well. Note that this description does not provide the instantaneous values for the position and the momentum of the particle; all one can say is that their ranges of variation are $\Delta x = L$ and $\Delta p = nh/L$.

This simple example points to a very interesting dynamical role played by the stationary modulation wave: it stabilizes the particle's motion around certain values of the velocity (or the energy), which in cases as the present one are determined by the geometry of the accessible space or the boundaries of the system. Of course from this kinematical approach one cannot tell how stable these stationary states are, nor does it provide any information on the dynamical process by which the particle *reaches* one such state starting from arbitrary initial conditions. What is remarkable, however, is that the particle's states of motion are determined (causally, of course) by nonlocal properties of the system, indirectly through the radiation field which acts as the intermediary. Since a stationary situation is being considered in which particle and field are already in equilibrium, the radiation field here is not simply the original, free zeropoint field, but the complete equilibrium field including radiation reaction, all subject to the appropriate boundary conditions.

A heuristic analysis similar to the above one can be performed in other simple cases in which the particle describes a closed orbit with constant velocity v. Take for instance the circular motion with radius r about a fixed axis. The periodicity condition on the wave reads, from equation (61), $2\pi r k_B = 2\pi n$, and with $k_B = mv/\hbar$ by equation (56) one gets the quantization of angular momentum for circular motion:

$$l = mvr = n\hbar \ . \tag{64}$$

The picture suggested by this calculation is not far from the popular image that has emerged from de Broglie's theory, of a standing wave filling the circumference of the orbit with exactly an integer number of half–waves.

Phase quantization appears whenever the particle performs a periodical motion in phase space, even if this motion proceeds with a variable velocity $v(x)$. For such cases one can attempt a generalization of the above results, by imposing the appropriate requisite of stationarity on the modulation. For motions with variable velocity one has $k_B = k_B(x)$, and the phase increment:

$$\Delta\phi = \int dx\, k_B = \frac{1}{\hbar}\int dx\, p = \frac{m}{\hbar}\int dx\, v \tag{65}$$

along one orbit is $\Delta\phi = \hbar^{-1}\oint dx\, p(x)$; hence the condition of quantization becomes, by equation (56),

$$\oint p\, dx = m\oint v\, dx = nh \ . \tag{66}$$

This formula coincides with the original Wilson–Sommerfeld quantization rule, which gives correct results only in some particular instances, notably when the particle is reflected head–on by sharp boundaries and v is constant along the trajectory. One should expect equation (66) to be a good approximation whenever the smoothed–out path of the particle coincides with the path of the associated wave. When the particle moves in a shallow potential well, things become complicated because the partial penetration of the modulated wave into a region where $p = \sqrt{2m(E-V)} < 0$ can contribute a considerable amount to its phase increment $\Delta\phi$, whereas it does not contribute to the (classical) phase integral $\int p\, dx$. Moreover, variations of $V(x)$ contribute also to alter the amplitude of the modulated wave, an effect that is not taken into account in writing equation (49), though it is of course properly described by equation (59).

For a general analysis of the quantization of a bounded, periodic or nonperiodic motion in more than one dimension, it would be necessary to consider the complete set of zeropoint waves interacting with the moving particle and to proceed with the corresponding derivation of the modulated wave and its eventual phase quantization. Unfortunately the problem becomes very involved and it is not possible to extract detailed conclusions without a close scrutiny, which would require a detailed development of the wave description. A very suggestive illustration of the potentialities of this approach, presented in a relativistic context, is given in Synge's monograph [17], where it is applied to the study of the Dirac equation.

Through the spatial extension of the modulation wave, the influence of a given setup on the associated particle will be very different in general from the influence this setup would have on the particle alone. We have just seen some examples of this difference, which refer to simple cases of stationary states for bound particles. More generally speaking, nonlocal effects arise, which become apparent through the influence of the modified wave on the particle. Consider, for instance, Young's double–slit experiment performed with electrons. The wave associated with each electron is diffracted and interferes with itself after passing through the slits. After having gone through either one of the slits, the particle enters the interference region, where it is influenced by the modified field, and finally lands at some specific point along the screen. If this experiment is performed with a collimated, monochromatic beam of (many noninteracting) electrons, sent either one by one or simultaneously, but without changing the setup, the coherence of the associated wave is not lost, and thus eventually the well–known electron interference pattern is built up on the screen. If one ignores the role played by the background field in these processes, one can be led to conclude that each electron 'interferes with itself'. Let us look at this kind of phenomena from the present point of view.

4.2 Interference of Particles

We shall refer to experiments made with a common type of particle interferometer, made of three slabs cut from a silicon monocrystal [18]-[20]. A beam of neutrons of $\sim 1\mathring{A}$ wavelength incident on the first slab is split by Bragg scattering into two coherent beams I and II; these beams are then redirected as they traverse the second slab, to recombine and interfere on the third slab. The phases of beams I and II can be changed by various means, thus affecting the interference pattern at D; the relative phase at D determines the counting rates at the detectors situated beyond the posterior face of the crystal. The neutrons have a relatively low velocity, of $\sim 10^5$ cm/sec, which makes it easy to control electronically any measurements that must be performed. Note, incidentally, that although the experiments are performed with *particles* (normally with neutrons), the wave terminology is freely used to describe what happens in the interferometer.

Originally the particles move between the slabs with an essentially constant velocity $v_0 = p_0/m = \sqrt{2E/m}$. Now assume that the upper beam is subject to an external field that varies locally within the interferometer, and call $V(x)$ the corresponding potential; according to equation (65) this produces a relative phase shift of the associated wave (whose wave front follows the trajectory of the particles), given by:

$$\Delta\phi = \frac{1}{\hbar}\int dx(p-p_0) = \frac{\sqrt{2m}}{\hbar}\int dx\,(\sqrt{E-V}-\sqrt{E}).$$

Under normal experimental conditions, $|V(x)| \ll E$ (typically $|V|/E \sim 10^{-7}$); hence one can altogether neglect the variation of the wave amplitude, and for the phase shift one gets

$$\Delta\phi \simeq -\frac{1}{\hbar}\int dx\,\frac{V(x)}{v(x)} = -\frac{1}{\hbar}\int dt\,V\,, \tag{67}$$

where the integration is of course performed along the path of the perturbed beam.

Several different experiments have been performed to which this result is applicable. For instance, with the interferometer in a vertical position, beam II is subject to a slightly smaller gravitational force; a difference in height of 2 cm between the two beams amounts to a difference in potential energy of about 10^{-9} eV. The direct mechanical effect of this difference on the neutrons is minute; but the phase shift that can be produced by it over a distance of about 10cm is of the order of 100 rad! [19]

In a different experiment [20] a uniform magnetic field **B** is placed in one of the beams, causing the neutron spin to precess around the direction $\hat{\mathbf{n}}$ of the field. The phase shift is in this case given by equation (67) with

$$V = -(e/mc)\mathbf{S}\cdot\mathbf{B} = -2\omega_L\mathbf{S}\cdot\hat{\mathbf{n}}\,, \tag{68}$$

where $\mathbf{S} = (\hbar/2)\hat{\sigma}$ is the spin angular momentum and $\omega_L = eB/2mc$ is the Larmor frequency; hence

$$\Delta\phi = \frac{2\omega_L}{\hbar}(\mathbf{S}\cdot\hat{\mathbf{n}})\,t = \omega_L\,(\hat{\sigma}\cdot\hat{\mathbf{n}})\,t\,. \tag{69}$$

On the other hand, during the same time interval t the neutrons rotate about the axis $\hat{\mathbf{n}}$ through an angle given by $\theta = \omega_B t$, where the frequency of rotation is $\omega_B = eB/mc = 2\omega_L$, hence equation (69) can be written as

$$\Delta\phi = \frac{1}{2}(\hat{\sigma}\cdot\hat{\mathbf{n}})\,\theta\,, \tag{70}$$

where $\hat{\sigma}\cdot\hat{\mathbf{n}} = \pm 1$. Note that the beam does not have to be spin–polarized for this shift to take place, since the spins are always set to precess around $\hat{\mathbf{n}}$ by the external magnetic field. An interesting feature of this phenomenon is that the phase shift is given by half the rotation angle; hence a spin rotation of 4π is needed to produce a phase shift of 2π (and thus recover the original interference pattern), as is well known from the (quantum) theory of spin.

A related effect that has been extensively discussed in the literature, and tested in the laboratory, is of course the Aharonov–Bohm effect [21]. In this case the perturbation is produced by a magnetic vector potential $\mathbf{A}(\mathbf{x})$, which can give rise to a phase shift even if the corresponding magnetic field is zero: $\nabla \times \mathbf{A} = 0$, so that there is no direct mechanical effect on the particles. The phase shift can again be obtained from equation (65), by considering that the velocity of the particles in terms of the linear momentum changes from $m\mathbf{v} = \mathbf{p}$ to $m\mathbf{v} = \mathbf{p} - (e/c)\mathbf{A}$, whence

$$\Delta\phi = -\frac{e}{\hbar c}\int d\mathbf{x} \cdot \mathbf{A} \tag{71}$$

If the two beams are in a region where $\mathbf{A}(\mathbf{x}) \neq 0$, the total relative phase shift is simply given by the integral over the closed circuit: $\Delta\phi = (-e/\hbar c)\oint d\mathbf{x} \cdot \mathbf{A}$. From this example, as from the previous ones, one sees that in an interference experiment the particles can be significantly deviated from their path by a mere phase shift, even when there are no external forces present, to the extent that a phase shift of 180° causes the particles to hit the screen on regions where originally no particles landed. This is a particularly impressive effect of the action of the background field on the particles.

4.3 Some Reflections on the Wave Function

Suppose that in the double–slit experiment, neutrons are being sent one at the time, and at some point one decides to intervene with the purpose of determining which slit a certain particle passed through [18]. If this cannot be done without affecting its state of motion, and consequently also its associated wave, this particular neutron will not contribute to the interference pattern that has been building up; this does not mean, however, that the whole of this pattern is destroyed. In fact, it shouldnt; if the experimenter has been careful enough not to upset the setup, he may abandon his pretension to determine the precise path of the particles and resume his original experiment with the neutrons contributing to the formation of the interference pattern. The pattern will be formed with those particles that belong to the coherent beam; particles hitting the screen without being part of the coherent beam will only add noise to the interference fringes and reduce their visibility.

If, however, it becomes possible to determine the path of every single particle without significantly altering its state of motion and destroying the coherence of the beam, an interference pattern with good fringe visibility can be obtained whereas at the same time it is clear that every neutron has travelled through only one of the two branches of the interferometer.

Let us now try to focus on the meaning of the wave function, drawing from all what has been said above. To make the discussion simple we refer here to one–particle stationary situations, such as those of the bound–particle problems or the conventional interference experiments just mentioned, for which ψ is the solution of the stationary Schrödinger equation (60) with appropriate boundary conditions. In previous paragraphs we have indistinctly referred to a representative particle or to an ensemble of equivalent particles, all subject to the same experimental conditions, as being associated with a modulated wave; a statistical meaning is of course assigned to the wave function. In addition, however, observe that ψ is not expected to refer to the particle (or the ensemble) as an abstraction, but always in a given setup or context. It would therefore be more correct to say that ψ refers to a *situation*; any particle (or any number of independent particles, whether a whole beam, or just one, or none at all) that is found in that situation will be correctly described by the same ψ —appropriately normalized—. Conversely, any particle that ceases to be in that situation (for instance, by being absorbed by a detector) does not form part any more of the ensemble considered; another ψ (an atomic wave function, say) will be necessary to describe its new situation.

All our analysis has been restricted here to systems of one particle or beams of equivalent, noninteracting particles. The extension to systems of two or more particles is not trivial, as is well known, and new phenomena appear, which are completely unknown in the classical domain, such as the phenomena related to the statistics of quantum particles and notably the Pauli exclusion principle. Since SED has basically not been applied to many–particle systems, such phenomena remain beyond the scope of the present work.

Acknowledgments

The authors wish to express their sincere thanks to P. Garbaczewski and other organizers of the 1995 Karpacz Winter School for their kind invitation and their warm hospitality, which made this meeting even more fruitful and agreeable.

References

1. L. de la Peña, in *Stochastic Processes Applied to Physics and Other Related Fields*, B. Gómez et al., eds. (World Scientific, Singapore, 1983).
2. T.W. Marshall and E. Santos, in *Problems in Quantum Physics: Gdansk '87*, L. Kostro et al., eds. (World Scientific, Singapore, 1988);
A. V. Barranco, S.A. Brunini and H.M. França, Phys. Rev. **A39**, 5492 (1989);
A.M. Cetto and L. de la Peña, Found. Phys. **19**, 419 (1989);
A. Rueda, Space Science Rev. **53**, 223 (1990); D.C. Cole, Found. Phys. **20**, 225 (1990).
3. L. de la Peña and A.M. Cetto, Rev. Mex. Fis. **39**, 653 (1993), Found Phys. **24**, 917 (1994), **25** (1995), in press.
4. A.F. Kracklauer, Phys. Essays **5**, 226 (1992).
5. L. de la Peña and A.M. Cetto, Ann. Fond. L. de Broglie (special issue) 311 (1993), Found. Phys. **24**, 753 (1994);
A.M. Cetto and L. de la Peña, in *Proceedings of the Oviedo Symposium on Fundamental Problems in Quantum Physics*, M. Ferrero and A. van der Merwe, eds (World Scientific, Singapore, 1995).
6. L. de Broglie, *Une Tentative d'Interprétation causale et non linéaire de la Mécanique Ondulatoire* (Gauthier–Villars, Paris, 1956); C.R.Acad. Sci. Paris

277B, 71 (1973). For a modern presentation of de Broglie's theory see F. Selleri, *Quantum Paradoxes and Physical Reality* (Kluwer, Netherlands, 1990).

7. T.W. Marshall, Nuovo Cim. **38**, 206 (1965); T.H. Boyer, Phys. Rev. **174**, 1764 (1968).
8. L. de la Peña and A.M. Cetto, book in preparation, chapter 3.
9. L. de la Peña, J.L. Jiménez and R. Montemayor, Nuovo Cim. **B 69**, 71 (1982).
10. G.H. Goedecke, Found. Phys. **13**, 1121 (1983), appendix B.
11. P.W. Milonni, *The Quantum Vacuum* (Academic Press, San Diego 1994).
12. See, e.g., H.C. Corben, *Classical and Quantum Theories of Spinning Particles* (Holden–Day, San Francisco, 1968) and references therein.
13. W.J. Lehr and J.L. Park, J. Math. Phys. **18**, 1235 (1977); B. Gaveau, T. Jacobson, M. Kac and L.S. Schulman, Phys. Rev. Lett. **53**, 419 (1984).
14. A. Rueda, Found. Phys. Lett. **6**, 75 (1993).
15. G. Cavalleri, Lett. Nuovo Cim. **43**, 285 (1985).
16. See, e.g., J.C.H. Spence, *Experimental High–Resolution Electron Microscopy*, 2nd. ed. (Oxford U.P., Oxford, 1988), chapter 4.
17. J.L. Synge, *Geometrical Mechanics and de Broglie Waves* (Cambridge U.P., Cambridge, 1954).
18. See e.g. D.M. Greenberger, Rev. Mod. Phys. **55**, 875 (1983).
19. R. Colella, A.W. Overhauser and S.A. Werner, Phys. Rev. Lett. **34**, 1472 (1975).
20. H. Rauch, A. Zeilinger, G. Badurek, A. Wilfing, W. Bauspiess and U. Bonse, Phys. Lett. **A54**, 425 (1975).
21. Y. Aharonov and D. Bohm, Phys. Rev. **115**, 485 (1959).

Lévy Processes and Relativistic Quantum Dynamics

*Piotr Garbaczewski**

Institute of Theoretical Physics, University of Wrocław,
PL–50 204 Wrocław, Poland

Abstract: The traditional Gaussian framework (Wiener process as the "free noise", with the Laplacian as noise generator) is extended to encompass any infinitely divisible probability law covered by the Lévy–Khintchine formula. It implies a family of random environment models (of the fluctuating medium) governed by the generally non–Gaussian "free noises". Since the so called relativistic Hamiltonians $|\nabla|$ and $\sqrt{-\Delta + m^2} - m$ are known to generate such laws, we focus on them for the analysis of probabilistic phenomena, which are shown to be associated with the relativistic quantum propagation once an analytic continuation in time of the corresponding holomorphic semigroup is accomplished. The pertinent stochastic processes are identified to be spatial jump processes.

The Schrödinger equation and the generalized heat equation are connected by analytic continuation in time, known to take the Feynman–Kac (holomorphic semigroup) kernel into the Green function of the corresponding quantum mechanical problem. For $V = V(x), x \in R$, bounded from below, the generator $H = -2mD^2\Delta + V$ is essentially selfadjoint on a natural dense subset of L^2, and the kernel $k(x, s, y, t) = [\exp[-(t - s)H]](x, y)$ of the related dynamical semigroup is strictly positive. The quantum unitary dynamics $\exp(-iHt)$ is a final result of the analytic continuation.

As repeatedly emphasized [1, 2, 3, 4, 5], any temporal evolution that is analyzable in terms of a probability measure may be interpreted as a stochastic process. In view of the Born statistical interpretation postulate for quantum mechanics, the analytic continuation in time induces a class of probability measures, namely, consider $\rho(x, t) = |\psi(x, t)|^2$ as the density of a probability measure associated with a given solution $\psi(x, t)$ of the Schrödinger equation. Then, it is possible to address the problem of that stochastic dynamics which would be either (i) measure preserving or (ii) induce the time evolution of the measure proper. Keep in mind that the Schrödinger equation itself *is not* a genuine partial differential equation of probability theory; rather it is the Born postulate which embeds the unitary evolution problem into the probabilistic framework.

* Supported by the KBN research grant No 2 P302 057 07

A simple illustration of the analytic continuation in time is provided by considering the force–free propagation, where the formal recipe gives rise to the equations of motion (one should be aware that to execute a mapping for concrete solutions, the proper adjustment of the time interval boundaries is indispensable):

$$i\partial_t \psi = -D\Delta\psi \longrightarrow \partial_t \theta_* = D\Delta\theta_* \ ,$$
$$i\partial_t \overline{\psi} = D\Delta\overline{\psi} \longrightarrow \partial_t \theta = -D\Delta\theta \ , \tag{1}$$
$$it \rightarrow t \ .$$

Then

$$\psi(x,t) = \left[\rho^{1/2}\exp(iS)\right](x,t) = \int dx' G(x-x',t)\psi(x',0) \ ,$$
$$G(x-x',t) = (4\pi i D t)^{-1/2} \exp\left[-\frac{(x-x')^2}{4iDt}\right] \ , \tag{2}$$
$$\theta_*(x,t) = \int dx' k(x-x',t)\theta_*(x',0) \ ,$$
$$k(x-x',t) = (4\pi D t)^{1/2} \exp\left[-\frac{(x-x')^2}{4Dt}\right] \ ,$$

The description in terms of the time adjoint pair of equations is not accidental and reflects the Markov property of probabilistic solutions of the associated Schrödinger problem: find an interpolation between the given pair of boundary (for the process on a finite fixed time interval) probability distributions.

Strictly positive semigroup kernels generated by Laplacians plus suitable potentials are very special examples in a surprisingly rich encompassing family. First of all, the concept of the "free noise", normally characterized by a Gaussian probability distribution appropriate to a Wiener process, can be extended to all infinitely divisible probability distributions via the Lévy–Khintchine formula. It expands our framework from continuous diffusion processes to jump or combined diffusion–jump propagation scenarios. All such (Lévy) processes are associated with strictly positive dynamical semigroup kernels.

Remark: Apart from the wealth of physical phenomena described in terms of Gaussian stochastic processes, there is a number of physical problems where the Gaussian tool–box proves to be insufficient to provide satisfactory probabilistic explanations. Non–Gaussian Lévy processes naturally appear in the study of transient random walks when long–tailed distributions arise [7, 8, 9]. They are also found necessary to analyze fractal random walks [10], intermittency phenomena, anomalous diffusions, and turbulence at high Reynolds numbers [7, 12, 11].

Let us consider Hamiltonians of the form $H = F(\hat{p})$, where $\hat{p} = -i\nabla$ stands for the momentum operator and for $-\infty < k < +\infty$, $F = F(k)$ is a real valued, bounded from below, locally integrable function. Then, $\exp(-tH) = \int_{-\infty}^{+\infty} \exp[-tF(k)]dE(k)$, $t \geq 0$, where $dE(k)$ is the spectral measure of $\hat{p}$.

Most of our discussion will pertain to processes in one spatial dimension, and let us specialize the issue accordingly. Because $(E(k)f)(x) = \frac{1}{\sqrt{2\pi}} \int_{-\infty}^{k} \exp(ipx)\hat{f}(p)dp$, where $\hat{f}$ is the Fourier transform of f, we learn that

$$[\exp(-tH)]\, f(x) = \left[\int_{-\infty}^{+\infty} \exp(-tF(k))dE(k)f\right](x) =$$

$$\frac{1}{\sqrt{2\pi}} \int_{-\infty}^{+\infty} \exp\left[-tF(k)\right] \frac{d}{dk} \left[\int_{-\infty}^{k} \exp(ipx)\hat{f}(p)dp\right] dk = \tag{3}$$

$$\frac{1}{\sqrt{2\pi}} \int_{-\infty}^{+\infty} \exp(-tF(k)) \exp(ikx)\hat{f}(k)dk = \left[\exp(-tF(p))\hat{f}(p)\right]^{\vee}(x)$$

where the superscript $\vee$ denotes the inverse Fourier transform.

Let us set $k_t = \frac{1}{\sqrt{2\pi}}[\exp(-tF(p))]^{\vee}$, then the action of $\exp(-tH)$ can be given in terms of a convolution: $\exp(-tH)f = f * k_t$, where $(f * g)(x) := \int_R g(x - z)f(z)dz$.

We shall restrict consideration only to those $F(p)$ which give rise to positivity preserving semigroups: if $F(p)$ satisfies the celebrated Lévy–Khintchine formula, then k_t is a positive measure for all $t \geq 0$. The most general case refers to a contribution from three types of processes: deterministic, Gaussian, and an exclusively jump process. We shall concentrate on the integral part of the Lévy–Khintchine formula, which is responsible for arbitrary stochastic jump features:

$$F(p) = -\int_{-\infty}^{+\infty} \left[\exp(ipy) - 1 - \frac{ipy}{1+y^2}\right] \nu(dy) \tag{4}$$

where $\nu(dy)$ stands for the so–called Lévy measure.

The disregarded Gaussian contribution would read $F(p) = p^2/2$. In this case we know in detail how the analytic continuation in time of the Laplacian generated holomorpic semigroup induces a mapping to a quantum mechanical (since the Schrödinger equation is involved) diffusion processes [5, 2, 3].

Our further attention will focus on two selected choices for the characteristic exponent $F(p)$, namely: $F_0(p) = |p|$ and $F_m(p) = \sqrt{p^2 + m^2} - m, m > 0$, where we have chosen suitable units so as to eliminate inessential parameters. (The relativistic Hamiltonian is better known in the form $\sqrt{m^2c^4 + c^2p^2} - mc^2$ where c is the velocity of light.)

The respective Hamiltonians (semigroup generators) H_0, H_m are pseudodifferential operators. The semigroup kernels k_t^0, k_t^m in view of the "free noise" restriction (no potentials, will be defined in below) are transition densities of the jump (Lévy) processes regulated by the corresponding Lévy measures $\nu_0(dy)$, $\nu_m(dy)$. It is instructive to notice that as in the case of Gaussian derivations

(1),(2), the pseudodifferential analog of the Fokker–Planck equation can be introduced. Namely, as a consequence of $[\exp(-tH)\overline{\rho}](x) = \overline{\rho}(x,t)$ and in view of the identification $F(p \to -i\nabla) := H$ we arrive at

$$F_0(p) \Longrightarrow \partial_t \overline{\rho}(x,t) = -|\nabla|\overline{\rho}(x,t) \tag{5}$$

or

$$F_m(p) \Longrightarrow \partial_t \overline{\rho}(x,t) = -\left[\sqrt{-\Delta + m^2} - m\right] \overline{\rho}(x,t) \tag{6}$$

respectively.

Although the pseudodifferential generator of the semigroup implies that the Fokker–Planck equation is no longer exclusively differential but an integro–differential equation, each solution $\overline{\rho}(x,t)$ in the present case is nevertheless a solution of a partial differential equation of higher order. Specifically, the respective partial differential equations are of the second order, see[4]. Our two semigroups are holomorphic, hence we can replace the time parameter t by a complex one $\sigma = t + is$, $t > 0$ so that $\exp(-\sigma H) = \int_R \exp(-\sigma F(k))\, dE(k)$. Its action is defined by

$$[\exp(-\sigma H)]\, f = \left[(\hat{f} \exp(-\sigma F)\right]^{\vee} = f * k_\sigma \ . \tag{7}$$

Here, the kernel reads $k_\sigma = \frac{1}{\sqrt{2\pi}}[\exp(-\sigma F)]^{\vee}$. Since H is selfadjoint, the limit $t \downarrow 0$ leaves us with the unitary group $\exp(-isH)$, acting in the same way: $[\exp(-isH)]f = [\hat{f} \exp(-isF)]^{\vee}$, except that now $k_{is} := \frac{1}{\sqrt{2\pi}}[\exp(-isF)]^{\vee}$ in general is *not* a measure. In view of unitarity, the unit ball in L^2 is an invariant of the dynamics. Hence density measures can be associated with solutions of the Schrödinger pseudodiferential equations:

$$F_0(p) \Longrightarrow i\partial_t \psi(x,t) = |\nabla|\psi(x,t) \tag{8}$$

or

$$F_m(p) \Longrightarrow i\partial_t \psi(x,t) = \left[\sqrt{-\Delta + m^2} - m\right] \psi(x,t) \tag{9}$$

provided with the appropriate initial data functions $\psi(x,0)$.

An obvious consequence is that the corresponding partial differential equation of the second order takes on a familiar *relativistic* form

$$F_0(p) \Longrightarrow \Box\psi(x,t) := (-\Delta + \Delta_t)\psi(x,t) = 0 \tag{10}$$

while after setting $\psi(x,t) = \tilde{\psi}(x,t)\exp(imt)$, we arrive at the Klein–Gordon equation:

$$F_m(p) \Longrightarrow (\Box + m^2)\tilde{\psi}(x,t) = 0 \tag{11}$$

where the D'Alembert operator $\Box = -\Delta + \Delta_t$ replaces its Euclidean counterpart $-\Box_E$.

We have thus reached a point, at which our major question can be precisely stated:

What are the stochastic processes consistent with the probability measure dynamics $\rho(x,t) = |\psi(x,t)|^2$, determined by the pseudodifferential equations (8) and (9)?

We have chosen two rather special pseudodifferential counterparts of the Laplacian guided by two reasons: (i) their similarity on analytic grounds (the same criteria [13] for the existence of the bound state spectrum if summed with suitable potentials, (ii) the claim of Ref. [14] that the pertinent stochastic process in the mass $m > 0$ case actually displays the Markov property.

If the Markov property would hold true for the relativistic Hamiltonian generated dynamics, we would be able to repeat almost all steps of the previous, Schrödinger picture, quantum dynamics analysis [1, 2, 3]. However, the situation is not that simple, and the argument of [4] excludes the Markov property, in all nonstationary situations, in a flat contradiction with general statements by De Angelis [14].

Let us introduce some probabilistic notions, which will tell us how to work with pseudodifferential operators. We shall notice that for explicit computational purposes, the Cauchy generator $|\nabla|$ is much more suited than the $m > 0$ relativistic Hamiltonian. It is a real disadvantage when dealing with Lévy processes that rather limited number of concrete examples is available, in contrast to the wealth of the general theory.

The Lévy–Khintchine formula tells us that the action of the Hamiltonian $H = F(-i\nabla)$ on a function in its domain can be represented as follows:

$$(H\,f)(x) = -\int_R \left[f(x+y) - f(x) - \frac{y\,\nabla f(x)}{1+y^2}\right] \nu(dy) \tag{12}$$

It is important to observe that for the "free noise" processes whose semigroup generators are $|\nabla|$ and $\sqrt{-\Delta + m^2} - m$ we do know explicitly their kernels (transition probability densities) and the involved Lévy measures, as well as about the extension of the Feyman–Kac *path integral* construction of the semigroup kernels to these particular Lévy processes [13], in case of arbitrary space dimensions. Therefore we feel free to use the Feynman–Kac kernel notion instead of the semigroup kernel.

For the Cauchy process, whose generator is $|\nabla|$, we deal with a probabilistic classics:

$$\overline{\rho}(x,t) = \frac{1}{\pi}\frac{t}{t^2+x^2} \Longrightarrow k^0(y,s,x,t) = \frac{1}{\pi}\frac{t-s}{(t-s)^2+(x-y)^2} \tag{13}$$

$$0 < s < t$$

$$\langle \exp\left[ipX(t)\right]\rangle := \int_R \exp(ipx)\overline{\rho}(x,t)\,dx = \exp\left[-tF_0(p)\right] = \exp(-|p|t)$$

The characteristic function of $k^0(y,s,x,t)$ for y, s fixed, reads $\exp[ipy - |p|(t-s)]$, and the Lévy measure needed to evaluate the Lévy–Khintchine integral reads:

$$\nu_0(dy) := \lim_{t\downarrow 0}\left[\frac{1}{t}k^0(0,0,y,t)\right] dy = \frac{dy}{\pi y^2} \tag{14}$$

In the case of the relativistic generator $\sqrt{-\Delta + m^2} - m$, formulas determining the stochastic jump process are much less appealing:

$$\langle \exp[ipX(t)] \rangle := \exp[-tF_m(p)] = \exp\left[-t(\sqrt{p^2+m^2} - m)\right]$$

$$\overline{\rho}(x,t) = \frac{m}{\pi}\frac{t\exp(mt)}{\sqrt{x^2+t^2}} K_1(m\sqrt{x^2+t^2}) \tag{15}$$

$$[\exp(-(t-s)F_m(-i\nabla))](x-y) = k^m(y,s,x,t) := \overline{\rho}(x-y, t-s)$$

$$\nu_m(dy) = \frac{m}{\pi|y|} K_1(m|y|)dy$$

where $K_1(z)$ is the modified Bessel function of the third kind of order 1.

We are interested in acting with the pseudodifferential generators $H = F(-i\nabla)$ on functions in the exponential form (recall the familiar Madelung procedure in the Gaussian case) $f(x,t) = \exp \Phi(x,t)$:

$$(H \exp \Phi)(x) = -\int_R \left[\exp \Phi(x+y) - \exp \Phi(x) - \frac{y\Phi'(x)\exp\Phi(x)}{1+y^2}\right] \nu(dy) =$$

$$= \exp\Phi(x) \int_R \left[\exp(\Phi(x+y) - \Phi(x)) - 1 - \frac{y\Phi'(x)}{1+y^2}\right] \nu(dy) \tag{16}$$

where $\Phi'(x) = \nabla\Phi(x)$. Since $(H\Phi)(x) = -\int_R [\Phi(x+y) - \Phi(x) - y\Phi'(x)/(1+y^2)]\nu(dy)$, we can make a safe rearrangement of (16):

$$(H\exp\Phi)(x) = \exp\Phi(x) \left[(H\Phi)(x) - \int_R (\exp\Phi_{xy} - 1 - \Phi_{xy})\nu(dy)\right] \tag{17}$$

$$\Phi_{xy} := \Phi(x+y) - \Phi(x)$$

In application to the pseudodifferential dynamics $i\partial_t\psi(x,t) = (H\psi)(x,t)$ with $\psi = \exp(R+iS)$, one easily derives [4] its implications for the real functions $\Theta = \exp(R+S)$ and $\Theta_* = \exp(R-S)$; plus a trivial extension from H to $H+V$ situations.

Remark : Experience [15, 2] with the Gaussian (standard Laplacian generated) noise proves that the Madelung substitution $\psi(x,t) = \exp[R(x,t)+iS(x,t)]$ would associate with the Schrödinger equation a pair of time adjoint generalised diffusion equations where the Feynman–Kac potential (time dependent in the general case) equals $\frac{1}{2mD}[2Q(x,t) - V(x)]$. Here $Q(x,t) = 2mD^2\frac{\Delta\rho^{1/2}}{\rho^{1/2}}(x,t)$ and $V(x)$ is taken as an external conservative force potential. Let us emphasize that $V(x)$ actually *was* the Feynman–Kac potential of the dynamical semigroup prior to the analytic continuation in time procedure. The mapping $V(x) \to 2Q(x,t) - V(x)$ is an effect of the analytic continuation in time, as manifested on the level of the associated [2, 3] Feynman–Kac kernels.

In view of (17), the pseudodifferential Schrödinger equation $i\partial_t\psi(x,t) = H\psi(x,t)$ implies the following time evolution of the Madelung exponents:

$$\partial_t R = HS - \int_R [\exp(R_{xy}) \sin S_{xy} - S_{xy}]\, d\nu(y)$$

$$\partial_t S = -HR + \int_R [\exp(R_{xy}) \cos S_{xy} - 1 - R_{xy}]\, d\nu(y) \tag{18}$$

where $H = F(-i\nabla)$.

By employing (17) with respect to $\rho^{1/2} = \exp(R)$, we arrive at:

$$Q := \frac{H\rho^{1/2}}{\rho^{1/2}} = HR - \int_R [\exp(R_{xy}) - 1 - R_{xy}]\, d\nu(y) \tag{19}$$

and hence:

$$\partial_t S = -Q + \int_R \exp(R_{xy})\, [\cos(S_{xy}) - 1]\, d\nu(y) \tag{20}$$

The same procedure can be repeated for $\Theta = \exp(R+S)$ and $\Theta_* = \exp(R-S)$, which implies:

$$\partial_t\Theta = H\Theta + \Theta\left[-2Q + \int_R \exp(R_{xy})[-\sin S_{xy} + \cos S_{xy} + \exp(S_{xy}) - 2]\, d\nu(y)\right] \tag{21}$$

$$\partial_t\Theta_* = -H\Theta_* + \Theta_*\left[2Q - \int_R \exp(R_{xy})\, [\sin S_{xy} + \cos S_{xy} + \exp(-S_{xy}) - 2]\, d\nu(y)\right]$$

In contrast to the Gaussian case [15, 2], equations (21) do not take the form of a time adjoint pair, unless some additional restrictions are imposed on the Madelung exponent $S(x,t)$ (notice that we have restored time dependence, skipped before for convenience). An obvious demand is $S(x+y,t) = S(x,t)$ for all y, t, and any fixed x. But then, equations (21) would manifestly refer to the *stationary* (measure preserving) random dynamics, governed by the pair of equations:

$$\partial_t\Theta = H\Theta - 2Q\Theta$$

$$\partial_t\Theta_* = -H\Theta_* + 2Q\Theta_* \tag{22}$$

which are mutually time adjoint. Hence they would fall into the Schrödinger problem framework[2, 4] with a trivial implication that the measure preserving process is Markovian. This however cannot be a property of the "free" dynamics since we need external potentials to secure stationarity. Let us therefore make an essential amelioration by performing the previous analysis for the case $i\partial_t\psi = (H+V)\psi$ with $V = V(x)$. Then, the stationary system of equations would take the form:

$$\partial_t\Theta = H\Theta - (2Q+V)\Theta \tag{23}$$

$$\partial_t \Theta_* = -H\Theta_* + (2Q + V)\Theta_*$$

which upon substituting $S(x,t) = Et$, where E is a constant, yields a pseudodifferential version of the Sturm–Liouville problem:

$$H\rho^{1/2}(x) - \left[2\frac{H\rho^{1/2}}{\rho^{1/2}} + V(x) - E\right]\rho^{1/2}(x) = 0 \tag{24}$$

$$\Downarrow$$

$$V(x) - E = -\frac{H\rho^{1/2}(x)}{\rho^{1/2}(x)}$$

to be solved (for a chosen value of E) with respect to the square root of the probability density $\rho(x)$, once the external force potential $V(x)$ is selected.

This problem has its Gaussian counterpart in the study of the measure preserving dynamics [2] , and in the present context it can be solved by invoking those potentials for the original pseudodifferential Schrödinger equation, for which the bound states (i.e., stationary solutions) have granted the existence status. The relevant analysis has been carried out in the studies of the relativistic stability of matter [13]. In addition we know that in the stationary case, the Feynman–Kac path integral generalization to Lévy semigroup kernels is available.

However, the Markov property cannot [4] automatically be attributed to the nonstationary dynamics, as described by (21).

The probability density $\rho(x,t)$ (respectively $\overline{\rho}(x,t)$) was a fundamental entity in our previous considerations: in fact, providing the time evolution of the probability measure for the whole time interval of interest, so that the transition probability densities could be sought for [4].

In the Gaussian case we dealt with the temporal evolution of the probability density given in its traditional Fokker–Planck form appropriate for Markov diffusion processes [2, 3]. In connection with the pseudodifferential ("free noise") dynamics, we address an obvious extension of the previous notion to a class of jump processes. We shall extend the usage of the name Fokker–Planck equation to any first order in time differential equation determining the space–time properties of $\rho(x,t)$ or $\overline{\rho}(x,t)$.

Let us investigate the time developement of $\overline{\rho}(x,t) = \theta(x,t)\theta_*(x,t)$, where $\theta(x,t)$, $\theta_*(x,t)$ come out as solutions of the temporally adjoint pair of equations of the form

$$\partial_t \theta = H\theta - V\theta \tag{25}$$

$$\partial_t \theta_* = -H\theta_* + V\theta_*$$

with a Feynman–Kac potential V. Then, in view of (17) and $\theta = \exp(R + S)$, $\theta_* = \exp(R - S)$, we get an evolution equation for the probability density:

$$\partial_t \overline{\rho}(x,t) = \theta_*(x,t)\,(H\theta)(x,t) \; - \; \theta(x,t)\,(H\theta_*)(x,t) = \tag{26}$$

$$\int_R \left[-\theta_*(x,t)\theta(x+y,t) + \theta(x,t)\theta_*(x+y,t) + 2\overline{\rho}(x,t)\nabla S(x,t)\frac{y}{1+y^2}\right] d\nu(y)$$

Following the traditional recipes when dealing with Lévy measures [6], let us consider an open neighborhood of the origin $|\epsilon| \ll 1$. Instead of integrating over all possible jump sizes, let us integrate over jumps of size $|y| > \epsilon > 0)$. The removal of this lower bound as $\epsilon \to 0$ will eventually amount to evaluating the principal value of the integral. In case $\epsilon > 0$, we can safely remove the compensating term including $y/(1+y^2)$ from the integral, and restrict considerations to the contribution from the first two terms only.

Our purpose is to establish a connection with the conventional theory of jump stochastic processes, as developed in [16]. Integrating over a Borel set $A \subset R$, $x \in A$ we get:

$$\int_A dx \int_{|y|>\epsilon} [-\theta_*(x,t)\theta(x+y,t) + \theta(x,t)\theta_*(x+y,t)]\, d\nu(y) =$$

$$\int_R dx \int_{|y|>\epsilon} \chi_A(x) \left[-\overline{\rho}(x,t)\frac{\theta(x+y,t)}{\theta(x,t)} + \overline{\rho}(x+y,t)\frac{\theta(x)}{\theta(x+y)}\right] d\nu(y) = \tag{27}$$

$$\int_R dx\, \overline{\rho}(x,t) \int_{|y|>\epsilon} \frac{\theta(x+y,t)}{\theta(x,t)} [\chi_A(x+y) - \chi_A(x)]\, d\nu(y)$$

where we interchanged the order of integrations, and made appropriate adjustments of integration variables ($x \to x - y$ and $y \to -y$), while exploiting the property $d\nu(-y) = -d\nu(y)$ of Lévy measures; $\chi_A(x)$ is an indicator function of the Borel set $A \subset R$, equal to 1 when $x \in A$ and 0 otherwise.

In the present case we deal with a Markov process with transition probability densities given for arbitrary time instants: $\overline{\rho}(x,t) = \int_R p(y,s,x,t)\overline{\rho}(y,s)dy$, $s < t$. By invoking the standard wisdom about jump Markov processes [16], and exploiting $\lim_{t\downarrow s} p(y,s,A,t) = \chi_A(y)$, for any Borel set $A \subset R$ away from $(-\epsilon, +\epsilon)$, we can define the jump process running with jumps of size $|y| > \epsilon > 0$. It should be viewed as an approximation of the original stochastic process governed by (26), with the initial data $\overline{\rho}(x,0)$ common for both:

$$\partial_t \overline{\rho}_\epsilon(A,t) = \int_R q(x,t,A)\, \overline{\rho}_\epsilon(x,t)\, dx + \langle v\rangle_A(t) \int_{|y|>\epsilon} \frac{y}{1+y^2} d\nu(y) \tag{28}$$

where

$$q(x,t,A) := \lim_{u\downarrow t} \frac{1}{u-t} [p(x,t,A,u) - \chi_A(x)] =$$

$$\int_{|y|>\epsilon} \frac{\theta(x+y,t)}{\theta(x,t)} [\chi_A(x+y) - \chi_A(x)]\, d\nu(y), \tag{29}$$

$$\langle v\rangle_A(t) := \int_A \overline{\rho}(x,t)\, [2\nabla S(x,t)]\, dx$$

Here $q(x,t,A) \geq 0$ for all x which are *not* in A, in agreement with [16]. We have also introduced a pseudodifferential counterpart of the current velocity field $v(x,t) = 2\nabla S(x,t)$, previously attributed to diffusion processes where the probability conservation law (a continuity equation in another lore) $\partial_t \rho = -\nabla(v\rho)$ plays the rôle of the Fokker–Planck equation.

Notice, that in the particular case of $\theta(x,t) \equiv 1$ for all x,t, and $V = 0$, Eq.(26) reduces to the "free noise" situation covered by the traditional Fokker–Planck equations. Then, $q(t,x,A) = \int_{|y|>\epsilon}[\chi_A(x+y) - \chi_A(x)]d\nu(y)$, while $R = -S$, $\overline{\rho} = \exp(2R) = \theta_*$ implies $\langle v\rangle_A(t) = -\overline{\rho}(x,t)|_a^b$ where $[a,b] := A \subset R$.

Now, let us address the Fokker–Planck equation for the pseudodifferential–Schrödinger dynamics case, which we consider in the form analogous to (26); see also (1) for comparison:

$$i\partial_t\psi = H\psi + V\psi \tag{30}$$
$$i\partial_t\overline{\psi} = -H\overline{\psi} - V\overline{\psi}$$

We re-emphasize that to define the probability density $\rho(x,t) = |\psi(x,t)|^2$ one actually employs solutions of the time adjoint pair of Schrödinger equations.

In view of (30), the pseudodifferential continuity equation follows:

$$\partial_t\rho(x,t) = -i\left[\overline{\psi}(x,t)(H\psi)(x,t) - \psi(x,t)(H\overline{\psi})(x,t)\right] = \tag{31}$$
$$-i\int_R \left[-\overline{\psi}(x,t)\psi(x+y,t) + \psi(x,t)\overline{\psi}(x+y,t) + 2i\rho(x,t)\nabla S(x,t)\frac{y}{1+y^2}\right] d\nu(y)$$

Our next step is a repetition of the procedures behind (27), which implies:

$$\partial_t\rho(x,t) = \int_R \left[2\mathcal{I}\left[\psi(x,t)\overline{\psi}(x+y,t)\right] + 2\rho(x,t)\nabla S(x,t)\frac{y}{1+y^2}\right] d\nu(y)$$
$$\Downarrow$$
$$\int_A dx \int_{|y|>\epsilon} 2\mathcal{I}\left[\psi(x,t)\overline{\psi}(x+y,t)\right] =$$
$$\int_R dx \int_{|y|>\epsilon} \chi_A(x) 2\rho^{1/2}(x,t)\rho^{1/2}(x+y,t)\sin\left[S(x,t) - S(x+y,t)\right] d\nu(y) =$$
$$\int_R \rho(x,t)dx \int_{|y|>\epsilon} \frac{\rho^{1/2}(x+y)}{\rho^{1/2}(x)}\sin\left[S(x+y,t) - S(x,t)\right]\left[\chi_A(x+y) - \chi_A(x)\right] d\nu(y) \tag{32}$$

where $\mathcal{I}[f(x,t)]$ stands for an imaginary part of a complex function $f(x,t)$. So, a counterpart of (28) reads:

$$\partial_t\rho_\epsilon(A,t) = \int_R q(x,t,A)\rho_\epsilon(x,t)dx + \langle v\rangle_A(t)\int_{|y|>\epsilon} \frac{y}{1+y^2}d\nu(y) \tag{33}$$

where, however

$$q(x,t,A) := \int_{|y|>\epsilon} \mathcal{I}\left[\frac{\psi(x+y,t)}{\psi(x,t)}\right] [\chi_A(x+y) - \chi_A(x)]\, d\nu(y) \tag{34}$$

no longer can be derived from transition probability densities of the process, as in the previous discussion, because in general our process is *not* Markovian [4]. At least in the case of nonstationary dynamics, the only transition probability density which is at our disposal connects an initial instant of the evolution with any later one. In fact, we might even not be sure that $q(x,t,A)$ is a well defined probabilistic object, because of the presence of $\sin[S(x+y,t) - S(x,t)]$ in the integrand. At this point an observation of [14] helps. Namely, in view of the identity:

$$\int_R dx \int_{|y|>\epsilon} |\psi(x+y,t)\psi(x,t)|\, [\chi_A(x+y) - \chi_A(x)]\, d\nu(y) = 0 \tag{35}$$

valid for Borel sets $A \subset R$, which are away from $(-\epsilon, +\epsilon)$, we can always pass from (32) to the rearranged form of (33):

$$q(x,t,A) = \int_{|y|>\epsilon} \left[\left|\frac{\psi(x+y,t)}{\psi(x,t)}\right| + \mathcal{I}\left[\frac{\psi(x+y,t)}{\psi(x,t)}\right]\right] [\chi_A(x+y) - \chi_A(x)]\; d\nu(y) \tag{36}$$

implying that $q(x,t,A)$ is positive for all x which are *not* in A, as should be the case [16].

In fact, our Fokker–Planck equations involve exclusively the integral term on their right–hand–side:

$$\partial_t \overline{\rho}_\epsilon(A,t) = \int_R \overline{q}(x,t,A)\overline{\rho}_\epsilon(x,t)dx \tag{37}$$

$$\partial_t \rho_\epsilon(A,t) = \int_R q(x,t,A)\rho_\epsilon(x,t)dx$$

where an overbar distinguishes between probabilistic quantities characterising different families of stochastic jump processes before and after an analytic continuation in time of the given holomorphic semigroup. respectively. Let us emphasize that the above simplification occurs only in the $|y| > \epsilon > 0$ jumping size regime. The real rôle of the two spurious, in the present regime, terms is to compensate the divergent contributions from the Lévy measure when the principal value integral $\epsilon \to 0$ limit is considered; then the *standard* jump process theory does not apply. Anyway, those two terms are irrelevant for any $\epsilon > 0$, irrespectively of how small ϵ is.

More detailed analysis and a number of extensions of the described formalism can be found in the original publication [4], while a discussion of the Gaussian (Wiener measure generated) case and this of the Schrödinger interpolation problem in Refs. [1, 2, 15, 3, 17].

References

1. J.C. Zambrini, J. Math. Phys. **27**, 3207 (1986).
2. Ph. Blanchard, P. Garbaczewski, Phys. Rev. **E49**, 3815 (1994).
3. P. Garbaczewski, R. Olkiewicz, "Why Quantum Dynamics can be Formulated as a Markov Process", Phys. Rev. **A**, in press (1995).
4. P. Garbaczewski, J.R. Klauder, R. Olkiewicz, "The Schrödinger Problem, Lévy Processes and Noise in Relativistic Quantum Mechanics", Phys. Rev. **E**, in press (1995).
5. E. Nelson, "Quantum Fluctuations", Princeton University Press, Princeton, 1985.
6. L. Breiman, "Probability", Addison–Wesley, Reading, 1968.
7. E.W. Montroll, B.J. West., in: "Fluctuation Phenomena", ed. by E.W. Montroll and J.L. Lebowitz, North–Holland, Amsterdam, 1987
8. A. Janicki, A. Weron, "Simulation and Chaotic Behaviour of α–stable Stochastic Processes", Marcel Dekker, New York, 1994.
9. H.C. Fogedby, Phys. Rev. Lett. **73**, 2517 (1994).
10. B.B. Mandelbrot, "The Fractal Geometry of Nature", W.H. Freeman, New York,1982.
11. J. Klafter, A. Blumen, M.F. Shlesinger, Phys. Rev. **A35**, 3081 (1987).
12. X.J. Wang, Phys. Rev. **A45**, 8407 (1992).
13. R. Carmona, W.C. Masters, B. Simon, Journ. Funct. Anal. **91**, 117 (1990).
14. G.F. De Angelis, J. Math. Phys. **31**, 1408 (1990).
15. P. Garbaczewski, Phys. Lett. **A172**, 208 (1993).
16. I.I. Gikhman, A.V. Skorokhod, "Introduction to the Theory of Random Processes", W.B. Saunders Comp., Philadelphia, 1969.
17. P. Garbaczewski and R. Olkiewicz, "Feynman–Kac Kernels in Markovian Representations of the Schrödinger Interpolating Dynamics", in preparation

Quantum Coherence and Decoherence in a Classically Chaotic Experimentally Accessible Quantum Optical System

Robert Graham

Fachbereich Physik, Universität-Gesamthochschule Essen
45117 Essen, Germany

Abstract: The quantum behavior of classically chaotic systems is dominated by coherence effects. Dynamical localization is the particular coherence effect associated with classical chaotic diffusion. This effect is considered for a system of laser-cooled two-level atoms subject to a standing wave laser field periodically modulated in time. If the field is only turned on during short periodic pulses, momentum is transferred from the field to the atoms only during short kicks and diffuses classically, but is dynamically localized quantally. Spontaneous emission from the two-level atoms perturbs the quantum coherence and resurrects the diffusion of the atomic momentum, albeit with a smaller diffusion constant. The theory of this effect based on a particular linear version of the stochastic Schrödinger equation is discussed.

1 Introduction

Chaos in classical Hamiltonian systems [1], like e.g. a periodically driven pendulum, is characterized by the existence of a positive Lyapunov exponent. It describes a non-zero positive average rate of initially small perturbations like small changes in the initial conditions. Correlation functions of observables have a spectrum with a continuous component, and the dynamics on long time-scales is unpredictable and random. Suppose one constructs such a system on a tiny scale, e.g. a tiny periodically driven pendulum. How this can be done in practice will be indicated below. What is the role, in such systems, of the quantum mechanical coherence of the wave-function? If a classical random process, or a classical chaotic process is coupled to an otherwise closed quantum system, then transitions with a rate γ_{ind} are induced in the quantum system which will destroy quantum coherence on a time-scale $T_{dec} = 1/\gamma_{ind}$. For example, a harmonic quantum oscillator in a superposition of 2 coherent states $|\alpha\rangle + |-\alpha\rangle$, if coupled to a thermal field at temperature T decoheres in a time

$$T_{dec} = \left[\gamma|\alpha|^2(1+(\exp\left(\frac{\hbar\omega}{k_BT}\right)-1)^{-1})\right]^{-1} \tag{1}$$

(for a detailed analysis see [2]). After the time $T_{\rm dec}$ quantum coherence ceases to play any role. However, what is the role of quantum coherence if it is the chaotic dynamics itself which is quantized? The résumée (see e.g. [3]) of much work of a great number of people over many years is: Quantum coherence effects dominate the physical properties of closed classically chaotic systems just as much as they do for classically regular systems. But there are some new unexpected quantum coherence effects. Let us consider some examples with increasing level of unexpectedness.

Time–dependent expectation values of dynamical variables A have discrete Fourier spectra as opposed to the continuous classical spectra of chaotic systems. This is readily seen from the general expression

$$\langle A(t)\rangle = \sum_{\mu,\lambda} C_\mu^* C_\lambda e^{-i(\omega_\lambda-\omega_\mu)t}\langle\varphi_\mu|A|\varphi_\lambda\rangle \tag{2}$$

where C_λ are the amplitudes of the energy (or quasi–energy) eigenstates $|\varphi_\lambda\rangle$ in the state vector. The products $C_\mu^* C_\lambda$ for $\mu \neq \lambda$ are the 'coherences' of the state vector, whose presence is necessary for the appearance of the discrete spectral lines.

Level repulsion between overlapping states sitting in the same chaotic component of phase space is also due to the presence of coherences. To determine two such levels one may imagine to pick two linearly independent, normalized test–wave functions $|t_{1,2}\rangle$, mutually orthogonal and also orthogonal to all other wavefunctions. Using the ansatz $|\psi\rangle = C_1|t_1\rangle + C_2|t_2\rangle$, $|C_1|^2 + |C_2|^2 = 1$ in Ritz's variational principle $\delta(|C_1|^2H_{11} + |C_2|^2H_{22} + C_1^*C_2H_{12} + C_1C_2^*H_{21}) = 0$ one readily obtains the energy eigenvalues $E = \frac{1}{2}(H_{11} + H_{22}) \pm \frac{1}{2}\Delta E$ with the spacing $\Delta E = \sqrt{(H_{11} - H_{12})^2 + 4|H_{12}|^2}$. There is level repulsion due to the appearance of $|H_{12}|^2$, which can be traced back to the cofactors of the coherences $C_1^*C_2$ of $|\psi\rangle$ in the basis of the test functions. There is level repulsion because wave–functions overlap and still must be orthogonal. If $|t_1\rangle$ and $|t_2\rangle$ are non–overlapping, e.g. because they happen to be localized in different parts of phase–space, their mutual coherence becomes irrelevant and level repulsion disappears.

Tunneling through chaotic domains of phase–space [4] is also dominated by coherence effects. In the mechanism called chaos–assisted tunneling degenerate states localized e.g. on different regular islands mix with nearly degenerate sufficiently delocalized states in the chaotic domain between the regular islands to first tunnel into these and then on further to the next regular island without necessarily ever appearing fully in the chaotic domain. Instead coherent higher–frequency oscillations are superimposed on the familiar tunneling oscillations with the basic splitting frequency of the tunneling doublet.

The (originally) most unexpected quantum coherence effect arises in the rather commonly occurring systems where the classical transport through the chaotic phase–space takes place in a local diffusive fashion [1] such that for an action variable p one finds classically $\langle(p(t) - p_0)^2\rangle = Dt$. Then quantum states over the same chaotic part of phase-space may turn out to be localized, i.e.

may be not all spread out over the entire chaotic domain, and not all appreciably overlapping. The physical reason is that there are *many* different paths between two states with sufficiently widely spaced classical actions which contribute appreciably to the Feynman path integral for the transition amplitude, but all contribute with different and mutually random phases. The net result is zero by destructive interference: because there are so *many* different paths they conspire coherently and mutually block each other, and the state is localized. (Unfortunately for the case of dynamical localization, nobody seems to be able to mathematically realize this appealing and physically very simple picture). This is exactly the mechanism of Anderson localization in spatially disordered systems, yet localization not in space but in an action variable, and disorder not provided by external conditions and statistically, but generated internally by the dynamics. Hence a new name: dynamical localization (for reviews see e.g.[3, 5, 6]).

In practice only the 1–dimensional case, i.e. the case of a single, classically diffusing action variable p is important, so far. Let p be quantized by $p = \hbar l$, l a quantum number. Then the localization length of p is written as $\hbar\xi$. An initial state with action p_0, say, overlaps only with ξ other localized (quasi–)energy states. Let the local density of states (# states per unit of (quasi–)energy and unit of l) be ρ_0. Then $\rho = \rho_0\xi$ is the density of states (# per unit of (quasi–)energy) overlapping with the initial state. As pointed out by Chirikov [6], this level density is effectively continuous for times $t < \hbar\rho$ and classical diffusion prevails, $\langle(p-p_0)^2\rangle = Dt$. At $t = 2\pi\alpha\hbar\rho$ where α is of order 1 the (on the average) exponentially localized state $|\psi|^2 \simeq \frac{1}{\xi}e^{-2(l-l_0)/\xi}$ is reached, with $\langle p^2\rangle = \frac{1}{2}\hbar^2\xi^2$. One obtains the important relation

$$\xi = 4\pi\alpha\rho_0 D/\hbar\,. \tag{3}$$

The most important application are periodically driven systems, where the local density of states is simply $\rho_0 = \frac{1}{\hbar\omega}$ due to Floquet's theorem. Hence $\xi = \frac{2\alpha D}{\hbar^2}T$. Numerically [7] and theoretically [8] $\alpha \simeq 0.5$ for kicked systems. In the following we shall be exclusively concerned with the phenomenon of dynamical localization.

2 Dynamical Localization

As an interesting physical system to consider in this field we proposed some time ago the center–of–mass dynamics of two–level atoms in a standing–wave laser field [9]. (The same model in a different physical context is treated in [10]). This proposal was motivated by the rapid experimental progress with matter waves in laser fields [11] and it led to more experiments [12, 13] in which indeed the predicted coherence effects were observed. Two–level atoms in a standing–plane–wave laser–field are governed by the Hamiltonian

$$H = -\frac{\hbar^2}{2M}\frac{\partial^2}{\partial x^2} + K\cos k_L x(|g\rangle\langle e| + |e\rangle\langle g|)f(t) + \hbar\delta_L|e\rangle\langle e| \tag{4}$$

where M is the total mass, x is the center–of–mass coordinate parallel to the plane–wave (the other two components can be separated off and are irrelevant), $|g\rangle$, $|e\rangle$ are the atomic ground state and excited state, respectively, $\delta_L = \omega_0 - \omega_L$ is the detuning between the atomic transition and the laser, which has frequency ω_L and wave–number k_L, $f(t) = f(t+T)$ is an externally imposed periodic modulation of the laser amplitude which is nonzero, $f(t) \neq 0$, only during a periodic sequence of short pulses of duration ΔT and period T. We assume $f(t)$ to be normalized by $\int_0^T f(t)dt = 1$. The coupling constant then is $K = dE_0 \Delta T$ (the product of dipole matrix–element d, laser amplitude E_0 and ΔT). In [9] we worked with a Hamiltonian, where $K \cos k_L x f(t)$ is replaced by $dE_0 \cos(k_L(x - \Delta L \sin \omega t))$ where $k_L \Delta L$ is the phase–modulation amplitude. It is this Hamiltonian that was realized in the experiments [12, 13] and we shall return to it in section IV.

The internal two–level degree of freedom may be separated out by the ansatz

$$|\psi_+(t)\rangle = \sum_{l=-\infty}^{+\infty} \int_0^{k_L} dk e^{ikx} \left(C_{2l}^{(+)}(k,t) e^{i2lk_L x} |g\rangle + C_{2l+1}^{(+)}(k,t) e^{i(2l+1)k_L x} |e\rangle \right) \quad (5)$$

$$|\psi_-(t)\rangle = \sum_{l=-\infty}^{+\infty} \int_0^{k_L} dk e^{ikx} \left(C_{2l+1}^{(-)}(k,t) e^{i(2l+1)k_L x} |g\rangle + C_{2l}^{(-)}(k,t) e^{i2lk_L x} |e\rangle \right) \quad (6)$$

and defining the 2π-periodic wave-functions

$$\psi^{(\pm)}(k, k_L x, t) = \sum_{l=-\infty}^{+\infty} C_l^{(\pm)}(k,t) e^{ilk_L x} . \quad (7)$$

The resulting Schrödinger equations are decoupled in k and in $(\pm)$ and read

$$i\hbar \dot{\psi}^{(\pm)}(k,q,t) = \frac{\hbar^2 k_L^2}{2M} \left(-i\frac{\partial}{\partial q} + \frac{k}{k_L} \right)^2 \psi^{(\pm)}(k,q,t)$$
$$+\frac{1}{2}\hbar\delta_L \left(\psi^{(\pm)}(k,q,t) \pm \psi^{(\pm)}(k,q+\pi,t) \right) + K \cos q f(t) \psi^{(\pm)}(k,q,t) \quad (8)$$

If ΔT satisfies $\delta_L \Delta T << 1$, $\langle |(p/m)| \rangle \Delta T << 2\pi/k_L$ then $f(t)$ may be considered as a periodic sequence of δ-functions

$$f(t) \to \sum_{n=-\infty}^{+\infty} \delta(t - nT) = \frac{\omega}{2\pi} \sum_{m=-\infty}^{\infty} e^{i\omega m t} \quad (9)$$

and the independent Schrödinger equations (8) are all of the form of that of a periodically kicked rotor.

Let us now consider this case of a rotor. It is well known, that in the angular momentum (l–) representation, its Schrödinger equation for the Floquet–states

$$\psi^{(\pm)}(t) = e^{-i\varepsilon t} U^{(\pm)}(t) ,$$
$$U^{(\pm)}(t+T) = U^{(\pm)}(t) \quad (10)$$

can be mapped on a 1–dimensional tight–binding solid–state Schrödinger equation [14]

$$T_l^{(\pm)}(k)\bar{U}_l^{(\pm)} + \sum_r W_r \bar{U}_{l-r}^{(\pm)} = 0 \tag{11}$$

where, in the present case

$$W_l = -\int_0^{2\pi} dq e^{-ilq} \tan\left(\frac{K}{2\hbar}\cos q\right)$$
$$T_l^{(\pm)}(k) = \tan\left[\frac{T}{2}\left(\varepsilon - \frac{\hbar(k_L l + k)^2}{2M} - \frac{\delta_L}{2}(1 \mp (-1)^l)\right)\right]$$
$$\bar{U}_l^{(\pm)} = \frac{1}{2}\int_0^{2\pi} dq e^{-ilq}\left(1 + e^{-i\frac{K}{\hbar}\cos q}\right) U^{(\pm)}(k,q,T-0)\,. \tag{12}$$

Anderson localization occurs in this tight–binding model [14], if $T_l^{(\pm)}(k)$ can be considered as random in l and W_l (which really is only well defined for $|K| < \pi\hbar$, but similar representations where this problem is avoided can also be given, see [15]) is sufficiently short–ranged. A proof has not been constructed, but it appears that both properties are satisfied. The parameters k (a shift of the origin of angular momentum within the interval $0 \leq k < k_L$) and δ_L (the detuning) only enter $T_l^{(\pm)}$. The presence of δ_L helps (but is not essential) to randomize $T_l^{(\pm)}$ but has otherwise little influence. One predicts no important influence of δ_L on the averaged localization length (as long as the underlying assumption $\delta_L \Delta T << 1$ remains valid). The parameter k is more important. If it vanishes, $k = 0$, the Schrödinger equations (8) have two time reversal symmetries [16].

$$\begin{aligned} T: &\quad T\psi^{(\pm)}(0,q,t) = \psi^{(\pm)*}(0,q,-t) \\ S: &\quad S\psi^{(\pm)}(0,q,t) = \psi^{(\pm)*}(0,-q,-t) \end{aligned} \tag{13}$$

which are connected by a parity transformation

$$R = ST: \qquad R\psi^{(\pm)}(0,q,t) = \psi^{(\pm)}(0,-q,t)\,. \tag{14}$$

If $k \neq 0$ then T and R are broken but S still remains as time–reversal symmetry. This is important for dynamical localization because the localization effect is strengthened by the so–called weak localization effect (constructive interference of the amplitude from each periodic orbit with its time–reversed [17]). The presence of R for $k = 0$ means that Floquet states are in that case either symmetric or anti–symmetric in q, or in angular–momentum if that representation is used, which gives the origin, $l = 0$, a special significance [16]. States localized around l_0 far from the origin necessarily have also a component localized around $-l_0$, which in such measures of localization, like the inverse participation ratio [18] or the spectral entropy [19] doubles their localization length [16]. If $k \neq 0$ is switched on the symmetry of $|U_l^{(\pm)}|^2$ with respect to $l \to -l$, and the special role of the origin $l = 0$ both disappear and the localization length is only determined by the local width of each Floquet state (compare figures 4a and 6a of [16]).

For the Anderson–type Hamiltonian (11) one expects and numerically finds exponentially localized states of the form

$$U_l = \frac{a(l)}{\sqrt{\xi(l_0)}} \exp\left[-\frac{|l-l_0|}{\xi(l_0)}\right] \tag{15}$$

with a rapidly fluctuating amplitude factor $a(l)$, normalized such that $\overline{|a(l)|^2} = 1$. The localization length $\xi(l_0)$ also fluctuates strongly from state to state. Numerically one finds that on the average $(\xi(l_0))^{-1} \simeq 2\xi^{-1}$ [20] were ξ is the localization length of a wave–packet, starting, at time 0, as an l-eigenstate as discussed in the preceding section. In the present case, with $D = (K^2 k_L^2/2T)$, $\rho_0 = T/2\pi\hbar$ eq. (3) gives

$$\xi = \frac{\alpha K^2}{\hbar^2}. \tag{16}$$

The value of α depends on whether the distribution of $T_l^{(\pm)}$, according to eq. (12) has a second moment (then $\alpha = 1$ [8]) or not. In the present case the second moment does not exist. According to eq. (12) the $T_l^{(\pm)}$ are Cauchy–distributed [14] for generic values of the T, k_L, M, δ_L,

$$P(T_l^{(\pm)}) = \frac{1}{\pi}\frac{1}{1+(T_l^{(\pm)})^2} \tag{17}$$

like in the Lloyd–model [14], and, like there one expects [8] and finds [7] $\alpha = 1/2$.

3 Decoherence from Spontaneous Emission

Because in our realization of the kicked rotor by two–level atoms the upper state is freely excited, we have to take into account spontaneous emission. Physically this amounts to a degradation of quantum coherence, and the question is how sensitive the particular coherence effect under study will be against this degradation. To describe spontaneous emission we use the technique of the stochastic Schrödinger equation based on the Wigner–Weißkopf theory of spontaneously decaying states [20]. We here use this method in the particular form of a *linear* stochastic Schrödinger equation discussed at length in recent papers [21, 2], where references to earlier work can also be found. (See also Belavkin's contribution to these proceedings). This linear stochastic Schrödinger equation is here taken in the form (see [22])

$$\begin{aligned} |\psi(t+dt)\rangle = {} & (1-dP(t))e^{\frac{\gamma}{2}dt}e^{(-\frac{i}{\hbar}H_0(t)-\frac{\gamma}{2}|e\rangle\langle e|)dt}|\psi(t)\rangle \\ & +dP(t)e^{-ik_Lxu}|g\rangle\langle e|\psi(t)\rangle\,. \end{aligned} \tag{18}$$

Here H_0 is the Hamiltonian of the preceding section without spontaneous emission. It should be noted, however, that the spacing $\hbar\omega_0$ of the atomic levels now includes the renormalization due to the coupling to the electromagnetic vacuum. γ is a spontaneous decay rate, which is given by the usual expression in terms of the square of the dipole matrix–element. $dP(t)$ is an uncorrelated Poisson process satisfying $dP^2(t) = dP(t)$, i.e. $dP(t) = 0, 1$, with $\langle dP(t)\rangle = \gamma dt$. We shall call the measure of this Poisson process $d\mu^P[P]$. We note that $|\psi(t)\rangle$, according to eq. (18), does not preserve normalization, but that

$$\int d\mu^P[P] \quad \langle\psi(t)|\psi(t)\rangle = 1\,. \tag{19}$$

To maintain this normalization is the raison d'être of the factor $e^{(\gamma/2)dt}$ in the first term of the right hand side of (18). This term describes the time–evolution, if there is *no* spontaneous emission event in dt, to which case the Poisson process assigns the probability $e^{-\gamma dt}$, *independent* of the occupation probabilities of the two atomic levels. Is this correct? Yes, if total probabilities are finally computed not from $d\mu^P[P]$ alone, but from the total (and then automatically normalized) measure

$$d\mu = d\mu^P[P]\langle\psi|\psi\rangle\,, \tag{20}$$

to which the no–spontaneous–emission–term of eq. (18) contributes with the probability (for $\gamma dt \to 0$)

$$e^{-\gamma dt}e^{\gamma dt}(1 - \gamma dt \int |\psi_e^{(t)}|^2 dq)\,. \tag{21}$$

This is indeed the correct probability that no spontaneous emission occurs in $(t, t+dt)$. To the second term on the right–hand–side of eq. (18), which describes a spontaneous decay into the ground state, the Poisson process assigns the tiny probability γdt, again independent of how the atomic levels are occupied: in the measure $d\mu$, eq. (20), the probability γdt *automatically* gets multiplied by the correct factor $\int |\psi_e|^2 dq$ to give the total probability for a spontaneous transition in $(t, t+dt)$, adding up to 1 with the complementary probability (21), as required by normalization. The phase–factor in the second term of (18) describes the atomic recoil, the parameter u is given by $u = \sin\Theta \sin\phi$ where Θ is the angle between the direction of polarization of the laser field, taken in the y–direction, and the momentum $\hbar\mathbf{k}_L$ of the emitted photon, and ϕ is the angle between the z–axis and the projection $\mathbf{k}_L$ into the (x, z)–plane. The distribution $W(u)$ of u is determined by that of Θ, ϕ, which is given by

$$P(\Theta, \phi)d\Theta d\phi = \frac{3}{8\pi}(1 - \cos^2\Theta)\sin\Theta d\Theta d\phi\,. \tag{22}$$

The nice feature of (18), which distinguishes it from other nonlinear versions (see e.g. [20, 23]), is its linearity, which preserves the superposition principle, and allows for analytical solutions at least in simple cases like the periodically kicked rotor, in which we are interested here. Numerical solutions of eq. (18) proceed as follows. The state–vector describing the initial state is specified in normalized form. At time $t = 0$, say, the Poisson process determines the probability that spontaneous emission first occurs after the interval t_s in dt_s as $p(t_s)dt_s = \gamma e^{-\gamma t_s}dt_s$. Hence, picking a random number $0 \le z_s \le 1$, the random time t_s is determined from $\int_0^{t_s} d\tau p(\tau) = 1 - e^{-\gamma t_s} = z_s$. Then Eq. (18)

is integrated from $t = 0$ to $t = t_s$ using the first term on the right hand side only. Then the second term is applied between t_s and $t_s + dt_s$. Next a new time–interval t'_s is determined by repeating this procedure with a new random number $0 \le z'_s \le 1$ and the unnormalized state–vector is propagated for t'_s etc. up to the desired total time–interval. This is repeated for many ($N1$) runs. The desired expectation values, e.g. $\langle\psi_j(t)|p^2|\psi_j(t)\rangle$ are computed during each run (labeled by j) using the *unnormalized* wave–function. Finally, the true and correctly normalized expectation value is estimated by taking the ensemble average over the N sample paths as, e.g. in $\langle p^2\rangle = \frac{1}{N}\sum_{j=1}^{N}\langle\psi_j(t)|p^2|\psi_j(t)\rangle$. The statistical operator of the atoms is given in terms of the stochastic $|\psi(t)\rangle$ as the average

$$\rho(t) = -\int |\psi(t)\rangle\langle\psi(t)|d\mu^P[P] \tag{23}$$

over the Poisson process. It is easy to show that, according to eq. (18), $\rho(t)$ satisfies the usual quantum optical master equation

$$\begin{aligned}\dot\rho(t) = &-\frac{i}{\hbar}\,[H_0(t),\rho(t)] - \frac{\gamma}{2}\,(|e\rangle\langle e|\rho(t) + \rho(t)|e\rangle\langle e|) \\ &+\gamma\int_{-1}^{1} du W(u) e^{-ik_Lxu}|g\rangle\langle e|\rho(t)|e\rangle\langle g|e^{ik_Lxu}\,.\end{aligned} \tag{24}$$

Turning to the special case of the kicked rotor it is possible to obtain a stochastically equivalent but simpler stochastic Schrödinger equation integrated for one period. Let us choose as a reference time t_0 within a given period the time immediately before one kick. Our aim is to incorporate the fact that, during each kicking period, there can be at most one spontaneous emission event. This can be described by a single Poisson variable $dP(T)$ for each period by putting

$$\begin{aligned}|\psi(t_0+T)\rangle = &\ (1-dP(T))e^{\gamma T/2}V(T)K_0|\psi(t_0)\rangle \\ &+dP(T)V(T-\tau_s)e^{-ik_Lxu}|g\rangle\langle e|e^{\gamma\tau_s/2}V(\tau_s)K_0|\psi(t_0)\rangle\end{aligned} \tag{25}$$

with the Poisson process $dP(T)$, statistically independent for different periods, defined by $dP^2(T) = dP(T)$, $\langle dP(T)\rangle = 1 - e^{-\gamma T}$. If $dP(T)$ is (randomly) determined to be non-zero (i.e. equal to 1) because a random number $0 \le z_s \le 1$ turns out to be $z_s \le 1 - e^{-\gamma T}$, then one sets $\tau_s = \gamma^{-1}|\ln(1-z_s)|$ in eq. (25) and similarly draws u randomly from its distribution

$$W_u(u) = \int P(\Theta,\phi)\delta(u - \sin\Theta\sin\phi)d\Theta d\phi\,. \tag{26}$$

Not surprisingly, after what we have already said, the time–evolution operator $V(\tau)$ is not hermitian. It is given by

$$V(\tau) = e^{-i\frac{\hbar l^2}{2M}\tau - i\delta_L\tau - \frac{\gamma}{2}\tau}|e\rangle\langle e| + e^{-i\frac{\hbar l^2}{2M}\tau}|g\rangle\langle g|\,. \tag{27}$$

The operator K_0 describes the kick applied immediately after the time $t = t_0$

$$K_0 = e^{-\frac{i}{\hbar}K\cos q}\,. \tag{28}$$

Numerical results obtained by the use of this stochastic description are reported in [24, 25]; restrictions of space do not permit us to discuss them here. The solution of the master–equation (24) for one kicking period, which also follows from eq. (25), is given by

$$\begin{aligned}\rho(t_0+T) &= V(T)K_0\rho(t_0)K_0^\dagger V^\dagger(T) \\ &+\gamma\int_{-1}^{1} duW_u(u)\int_0^T d\tau_s e^{-\gamma\tau_s}V(T-\tau_s)e^{-ik_Lux}|g\rangle\langle e|e^{\gamma\tau_s/2}V(\tau_s)K_0 \\ &\quad \rho(t_0)K_0^\dagger V^\dagger(\tau_s)e^{\gamma\tau_s/2}|e\rangle\langle g|e^{ik_Lux}V^\dagger(T-\tau_s)\,. \end{aligned} \tag{29}$$

It is less convenient for numerical work than the stochastic Schrödinger equation (because of the problems connected with storing and evolving in time a large density matrix). However, eq. (29) is useful for finding, in an approximate analytical way, the influence of spontaneous emission on dynamical localization.

The interesting case is $\gamma T << 1$, which we now assume. In fact, we shall even make the stronger assumption $\gamma T\xi << 1$. We consider first the expectation value $\langle p^2(t)\rangle = Tr\rho(t)p^2$ for a large time $t = NT\gamma^{-1}$. It is useful to decompose $\langle p^2(t)\rangle$, according to eq. (29), into the different contributions with a fixed number n, $0 \le n \le N$, of spontaneous decay events, $\langle p^2(t)\rangle = \sum_{n=0}^N \langle p^2(t)\rangle^{(n)}$. It follows from the structure of the equation (and is physically obvious) that these contributions are superposed incoherently (i.e. in their probabilities or averages, not in their probability amplitudes). If there are n decays (in n out of N periods), $\langle p^2(t)\rangle^{(n)}$, according to the second term in (29), contains a statistical factor $(\gamma T)^n$; a combinatorial factor $\binom{N}{n}$; a further factor $(\frac{1}{2})^n$ because the upper state, on the average, is occupied with probability 1/2; a factor $e^{-\frac{\gamma}{2}NT}$ because $V(T)V^\dagger(T)$ decays, on the average, with a factor $e^{-\frac{\gamma}{2}T}$; and, from the second moment of p^2, a factor $(n+1)\hbar^2k_L^2\xi^2/2$. The latter factor certainly is the correct value of $\langle p^2\rangle$ in the localized state if $n = 0$, and each spontaneous decay adds a further spreading by the same amount [26, 27] provided $\gamma\xi T << 1$. All other details contained in eq. (29) will be averaged out on long time–scales. Putting this together we obtain

$$\langle p^2(t)\rangle = \sum_{n=0}^{N}(\gamma T)^n\binom{N}{n}\left(\frac{1}{2}\right)^n(n+1)\frac{\hbar^2k_L^2\xi^2}{2}e^{-\frac{\gamma}{2}NT} \tag{30}$$

which can be summed and gives, for $\gamma T << 1$,

$$\langle p^2(t)\rangle = \frac{\hbar^2k_L^2\xi^2}{2} + \frac{\gamma t}{2}\frac{\hbar^2k_L^2\xi^2}{2}\,. \tag{31}$$

Thus diffusion of p is resurrected by spontaneous emission, however with a new diffusion constant, proportional to γ, which under our assumptions turns out to be much smaller than the classical diffusion constant. This result is qualitatively similar to, but, due to considerable differences in the models, quantitatively different from results obtained before [27]. We note that in (31) there does *not* occur the usual enhancement of decoherence (shortening of decoherence times like in eq. (1)) characteristic of noise or dissipation in collective bosonic degrees of freedom; the reason is that here noise is connected to a fermionic (two–level) degree of freedom.

A very similar calculation can be performed to obtain the time-dependent occupation probability of angular momentum states $\overline{|\psi(l)|^2} = \int d\mu^P[P]\langle l|\psi(t)\rangle\langle\psi(t)|l\rangle$. Again, we decompose in contributions from a fixed number $n(0 \leq n \leq N)$ of spontaneous decays. The statistical, combinatorial, and decay factors for each contribution is the same as before, only, instead of the contribution from $\langle p^2\rangle$ we have now to insert the contribution from $\langle|l\rangle\langle l|\rangle$. Each spontaneous decay folds the distribution preceding it with $\xi^{-1}\exp(-2|l|/\xi)$. This is most conveniently handled in Fourier space where this folding simply becomes multiplication with $(1+v^2\xi^2/4)^{-1}$ where v is the Fourier–conjugate of l (taken as continuous variable, for simplicity). Hence, we obtain, transforming back to l–space

$$\overline{|\psi(l)|^2} = \int \frac{dv}{2\pi} e^{-ilv} \sum_{n=0}^{N} (\gamma T)^n \binom{N}{n} \left(\frac{1}{2}\right)^n \left(1+\frac{v^2\xi^2}{4}\right)^{-n-1} e^{-\frac{\gamma T N}{2}} \qquad (32)$$

which can be summed and simplified for $\gamma T << 1$ to give

$$\overline{|\psi(l)|^2} = \int \frac{dv}{2\pi} e^{-ilv} \frac{\exp\left(-\frac{\gamma t}{2}\frac{v^2\xi^2/4}{1+v^2\xi^2/4}\right)}{1+v^2\xi^2/4}. \qquad (33)$$

It can be seen that the exponential shape of the probability distribution persists for $l < \xi$ (corresponding to $v\xi/21$), while for $l\xi$ the distribution becomes approximately Gaussian. As eq. (33) gives the characteristic function of $\overline{|\psi(l)|^2}$ in a simple closed form, all moments of this distribution can also be written down explicitly. The result (31) can also be rederived this way.

4 De Broglie Waves in a Phase–Modulated Standing Wave Laser–Field

The original proposal of [9] used a phase–modulated standing–wave field which seems less idealized than the periodic standing–wave kicks assumed in the preceding sections. In fact, such experiments have recently been realized [12, 13] and a partial comparison with theory became possible. Therefore, we now consider this case in some more detail. As was already mentioned after eq. (4) the Hamiltonian is still of the form (4), but the interaction term has to be replaced by the phase–modulated term, also given in the discussion following eq. (4). Both, theory and experiment were so far done for the strongly detuned case, where the occupation probability of the excited atomic state is strongly suppressed and spontaneous emission becomes weak. Then, it becomes possible to eliminate the excited atomic state adiabatically. In the simplest approximation [9] (for improvements see [25]) we get the new effective Schrödinger equation

$$i\hbar\frac{\partial\psi_g}{\partial t} = -\frac{\hbar^2}{2M}\frac{\partial^2\psi_g}{\partial x^2} - \frac{\hbar\Omega_{\text{eff}}}{8}\cos(2k_L(x-\Delta L\sin\omega t))\psi_g \qquad (34)$$

for the center of mass–motion of the atom in its ground–state, with $\Omega_{\text{eff}} = \frac{4(dE_0)^2}{\hbar^2\delta_L}$. Basically, this Hamiltonian describes again a periodically kicked particle, but now the kick has a finite extension ΔT in time with $\Delta T \simeq \pi/(2k_L\Delta L\omega)$. One way to see this [28] is by Fourier–expanding the interaction term, $\cos(2k_L(x - \Delta L\sin\omega t)) = \sum_m J_m(2k_L\Delta L)\cos(2k_Lx - m\omega t)$ and noting that the intensity spectrum of this expansion, given by $|J_m(2k_L\Delta L)|^2$ is on the average $\langle|J_m(2k_L\Delta L)|^2\rangle \sim \frac{1}{\pi 2k_L\Delta L}$ independent of m for $|m|\lesssim 2k_L\Delta L$ and falls off exponentially for larger $|m|$. Hence ψ_g roughly behaves like in a kicked rotor with Hamiltonian [9]

$$H = \frac{p^2}{2M} + \frac{2\pi\hbar\Omega_{\text{eff}}}{8\omega\sqrt{2\pi k_L\Delta L}}\cos(2k_Lx)\sum_{n=-\infty}^{+\infty}\delta_{\Delta T}\left(t - n\frac{2\pi}{\omega}\right) \tag{35}$$

where $\delta_{\Delta T}$ has a (small) finite width $\Delta T = \pi/(2k_L\Delta L\omega)$. Neglecting ΔT we can read off the condition [9]

$$\frac{2\pi\hbar\Omega_{\text{eff}}}{8\omega\sqrt{2\pi k_L\Delta L}}\frac{2\pi}{\omega}\frac{4k_L^2}{M} \gtrsim 1 \tag{36}$$

for the occurrence of global classical chaos in the domain $|p| < M\Delta L\omega$; the diffusion constant D of classical momentum diffusion $\langle(p - p_0)^2\rangle = Dt$ as [9]

$$D = \frac{\hbar^2\Omega_{\text{eff}}^2 k_L}{32\omega\Delta L}; \tag{37}$$

and the localization length ξ of the localized momentum $p = 2\hbar k_L\xi$ in the quantized case [9]

$$\begin{aligned}\langle(p - p_0)^2\rangle &= \frac{1}{2}(2\hbar k_L)^2\xi^2 ,\\ \xi &= (2\pi\alpha/k_L\Delta L)(\Omega_{\text{eff}}/8\omega)^2 .\end{aligned} \tag{38}$$

Classically, the momentum would spread diffusively over the whole classically chaotic domain $|p|\lesssim M\Delta L\omega$, i.e. one would obtain $\sqrt{\langle p^2\rangle} \simeq \frac{1}{\sqrt{3}}M\Delta L\omega$. Quantum mechanically, starting initially near $\langle p^2\rangle = 0$, one predicts, with $\alpha = 1/2$

$$\sqrt{\langle p^2\rangle} \simeq \frac{1}{\sqrt{2}}2\hbar k_L\,(\pi/k_L\Delta L)\,(\Omega_{\text{eff}}/8\omega)^2 . \tag{39}$$

Thus, increasing the modulation amplitude ΔL and looking at the system in its steady state one predicts [9] first a (classical) rise of $\sqrt{\langle p^2\rangle}$ proportional to ΔL, as long as the size of the chaotic domain is smaller than the localization length, then a cross–over to the quantum regime at $\Delta L \simeq (\sqrt{6}\pi\hbar/M\omega)^{1/2}(\Omega_{\text{eff}}/8\omega)$, and finally a *decrease* of $\sqrt{\langle p^2\rangle}$ proportional to ΔL^{-1} for $\Delta L \Delta L_c$. This predicted behavior has indeed been observed experimentally [12] and a quantitative comparison with these estimates and with more detailed numerical simulations based on eq. (35) have been made.

It is also possible to include spontaneous emission in this reduced description. In place of eq. (34) we then obtain

$$\begin{aligned} |\psi_g(t+dt)\rangle = & (1-dP(t))e^{\frac{\gamma}{2}dt}e^{-\frac{i}{\hbar}\tilde{H}(t)dt}|\psi_g(t)\rangle \\ & +dP(t)\frac{\Omega}{2\delta_L}e^{-ik_Lxu}\cos[k_L(x-\Delta L\sin\omega t)]|\psi_g(t)\rangle \end{aligned} \tag{40}$$

where

$$\tilde{H} = -\frac{\hbar^2}{2M}\frac{\partial^2}{\partial x^2} - \frac{\hbar\Omega_{\rm eff}}{4}\left(1+\frac{i\gamma}{2\delta_L}\right)\cos^2(k_L(x-\Delta L\sin\omega t)) \tag{41}$$

is again non–hermitian. The factor $(1+\frac{i\gamma}{2\delta_L})$ has its origin in the replacement $\delta_L \to \delta_L - i\frac{\gamma}{2}$, which, to first order in $\frac{\gamma}{\delta_L}$, leads to a renormalization of $\Omega_{\rm eff}$ by this factor. Eq. (40), like all linear stochastic Schrödinger equations, does not preserve the norm $\langle\psi_g(t)|\psi_g(t)\rangle$ but it preserves $\int d\mu^P[P]\langle\psi_g|\psi_g\rangle = 1$. The calculations of momentum diffusion and wave–packet diffusion leading to eqs. (31), (33) can now be repeated with eq. (40) and we obtain

$$\begin{aligned} \langle p^2(t)\rangle &= \frac{1}{2}(2\hbar k_L)^2\xi^2\left(1+\frac{\gamma}{2}\left(\frac{\Omega}{2\delta_L}\right)^2 t\right) \\ \overline{|\psi_g(l)|^2} &= \int\frac{dv}{2\pi}e^{-ik_Lv}\frac{\exp\left(-\frac{\gamma}{2}\left(\frac{\Omega}{2\delta_L}\right)^2 t\frac{v^2\xi^2/4}{1+v^2\xi^2/4}\right)}{1+v^2\xi^2/4}\,. \end{aligned} \tag{42}$$

with ξ given by eq. (38). For the experiments [12, 13] spontaneous emission has not yet been analyzed. In [12, 13] the values of the parameter $\lambda = 2k_L\Delta L$ are smaller than 10. Even though these values reach well beyond the classical to quantum cross–over point (which is near $\lambda = 2.5$ in the experiments) they are not yet as large as one would like them to be on theoretical grounds, in order to have a rather large classical chaotic domain containing many overlapping resonances. For the experimentally realized values of λ individual regular islands within the chaotic domain are still of major importance and their influence has to be carefully assessed [13, 29] in order to nail down the unequivocal observation of dynamical localization in this system. Fortunately, many parameters of this system are at the experimenters disposal and I believe that the prospects of achieving this goal in the present system are much better than in the competing experiments on Rydberg atoms in microwave fields [30]–[32] where similar problems – the existence of effects obfuscating dynamical localization – also exist, but only very few parameters are easily varied. It would e.g. be very interesting to gradually tune the laser from the off–resonance case, described above, towards resonance, thereby gradually turning on spontaneous emission. Exactly on resonance the system again decouples into two phase–modulated quantized pendula with different parameters but otherwise similar behavior to the one described above, except for the importance of spontaneous emission near resonance. Clearly by observing the change of the localization length with the detuning, its intricate interplay with the underlying nonlinear dynamics and with spontaneous emission, we could still learn a lot about this subtlest of all coherence effects in the field of quantum chaos.

Acknowledgments

I am indebted to the organizers of this winter school for the opportunity to present our work in a very stimulating atmosphere. The present work was supported by the Deutsche Forschungsgemeinschaft through the Sonderforschungsbereich "Unordnung und große Fluktuationen".

References

1. A.J. Lichtenberg and M.A. Lieberman, 'Regular and Stochastic Motion', (Springer, Berlin 1983);
   R.S. MacKay and J.D. Meiss (eds.), 'Hamitonian Dynamical Systems' (Hilger, Bristol 1987).
2. P. Goetsch, R. Graham and F. Haake, Phys. Rev. **A51**, 136 (1995); ibid. to appear.
3. G. Casati and B.V. Chirikov (eds.), 'Quantum Chaos: Between Order and Disorder', (Cambridge Univ. Press, Cambridge 1995).
4. O. Bohigas, S. Tomsovic and D. Ullmo, Phys. Rev. Lett. **64**, 1479 (1990); Phys. Rev. Lett. **65**, 5 (1990); Phys. Reports **223**, 43 (1993).
5. S. Fishman, in 'Quantum Chaos', G. Casati, I. Guarneri and U. Smilansky (eds.), (North Holland, Amsterdam 1993);
   R.E. Prange, in 'Chaos and Quantum Chaos', W.D. Heiss (ed.), (Springer, Berlin 1992).
6. See e.g. B.V. Chirikov, in 'Chaos and Quantum Chaos', W.D. Heiss (ed.), (Springer, Berlin 1992) and references given there.
7. D.L. Shepelyansky, Physica **D28**, 103 (1987).
8. S. Fishman, R.E. Prange and M. Griniasty, Phys. Rev. **A39**, 1628 (1989).
9. R. Graham, M. Schlautmann and P. Zoller, Phys. Rev. **A45**, R19 (1992).
10. R. Graham, M. Schlautmann and D. L. Shepelyansky, Phys. Rev. Lett. **67**, 255 (1991).
11. P.L. Gould et al, Phys. Rev. Lett. **56**, 827 (1986);
    C. Salomon et al, Phys. Rev. Lett. **59**, 1659 (1987);
    P.J. Martin et al, Phys. Rev. **A36**, 2495 (1987);
    P.J. Martin et al, Phys. Rev. Lett. **60**, 515 (1988).
12. F.L. Moore, J.C. Robinson, C. Bharucha, P.E. Williams and M.G. Raizen, Phys. Rev. Lett. **73**, 2974 (1994).
13. J.C. Robinson, C. Bharucha, F.L. Moore, R. Jahnke, G.A. Georgakis, Q. Niu and M.G. Raizen, Phys. Rev. Lett., to appear.
14. S. Fishman, D.R. Grempel and R.E. Prange, Phys. Rev. Lett. **49**, 509 (1982);
    D.R. Grempel, R.E. Prange and S. Fishman, Phys. Rev. **A29**, 1639 (1984).
15. D.L. Shepelyansky, Phys. Rev. Lett. **56**, 677 (1986).
16. G. Casati, R. Graham, I. Guarneri and F. Izrailev, Phys. Lett. **A190**, 159 (1994).
17. T. Dittrich and U. Smilansky, Nonlinearity **4**, 59 (1991);
    R. Blümel and U. Smilansky, Phys. Rev. Lett.**69**, 217 (1992).
18. D. Weaire and V. Srivasta, J. Phys. **C10**, 4309 (1977).

19. R. Blümel and U. Smilansky, Phys. Rev. **A30**, 1040 (1984); F.M. Izrailev, J. Phys. **A22**, 865 (1989).
20. K. Mølmer, Y. Castin and J. Dalibard, J. Opt. Soc. Am. **B10**, 524 (1993); R. Dum, P. Zoller and H. Ritsch, Phys. Rev. **A45**, 4879 (1992).
21. P. Goetsch and R. Graham, Phys. Rev. **A50**, 5242 (1994).
22. A. Barchielli, preprint IFUM 449/FT (1993).
23. H.M. Wiseman and G.J. Milburn, Phys. Rev. **A47**, 1652 (1993); S. Dyrting and G.J. Milburn, Phys. Rev. A, to appear.
24. M. Schlautmann and R. Graham, Phys. Rev. **E**, to appear.
25. R. Graham and S. Miyazaki, 'Dynamical localization of atomic De Broglie waves: the influence of spontaneous emission', preprint 1995.
26. E. Ott, T.M. Antonsen and J.D. Hanson, Phys. Rev. Lett. **53**, 2187 (1984).
27. T. Dittrich and R. Graham, Ann. Phys. (N.Y.) **200**, 363 (1990); D. Cohen, Phys. Rev. **A43**, 639 (1991); **44**, 2292 (1991).
28. B.V. Chirikov and D.L. Shepelyansky, Sov. Phys. Tech. Phys. **27**, 156 (1982).
29. J.P. Bardroff, I. Bialynicki-Birula, D.S. Krähmer, G. Kurizki, E. Mayr, P. Stifter and W.P. Schleich, preprint 1995.
30. E.J. Galvez, B.E. Sauer, L. Moorman, P.M. Koch and D. Richards, Phys. Rev. Lett. **61**, 2011 (1988).
31. J.E. Bayfield, G. Casati, I. Guarneri and D.W. Sokol, Phys. Rev. Lett. **63**, 364 (1989).
32. R. Blümel, R. Graham, L. Sirko, U. Smilansky, H. Walther and K. Yamada, Phys. Rev. Lett. **62**, 341 (1989).

Anomalous Diffusion, Spontaneous Localizations and the Correspondence Principle

Paolo Grigolini

[1] Istituto di Biofisica del Consiglio Nazionale delle Ricerche,
Via S. Lorenzo 26, 56127 Pisa Italy
[2] Department of Physics,
University of North Texas, P.O. Box 5368, Denton, Texas 76203

Abstract: We show that the equivalence between quantum and classical physics can be obtained, in the proper limit, by using arguments of ordinary statistical mechanics. It is argued that conditions of statistical mechanics as non ordinary as anomalous diffusion might produce the breakdown of this equivalence

1 Introduction

Quantum Mechanics rests on the joint use of two main ingredients. Borrowing the terminology developed by Penrose [1] we can say that it consists of the joint action of a unitary evolution, the **U** element of Penrose, and of a stochastic process, related to the measurement action, called by Penrose the **R** element .

In other words, the key quantum mechanical ingredient is the Schrödinger equation

$$ih\frac{\partial\,|\psi(t)>}{\partial t} = \hat{H}|\psi(t)> \ , \tag{1}$$

which leads to the unitary time evolution

$$|\psi(t)>= \exp(-\frac{i}{\hbar}\hat{H}t)|\psi(0)> \ . \tag{2}$$

This is the "deterministic" ingredient of quantum mechanics, the **U** element of Penrose [1]. In spite of the fact that quantum mechanics, with the Heisenberg uncertainty principle,

$$\Delta p \Delta q \geq \hbar/2 \tag{3}$$

is regarded as being the kingdom of uncertainty and indeterminacy, the adoption of the Schrödinger equation alone would make it the kingdom of the Universal Determinism and of Cosmic Certainty. This is so because in quantum mechanics, as well known, also two non interacting systems must maintain unchanged their original correlations. The so called Einstein, Podolsky, Rosen (EPR) effect [2], and the so called EPR correlations as well [3], are manifestations of the tendency of Quantum Mechanics to extend the range of correlations beyond any spatial limit. A specially relevant example can be illustrated by means of the Hamiltonian

$$\hat{H}_T = \hat{V} + \hat{H}_\mathrm{B} \ , \tag{4}$$

which has the following physical meaning. We focus our attention on a two–level system with the degenerate states $|+\rangle$ and $|-\rangle$, which are orthogonal to one another, and might refer to distinct positions. This quantum mechanical system interacts with its environment, described by the Hamiltonian $\hat{H}_B$, by means of the interaction term $\hat{V}$, which is given the following form

$$\hat{V} = g\hat{s}_z\hat{y} \tag{5}$$

where g is the coupling strength, $\hat{s}_Z$ is the z–component of the Pauli vectorial matrix, and $\hat{y}$ is an operator belonging to the Hilbert space of the environment. This is an extremely simplified model of interaction, which, nevertheless, serves very well the purpose of illustrating a striking property of quantum mechanics. Let us imagine that the whole Universe starts from the initial condition

$$|\psi_T(0)>= (c_+|+> +c_-|->)|E> \tag{6}$$

It is a straightforward exercise of quantum mechanics to prove that the time evolution of the system is described by

$$|\psi_T(t)>= c_+|+> |E_+(t)> +c_-|-> |E_+(t)> \ , \tag{7}$$

where

$$|E_\pm(t)>= \exp\left(-\frac{i}{\hbar}\hat{H}_\mathrm{B}t\right)\left(-\frac{i}{\hbar}\int_0^t (\pm)\,\hat{y}(t')dt'\right)|E> \ . \tag{8}$$

It must be recalled that the states $|E\rangle$, $|E_+\rangle$ and $|E_-\rangle$ are states of the environment, thereby involving an extremely large number of degrees of freedom, regardless the system of interest is microscopic or macroscopic. It is evident that in both cases the quantum correlation is transferred to a system of dimensions larger than the original system with a physical mechanism that at first sight seems to contradict physical intuition, but which is a correct quantum mechanical transport or enlargement of quantum correlations, taking place with no need of energy exchange between the system of interest and the environment.

It must be remarked that, if the system of interest, the system with the eigenstates $|+\rangle$ and $|-\rangle$ is already macroscopic, and $|+\rangle$ and $|-\rangle$ refer to macroscopically distinct positions, then the entanglement property, which, in a sense, makes quantum mechanics still more paradoxical, is the key ingredient to prove, in accordance with the daily experience, that the detections of correlations among spatially distinct components of a single state of a macroscopic system is impossible. This is so because after a very short time τ_D the two states $|E\rangle$ and $|E_-\rangle$ become orthogonal,

$$< E_-(\tau_D)|E_+(\tau_D) >\approx 0 \ , \tag{9}$$

thereby making the communication between the two branches of (7) impossible.

The statistical consequence of the orthogonality condition (9) can be properly considered by using the reduced density equation

$$\rho \equiv Tr_E \rho_T \quad , \tag{10}$$

which reads

$$\begin{aligned}\rho(t) = &\, c_+ c_+^* \left(|+><+|\right) < E_+(t)|E_+(t)> \\ &+ c_- c_-^* \left(|-><-|\right) < E_-(t)|E_-(t)> \\ &+ c_+ c_-^* \left(|+><-|\right) < E_-(t)|E_+(t)> \\ &+ c_- c_+^* \left(|-><+|\right) < E_+(t)|E_-(t)>\end{aligned} \tag{11}$$

and, due to the orthogonality condition (9), in a time τ_D whose order of magnitude will be assessed later, results in the statistical mixture

$$\rho(\infty) = |c_+|^2 \, |+><+| + |c_-|^2 \, |-><-| \ . \tag{12}$$

This statistical mixture is indistinguishable from the result which would be produced by the occurrence of real wave function collapses according to the measurement postulate of ordinary quantum mechanics. For this reason, as stressed in the work of Zurek [4, 5, 6] and in the earlier work of other scientists (see, for instance, Ref. [7]), it is thought that the decoherence processes caused by the unavoidable interaction between a system of interest and its environment has the key role of realizing, from within ordinary quantum mechanics, the process of the wave–function collapse, which is indispensable to give a physical meaning to the so called quantum mechanical mean value of an observable. Throughout this paper we shall be referring to this as Decoherence Theory (DT).

It must be stressed that the short, but finite, time τ_D can be evaluated by using a key ingredient of ordinary statistical mechanics. We see indeed from (8) that

$$\Phi(t) \equiv < E_+(t)|E_-(t)> = < E| \left(\frac{2i}{\hbar} \int_0^t \hat{y}(t')dt' \right) |E> \equiv < E| \left(\frac{2i}{\hbar} \hat{x}(t) \right) |E> \ . \tag{13}$$

If we take into account that the environment is characterized by a large number of degrees of freedom, the "relaxation" function $\Phi(t)$ can be dealt with adopting the standard arguments of statistical mechanics concerning Gaussian statistical processes. This means that (13) can be written [8] under the form

$$\Phi(t) = \exp\left\{ -\frac{4}{\hbar^2} \left\langle \hat{x}^2(t) \right\rangle \right\} \ . \tag{14}$$

By making the assumption that the "fast" fluctuation $\hat{y}(s)$ results in ordinary diffusion (see also Section 2), we are immediately led to an exponential decay with lifetime τ_D given by

$$\tau_D = \frac{\hbar^2}{4g^2 < \hat{y}^2(t) >_{\rm eq} T_{\rm relax}} \ , \tag{15}$$

where T_{relax} is the correlation time of the the variable $\hat{y}(s)$, namely, the lifetime of the corresponding correlation function. Notice that this is the ratio of a microscopic ($\hbar^2$) to a macroscopic (the denominator of (15)) quantity, thereby leading to the extremely low value of τ_D.

Is it really possible to describe all the relevant dynamical properties, including the transition from a pure state to the statistical density matrix with the use of (1) alone? This would make it unnecessary to supplement quantum mechanics with the randomness element **R** of Penrose [1]. To explain what the element **R** is all about, it is convenient to remind the reader about the fundamental postulate of quantum mechanics, and some proper comments concerning the meaning of quantum mean value are also in order. Let us consider an observable $\hat{A}$, characterized by the eigenvalues ϵ_i and the eigenvectors $|u_i\rangle$,

$$\hat{A}|u_i\rangle = \epsilon_i|u_i\rangle \ . \tag{16}$$

We are now ready for a general discussion, including also the earlier discussed case of the two–value observable $\hat{\sigma}_i$. Let us assume that the system at time t is expressed by the wave–function $|\psi(t)\rangle$ and that this is conveniently expressed in terms of the eigenstates $|u_i\rangle$ as follows:

$$|\psi(t)>=\sum_{i=1}^{\infty} c_i(t)|u_i> \ . \tag{17}$$

According to the formal prescriptions of quantum mechanics the quantum mechanical value of the observable $\hat{A}$, given the condition of the system in the state $|\psi(t)\rangle$, is defined by

$$<\hat{A}>=\sum_{i=1}^{\infty} |c_i(t)|^2 \varepsilon_i \ . \tag{18}$$

How to connect this theoretical prediction to the results of real experiments? This is where the measurement postulate of quantum mechanics comes into play. If a measurement process is carried out on the system to determine the value of the observable $\hat{A}$, then the system is assumed to collapse into one of the eigenstates of this observable, $|u_i\rangle$, with a probability p_i. If the experiment is repeated in the same conditions, or equivalently if a Gibbs ensemble of systems prepared in the same state is used, then a distribution of possible values with the probability p_i is obtained, thereby resulting in the "experimental" mean value

$$<\hat{A}(t)>_{\text{exp}}=\sum_{i=1}^{\infty} p_i(t)\varepsilon_i \ . \tag{19}$$

The main tenet of the quantum mechanical postulate is that $p_i(t) = |c_i(t)|^2$ so that the "experimental" mean value (19) equal the "theoretical" prediction (18). It is evident that this postulate introduces an element of randomness within quantum mechanics and consequently the need of adopting a statistical view as conveniently expressed by the statistical density matrix ρ defined by

$$\rho(t) = \sum_{j=1}^{N} \frac{|\psi^{(j)}(t) >< \psi^{(j)}(t)|}{N} \, , \tag{20}$$

which implies a proper average over a Gibbs ensemble of systems. Note that even if all the systems are described by the same wavefunction $|\psi(t)\rangle$ and consequently the density matrix is that of a pure state $|\psi(t)\rangle\langle\psi(t)|$ yet the statistical density matrix retains its original statistical meaning, of an average over a Gibbs ensemble, due to the measurement postulate, according to which indeed the measurement act results in the following transition

$$\rho(t) = |\psi(t) >< \psi(t)| \Rightarrow \sum_{i=1}^{\infty} p_i(t)|u_i >< u_i| \; . \tag{21}$$

In other words, while in classical statistical mechanics randomness results from the uncertainty over the initial conditions, in quantum mechanics even if all the systems are identically prepared, randomness, and consequently, the adoption of a statistical treatment, is dictated by the measurement postulate itself. This means that the probabilistic character of quantum mechanics, stressed by the uncertainty relation (3), is closely connected with the statistical nature of the measurement process, as it is shown by noticing that the variances defining (3) imply the occurrence of the random process associated to the measurement and leading to the transition described by (21).

The statistical density matrix resulting from the transition (21) can be derived, as shown earlier in the two– dimensional case, also by taking the influence of the environment into account. However, the definition itself of reduced density matrix (10) is subtly related to the conventional definition of quantum mean value of an observable $\hat{A}$. The definition (19) can be written also as $\langle\hat{A}\rangle = Tr\rho\hat{A}$. If this observable belongs to a subsystem S of the whole Universe U the environment E being the remainder part of the Universe, the quantum mechanical mean value is written as $\langle\hat{A}\rangle = Tr\rho_T\hat{A}$, which is immediately proved to be equivalent to $\langle\hat{A}\rangle = Tr\rho\hat{A}$, where ρ is the reduced density matrix (10). This shows that the definition of reduced density matrix is subtly related to the measurement postulate and corresponds in a sense to measuring the environment of the system of interest [11].

All this shows that the element of randomness **R** associated with the measurement act does not need to be explicitly adopted and that the border between quantum and classical mechanics can be crossed without any risk if the Gibbs picture is suitably adopted. This paper is devoted to assessing whether special physical conditions exist which might invalidate this unified representation of quantum and classical mechanics.

In the case of regular classical motion the spreading of the wave function is extremely slow and this leads to a satisfactory derivation of classical from quantum physics, once initial conditions corresponding to a single trajectory are created. In the opposite limit of a totally chaotic classical system, although the spreading of the wave functions is extremely fast, the quantum–classical correspondence is recovered if a proper use of the statistical nature of quantum mechanics is made. This is so because in the chaotic case classical as well as quantum mechanics has statistical properties independent of the properties of single trajectories. Thermodynamics and ordinary statistical mechanics seem to be generated by physical conditions where a time scale separation between macroscopic and macroscopic processes is possible. If our attention is shifted from chaotic processes which are the source of ordinary diffusion to those without a defined time scale, which are responsible for anomalous diffusion, it is not yet clear which kind of statistical mechanics will be generated. In addition to that, we think that anomalous diffusion also violates the conditions necessary to recover in the proper limit classical from quantum mechanics. We see, for instance, that (14) and the corresponding exponential behavior of the "relaxation" function $\Phi(t)$, rests on conditions of "normal diffusion" (see Section 2), and a pathological diffusion condition might maintain the quantum coherence for a period of time much more extended than that predicted by ordinary quantum mechanics. We shall explore the quantum behavior of systems which in classical physics would exhibit anomalous diffusion in Section 3.

On the other hand, the power of the statistical interpretation of quantum mechanics, according to which classical mechanics can be derived from quantum mechanics if a proper use is made of the fact that no isolated systems exists, rests on the fact that the experimental observations are usually made adopting a Gibbs ensemble picture. In the last few years, on the contrary, it was shown that individual system observations are possible [9, 10], and for this reason we plan to discuss in this paper also the problems raised by this new perspective. In Section 4 we shall illustrate some attractive aspects of a recent proposal resting on a wise modification of quantum mechanics and including the **R** element as a substantial part of dynamics, rather than as a measurement postulate.

Finally, in Section 5 we plan to discuss also the problem of how to derive thermodynamics from a microscopic picture and to show that the improved interpretation of quantum mechanics, and especially the generalization of quantum mechanics resting on the dynamical role of spontaneous localizations, lead to a more satisfactory treatment of this difficult problem.

2 Dynamic Approach to Anomalous Diffusion

In classical physics diffusion processes stem from the integration of a remarkably simple equation, which, in the one–dimensional case, reads as follows:

$$dx/dt = \xi(t) \tag{22}$$

where $x(t)$ is the diffusing variable and $\xi(t)$ is an erratic process. From the formal integration of this equation we get:

$$x(t) = \int_0^t dt'\xi(t') + x(0) \ . \tag{23}$$

Since $\xi(t)$ is an erratic process, we are naturally led to derive from (23) only statistical information. We assume the diffusion process to be unbiased, $\langle \xi(t) \rangle = 0$ and consequently the simplest statistical information is that provided by the second moment of the diffusing variable $x(t)$. By evaluating the square value of (23) and averaging over several initial conditions we get:

$$< x^2(t) >= \int_0^t dt' \int_0^t dt'' < \xi(t')\xi(t'') > +2 \int_0^t dt' < \xi(t')x(0) > + < x^2(0) > \ . \tag{24}$$

The time evolution of $\xi(t)$ is independent of the arbitrarily chosen initial value $x(0)$. Consequently, $\langle \xi(t)x(0) \rangle = \langle \xi(t) \rangle \langle x(0) \rangle = 0$. Let us now make the assumption that the statistical process $\xi(t)$ is stationary, thereby implying

$$< \xi(t)\xi(t') >=< \xi(|t - t'|)\xi(0) > \ . \tag{25}$$

This leads us to

$$< x^2(t) >= 2 < \xi^2 > \int_0^t dt' \int_0^{t'} dt'' \Phi_\xi(t'') + < x^2(0) > \ , \tag{26}$$

where

$$\Phi_\xi(t) \equiv < \xi(0)\xi(t) > / < \xi^2 > \ . \tag{27}$$

Ordinary statistical mechanics is the kingdom of the time scale separation between microscopic and macroscopic processes, and ordinary diffusion stems naturally from this condition. The "microscopic" time scale is defined by the correlation time τ given by

$$\tau \equiv \int_0^\infty \Phi_\xi(t')dt' \ . \tag{28}$$

It is evident from (26) that if the long–time region $t \gg \tau$ is explored, then (26) becomes

$$< x^2(t) >= 2Dt + < x^2(0) > \ , \tag{29}$$

where D is the standard diffusion coefficient defined by

$$D =< \xi^2 > \tau \ . \tag{30}$$

At the same time the generalized version of the central limit theorem leads us to the conclusion that the distribution of the variable x is a Gaussian function whose width is the linearly increasing function of time given by (29).

This result illustrates the essence of ordinary statistical mechanics. It has been recently proved [12] that there is a close connection between the Kubo linear response theory and ordinary statistical mechanics. This, in turn, implies that it is possible to define a microscopic time scale and that it is possible to establish a neat separation between microscopic and macroscopic dynamics. Standard diffusion is an attractively simple way of illustrating this property. We have seen indeed that the existence of a finite time scale, (28), implies the possibility of defining a much larger time scale, where microscopic processes are perceived always as the source of gaussian fluctuations.

What about the case where it is not possible to define the finite time scale (28)? Before addressing this intriguing problem, we have to solve the equally intriguing problem of assessing whether or not processes with no finite time scale exist. Obviously, a process characterized by the correlation function

$$\lim_{t\to\infty} \Phi_\xi(t) \propto \frac{1}{t^\beta} \tag{31}$$

with $0 < \beta < 1$, would be incompatible with the condition of resulting in the finite time scale of (28) and, as it can be easily proved by differentiating twice (24) with respect to time, would result in $< x^2(t) > \propto t^{2H}$ with $H = 1 - \beta/2$ and $1 > H > 1/2$. Does an inverse–power law like that of (31) have a physical significance? The solution of the problem can be postponed by establishing a connection between the correlation function $\Phi_\xi(t)$ and another statistical property, the sojourn time distribution $\psi(t)$.

Let us imagine that the variable ξ fluctuates between two distinct values, a and $-a$, with a statistical weight p, so that $\langle\xi\rangle_{\mathrm{eq}} = 0$ and $\langle\xi^2\rangle_{\mathrm{eq}} = 2pa^2$. The sojourn time distribution $\psi(t)$ refers to the times of sojourn in one of these two states. It is possible to prove that the correlation function $\Phi_\xi(t)$ is related to the sojourn time distribution $\psi(t)$ by an important relation established by Geisel et al. [13] on the basis of the renewal theory. This is

$$\Phi_\xi(t) = \frac{1}{<\xi^2>}\frac{1}{<t>}\int_t^\infty (t'-t)\psi(t')dt' \ , \tag{32}$$

where $\langle t\rangle$ denotes the average sojourn time. From the asymptotic analysis of (32) we find that the condition (31) is realized if

$$\lim_{t\to\infty} \psi(t) \propto \frac{1}{t^\mu} \tag{33}$$

and the condition: $2 < \mu < 3$ holds. This is so since from the asymptotic analysis of (32) we derive: $\beta = \mu - 2$.

The problem of the physical origin of (31) is thus traced back to that of the physical origin of the inverse power law for the distribution time $\psi(t)$. An illuminating example of this property is given by the standard map [14], which reads

$$\begin{aligned} p_{n+1} &= p_n + K\sin\theta_n \\ \theta_{n+1} &= \theta_n + p_n, \ \ (\mathrm{mod}\ 2\pi) \ . \end{aligned} \tag{34}$$

Under specific conditions on the strength K physical conditions exist, where the (p, θ) phase–space of the system is characterized by acceleration islands surrounded by a chaotic sea [15]. When the phase space point sticks to the border with the acceleration island it tends to sojourn there with a waiting time distribution characterized by an inverse power–law distribution. There are several theoretical explanations of this behavior, resting on the fact that the region at the border between the chaotic sea and the acceleration islands is fractal and the particle tends to imbed itself into regions of smaller and smaller dimensions [15].

At this stage, the possibility of deriving from a completely dynamical, and Hamiltonian, picture a process of anomalous diffusion is almost completely accounted for theoretically. To stress the departure of the process of anomalous diffusion from conventional statistical mechanics, it is convenient to establish a connection with the Lévy processes [16], which are an exact solution of the Chapmann–Kolmogoroff equation not fulfilling the standard Gaussian prescription. The Fourier–Laplace transform of the Lévy distribution $P(x, t)$, denoted by $\hat{P}(k, s)$, has the explicit form

$$\hat{P}(k, s) = \frac{1}{s + b\,|k|^{\alpha}} \tag{35}$$

with

$$0 \leq \alpha \leq 2 \ . \tag{36}$$

The Lévy processes represent a condition of anomalous statistical mechanics, with distributions characterized by tails slower than the conventional Gaussian distributions, which are now the subject of investigation by many groups. The main reason why we mention here the Lévy processes is because they are proved [17, 18]to be naturally produced by the Hamiltonian physical conditions leading to (33). It was recently shown [19] that if the sojourn time distribution is expressed in a form valid also in the short–time region, and precisely,

$$\psi(t) = \frac{A}{(B + t)^{\mu}} \tag{37}$$

with A connected to B by the normalization condition,

$$A = (\mu - 1)B^{\mu - 1} \ , \tag{38}$$

then to a good approximation we get (35) for the Fourier–Laplace transform of $P(x, t)$, with b defined by

$$b = \frac{B^{\mu-1}\Gamma(2 - \mu)}{T} \cos\left(\frac{\pi}{2}(\mu - 1)\right) \tag{39}$$

and

$$\mu = 1 + \alpha \ . \tag{40}$$

It has to be pointed out that if an one–dimensional map, responsible for anomalous is perturbed by making its dynamics also depend on the addiction of a noise of intensity D, then the asymptotic limit of the waiting time distribution is proved [20] to be given by

$$\psi(t) \approx \exp(-D^{1/(2\mu-1)}t) \ , \tag{41}$$

which suggests that normal diffusion is recovered at a time scale which becomes increasingly large upon decrease of the noise intensity. It must be pointed out that when normal diffusion is recovered the time evolution of the second moment of the distribution is shown [21] to be given by

$$< x^2(t) > \approx D^{[(\mu-3)/(2\mu-1)]}t \ , \tag{42}$$

which means that the diffusion coefficient of the process becomes extremely large if the noise is very weak. This has to be contrasted with the negligible influence of a weak fluctuation on normal diffusion. The essential statistical equivalence of classical and quantum mechanics in the case of normal diffusion might be due to the fact that a single trajectory can be significantly perturbed by a weak noise with no essential change, though, on the statistics of the process. Anomalous diffusion, on the contrary, exhibits a much stronger sensitivity to external perturbations. Thus we make the plausible conjecture that anomalous diffusion is also sensitive to quantum fluctuations, thereby leading to the interesting consequence that a departure of quantum from classical mechanics can be detected even if a statistical treatment is adopted. This issue will be discussed in Section 3.

3 The Wigner Method, the Correspondence Principle and the Quantum Mechanical Manifestations of Anomalous Diffusion

One of the most striking aspects of the quantum systems which would be classically chaotic is the discovery of a new possible channel for the macroscopic manifestation of quantum mechanics. The majority of authors [22, 23, 24, 25] with but a few exceptions [26] agree on this issue with different arguments leading however to equivalent conclusions. We think that probably the most intuitive way to reach this conclusion is given by the picture recently proposed by Zurek an Paz [25]. Following Zurek and Paz [25] let us consider a quantum system in the physical condition where, according to a traditional wisdom quantum dynamics is supposed to be virtually coincident with classical mechanics. Let us consider, for instance, a driven one–dimensional system, namely a particle moving in a time dependent potential $V(q,t)$. We adopt the Wigner formalism [27] leading to the following equation of motion for the quasiprobability $\rho_W(q,p;t)$

$$\frac{\partial}{\partial t}\rho_W(q,p;t) = \left(\hat{L}_{\text{class}} + \hat{L}_Q\right)\rho_W(q,p;t) \ , \tag{43}$$

where $\hat{L}_{\rm class}$ coincides with the classical Poisson bracket and $\hat{L}_Q$ is the quantum contribution reading

$$\hat{L}_Q = \sum_{r=3}^{\infty} \left(\frac{\partial^r}{\partial q^r} V(q,t) \right) \left(\frac{\hbar}{2i} \right)^{r-1} \frac{1}{r!} \frac{\partial^r}{\partial q^r} \; . \tag{44}$$

Note that, even if $\hat{L}_Q$ is disregarded the resulting equation is linear. Therefore, in spite of the exact statistical agreement that would be obtained in that case between classical and quantum mechanics, it would be impossible to rule out the linear combinations of distinct positions that are possible in quantum mechanics and are incompatible with classical mechanics. The operator $\hat{L}_Q$ has the role of signaling at a statistical level the effects that such superpositions, if any, might have on the time evolution of the systems. Let us consider the case where the Planck constant $\hbar$ is extremely small compared to the typical macroscopic action ΔI namely the volume of the phase space available to the system within the classical representation. Under this condition, where the classical picture is expected to hold, we adopt an initial distribution given by a smooth Wigner density $\rho_W(q,p;0)$. The classical evolution of the system is given by

$$\rho_L(q,p;t) = \exp(\hat{L}_{\rm class} t) \rho_L(q,p;0) \; . \tag{45}$$

The quantum evolution of the system is described by

$$\rho_W(q,p;t) = \exp\left[(\hat{L}_{\rm class} + \hat{L}_{\rm class}) t \right] \rho \;\cdot_L(q,p;0) \; . \tag{46}$$

The validity of the classical picture should be broken when $\rho_W(q,p;t)$ significantly departs from $\rho_L(q,p;t)$. In the case of ordinary classical motion the time required for this breakdown to take place is inversely proportional to the Planck constant to a power of the order of unity, and consequently a time so astronomically large as to result in a complete fulfillment of the expectations of the correspondence principle [28].

In the case where the classical picture is fully chaotic completely different situation would occur. Let us see why. The Liouville density undergoes a rapid process of fragmentation and the originally smooth distribution quickly develops whorls and tendrils, and it becomes more and more finely fragmented with increasing time. This process of increasing fragmentation has the effect of enhancing the role of the operator L_Q. This operator would rigorously vanish in the case of linear systems, thereby making unlimited the regime of validity of the correspondence principle. Even in the nonlinear case this correction can be neglected if the Liouville density (and consequently the Liouville distribution) remained smooth. However, when the Wigner distribution, originally identical to the Liouville density, becomes extremely fragmented, the differential operator, applied to the quasi– distribution, yields terms comparable to that of the classical evolution, thereby resulting in a departure of the Liouville density from the Wigner quasi–distribution. Zurek and Paz [25] argue that this departure of the classical from the quantum description takes place at a time t_χ given approximately by

$$t_\chi \approx \frac{1}{\lambda} \ln \left(\frac{1}{\hbar} \right) , \tag{47}$$

where λ is the largest Lyapunov coefficient. The time required for the quantum–classical breakdown to take place is proportional to the logarithm of the inverse of the Planck constant. This is a relatively short time, which would allow, also for macroscopic systems, quantum effects to show up in experimentally observable times. This would raise the serious problem of the validity of the correspondence principle in classically chaotic systems.

Now we have to answer the question of why, in spite of this widely accepted conclusion, there seems to be no deep concern, among the physicists, on the inconsistency of quantum mechanics as "universal" theory from which classical mechanics can be naturally derived. This depends on the fact that all the authors working in this area have explicitly, or implicitly, in mind a definition of the correspondence principle, which does not correspond to the naive formulation resting on the existence of the single classical trajectories, thought of as the motion of narrow packets (naive definition). The postulate of the theory of measurement illustrated in Section 1 implies that the quantum expectation value is experimentally reproduced making an average over a set of independent measurements, or, equivalently, over a Gibbs ensemble. A proper formal definition of correspondence principle, in line with this perspective, can be given by using the Wigner formalism. According to the Wigner formalism, the mean value of a quantum mechanical observable $\hat{A}$ can be expressed by

$$< \psi(t)|\hat{A}|\psi(t) >= \int dpdqA_W(q,p)\rho_W(q,p;t) , \tag{48}$$

where $A_W(q,p)$ is the classical variable corresponding to the quantum observable within the Weyl picture. It is thus clear that the adoption of a Gibbs ensemble is necessary even if only one wave function (and not a statistical mixture) is considered. Within this interpretation scheme the classical behavior of our system (and consequently the correspondence principle) holds true, if in (48) $\rho_W(q,p;t)$ can be safely replaced by the Liouville density $\rho_L(q,p;t)$. This statistical definition of the correspondence principle is the one which must be adopted to be in line with the spirit of ordinary quantum mechanics.

What about the quantum treatment of processes which would exhibit anomalous diffusion in classical mechanics? This is still the object of investigation and attempts are being made to prove that in that case a breakdown of the correspondence principle occurs in logarithmic times even if the statistical definition of the correspondence principle is applied. There seems to be numerical evidence of this property [29], and from an intuitive point of view the physical reason of that seems to be the fact that anomalous diffusion is much more sensitive than normal diffusion to the influence of fluctuations. In other words, we make the plausible conjecture that anomalous diffusion is sensitive to quantum fluctuations as it is, as illustrated in Section 2, to thermal fluctuations.

4 From Statistical to Individual–System Pictures

In Section 1 we have mentioned that the dephasing processes caused by the interaction between a system of interest and its environment result in a statistical density matrix indistinguishable from that which according to the measurement postulate of quantum mechanics would be produced by a measurement process. Let us consider now a quantum oscillator, interacting with an environment of harmonic oscillators, so that its motion is equivalent to that of a classical oscillator with damping Γ. Let us consider the coherent states $|\alpha\rangle$ and $|-\alpha\rangle$, obtained by shifting the ground state of the harmonic oscillator by the quantity $\Delta Q/2$ in the positive and negative direction, respectively. The motion of $|\alpha\rangle(|-\alpha\rangle)$ would be indistinguishable from that of a classical particle with mass M with the initial condition given by $q = \Delta Q/2(q = -\Delta Q/2)$, $p = 0$. The striking quantum mechanical aspect is that a linear superposition of both trajectories is possible, with the "classical" particle being simultaneously present in two distinct positions at a distance ΔQ from one another. This leads us to the same initial condition as (6) with $|+\rangle$ replaced by $|\alpha\rangle$ and $|-\rangle$replaced by $|-\alpha\rangle$. The environment is split into two orthogonal states in a time τ_D, whose explicit expression in this case is:

$$\tau_D = \frac{\hbar^2}{2\Gamma M k_{\mathrm{B}} T (\Delta Q)^2} , \tag{49}$$

as it can be proved by adapting to this specific case the calculations illustrated in Section 1 (see Eq. (15)).

The DT illustrated in Section 1 and resulting in (15) is probably the most popular interpretation of quantum mechanics, and it is based on the conviction that the problems posed by quantum mechanics can be settled without changing this theory. We cannot disregard, however, the point of view of another group of physicists who, on the contrary, believe that to derive classical from quantum mechanics it is necessary to change or generalize quantum mechanics. We essentially adopt the theoretical proposal made by Ghirardi, Rimini and Weber [30] and shall be referring to this theory as the Spontaneous Localization Theory (SLT). According to these authors the wave function collapses already takes place at the level of microscopic systems without resulting, though, in significant dynamical corrections to the prescriptions of ordinary quantum mechanics. This is so for the following reason. Let us imagine that the linear superposition $|\psi(t)\rangle = |\psi_1(t)\rangle + |\psi_2(t)\rangle$ refers to a microscopic body, e.g. a proton, and that the distance between the two distinct positions is ΔQ. The process of direct collapse affecting an individual microscopic constituent is characterized by the two key parameters λ and $1/\sqrt{\alpha}$. The first is the rate of occurrence of the process of wave function collapse, and its value is set [30] $\lambda = 10^{-17}\mathrm{sec}^{-1}$, thereby implying one collapse to occur in 10^8 years. The value of the second parameter is assumed [30] to be $1/\sqrt{\alpha} = 10^{-5}\mathrm{cm}$ and represents the distance at which two distinct spatial components of the same state must be found for a collapse to really occur. In other words, if $\Delta Q \ll 1/\sqrt{\alpha}$ the collapse into one of the two distinct positions never occurs and if $\Delta Q \sim 1/\sqrt{\alpha}$ or larger, it occurs in 10^8 years. It is evident that with these new dynamical rules the motion of a microscopic body fits almost exactly the predictions of ordinary quantum mechanics. It is evident that if we apply this theory to the case of the oscillator earlier mentioned in this Section, and this oscillator does not interact with the environment, no substantial modification from the predictions of ordinary quantum mechanics occurs.

What about the case when the oscillator interacts with a bath? The case of the oscillator interacting with a bath is similar to that of $N+1$ particles, interacting with one another. Let us consider, first, a physical system consisting of $N+1$ free particles coupled to one another by means of harmonic interactions. We notice that in this special case it is convenient to express the SLT theory by means of the Wigner formalism already used in Section 3, which makes the standard quantum mechanical contribution formally equivalent to the predictions of classical statistical physics. In the special case under discussion the quantum Liouville equation corrected by the occurrence of the processes of spontaneous localization is shown to read:

$$\frac{\partial \rho_W}{\partial t} = \{H, \rho_W\}_{PB} + \sum_{i=0}^{N} \lambda_i \left[\exp\left(\frac{\alpha h^2}{4}\frac{\partial^2}{\partial p_i^2}\right) - 1\right] \rho_W \ , \tag{50}$$

where the first term on the r.h.s. of Eq. (50) is the standard quantum mechanical term (which, thank to the harmonic nature of the Hamiltonian within the Wigner formalism turns out to be indistinguishable from the classical Poisson bracket) and the second term on the r.h.s. of (50) refers to the dynamical corrections resulting from the process of spontaneous localization. The p_i's denote the classical momenta of the particles of the system under study. Notice that it is plausible that the parameters λ_i have the same order of magnitude, namely, 10^{-17}sec^{-1}.

It is evident that this modification of the dynamical rules of ordinary quantum mechanics does not result in any significant modification if microscopic variables are considered. However, the dynamics of the center of gravity of a macroscopic body is strongly affected by the accumulation of these processes, thereby resulting in a motion virtually indistinguishable from that of ordinary classical mechanics. This is easily proved by focusing our attention on the reduced probability distribution concerning the center of gravity of the system, denoted by the symbol ρ_{WB}. With simple calculations [30, 31] we derive from (50) the following equation of motion for ρ_{WB}

$$\frac{\partial \rho_{WB}}{\partial t} = \{H_{\mathrm{B}}, \rho_{WB}\}_{PB} + \left(\sum_{i=0}^{N} \lambda_i\right) \left[\exp\left(\frac{\alpha h^2}{4}\frac{\partial^2}{\partial p_{\mathrm{B}}^2}\right) - 1\right] \rho_{WB} \ , \tag{51}$$

where H_B denotes the classical Hamiltonian of a free particle with the total mass $M = \sum_{i=0}^{N} m_i$ and the momentum $p_{\mathrm{B}} = \sum_{i=0}^{N} p_i$. We see from (51) that the quasiprobability, and hence the wave function, of the center of gravity of this macroscopic body is driven by an equation which has the same structure as (50), namely by a term corresponding to the ordinary Schrödinger equation, plus a correction with the same structure as that concerning a single particle but with an enhanced rate of wave function collapse given by $\lambda_{tot} = N\lambda$. It is straightforward to prove [30] that in the case of a macroscopic body of one gram a quantum state is left in a superposition of two distinct positions corresponding to the distance $\Delta Q = 1\text{cm}$ for a time of the order of 10^{-7} seconds. This means that the quantum mechanical coherence survives for times much shorter than the time required for us to perceive the motion of macroscopic bodies.

The center of gravity is an illuminating example of macroscopic variable subject to a significant process of accumulation of the individual spontaneous localizations. However, in the field of the condensed matter we cannot ignore considering the fluctuation–dissipation processes as a possible source of accumulation of the CSL processes. This process of accumulation is especially relevant because it is triggered by the same key mechanism as that resulting in the decoherence process. The entangled condition corresponding to that of (7) means that the individual i-th oscillator of the bath, with the frequency ω_i and the coupling strength ϵ_i, tends to set itself in a state which is a linear superposition of two wave–packets localized around two distinct spatial positions at a distance

$$\Delta L_i = \frac{\varepsilon_i \Delta Q}{m_i \omega_i^2} \tag{52}$$

from one another. This is easily shown by proving that the coupling between the oscillator of interest and the i-th bath oscillator shifts the origin of the harmonic potential of the latter oscillator by the quantity (52) in the positive or in the negative direction, according to whether the oscillator of interest is shifted by the quantity ΔQ in the positive or the negative direction, respectively. In the quantum case that we are considering the oscillator of interest is initially placed in a superposition of these two distinct positions, thereby forcing the bath oscillator into a superposition of two distinct positions shifted from the origin by the quantity ΔL_i of (52).

The direct CSL process acting on this "bath" oscillator causes a reduction of its wave–function and hence of the whole entangled state (51) with increasing efficiency as ΔL_i widens, till to fulfill the condition

$$\frac{1}{\sqrt{\alpha}} << \Delta L_i \ , \tag{53}$$

which ensures the complete efficiency of the CSL reduction process: The larger $\Delta Q_i$the more numerous the bath oscillators satisfying (53) are, thereby resulting in an increased rate for the collapse of the wave–function of the central oscillator.

After some calculations, described in detail in Ref. [31] we prove that the rate of decoherence process affecting the off–diagonal density matrix element is given by

$$R_T = R_S + R_C \ , \tag{54}$$

where $R_S = 1/\tau_D$ is, with τ_D given by (49) is the prediction of ordinary quantum mechanics and $R_c = 1/\tau_C$ is the correction resulting from the SLT. The explicit expression for this correction is

$$\frac{1}{\tau_C} = \frac{\sqrt{\lambda \alpha \Gamma \Delta Q^2}}{2} \; . \tag{55}$$

In general we have the condition

$$\tau_D \ll \tau_C \tag{56}$$

which implies a statistical equivalence between the SLT and standard quantum mechanics. In Section 3 we have been emphasizing that, although the connection between quantum mechanics and classical mechanics is difficult to make in terms of single trajectories, which might not be recovered from within quantum mechanics, when the comparison between quantum and classical physics is made adopting a statistical view, then the quantum predictions agree with the classical predictions. This happens not only in the case when the conditions are considered for the fulfillment of the correspondence principle, but it applies also to the case when the initial conditions are essentially quantum. The case of the ionization of the hydrogen atom is a relevant example of this intriguing aspect [32]. Consequently the fact that the SLT leads to an almost exact statistical agreement with ordinary quantum mechanics must be considered as a element of strength of the new theory.

An intriguing issue to discuss is whether or not an individual–system experiment might lead us to assess whether the prediction of the DT or that of the SLT is correct. In principle this seems to be possible, but the nature itself of the quantum jumps experiments, resting only indirectly on the collapse processes [9], makes it extremely difficult to establish whether the DT or the SLT is the correct theory.

5 From the Processes of Spontaneous Localization to the Boltzmann Dream Becoming True and Conclusions

It must be pointed out that in spite of the philosophical discordance, the DT and the SLT leads to proposals for the derivation of thermodynamics which are formally equivalent. Notice that if only the second moment is considered the proposal of the DT theory

$$\begin{aligned} \frac{\partial}{\partial t}\rho_W(q,p;t) &= -\left[\frac{p}{m}\frac{\partial}{\partial q} + \left(\frac{\partial V}{\partial q}\right)\frac{\partial}{\partial p}\right]\rho_W(q,p;t) \\ &+ \left[\sum_{n=1}^{\infty}\frac{1}{(2n+1)!}\left(\frac{\hbar}{2i}\right)^{2n} V^{(2n+1)}\frac{\partial^{2n+1}}{\partial p^{2n+1}}\right]\rho_W(q,p;t) \\ &+ \left[\Gamma\frac{\partial}{\partial p}p + D\frac{\partial^2}{\partial p^2}\right]\rho_W(q,p;t) \end{aligned} \tag{57}$$

is essentially indistinguishable from that of the SLT, which would lead to

$$\frac{\partial}{\partial t}\rho_W(q,p;t) = -\left[\frac{p}{m}\frac{\partial}{\partial q} + \left(\frac{\partial V}{\partial q}\right)\frac{\partial}{\partial p}\right]\rho_W(q,p;t)$$
$$+\left[\sum_{n=1}^{\infty}\frac{1}{(2n+1)!}\left(\frac{\hbar}{2i}\right)^{2n} V^{(2n+1)}\frac{\partial^{2n+1}}{\partial p^{2n+1}}\right]\rho_W(q,p;t)$$
$$+\lambda\left[\exp\left(\frac{\alpha h^2}{4}\frac{\partial^2}{\partial p_{\mathrm{B}}^2}\right)-1\right]\rho_W(q,p;t) \ . \tag{58}$$

Let us focus our attention first on the physical predictions stemming from (57). The issue to discuss is that concerning the possibility of realizing, through mixing, the condition of microcanonical distribution, starting from any initial condition for the distribution $\rho_W(q,p;0) = \rho_L(q,p;0)$. This is made possible by the fact that even if the volume of the phase space corresponding to a given energy E must remain constant, mixing results in a progressive fragmentation of the distribution, thereby making it necessary to adopt a coarse–graining picture, which changes the Gibbs entropy, which would be time independent, into an increasing function of time coinciding with the Boltzmann entropy.

Zurek and Paz [25] noticed that the diffusion term does not compete with the process of dilatation associated to mixing, but it does with the process of contraction, thereby reaching in a time much shorter than the time τ_c, Eq. (47), a balance with it. As a result of that, after that very short time, depending logarithmically on the noise intensity, and consequently essentially independent of the noise intensity, the Wigner distribution, which is still indistinguishable from the Liouville density, becomes a smooth and compact distribution. This is equivalent to the coarse–graining process earlier described, and leads naturally to a significant increase of the entropy with no need of adopting a coarse–graining picture.

It must be pointed out that these interesting results concern essentially the time region $t \ll 1/\gamma$. Within this "short–time region" a sort of genuinely mechanical thermodynamics is produced by the key action of chaos. In the long time region $t \gg 1/\gamma$ the system gets the temperature and all the thermodynamical properties of its environment. However all this takes place in times which are so astronomically long as to make this condition void of interest.

Due to the "equivalence" of (58) with (57) we can reach the same conclusion starting from (58) rather than from (57). In this case we have to take into account that the SLT results in a temperature increase of 10^{-15} ^{0}K per year. This leads us to the same conclusion as (58): If we confine our attention to the regime of "short times", where the temperature increase can be neglected, we obtain a sort of thermodynamics, derived from mechanics, with no need of using any coarse–graining procedure. From a conceptual point of view, however, (58) is still more satisfactory than (57). This is so because from within the ordinary classical or quantum physics the derivation of the Fokker–Planck equation cannot be made without adopting subtle subjective assumptions. On the contrary the SLT is already irreversible, and only approximately reversible when the proper conditions for ordinary classical and quantum mechanics apply. Consequently within the SLT the second principle of thermodynamics can be derived from mechanics with no ad hoc assumption whatsoever. In conclusion, the SLT makes it possible to establish a connection between mechanics and thermodynamics free from any subjective assumption. If we disregard this aspect, choosing between the DT and the SLT seems to be essentially a question of tastes.

Is there any real possibility to distinguish experimentally between the two theories? In principle, as shown in Section 4, the experimental observation of a individual system should make this possible, since the two predictions depart from one another by many orders of magnitude [31]. However, the current techniques of quantum jump experiments are based on the indirect effects of quantum jumps rather than on the direct observation of the elementary jumps [9]. This seems to make extremely difficult to distinguish the predictions of the DT from those of the SLT. We have also seen that within ordinary statistical mechanics the SLT leads to the same predictions as the DT theory. This is so because the key ingredient of the DT is indeed ordinary statistical mechanics. If move from ordinary statistical mechanics to anomalous statistical mechanics, the situation changes drastically, and the equivalence between classical and quantum statistical mechanics is lost. This is where we expect to see the the statistical predictions of the SLT to depart from those of the DT theory.

References

1. R. Penrose, *Shadows of the Mind, A Search for the Missing Science of Consciousness*, Oxford University Press, Oxford (1994).
2. A. Einstein, B. Podolsky, and N. Rosen, Phys. Rev. **47**, 777 (1935).
3. H. Primas, *Chemistry, Quantum Mechanics and Reductionism*, Springer– Verlag, Berlin, 1983.
4. W.H. Zurek, Phys. Rev. **D24** 1516 (1981); **26** 1862 (1982).
5. W.H. Zurek, in *Frontiers of Nonequilibrium Statistical Physics*, NATO ASI series. Series B, Physics; vol. **135**, Plenum Press, New York (1986) p. 145.
6. W.H. Zurek, Physics Today **44**, 36 (1991).
7. N.G. Van Kampen, Physica **20**, 603 (1954).
8. S.–K.Ma, *Statistical Mechanics*, World Scientific, Singapore, 1985.
9. R.J. Cook, Prog. in Optics XXVIII, Elsevier 1990, p. 361.
10. H. Dehmelt, Rev. Mod. Phys. **62**, 525 (1990).
11. V.B. Braginsky, and F. Ya. Khalili, *Quantum Measurement*, Cambridge University Press, Cambridge, 1992.
12. M. Bianucci, R. Mannella, B.J. West and P. Grigolini, Phys. Rev. **E50**, 2630 (1994).
13. T. Geisel, J. Nierwelberg, and A. Zacherl, Phys. Rev. Lett. **54**, 616 (1985).
14. M. Tabor, *Chaos and Integrability in Nonlinear Dynamics*, John Wiley & Sons, New York (1989).
15. R. Ishizaki, H. Hata, T. Horita, and H. Mori, Prog. Theor. Phys. **84**, 179 (1990); R. Ishizaki, T. Horita, T. Kobayashi, and H. Mori, Prog. Theor. Phys. **85**, 1013 (1991).

16. E.W. Montroll and B.J. West, in *Fluctuation Phenomena*, 2nd ed. edited by E.W. Montroll and J.L. Lebowitz, Studies in Statistical Mechanics vol 7, North– Holland, Amsterdam, 1987.
17. G. Zumofen and J. Klafter, Phys. Rev. E **47**, 851 (1993).
18. G. Zumofen and J. Klafter, Europhysics Letters **25**, 565 (1994).
19. G. Trefán, E. Floriani, B.J. West, an P. Grigolini, Phys. Rev. E **50**, 2564 (1994).
20. E. Floriani, R. Mannella, P. Grigolini, in preparation.
21. R. Bettin, R. Mannella, B.J. West, P. Grigolini, Phys. Rev. **E51**, 212 (1995).
22. M. V. Berry, Ann. Phys. **131**, 163 (1981).
23. G.P. Berman and G.M. Zaslavsky, Physica **A91**, 450 (1978).
24. G.M. Zaslavsky, Phys. Rep. **80**, 157 (1981).
25. W.H. Zurek and J.P. Paz, Phys. Rev. Lett. **72**, 2508 (1994).
26. B.S. Helmkamp and D.A.Brown, Phys. Rev. **E49**, 1831 (1994).
27. M. Hillery, R.F. O'Connell, M.O. Schully and E.P. Wigner, Phys. Rep. **106**, 122 (1984).
28. P. Grigolini, *Quantum Irreversibility and Measurement*, World Scientific, New York, 1993.
29. R. Roncaglia, L. Bonci, B.J. West and P. Grigolini, submitted to Phys. Rev. **E**.
30. G.C. Ghirardi, A. Rimini, and T. Weber, Phys. Rev. **D34**, 470 (1986).
31. L. Tessieri, D.Vitali, P. Grigolini, submitted to Phys. Rev. **E**.
32. J. Bayfield and P. Koch, Phys. Rev. Lett. **33**, 258 (1974).

Quantum Open Systems as Random Classical Dynamical Systems

Z. Haba

Institute of Theoretical Physics, University of Wrocław
PL–50–204 Wrocław, Poland

Abstract: We discuss an interaction of a quantum oscillator with a heat bath of harmonic oscillators. We define a classical Markov process whose correlations coincide with correlations of the quantum oscillator in a heat bath. In this way we replace a quantum non-commutative problem by a classical commutative one. In this framework we discuss briefly dissipation, noise,spontaneous localization and decoherence in quantum mechanics.

1 Introduction

The mathematical theory of diffusion processes originates from the physical theory of Brownian motion. A test particle placed in a fluid (or a gas) at temperature T undergoes a random motion as a result of a stochastic bombardment of molecules moving chaotically with a mean velocity proportional to $\sqrt{T}$. The question what is the quantum analog of the classical Brownian motion occupied mathematicians as well as physicists [1] [2] [4] [5] [3] [6] [7].It has been argued that an operator-valued process has to be introduced in order to describe the physical process of an interaction of a quantum particle with an enviroment.Operator-valued processes have been applied to laser physics and non-linear optics as solutions of quantum Langevin equations. However, as emphasized by some authors [1] [8] the operator Langevin equations encounter serious difficulties, because the noise entering these equations cannot be Gaussian, is a priori unknown and difficult to determine from the underlying physical model. This situation is in a sharp contrast with the classical diffusion where we know the form of the Fokker-Planck equation and subsequently the corresponding Langevin equation.

We have shown in our earlier papers [9] [10] that quantum dynamics can be expressed in terms of conventional diffusion processes (time-ordering replaces the non-commutativity). In this paper we discuss an interaction with a heat bath. We treat in detail the well-known soluble model of an interaction with an environment consisting of harmonic oscillators kept at fixed temperature T. There is a revival of an interest recently in this model [13] [14], because of its relevance for understanding of the foundations of quantum mechanics. Our motivation is to show on this text-book example the relevance of different sources of noise and the distinction of a classical and quantum noise (discussed recently by Senitzky [15].

2 Stochastic Representation of Quantum Dynamics in a Pure State

We review in this section our approach to quantum dynamics based on stochastic differential equations [9][10][11]. It can be considered as a mathematical version of the Feynman approach via sum over classical trajectories.Our motivation is just a technical one: analogously as in the theory of diffusions we would like to have at our disposal in addition to the semigroup also the stochastic differential equations and the path integral. These tools are well appreciated in the theory of diffusion. We would like to point out their usefulness in the study of quantum dynamics.

We consider a unitary evolution U_t generated by the Hamiltonian

$$H = -\frac{\hbar^2}{2m}\Delta + V \tag{1}$$

Assume that χ_t is a solution of the Schrödinger equation $0 \leq t \leq s$

$$i\hbar\partial_t\chi_t(x) = H\chi_t(x) \tag{2}$$

with the initial condition (this should rather be called a final condition) $\chi_{t=s}(x) = \chi(x)$.Let ϕ_t be the solution of the equation

$$\partial_t\phi_t(x) = \frac{i\hbar}{2m}\Delta + i\hbar(\chi_t(x))^{-1}\nabla\chi_t(x)\nabla\phi_t(x) \tag{3}$$

with the final condition ϕ. Then it is easy to check through a direct calculation that $\psi_t = \chi_t\phi_t$ is the solution of the Schrödinger equation with the final condition $\psi = \chi\phi$.

Eq.(3) resembles a diffusion equation but now the diffusion constant is imaginary. One can easily see that the standard arguments [16] connecting stochastic differential equations to partial differential equations still apply if the final condition ϕ is an analytic function.So, let us consider a stochastic equation in the integral form of Gikhman and Skorohod [16]

$$q_{t,x}(s) = x + i\hbar\int_t^s (\chi_\tau(q_{t,x}(\tau))^{-1}\nabla\chi_\tau(q_{t,x}(\tau))d\tau + \overline{\lambda}\sigma(b(s) - b(t)) \tag{4}$$

where $b(t)$ $(t \geq 0)$is the Brownian motion i.e. a Gaussian process with independent increments and the variance

$$E[b(t)^2] = t$$

and

$$\lambda = \frac{(1+i)}{\sqrt{2}}$$

$$\sigma = \sqrt{\frac{\hbar}{m}}$$

The proof of Gikhman and Skorohod can now be applied to show that the solution of eq.(3) with the final condition $\phi_{t=s} = \phi$ can be expressed in the form

$$\phi_t(x) = E[\phi(q_{t,x}(s))] \tag{5}$$

for $0 \leq t \leq s$.In order to express the solution of the Schrödinger equation for negative time we may use the time reflection invariance

$$(U_t\psi)(x) = \overline{U_{-t}\overline{\psi}(x)}$$

In mathematical physics it is customary to work with the initial condtion at $s \leq t$.In order to rewrite eqs.(4)-(5) with this initial condition we use the invariance of the Brownian paths (hence also of q) under the inversion of time $\tau \to t - \tau$ (i.e. $b(\tau) \to b(t-\tau) - b(\tau)$).After such an inversion eq.(4) can be replaced by (see ref.[17])

$$q_{s,x}(t) = x + i\hbar\int_s^t (\chi_{t-\tau}(q_{s,x}(\tau)))^{-1}\nabla\chi_{t-\tau}(q_{s,x}(\tau))d\tau + \lambda\sigma(b(t) - b(s)) \tag{6}$$

now χ_t, where $t \geq s$, is the solution of the Schrödinger equation with the initial condition χ.The solution of the Schrödinger equation with the initial condition $\psi = \chi\phi$ takes the form

$$\psi_t(x) = \chi_t(x)E[\phi(q_{s,x}(t))] \tag{7}$$

We were so detailed about initial conditions for time-dependent wave functions χ, because they will play an important role in subsequent sections.

It can be seen from eqs.(6)-(7) that if χ_t is of the form $\chi_t = exp(\frac{i}{\hbar}W_t)$ (W_t may be complex), where W_t is regular in $\hbar$ then as $\hbar \to 0$

$$(\chi_t(x))^{-1}\psi_t(x) \to \phi(x_t(x)) \tag{8}$$

where $x_t(x)$ is the solution (in general complex) of eq.(6) at $\hbar = 0$.We gave a complete proof of this statement in ref.[12] and explained its relevance to a classical behaviour of a coherent state.

3 Stochastic Description of the Quantum Dynamics in a Mixed State

Let ρ be the density matrix i.e. a positive operator of trace 1. Taking a complete basis $\mid n >$ in the Hilbert space we can write

$$\rho = \sum_n \mid n >< m \mid \rho_{nm} \tag{9}$$

Because ρ is positive and the matrix ρ_{nm} is positive definite there exist random variables α_n such that the matrix elements ρ_{nm} can be expressed as an expectation value

$$E_\rho[\bar{\alpha}_n \alpha_m] = \rho_{nm} \tag{10}$$

Then, let us define

$$\mid \chi >= \sum_n \alpha_n \mid n > \tag{11}$$

Now,we can express the density matrix in the form

$$\rho = E_\rho[\mid \chi >< \chi \mid] \tag{12}$$

The formula (10) comes out in an explicit form in many models of the density matrix; in particular, if ρ results from a pure state

$$\psi(x, y) = \sum_{nm} c_{nm} \phi_n(y) \psi_m(x)$$

by an averaging over the coordinates y of the environment i.e.

$$\rho(x, x') = \int dy \overline{\psi(x, y)} \psi(x', y)$$

In subsequent sections we discuss a decomposition of ρ in coherent states

$$\rho = \int d\alpha P(\alpha) \mid \alpha >< \alpha \mid \tag{13}$$

then the measure $d\alpha P(\alpha)$ gives a probability distribution of α.

Let U_t be a unitary Schrödinger evolution determined be the Hamiltonian

$$H = -\hbar^2 \frac{1}{2m} \Delta + V$$

Assume we know $|\chi >_t = U_t |\chi >$ (where χ is defined in eq.(11)). Let $\psi(x) = \chi(x)\phi(x)$ where ϕ is an analytic function then from eq.(7) (we put $s = 0$ here and simplify the notation)

$$(U_t\psi)(x) = \chi_t(x) E[\phi(q_t(x)] \tag{14}$$

where $q_t(x)$ is a complex diffusion process starting at $\tau = 0$ from x solving the stochastic differential equation

$$dq_\tau = i\hbar \chi_{t-\tau}^{-1} \nabla \chi_{t-\tau} d\tau + \lambda \sigma db_\tau \tag{15}$$

We can express arbitrary operators as well as correlation functions as expectation values with respect to the Brownian motion.Owing to the formula (12) in order to obtain the correlation functions we need to take the average $E_\rho[..]$in addition to the expectation value over the Brownian motion discussed in [9].So,the correlation function in a mixed state ρ is expressed as follows

$$< F(x_t)G(x) >_\rho = Tr(\rho U_t^+ F(x)U_t G(x))$$
$$= E_\rho E[\int dx|\chi_t(x)|^2 F(x)G(q_t(x))] \tag{16}$$

4 Example: Independent Oscillators at Finite Temperature

Let us first consider as a simple exercise a model of independent oscillators.

$$H_0 = -\frac{\hbar^2}{2}\sum_k \frac{\partial^2}{\partial x_k^2} + \sum_k \frac{1}{2}\omega_k^2 x_k^2 \tag{17}$$

The ground state solution for the Schrödinger equation is

$$\chi_0(x) = exp(-\frac{1}{2}\sum_k \omega_k x_k^2) \tag{18}$$

The stochastic equation (15) reads

$$dq_k = -i\omega_k q_k dt + \lambda\sigma db_k \tag{19}$$

As an example of a mixed state let us consider the canonical ensamble

$$\rho_0 = \frac{exp(-\beta H_0)}{Tr(exp(-\beta H_0))} \tag{20}$$

In order to express the density matrix as a random pure state (12) we express ρ_0 in terms of the coherent states $\mid \alpha >$. We can see that α_k are complex independent Gaussian random variables with the correlations

$$E_\rho[\overline{\alpha_k^n}\alpha_r^p] = \delta_{np}\delta_{kr} n! exp(-n\beta\hbar\omega_k)\prod_m(1 - exp(-\beta\hbar\omega_m)) \tag{21}$$

Now,

$$< x \mid U_t|\chi_\alpha >= exp(-\frac{1}{2\hbar}\sum_k \omega_k x_k^2 + \frac{1}{\hbar}\sum_k exp(-i\omega_k t)\alpha_k x_k + r_t(\alpha)) \tag{22}$$

where r is independent of x. Hence, the stochastic process in a mixed state is a solution of the equation

$$dq_k(\tau) = -i\omega_k q_k(\tau)d\tau + iexp(-i\omega_k(t-\tau))\alpha_k d\tau + \lambda\sigma db_k(\tau) \tag{23}$$

It is instructive to differentiate eq.(23) once more.We get

$$\frac{d^2 q_k}{dt^2} + {\omega_k}^2 q_k = \\ -i\omega_k \lambda\sigma \frac{db_k}{dt} + \lambda\sigma \frac{d^2 b_k}{dt^2} \tag{24}$$

So,the random force does not depend on the environment at all.The solution $q(t)$ depends on α only through the boundary conditions

$$q_k(\tau = 0) = x_k$$

$$\frac{dq_k}{d\tau}(\tau = 0) = i(\alpha_k exp(-i\omega_k t) - \omega_k x_k) + \lambda\sigma \frac{db_k}{d\tau}(\tau = 0)$$

The solution of eq.(23) with the initial condition x reads

$$q_k(s;\alpha) = exp(-i\omega_k s)x_k + \frac{\alpha_k}{2\omega_k}(exp(-i\omega_k(t-s)) - exp(-2i\omega_k(t+s))) + \\ +\lambda\sigma \int_0^s exp(-i\omega_k(s-\tau))db_k(\tau) \tag{25}$$

We can now insert the solution (25) into eq.(16). A simple computation of the Gaussian integral leads to the well-known result

$$< x_k(t)x_r(0) >_\rho = \delta_{kr}\omega_k^{-1} cth(\frac{1}{2}\beta\omega_k) exp(-i\omega_k t) \tag{26}$$

5 Interaction with an Environment

We discuss now a standard model of an interaction with an environment (see refs.[1][2][18])

$$H = \hbar\omega_0 A^+ A + \sum_k \hbar\omega_k a_k^+ a_k + \sum_k g_k A^+ a_k + \overline{g_k} a_k^+ A \tag{27}$$

Using the Schrödinger representation of the creation and anihilation operators

$$a = \sqrt{\frac{\hbar}{2\omega}} \frac{\partial}{\partial x} + \sqrt{\frac{\omega}{2\hbar}} x$$

we express H in the form (1). It is clear that the product of oscillator ground states is the ground state for H. Then,taking the ground state as χ in eq.(7) we get the following equation for the stochastic process at zero temperature

$$dq_k = -i\omega_k q_k dt + v_k \sqrt{\frac{\omega_0}{\omega_k}} Q dt + \lambda\sigma\gamma_k db_k + \lambda\sigma \Gamma_k db \tag{28}$$

$$dQ = -i\omega_0 Q dt - \sum_k v_k \sqrt{\frac{\omega_k}{\omega_0}} q_k dt + \lambda\sigma \sum_k \delta_k db_k + \lambda\sigma\Gamma db$$

Here,we denoted the coordinates of the bath by small letters and coordinates of a test particle by capital letters. The constants entering the equations are defined as follows

$$g_k = \frac{1}{2}(u_k + iv_k)$$

whereas γ, Γ and δ are arbitrary solutions of the equations

$$\gamma_k^2 + \Gamma_k^2 = 1$$

$$\sum_k \delta_k^2 + \Gamma^2 = 1$$

$$\delta_k \gamma_k + \Gamma_k \Gamma = \sqrt{\frac{1}{4\omega_k\omega_0}} u_k$$

Eqs.(28) are particularly simple if g_k are purely imaginary (i.e.$u_k = 0$).Then

$$\Gamma_k = \delta_k = 0$$

and

$$\gamma_k = \Gamma = 1$$

Although eqs.(28) are linear it is still not easy to obtain an explicit solution.We shall write down an explicit solution in a simplified case (this is the model discussed by van Kampen [7])of a purely imaginary

$$g_k = \frac{i}{2}\mu_k\sqrt{\omega_0}$$

where

$$\omega_0 = \frac{2\pi}{\tau_0} \rightarrow 0$$

,where τ_0 is the period of the oscillator.This is the limit of a weak coupling between the particle and the heat bath and slow oscillation (in comparison with the frequencies of the bath). Now, eqs.(28) simplify

$$dq_k = -i\omega_k q_k dt + \lambda\sigma db_k \tag{29}$$

$$dQ = -\mu_k\sqrt{\omega_k} q_k dt + \lambda\sigma db \tag{30}$$

These equations are easy to solve. So,$q_k(t)$ is the same as for independent oscillators and expressed by $q_k(t; \alpha = 0)$ in eq.(25),whereas

$$Q_t = X - \int_0^t \sum_k \mu_k \sqrt{\omega_k} q_k(s) ds + \lambda\sigma b_t \tag{31}$$

The heat bath is usually defined by

$$\rho = exp(-\beta H_0) \tag{32}$$

where H_0 is the same as in eq.(17). In such a case the random variables α_k are defined in the same way as for independent oscillators. However,the dynamics is now different because the oscillators are coupled.In order to calculate $U_t \mid \chi_\alpha >$ we may use the formula (7). So,

$$< x, X \mid U_t \mid \chi_\alpha >=< x, X \mid \chi_0 > E[exp(\hbar^{-1} \sum_k \alpha_k q_k(t; x, X))] \quad (33)$$

In order to get an explicit formula we would need an explicit solution for q_t.We shall again restrict ourselves to the limit $\omega_0 \to 0$ (eqs.(29)-(30)). In this limit $U_t \mid \chi_\alpha >$ is expressed by eq.(22).The stochastic process (15) for a quantum particle interacting with a heat bath is a solution of the stochastic equation

$$dq_k = -i\omega_k q_k d\tau + iexp(-i\omega_k(t-\tau))\alpha_k d\tau + \lambda\sigma db_k \quad (34)$$

$$dQ = -\sum_k \mu_k \sqrt{\omega_k} q_k d\tau + \lambda\sigma db \quad (35)$$

The solution of eq.(34) for $q_k(t;\alpha)$ is given by eq.(22) whereas for $Q(t)$ by eq.(31) with $q_k(s)$ replaced by $q_k(s;\alpha)$.

We would like to mention a model of an interaction with a heat bath defined by the ground state which is not of the Fock type. Let us consider the ground state

$$\chi = exp(-\frac{S}{\hbar}) \quad (36)$$

with

$$S(x, X) = \frac{1}{2}\omega_0 X^2 + X \sum_k c_k x_k + \frac{1}{2} \sum_k \omega_k x_k^2 \quad (37)$$

Then, the stochastic equations at zero temperature read

$$dq_k = -i\omega_k q_k dt - ic_k Q dt + \lambda\sigma db_k \quad (38)$$

$$dQ = -i\omega_0 Q dt - i \sum_k c_k q_k dt + \lambda\sigma db \quad (39)$$

6 Dissipation and Localization in Quantum Mechanics

Using the framework of preceding sections we discuss now the dynamics of a subsystem.First, consider the formula (27) with g_k purely imaginary. Let us solve eqs.(28) for q_k. We get

$$q_k(t) = exp(-i\omega_k t)x_k + v_k \sqrt{\frac{\omega_0}{\omega_k}} \int_0^t exp(-i\omega_k(t-s))Q(s)ds + \\ +\lambda\sigma \int_0^t exp(-i\omega_k(t-s))db_k(s) \quad (40)$$

Let us introduce the notation

$$N_D(s) = \sum_k v_k x_k \sqrt{\frac{\omega_k}{\omega_0}} exp(-i\omega_k s) \tag{41}$$

and

$$N_R(s) = \sum_k v_k \sqrt{\frac{\omega_k}{\omega_0}} \int_0^s exp(-i\omega_k(s-\tau))db_k(\tau) \tag{42}$$

corresponding to a "deterministic" and random noise. Next let us denote

$$K(s) = \sum_k v_k^2 exp(-i\omega_k s) \tag{43}$$

Then the equation for Q (28) reads

$$dQ = -i\omega_0 Q dt - \int_0^t K(t-s)Q(s)dsdt - N_D(t)dt - \lambda\sigma N_R dt + \lambda\sigma db \tag{44}$$

K(s) is positive for small s.Hence, the term KQ describes a dissipation. N_D,N_R and b constitute a random external force. In computations with the ground state of independent oscillators(18) x_k are independent random variables with the variance

$$E[x_k x_r] = \frac{\hbar}{\omega_k}\delta_{kr} \tag{45}$$

Hence, the noise N_D is a Gaussian complex stochastic process with the covariance

$$E[N_D(s)N_D(t)] = \hbar\omega_0^{-1} \sum_k v_k^2 exp(-i\omega_k(t+s)) \tag{46}$$

$$E[\overline{N_D(s)}N_D(t)] = \hbar\omega_0^{-1} \sum_k v_k^2 exp(-i\omega_k(t-s))$$

In general, it is not easy to derive an explicit solution of eq.(44). We restrict ourselves to the "Ohm approximation"

$$K(s) = a\delta(s) \tag{47}$$

where a is a positive constant.One can justify the approximation (47) (see ref.[2]) if $\omega_k = k\omega$ where $k \in [0,\infty]$ and $v_k^2 = \frac{a}{\pi}\omega\Delta k$. In such a case the sum over infinitesimal Δk can be replaced by an integral

$$\int_0^\infty dk exp(-ik\omega s) = \frac{\pi}{\omega}\delta(s) - \frac{i}{s\omega}$$

Then, the term s^{-1} is considered as small in comparison to $\delta(s)$. Inserting the approximation (47) for K into eq.(44) we get an equation of motion of a test particle (oscillator) Q in an environment of oscillators at zero temperature

$$dQ = -i\omega_0 Q ds - aQds - N_D ds - \lambda\sigma N_R ds + \lambda\sigma db \tag{48}$$

At positive temperature we should add a thermal noise N_T to eq.(48). Its form can be found from eqs.(23) and (28)

$$N_T(s) = -\sum_k v_k \frac{\alpha_k}{2\sqrt{\omega_0\omega_k}}(exp(-i\omega_k(t-s) - exp(-i\omega_k(t+s))) \tag{49}$$

where α_k are the Gaussian random variables (21) (the dependence of N_T on t has its origin in eq.(6)). The solution of eq.(48) (which appears in the formula (7)) is (with the thermal noise added)

$$Q_t = exp(-at - i\omega_0 t)X+$$

$$+\int_0^t exp(-(t-s)(a+i\omega_0))(N_T(s)ds - N_D(s)ds - \lambda\sigma N_R(s)ds + \lambda\sigma db(s)) \tag{50}$$

The behaviour of a subsystem is determined by an average over the environment.The stochastic processes introduced in preceding sections can be used as an efficient tool in computations of the expectation values over the environment. We define the density matrix for a subsystem (with coordinates X) as

$$\rho_P(X,X') = \int \prod_k dx_k \rho(X,x;x,X') =$$
$$\int \prod_k dx_k E_\rho[< x,X|\chi_\alpha >< \chi_\alpha|x,X' >] \tag{51}$$

Let us first make the usual assumption that initially the system and the environment are separated i.e. that ρ is of the product form. We shall consider as a system an oscillator with a frequency ω_0. So, we assume

$$\rho(X,x;x,X') = exp(-\frac{\omega_0}{2\hbar}(X^2+X'^2))\overline{\nu(X)}\nu(X')\rho_0(x,x) \tag{52}$$

We take ρ_0 of the form (20) with H_0 defined in eq.(17) i.e.the heat bath consists of independent oscillators. If (to simplify the matters) we additionally require that the ground state of the free system coincides with the ground state of an interacting system (as it was in the model (27)) then we can express in a simple way the evolution of system's density matrix in terms of the Brownian motion

$$\rho_P(t;X,X') =$$

$$E_\rho[\int \prod_k dx_k exp(-\tfrac{\omega_0}{\hbar}(X^2+X'^2))|\chi_0(x)|^2 \overline{E[\nu(Q_t(X,x))exp(\tfrac{1}{\hbar}\sum_k \alpha_k q_k(t;x,X))]}$$

$$E[\nu(Q_t(X',x))exp(\frac{1}{\hbar}\sum_k \alpha_k q_k(t;x,X'))]] \tag{53}$$

For the model (27) the stochastic process (Q, q_k) is the solution of the stochastic equation (28).

We are interested whether an interaction with the heat bath can induce a "decoherence" [13] in the sense that the off-diagonal elements $\rho_P(X, X')$ decrease fast for a large time when $|X - X'| \neq 0$ (we can discuss in a similar framework any other observable and decoherence with respect to this observable).These problems have been intensively studied recently [13][14][20]. We would like to see in a general framework when and how the decoherence is possible. The model (27) is soluble, so the questions are reduced to analysis of linear problems. These problems can be formulated quite explicitly in the ohmic approximation (47). We can compute explicitely the expectation value in eq.(53) because the integral is Gaussian. We get

$$\rho_P(t; X, X') = C(X, X')A(t, X + X')exp(-f(t)(X - X')^2) \tag{54}$$

with

$$f(t) = \sum_k \gamma_k(1 + exp(-2at) - 2exp(-at)cos((\omega_0 - \omega_k)t) \tag{55}$$

where γ_k depend on v_k in eq.(28). It can be seen from eq.(55) that for the decoherence (at large t) we must set $t = 0$. Then, the increase of $f(t)$ depends on the spectrum of the frequencies ω_k and on v_k.

We can discuss now the problem of decoherence in general. First, let us consider eq.(53).Eq.(53) gives a formula for the density matrix of a general Hamiltonian system under an assumption that the ground state of the environment coincides with the ground state of independent harmonic oscillators.If in eq.(53) we first take the expectation value over the random variables α,then (assuming that they are Gaussian) we get

$$\begin{aligned}\rho_P(t; X, X') = \int \prod_k dx_k E[w(q, \tilde{q}, Q, \tilde{Q})\\ exp(\sum_{k,r} E[\overline{\alpha_k}\alpha_r]q_k(t, x, X)\tilde{q}_r(t, x, X'))]\end{aligned} \tag{56}$$

where w is a certain weight function, which depends on the model,q and $\tilde{q}$ are independent realizations of the same stochastic process. If now $E[\overline{\alpha}_k\alpha_r] = p_k\delta_{kr}$ and q_k has a term of the form

$$q_k \sim Xm_k \int_0^t exp(-i\omega_k s)ds$$

(such a term comes from a linear part of eq.(28)) then from eq.(56) we get the formula (54) with

$$f(t) = \sum_k m_k^2 p_k(1 - cos\omega_k t) \tag{57}$$

In the model (28) we have got the decoherence from an average over the random variables α describing the heat bath. We would like to point out that the positive temperature is not necessary for this effect. There is still another source of noise and another cause of decoherence. Let us consider the model (36). Its simplest version is described by eq.(37). If in $S(q,Q)$ there is a coupling between the system and the environment of the Leggett form [19] $X\sum_k c_k x_k$ then the approximation

$$q_k \sim x_k exp(-i\omega_k t)$$

together with an approximation of the x-integral in eq.(51) by a Gaussian integral leads to the result (54) with

$$f(t) \approx \sum_k \frac{c_k^2}{\omega_k} sin^2 \omega_k t \tag{58}$$

Summarizing the behaviour of the reduced density matrix defined in eqs.(51) and (53) we can see that the crucial property responsible for the decoherence is the Leggett-type interaction where the particle coordinate X interacts with a collection of heat bath coordinates $\sum_k c_k x_k$.

There is another theory explaining the decoherence.The authors [21][22] explain the appearance of classical properties through an addition of some stochastic terms to the Schrödinger equation. We would like to indicate that there may be a relation between these two approaches.For simplicity of arguments let us neglect the termal noise (zero temperature) and (as a noise coming from the environment) consider only N_D in eq.(48) corresponding to the model (27).We make the assumption of the Ohm dissipation (47). The corresponding choice of v_k^2 (discussed below eq.(51)) leads to the following approximation to N_D

$$E[N_D(s)N_D(t)] = 0 \tag{59}$$
$$E[\overline{N_D(s)}N_D(t)] = \hbar\epsilon\delta(t-s)$$

where

$$\epsilon = \frac{a}{\omega_0}$$

So,N_D is the white noise. We denote the corresponding Brownian motion by B_D i.e.

$$\frac{dB_D}{ds} = N_D$$

Now,eq.(48) reads

$$dQ = -i\omega_0 Q ds - aQds - dB_D(s) + \lambda\sigma db(s) \tag{60}$$

whereas eq.(53) (with $\alpha = 0$)

$$\rho_P(t; X, X') =$$

$$E_\rho[\int \prod_k dx_k exp(-\frac{\omega_0}{\hbar}(X^2 + X'^2))|\chi_0(x)|^2 \overline{E[\nu(Q_t(X))]} E[\nu(Q_t(X'))] \tag{61}$$

$$\simeq exp(-\frac{\omega_0}{2\hbar}(X^2 + X'^2)) E_N[\overline{E[\nu(Q_t(X))]} E[\nu(Q_t(X'))]]$$

where E_N means that the integration over x is replaced by an expectation value with respect to the complex white noise N_D. An application of the standard stochastic calculus shows that $\nu(Q_t)$ fulfills the equation

$$\begin{aligned} d\nu = & \frac{\partial \nu}{\partial Q}(-i\omega_0 Q ds - aQds - dB_D(s) + \lambda\sigma db) + \\ & + \frac{\hbar}{2}\frac{\partial^2 \nu}{\partial Q^2} ids \end{aligned} \tag{62}$$

The wave function $\psi(X) = exp(-\frac{\omega_0 X^2}{2\hbar})\nu(X)$ satisfies the stochastic Schrödinger equation (which is a stochastic generalization of the Schrödinger equation for an oscillator).Then, as a consequence we obtain a Lindblad-type master equation for the density matrix ρ_P (eq.(61)). This equation, when expressed in the momentum representation, in the limit of small ω_0 takes the form

$$\partial_t < p|\rho_P|p' >= -\frac{\epsilon}{2}(p-p')^2 < p|\rho_P|p' > \tag{63}$$

Eq.(63) means a localization in momentum.

Summarizing, we can say that models of an interaction with an environment lead to an appearance of some classical properties of quantum systems. So, the Wigner-Weisskopf-Louisell model of radiation shows a continuous spontaneous localization of the momentum and to some extent also a decoherence (54)-(55) in space. The model (36)-(39), which is a modification of the Leggett model, leads to a quite efficient localization in space with a proper choice of frequencies and amplitudes in eq.(58).

References

1. W.H. Louisell, Quantum Statistical Properties of Radiation, Wiley, New York, 1973.
2. H. Haken, Light, waves, photons, atoms, Springer, 1986.
3. G.S. Agarwal, Quantum Statistical Theories of Spontaneous Emission and Their Relation to Other Approaches, Springer, 1974.
4. L. Accardi and Y.G. Lu, in Quantum Probability VII, Springer Lecture Notes in Math., 1992.
5. R. Benguria and M. Kac, Phys. Rev. Lett. **46**, 1(1981).
6. I.R. Senitzky, Phys. Rev. **119**, 670(1960)**124**, 642(1961).
7. N.G. van Kampen, in Stochasticity and Quantum Chaos, ed. by Z. Haba et al, Kluwer, 1995, see also J. Stat. Phys. **78**, 299(1995).
8. M. Sargent III, M.O. Scully and W.E. Lamb, Laser Physics, Addison-Wesley, 1974.
9. Z. Haba, Phys. Lett. **175A**, 371(1993).
10. Z. Haba, Phys. Lett. **189A** , 261(1994).
11. Z. Haba, J. Phys. **A27**, 6457(1994).
12. Z. Haba, Coherent States and Quantum Dynamics of Non-linear Systems, subm. for publication.
13. W.G. Unruh and W.H. Zurek, Phys. Rev. **D40**, 1071(1989);
    M. Gell-Mann and J.B. Hartle, in Complexity, Entropy and the Physics of Information, ed. W.H. Zurek, Addison-Wesley, 1990.

14. J.P. Paz, S. Habib and W.H. Zurek, Phys. Rev. **D47**, 488(1993).
15. I.R. Senitzky, Phys. Rev. **A48**, 4629(1993).
16. I.I. Gikhmana and A.V. Skorohod, Stochastic Differential Equations, Springer, New York, 1972.
17. M. Freidlin, Functional Integration and Partial Differential Equations, Princeton, 1985.
18. G.W. Ford, J.T. Lewis and R.F. O'Connell, Phys. Rev. **A37**, 4419(1988).
19. A.O. Caldeira and A.J. Leggett, Physica **A121**, 587(1983).
20. L.E. Ballentine, Phys. Rev. **A43**, 9(1991);
E. Joos and H.D. Zeh, Z. Phys. **B59**, 223(1985);
L. Diosi, Phys. Rev. **A40**, 1165(1985).
21. G.C. Ghirardi, P. Pearl and A. Rimini, Phys. Rev. **A42**, 78(1990).
22. N. Gisin and I.C. Percival, J. Phys. **A25**, 5677(1992).

Large–Scale Structure of the Universe and Asymptotics of Burgers' Turbulence with Heavy–tailed Dependent Data

Yiming Hu[2] *and W.A. Woyczynski*[1,2]

[1] Department of Statistics
[2] Center for Stochastic and Chaotic Processes in Science and Technology
Case Western Reserve University
Cleveland, Ohio 44106

Abstract: Large-time asymptotics of the statistical solutions $v(t,x)$ of the Burgers' equation $v_t + vv_x = \nu v_{xx}$ is considered. The initial velocity potential is assumed to be of the shot noise type with dependent amplitudes and heavy tails. The problem arises naturally in the adhesion model of the large-scale distribution of matter in the Universe. As $t \to \infty$, the random field $v(t,x)$ becomes stochastically relatively asymptotically equivalent to a field with "saw-tooth" trajectories. The intermittent shocks of the velocity field correspond then to the regions of high density in the coupled passive tracer density field. This paper extends a result of S.Albeverio, S.A.Molchanov and D. Surgailis (1992), obtained for the case of independent amplitudes.

1 Introduction

It is a well known observational fact that the matter in the Universe is distributed in clusters and superclusters of galaxies, with giant voids between them. At this late epoch of the formation of the large scale structure,

- the dark (nonluminous) matter dominates;
- it acts as colisionless dustlike particles;
- no pressure effects need to be taken into account, with the Newtonian gravity being the only force of consequence;
- the radiative and gas dynamics effects are short range.

Assuming the flat, expanding universe, with the scale factor

$$a(t) = t^{2/3} ,$$

and the mean density

$$\bar{\rho} \propto a^{-3} ,$$

the evolution of the matter density $\rho = \rho(\mathbf{x}, t)$ is usually (see e.g. Peebles (1980), Kofman et al. (1992)) described by the system of three coupled partial differential equations

$$\frac{\partial \rho}{\partial t} + 3H\rho + \frac{1}{a}\nabla \cdot (\rho v) = 0 , \tag{1.1}$$

$$\frac{dv}{dt} + Hv = -\frac{1}{a}\nabla\varphi , \tag{1.2}$$

$$\nabla^2\varphi = 4\pi G a^2(\rho - \bar{\rho}) , \tag{1.3}$$

where v is the local velocity, φ is the gravitational potential, and H and G are, respectively, the Hubble and the gravitational constants. The three equations are, of course, the continuity equation, the Euler equation and the Poisson equation.

This system is not easy to analyze and several attempts have been made at simplifying it, while preserving the predictive ability of the reduced models. Introducing the velocity $u = dx/da$ in the coordinates comoving with the expanding Universe, the Euler equation (1.2) is transformed into equation

$$\frac{\partial u}{\partial a} + (u \cdot \nabla)u = -\frac{3}{2a}(u + A\nabla\varphi) , \tag{1.4}$$

with

$$A = \left(\frac{3}{2}H^2a^3\right)^{-1} = const,$$

where the right-hand side represents, in the Lagrangian approach, the force acting on the particle. It is still a nonlocal operator so, in (1970), Zeldovich proposed a model in which it was assumed to be 0. This gives a clear Lagrangian picture as (1.4) becomes then the classical Riemann equation, which could explain formation of the pancake structures. This model has been adjusted by Gurbatov, Saichev and Shandarin (1985) (see also Shandarin and Zeldovich (1989), Weinberg, Gunn (1990)) who replaced the nonlocal term on the right-hand side of (1.4) by the Laplacian, to yield the Burgers equation

$$\frac{\partial u}{\partial a} + (u \cdot \nabla)u = \nu\nabla^2 u , \tag{1.5}$$

where the viscosity term is supposed to mimick the gravitation adhesion. The constant ν should be small so that the viscosity effects do not affect the motion of the matter outside clusters. This adhesion model of the large structure of the Universe has been extensively studied in the astrophysical literature and satisfactorily tested against high resolution (512×512) N-body simulations (Kofman et al (1992)).

In this paper we consider the Cauchy problem for the one-dimensional Burgers' equation

$$v_t + vv_x = \nu v_{xx} ,$$

with the shot noise initial velocity potential $V(x) = \int_0^x v(0, y)\, dy$ of the form

$$V(x) = \sum_{i=-\infty}^{\infty} e^{\xi_i}\delta(x - x_i) , \tag{1.6}$$

where $\{x_i\}$ is the standard Poisson point process on the real line and random variables

$$\xi_i = \sum_{j=0}^{n} c_j\eta_{i+1-j}, \quad i \in \mathbf{N} ,$$

form a strictly stationary sequence of moving averages of independent and identically distributed symmetric α-stable random variables η_i, $i \in \mathbf{N}$, with $1/2 < \alpha < 2$. Recall that $E\exp(is\eta_i) = \exp(-|s|^\alpha)$. The positive integer n will be fixed throughout this paper.

The solution random field of the Burgers' equation with these initial data can be written in the form

$$v(t,x) = -2\nu \frac{\sum_i (x - x_i) g(t,x,x_i) e^{\xi_i}}{t \sum_i g(t,x,x_i) e^{\xi_i}} \tag{1.7}$$

where $g(t,x,y) = (4\pi\nu t)^{-1/2} \exp[(x-y)^2/4\nu t]$ is the usual Gaussian kernel (see e.g. a survey article by Woyczynski (1992)). Without loss of generality, we let $\nu = 1/2$.

Our main objective is to study the asymptotic behavior of solutions (1.7) for large times t. The asymptotic behavior of solutions $v(t,x)$ will be described in terms of what can be called a *stochastic relative asymptotic equivalence* of processes, the notion introduced and studied in the Burgers equation context by Albeverio, Molchanov and Surgailis (1995). Formally, for two processes X_t and Y_t we will write that

$$X_t \overset{P}{\sim} Y_t \qquad (t \to \infty) ,$$

whenever

$$\frac{|X_t - Y_t|}{|Y_t|} \xrightarrow{P} 0 \qquad \text{as} \qquad t \to \infty , \tag{1.8}$$

where $\xrightarrow{P}$ stands for convergence in probability. To simplify the presentation we write and prove our results for the simplest, pair-dependent case ($n = 1$) of shot noise stresses, but exactly the same method works for the general finite moving averages. In this context our basic result can be formulated as follows:

Theorem 1.1. *Let the initial velocity potential*

$$\int_{-\infty}^{x} v(0,y)dy = \sum_i e^{\xi_i} \delta(x - x_i) ,$$

where $\{x_i\}$ is the standard Poisson processes on R, and $\{\xi_i\} = \{c_1\eta_i + c_0\eta_{i+1}\}$ with $\{\eta_i\}$ being i.i.d symmetric α-stable random variables $(1/2 < \alpha < 2)$ and independent of the Poisson ensemble. Then, for each $x \in R$, the solution random field v of the Burgers' equation

$$v_t + vv_x = \nu v_{xx}$$

satisfies the asymptotic condition

$$v(t,x) \overset{P}{\sim} -\frac{x - x_{i^*}}{t}, \qquad (t \to \infty) \tag{1.9}$$

where $x_{i^*} = x_{i^*(x,t)}$ is the point where the random field $g(t,x,x_i)e^{\xi_i}$ attains its maximum, i.e.

$$g(t,x,x_{i^*})e^{\xi_{i^*}} = \max_i g(t,x,x_i)e^{\xi_i} . \tag{1.10}$$

The proof of this theorem will appear in Section 4 and it will rely on several auxiliary results that will be discussed in Sections 2 and 3.

There are several other recent papers in this area, some of them surveyed in the already quoted paper by the second-named author. In particular, our result is an extension of a theorem of Albeverio, Molchanov and Surgailis (1995), where independence of the shot noise's amplitudes is assumed. Other asymptotic results for the statistical solutions of Burgers equation that permit dependent structure in the initial data can be found in Funaki, Surgailis and Woyczynski (1995), Hu and Woyczynski (1994, 1995), Surgailis, Woyczynski (1993) and Molchanov, Surgailis, Woyczynski (1995).

2 Asymptotic Equivalence of Sums and Maxima

Before we formulate our first auxiliary proposition a few simple observations are in order.

1°. $\{\xi_i\} = \{c_1\eta_i + c_0\eta_{i+1}\}$ is a strictly stationary sequence of α-stable random variables.

2°. Functions

$$H_1(a) = H_\eta(a) = P\{e^{\eta_0} > a\}, \qquad \text{and} \qquad H_2(a) = H_\xi(a) = P\{e^{\xi_0} > a\} ,$$

are slowly varying as $a \to \infty$.

3°. For $k = 1, 2$,

$$\lim_{a\to\infty} \frac{H_k(a(\log a)^{-1/2})}{H_k(a)} = 1 . \tag{2.1}$$

4°. For $k = 1, 2$,

$$\int_R H_k(e^{x^2})dx < \infty . \tag{2.2}$$

Properties 3° and 4° follow directly from property 2°. The need to secure condition (2.2) was the main reason why the range of the parameter α was restricted to the interval $(1/2, 2)$.

The asymptotic equivalence of sums and maxima is a well known phenomenon for independent random variables (see e.g. Kwapien and Woyczynski (1992)). Here, we will need it for dependent random variables of the form

$$X_{i,L} = g_L(x_i)e^{\xi_i} , \tag{2.3}$$

where (x_i) and ξ_i are as in Section 1, and the kernel $g_L(x) \geq 0$ satisfies the following three conditions:

$$\int_R H_k(g_L(x)^{-1})dx < \infty , \tag{2.4}$$

$$\sup_{L>0}\ \sup_{x\in R} g_L(x) < \infty , \tag{2.5}$$

and

$$|\{x \in R : g_L(x) > c\}| \to \infty \qquad \text{as} \qquad L \to \infty\,, \tag{2.6}$$

where $c > 0$ is a constant and $|\cdot|$ denotes the Lebesgue measure. Under these circumstances we have the following

Proposition 2.1. *Let*

$$X_L^* = \sup_i X_{i,L}, \qquad \text{and} \qquad S_L = \sum_i X_{i,L}\,. \tag{2.7}$$

Then, for any $L > 0$, the series S_L converges a.s. , and

$$\frac{S_L}{X_L^*} \xrightarrow{P} 1 \qquad \text{as} \qquad L \to \infty\,. \tag{2.8}$$

The proof of the proposition will be preceded by a combinatorial lemma.

Lemma 2.1. *Let X_i be a sequence of random variables and A_{ij} be a finite set of positive integers such that $\#A_{ij} = s < \infty$, and $i, j \notin A_{ij}$. Then*

$$\sum_{i \neq j} P\Big\{X_j > X_i > X_k;\ k \in \mathbf{N} \setminus (A_{ij} \cup \{i,j\})\Big\} \leq (s+2)!\,. \tag{2.9}$$

Proof. First, observe that for fixed $k < \infty$, $k \in N$, we have

$$1 \geq P\{X_{(1)} > X_{(2)} > \ldots > X_{(k+1)}\}$$

$$= \sum_{i_1} \sum_{i_2 \neq i_1} \sum_{i_3 \neq \{i_1,i_2\}} \cdots \sum_{i_k \neq \{i_1,\ldots,i_{k-1}\}}$$

$$P\Big\{X_{i_1} > X_{i_2} > \ldots > X_{i_k} > X_{(k+1)}, i_1 = (1), i_2 = (2), \ldots, i_k = (k)\Big\}\,,$$

where

$$X_{(1)} = \max\{X_i\},\ X_{(2)} = \max\{\{X_i\} \setminus X_{(1)}\},$$
$$X_{(3)} = \max\{\{X_i\} \setminus \{X_{(1)}, X_{(2)}\}\}, \ldots$$

is the order statistics of the random sequence $X_1, X_2, \ldots$. Therefore,

$$\sum_{i_1 \neq i_2 \neq \ldots \neq i_k} P\Big\{X_{i_1} > X_{i_2} > \ldots > \max\{X_l : l \in \mathbf{Z} \setminus \{i_1, i_2, \ldots, i_k\}\Big\} \leq 1\,. \tag{2.10}$$

Next, we notice that for fixed $\{i,j\}$ and $i_k \in A_{ij}$, there are

$$\sum_{s_1+s_2 \leq s} s_1! \binom{s}{s_1} s_2! \binom{s-s_1}{s_2} \tag{2.11}$$

ways to put elements $\{X_{i_1}, X_{i_2}, \ldots, X_{i_s}\}$ into intervals $(-\infty, X_i)$, (X_i, X_j) and (X_j, ∞), distinguishing order of elements in (X_i, X_j) and (X_j, ∞), and disregarding it for elements falling into the interval $(-\infty, X_i)$. For each particular ordering

$$X_{k_1} > X_{k_2} > \ldots X_{k_{s_1}} > X_j > X_{k_{s_1+1}} > \ldots > X_{k_{s_1+s_2}} > X_i$$

$$> \max\Big(X_k : k \in \mathbf{Z} \setminus \{i, j, k_1, \ldots, k_{s_1+s_2}\}\Big) ,$$

we have

$$\sum_{i \neq j} P\Big\{X_{k_1} > \ldots > X_{k_{s_1}} > X_j > X_{k_{s_1+1}} > \ldots$$

$$> X_{k_{s_1+s_2}} > X_i > \max\Big(X_k : k \in \mathbf{Z} \setminus \{i, j, k_1, \ldots, k_{s_1+s_2}\}\Big)\Big\}$$

$$\leq \sum_{i_1 \neq i_2 \neq \ldots \neq i_{s_1+s_2+2}} P\Big\{X_{i_1} > X_{i_2} > \ldots > X_{i_{s_1+s_2+2}}$$

$$> \max\Big(X_{i_k} : k \in \mathbf{Z} \setminus \{i_1, i_2, \ldots, i_{s_1+s_2+2}\}\Big)\Big\} ,$$

and, by (2.10) and (2.11), we obtain that the left hand side of (2.9) is less than or equal to

$$\sum_{s_1+s_2 \leq s} s_1! \binom{s}{s_1} s_2! \binom{s-s_1}{s_2} = \sum_{s_1+s_2 \leq s} \frac{s!}{(s-s_1-s_2)!} \leq s^2 s! \leq (s+2)!$$

Therefore, Lemma 1 follows.

Now, we return to

The proof of Proposition 2.1. Proposition 2.1 follows immediately if we establish that

$$\lim_{L \to \infty} P\{X_L^* < K\} = 0, \qquad (\forall K < \infty) \tag{2.12}$$

and

$$\lim_{K \to \infty} \sup_{L > 0} P\left\{\frac{S_L^-}{X_L^*} > \epsilon, X_L^* > K\right\} = 0, \qquad (\forall K < \infty,\ \forall \epsilon > 0) \tag{2.13}$$

where $S_L^- = S_L - X_L^*$. For simplicity of notation let

$$\varphi_i = e^{\xi_i} . \tag{2.14}$$

We shall prove (2.12) first. Let

$$G_L = \{x \in R : g_L(x) \geq c\}, \qquad N = \#\{x_i \in G_L\} .$$

Then we have

$$P\{X_L^* \le K\} \le P\{g_L(x_i)\varphi_i \le K, x_i \in G_L\} \le P\{\varphi_i \le K/c, x_i \in G_L\}$$

$$= \sum_{n=0}^{\infty} P\{N=n\} P\Big\{\max\{\varphi_1,\dots,\varphi_n\} \le K/c\Big\}$$

$$\le \sum_{n=0}^{\infty} P\{N=n\} P\Big\{\max\{\varphi_1,\varphi_3\ ,\dots,\varphi_{2[n/2]-1}\} \le K/c\Big\}$$

$$\sim \sum_{n=0}^{\infty} \frac{|G_L|^n e^{-|G_L|}}{n!} F_\xi^{n/2}(K/c) = \exp\Big\{-|G_L|\Big((1-F_\xi^{1/2}(K/c)\Big)\Big\} \to 0\ .$$

Next, we shall prove (2.13). Let $\chi\{A\}$ be the indicator function of set A, and i_L^* denote the position of the maximum, that is $X_{i_L^*} = \max_i g_L(x_i)e^{\xi_i}$. Then

$$P\{S_L^-/X_L^* > \epsilon, X_L^* > K\} \le \frac{1}{\epsilon} E\Big[E[S_L^-\chi\{X_L^* > K\}/X_L^*] \,\Big|\, \{x_k\}\Big]$$

$$= \frac{1}{\epsilon}\sum_i E\Big[\sum_{j:j\neq i} E\Big(\frac{X_{j,L}}{X_{i,L}}\chi\{X_{i,L} > K, i_L^* = i\} \,\Big|\, \{x_k\}\Big)\Big]$$

$$\begin{aligned} &= \frac{1}{\epsilon}\sum_i E\Big[\Big\{\sum_{j:|j-i|>1} E\Big(\frac{X_{j,L}}{X_{i,L}}\chi\{X_{i,L} > K, i_L^* = i\} \,\Big|\, \{x_k\}\Big)\Big\} \\ &\qquad + E\Big(\frac{X_{i+1,L}}{X_{i,L}}\chi\{X_{i,L} > K, i_L^* = i\} \,\Big|\, \{x_k\}\Big) \\ &\qquad + E\Big(\frac{X_{i-1,L}}{X_{i,L}}\chi\{X_{i,L} > K, i_L^* = i\} \,\Big|\, \{x_k\}\Big)\Big] \end{aligned} \tag{2.15}$$

$$= \frac{1}{\epsilon}\sum_i E\Big(A_i + B_i + C_i\Big)\ .$$

We shall consider separately terms A_i and, B_i and C_i.

For A_i, we have that $|i-j| > 1$ for each term, and

$$E\Big[\frac{X_{j,L}}{X_{i,L}}\chi\{X_{i,L} > K, i_L^* = i\} \,\Big|\, \{x_k\}\Big]$$

$$\le E\Big[\frac{X_{j,L}}{X_{i,L}}\chi\Big\{X_{i,L} > K, X_{k,L} < X_{i,L}, k \in \mathbf{N}/\{j-1, j+1\}\Big\} \,\Big|\, \{x_k\}\Big]$$

$$= \int_{Kg(x_i)^{-1}}^{\infty} \Big[\frac{g(x_j)}{g(x_i)a} \int_0^a bE\Big(\chi\Big\{g(x_k)\varphi_k < g(x_i)a;$$

$$k \in \mathbf{N}/\{j-1, j+1\}\Big\} \,\Big|\, \varphi_i = a, \varphi_j = b\Big) dH_\xi(b)\Big] dH_\xi(a)$$

$$= -\int_{Kg(x_i)^{-1}}^{\infty} H_\xi\Big(g(x_j)^{-1}g(x_i)a\Big) \int_0^1 \Big(\frac{H_\xi(u(g(x_j)^{-1}g(x_i)))}{H_\xi(a(g(x_j)^{-1}g(x_i)))} - 1\Big) du$$

$$\times E\Big(\chi\Big\{X_{k,L} < a, k \in \mathbf{N}/\{j-1, j+1\}\Big\} \,\Big|\, \varphi_i = a\Big) dH_\xi(a) \tag{2.16}$$

$$\leq -\delta \int_0^{\infty} H_\xi(g(x_j)^{-1}g(x_i)a) E\Big(\chi\Big\{X_{k,L} < a,$$

$$k \in \mathbf{N}/\{j-1, j+1\}\Big\} \,\Big|\, X_{i,L} = a\Big) dH_\xi(a). \tag{2.17}$$

$$= \delta P\Big\{X_j > X_i > X_k, k \in \mathbf{N}/\{i, j-1, j, j+1\}\Big\}$$

In (2.16) we used the fact that, for $k \in \mathbf{N}/\{j-1, j+1\}$, $\Big\{\{X_{k,L}\}, X_{i,L}\Big\}$ and $X_{j,L}$ are independent, and in (2.17) the fact that $H_\xi(a)$ is slowly varying as a function of $a \to \infty$, as $L \to \infty$, played a role. Therefore, by (2.9), we have

$$\tfrac{1}{\epsilon} \textstyle\sum_i E(A_i) \leq$$

$$\tfrac{1}{\epsilon}\delta \textstyle\sum_i E\Big[\sum_{j \neq i}\Big(P\Big\{X_j > X_i > X_k, k \in \mathbf{N}/\{i, j-1, j, j+1\}\Big\}\Big)\Big] \leq 4! \tfrac{1}{\epsilon}\delta \, .$$

Now, consider terms B_i, where $j = i+1$. Then,

$$B_i = E\Big[\frac{X_{i+1,L}}{X_{i,L}} \chi\{X_{i,L} > K;\ i_L^* = i\} \,\Big|\, \{x_k\}\Big]$$

$$\leq E\Big[\frac{g_L(x_{i+1}) e^{c_1\eta_{i+1}} e^{c_0\eta_{i+2}}}{g_L(x_i) e^{c_1\eta_i} e^{c_0\eta_{i+1}}} \chi\Big\{g_L(x_i) e^{c_1\eta_i} e^{c_0\eta_{i+1}} > K;$$

$$g_L(x_k) e^{c_1\eta_k} e^{c_0\eta_{k+1}} < g_L(x_i) e^{c_1\eta_i} e^{c_0\eta_{i+1}}, k \in \mathbf{N}/\{i, i+1, i+2,\}\Big\}\Big]$$

$$= -\int \int_{(Kg_L(x_i)^{-1}a^{-c_0})^{-1}}^{\infty} \Big[\frac{g_L(x_{i+1})}{g_L(x_i)} \frac{a^{c_1}}{b^{c_1}a^{c_0}} \int_0^{(g_L(x_{i+1})^{-1} g_L(x_i) b^{c_1} a^{c_0} a^{-c_1})^{1/c_0}} c^{c_0}$$

$$\times E\Big(\chi\Big\{g_L(x_k) e^{c_1\eta_k} e^{c_0\eta_{k+1}} < g_L(x_i) b^{c_1} a^{c_0}; k \in \mathbf{N}/\{i, i+1, i+2\}\Big\}\Big|\, e^{\eta_i+1} = a \, ,$$

$$e^{\eta_i} = b, e^{\eta_{i+2}} = c\Big) dH_{\eta_{i+2}}(c)\Big] dH_{\eta_i}(b) dH_{\eta_{i+1}}(a)$$

For $k \in \mathbf{N}/\{i, i+1, i+2\}$,

$$E\Big(\chi\Big\{g_L(x_k)e^{c_1\eta_k}e^{c_0\eta_{k+1}} < g_L(x_i)b^{c_1}a^{c_0}\ ,$$

$$k \in \mathbf{N}/\{i, i+1, i+2\}\Big\}\ \Big|\ e^{\eta_{i+1}} = a, e^{\eta_i} = b, e^{\eta_{i+2}} = c\Big)$$

is independent of c.

Since $H_\eta(a)$ is slowly varying, we have

$$-\frac{g_L(x_{i+1})}{g_L(x_i)}\frac{a^{c_1}}{b^{c_1}a^{c_0}}\int_0^{(g_L(x_{i+1})^{-1}g_L(x_i)b^{c_1}a^{c_0}a^{-c_1})^{1/c_0}} c^{c_0}dH_\eta(c)$$

$$= H_\eta\Big((g_L(x_{i+1})^{-1}g_L(x_i)b^{c_1}a^{c_0}a^{-c_1})^{1/c_0}\Big)$$

$$\times \int_0^1 \Big(\frac{H_\eta((g_L(x_{i+1})^{-1}g_L(x_i)b^{c_1}a^{c_0}a^{-c_1})^{1/c_0}u)}{H_\eta((g_L(x_{i+1})^{-1}g_L(x_i)b^{c_1}a^{c_0}a^{-c_1})^{1/c_0})} - 1\Big)du^{c_0}$$

$$\leq \delta H_\eta\Big((g_L(x_{i+1})^{-1}g_L(x_i)b^{c_1}a^{c_0}a^{-c_1})^{1/c_0}\Big)\ .$$

Using the above two facts, we get

$$B_i \leq \delta \int\int H_\eta\Big((g_L(x_{i+1})^{-1}g_L(x_i)b^{c_1}a^{c_0}a^{-c_1})^{1/c_0}\Big) E\Big(\chi\Big\{g_L(x_k)e^{c_1\eta_k}e^{c_0\eta_{k+1}}$$

$$< g_L(x_i)b^{c_1}a^{c_0}, k \in \mathbf{N}/\{i+2\}\Big\}\ \Big|\ e^{\eta_{i+1}} = a, e^{\eta_i} = b\Big)dH_\eta(b)dH_\eta(a)$$

$$= \quad \delta P\Big\{X_{i+1,L} > X_{i,L} > X_{k,L}\ , k \in \mathbf{N}/\{i, i+1, i+2\}\Big\}\ .$$

Therefore

$$\frac{1}{\epsilon}\sum_i E(B_i) \leq 3!\frac{1}{\epsilon}\delta\ .$$

For terms C_i's, we can proceed as in the case of B_i's, to obtain a similar estimate. Combining the above three estimates, we obtain that

$$P\Big\{S^-/X_L^* > \epsilon, X_L^* > K\Big\} \leq (4! + 2\times 3!)\frac{1}{\epsilon}\delta$$

for sufficiently large L and K. Since δ is arbitrary, the proof of Proposition 2.1 is complete.

3 Asymptotic Boundedness in Probability

Let

$$g_t(x) = t^{1/2} g(t,0,x) ,$$

and

$$g_{1,t}(x) = x g_t(x) .$$

Since $H_k(a), k = 1,2$, satisfy conditions 1°-4° of section 2, for any $\delta > 0$, $\sigma > 0$, there is $K < \infty$ such that the inequality

$$\left[H_i\left(a\frac{g_{1,t}(x)}{g_{1,t}(y)}\right) - H_i\left(a\frac{g_t(x)}{g_t(y)}\right)\right]_+ < \sigma H_i\left(a\frac{g_t(x)}{g_t(y)}\right) \tag{3.1}$$

holds for all $a > 1$, $i = 1,2$, $x, y \in R$, $|y| > K|x|$, $|x| > \delta\sqrt{t}$, $t > 1$. This result can be found in Albeverio, Molchanov and Surgailis (1995), and it will be used to establish the asymptotic boundedness in probability of the ratio $x_{i_1^*}/x_{i^*}$, where

$$g_t(x_{i^*})e^{\xi_{i^*}} = \max_i\left(g_t(x_i)e^{\xi_i}\right), \quad \text{and} \quad g_{1,t}(x_{i_1^*})e^{\xi_{i_1^*}} = \max_i\left(g_{1,t}(x_i)e^{\xi_i}\right) .$$

Lemma 3.1. *Under the above notation, we have*

$$\lim_{K\to\infty} \limsup_{t\to\infty} P\left\{|x_{i_1^*}| > K|x_{i^*}|]\right\} = 0 . \tag{3.2}$$

Proof. Let us first estimate the probability of "small" values of $|x_{i^*}|$. Namely, we claim that for any $\delta > 0$

$$P\left\{|x_{i^*}| < \delta\sqrt{t}\right\} < C\delta^3 , \tag{3.3}$$

where $C < \infty$ is independent of t and δ. Indeed,

$$P\left\{|x_{i^*}| < \delta\sqrt{t}\right\} = E \sum_{i:|x_i|<\delta\sqrt{t}} P[X_p < X_i, p \neq i \,\Big|\, \{x_s\}]$$

$$\leq E \sum_{i:|x_i|<\delta\sqrt{t}} P\left\{X_p < X_i, p = i+2k, k \in \mathbf{N}/\{0\} \,\Big|\, \{x_s\}\right\}$$

$$= E \sum_{i:|x_i|<\delta\sqrt{t}} \int_0^\infty \prod_{k\in\mathbf{N}/\{0\}} F_\xi\left(a\frac{g_t(x_i)}{g_t(x_{i+2k})}\right) dF_\xi(a)$$

$$\cong E\frac{1}{2}N(\delta\sqrt{t}) \int_0^\infty F_\xi(ca)^{N(\sqrt{t})/2-1} dF_\xi(a) ,$$

where $N(r) := \#\{i : |x_i| < r\}$ and

$$c = \sup\{g_t(x)/g_t(y) : |x| < \sqrt{t}, |y| < \sqrt{t}\} \leq e^{1/2} .$$

Next

$$\int_0^{\infty} F_\xi(ca)^{N/2-1} dF_\xi(a) = \int_0^{\infty} H_\xi(a/c) dF_\xi^{N/2-1}(a)$$

$$\leq c_1 \int_0^{\infty} H_\xi(a) dF_\xi^{N/2-1}(a) = c_1 (N/2)^{-1} ,$$

as $H_\xi(a/c) \leq c_1 H_\xi(a)$ for all $a > 0$, where c_1 is a constant, which follows from the monotonicity and slow variation of $H_\xi(a)$. Consequently, the right hand side of (3.2) does not exceed $c_1 E[N(\delta\sqrt{t})/N(\sqrt{t})] = c_1\delta^3$, which was to be proved.

Now, using an easy observation to the effect that $\varphi_{i^*} \to \infty$, as $t \to \infty$, the lemma follows from the fact that

$$\lim_{K\to\infty} \limsup_{t\to\infty} P\Big\{|x_{i_1^*}| > K|x_{i^*}|, |x_{i^*}| > \delta\sqrt{t}, \varphi_{i^*} > 1\Big\} = 0 . \tag{3.4}$$

The latter statement can be demonstrated as follows: Denoting the probability in (3.4) by $P_{K,t}$, one can rewrite it as

$$P_{K,t} = E \sum_{|x_i|>\delta, |x_j|>K|x_i|} P_{K,t}^{i,j} ,$$

where

$$P_{K,t}^{i,j} = P\Big\{X_p < X_i, p \neq i; X_{1,q} < X_{1,j}, q \neq j; \varphi_i > 1 \Big| \{x_s\}\Big\}$$

$$= E \sum_{|x_i|>\delta, |x_j|>K|x_i|} P\Big\{\varphi_p < \min\Big(g_t(x_i)g_t(x_p)^{-1}\varphi_i, g_{1,t}(x_j)g_{1,t}(x_p)^{-1}\varphi_j\Big);$$

$$g_{1,t}(x_i)g_{1,t}(x_j)^{-1}\varphi_i < \varphi_j < g_t(x_i)g_t(x_j)^{-1}\varphi_i,\ p \in \mathbf{N}/\{i,j\}, \varphi_i > 1 \Big| \{x_s\}\Big\} .$$

Consider separately the terms in the above sum corresponding to $|i-j| > 1$ and $|i-j| = 1$.

For the case $|i-j| > 1$,

$$P_{K,t}^{i,j} \leq$$

$$P\Big\{\varphi_p < \min\Big(g_t(x_i)g_t(x_p)^{-1}\varphi_i, g_{1,t}(x_j)g_{1,t}(x_p)^{-1}\varphi_j\Big); g_{1,t}(x_i)g_{1,t}(x_j)^{-1}\varphi_i$$

$$< \varphi_j < g_t(x_i)g_t(x_j)^{-1}\varphi_i,\ p \in \mathbf{N}/\{i-1, i, i+1, j-1, j, j+1\} \Big| \{x_s\}\Big\}$$

$$\leq \int_1^{\infty} dF_\xi(a) \Big[H_\xi\Big(a\frac{g_{1,t}(x_i)}{g_{1,t}(x_j)}\Big) - H_\xi\Big(a\frac{g_t(x_i)}{g_t(x_j)}\Big)\Big]_+$$

$$\times P\Big\{\varphi_p < g_t(x_i)g(x_p)^{-1}\varphi_i\ p \in \mathbf{N}/\{i-1, i, i+1, j-1, j, j+1\} \Big| \{x_s\}\Big\}$$

(using (3.1))

$$\leq \sigma E \int_0^\infty H_\xi\Big(a\frac{g_t(x_i)}{g_t(x_j)}\Big) P\Big\{\varphi_p < g_t(x_i)g_t(x_p)^{-1}\varphi_i \,,$$

$$p \in \mathbf{N}/\{i-1, i, i+1, j-1, j, j+1\}\Big\} dF_\xi(a)$$

$$\leq \sigma P\Big\{g_t(x_j)\varphi_j > g_t(x_i)\varphi_i > g_t(x_p)\varphi_p;\ p \in \mathbf{N}/\{i-1, i, i+1, j-1, j, j+1\}\Big\}\,.$$

For the case $j = i+1$, for a given $\{x_s\}$,

$$P^{i,i+1}_{K,t} \leq P\Big\{\varphi_p < g_t(x_i)g_t(x_p)^{-1}\varphi_i;\, g_{1,t}(x_i)g_{1,t}(x_{i+1})^{-1}\varphi_i < \varphi_{i+1}$$

$$< g_t(x_i)g_t(x_{i+1})^{-1}\varphi_i,\ \ p \in \mathbf{N}/\{i-1, i, i+1, i+2\}, \varphi_i > 1\Big\}\,.$$

Let $F(a) = 1 - H_\eta(a)$. Then,

$$P^{i,j}_{K,t} = P\Big\{\varphi_p < g_t(x_i)g_t(x_p)^{-1}e^{c_1\eta_i + c_0\eta_{i+1}};\, g_{1,t}(x_i)g_{1,t}(x_{i+1})^{-1}e^{c_1\eta_i + c_0\eta_{i+1}}$$

$$< e^{c_1\eta_{i+1} + c_0\eta_{i+2}} < g_t(x_i)g_t(x_{i+1})^{-1}e^{c_0\eta_i + c_1\eta_{i+1}},$$

$$p \in \mathbf{N}/\{i-1, i, i+1, i+2\}, \varphi_i > 1\Big\}$$

$$= \int \int_{\{b^{c_1}a^{c_0} > 1\}} P\Big\{\varphi_p < g_t(x_i)g_t(x_p)^{-1}b^{c_1}a^{c_0},$$

$$g_{1,t}(x_i)g_{1,t}(x_{i+1})^{-1}b^{c_1}a^{c_0} < a^{c_1}e^{c_0\eta_{i+2}} < g_t(x_i)g_t(x_{i+1})^{-1}b^{c_1}a^{c_0},$$

$$p \in \mathbf{N}/\{i-1, i, i+1, i+2\} \,\Big|\, e^{\eta_i} = b, e^{\eta_{i+1}} = a\Big\} dF(b)dF(a)$$

$$\leq \int \int_{\{b^{c_1}a^{c_0} > 1\}} P\Big\{\varphi_p < g_t(x_i)g_t(x_p)^{-1}b^{c_1}a^{c_0}\Big\}$$

$$\Big[H_{\eta_{i+2}}\Big(g_{1,t}(x_i)g_{1,t}(x_{i+1})^{-1}b^{c_1}a^{c_0}\Big)$$

$$-H_{\eta_{i+2}}\Big(g_t(x_i)g_t(x_{i+1})^{-1}b^{c_1}a^{c_0}\Big)\Big]_+ dF(b)dF(a)\,.$$

Once again, using (3.1), we will have

$$P_{K,t}^{i,i+1} \le \sigma \int\int H_{\eta_{i+2}}\Big(g_t(x_i)g_t(x_{i+1})^{-1}b^{c_1}a^{c_0}\Big)$$

$$\times P\Big\{\varphi_p < g_t(x_i)g_t(x_p)^{-1}b^{c_1}a^{c_0}\Big\}dF(b)dF(a)$$

$$= \sigma P\Big\{g_t(x_{i+1})\varphi_{i+1} > g_t(x_i)\varphi_i > g_t(x_p)\varphi_p;\ \ p \in \mathbf{N}/\{i-1, i, i+1, i+2\}\Big\}.$$

For the case $j = i - 1$, similarly as in the case $j = i + 1$, we have

$$P_{K,t}^{i,i-1} \le \sigma P\Big\{g_t(x_{i+1})\varphi_{i+1} > g_t(x_i)\varphi_i > g_t(x_p)\varphi_p;\ \ p \in \mathbf{N}/\{i-1, i, i+1, i+2\}\Big\}.$$

Combining the above estimates, we have

$$P_{K,t} \le \sigma \sum_{i \neq j}\Big[P\Big\{g_t(x_j)\varphi_j > g_t(x_i)\varphi_i > g_t(x_p)\varphi_p;$$

$$p \in \mathbf{N}/\{i-1, i, i+1, j-1, j, j+1\}\Big\}$$

$$+2P\Big\{g_t(x_j)\varphi_j > g_t(x_i)\varphi_i > g_t(x_p)\varphi_p; p \in \mathbf{N}/\{i-1, i, j, i+1\}\Big\}\Big]$$

$$\le (6! + 2 \times 5!)\sigma\ .$$

Since $K \to \infty$, $t \to \infty$, and σ can be arbitrarily small, Lemma 3.1 has been proved.

4 Proof of Theorem 1.1.

We shall consider $x = 0$ only and assume, similarly as in the proof of Proposition 2.1, that the distribution of $\varphi_i = e^{\xi_i}$ is continuous. Let us introduce the notation

$$u(t) = \sum_i X_i = \sum_i g_t(x_i)\varphi_i\ , \tag{4.1}$$

$$v_1(t) = \sum_i X_{1,i} = \sum_i g_{1,t}(x_i)\varphi_i\ . \tag{4.2}$$

It is easy to check that, if $t > (4\nu)^{-1}$, then functions $g_t(x)$ and $g_{1,t}(x)$ satisfy conditions (2.4), (2.5) and (2.6). In such a case, $v(t) = v(t, 0)$ can be written as

$$v(t) = \frac{v_1(t)}{tu(t)}\ , \tag{4.3}$$

and the relation (1.9) becomes

$$v(t) \overset{P}{\sim} t^{-1}x_{i^*}, \qquad (t \to \infty) \tag{4.4}$$

which is equivalent to the condition

$$\theta_t = \frac{1}{|x_{i^*}|}\left|\frac{v_1(t)}{u(t)} - x_{i^*}\right| \xrightarrow{P} 0, \quad (t \to \infty) .$$

Hence, by Proposition 2.1,

$$\frac{u(t)}{X_{i^*}} \xrightarrow{P} 1 \tag{4.5}$$

and

$$\frac{\sum_{i\neq i_1^*} X_{1,i}}{X_{1,i_1^*}} \xrightarrow{P} 0 \tag{4.6}$$

as $t \to \infty$, where $X_{1,i_1^*} = \max_i X_{1,t}$. Now

$$\theta_t \leq \left| \frac{X_{i^*}}{u(t)} - 1 \right| + \frac{1}{|x_{i^*}|}\sum_{i=i^*} \frac{X_{1,i}}{X_{i^*}} = \theta_t' + \theta_t'' ,$$

where $\theta_t' \to 0$ according to (4.5), while

$$\theta_t'' \leq \frac{|x_{i_1^*}|}{|x_{i^*}|}\left[\frac{X_{i_1^*}}{X_{i^*}} 1(i_1^* \neq i^*) + \sum_{i\neq i_1^*} \frac{X_{1,i}}{X_{1,i_1^*}}\right] .$$

The two terms in the square brackets vanish in probability as $t \to \infty$ because of (4.5) and (4.6), respectively. The ratio $|x_{i_1^*}|/|x_{i^*}|$ is asymptotically bounded in probability by Lemma 3.1. Thus the proof of Theorem 1.1 is complete.

5 Conclusions

We have provided a rigorous proof of the fact that for large times the statistical solutions of the Burgers' equation with the initial velocity potential data of the shot noise type, with dependent amplitudes and heavy tails, converge (in the sense of the stochastic relative asymptotic equivalence) to a random field with with "saw-tooth" trajectories. The location of the shocks is governed by a standard Poisson process location of the maxima of the initial shot noise and by their amplitudes.

If the above random field of velocities is coupled with the continuity equation, the resulting density field will display the characteristic intermittencies, with density maxima corresponding to the location of velocity shocks. This establishes analytically the existence of a one-dimensional version of the netlike-structures in the adhesion model of matter distribution in the Universe.

In the case of the initial shot noise velocity potential with independent amplitudes, Hu and Woyczynski (1995a) have established a power law for the shock density. However, at this point it is not clear if that result can be extended to the more general framework considered in the present paper.

Acknowledgments

This work was partially supported by the U.S. Office of Naval Research and the National Science Foundation.

References

1. S. Albeverio, S.A. Molchanov and D. Surgailis, Stratified structure of the Universe and Burgers' equation - a probabilistic approach, Prob. Theory Rel. Fields **100**, (1994) to appear.
2. T. Funaki, D. Surgailis and W.A. Woyczynski, Gibbs-Cox random fields and Burgers' turbulence, Ann. Appl. Probability, **5**, 701–735 (1995).
3. S.N. Gurbatov, A.I. Saichev, D.F. Shandarin, Soviet Phys. Dokl., **30**, 921 (1985).
4. Y. Hu and W.A. Woyczynski, An extremal rearrangement property of statistical solutions of Burgers' equation, Annals of Applied Probability, **4**, No.3, 838–858 (1994).
5. Y. Hu and W.A. Woyczynski, Limit behavior of quadratic forms of moving averages of i.i.d. random variables and statistical solutions of the Burgers' equation. J. Multivariate Anal., **52**, 15–44 (1995).
6. Y. Hu and W.A. Woyczynski, Shock density in Burgers' turbulence, in *Nonlinear Stochastic PDE's: Burgers' Turbulence and Hydrodynamic Limit*, IMA Volumes, Springer-Verlag, pp. 211-226 (1995).
7. L. Kofman, D. Pogosyan, S.F. Shandarin and A.L. Mellot, Coherent structures in the Universe and the adhesion model, Astrophys. J., **393**, 437–449 (1992).
8. S. Kwapien and W.A. Woyczynski, *Random Series and Stochastic Integrals: Single and Multiple.* Birkhäuser, Boston (1992).
9. S.A. Molchanov, D. Surgailis and W.A. Woyczynski, Hyperbolic asymptotics in Burgers Turbulence and extremal processes, Comm. Math. Phys., 1–27 (1994).
10. P.J.E. Peebles, *The Large-Scale Structure of the Universe*, (1980) Princeton University Press, Princeton, N.J.
11. S.F. Shandarin and B.Z. Zeldovich, Turbulence (1989), intermittency, structures in a self-gravitating medium: the large scale structure of the universe, Rev. Modern Phys., **61**, 185–220.
12. Surgailis D., Woyczynski W.A. (1993), Long range predictions and scaling limit for statistical solutions of the Burgers' equation, in *Nonlinear Waves and Weak Turbulence*, Birkhäuser, Boston, pp. 313-338.
13. Surgailis D., Woyczynski W.A. (1994), Scaling limits of the solution of the Burgers' equation with singular Gaussian initial data, in *Chaos Expansions, Multiple Wiener-Ito Integrals and Their Applications*, C. Houdre and Perez-Abren, Eds. CRC Press, pp. 145-162.
14. Vergassola M., Dubruille B., Frisch U. and Noulles A. (1994), Burgers' equation, devil's staircases and the mass distribution for the large-scale structures, Astron. et Astrophys., to appear.
15. Weinberg D.H., Gunn J.E. (1990), Large-scale structure and the adhesion approximation, Monthly Not. Royal Astronom. Soc. 247, 260-286.
16. Woyczynski W.A., Stochastic Burgers' flows, in *Nonlinear Waves and Weak Turbulence with Applications in Oceanography and Condensed Matter Physics*, N. Fitzmaurice et al., eds. 279-312. Birkhauser, Boston (1993).

Convergence of Iterative Methods in Perturbation Theory

H.R. Jauslin[1], *M. Govin*[1] *and M. Cibils*[2]

[1] Laboratoire de Physique de l'Université de Bourgogne
21000 Dijon, France.
e-mail: jauslin at satie.u-bourgogne.fr

[2] Institut de Physique Théorique, EPFL, 1015 Lausanne, Switzerland.
e-mail: cibils at eldp.epfl.ch

Abstract: We discuss iterative KAM type methods for eigenvalue problems in finite dimensions. We compare their convergence properties with those of straight forward power series expansions.

1 Introduction

The KAM techniques were developed to treat the problem of the persistence under perturbations of invariant tori in classical mechanics [1] [2]. Similar ideas have been also applied in quantum mechanics, e.g. to prove the persistence of point spectrum under perturbations [3]–[6]. The method is based on an iteration scheme, which under suitable restrictions on the characteristic frequencies of the system is shown to converge to an attractive fixed point, provided that the perturbation is sufficiently small.

In this note we first discuss how the classical and quantum KAM problems can be stated in a formally similar setting. We then address the question of the convergence of iterative KAM schemes in simple finite dimensional eigenvalue problems for perturbations of arbitrarily large intensity. We compare with the radius of convergence of traditional power series expansions.

The general idea can be formulated as follows. The system is described by a Hamiltonian of the form $H = H_0 + \varepsilon V_1$ where H_0 is meant to be a simple system whose solution is explicitly known. In classical mechanics it is an integrable system, which in the in action-angle representation depends only on the action variables. In quantum mechanics H_0 has usually pure point spectrum and the eigenvalues and eigenvectors are explicitly known. They are usually obtained through the use of the symmetry properties of the system. In this basis H_0 is diagonal.

The goal, which in general cannot be easily attained, can be thought of as finding a transformation of coordinates or of basis, such that in the new representation the complete Hamiltonian H is simple (integrable or diagonal). The transformation should maintain the basic Hamiltonian structure of the problem.

This means that classically one is restricted to canonical transformations, while in quantum mechanics one uses unitary transformations.

A first approach to the construction of this transformation is a development in power series in ε. In classical mechanics this approach encounters difficulties due to the presence of resonances and small denominators. These difficulties were overcome by the methods of KAM theory, which use an iterative scheme instead of the expansion in power series. We remark that recently some results of KAM theory have been recovered by resumation techniques from the power series expansion [7] [8].

2 Formulation as an Eigenvalue Problem

The formalisms of classical and quantum transformations can be set in a framework of Lie groups and Lie algebras, which allows to treat them in a formally similar manner.

A] *Unitary transformations in quantum mechanics:* Unitary transformations U that are homotopic to the identity (i.e. that can be embedded into a family of unitary operators U_κ, depending continuously on the parameter $\kappa \in \mathbf{R}$, such that $U_{\kappa=0} = \mathbf{1}$ and $U_{\kappa=1} = U$) can be represented as $U = e^W$, where the generator W is antihermitian $W^\dagger = -W$. The unitary transformation acts on the Hamiltonian as

$$H \mapsto H' = e^W H e^{-W} = H + [W, H] + \frac{1}{2!}[W, \ [W, H]] + \frac{1}{3!}[W, \ [W, \ [W, H]]] + \dots$$

B] *Canonical transformations in classical mechanics:* In classical mechanics one can use a similar construction by working with Poisson brackets [9]. For two functions on phase space $f(p, q), g(p, q)$

$$\{f, g\} = \sum_j \frac{\partial f}{\partial p_j}\frac{\partial g}{\partial q_j} - \frac{\partial g}{\partial p_j}\frac{\partial f}{\partial q_j}$$

The functions on phase space have the structure of a Lie algebra with respect to the Poisson bracket.

To any function $f(p, q)$ one can associate a linear operator $\hat{f}$ defined by $g \mapsto \hat{f}(g) = \{f, g\}$, that is called the Lie operator associated to f. The Lie operators form a Lie algebra with respect to their commutator $[\hat{f}_1, \hat{f}_2](g) = \hat{f}_1(\hat{f}_2(g)) - \hat{f}_1(\hat{f}_2((g))$. The relation between the Lie algebra of phase space functions and the Lie algebra of Lie operators is given by $\{f, g\} = h \iff [\hat{f}, \hat{g}] = \hat{h}$. The exponential of a Lie operator is defined by $\exp(\hat{f}) = \sum_{n=0}^{\infty} \hat{f}^n/n!$ where $\hat{f}^n$ denotes the operator $\hat{f}$ composed n times. Canonical transformations can be represented by exponentials of Lie operators by the following properties [9]:

a) For any function $S(p, q)$ on phase space, if $\exp(\hat{S})$ converges, the transformation

$$p \mapsto p' = e^{\hat{S}} p, \qquad q \mapsto q' = e^{\hat{S}} q \tag{2.1}$$

is a canonical transformation.

b) Every canonical transformation that is homotopic to the identity can be represented as an infinite product of exponentials of Lie operators (2.1). The operator $\hat{S}$, or by extension the function $S(p,q)$ is called the Lie-generator of the canonical transformation.

Under a canonical transformation, the Lie operator associated with the Hamilton function transforms as

$$\hat{H} \mapsto \hat{H}' = e^{\hat{S}}\hat{H}e^{-\hat{S}} = \hat{H} + [\hat{S},\hat{H}] + \frac{1}{2!}[\hat{S},[\hat{S},\hat{H}]] + \frac{1}{3!}[\hat{S},[\hat{S},[\hat{S},\hat{H}]]] + \ldots$$

the corresponding Hamilton function as

$$H \mapsto H' = e^{\hat{S}}H = H + \{S,H\} + \frac{1}{2!}\{S,\{S,H\}\} + \frac{1}{3!}\{S,\{S,\{S\ ,H\}\}\} + \ldots$$

We consider now an integrable classical system with N degrees of freedom described by a Hamilton function H_0, which expressed in action ($\underline{A} \in \mathbf{R}^N$) and angle ($\underline{\varphi} = (\varphi_1, \varphi_2 \ldots \varphi_N) \in \mathbf{T^N}$) variables depends only on the actions $H_0 = H_0(\underline{A})$. To follow the analogy with quantum mechanics we determine the eigenvalues $i\lambda$ and eigenfunctions $\psi(\underline{A},\underline{\varphi})$ of the corresponding Lie operator

$$\hat{H}_0\psi = i\lambda\psi, \quad i.e. \quad \{H_0,\psi\} = i\lambda\psi$$

which taking into account the independence of H_0 on φ can be written as

$$\underline{\omega}(\underline{A}) \cdot \frac{\partial}{\partial\varphi}\psi = i\lambda\psi \tag{2.2}$$

with $\underline{\omega}(\underline{A}) = \partial H_0/\partial\underline{A}$. The solution of (2.2) can be written as

$$\psi(\underline{A},\underline{\varphi}) = e^{i\underline{n}\cdot\underline{\varphi}}\,\delta(\lambda - \underline{n}\cdot\underline{\omega}(\underline{A}))$$

with $\underline{n} \in \mathbf{Z}$. If we restrict to the invariant torus corresponding to a fixed value of the frequency vector $\underline{\omega}$, the spectrum is pure point with purely imaginary eigenvalues $i\lambda = i\underline{\omega}\cdot\underline{n}$, and eigenfunctions $\phi(\underline{\varphi}) = \exp(i\underline{n}\cdot\underline{\varphi})$, with $\underline{n} \in \mathbf{Z}^N$. We remark that if two components of $\underline{\omega}$ are incommensurate, the point spectrum is dense. We remark that the restriction to a fixed frequency vector is what is appropriate in KAM theory. Thus in classical mechanics the question of persistence upon perturbation of invariant tori of a given frequency can be interpreted as the persistence of dense point spectrum under perturbation. The break up of an invariant torus is signaled by the appearance of continuous spectrum.

3 KAM Type Iterations for Eigenvalue Problems

We consider a quantum Hamiltonian of the form

$$H_1 = D_1 + V_1$$

where $D_1 \equiv H_0$ is diagonal in the eigenbasis $\{\phi_n\}$ with corresponding eigenvalues d_n^1, and V_1 is of order $\mathcal{O}(\varepsilon)$ has zero diagonal elements (in the basis $\{\phi_n\}$). The idea is to find a unitary transformation U_1, such that

$$U_1(D_1 + V_1)U_1^{-1} = D_2 + V_2 \tag{3.1}$$

where D_2 is diagonal and V_2 is of $\mathcal{O}(\varepsilon^2)$ and with zero diagonal elements (always in the original basis $\{\phi_n\}$).
If we iterate this procedure k times we will get

$$U_k(D_k + V_k)U_k^{-1} = D_{k+1} + V_{k+1}$$

the non diagonal part becomes of $\mathcal{O}(\varepsilon^{2^k})$. For small ε, when the iteration converges to a diagonal D_∞ the approach will be exponential.
The unitary transformation U_k required at the k-th step can be constructed as follows (to keep the notation simple we write the construction for k=1; the construction is the same for any k). We can represent U_1 as

$$U_1 = e^{W_1}$$

where $W_1^\dagger = -W_1$ is antihermitian and of order $\mathcal{O}(\varepsilon)$. We insert it into (3.1), develop the exponentials, and separate the terms of $\mathcal{O}(\varepsilon)$ from the higher order terms to obtain

$$U_1(D_1 + V_1)U_1^{-1} = D_1 + V_1 + [W_1, D_1] + \tilde{V}_2,$$

$$\tilde{V}_2 = [W_1, V_1] + 1/2!\,[W_1, [W_1, D_1 + V_1]] + +1/3!\,[W_1, [W_1, [W_1, H_1]]] \ldots = \mathcal{O}(\varepsilon^2). \tag{3.2}$$

The condition to eliminate the non diagonal terms of order ε is thus

$$V_1 + [W_1, D_1] = 0 \tag{3.3}$$

The equation (3.3) can be solved by expressing it as a matrix equation in the basis $\{\phi_n\}$:

$$(D_1)_{jk} = <\phi_j \mid D_1 \mid \phi_k> = d_j^1 \delta_{j,k}$$

$$(V_1)_{jk} = <\phi_j \mid V_1 \mid \phi_k>$$

$$(W_1)_{jk} = <\phi_j \mid W_1 \mid \phi_k>$$

to obtain

$$(W_1)_{jk} = \begin{cases} (V_1)_{jk}/(d_j^1 - d_k^1), & \text{for } j \neq k \\ 0 & \text{for } j = k \end{cases} \tag{3.4}$$

We remark that the $(W_1)_{jj} = 0$ is not imposed by (3.3), and could be chosen differently.

The corresponding $\tilde{V}_2$ constructed with (3.2) will in general have non zero diagonal elements, which we group together with the initial diagonal part:

$$(V_2)_{jk} = (\tilde{V}_2)_{jk} - \delta_{j,k}(\tilde{V}_2)_{jj}$$

$$(D_2)_{jk} = \big((D_1)_{jj} + (\tilde{V}_2)_{jj}\big)\delta_{jk}$$

This can be written equivalently as

$$D_2 = \quad \textit{diagonal part of } U_1(D_1 + V_1)U_1^{-1}$$

$$V_2 = \textit{non} - \textit{diagonal part of } U_1(D_1 + V_1)U_1^{-1}.$$

We can interpret the procedure as the iteration of a map T in the space of hermitian matrices $T : H \mapsto H'$. If we denote by D and V the diagonal and non diagonal parts of H, we can write T symbolically as

$$D' = F(D, V)$$

$$V' = G(D, V)$$

where F and G are the nonlinear maps on the space of matrices defined by the construction above.

4 Convergence of the Iteration, Comparison with Power Series

Provided that one avoids resonances and small denominators in (3.4) it can be shown that the above iteration converges for small enough perturbations. The fact that the iteration does not converge when there are resonances or too small denominators has a physical interpretation: in those situations the perturbed system has a behavior that is qualitatively different from the unperturbed one. Typically the unperturbed problem has a quasi-periodic character which is destroyed by the perturbation. Therefore the perturbed system cannot be transformed by a regular change of coordinates into one with the same form as the unperturbed one.

A natural question in this context is to ask whether the limits of the region of convergence of the iterative procedure are the intrinsic ones imposed by the qualitative change in behavior at some critical intensity of the perturbation ε_{crit}, or whether the procedure ceases to converge for perturbations smaller than ε_{crit}, which should be interpreted as an artifact produced by the chosen algorithm.

A related question is the comparison of the domains of convergence of the iterative method with the region of convergence of a standard power series expansion. The answer to these questions is not known for classical systems nor for quantum systems in infinite dimensional Hilbert spaces. Our aim is to get some preliminary intuition of what can be expected by analysing the question in quantum systems with finite dimensional Hilbert space, like e.g. spin degrees of freedom. By construction these examples do not have any intrinsic qualitative change of behavior, so any limitation of convergence is to be attributed to the chosen method.

The simplest illustration is the case where H is a symmetric 2×2 matrix, which we can take of trace zero:

$$H = \begin{pmatrix} d & v \\ v & -d \end{pmatrix}$$

The transformation is given by $U = \exp W$, $W = \begin{pmatrix} 0 & v/(2d) \\ -v/(2d) & 0 \end{pmatrix}$ which yields

$$U = \begin{pmatrix} \cos v/(2d) & \sin v/(2d) \\ -\sin v/(2d) & \cos v/(2d) \end{pmatrix}$$

The space of symmetric 2×2 matrices with trace zero is two dimensional and can be parametrized by the vectors (d, v). In this representation the transformation T is given by

$$\begin{pmatrix} d' \\ v' \end{pmatrix} = \begin{pmatrix} \cos v/d & \sin v/d \\ -\sin v/d & \cos v/d \end{pmatrix} \begin{pmatrix} d \\ v \end{pmatrix}$$

The transformation can be thus interpreted geometrically as a rotation by an angle $\alpha = v/d$. It is therefore natural to use polar coordinates (r, ϑ) defined by $r^2 = d^2 + v^2, \quad \tan(\vartheta) = v/d$. The radius is invariant, and the transformation is described by

$$T: \quad \vartheta \mapsto \vartheta' = \vartheta - \tan(\vartheta)$$

In this representation we can analyse some of the convergence properties easily.

a) There are two attracting fixed points ($P_1^* : \vartheta = 0$ and $P_2^* : \vartheta = \pi$). They correspond to the two possible permutations of the eigenvalues in the diagonalized matrix:

$$\begin{pmatrix} \lambda_+ & 0 \\ 0 & \lambda_- \end{pmatrix} \quad \text{and} \quad \begin{pmatrix} \lambda_- & 0 \\ 0 & \lambda_+ \end{pmatrix}$$

b) There are infinitely many other fixed points, periodic orbits of all periods, and chaotic orbits, but they are all unstable.

c) We can determine the immediate domain of attraction of the two stable fixed points (i.e. the region around the fixed point on which the application is contractive). For P_1^* the condition of contractivity is $|T\vartheta| < |\vartheta|$. Since for $|\vartheta| < \pi/2$, we have $|tan(\vartheta)| > |\vartheta|$, the border of the immediate domain of attraction is given by $|\vartheta| < |\vartheta_c|$ where ϑ_c is defined by $\vartheta_c - \tan(\vartheta_c) = -\vartheta_c$ which can be evaluated as $\vartheta_c \simeq 1.1655...$ which corresponds to $\alpha_c = \tan(\vartheta_c) = (v/d)_{c(iteration)} =\simeq$

2.3311.... The border of the immediate domain of attraction is thus an unstable orbit of period 2.

We can compare this result with the radius of convergence of the power series expansion (i.e. the usual Rayleigh-Schrödinger perturbation theory). The radius of convergence can be read e.g. from the expansion of the eigenvalues: $\lambda_{\pm} = \pm d\sqrt{1+(v/d)^2} = \pm d \sum_k b_k (v/d)^{2k}$, $b_0 = 1, b_1 = 1/2, b_2 = -(1)/(2\cdot 4)$, $b_3 = (1\cdot 3)/((2\cdot 4\cdot 6), \ldots$ Taking $\varepsilon \equiv v/d$ as the expansion parameter, the radius of convergence is given by $(v/d)_{c(series)} = 1/(\limsup (b_k)^{1/(2k)}) = 1$. Thus, the iterative method converges monotonically in a region that goes way beyond the radius of convergence of the power series.

We have seen that the iterative procedure does not converge for the value ϑ_c. The next question is if it converges for the remaining values $|\vartheta| \in (\vartheta_c, \pi/2)$. The convergence in this region can be explored numerically. Since it is not in the immediate domain of attraction the angle ϑ can grow at first, but it can end up falling into the immediate domain of attraction. The object to study are the preimages of the immediate domain of attraction. The picture that emerges is that all points except for a Cantor set converge to P_1^* or to P_2^*. We have observed the same type of behavior for $N \times N$ matrices with large N [10].

The Cantor set on which the iteration does not converge reflects a property of the iteration scheme, not an intrinsic property of the system. This point can be further illustrated by comparing with different schemes of iteration. A natural variation of the scheme we have started with is to use the Cayley representation of the unitary matrix U instead of the exponential:

$$U = \frac{1+W/2}{1-W/2}$$

This leads to exactly the same equations (3.3) to determine W, but the higher order terms V_2 will be different, and thus the iteration and its convergence properties will be different. In this version, the iteration is equivalent to the iteration of the following rational function:

$$z \mapsto z' = \frac{z(z^4+50z-15)}{(1+50z-15z^4)}$$

where the variable z is related to v/d by $v/d = (i/2)(z+1/z)$. We can study this iteration by continuing it into the complex plane. In this representation the original real axes corresponds to the imaginary axis, and the two stable fixed points are at $z = \pm i$. The set of points that do not converge to one of the two attracting fixed points is a Julia set with an intricate self similar structure in the complex plane, as shown in the figure. The white and the black regions correspond respectively to points attracted to one or the other fixed point. The Julia set is the boundary of the two domains of attraction. Figure b is an enlargement of the center of figure a.

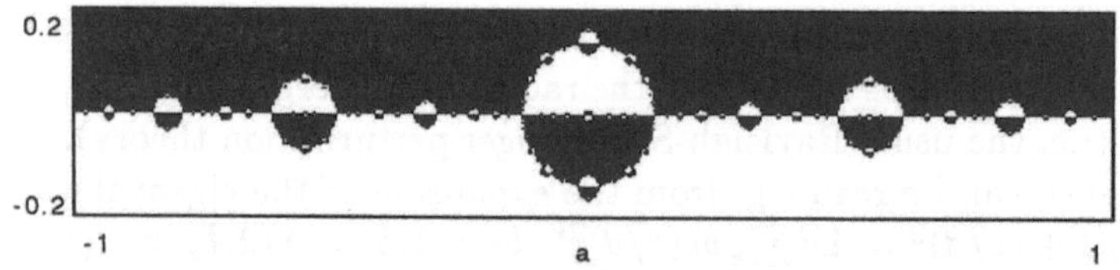

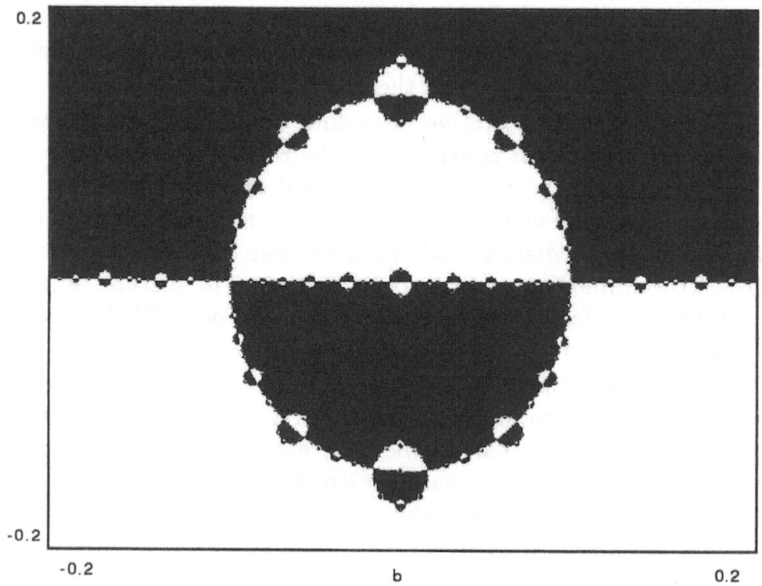

Acknowledgments

Support from the EC contract ERBCHRXCT94-0460 for the project "Stability and universality in classical mechanics" is acknowledged.

References

1. V.I. Arnold; *Mathematical Methods of Classical Mechanics*, Springer Vlg., 1978.
2. J. Moser; *Stable and random motions in dynamical systems*, Princeton U. Press, 1973.
3. J. Bellissard; in *Trends in the Eighties*, edited by S. Albeverio and Ph. Blanchard, World Scientific, Singapore 1985.
4. M. Combescure; Ann. Inst. H.Poincaré **47** (1987) 63; Ann. Phys. **185**, 86(1988).
5. P. Blekher, H.R. Jauslin, J.L. Lebowitz; Floquet spectrum for two-level systems in quasiperiodic time dependent fields, J. Stat. Phys. **68**, 271(1992).

6. H.R. Jauslin; Small divisors in driven quantum systems, in *Stochasticity and Quantum Chaos*, Z. Haba, W. Cegla, L. Jakóbczyk (eds.), Kluwer Publ. 1995.
7. L.H. Eliasson; *Absolutely convergent series expansions for quasiperiodic motions*, report 2-88, Dept. of Mathematics, University of Stokholm, 1988.
8. G. Gallavotti; Commun. Math, Phys. **164**, 145(1994).
9. A.J. Dragt, J.M. Finn; J. Math. Phys. **17**, 2215(1976).
10. M. Govin, H.R. Jauslin, M. Cibils; in preparation.

[illegible]

Eigenfunction Expansions for Time Dependent Hamiltonians

H.R. Jauslin, S. Guerin and A. Deroussiaux

Laboratoire de Physique de l'Université de Bourgogne
21000 Dijon, France
e-mail: jauslin at satie.u-bourgogne.fr

Abstract: We describe a generalization of Floquet theory for non periodic time dependent Hamiltonians. It allows to express the time evolution in terms of an expansion in eigenfunctions of a generalized quasienergy operator. We discuss a conjecture on the extension of the adiabatic theorem to this type of systems, which gives a procedure for the physical preparation of Floquet states.

1 Introduction

We consider a quantum mechanical system subject to an explicitly time dependent external field, described a Hamiltonian of the form

$$H = H_0 + \hat{V}(x,t)$$

acting on a Hilbert space $\mathcal{H}$. We assume that the time dependence is introduced through a classical dynamical system, i.e.

$$\hat{V}(x,t) = V(x, \underline{\theta}(t))$$

where $\underline{\theta}(t) = \Phi_t \underline{\theta}$ is a classical flow on a domain Ω, that is ergodic with respect to an invariant measure $d\mu$. Some special examples are periodic forces with frequency ω (Ω = unit circle $S^1, d\mu = d\theta, \Phi_t\ \theta = \theta + \omega t$), or quasiperiodic forces with N frequencies $\underline{\omega} = (\omega_1, \ldots, \omega_N)$, ($\Omega$ = N – dimensional torus $\mathbf{T}^N \equiv S^1 \times S^1 \ldots \times S^1, d\mu = d\theta_1 \ldots d\theta_N, \Phi_t\ \underline{\theta} = \underline{\theta} + \underline{\omega} t$).
We consider systems where H_0 has discrete spectrum with eigenvalues E_n and eigenfunctions φ_n. We characterize the driving flow Φ_t in terms of the associated Koopman operator $\mathcal{T}_t$, defined on functions $\xi(\underline{\theta}) \in L_2(\Omega, d\mu)$ by

$$(\mathcal{T}_t \xi)(\underline{\theta}) = \xi(\Phi_t \underline{\theta}) \equiv \xi(\underline{\theta}(t)) \tag{1.1}$$

We consider here driving flows with pure point Koopman spectrum. Since $\mathcal{T}_t$ is a continuous family of unitary operators it has a complete basis of orthogonal eigenvectors $\xi_k(\underline{\theta})$ with eigenvalues of absolute value equal to one:

$$\mathcal{T}_t \xi_k = e^{-i\nu_k t} \xi_k \tag{1.2}$$

In absence of the perturbation V the solution of the Schrödinger equation $i\partial\phi^0/\partial t = H_0\phi^0$ can be expressed in terms of the eigenfunction expansion

$$\phi^0(t) = \sum_{n=0}^{\infty} e^{-iE_n t} c_n \varphi_n \tag{1.3}$$

where the coefficients c_n are determined from the initial condition $c_n = < \varphi_n, \phi^0(t=0) >$. This representation shows immediately that the time evolution is almost periodic. This means in particular that the trajectories $\phi^0(t)$ in Hilbert space are precompact, which means that the evolution happens essentially in a finite dimensional subspace of $\mathcal{H}$: If at time $t = 0$ we make an approximation of the initial condition $\phi^0(t=0) = \sum_{n=0}^{N-1} c_n\varphi_n + r_N(0)$ such that the error is $|r_N(0)| < \epsilon$, then the time evolution can be approximated by the N dimensional truncation

$$\phi^0(t) = \sum_{n=0}^{N-1} e^{-iE_n t} c_n \varphi_n + r_N(t)$$

The error made will have for all times the same bound as initially $|r_N(t)| < \epsilon$. This property is important for the numerical solution of the Schrödinger equation, since it justifies the use of a finite dimensional truncation.

We will describe a method to extend this kind of eigenfunction expansion to time dependent Hamiltonians [1] –[5], which is the natural extension of the Floquet or quasienergy techniques that are widely used for periodic time dependent forces [6]–[10]. The idea is to construct a time independent operator K in an enlarged Hilbert space $\mathcal{K} = \mathcal{H} \otimes L_2(\Omega, d\mu)$. We consider the Hamiltonian $H = H_0 + V(\underline{\theta}(t)) = H_0 + V(\Phi_t\underline{\theta})$ as a family of Hamiltonians parametrized by the initial phase $\underline{\theta} \in \Omega$. We denote by $U(t, t_0\, ;\, \underline{\theta})$ the propagator of the corresponding Schrödinger evolution, which satisfies $U(t+a, t_0+a\, ;\, \underline{\theta}) = U(t, t_0\, ;\, \Phi_a\underline{\theta})$. By composition of the propagator with the Koopman operator $\mathcal{T}_t$ defined in (1.1) one can define

$$W_{t-t_0} = \mathcal{T}_{-t} \circ U(t, t_0\, ;\, \underline{\theta}) \circ \mathcal{T}_{t_0} \equiv \mathcal{T}_{-(t-t_0)} \circ U(t - t_0, 0\, ;\, \underline{\theta}) \tag{1.4}$$

which is a one parameter $(t - t_0)$ family of unitary operators acting on the enlarged Hilbert space $\mathcal{K}$. By Stone's theorem this defines a self-adjoint operator K by

$$e^{-iK(t-t_0)} = W_{t-t_0} \tag{1.5}$$

By differentiation with respect to t and evaluating at $t = t_0$, K can be written acting on $\psi(\underline{\theta}) \in \mathcal{K}$ as

$$(K\psi)(\underline{\theta}) = \left[-i\underline{g}(\underline{\theta}) \cdot \frac{\partial}{\partial\underline{\theta}} + H(\underline{\theta})\right]\psi(\underline{\theta})$$

where $H(\underline{\theta}) = H_0 + V(\underline{\theta})$ and $\underline{g}(\underline{\theta})$ is the vector field of velocities associated to the flow Φ_t:

$$\underline{g}(\underline{\theta}) = \frac{d}{dt}\Phi_t\underline{\theta}\Big|_{t=0}, \quad \text{i.e.} \quad \frac{d}{dt}\mathcal{T}_t = \underline{g}(\underline{\theta}(t)) \cdot \frac{\partial}{\partial\underline{\theta}(t)}\mathcal{T}_t \equiv \mathcal{T}_t\underline{g}(\underline{\theta}) \cdot \frac{\partial}{\partial\underline{\theta}} \tag{1.6}$$

Equations (1.4) (1.5) allow to relate the time evolution of the system to the spectral properties of the quasienergy operator K. If K has pure point spectrum

$$K\psi_m = \lambda_m \psi_m,$$

then, using the inverse of (1.4), the time evolution can be expressed by the eigenfunction expansion

$$\begin{aligned}\phi(t) = & \ U(t,0\,;\,\underline{\theta})\phi(0) = \mathcal{T}_t e^{-iKt}\phi(0) \\ = & \ \mathcal{T}_t \sum_m c_m e^{-i\lambda_m t}\psi_m(x,\underline{\theta})\end{aligned} \tag{1.7}$$

where the coefficients c_m are determined by the scalar product (in $\mathcal{K}$) $c_m = <\psi_m, \phi(0)\otimes 1>_{\mathcal{K}}$. We can further develop the eigenfunctions $\psi_m(x,\underline{\theta})$ with respect to the eigenfunctions $\xi_k(\underline{\theta})$ of the Koopman operator:

$$\psi_m(x,\underline{\theta}) = \sum_k \tilde{\psi}_m(x,k)\xi_k(\underline{\theta})$$

Using (1.2) we can write

$$\phi(t) = \sum_{m,k} c_m \tilde{\psi}_m(x,k)\xi_k(\underline{\theta})e^{-i(\nu_k+\lambda_m)t}$$

As in the time independent case (1.3), this expansion implies that the time evolution is almost periodic and can be described essentially in an finite dimensional space, with an error that is bounded uniformly in time, i.e. that is determined by the chosen approximation for the initial condition. This property is important for the numerical treatment of these models. We remark however, that the quasienergy operator can have continuous spectrum, which is the signature of an instability in the system, often due to some resonance phenomenon [8] [4].

2 Structure of the Floquet Eigenfunctions

We first remark that if $\psi_m(x,\underline{\theta})$ is a Floquet eigenfunction with eigenvalue λ_m, then $\xi_k(\underline{\theta})\psi_m(x,\underline{\theta})$ will also be an eigenfunction corresponding to the eigenvalue $\nu_k + \lambda_m$. The eigenfunctions and eigenvalues can thus be grouped into infinite families. As a consequence, the Floquet eigenvalues form in general a dense point spectrum that makes perturbation theory difficult. In order to get some insight into the structure of these families, we restate the problem in a different but essentially equivalent way. We ask if the Schrödinger equation

$$i\frac{\partial}{\partial t}\phi = H(\underline{\theta}(t))\phi$$

can be transformed by a unitary transformation $\phi = C\bar{\phi}$ into a Schrödinger equation with some time independent Hamiltonian B:

$$i\frac{\partial}{\partial t}\bar{\phi} = B\bar{\phi}$$

This can be expressed in terms of the propagators as

$$U(t,0\,;\,\underline{\theta}) = C(\underline{\theta}(t))e^{-iBt}C^{-1}(\underline{\theta}) \tag{2.1}$$

There always exists a time dependent transformation C that does this, but we impose the stronger requirement that the time dependence of C is the same one as that of the driving force, i.e.

$$C = C(\underline{\theta}(t))$$

The relation between the original and the transformed Hamiltonian is

$$\begin{aligned} B = {} & C^{-1}HC - iC^{-1}\frac{dC}{dt} \\ = {} & C^{-1}(\underline{\theta}(t))H(\underline{\theta}(t))C(\underline{\theta}(t)) - iC^{-1}(\underline{\theta}(t))\frac{\partial C(\underline{\theta}(t))}{\partial \underline{\theta}(t)} \cdot \underline{g}(\underline{\theta}(t)) \end{aligned}$$

Embedding eq. (2.1) into the enlarged space $\mathcal{K}$ and acting from the left by $\mathcal{T}_{-t}$ we obtain the relation

$$e^{-iK(\underline{\theta})t} = C(\underline{\theta})\; \mathcal{T}_{-t}e^{-iBt}\; C^{-1}(\underline{\theta})$$

The question is thus equivalent to showing that the quasienergy operator K is unitarily equivalent to a trivial one $K_B = -i\underline{g}(\underline{\theta}) \cdot \partial/\partial\underline{\theta} + B$ corresponding to a time independent Hamiltonian B. Since the eigenfunctions of K_B are $\xi_k(\underline{\theta}) \otimes \varphi_n^B(x)$, with $B\varphi_n^B = E_n^B\varphi_n^B$, the eigenfunctions of K can be written as

$$\psi_{n,k}(x,\underline{\theta}) = C(x,\underline{\theta})\left[\xi_k(\underline{\theta}) \otimes \varphi_n^B(x)\right] \tag{2.2}$$

with eigenvalues $\lambda_{n,k} = E_n^B - \nu_k$.

We remark that inserting (2.2) into the expansion (1.7) we can write

$$\phi(t) = C(x,\underline{\theta}(t))\sum_{n,k} c_{n,k}e^{-iE_n^Bt}\xi_k(\underline{\theta})\varphi_n^B(x)$$

3 Preparation of Floquet States, Adiabatic Conjecture

The eigenstates of the quasienergy operator will be called Floquet states. Their time evolution is particularly simple:
if $\phi(0) = \psi_m(x, \underline{\theta})$ then $\phi(t) = \exp(-i\lambda_m t)\psi_m(x, \underline{\theta}(t))$. They play a role similar to the one of stationary states in time independent systems.
We will next discuss how the Floquet states can play a role beyond the mathematical tool for formal expansions. We will discuss a procedure to physically prepare a system into a Floquet state. The procedure is based on the adiabatic switching of the time dependent field, as described e.g. By the Hamiltonian

$$H = H(\alpha(t), \underline{\theta}(t)) = H_0 + \alpha_\tau(t)V(\underline{\theta}(t))$$

with

$$\alpha(t) = \begin{cases} 0 & \text{if } t \le -\tau \\ (t+\tau)/\tau & \text{if } -\tau < t < 0 \\ 1 & \text{if } t \ge 0 \end{cases} \tag{3.1}$$

In order to treat this situation we have to generalize somewhat the Floquet formalism. Besides the dependence on $\underline{\theta}(t)$, the Hamiltonian depends on time also through another variable $\alpha(t)$. On the enlarged Hilbert space $\mathcal{K} = \mathcal{H} \otimes L_2(\Omega, d\mu(\underline{\theta}))$ we define now a time dependent self-adjoint operator

$$K(t) = -i\underline{g}(\underline{\theta}) \cdot \frac{\partial}{\partial \underline{\theta}} + H(\alpha(t), \underline{\theta}) \tag{3.2}$$

If we define $U^K(t, t_0)$ as the propagator associated to $K(t)$

$$i\frac{\partial}{\partial t}U^K(t,t_0) = K(t)U^K(t,t_0); \qquad U^K(t,t) = \mathbf{1} \tag{3.3}$$

then one obtains the following relation with the propagator of the original Hamiltonian

$$U(t,t_0\ ;\ \underline{\theta}) = \mathcal{T}_t U^K(t,t_0)\mathcal{T}_{-t_0} \quad \text{i.e.} \quad U^K(t,t_0) = \mathcal{T}_{-t}U(t,t_0\ ;\ \underline{\theta})\mathcal{T}_{t_0} \tag{3.4}$$

This relation can be verified by differentiation, which using equation (1.6) yields

$$i\frac{\partial}{\partial t}U = i\Big(\frac{\partial}{\partial t}\mathcal{T}_t\Big)U^K\mathcal{T}_{-t_0} + i\mathcal{T}_t\frac{\partial}{\partial t}U^K\mathcal{T}_{-t_0} = i\mathcal{T}_t\underline{g}(\underline{\theta}) \cdot \frac{\partial}{\partial \underline{\theta}}U^K\mathcal{T}_{-t_0} + \mathcal{T}_t K U^K \mathcal{T}_{-t_0}$$

and substituting (3.2) in the last term we obtain

$$\begin{aligned} i\frac{\partial}{\partial t}U(t,t_0\ ;\ \underline{\theta}) &= \\ &= i\mathcal{T}_t\underline{g}(\underline{\theta}) \cdot \frac{\partial}{\partial \underline{\theta}}\mathcal{T}_t U^K\mathcal{T}_{-t_0} - \mathcal{T}_t i\underline{g}(\underline{\theta}) \cdot \frac{\partial}{\partial \underline{\theta}}U^K + \mathcal{T}_t H(\alpha(t),\underline{\theta})U^K\mathcal{T}_{-t_0} \\ &= \mathcal{T}_t H(\alpha(t),\underline{\theta})U^K\mathcal{T}_{-t_0} = \mathcal{T}_t H(\alpha(t),\underline{\theta})\mathcal{T}_{-t}U(t,t_0\ ;\ \underline{\theta}) \\ &= H(\alpha(t),\underline{\theta}(t))U(t,t_0\ ;\ \underline{\theta}) \end{aligned}$$

which is the defining equation of $U(t, t_0 \,;\, \underline{\theta})$.
We can now treat equation (3.3) by adiabatic techniques. In the case of periodic forces this has been discussed e.g. in Refs. [11][14]. The usual adiabatic theorem cannot be applied directly since in this type of problem the instantaneous point spectrum is dense. Under suitable conditions the following statement can be conjectured:

Adiabatic conjecture: Denoting by $P(t)$ the projection in $\mathcal{K}$ on an instantaneous eigenspace of $K(t)$,

$$\lim_{\tau\to\infty} U^K(t, t_0)P(t_0) = \lim_{\tau\to\infty} P(t)U^K(t, t_0). \tag{3.5}$$

Assuming this conjecture, by the relation (3.4) we obtain for the original problem

$$\lim_{\tau\to\infty} U(t, t_0 \,;\, \underline{\theta})\mathcal{T}_{t_0}P(t_0)\mathcal{T}_{-t_0} = \lim_{\tau\to\infty} \mathcal{T}_t P(t)\mathcal{T}_{-t}U(t, t_0 \,;\, \underline{\theta}) \tag{3.6}$$

We remark that this is a relation in the enlarged space $\mathcal{K}$, i.e. $U(t, t_0 \,;\, \underline{\theta})$ is considered as an operator in $\mathcal{K}$.

We can give a more concrete picture by expressing this relation in terms of the eigenfunctions: We denote by $\psi_m^{\alpha(t)}(\underline{\theta})$ and $\lambda_m^{\alpha(t)}$ the instantaneous eigenfunctions and eigenvalues of the operator $K(t)$, and take in (3.6) $P(t) = |\psi_m^{\alpha(t)}><\psi_m^{\alpha(t)}|$. If at time t_0 the system is in an instantaneous Floquet state $\phi(t_0) = \psi_m^{\alpha(t_0)}(\underline{\theta}(t_0))$ then in the adiabatic limit $\tau \to \infty$ the time evolution determined by the Schrödinger equation $i\partial\phi/\partial t = H(\alpha(t), \underline{\theta}(t))\phi$ is such that $\phi(t)$ stays for all t an instantaneous Floquet eigenstate, up to a phase:

$$\phi(t) = e^{i\delta(t,t_0)}\psi_m^{\alpha(t)}(\underline{\theta}(t)). \tag{3.7}$$

In the case of adiabatic switching (3.1) that we are considering, the instantaneous eigenstates for $t < -\tau$ are of the form $\varphi_n \otimes 1$, with $\varphi_n \in \mathcal{H}$ a stationary state of the unperturbed problem. Thus if we prepare the system in a stationary state of H_0, the adiabatic switching will move the system into a Floquet state. The fact that equation (3.7) is a consequence of the relation (3.6) can be seen as follows. First we express the initial condition as $\phi(t_0) = \psi_m^{\alpha(t_0)}(\underline{\theta}(t_0)) \equiv \mathcal{T}_{t_0}|\psi_m^{\alpha(t_0)}>$. The left hand side of (3.6) applied to this $\phi(t_0)$ is just $\phi(t)$:

$$\begin{aligned} &U(t, t_0 \,;\, \underline{\theta})\mathcal{T}_{t_0}P(t_0)\mathcal{T}_{-t_0}\phi(t_0) \;= \\ &= \; U(t, t_0 \,;\, \underline{\theta})\mathcal{T}_{t_0}|\psi_m^{\alpha(t_0)}><\psi_m^{\alpha(t_0)}|\mathcal{T}_{-t_0}\mathcal{T}_{t_0}|\psi_m^{\alpha(t_0)}> \\ &= \; U(t, t_0 \,;\, \underline{\theta})\mathcal{T}_{t_0}|\psi_m^{\alpha(t_0)}> \equiv \phi(t) \end{aligned}$$

Applying the right hand side of (3.6) to $\phi(t_0)$ we can write thus

$$\phi(t) = \mathcal{T}_t P(t)\mathcal{T}_{-t}U(t, t_0 \,;\, \underline{\theta})\mathcal{T}_{t_0}|\psi_m^{\alpha(t_0)}> \tag{3.8}$$

The set of functions defined by $\mathcal{T}_t|\psi_n^{\alpha(t)}>$ form for fixed t a basis of the Hilbert space $\mathcal{K}$, because the eigenfunctions form a basis and $\mathcal{T}_t$ is unitary. We can therefore make the following decomposition

$$U(t,t_0\ ;\ \underline{\theta})\mathcal{T}_{t_0}|\psi_m^{\alpha(t_0)}>=\sum_n c_{m,n}(t,t_0)\mathcal{T}_t|\psi_n^{\alpha(t)}>$$

with $c_{m,n}(t,t_0)=<\psi_n^{\alpha(t)}|\mathcal{T}_{-t}U(t,t_0;\underline{\theta})\mathcal{T}_{t_0}|\psi_m^{\alpha(t_0)}>\equiv<\psi_n^{\alpha(t)}|U^K(t,t_0)|\psi_m^{\alpha(t_0)}>$. Inserting into (3.8) we can write

$$\begin{aligned}\phi(t) &= \mathcal{T}_tP(t)\mathcal{T}_{-t}\sum_n c_{m,n}(t,t_0)\mathcal{T}_t|\psi_n^{\alpha(t)}> \\ &= \mathcal{T}_t\sum_n c_{m,n}(t,t_0)|\psi_m^{\alpha(t)}><\psi_m^{\alpha(t)}\mid\psi_n^{\alpha(t)}>=c_{m,m}(t,t_0)\mathcal{T}_t|\psi_m^{\alpha(t)}>\ .\end{aligned}$$

Putting the two sides together we obtain thus

$$\phi(t)\equiv U(t,t_0\ ;\ \underline{\theta})\mathcal{T}_{t_0}|\psi_m^{\alpha(t_0)}>=c_{m,m}(t,t_0)\mathcal{T}_t|\psi_m^{\alpha(t)}>\ .$$

Since all the operators in this equation are unitary we can identify $c_{m,m}(t,t_0)=e^{i\delta(t,t_0)}$ and obtain (3.7).

Acknowledgments

Support from the EC contract ERBCHRXCT94-0460 for the project "Stability and universality in classical mechanics" is acknowledged.

References

1. J. Bellissard, in *Stochastic Processes in Classical and Quantum Systems* edited by S. Albeverio, G. Casati and D. Merlini, Springer 1986.
2. T.S. Ho, S.I. Chu; J. Phys. **B17**, 2101 (1984).
3. H.R. Jauslin, J.L. Lebowitz; Chaos **1**, 114 (1991).
4. P. Blekher,H.R. Jauslin, J.L. Lebowitz; Floquet spectrum for two-level systems in quasiperiodic time dependent fields, J. Stat. Phys. **68**, 271 (1992).
5. H.R. Jauslin; Stability and chaos in classical and quantum Hamiltonian systems; in P.L. Garrido, J. Marro (eds.), *II Granada Lectures in Computational Physics*, World Scientific, Singapore 1993.
6. J.S. Howland; Ann. Inst. H. Poincaré, **49** (1989) 309 and **49**, 325 (1989).
7. K. Yajima; Comm. Math. Phys. **87**, 331 (1982).
8. J. Bellissard; in *Trends in the Eighties*, edited by S. Albeverio and Ph. Blanchard, World Scientific, Singapore 1985.
9. M. Combescure; Ann. Inst. H. Poincaré **44**, 293 (1986), and Ann. Phys. **173**, 210 (1987).
10. G. Casati, L. Molinari; Progr. Theo. Phys. Suppl. **98**, 287 (1989).
11. H.P. Breuer, M. Holthaus; Phys. Lett. **A140**, 507 (1989).

12. H.P. Breuer, K. Dietz, M. Holthaus; J.Phys. **B 24**, 1343 (1991).
13. D.J. Moore; Phys. Rep. **210**, 1 (1991).
14. R.H. Young, W.J. Deal, Jr.; J. Math. Phys. **11**, 3298 (1970).

Strange Attractors in Higher–Dimensional Phase Space

Tomasz Kapitaniak and Jerzy Wojewoda

Division of Dynamics, Technical University of Łódz
PL–90–924 Łódz, Poland

Abstract: We discuss properties of higher dimensional strange attractors of dissipative dynamical systems in the neighbourhood of chaos-hyperchaos transition. We show new types of bifurcations which are typical for coupled chaotic systems. Particularly we introduce the concept of monotonic stability for which all perturbations decay monotonically and describe transition from monotonically stable to asymptotically stable chaotic attractor.

1 Introduction

In last two decades it has been shown that chaotic behavior is typical for three dimensional dissipative dynamical systems. The chaotic attractor is characterized by one positive Lyapunov exponent indicating sensitive dependence on initial conditions (exponential spreading within the attractor in the direction transverse to the flow) and its information dimension is larger than 2 but smaller than 3. In higher (at least four) dimensional systems beside chaotic attractors it is possible to find hyperchaotic attractors with two positive Lyapunov exponents [2-9]. Such attractors involve two directions of spreading within the attractor and their information dimension is larger than 3. General classification of attractors in four- and five-dimensional dissipative dynamical systems is given in [5]. Recently, it has been shown that chaos-hyperchaos transition is strictly connected with the chaos synchronization problems [2,3,8,9,15,16].

The problem of synchronization of chaotic systems can be understood as a problem of stability of n-dimensional chaotic attractor in m-dimensional phase space $m > n$. Let $\mathbf{A}$ be a chaotic attractor. The basin of attraction $\beta(\mathbf{A})$ is the set of points whose ω -limit set is contained in $\mathbf{A}$. According to Milnor [1] we say that $\mathbf{A}$ is an asymptotically stable attractor if it is Lyapunov stable ($\beta(\mathbf{A})$ has positive Lebesque measure) and $\beta(\mathbf{A})$ contains neighbourhood of $\mathbf{A}$.

In this paper we investigate stability of the chaotic attractor in the neighbourhood of the transition from chaos to hyperchaos for two coupled systems. Additionally we define monotonic stability of the attractor as a special case of asymptotic stability and describe a new type of bifurcation which is characteristic for coupled systems.

2 Coupled Systems

Consider a system

$$\dot{z} = f(z) \quad (1)$$

consisting of two coupled identical subsystems governed by

$$\dot{x} = f(x) + D(y - x)\dot{y} = f(y) - D(x - y) \quad (2)$$

where: $(x, y) \in \mathbf{R^n}, \mathbf{n} \geq \mathbf{3}, \mathbf{D} \in \mathbf{R}$. Assume that $x = f(x)$ and $y = f(y)$ have asymptotically stable chaotic attractor **A** in invariant manifold $\mathbf{R^n} = \mathbf{M}$.]

For the simplified analysis of the stability of chaotic attractor **A** let us introduce new variable

$$e(t) = x(t) - y(t) \quad (3)$$

With this transformation one replaces system (1) with the equivalent system

$$\dot{x} = f(x) - De \quad (4)$$

and

$$\dot{e} = f(x) - f(x - e) - 2De = g(x, e) \quad (5)$$

The variable $e(t)$ (eq.(5)) describes the evolution transverse to the n-dimensional invariant manifold **M** while the first equation (4) describes the evolution on the manifold **M**. The spectrum of Lyapunov exponents of eq.(4,5) can be easily divided into two subsets $\lambda^{(1)}$ associated with the evolution of $x(t)$ describing dynamics on the manifold **M** while the other set $\lambda^{(2)}$ describes propagation of perturbation normal to **M**.

In terms of $(e(t))$ we can define synchronization in the following way. Two chaotic systems $x = f(x)$ and $y = f(y)$ are synchronized if

$$lim_{t \to \infty} e(t) = 0 \quad (6)$$

If the limit (6) is fulfilled for all initial values in the neighbourhood of $(e(t) = 0)$, then the attractor **A** is *asymptotically stable*.

If additionally the following inequality holds

$$\frac{d|e(t)|}{dt} < 0 \quad (7)$$

then the synchronization is monotonic, i.e. after each perturbation the distance between the actual trajectory and the attractor is a decreasing function of time. In the case of monotonic synchronization for all initial values in the neighbourhood of $(e(t) = 0)$ we refer to the chaotic attractor **A** as *monotonically stable*. Of course, the monotonic stability is a special case of asymptotic stability. It should be noted here that monotonic stability depends on a distance metric. In this sense monotonicity is a quantitative property of the attractor and may depend on the observables.

Let us next describe various stability transitions of the chaotic attractor **A** in terms of the control parameter D. For sufficiently large D, say $D > D_3$, the chaotic attractor **A** is monotonically stable, the strongest type of stability discussed here.

Most analyses, however have in the present context been dealing with asymptotic stability. It can be shown [3] that the chaotic attractor **A** is asymptotically stable in $\mathbf{R}^{2n}$ if $D > D_2 = \lambda/2$ where λ is the largest Lyapunov exponent of chaotic state. At $D_2 = \lambda/2$ we have a bifurcation where chaotic attractor **A** loses its asymptotic stability. For $D < D_2$ the attractor **A** is still stable (there exist a $2n$-dimensional basin $\beta(\mathbf{A})$ with a positive Lebesque measure). However, it is not asymptotically stable as $\beta(\mathbf{A})$ does not contain whole neighbourhood of **A**. In this case, for a typical trajectory of (4,5), all transverse Lyapunov exponents $\lambda^{(2)}$ are negative. However, there are still initial conditions dense in the attractor for which one of the transverse exponents is positive. In the region $D < D_2$, the chaotic attractor **A** can also have a riddled basin [2,4,5], if there is an $\epsilon > 0$ such that for every point $x \in \beta(\mathbf{A})$ any arbitrarily small ball centered on x contains a set of points of positive measure whose orbits exceed a prescribed distance ϵ from **A**. In our case of two coupled systems there would be riddled basins of at least two different attractors.

With a further decrease of the control parameter D our system undergoes a blowout bifurcation (chaos-hyperchaos transition [2,4,6,8]) at, say $D = D_1$. After this bifurcation, one of the transverse Lyapunov exponents for a typical orbit on the attractor will always be positive. The chaotic attractor **A** becomes a chaotic saddle. A chaotic invariant set **A** having a dense orbit is a chaotic saddle [2] if there is a neighbourhood **U** of **A** such that $\beta(\mathbf{A}) \cap \mathbf{U}$ is greater than **A** but has zero Lebesque measure. Here: $D < D_1$, we observe the phenomenon of on-off intermittency (chaos-hyperchaos intermittency) [2] in which a typical phase space trajectory spends some of the time in the neighbourhood of the attractor **A** and occasionally burst away from it. For on-off (chaos-hyperchaos) intermittency the largest transverse Lyapunov exponent is positive but small. Due to the finite fluctuations of it there are stretches of time where the orbit is attracted to the invariant manifold **M** (i.e. the fluctuations may permit all transient Lyapunov exponents negative in this stretches). For $D < D_0$ largest transverse Lyapunov exponent is sufficiently large and an escape to a completely different attractor is possible. In this case the chaotic attractor **A** becomes a normally repelling chaotic saddle, i.e. there is an attractor in the invariant manifold **M**, but all points which are not lying on this manifold eventually leave a neighbourhood of **A**.

The typical bifurcations of system (2) are summarized in Figure 1. Each of these bifurcations can be seen as supercritical (subcritical) according to the creation of nearby invariant sets as D increases (decreases) through the bifurcation point. In Figure 1 bifurcations are shown as subcritical).

3 Transverse Linearization

Let us linearize eq.(5) in the neighbourhood of the attractor **A**, i.e. in the neighbourhood of the fixed point $e = 0$ with the condition $x \in \mathbf{A}$. In this case one

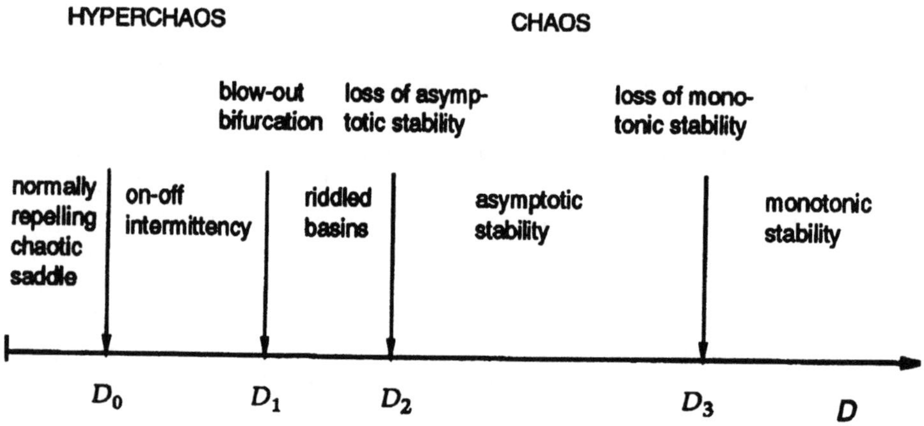

Fig. 1. Sequence of bifurcations for $D > D_0$ which are characteristic for system (1).

obtains equation

$$\dot{e} = B(x, D)e \tag{8}$$

where

$$B(x, D) = \left.\frac{\partial g(\bar{x})}{\partial \bar{x}}\right|_{x \in \mathbf{A}, \mathbf{e}=\mathbf{0}} \tag{9}$$

and $x = [x, e]^T$.

The introduced concept of linearization in the neighbourhood of the attractor allows us to reduce the problem of analysis of the stability of the attractor to the problem of investigation of the stability of the fixed point $e = 0$ of the eq.(4,5). Matrix $\mathbf{B(x, D)}$ is defined at given values of $x(t)$ which represents solution of eq.(4). Eigenvalues of $\mathbf{B(x, D)}$ can be considered only as functions of $x^{(i)}$, where $x^{(i)}$ is a discrete series such as $x^{(i)} \in x(t)$, i.e. solution $x(t)$ has been idiscretased. We have the following result

Proposition :
The chaotic attractor $\mathbf{A}$ is asymptotically and monotonically stable in $\mathbf{R}^{2n}$ if for all $x \in \mathbf{A}$, $\mathbf{e} = \mathbf{0}$ is the asymptotically stable fixed point of eq.(8).

Proof of the above proposition is based on the fundamental results of the linear stability. It will be given elsewhere [14].

4 Example

As an example let us consider two coupled Rossler systems

$$\dot{x} = -(y + z) + D(u - x)$$

$$\begin{aligned} \dot{y} &= x + ay + D(v - y) \\ \dot{z} &= b + z(x - c) + D(w - z) \end{aligned} \tag{10}$$

$$\begin{aligned} \dot{u} &= -(w + v) + D(x - u) \\ \dot{v} &= u + av + D(y - v) \\ \dot{w} &= b + w(u - c) + D(z - w) \end{aligned} \tag{11}$$

where a, b and c are constant. In our numerical investigation we considered the following parameter values: a=0.15, b=0.20, c=10.0 and D was taken as a control parameter. In the case of $D = 0$ (no coupling) the dynamics of both Rossler systems evolve along the chaotic attractor **A** [12] and typical trajectories are characterized by the following Lyapunov exponents: $\lambda_1 = 0.13, \lambda_2 = 0, \lambda_3 = -14.1$.

For this example eq.(5) describing transverse flow in the neighbourhood of chaotic attractor is given by

$$\begin{aligned} \dot{e}_1 &= -2De_1 - e_1 - e_3 \\ \dot{e}_2 &= -2De_2 + e_1 + ae_2 \\ \dot{e}_3 &= -2De_3 + ze_1 + (x - c)e_3 \end{aligned} \tag{12}$$

and is linear in e variable.

The exact system of equations (10) and (11) allows for a direct numerical analysis of the transverse behaviour. In Figure 2(a,b) we illustrate the typical evolution of the transverse flow corresponding to each characteristic behaviour indicated in Figure 1. These figures show the 3D-trajectories of $\mathbf{e(t)}$. In Figure 2(a) the chaotic attractor **A** represents a monotonically asymptotically stable one, and we can observe that the distance between points on the transverse trajectory $\mathbf{e(t)}$ and the attractor **A** seems to decrease monotonically. Asymptotic stability which is not monotonic is described in Figure 2(b). Transverse evolution $\mathbf{e(t)}$ approaches the attractor **A** and finally finishes on it, but its distance from the attractor may sometimes increase slightly. In Figure 2(c) we observe an on-off (chaos-hyperchaos) intermittency. The transverse trajectory $\mathbf{e(t)}$ spends some time in the neighbourhood of the chaotic attractor **A**, but occasionally bursts away from it. Finally in Figure 2(d) the periods of evolution in the neighbourhood of **A** are not visible anymore and the chaotic attractor is a normally repelling chaotic saddle. As it was pointed out in Figure 1 for $D \in [D_1, D_2]$ one can expect riddled behaviour. In this interval we observed coexistence of several different types of attractors: hyperchaotic (no synchronization), chaotic (synchronization), periodic of different periods (no synchronization). Small differences in the initial conditions can move system evolution from one attractor to another. The detailed study of this behaviour will be given in [17].

For this example the transverse linearized flow in the neighbourhood of the chaotic attractor **A** can give an adiabatic interpretation of its local stability, parametrized in the x and z components of the attractor. In Figure 3(a,b) we show the characteristic types of eigenvalues of the matrix $\mathbf{B(x, z, D)}$ for x and

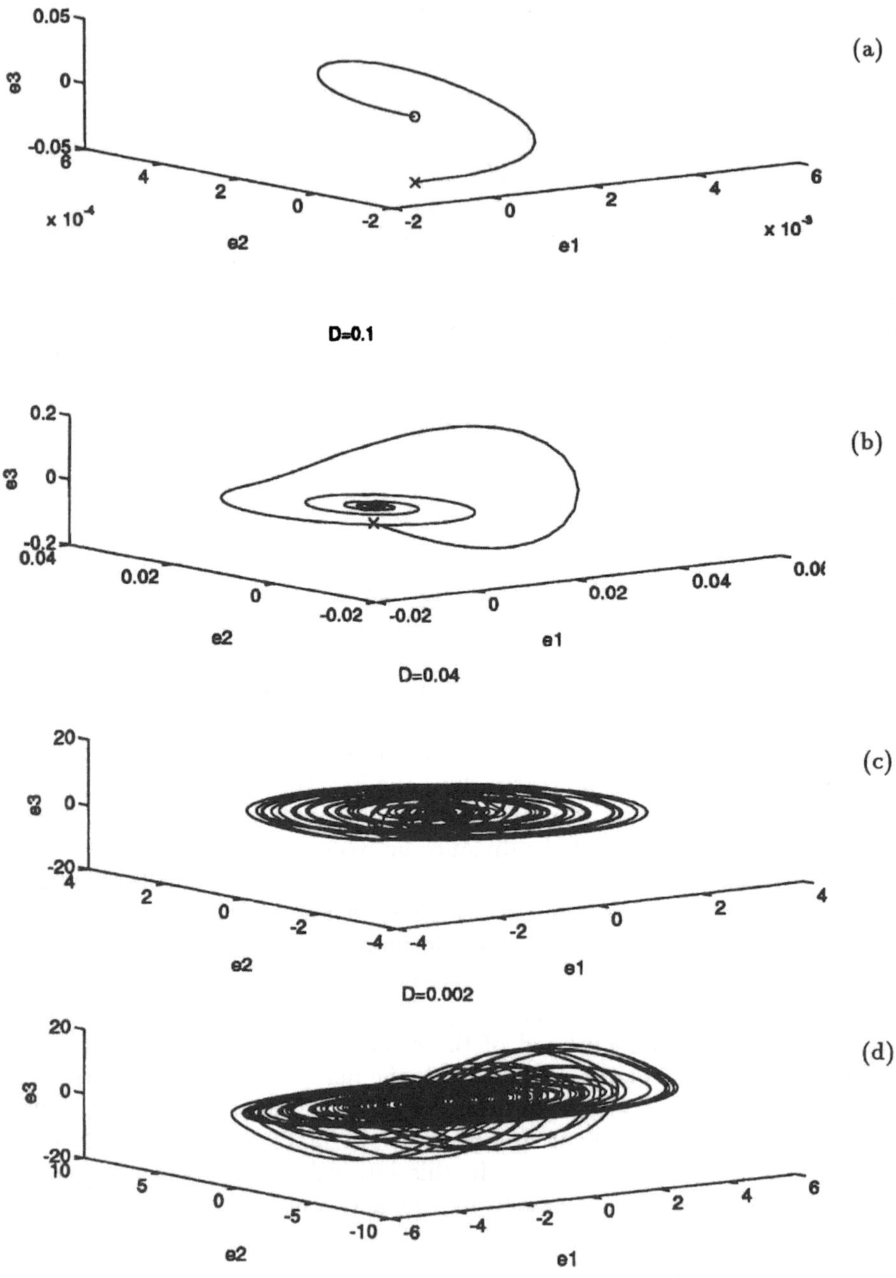

Fig. 2. Typical behaviours of the transverse motion: (a) monotonic, asymptotic stability, (b) asymptotic stability, (c) on-off intermittency, (d) normally repelling chaotic saddle.

z in the ranges [-20.0,20.0] and [-1.0,40.0] respectively. In the gray regions all eigenvalues are either real and negative or complex with negative real parts, while in the white regions at least one real eigenvalue is positive or a pair of complex eigenvalues has positive real parts. Figure 3(a) illustrates that, for $D = 3.5$ all $(x, z) \in \mathbf{A}$ are in gray region and the eigenvalues of the matrix $\mathbf{B}(\mathbf{x}, \mathbf{z}, \mathbf{D})$ have all negative real parts, i.e. the fixed point $\mathbf{e}(\mathbf{t}) = \mathbf{0}$ is asymptotically stable for all points on the attractor $\mathbf{A}$ and we conclude that the chaotic attractor is monotonically stable. We identified monotonic bifurcation point for $D = D_2 = 3.49$.

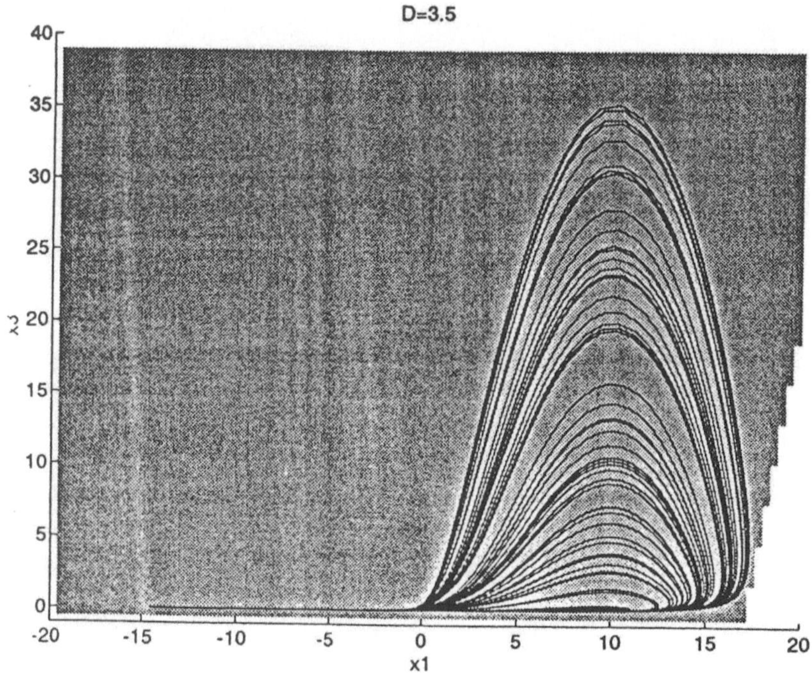

Fig. 3 a. Illustration of the local stability (gray region $\mathbf{e}(\mathbf{t})$ locally asymptotically stable, white region $e(t) = 0$ locally unstable: (a) monotonic, asymptotic stability

For smaller values of D, part of the attractor is in the grey region and part of it is in the white one as shown in Figure 3(b). This means that in one part of the attractor $\mathbf{e}(\mathbf{t})$ is a locally stable fixed point, while in the other part it is unstable. From adiabatic point of view it would be possible to investigate the asymptotic stability further by integrating the leading local exponent of the transverse perturbation along with the attractor and see if the time averaged value stabilizes as negative or not. However, such a calculation would be similar to the analysis of Lyapunov exponents numerically. In this study we have confirmed in all our calculations that $D_1 > D > 0.065(= \lambda_1/2)$ stabilizes the synchronized attractor, but not monotonically.

All our numerical computations have been performed using the software INSITE [13].

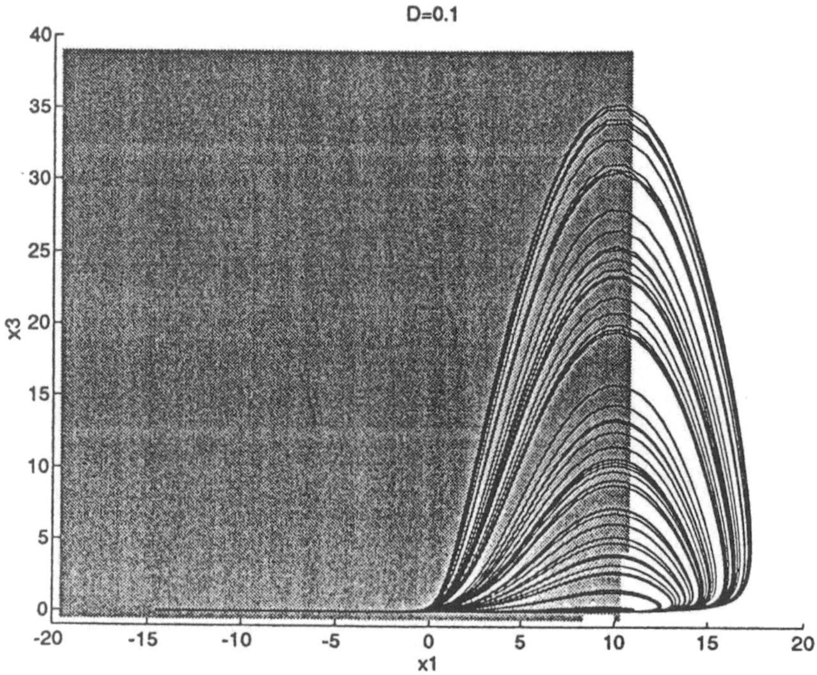

Fig. 3 b. Illustration of the local stability (grey region e(t) locally asymptotically stable, white region $e(t) = 0$ locally unstable: (b) asymptotic stability.

5 Conclusions

To summarize we described typical bifurcations of strange attractors in higher-dimensional phase space associated with the change of their stability properties. We have defined a monotonic stability of strange chaotic attractor and analyzed the bifurcation from asymptotic stability to monotonic stability. Associated with monotonic stability monotonic chaos synchronization is characterized by short synchronization time and may have practical applications. The introduced concept of linearization in the neighbourhood of the attractor allows us to describe this bifurcation in three- dimensional subspace transverse to the attractor.

References

1. J. Milnor, Commun. Math. Phys., **99**, 177 (1985).
2. P. Ashwin, J. Buescu and I. Stewart, Phys. Lett. **A193**, 126 (1994).
3. T. Yamada and H. Fujisaka, Prog. Theor. Phys., **70**, 1240 (1983).
4. E. Ott, J.C. Sommerer, Phys. Lett. **A188**, 39 (1994).
5. E. Ott, J.C. Sommerer, J.C. Alexander, I. Kan and J.A. Yorke, Physica **D76**, 384 (1994).
6. T. Kapitaniak, and W.-H. Steeb, Phys. Lett. **152A**, 33 (1991).
7. T. Kapitaniak, Chaotic Oscillations in Mechanical Systems, Manchester University Press: Manchester (1991).
8. T. Kapitaniak, Phys. Rev. **47E**, R2975 (1993).
9. T. Kapitaniak and L.O. Chua, Int. J. Bif. Chaos, 477 (1994).
10. V.S. Anishchenko, T. Kapitaniak, M.A. Safonova and O.V. Sosnovzeva, Phys. Lett. **A192**, 207 (1994).
11. J.C. Alexander, I. Kan, J.A. Yorke and Z. You, Int. J. Bif. Chaos, **2**, 795 (1992).
12. O. E. Rossler, Phys. Lett., **57A**, 397 (1976).
13. T.S. Parker, and L.O. Chua, Practical Numerical Algorithms for Chaotic Systems, Springer: New York (1989).
14. T. Kapitaniak and K.-E. Thylwe (in preparation).
15. L.M. Pecora and T.L. Carroll, Phys. Rev. Lett. **64**, 821 (1990).
16. M. de Sousa Viera, A.J. Lichtenberg, and M.A. Lieberman, Phys. Rev. **A46**, 7359 (1992).
17. T. Kapitaniak and J. Wojewoda (in preparation).

Mathematical Classification of Complete Chaos

Michael Keane

[1] CWI, P. O. Box 94079,
1090 GB Amsterdam, The Netherlands
[2] Delft University of Technology
2628 CD Delft, The Netherlands

In this essay, we attempt to give a brief survey of the ideas developed over the past years by mathematicians in an attempt to explain the apparent appearance of randomness in deterministic situations. The concepts discussed have a long history, and a detailed investigation would be necessary to give the exact origins, which will not be attempted here. Also no proofs will be given, as there are sufficient expositions at many levels in the literature.

Let us begin by explaining the elementary ideas behind the apparent appearance of randomness. Consider a very simple dynamical system, whose states are the complex numbers of modulus one, and in which the discrete passage of time is given by moving from the state z to the state z^2. In addition, we assume that the system is in statistical equilibrium, given by the uniform probability distribution on the set of states of the system. This probability distribution is easily seen to be invariant with respect to the passage of time. This simple dynamical system is clearly deterministic, in that complete knowledge of the state of the system at a given time t implies complete knowledge of its state at any time in the future. Nevertheless, there is an element of randomness if complete knowledge of the state is not assumed. Let us suppose that the system is hidden from the observer, who is told at a given time only whether the state z is in the right half or in the left half of the complex plane. Then the observer sees a sequence of zeroes and ones (say, zero for right half and one for left half), and thus progressively gathers information about the beginning state. However, at each observation, the probability of seeing a zero or a one is exactly one-half, independent of the previous observations, so that the observer's knowledge of the future remains completely obscured. The sequence of zeroes and ones which is seen has the same statistical properties as a sequence of random tosses of a fair coin, making the system appear completely random. It is this situation which has been studied mathematically, and we can formalize it as follows.

A *dynamical system* consists of a compact metric space X, together with a homeomorphism S from X to itself and a S-invariant probability measure μ

defined on (the Borel subsets of) the space X. For simplicity we assume that S is invertible – in the above example, this is not satisfied, but this should be considered as a technical detail. Thus by definition, a dynamical system is deterministic, and complete knowledge of the state (a point x of X) at a given time entails knowledge of the state at any time in the future or past, namely, at a given time n the system is in the state $S^n(x)$, if it is in state x at time zero.

Next let us introduce the dynamical systems corresponding to complete randomness, which are called *Bernoulli schemes.* Suppose that $p = (p_1, \cdots, p_n)$ is a probability vector, and let Y be the space of all doubly infinite sequences whose elements are taken from the finite alphabet consisting of the symbols 1 to n, n being a fixed positive integer. The left shift operator T acts naturally on Y, by shifting each coordinate of a point one position to the left, and the probability measure ν defined on Y to be the infinite product of the probability p on each coordinate, is S-invariant. Thus we have constructed in a very simple manner a dynamical system in the sense of the above definition, which we recognize as a model for independent repetitions of a single experiment with n outcomes whose corresponding probabilities are given by the vector p.

From the example given at the beginning it should be clear that the dynamical system whose states are complex numbers of modulus one is essentially the same system as a (one-sided) Bernoulli scheme. We now give a mathematical meaning to this notion of similarity. Two dynamical systems are said to be *metrically isomorphic* if there exists a measurable isomorphism between their state spaces which commutes with the corresponding homeomorphisms denoting the passage of time, with probability one. In this definition, the state spaces are considered to be probability spaces, so that *measurable isomorphism* means that there is a probability preserving measurable map (defined almost everywhere), invertible almost everywhere, between them. Thus if two systems are metrically isomorphic, then by relabeling the points of one of the systems (using as labels the points of the other system), the systems become indistinguishable with probability one.

The concept of metric isomorphism has turned out to be very useful in the study of dynamical systems. Only very recently has it been shown by Hungarian mathematicians that the systems consisting of finitely many hard spheres in a box (discretized in time) are metrically isomorphic to Bernoulli schemes. In general, producing metric isomorphisms is a very difficult task, and the constructions available seem to give little insight into the physical nature of the correspondances. The concept is considered by many to be too general to be applicable in real world situations, as it does not take into account the "nearness" of states – a measurable mapping, even if it is measure preserving, has nothing to say about the underlying distances between states. For these reasons, it seems to be desirable to introduce a more stringent notion of identification of systems, which we now discuss.

At first glance, one might expect that the proper concept to require for classification purposes is the continuity of the identification mapping. However, this turns out to be too strong for our purpose, and practically all systems are different if we require continuous isomorphisms. The proper concept seems to be that of *finitary isomorphism*, in which we require that the identification mapping is continuous only after removal of sets of measure zero in each of the state spaces. If we consider again the initial example, then we can see that the identification mapping is indeed not continuous, but only finitary. In fact, if z is a complex number of modulus one, then it can be written as $\exp(2\pi i x)$, with x in the unit interval, and then the sequence of zeroes and ones corresponding to z is just given by the binary expansion of x. The mapping which associates to each x in the unit interval its binary expansion is neither continuous nor invertible, essentially because each dyadic rational number has two different expansions. However, if we remove the dyadic rationals from the unit interval, and the corresponding sequences (which are eventually either all zeroes or all ones) from the sequence space of zeroes and ones, then this mapping becomes continuous, so that here we indeed have a finitary isomorphism. This proves also the utility of the finitary notion, at least by example, since we must admit that binary expansions are useful descriptions of real numbers, even if some continuity is lacking.

Next we would like to discuss the problem of the title. Given the definitions above, it seems clear that the Bernoulli schemes are suitable models for complete chaos. But there are very many Bernoulli schemes, one for each choice of the probability vector p. Let us introduce the notation $BS(p)$ to designate the Bernoulli scheme constructed using the probability vector p. We can then raise the following

QUESTION: When are the Bernoulli schemes $BS(p)$ and $BS(q)$ finitarily isomorphic?

It turns out that this question has a simple answer. To each probability vector p we can associate a number, $h(p)$, called the ***entropy*** of p, defined by

$$h(p) = -\sum_{i=1}^{n} p_i \log p_i ,$$

and then we can prove that $BS(p)$ and $BS(q)$ are finitarily isomorphic if and only if $h(p) = h(q)$. Intuitively, the entropy, which can more generally be defined for any dynamical system as defined above, measures the amount of information per unit time obtained by observing the system through a finite partition of the state space, so that it becomes an ***isomorphism invariant*** for dynamical systems. For the systems which are completely random, it is the only number needed to characterize them.

A complete mathematical proof of the above statement is beyond the scope of this exposition, although it is considerably simpler than the proofs of metrical isomorphism and gives more information. However, we can try to convey some of the flavor of the proof as follows. Suppose that p and q are two probability vectors with the same entropy, and denote by X and Y the sequence spaces of the corresponding Bernoulli schemes $BS(p)$ and $BS(q)$. Suppose also for simplicity that the symbol 1 possesses the same probability in both schemes: $p_1 = q_1$. We want to describe the renaming procedure, which takes a point x in X and turns it into a point y of Y. The first thing to do is to look at the infinite sequence x and copy all occurrences of the symbol 1 to the corresponding places in the point y we are looking for. With probability one, this now divides the remaining parts of x into finite strings of the remaining symbols. Using concepts of information theory, it is then possible to provide a sequential procedure to successively copy larger and larger blocks of these finite strings to the corresponding places in y, taking care that shifted objects are copied in the same manner.

Finally, we remark that the notion of finitary isomorphism is closely connected with the classes of observables which we consider. For this purpose, an *observable* should be thought of as a function defined on the state space of the dynamical system. In metric isomorphisms, the natural class of observables is the set of measurable functions, whereas in continuous isomorphisms the natural class of observables would be the continuous functions. For finitary isomorphisms, the natural class of observables is the set of functions which are continuous after removal of a set of measure zero, allowing many indicator functions of sets (those which have boundary measure zero) but still preserving the physical nature of closeness with probability one. Under such an isomorphism, these "almost continuous" functions are in one-to-one correspondence and hence their statistics behave in the same way in isomorphic dynamical systems. This property emphasizes the physical nature of the underlying systems.

Anomalous Diffusion and Lévy Statistics in Intermittent Chaotic Systems

J. Klafter[1], *G. Zumofen*[2] *and M.F. Shlesinger*[3]

[1] School of Chemistry, Tel-Aviv University, Tel-Aviv, 69978 Israel
[2] Physical Chemistry Laboratory, ETH-Zentrum, CH-8092 Zürich, Switzerland
[3] Physics Division, Office of Naval Research, Arlington, Virginia 22217-5660

Abstract: Deviations from simple Brownian motion have been observed in intermittent chaotic systems; of particular interest has been the case of enhanced diffusion. We review an approach to this anomalous behavior based on Lévy scale-invariant distributions to describe transport in such systems. We introduce the basic ingredients that make the approach useful in describing the non-Brownian behavior and demonstrate the applicability in the cases of the standard map, "egg-crate" potential and a one-dimensional iterated map which shows a combined laminar and dispersive motion.

1 Introduction

A great deal of interest has been devoted to the understanding of stochastic processes generated by deterministic systems with emphasis on the nature of particles' trajectories in such systems. In particular it has been observed that particles do not necessarily follow simple Brownian behavior and may even exhibit "strange kinetics" [1-14], namely dispersive or enhanced diffusional motion. The study of chaotic systems has opened therefore new challenges for theories of stochastic processes which reach beyond Brownian motion.

Examples of anomalous diffusion in dynamical systems cover both dissipative and Hamiltonian systems. One finds diffusion anomalies in numerical studies of one-dimensional maps [5, 6, 11, 15], of the Chirikov-Taylor standard map [1-4], of stochastic webs [7, 8] and in experiments on tracer diffusion in flow systems [16, 17, 18], where enhanced diffusion has been recently directly observed [18].

A new statistical description of anomalous diffusion has been recently introduced based on Lévy stable distributions which generalize the central limit theorem [10, 11, 19, 20]. This statistical description which we call the Lévy-walk approach introduces space-time coupling which, when combined with the Lévy distributions, leads to a useful description of anomalous transport and enables one to derive the corresponding propagator $P(r,t)$, the probability density to be at location r at time t. The Lévy-walk framework for enhanced diffusion has been implemented in a number of cases such as stochastic webs [7, 8], one-dimensional iterated maps [5, 11], hexagonal flows [21] and has also been introduced into the calculations of turbulent flows [22, 23]. More recently Lévy walks have been used in analyzing diffusion at solid-liquid interfaces [24], predicting a regime dominated by enhanced diffusion. The well known Cauchy distribution

$$\psi(x) = \frac{1}{\pi} \frac{1}{1+x^2} \ , \tag{1}$$

is an example of a probability distribution whose first moment is zero, but whose second and higher even moments are infinite. Furthermore, if n identically distributed random variables, each with a Cauchy distribution, are added, the sum $x_1 + x_2 + \ldots + x_n$ also has a Cauchy distribution, $P_n(x)$ with

$$P_n(x) = \frac{1}{\pi n} \frac{1}{1+(x/n)^2} \ , \tag{2}$$

where $P_n(x)$ and $\psi(x)$ are related by a change of scale from x to x/n. This is the hallmark of fractals, where the whole $P_n(x)$ (the distribution of the sum) looks like $\psi(x)$ (the distribution of a single variable). In viewing the sum of random variables as adding random walk steps this scaling is equivalent to finding the distribution for the position of a random walker after one step and after n steps, i.e.

$$\psi(x) = n^{\alpha} P_n(x n^{\beta}) \ , \tag{3}$$

with $\alpha = 1$ and $\beta = 1$. P. Lévy in the 1920s attacked more generally the problem of when a random walk distribution is independent (up to a scalefactor) of the number of steps taken. For an unbiased one-dimensional random walk he found that

$$P_n(k) = \int_{-\infty}^{\infty} P_n(x) e^{ikx} dx = \exp(-n|k|^{\gamma}) \ , \ 0 < \gamma \leq 2 \ . \tag{4}$$

Here and in what follows the argument indicates in which space the function is meant to hold, e.g. $P_n(k)$ is the Fourier ($x \to k$) transform of $P_n(x)$. Eq. (4) implies the scaling equation

$$\psi(x) = n^{1/\gamma} P_n(x n^{1/\gamma}) \ . \tag{5}$$

When $\gamma < 2$ the second moment of the probability diverges, so no proper characteristic scale exists and both $\psi(x)$ and $P_n(x)$ behave asymptotically as a power law $\sim |x|^{-\gamma-1}$. Again this is the signature of fractal behavior. The random-walk process which follows Eqs. (4) and (5) is known as the Lévy flight [25].

2 Lévy Flights and Lévy Walks

In this Section we look in more detail into the nature of Lévy flights and their relationship to anomalous diffusion. As mentioned, Lévy [25, 26, 27] considered a set $\{x_i\}$ of identically distributed random variables each governed by the probability density $\psi(x_i)$. He asked the question of when the new random variable x_0, given by

$$c_0 x_0 = c_1 x_1 + \ldots + c_n x_n \quad , \tag{6}$$

with c_i values being constants related by an auxiliary condition

$$c_0^\gamma = c_1^\gamma + \ldots + c_n^\gamma \quad , \tag{7}$$

has the same probability density as the x_i. For instance, for $i > 0$ choose $\langle x_i \rangle = 0$, $c_i = 1$, and $\gamma = 2$. Then $x_0 = n^{-1/2} \sum_{i=1}^n x_i$. This sum of n random variables properly normalized by the coefficient $n^{-1/2}$, has the Gaussian density, or $\psi(k) = \exp(-Dk^2)$, with D a constant. For $\gamma < 2$, Lévy showed that the solution is

$$\psi(k) = \exp(-D|k|^\gamma) \quad . \tag{8}$$

This corresponds to

$$\psi(x) \sim |x|^{-\gamma-1} \quad , \quad |x| \to \infty \quad , \tag{9}$$

that is the moments $\langle |x|^\mu \rangle$ are finite for $\mu < \gamma$ and are infinite for $\mu \geq \gamma$; in particular the variance diverges. These processes represent a random walker (with step i given by x_i) that visits a disconnected self-similar set of points for which the mean-squared displacement $\langle x^2 \rangle$ diverges.

We now introduce a discrete space Lévy flight to demonstrate, how the Lévy process results naturally from a scaling argument. Let us consider a random walk in a one-dimensional space and let $p(x)$ be the probability for a random-walk jump of displacement x. We choose [28, 29]

$$p(x) = \frac{N-1}{2N} \sum_{j=0}^{\infty} N^{-j} (\delta_{x,b^j} + \delta_{-x,b^j}) \quad , \tag{10}$$

with $N, b > 1$. Jumps of all orders of magnitude can occur in base b, but each successive order of magnitude in displacement occurs with an order of magnitude less probability in base N. The walker makes about N jumps of unit length before a jump of length b occurs, and a new cluster of sites visited starts to be generated. Eventually a fractal set of points is visited. To see this better, consider

$$p(k) = \frac{N-1}{N} \sum_{j=0}^{\infty} N^{-j} \cos(b^j k) \quad , \tag{11}$$

which is the Weierstrass example of a continuous but everywhere non-differentiable function for $b > N$. For small k,

$$p(k) = 1 - \frac{1}{2} k^2 \langle x^2 \rangle \quad , \tag{12}$$

if $\langle x^2\rangle = \sum x^2 p(x)$ is finite. If $b^2 > N$, then

$$\langle x^2\rangle = \sum_{j=0}^{\infty}(b/N)^j \quad , \tag{13}$$

diverges and Eq. (12) cannot be used. Instead note that

$$p(bk) = Np(k) - (N-1)\cos k \quad , \tag{14}$$

which has the following small k behavior

$$p(k) \simeq 1 - |k|^\gamma \simeq \exp(-|k|^\gamma) \quad , \tag{15}$$

with

$$\gamma = \frac{\ln N}{\ln b} \quad . \tag{16}$$

The exponent γ now appears in the form of a fractal dimension. For $\gamma = 2$, as in Eq. (12), the Gaussian behavior results while for $\gamma < 2$, $\langle x^2\rangle = \infty$ and the Lévy flights are obtained.

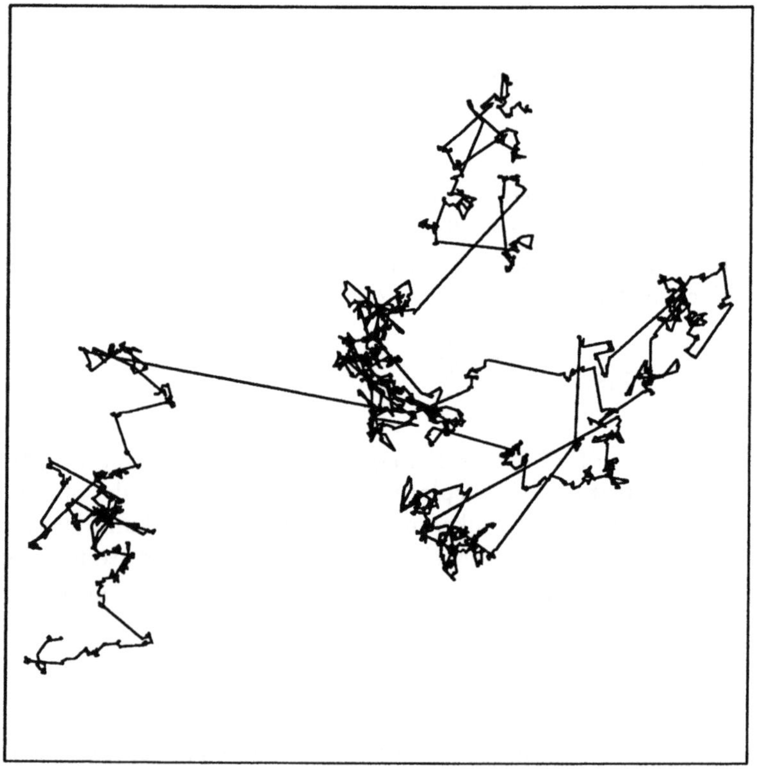

Fig. 1. The "Lévy flight": a typical trajectory of 10^4 jumps is shown with jump distances r chosen randomly from the power-law distribution $\psi(r)$, Eq. (21), with $\gamma = 1.5$. The turning points of the walk form a fractal structure of dimension γ.

A particle performing a Lévy flight jumps between sites, however distant, which leads to a divergence of the mean-squared displacement. One is interested in modifying the Lévy-flight law so that the motion still follows Eqs. (8) or (9) but with a finite mean-squared displacement. We shall therefore explore a stochastic process which visits the same sites as in the Lévy flight, but with a 'time cost' which depends on the distance. For the flight we need only to specify $\psi(x)$, the probability that a flight event takes place over distance x according to Eqs. (8) and (9). In the framework of the continuous time random walks (CTRW) [19, 20] we shall introduce a probability distribution $\psi(x,t)$ such that

$$\psi(x,t) = p(x|t)\psi(x) \tag{17a}$$
$$p(x|t) = \delta(|x| - t^{\nu}) \ , \tag{17b}$$

where $p(x|t)$ is the conditional probability to move a distance x in time t^{ν}. ν is an exponent which may take values smaller or larger than 1 and can be considered as a further parameter for the description of complex anomalous diffusion behavior [30]. Eqs. (17), within the CTRW approach, define Lévy walks which result in finite mean-squared displacements and account for the possibility of enhancement [20, 28, 31]. Alternatively, one can write Eq. (17a) as

$$\psi(x,t) = \frac{1}{2}\delta(|x| - t^{\nu})\psi(t) \ , \tag{18}$$

where now $\psi(t)$ is a time-dependent power-law. Other forms of conditional probability densities $p(x|t)$ were discussed in Ref. [19]. Lévy walks, as we see later, are characterized by a rich dynamical behavior depending on the power-law exponents of $\psi(x)$ or $\psi(t)$.

So far we considered one-dimensional flights and walks and also in the following Sections we will concentrate primarily on the one-dimensional case. Nevertheless, the extension to higher dimensions is possible [32]. Starting from Eq. (8) we may assume that $\mathbf{k}$ is a vector in d-dimensional space so that $|\mathbf{k}| = (k_x^2 + k_y^2 + ...)^{1/2}$. Assuming spherical symmetry we thus write for the distribution to move a distance r in Fourier space

$$\psi(\mathbf{k}) = exp(-D|\mathbf{k}|^{\gamma^*}) \ , \tag{19}$$

where γ^* is related to γ as we will show immediately. Eq. (19) is thus the extension of the one-dimensional distribution of Eq. (8) to the spherical symmetric case in d dimensions with $0 < \gamma^* \leq 2$. From the Fourier-back transformation we have for the radially symmetric distributions

$$\psi(r) = \begin{cases} \frac{1}{2\pi}\int\limits_{-\infty}^{\infty} exp(-D|k|^{\gamma^*} - ikr)dk & , \quad d=1 \\ \frac{1}{2\pi}\int\limits_{0}^{\infty} exp(-Dk^{\gamma^*})J_0(kr)dk & , \quad d=2 \\ \frac{1}{2\pi^2 r}\int\limits_{0}^{\infty} exp(-Dk^{\gamma^*})sin(kr)dk & , \quad d=3 \end{cases} \ , \tag{20}$$

which for large $|r|$ gives

$$\psi(r) \sim |r|^{-d-\gamma^*} \quad 0 < \gamma^* \leq 2 \ . \tag{21}$$

Comparing the exponent in this expression with that in Eq. (9) we can equate $\gamma^* = \gamma - d + 1$. A typical trajectory of a Lévy walker in two dimensions is displayed in Fig. 1.

In the next Section we introduce the Lévy walk formulation within the velocity model [11]. It should be emphasized that the CTRW formulation allows also for the description of dispersive transport and provides therefore a unified approach for both enhanced and dispersive motions. We combine these two behaviors in Section 5.

3 Lévy Walks: the Velocity Model

We now briefly outline the main ingredients in the Lévy-walk process. We choose the velocity picture in which the particle moves continuously at a constant velocity and changes directions at random [11]. Let $\mathbf{v}(t)$ denote a stochastic time dependent function such that the motion takes place at a constant velocity v for some time after which the direction and the length of the next motion event is chosen randomly but at the same velocity. The motion events are uncorrelated and the duration times of the motion events follow a probability distribution $\psi(t)$, the waiting time distribution. This idealized picture of the motion defines the probabilistic description of the stochastic process and was introduced in the CTRW approach [11, 19, 20]. The approach is based on a space-time coupled memory function $\psi(\mathbf{r}, t)$, the probability to move a displacement $\mathbf{r}$ in time t in a single motion event and to stop at $\mathbf{r}$ for initiating a new motion event at random. We use

$$\psi(\mathbf{r}, t) = \frac{1}{2}\delta(|\mathbf{r}| - t)\psi(t) \ , \tag{22}$$

where the delta-function accounts for the motion at a constant velocity and where length and time are given in dimension-less units. For $\psi(t)$ we consider a power law

$$\psi(t) \sim t^{-\gamma-1} \ , \quad \gamma > 0 \ , \tag{23}$$

We also introduce $\Psi(\mathbf{r}, t)$, the probability density to move a displacement $\mathbf{r}$ in time t in a single motion event and not necessarily to stop at $\mathbf{r}$

$$\Psi(\mathbf{r}, t) = \frac{1}{2}\delta(|\mathbf{r}| - t) \int_t^\infty \psi(\tau) d\tau \ . \tag{24}$$

$\psi(\mathbf{r}, t)$ and $\Psi(\mathbf{r}, t)$ are the relevant quantities for characterizing the motion. The motion consists of a sequence of these events and thus the propagator $P(\mathbf{r}, t)$ can be cast in the following way

$$P(\mathbf{r}, t) = \Psi(\mathbf{r}, t) + \int_{-\infty}^{\infty} \int_0^t \psi(\mathbf{r}', t')\Psi(\mathbf{r} - \mathbf{r}', t - t') d\mathbf{r}' dt' + \ldots \ . \tag{25}$$

The first term denotes the probability to reach location $\mathbf{r}$ in time t in a single motion event. The second term is the probability to reach $\mathbf{r}$ at time t with one stop and so on to include all combinations of motion events. In the Fourier-Laplace space ($\mathbf{r} \to \mathbf{k}, t \to u$) the convolution integrals simplify and the series in Eq. (25) can be given in a closed form as

$$P(\mathbf{k}, u) = \frac{\Psi(\mathbf{k}, u)}{1 - \psi(\mathbf{k}, u)} \quad . \tag{26}$$

A more detailed derivation of $P(\mathbf{r}, t)$ including dependence on initial conditions was reported in Refs. [11, 33].

From the propagator we calculate the time evolution of the mean-squared displacements

$$\langle r^2(t) \rangle = -\frac{\partial^2}{\partial \mathbf{k}^2} P(\mathbf{k}, u)\Big|_{\mathbf{k}=0} \quad . \tag{27}$$

Depending on γ which governs the decay of the tails of $\psi(t)$ we distinguish among three characteristic motion regimes: the ballistic-type, the intermediate-enhanced and the regular Brownian-type diffusion

$$\langle r^2(t) \rangle \sim \begin{cases} t^2 & , \quad 0 < \gamma^* < 1 \\ t^{3-\gamma^*} & , \quad 1 < \gamma^* < 2 \\ t & , \quad 2 < \gamma^* \end{cases} \quad . \tag{28}$$

Thus the Brownian motion is a special case in this Lévy-walk description. Here we concentrate on the intermediate enhanced diffusion regime $1 < \gamma^* < 2$ for which the asymptotic long-time behavior of the propagator has been shown to obey scaling [11]. Here for the spherical symmetric representation we have

$$P(r, t) \sim f(\xi)/t^{1/\gamma^*} \quad , \tag{29}$$

where ξ is the scaling variable, $\xi = |\mathbf{r}|/t^{1/\gamma^*}$, and where $f(\xi)$ is the scaling function. This function follows a Lévy distribution which to lowest order in small and large ξ values is given by

$$f(\xi) \simeq \begin{cases} \exp(-c\xi^2) & , \quad \xi < 1 \\ \xi^{-\gamma-1} & , \quad 1 < \xi \, , \; |r| \lesssim t \\ 0 & , \quad |r| > t \end{cases} \quad . \tag{30}$$

where the cutoff at $|r| = t$ is a consequence of the constant velocity. The fact that $f(\xi)$ is characterized by a Lévy distribution for $|r| < t$ is a manifestation of the generalization of the central limit theorem. For the propagator corresponding to $0 < \gamma^* < 1$ the reader is referred to Ref. [11]. We recall that in this Section we extended the considerations to higher dimensions with $\gamma^* = \gamma - d + 1$; in the next two Sections we will concentrate on the one-dimensional case so that $\gamma^* = \gamma$.

4 Lévy Walks in the Standard Map and in the Egg-Crate Potential

We now show that diffusion generated in the standard map provides an instructive example for the Lévy-walk statistics in conservative systems. The standard map which can be regarded as a discretized Hamiltonian for the kicked-rotor is a two-dimensional map of the form [2, 34]

$$x_{n+1} = x_n + K sin(2\pi\theta_n) \ , \ \ \theta_{n+1} = \theta_n + x_{n+1} \bmod 1 \ . \tag{31}$$

This map has been shown to have ranges of the stochasticity parameter K for which the motion along the x-direction is enhanced [35, 36]. This is due to the accelerating modes of various periods which give rise to islands of stability. These islands are encircled by cantori which separate areas of quasi- stability. Islands and cantori show some hierarchical self-similar structure [1-5, 35-37]. Orbits approaching these islands tend to stick to the critical tori and become quasi stable for some time such that laminar phases result. Examples of chaotic motion and of laminar phase motion are presented in Fig. 2. Due to repeated visits to the self-similar areas of quasi-stability an intermittent behavior results which is characterized by broad distributions of sticking times [1-4] which correspond to the previously introduced $\psi(t)$, Eq. (23). These broad distributions cause the motion to be anomalous and to be a promising candidate to be described in terms of Lévy walks.

Enhanced diffusion in the standard map is observed to be dominated by accelerating modes of period p [35, 36]. The acceleration is due to performing ℓ step units along x in p iterations. For the K values studied here, enhanced diffusion results from the fundamental modes for which $|\ell| = p$. This gives rise to a constant velocity $|v| \simeq |\ell|/p = 1$. Therefore sticking to the islands means being locked in a mode of constant velocity. Thus the motion is characterized by laminar phases of constant velocity interrupted by short periods of chaotic motion in the chaotic sea, which makes the space-time coupled waiting time distribution apply. The self-similar properties of the trajectory are demonstrated in Fig. 3.

In our numerical investigations we considered the $p = 3$ and $p = 5$ (period three and period five fundamental) accelerating modes with K values 1.1 and 1.03, respectively [14, 36]. Our analysis of the waiting time distribution $\psi(t)$ calculated for orbits starting from the chaotic sea revealed that the sticking time distributions follow approximately power laws and that the asymptotic behavior starts to hold for times $t \geq 10^2$, as displayed in Fig. 4, for both $p = 3$ and $p = 5$. In the figure plotted are $\psi(t)$ and $\psi(r)$ which were calculated independently from the segments of laminarity in orbits of typical length of 10^{11} number of iterations. Here r denotes the displacement between two locations: $r(t; n) = x_{n+t} - x_n$. The statistics obtained from a few long orbits or from many shorter orbits did not significantly differ. From the figure it is evident that $\psi(t)$ and $\psi(r)$ fall on top of each other for times $t \geq 10^2$. This collapse coincides with the inset of the power-law decays of the distributions. For times $t \geq 10^2$ we observe exponents

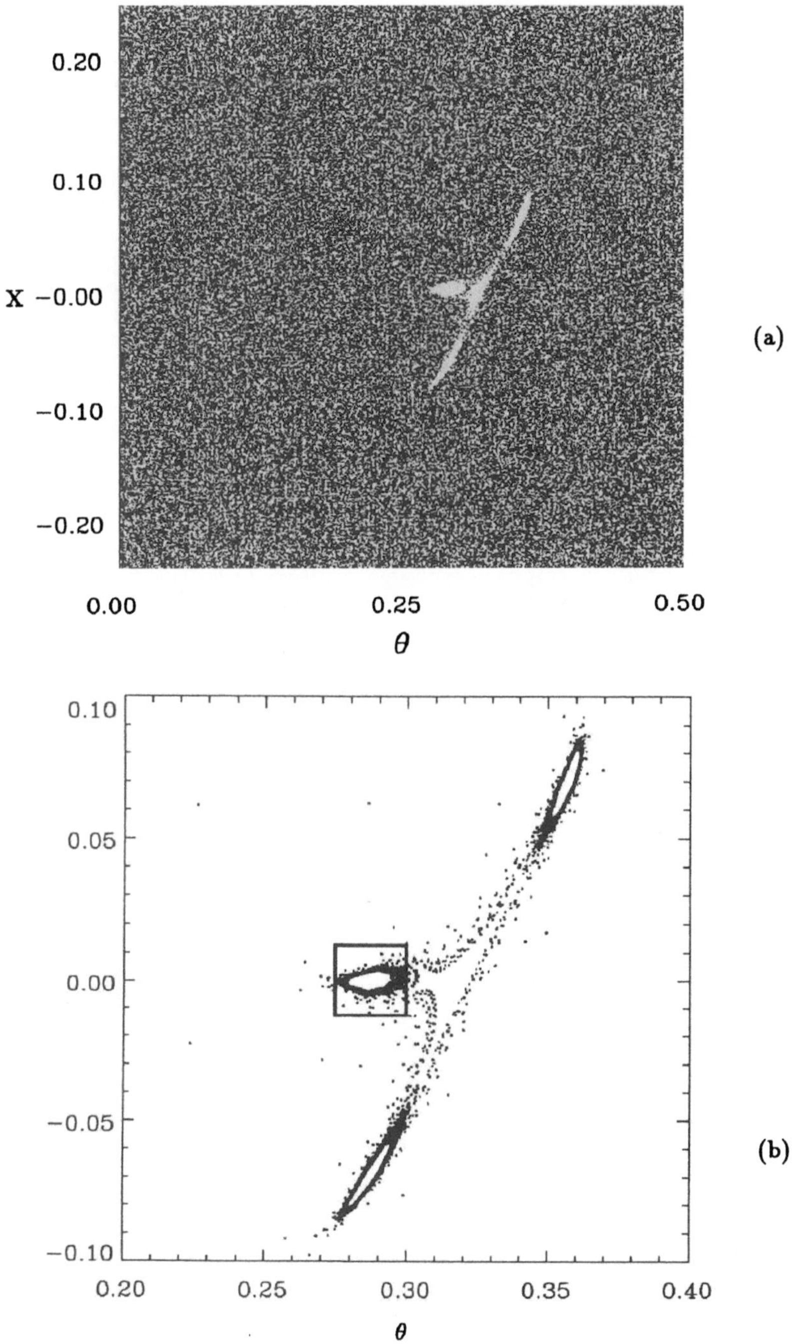

Fig. 2a,b. Trajectories of the iterated standard map for the $p = 3$ mode with $K = 1.1$. In the upper shown is the trajectory for motion in the chaotic sea. The bright area indicates the unreachable area of the islands of stability. The next figures: 2b,c,d display a trajectory of laminar motion for 10^5 iteration steps.

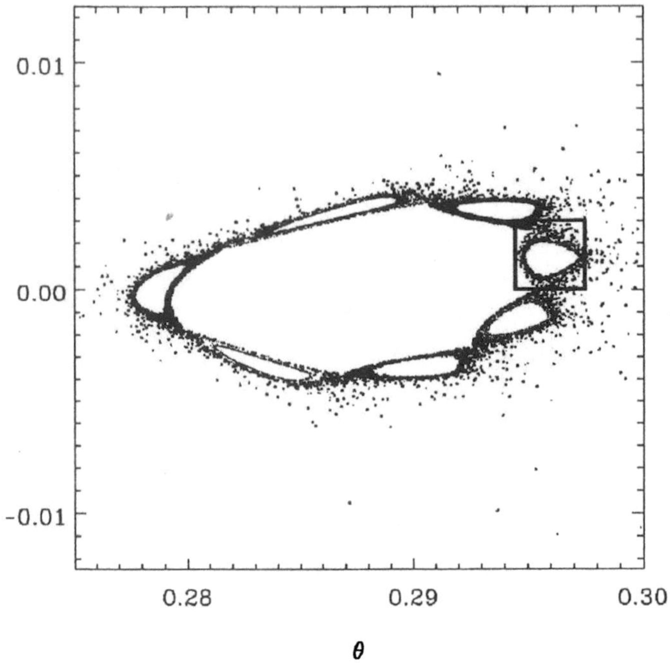

(c)

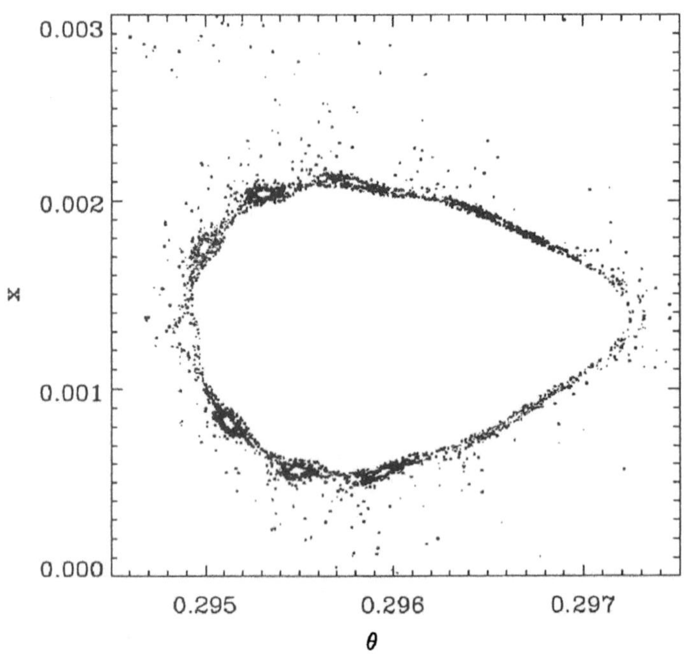

(d)

Fig. 2c,d. In fig. 2b,c,d three generations of islands of stability are shown. The inserted frames indicate the window of the next step of zooming-in.

which correspond to $\gamma \simeq 1.8$ for $p = 3$ and $\gamma \simeq 1.2$ for $p = 5$. If one chooses different time-windows different exponents of values around 1.75 are obtained. Exponents in the same range were obtained in Refs. [2, 3, 4] where the Poincaré recurrences were studied.

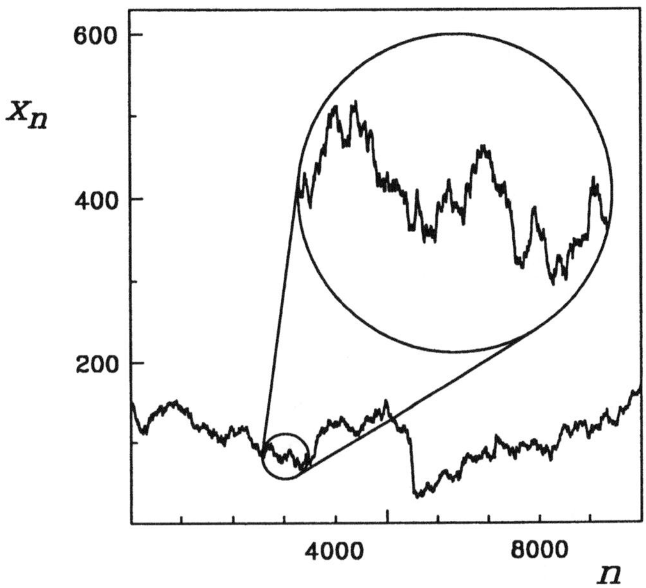

Fig. 3. Typical trajectory obtained from the iterated standard map, Eqs. (31) for the period-five mode. Plotted are x_n for 1000 iteration steps. The stochasticity parameter was set to $K = 1.04$ so that the motion is governed by the fundamental $p = |\ell| = 5$ mode. The self-similar property of the trajectory is obvious from the insert.

We point out that the collapse of $\psi(t)$ and $\psi(r)$ for $t \geq 10^2$ confirms the assumption that the motion is governed by fundamental accelerating modes at longer times and that the motion during the laminar phases takes place at a constant velocity. This in turn justifies the δ-function approach in Eqs. (22). In order to further investigate the sticking properties of the orbits near the islands of stability we carried out detailed computer calculations of the distribution $\phi(t)$ of exit times. $\phi(t)$ is the distribution of times it takes an orbit with initial coordinates placed close to the islands of stability to reach the chaotic sea. It turns out that $\phi(t)$ also follows a power-law, however, with an exponent approximately equal to γ being smaller than that of the corresponding $\psi(t)$ distribution by one [36].

An intuitive argument that leads to power-law behaviors of $\psi(t)$ and $\phi(t)$ can be given by following Hanson et al. [37] who make use of a self-similar system of rate equations for the Markovian motion in the hierarchical structure of belts, i.e. of zones of quasi-stability. In a simplified way we write

$$\dot{N}_j = p_{j-1\to j}N_{j-1} + p_{j+1\to j}N_{j+1} - (p_{j\to j-1} + p_{j\to j+1})N_j \ , \tag{32}$$

where N_j is the occupation probability of belt j and $p_{j\to j'}$ is the probability for the transition to take place from belt j to j'. Assuming a hierarchy in the probabilities p and also a hierarchy of belt areas one has

$$\frac{p_{j\to j+1}}{p_{j-1\to j}} = w \ , \ \frac{p_{j\to j+1}}{p_{j+1\to j}} = a \ , \tag{33}$$

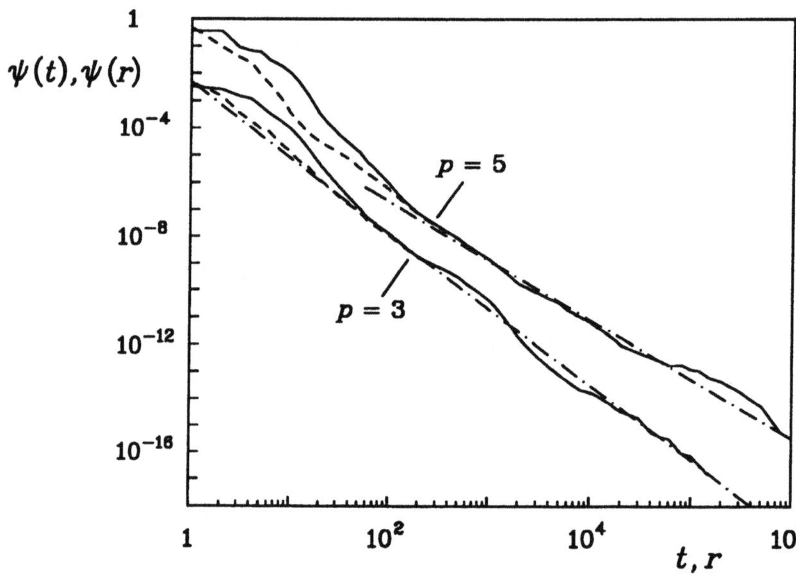

Fig. 4. Probability distributions for the standard map, Eqs. (31). Solid lines give $\psi(t)$, dashed lines are $\psi(r)$ and dash-dotted lines indicate the power-law behavior with exponents $\gamma = 1.8$ for $p = 3$ and $\gamma = 1.2$ for $p = 5$.

where w and a are the scaling parameters for the rates and for the areas, respectively. Introducing a hierarchy of time intervals one can approximate the probability density $\psi(t)$ of a trajectory to enter the system of belts from the chaotic sea and to stay in the system for the time t before returning to the chaotic sea by a sum of exponentials

$$\psi(t) \sim \sum_{j=1}^{\infty} (aw^2)^{-j} \exp(-w^j t) \ . \tag{34}$$

The overall behavior of this sum is known to follow a power-law, $\psi(t) \sim t^{-\gamma-1}$ with the exponent γ being related to a and w by $\gamma = 1 + \ln(a)/\ln(w)$ [36, 37].

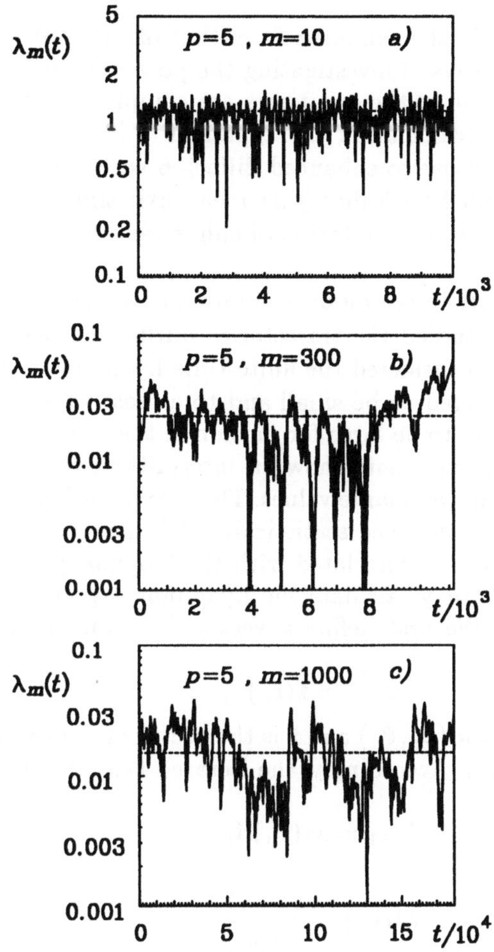

Fig. 5. The finite-time Lyapunov exponents for the standard map. Plotted are $\lambda_m(t)$ as full lines obtained for the stochasticity parameter equal to $K = 1.03$ and for various m values as indicated. Fig. 5a corresponds to the situation of chaotic motion, Figs. 5b and 5c to the motion in laminar phases of 10^4 and 10^5 iterations steps, respectively. The dashed lines give the average values $\langle\lambda\rangle$ with the average taken over the time window presented in the figures. Note the logarithmic vertical scales and also the changes in the vertical scale from Fig. 5a to 5b and in the horizontal scale from Fig. 5b to 5c.

Obviously, Equation (34) is a first step approximation, a more involved approach is required for a detailed analysis [6, 21, 37].

We now consider the exit times, for which the initial coordinate of the trajectory is chosen randomly in space. The corresponding initial condition in the system of cantori is given by a random choice of the belts. The approach, based on the sum of exponentials, has to be modified accordingly, namely the weights $(aw^2)^j$ in Eq. (34) have to be replaced by $(aw)^j$. This leads again to a power-law,

$\phi(t) \sim t^{-\gamma}$, which differs in the exponent from that of the waiting times by the amount of one. The two ways of investigating the power-law behavior, based on times of laminar phases and of the exit times are complementary and represent independent methods for calculating the exponent γ [36]. We point out that laminar phases which give rise to enhanced diffusion with γ-values in the range of $1 < \gamma < 2$ are associated with diverging mean exit times. This might have some implications for the in interpretation of enhanced diffusion in conservative systems.

In order to give support to the model of motion in a system of hierarchically nested cantori, we have studied the trajectories during the periods of laminar phases. For this study we considered the finite-time Lyapunov exponents $\lambda_m(t)$. These exponents are expected to be small and to almost vanish during the period of laminar motion and to be large for motion in the chaotic sea. As a result the finite-time Lyapunov exponents show an intermittent behavior of jumping stochastically between small and large values. The stochastic behavior of the Lyapunov exponents has been analyzed statistically [39], here we only demonstrate that the stochastic behavior is correlated with the laminarity of the motion.

For iterated maps one may calculate the Lyapunov exponents from the tangential vector iteration. We first define a vector function for the iteration so that

$$\mathbf{r}_{n+1} = \mathbf{f}(\mathbf{r}_n) \quad , \tag{35}$$

where $\mathbf{r}_n$ denotes the vector (x_n, θ_n) and $\mathbf{f}$ is the vector of the two corresponding map functions in Eqs. (31). Accordingly, the iterated tangent vector $\mathbf{d}_n$ is given by

$$\mathbf{d}_{n+1} = \mathbf{M}(\mathbf{r}_n)\mathbf{d}_n \quad , \tag{36}$$

with

$$\mathbf{M}(\mathbf{r}_n) = \frac{\partial \mathbf{f}(\mathbf{r}_n)}{\partial \mathbf{r}_n} \quad , \tag{37}$$

From the tangent vectors $\mathbf{d}_n$ the Lyapunov exponents are calculated as

$$\lambda(t) = \ln \frac{|\mathbf{d}_{t+1}|}{|\mathbf{d}_t|} \quad , \tag{38}$$

which denotes the finite-time Lyapunov exponent at time t for one single iteration step. For m iterations we have

$$\lambda_m(t) = \frac{1}{m} \ln \frac{|\mathbf{d}_{t+m}|}{|\mathbf{d}_t|} = \frac{1}{m} \sum_{j=0}^{m-1} \lambda(t+j) \quad . \tag{39}$$

The asymptotic long-time Lyapunov exponent follows from $\lambda_\infty = \lim_{m\to\infty} \lambda_m(t)$. In non-contracting conservative system $\lambda_\infty > 0$ for orbits starting in the chaotic sea and $\lambda_\infty = 0$ for trajectories initiated in the areas of stability.

In Figs. 5 shown are the finite-time Lyapunov exponents $\lambda_m(t)$ for three different situations. In Fig. 5a a typical behavior is shown for an arbitrarily chosen time window. In Fig. 5b and 5c the behavior is plotted for laminar phases of 10^4 and 10^5 iteration steps. The horizontal dashed lines indicate the average

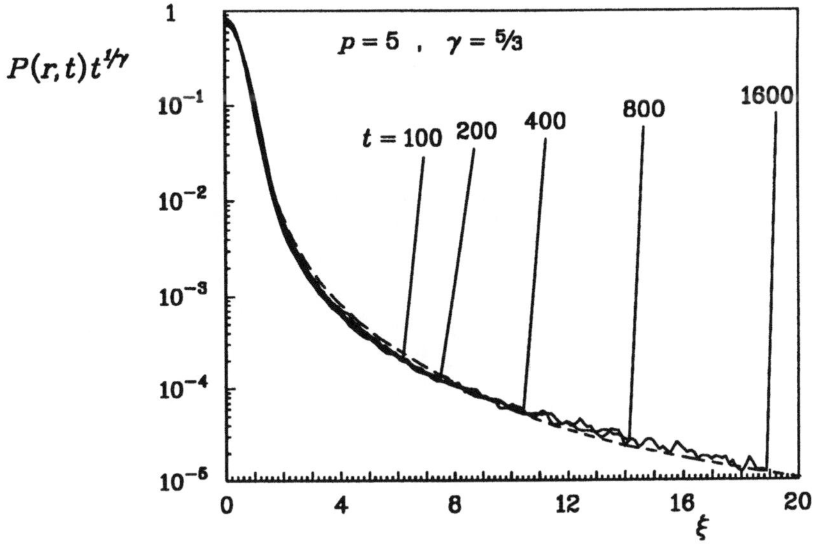

Fig. 6. The propagator for the standard map. Plotted are as solid lines the simulation results in the scaling representation for times as indicated. The stochasticity parameter was set to K=1.03 so that the $p = |\ell| = 5$ mode governs the motion. The dashed line is the stable law $cL_\gamma(cr)$ with $\gamma = 5/3$ and with c introduced as an adjustable parameter. The sharp peaks at the outermost wings are due to stationary conditions considered in the averaging procedure [33].

value $\langle\lambda\rangle$ with the average taken over the time period shown in the figure, i.e. $\langle\lambda\rangle = \lambda_{\Delta t}(t)$ with Δt equal to 10^3, 10^3, and 10^5 in Figs. 5a, 5b, and 5c, respectively. We notice that for the chaotic motion in Fig. 5a $\langle\lambda\rangle \simeq 1$, while in Figs. 5b and 5c $\langle\lambda\rangle$ is smaller by two orders of magnitude. The m-values in the Figs. 5 were chosen so that unnecessary noise is reduced but the nature of the stochastic behavior is still visible.

In all three cases shown in Fig. 5 the behavior is characterized by short periods of small and very small λ values. With increasing length of the laminar phase the jumps to small values are more pronounced. We interpret this jumps as resulting from visits to belts closer to the stability islands and we argue that the behavior of the finite-time Lyapunov exponents supports the picture of motion in a system of hierarchically organized cantori, as considered in Eqs. (32) and (33).

We continue by discussing the propagator $P(r,t)$ which was calculated from $P(r,t) = \langle\delta(r - x_t)\rangle$ with the average taken over trajectories of 10^{11} steps. In Fig. 6 the propagator $P(r,t)$ is shown for the $p = 5$ period mode in the scaling representation for the time regime $10^2 < t < 2 \times 10^3$. The numerical data are compared with the theoretical $P(r,t)$ of Eqs. (29) and (30); for $\gamma = 5/3$ a reasonable fit is achieved. We notice that the curves for different times fall on top of each other; this data collapse indicates that scaling holds at least in this

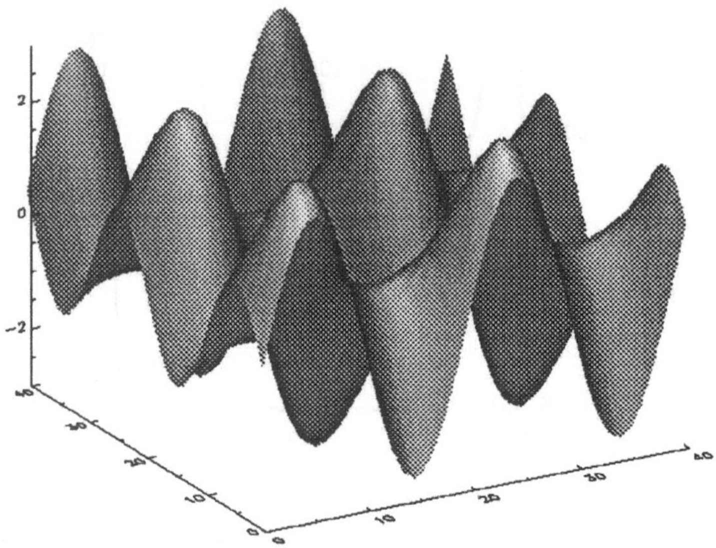

Fig. 7. The egg-crate potential. Plotted is the potential $V(x,y)$, Eq. (40), as a function of x and y for $A=2.5, B=1.5$, and $C=0.5$.

intermediate time regime. Furthermore, we observe that the decay in the wings follows a power-law, thus demonstrating the Lévy behavior. We thus conclude that the chaotic dynamics in the standard map is intimately related to the Lévy processes.

Another example of anomalous diffusion which follows the behavior of Lévy walks is given by the friction free motion subject to an egg-crate potential in two-dimensions

$$V(x,y) = A + B(\cos x + \cos y) + C \cos x \cos y \ . \tag{40}$$

In this potential the third term is responsible for the non-integrability of the corresponding Hamiltonian. In Fig. 7 the egg-crate potential is shown in a three-dimensional presentation. This potential can bee considered for modeling the persistent diffusion behavior observed for atoms or molecules on crystal surfaces [40].

The motion in the egg-crate potential has been studied by Geisel et al. [6] and by Chernikov et al. [7] and has been shown to exhibit both regular and enhanced diffusion, namely the mean-squared displacements which grow as $\langle r^2(t)\rangle \sim t^\alpha$, $1 \leq \alpha < 2$. It has been shown that the motion in the above potential also belongs to the class of Lévy walks and can be studied in terms of the CTRW

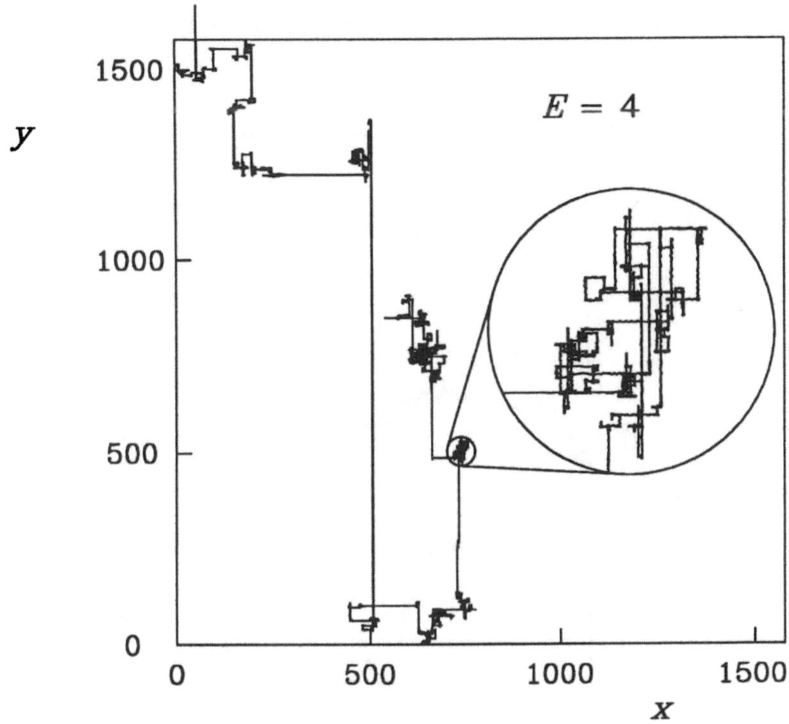

Fig. 8. The trajectory in the egg-crate potential. Plotted is an example of the two-dimensional trajectory $\mathbf{r}(t)$ obtained from the equations of motion for $E = 4$ and $t = 10^5$. The insert shows a part of the trajectory on an enlarged scale to strengthen the impression of the self-similarity.

formalism [6, 41]. The numerical investigation was based on solving numerically the equation of motion for different energies E; for details see Ref. [41].

In Fig. 8 we present a typical trajectory obtained from the numerical solution of the equations of motion for the constants being $A = 2.5$, $B = 1.5$, $C = 0.5$ and the energy $E = 4$. We notice the self-similar nature of the trajectory with laminar phases on all scales. The insert supports the impression and also shows that on the scale of a unit cell the trajectory follows a wiggly curve.

Another interesting property of the trajectories resulting from the particular shape of the egg-crate potential is the coexistence of phases of laminar motion with phases of localization. This aspect is better visible in Fig. 9 where the two-dimensional trajectory $\mathbf{r}(t)$ is plotted as a one-dimensional trajectory $x(t) + y(t)$ for two different energies. This presentation is reasonable because the motion is predominantly coaxial, that is, the motion occurs either parallel to the x or to y axis. From the figure we notice that for $E = 2.5$ phases of laminar motion are interrupted by periods of localization while for $E = 4$ these periods are hardly visible. The periods of localization are the subject of the next Section.

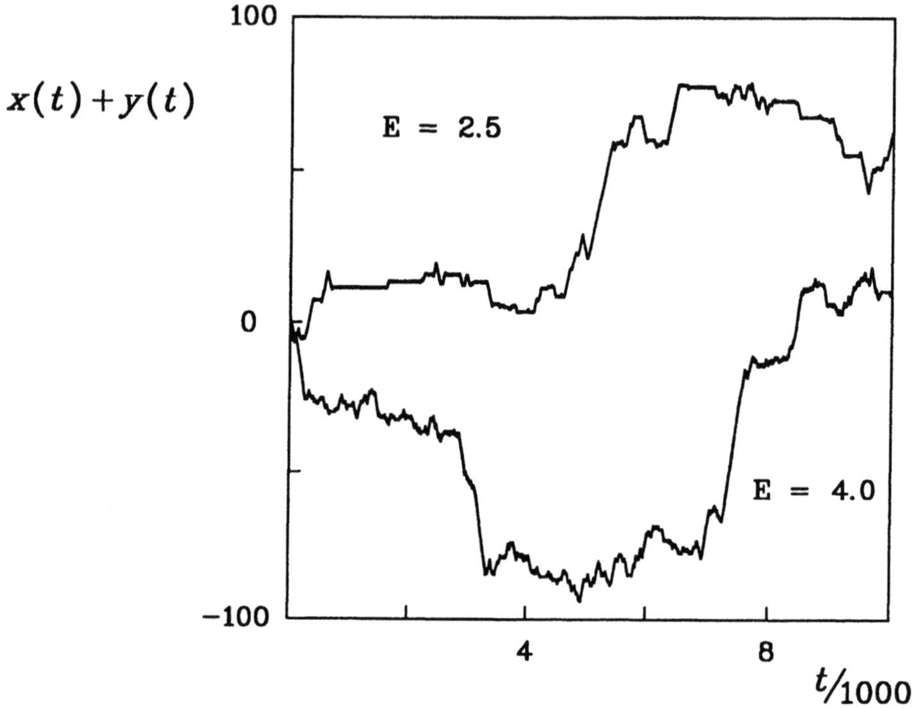

Fig. 9. Two typical one-dimensional trajectories $x(t)+y(t)$ in the egg-crate potential for energies as indicated.

5 Coexistence of Laminar and Dispersive Motions

We have concentrated hitherto on one aspect of diffusion in dynamical systems, that of enhanced motion. It has been recognized however, that dispersive transport for which the mean-squared displacement grows sub-linearly with time $\langle r^2(t)\rangle \sim t^\alpha$, $0 < \alpha < 1$, is also generated by some dynamical systems [11, 42]. Also dispersive transport can be described within the CTRW framework [11, 43, 44], however, with a broad distribution of waiting times which correspond to time spent at spatial locations in contrast to the broad distributions of time being locked in laminar phases, Eqs. (17).

An interesting situation occurs when there is a coexistence of laminar and dispersive (localized) phases. The trajectories in such cases demonstrate the interplay between the two competing motion modes and can be analyzed in terms of Lévy statistics. In a recent experiment by Solomon et al. [18] on tracer particles in a two-dimensional rotating flow it has been observed that dispersive and enhanced modes of motion can coexist in a way that a particle may perform long flights and be also intermittently trapped in space. This behavior is reminiscent of the observations for two-dimensional Hamiltonian systems discussed in the previous Sections namely: motion in the Chirikov-Taylor map and in the egg-crate potential. The coexistence of laminar and localized modes has been addressed by Chaikovosky and Zaslavsky [8] and Zaslavsky [45] using fractional Fokker-Planck equations.

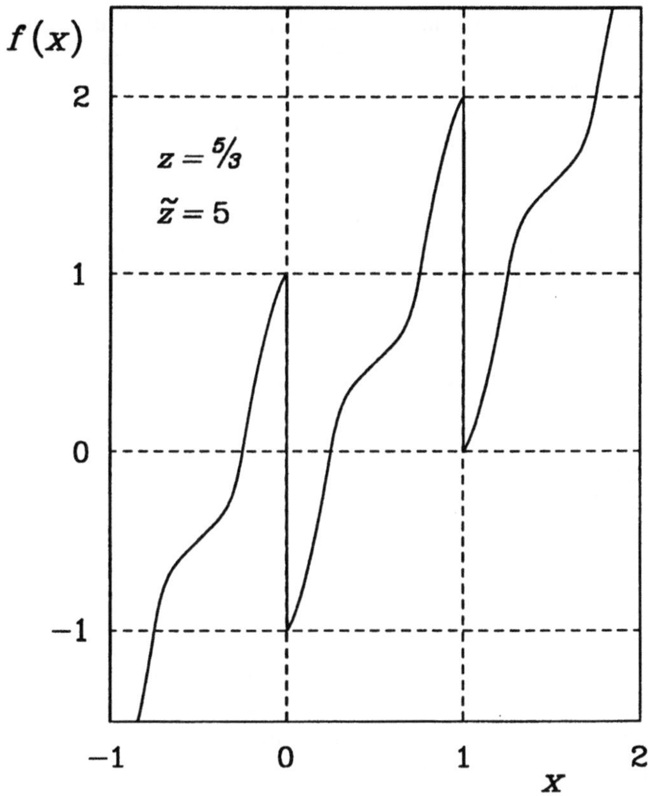

Fig. 10. The map function f(x) for the one-dimensional combined map, Eq. (41), for $z = 5/3$ and $\tilde{z} = 5$, as indicated corresponding to the exponents $\gamma = 3/2$ and $\tilde{\gamma} = 1/4$, respectively.

Here we introduce a one-dimensional map, which is characterized by intermittent chaotic motion with coexisting dispersive and laminar motion events [46] and we demonstrate the applicability of the random-walk scheme with Lévy stable-law distributions in analyzing the motion generated by this map. We show how the competing trends of laminar and localized phases lead to diffusional behavior and cover the whole range of dispersive, regular and enhanced behaviors.

Fig. 10. shows an example of the one-dimensional map which is defined as

$$f(x) = \begin{cases} (1+\epsilon)x + ax^z - 1 & , \quad 0 \le x \le \frac{1}{4} \\ (1+\tilde{\epsilon})x - \tilde{a}(\frac{1}{2} - x)^{\tilde{z}} & , \quad \frac{1}{4} \le x \le \frac{1}{2} \end{cases} . \tag{41}$$

Translational and inversion symmetries are inferred so that diffusional motion from one to the other unit cell is enabled. This map represents a combination of two maps introduced previously [5, 11, 42]. The map is discontinuous at the boundaries of each box but is continuous otherwise. For $\epsilon = \tilde{\epsilon} = 0$ the map shows marginal stable fixed points at $x = 0$ and $x = \frac{1}{2}$. The neighborhood of the fixed point at $x = 0$ is responsible for the laminarity, while the fixed point at $x = \frac{1}{2}$ gives rise to localization. The exponents z and $\tilde{z}$ determine the characteristic behaviors of the two types of events, free flights and localization, respectively. The prefactors a and $\tilde{a}$ are weights and are chosen so that the map $f(x)$ is continuous at $x = \frac{1}{4}$ including the first derivative. For $\epsilon = \tilde{\epsilon} = 0$ we used $a = 4^z\tilde{z}/(z+\tilde{z})$ and $\tilde{a} = 4^{\tilde{z}}z/(z+\tilde{z})$. ϵ and $\tilde{\epsilon}$ were considered to be small quantities and to differ from zero in the cases $z > 2$ and $\tilde{z} > 2$, respectively, in order to avoid problems in the numerical realization of the statistical analyses. This relatively simple map displays a rich spectrum of behaviors and a unique interplay of the two modes of motion.

In Fig. 11 we present two typical trajectories generated by the map function Eq. (41). Again r denotes the displacement between two locations: $r(t;n) = x_{n+t} - x_n$. The interchange of laminar and dispersive (localization, no motion) behavior is evident and the behavior is reminiscent for the trajectories in the egg-crate potential but also for the observations, although in a completely different type system, that have been reported by Solomon et al. [18]. In this respect our map generates statistical properties that are amenable to experimental observations. What relate the different cases to each other are the underlying Lévy stochastic processes.

The interchange between laminar and localizing periods is also visible in Fig. 12 where the velocities are shown as a function of time. The intermittent characteristics is very prominent in this figure and the observation of this figure can be compared with similar results of other systems [47]. In the lower part of the figure shown is the analog of an idealized picture where the velocity takes the values −1, 0, and 1. The idealized picture serves as a basis for the CTRW description.

In Fig. 13 the distributions of flight times and of localizing times obtained from the simulation calculations are shown. They follow reasonably well power-laws

$$\psi(t) \sim t^{-\gamma-1} \text{ and } \tilde{\psi}(t) \sim t^{-\tilde{\gamma}-1} , \tag{42}$$

respectively, with the exponents $\gamma = (z-1)^{-1}$ and $\tilde{\gamma} = (\tilde{z}-1)^{-1}$. The tilde is considered to mark the localization events.

For the analysis we consider the velocity picture introduced above and extend it accounting for phases of spatial localization. This means that the particle does not move at a constant velocity at all times but that the phases of laminar motion are intermittently interrupted by periods of no motion on the scale of typically

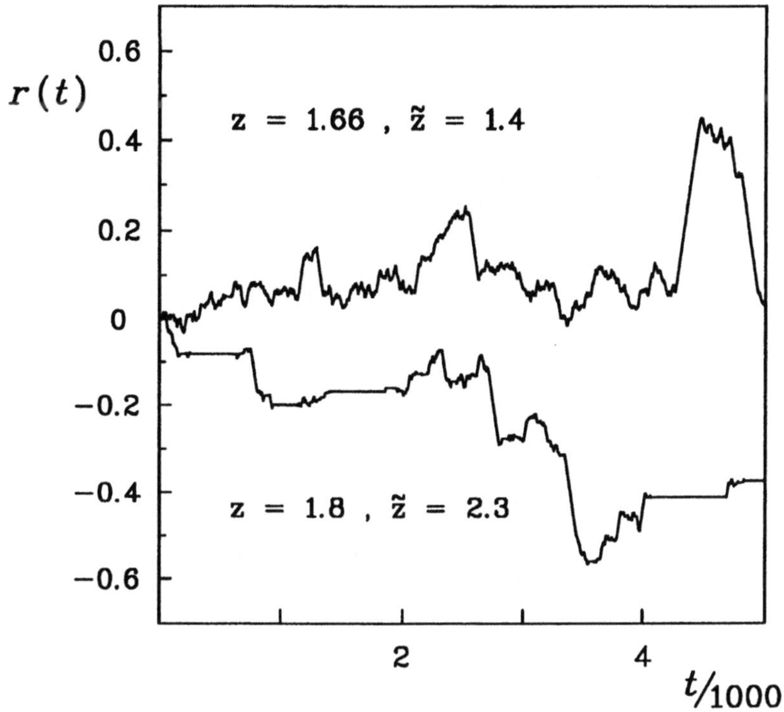

Fig. 11. Two typical trajectories from the iterated combined map, Eq. (41). The upper trajectory is obtained for values $z = 1.66$ ($\gamma = 1.5$) and $\tilde{z} = 1.4$ ($\tilde{\gamma} = 2.5$) which give rise to enhanced diffusion with $\alpha = 1.5$. The lower trajectory is obtained for values $z = 1.8$ ($\gamma = 1.25$) and $\tilde{z} = 2.33$) ($\tilde{\gamma} = 0.75$) giving rise to enhanced diffusion again with $\alpha = 1.5$.

one box. The probability distributions $\psi(r,t)$, Eq. (22) and $\Psi(r,t)$, Eq. (24) introduced for the velocity model, are complemented by $\tilde{\psi}(t)$ the probability distribution of times of no motion and by

$$\tilde{\Psi}(t) = \int_t^\infty \tilde{\psi}(\tau) d\tau \tag{43}$$

the probability for not having moved at all until time t. In the description of the propagator $P(r,t)$, the probability density to be at location r at time t, we assume that the observation starts with an event of motion at constant velocity; we obtain

$$P(r,t) = \Psi(r,t) + \int_0^t \psi(r,t')\tilde{\Psi}(t-t')dt'$$

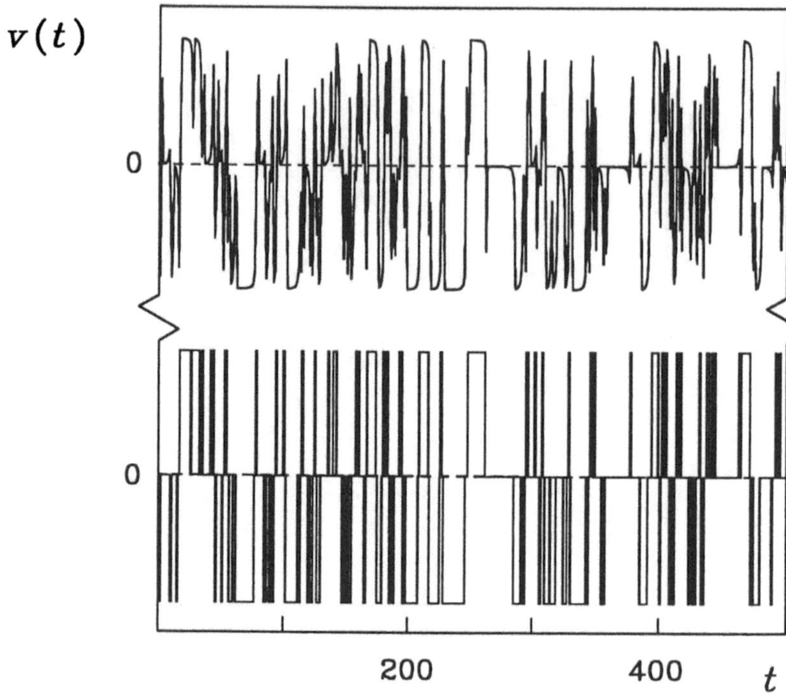

Fig. 12. Typical time dependent velocities $v(t) = x_{t+1} - x_t$ for the combined map of Eq. (41). In the upper part the results of the iteration is shown for $z = \tilde{z} = 1.8$ corresponding to $\gamma = \tilde{\gamma} = 1.25$. In the lower part the idealized motion is depicted which is considered in the CTRW analysis. The parameters chosen give rise for an enhanced diffusion with exponent $\alpha = 1.75$.

$$+ \int_{-\infty}^{\infty} dr' \int_0^t dt' \int_0^{t'} dt'' \psi(r', t'') \tilde{\psi}(t' - t'') \Psi(r - r', t - t') + \dots , \tag{44}$$

where the first term denotes the probability to reach location r in time t in a single motion event. The second term is the probability to reach r at an earlier time and to stay localized until time t. The third term is the probability to reach r in time t in two motion events interrupted by one period of localization. The sum has to be extended over all possible combinations of motion events interrupted by periods of localization. Taking the Fourier-Laplace transform and summing over even and odd terms independently we obtain

$$P(k, u) = \frac{\Psi(k, u) + \tilde{\Psi}(u) \psi(k, u)}{1 - \tilde{\psi}(u) \psi(k, u)} \quad . \tag{45}$$

A similar expression is obtained when the walks are initiated by a localization event followed by motion at constant velocity. For the description of the mean-

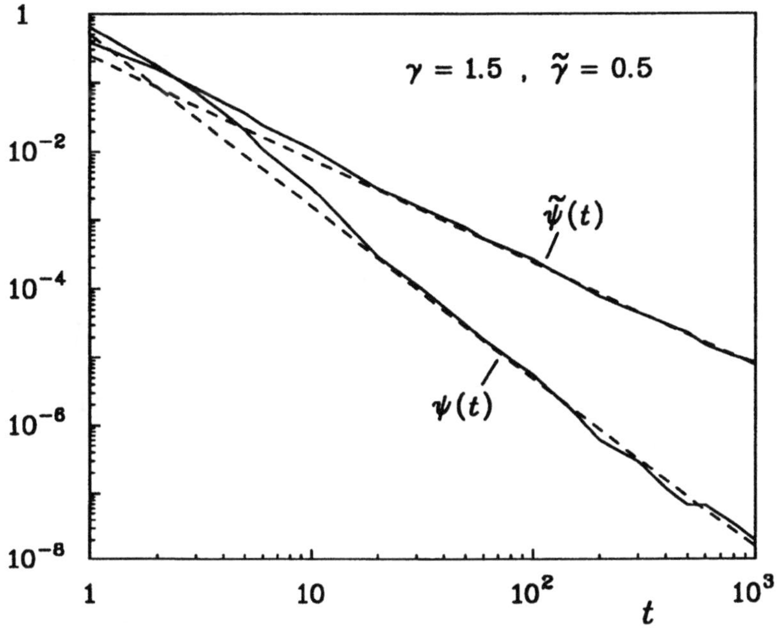

Fig. 13. The probability distributions $\psi(t)$ and $\tilde{\psi}(t)$ for $z = 5/3$ and $\tilde{z} = 3$ corresponding to $\gamma = 3/2$ and $\tilde{\gamma} = 1/2$, as indicated. Simulations results are given by solid lines, the dashed lines indicate the power-laws considered in Eqs. (42).

squared displacement we make use of Eq. (27) and obtain for the leading term in the asymptotic behavior a power law $\langle r^2(t)\rangle \sim t^{\alpha}$, with the exponent α depending on γ and $\tilde{\gamma}$ as

$$\alpha = \begin{cases} 2 + \min\{\tilde{\gamma}, 1\} - \min\{2, \gamma\} & , \quad \gamma > 1 \\ 2 + \min\{\tilde{\gamma}, \gamma\} - \gamma & , \quad 0 < \gamma < 1 \end{cases} . \tag{46}$$

For clarity, we concentrate on the range $1 < \gamma < 2$ which corresponds to the intermediate enhanced diffusion regime if localization is disregarded [11]. When localization is included the combined effects lead to

$$\langle r^2(t)\rangle \sim \begin{cases} t^{2+\tilde{\gamma}-\gamma} & , \quad 1 < \gamma < 2 \quad , \quad \tilde{\gamma} < 1 \\ t^{3-\gamma} & , \quad 1 < \gamma < 2 \quad , \quad \tilde{\gamma} > 1 \end{cases} . \tag{47}$$

Equation (46) indicates that, depending on the two exponents γ and $\tilde{\gamma}$, the motion shows enhanced, regular, or dispersive behavior. We recall that the exponents γ and $\tilde{\gamma}$ are related to those of the combined map by $\gamma = (z-1)^{-1}$ and $\tilde{\gamma} = (\tilde{z}-1)^{-1}$. It should be noted that for $\tilde{\gamma} > 1$, for which $\tilde{\psi}(t)$ has a finite first moment, the enhancement reduces to the result obtained by the original Lévy-walk scheme [11, 31]. In Fig. 14 we show some numerical results for the time evolution of the mean-squared displacements. Plotted is the ratio $\langle r^2(t)\rangle/t$ for various γ and $\tilde{\gamma}$ values. The denominator t has been chosen to strengthen

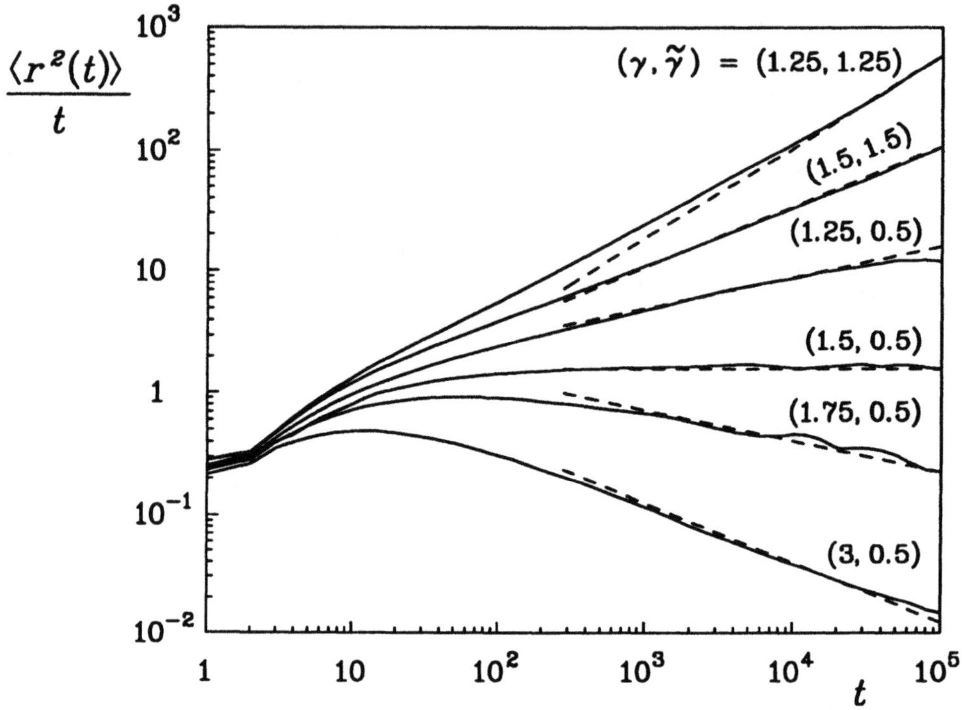

Fig. 14. The time evolution of the mean-squared displacement obtained from the iterated combined map, Eq. (41) for various enhanced and dispersive cases. Plotted are the ratios $\langle r^2(t)\rangle/t$ as solid lines for γ and $\tilde{\gamma}$ as indicated. The dashed lines give the predicted slopes according to Eqs. (46) and (47).

the impression of the deviation from simple Brownian motion. The results follow reasonably the predicted power-law behaviors.

For the tracer motion in a rotating annulus, mentioned above [18], the exponents $\mu = 2.3 \pm 0.2$ for the flights, $\nu = 1.6 \pm 0.3$ for the sticking and $\alpha = 1.65 \pm 0.15$ for the mean-squared displacement are reported which were determined from independent measurements. According to the present convention we have $\gamma = \mu - 1 = 1.3 \pm 0.2$ and $\tilde{\gamma} = \nu - 1 = 0.6 \pm 0.3$ so that Eq. (46) applies and therefore $\alpha = 1.3 \pm 0.5$ which is consistent with the experimental value.

We also discuss the propagator for the intermediate-enhanced diffusion regime, $1 < \gamma < 2$. We distinguish between the two cases $\tilde{\gamma} > 1$ and $\tilde{\gamma} < 1$. In the former case the small (k, u) expansion of the propagator gives to lowest order

$$P(k,u) \sim \frac{1}{u + c|k|^{\gamma}} \ , \tag{48}$$

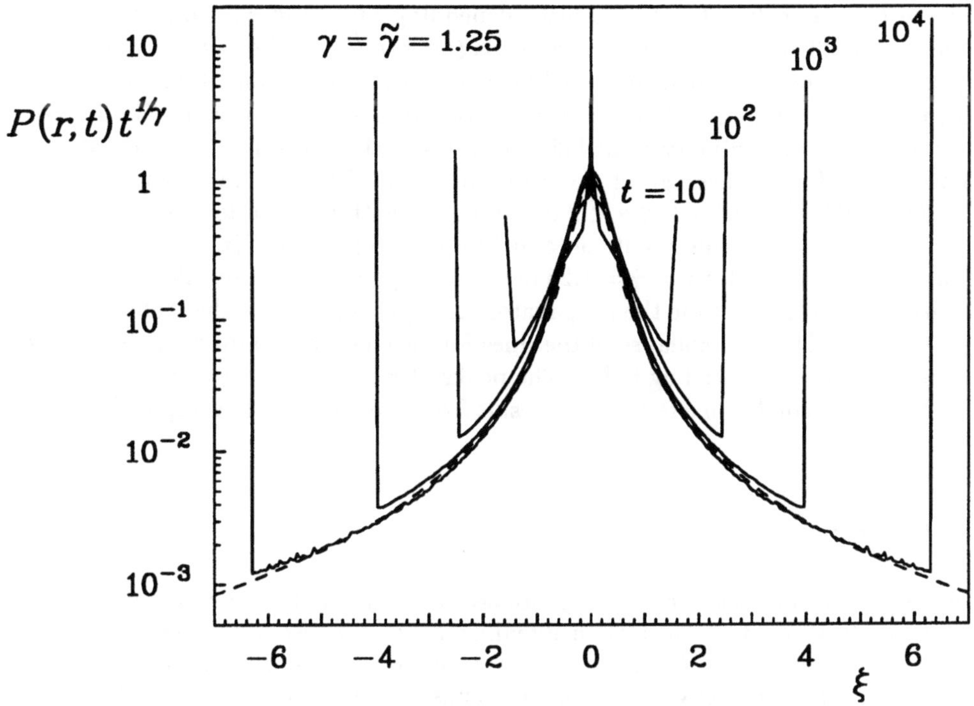

Fig. 15. The propagator $P(r,t)$ for the iterated combined map of Eq. (41). The numerical results are given as solid lines for times t and γ and $\tilde{\gamma}$ values, as indicated, in the scaling representation with the scaling variable $\xi = r/^{1/\gamma}$. The dashed line is the stable law $cL_\gamma(cr)$ with c introduced as an adjustable parameter. The sharp peaks are due to stationary conditions considered in the averaging procedure.

which yields the known Lévy stable distribution [11]

$$P(r,t) \sim \begin{cases} t^{-1/\gamma} L_\gamma(\xi) & , \quad r < t \\ 0 & , \quad r > t \end{cases} , \tag{49}$$

where ξ is the scaling variable $\xi = cr/t^{1/\gamma}$ and the cutoff is due to the δ-function correlation in Eq. (22). Here the laminar phase dominates the transport properties.

A different behavior is derived for $1 < \gamma < 2$ and $\tilde{\gamma} < 1$, where the small (k,u) expansion of the propagator gives

$$P(k,u) \sim \frac{1}{u^{1-\tilde{\gamma}}} \frac{1}{u^{\tilde{\gamma}} + c|k|^\gamma} . \tag{50}$$

This leads to the approximate scaling form

$$P(r,t) \sim \begin{cases} t^{-\tilde{\gamma}/\gamma} Q(\xi) & , \quad \xi \to 0 \\ t^{-\tilde{\gamma}/\gamma} \xi^{-\gamma-1} & , \quad r \lesssim t \\ 0 & , \quad r > t \end{cases} , \tag{51}$$

where Q is a scaling function which, depending on γ and $\tilde{\gamma}$, shows a cusp at the origin [33, 11]. The scaling variable is $\xi = r/t^{\tilde{\gamma}/\gamma}$. In the case of $\gamma > 2$ and $\tilde{\gamma} < 1$ we recover the dispersive behavior treated in Ref. [43]. Here the role of localization is pronounced and may even dominate, a situation which we did not observe in Hamiltonian systems [43]. From the scaling in Eqs. (50) and (51) we notice that for the average of the absolute value of the displacement we have $\langle |r(t)| \rangle \sim t^{\min\{1,\tilde{\gamma}\}/\gamma}$ for $1 < \gamma < 2$ in agreement with the results in Ref. [21, 45].

In Fig. 15 we present results obtained for the propagator $P(r,t)$. The numerical data are plotted for $\gamma = \tilde{\gamma} = 1.25$ in the scaling representation. Because of the finite mean trapping time the propagator, as expected, follows the stable law for $|r| < t$, Eq. (51). The collapse of the lines for various times onto a single master curve indicates that scaling holds. The peaks at $r \simeq 0$ and $|r| \simeq t$ are attributed to stationary condition effects which result from the method of averaging [11, 33].

6 Conclusions

We have reviewed the stochastic processes of Lévy flights and Lévy walks and their applications to transport in dynamical systems. We have introduced the idea of the 'time cost' for a particle to move over distance r in time t which is essential for the Lévy walk process. Thus the maximum possible displacement grows in time but remains finite and accordingly, also the mean-squared displacement remains finite.

A case, recently analyzed in terms of Lévy-flights, concerns the fluctuations of transition frequencies observed for single molecules in solid environments [48]. Due to the stochastic motion in the environment, which may arise from thermal and relaxational processes, the probe molecule is exposed to fluctuations in local stresses and electrostatic fields. A two-level system model has been proposed to account for the fluctuations due to long-range interactions between the two-level systems and the probe molecule. In the particular case of dipolar interactions in a three-dimensional space the fluctuations have been shown to follow a Cauchy random walk. It is concluded that within some assumptions the fluctuations in the spectra of single molecules embedded in solids can be described by Lévy flights.

From the distinction between flights and walks the following aspect arises which is merely a theoretical point rather than is of experimental consequences. Lévy flights are Markovian processes because at each time instance steps of any distance are possible [49]. Lévy walks are non-Markovian because the motion events last for some time on scales of the mean event time, if finite. Thus at least on these scales deviations between flights and walks are expected. For γ-values $\gamma < 1$ the deviations are more involved.

In summary chaotic dynamics in various dynamical systems have been shown to be intimately related to the Lévy processes. We have analyzed the transport in the Lévy walk framework which provides a useful random-walk description of enhanced as well as other diffusional behaviors. Lévy processes have been reported in other situations different from those mentioned above which leads one to believe that this generalization of Brownian motion may become as ubiquitous as its predecessor.

Acknowledgments

We thank F. Weber for technical assistance. A grant of computer time from the Rechenzentrum der ETH-Zürich is gratefully acknowledged.

References

1. C.F.F. Karney, Physica D **8**, 360 (1983).
2. B.V. Chirikov and D.L. Shepelyansky, Physica D **13**, 394 (1984).
3. B.V. Chirikov and D.L. Shepelyansky, in *"Renormalization Group"*, World Scientific (1988).
4. B.V. Chirikov and D.L. Shepelyansky. Phys. Rev. Lett. **61**, 1039 (1988).
5. T. Geisel, J. Nierwetberg, and A. Zacherl, Phys. Rev. Lett. **54**, 616 (1985).
6. T. Geisel, A. Zacherl, and G. Radons, Z. Phys. **B71**, 117 (1988).
7. A.A. Chernikov, B.A. Petrovichev, A.V. Rogalsky, R.Z. Sagdeev, and G.M. Zaslavsky, Phys. Lett. A**144**, 127 (1990).
8. D.K. Chaikovosky and G.M. Zaslavsky, Chaos **1**, 463 (1991).
9. X.-J. Wang, Phys. Rev. A **45**, 8407 (1992).
10. M.F. Shlesinger, G.M. Zaslavsky, and J. Klafter, Nature **263**, 31 (1993).
11. G. Zumofen and J. Klafter, Phys. Rev. E**47**, 851 (1993).
12. A.N. Jannacopoulos and G. Rowlands, J. Phys. A**26**, 6231 (1993).
13. J.-P. Bouchaud and A. Georges,Phys. Rep. **195**, 127 (1990).
14. R. Ishizaki, H. Hata, T. Horita, and H. Mori, Prog. Theor. Phys. **84**, 179 (1990); R. Ishizaki, T. Horita, T. Kobayashi, and H. Mori, Prog. Theor. Phys. **85**, 1013 (1991).
15. G. Zumofen, J. Klafter, and A. Blumen, Phys. Rev. E **47**, 2183 (1993).
16. O. Cardoso and P. Tabeling, Europhys. Lett. **7**, 225 (1988).
17. R. Ramshankhar, D. Berlin, and J.P. Gollub, Phys. Fluids A**2**, 1955 (1990).
18. T.H. Solomon, E.R. Weeks, and H.L. Swinney, Phys. Rev. Lett. **71**, 3975 (1993).
19. M.F. Shlesinger, J. Klafter, and Y.M. Wong, J. Stat. Phys. **27**, 499 (1982).
20. M.F. Shlesinger and J. Klafter, Phys. Rev. Lett. **54**, 2551 (1985).
21. G.M. Zaslavsky, D. Stevens, and H. Weitzner, Phys. Rev. E**48**,1683 (1993).
22. F. Hayot, Phys. Rev. A **43**, 806 (1991).
23. J.A. Viecelli, Phys. Fluids A**2**, 2036 (1990).
24. O.V. Bychuck and B. O'Shaugnessy, Phys. Rev. Lett. **74**, 1795 (1995).
25. B. Mandelbrot, *"The Fractal Geometry in Nature"* (Freeman, San Francisco, 1982).
26. P. Lévy, "Théorie de l'addition des variables aléatoires", Gauthier Villars, Paris 1937).
27. A. Janicki and A. Weron, *"Simulation of Chaotic Behavior of α-Stable Stochastic Processes"*, (Dekker, New York, 1994).
28. M.F. Shlesinger and J. Klafter, in *"On Growth and Form"*, (ed. H.E. Stanley and N. Ostrowsky, Kluwer, Boston, 1986).

29. E.W. Montroll and M.F. Shlesinger, in *"Studies in Statistical Mechanics"*, XI, (ed. J.L. Lebowitz and E.W. Montroll, North Holland, Amsterdam, 1984), p1.
30. J. Klafter, A. Blumen, and M.F. Shlesinger, Phys. Rev. A **35**, 3081 (1987).
31. M.F. Shlesinger, B.J. West, and J. Klafter, Phys. Rev. Lett. **58**, 1100 (1987).
32. V. Seshadri and B.J. West, Proc. Natl. Acad. Sci. USA **79**, 4501 (1982).
33. J. Klafter and G. Zumofen, Physica A**196**, 102 (1993).
34. B.V. Chirikov, Phys. Rep. **52**, 264 (1979).
35. Y.H. Ichikawa, T. Kamimura, and T. Hatori, Physica D**29**, 247 (1987).
36. G. Zumofen and J. Klafter, Europhys. Lett. **25**, 565 (1994).
37. J.D. Hanson, J.R. Cary, and J.D. Meiss, J. Stat. Phys. **39**, 27 (1985).
38. J. Klafter and G. Zumofen, (unpublished).
39. Y. Kikuchi and Y. Aizawa, Progr. Theor. Phys. **84** , 563 (1990).
40. D.S. Sholl and R.T. Skodje, Physica D**71**, 168 (1994).
41. J. Klafter and G. Zumofen, Phys. Rev. E**49**, 4873 (1994).
42. T. Geisel, S. Thomae, Phys. Rev. Lett. **52**, 1936 (1984).
43. J. Klafter and G. Zumofen, J. Phys. Chem. **98**, 7366 (1994).
44. A. Blumen, J. Klafter, and G. Zumofen, in *"Optical Spectroscopy of Glasses"*, ed. I. Zschokke (Reidel, Dordrecht, 1986), p.199.
45. G.M. Zaslavsky, Chaos, **4**, 1 (1994).
46. G. Zumofen and J. Klafter, Phys. Rev. E, **51**, xxx (1995).
47. P. Manneville, J. Phys. (Paris) **41**, 1235 (1980).
48. G. Zumofen and J. Klafter, Chem. Phys. Lett. **219**, 303 (1994).
49. G. Trefán, E. Floriani, B.J. West, and P. Grigolini, Phys. Rev. E **50**, 2564 (1994).
50. G. Zumofen and J. Klafter, Phys. Rev. E **50**, 5119 (1995).
51. G. Zumofen and J. Klafter, Phys. Rev. E **51**, yyy (1995).
52. W. Feller, *"An Introduction to Probability Theory and its Applications"* (Wiley, New York, 1971).

Classical and Quantum Chaotic Scattering

Ying-Cheng Lai

Department of Physics and Astronomy
Department of Mathematics
Kansas Institute for Theoretical and Computational Science
The University of Kansas
Lawrence, Kansas 66045, USA

Abstract: In this paper, several recent results on classical and quantum chaotic scattering by the author and collaborators are reviewed. New phenomena regarding qualitative changes of scattering characteristics as parameter changes in the classical context and quantum manifestation of chaotic scattering are discussed.

1 Introduction

Scattering is a fundamental tool to study many physical and chemical systems. In a scattering experiment, particles with different initial states interact with the system and the characteristics of the particles after the scattering are recorded. From such records, certain aspects of the system can be studied. Usually, one plots an output variable characterizing the particle after the scattering versus an input variable characterizing the particle before the scattering. Such a plot is called a scattering function. It may happen that there are an infinite number of singularities in the scattering function. Near these singularities, arbitrarily small change in the input variable can cause large changes in the output variable. This is a sensitive dependence on initial conditions, which signifies the appearance of chaos. The scattering in this case is chaotic [1].

Chaotic scattering occurs commonly in open physical systems. In the past decade, chaotic scattering has been studied extensively due to its potential relevance to disciplines such as astrophysics [2], fluid mechanics [3], chemistry [4] and solid state physics [5]. Previous studies have established that chaotic scattering is due to the existence of nonattracting chaotic invariant sets in the phase space that contains an infinite number of unstable periodic orbits [1, 6, 7]. These unstable periodic orbits are intimately related to the set of singularities on a line intersecting the closure of the stable manifold of the periodic orbits embedded in the chaotic invariant set and, hence, they are also closely related to the set of singularities in the scattering function. Studies have also revealed that there are two basic routes to chaotic scattering, one is the so-called 'abrupt' bifurcation in which a chaotic invariant set is suddenly created as a system parameter varies through a critical value [6], another is the saddle-center bifurcation followed by a period-doubling cascade [7], which is similar to the well known period-doubling bifurcation route to chaos in dissipative dynamical systems [8]. After the onset of chaotic scattering, further qualitative changes of the chaotic invariant set are possible as a system parameter changes [9, 10].

In terms of the dynamical properties of the chaotic invariant set, the dynamics of the chaotic scattering may be characterized as either *hyperbolic* [6] or *nonhyperbolic* [7, 11, 12]. In hyperbolic chaotic scattering, all the periodic orbits are unstable and there are no Kolmogorov-Arnol'd-Moser (KAM) surfaces in the scattering region. In this case, the survival probability of a particle in the scattering region typically decays exponentially with time:

$$P(t) \sim \exp(-\gamma t), \tag{1}$$

where t denotes time and γ is the rate of the exponential decay. In nonhyperbolic chaotic scattering, there are both KAM surfaces and chaotic invariant sets in the phase space [7, 11, 12]. A particle initialized in a region that contains the chaotic invariant sets can spend a long time in the vicinity of KAM surfaces. In this case, the survival probability of a particle decays algebraically with time:

$$P(t) \sim t^{-z}, for\ large\ t, \tag{2}$$

where z is the algebraic decay exponent.

In this paper, new results on both hyperbolic and nonhyperbolic chaotic scattering will be presented. Specifically, for *hyperbolic chaotic scattering*, we review a new bifurcation called the "crisis in scattering" [10]. Briefly, chaotic scattering occurs on both sides of the crisis. Before the crisis, there exist two topologically and dynamically isolated chaotic invariant sets in the phase space. At the crisis, the chaotic invariant sets collide with each other via an uncountably infinite number of heteroclinic tangencies. Consequently, an uncountably infinite number of new periodic and chaotic trajectories are suddenly created once the crisis occurs and, the fractal dimension of the set of singularities in the scattering function increases during the crisis, giving rise to an enhancement of chaotic scattering. For *nonhyperbolic chaotic scattering*, we focus on the algebraic decay process [Eq. (2)]. Numerical experiments reveal that, as a system parameter changes, the algebraic decay exponent z may exhibit rather large fluctuations. Our studies indicate that the fluctuating behavior of z, as a function of the parameter, can be attributed to the continual process of breakup of KAM surfaces in the phase space. A particularly interesting finding is that, when the outermost KAM surfaces enclosing some central island undergo transformations from being absolute barriers to partial barriers (Cantori), the survival probability $P(t)$ displays two different regions of scaling behavior with different decay exponents. The time scale where this crossover takes place is found to coincide with the typical time for a particle to penetrate the newly created Cantori [12].

An important direction of research in chaotic scattering is to study the quantum mechanics of systems which, when studied classically, exhibit chaotic scattering. This is particularly relevant to problems such as the ballistic electron transport in mesoscaopic junction systems in solid state physics [5]. Studies on quantum manifestation of chaotic scattering [13, 14, 5] show that the elements $S_{ij}(E)$ of the quantum scattering matrix (S-matrix) exhibit fluctuations in their dependence on the energy. Here we consider both hyperbolic and nonhyperbolic chaotic scattering and show that *the fine scale semiclassical quantum fluctuations of the S-matrix with energy are characteristically greatly enhanced in the nonhyperbolic case as compared with the hyperbolic case* [14].

To describe our results, it is convenient to use some simple scattering system as a model. We shall utilize the Troll-Smilansky scattering system [15] (to be described in Sec. 2) in this paper. We emphasize, however, that the occurrence of the phenomena described in this paper is not unique to the Troll-Smilansky system. In fact, we shall argue that these phenomena are expected to occur commonly in chaotic scattering systems. This paper is thus organized as follows. In Sec. 2, we introduce the Troll-Smilansky scattering system. In Sec. 3, we describe crisis in hyperbolic chaotic scattering. In Sec. 4, we study nonhyperbolic scattering and finite-time algebraic decay exponent fluctuations. In Sec. 5, we present results on quantum manifestations of both hyperbolic and nonhyperbolic scattering.

2 The Troll–Smilansky Scattering System

Troll and Smilansky [15] introduced an open Hamiltonian system to illustrate many common features of chaotic scattering. The system is shown schematically in Fig. 1, where a point particle is scattered by an infinite array of two-dimensional elastic scatterers. These scatterers are placed in the plane at constant intervals D along the y-axis. Each scatterer is represented by a circular attractive potential $V(r) < 0$ that becomes negligible for $r > R$, where $R < D/2$ (non-overlapping condition). The effect of each individual scatterer on a scattering trajectory can be characterized by the elastic deflection angle $\theta(l)$, which is a function of the angular momentum l, as shown in Fig. 1. Simple classical mechanics gives

$$\theta(l) = \pi - \sqrt{2}l \int_{r_t}^{\infty} \frac{dr}{r^2\sqrt{E - V(r) - l^2/2r^2}}, \tag{3}$$

where E is the particle energy. The mass of the particle is assumed to be unit, and r_t is the radius at the turning point of scattering trajectories determined by the following equation,

$$E - V(r_t) - \frac{l^2}{2r_t^2} = 0. \tag{4}$$

Note that $\theta(l) < 0$ for $l > 0$. Since the system is invariant under time reversal, we have $\theta(-l) = -\theta(l)\ mod(2\pi)$. Due to the finite range of each scatterer, $\theta(l) = 0$ for $l > l_{max} = uR$, where u is the particle velocity in the region where the potential is negligible. For a particle moving towards a scatterer with angular momentum l and velocity $\mathbf{u}$, its new velocity $\mathbf{u}'$ after scattering has the same

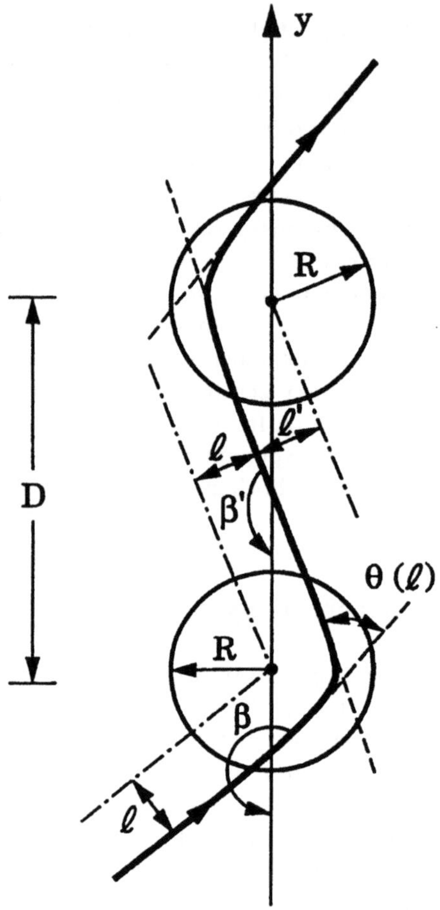

Fig. 1. The Troll and Smilansky scattering system in which particles interact with an infinite array of two-dimensional, non-overlapping and circular potential wells. The dynamical variables are the angular momentum l and the angle β

magnitude as $\mathbf{u}$ but has a different direction. Let β and β' be the angles measured counterclockwisely that $\mathbf{u}$ and $\mathbf{u}'$ make with respect to the $-y$-axis, respectively. Then from Fig. 1 we have

$$\beta' = \beta + \theta(l). \tag{5}$$

After this deflection, the particle may either collide with the scatterer above (if $u'_y = -\cos\beta' > 0$) or enter the scatterer below (if $u'_y < 0$). In either case, scattering is determined by the value of angular momentum l' relative to the new scatterer. Simple geometrical argument from Fig. 1 leads to the following expression for l',

$$l' = l + (Du)sgn(\cos\beta')\sin\beta'. \tag{6}$$

If $|l'| \geq l_{max}$, the particle continues to move along a straight line trajectory leaving the array of scatterers and never returns. Such particles will be regarded as having escaped from the scattering region. New scattering occurs only if $|l'| < l_{max}$. Following Troll and Smilansky [15], we choose β and l as our dynamical variables. Thus, the two-dimensional map Eqs. (5) and (6) can be symbolically represented as

$$(l', \beta') = M(l, \beta). \tag{7}$$

The phase space for M is defined by the domain $[0, 2\pi) \times [-l_{max}, l_{max}]$, which is a cylinder. It can be easily verified that M is area preserving.

3 Crisis in Hyperbolic Chaotic Scattering

The term "crisis" was first introduced by Grebogi, Ott and Yorke to describe sudden qualitative changes of a chaotic attractor as a parameter varies through a critical value in dissipative dynamical systems [16]. We use the term "crisis" in the context of chaotic scattering to refer to qualitative changes in the chaotic invariant sets and the scattering characteristics as a parameter varies. To illustrate crisis in chaotic scattering, we choose $V(r)$ to be the Woods-Saxon potential which is often used in the context of nuclear physics [17],

$$V(r) = -V_0/\{1 + \exp[(r - R_0)/\alpha]\}, \tag{8}$$

where $V_0 > 0$, and R_0 and α are constants. At large distance r, $V(r)$ vanishes exponentially. The crisis studied in this paper occurs in an energy range where the scattering is hyperbolic [10]. In the numerical experiments, we fix $V_0 = 10$, $R_0 = 0.5$, $\alpha = 0.1$, $D = 4$ and $R = 1.4$. Thus $V(r = R)/V_0 \sim 10^{-4}$ so that adjacent potentials do not appreciably overlap each other. Numerically, we find that for $E < 10$, there are only chaotic invariant sets in the phase space and the scattering is hyperbolic.

The chaotic invariant sets lie in the closure of the intersection of the stable and unstable manifolds. The map has two unstable fixed points: $(0, 0)$ (corresponding to a straight trajectory along the $-y$ axis) and $(0, \pi)$ (corresponding to a straight trajectory along $+y$ axis). For $E > E_c \approx 4.4$, there exist two *topologically and dynamically isolated* chaotic invariant sets associated with the unstable fixed points. Note that the two chaotic sets must be identical due to the symmetry of the system with respect to $\beta = 0$ (or 2π) and $\beta = \pi$. At the energy value $E_f \approx 4.4$, the stable manifold of one chaotic set becomes heteroclinically tangent to the unstable manifold of the other chaotic set, as shown in Fig. 2(a). At this crisis point, both chaotic sets are dynamically linked and particles initiated near one chaotic set can reach and exit along the unstable manifold of the other chaotic set. As E decreases passing through E_f, the closures of the stable and unstable manifolds of the two chaotic sets heteroclinically cross each other forming additional chaotic sets at the intersection. Since both the stable and unstable foliations of the chaotic sets have Cantor-like structures before the crisis, there must be an uncountably infinite number of tangencies between E_f and $E_l \approx 4.1$ [corresponding to the last tangency, as shown in Fig. 2(b)].

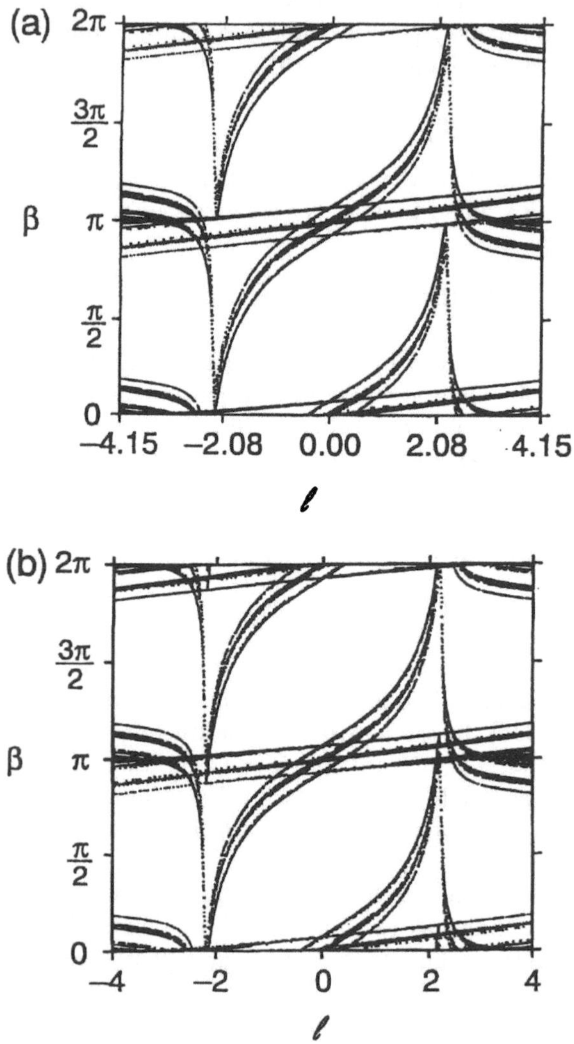

Fig. 2. Stable and unstable foliations for (a) $E = 4.4$ (near the crisis) and (b) $E = 4.1$ (near the last tangency). The unstable foliation is nearly horizontal

Physically, the occurrence of crisis for this particular system can be understood as follows. For a fixed particle energy, $\theta(l)$ is a function of the angular momentum l. At some $l = l_c$, $\theta(l)$ attains its maximum value θ_{max}. The locations where the tips of the stable manifolds cross the unstable manifolds in Figs. 2(a) and 2(b) correspond precisely to $\pm l_c$. This maximum deflection angle determines the extent to which a particle trajectory can turn over. The occurrence of crisis entails that a particle trajectory going upward can exit the system downward, and vice versa. This indicates that $|\theta_{max}|$ is close to π when the crisis occurs. Indeed, as E decreases, $|\theta_{max}|$ keeps increasing, as can be verified numerically. For $E \leq E^* \approx 3.2$, $|\theta_{max}| = \infty$. The situation where $|\theta_{max}| = \infty$ is referred to as "orbiting" [10].

Chaotic invariant sets in the phase space give rise to a set of an uncountably infinite number of singularities in the scattering function. This set of singularities can be conveniently characterized by its fractal dimension. To see the physical consequence of the crisis in chaotic scattering, we compute this fractal dimension d as the energy changes through the crisis. We use the uncertainty algorithm [18] to compute the fractal dimension. It can be shown that the uncertainty dimension is smaller than or equal to the box-counting dimension, and it has been proved that both dimensions are equal for nonattracting hyperbolic chaotic sets of typical dynamical systems [19]. Figure 3 shows the uncertainty dimension d of a Cantor set, obtained by fixing $\beta_0 = 2.55$, versus the particle energy E curve. As can be seen from Fig. 3, $d \approx 0.486$ for $E > E_f$. It increases as E decreases from E_f to $E_l \approx 4.1$. For $E < 4.1$, $d \approx 0.560$. The increase in the fractal dimension indicates a proliferation of singularities in the scattering function. Consequently, chaotic scattering is enhanced after the crisis. Changing β_0 does not significantly change the dimension values in Fig. 3, as long as the line cuts through all components of the stable foliations. This has been verified by computing the dimension at various energy values with slightly different β_0. The overall features of Fig. 3 appear therefore to be robust.

The dynamical process observed in our scattering model, including the increase of the fractal dimension of the invariant set after the crisis, can be understood by considering the following piecewise linear one-dimensional model [10],

$$
\begin{aligned}
f_\lambda(x) &= -s|x+1| + \lambda, \quad x < 0, \\
f_\lambda(x) &= s|x-1| + \lambda, \quad x \geq 0,
\end{aligned} \tag{9}
$$

where $s > \frac{1}{2}[3 + \sqrt{9 + 4(\lambda - 1)}]$ and $\lambda > -1$. For $-1 < \lambda < \lambda_f = s/(s+2)$, there are two isolated invariant sets. These are the "middle α" Cantor sets with $\alpha = (s-2)/s$, so the box-counting dimension of each one of these isolated invariant sets is $\ln 2/\ln s$. At $\lambda = \lambda_f$, the first tangency occurs, as shown in Fig. 4(a), which provides a link between both Cantor sets. Hence, for $\lambda > \lambda_f$, the invariant set of the map is the union of the two previously isolated Cantor sets plus the set created due to the linking of the two Cantor sets. At $\lambda_l = s/(s-2)$, the last tangency occurs, as shown in Fig. 4(b). For $\lambda > \lambda_l$, the large invariant Cantor set is self-similar and has box-counting dimension $2\ln 2/\ln s$. Hence, the graph of the fractal dimension of the invariant set has two constant pieces and the region $\lambda_f \leq \lambda \leq \lambda_l$ where the dimension increases from the lower constant

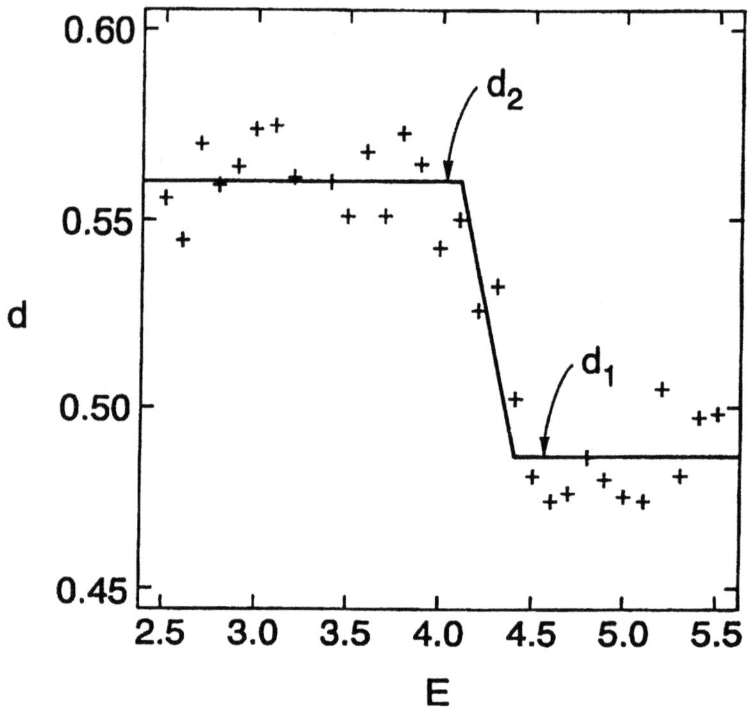

Fig. 3. Fractal dimension of the set of singularities of the scattering function as a function of E obtained by fixing $\beta_0 = 2.55$

value to the higher constant value. In fact, in this transition region the dimension can be shown to be non-decreasing.

Although the map for the scattering system is smooth and two-dimensional, while Eq. (9) has discontinuities in the derivative and is one-dimensional, Eq. (9) captures the essential features of the scattering system. For instance, although the tangencies are quadratic for the scattering system, the stable manifold exhibits a very large curvature at the tangency. Both stable and unstable manifolds away from the tangency points are almost straight lines. Therefore our one-dimensional model is a good dynamic representation of the scattering system at the heteroclinic crossing. The factor of 2 increase in the dimension could be decreased by changing slightly the local slope of the map, and thus, the analytic results would be even closer to the numerical results in Fig. 3. In our scattering system, Fig. 3 indicates that the d versus E curve during the crisis is non-decreasing. It has been conjectured that the d versus E curve during the crisis is a *devil staircase* based on a more detailed study of the one-dimensional model Eq. (9) [10].

Although the stable and unstable unstable manifolds associated with the

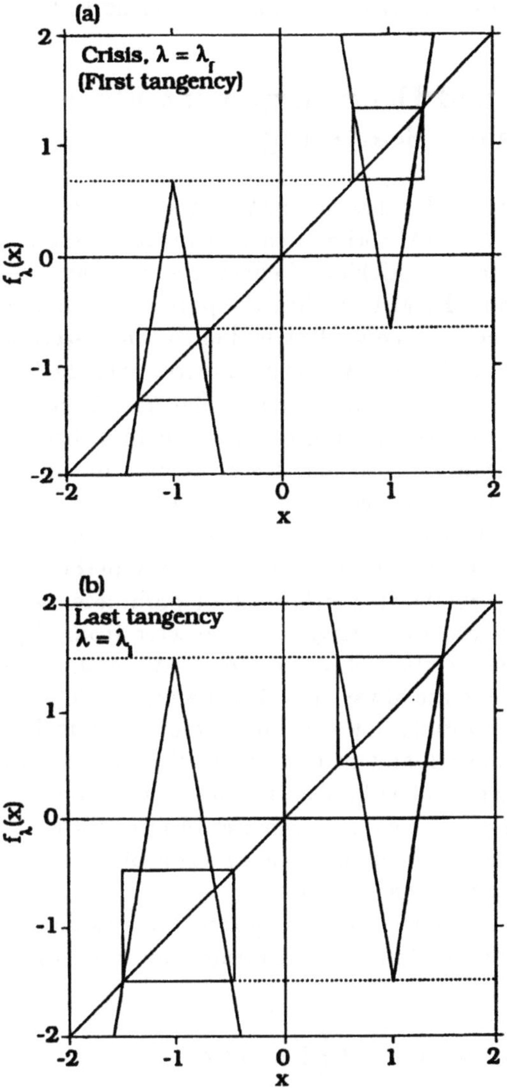

Fig. 4. The one-dimensional model Eq. (9), (a) the crisis; and (b) the last tangency

chaotic invariant sets in the Troll-Smilansky system with the Woods-Saxon potential look rather special, namely, the stable manifolds have a sharp bend and the unstable manifolds are almost straight [Figs. 2(a) and 2(b)], the condition for a crisis to occur is independent of these specific geometrical shapes of the manifolds. As we have discussed in the simple one-dimensional model, crisis and the subsequent increase in the fractal dimension *depend only on the occurrence of an infinite number of heteroclinic tangencies of the stable and unstable foliations.* Since fractal stable and unstable foliations are a general feature of chaotic scattering systems [1], we expect crisis in chaotic scattering to be typical.

4 Algebraic Decay and Decay Exponent Fluctuations in Nonhyperbolic Chaotic Scattering

The phase space of nonhyperbolic chaotic scattering system is characterized by the coexistence of families of KAM tori and chaotic invariant sets. The presence of the KAM tori gives rise to the "stickiness" effect, *i.e.*, if a particle is initialized in a chaotic region near some KAM tori, then the particle wanders close to that surface for a long time. The stickiness effect is due to a near-zero stretching of chaotic trajectory near the KAM tori. As a consequence, particle decay processes in a nonhyperbolic chaotic scattering are algebraic [20, 21, 22, 23]. That is, for a fixed set of parameter values, if one initializes a particle in a chaotic region near some KAM tori, the probability for this particle to remain in the region at time t decays with time algebraically, as described by Eq. (2).

As a system parameter changes, we expect the decay characteristics to change accordingly. Usually, a chaotic scattering system is nonhyperbolic at one set of parameter values and hyperbolic at another set of parameter values. In such systems the particle decay process changes its character from algebraic to exponential with the smooth variation of a parameter. We shall concentrate in a parameter regime where there are always KAM tori and, the decay process is always algebraic. The goal is to study how the finite-time algebraic decay exponent changes as a parameter changes and to investigate the mechanism underlying the change of the decay exponent [12]. We find that the exponent z fluctuates substantially as a function of the control parameter. As we will see in the sequel, the origin of such fluctuation can be understood in terms of the continual process of KAM surface breakup in the phase space.

To illustrate fluctuations of the decay exponent, we again use the Troll-Smilansky system, but with the following quadratic potential function for numerical convenience,

$$V(r) = -V_0[1-(\frac{r}{R})^2], \text{ for } r \leq R \tag{10}$$

and $V(r) = 0$ for $r > R$, where $V_0 \geq 0$ measures the depth of the potential well. Using simple classical mechanics we obtain for the deflection angle

$$\theta(l) = 2\sin^{-1}\frac{l}{\sqrt{2E}R} - \frac{l}{|l|}[\frac{\pi}{2} + \sin^{-1}\frac{l^2/R^2-(E+V_0)}{\sqrt{(E+V_0)^2-2l^2V_0/R^2}}], \tag{11}$$

where E is the particle energy and the particle mass is chosen to be 1. In our subsequent calculations, we fix all the parameters in Eq. (11): $V_0 = 0.2$, $E = 0.5$ and $R = 1.0$ (so $l_{max} = 1$), and use the distance D between adjacent potentials as the control parameter. For small D values, the map Eqs. (5) and (6), with $\theta(l)$ defined by Eq. (11), displays a mixture of regular motion on KAM surfaces and chaotic motion between these surfaces. For larger values of D, we did not find any KAM surfaces and particles appear to escape from the scattering region exponentially. Below we restrict our attention to the former range of D values.

To study the escape of particles, we choose a subregion in the chaotic region which apparently does not intersect any KAM surfaces. We then initialize $N(0)$ particles uniformly in that subregion, and count the number of particles $N(t)$ that have not yet escaped the system at time t. From these data we extract the algebraic decay behavior by plotting $\log_{10} N(t)$ versus $\log_{10} t$. It can be seen that for $D = 3.70$, the log – log plot shown in Fig. 5 is reasonably well fit by a straight line that extends over a few decades. The slope of the straigt line fit yields the decay exponent which, for the case in Fig. 5, is $z \approx 1.32$. The initial conditions used to generate Fig. 5 are placed at the lattice sites of a 2000×2000 uniform grid on the subregion: $3.25 \leq \beta \leq 3.30$ and $0.95 \leq l \leq 1.00$. For particles initialized in the chaotic region, their escape process will take place on a wide range of time scales. More specifically, based on the respective dynamical behavior in the phase space, the entire time interval can be divided into two subintervals, namely, $0 < t < t_1$ and $t > t_1$. For those particles that escape before $t = t_1$, it is likely that their motions are not impacted by any KAM surfaces along their trajectories. Thus the portion of the $N(t)$ versus t curve for $t < t_1$ is exponential rather than algebraic. For the particles which stay longer than $t = t_1$ in the region, their corresponding trajectories may spend substantial amount of time near some accessible KAM surfaces. Hence the curve $N(t)$ versus t is algebraic for $t > t_1$.

As the parameter D varies, the decay exponent z fluctuates as shown in Fig. 6. In what follows we carry out a sequence of numerical experiments addressing the origin of this fluctuation. More specifically, we examine the phase space structure and its associated dynamics for two pairs of D values, $D_1 = 3.60$, $D_2 = 3.62$, and $D_3 = 3.70$, $D_4 = 3.72$, respectively (see Fig. 6). Their corresponding decay exponents are denoted by z_1, z_2, z_3 and z_4, respectively. For the first pair of D values we have $z_2 > z_1$, while for the second pair we have $z_4 < z_3$. Thus, for increasing D values, the decay exponents increase for the first pair of D values. They decrease for the second pair. This is the reason for choosing the particular D values quoted above.

The phase-space structures corresponding to the first pair of D values, D_1 and D_2, are shown in Fig. 7(a) and Fig. 7(b), where we uniformly choose 144 initial conditions in the region $3.1 < \beta < 3.6$ and $0.87 < l < 0.92$. (Note that only parts of the phase space are shown in the figure highlighting the regions relevant to our discussion.) Some of the initial conditions lead to trajectories that lie on KAM surfaces, while others wander in the region of chaos. We have chosen to plot only the trajectories on the KAM surfaces. A noticeable feature common to both parameter values is that the phase space is divided into two distinct regions by KAM surfaces. While the region denoted by B is enclosed within the KAM surfaces, the other region, denoted by A, lies outside the surfaces. As a result of this space partitioning, particles launched from outside visit only region A before they exit the potential. The particle decay dynamics for both D_1 and D_2 is thus determined by the structures present in region A. As D increases from

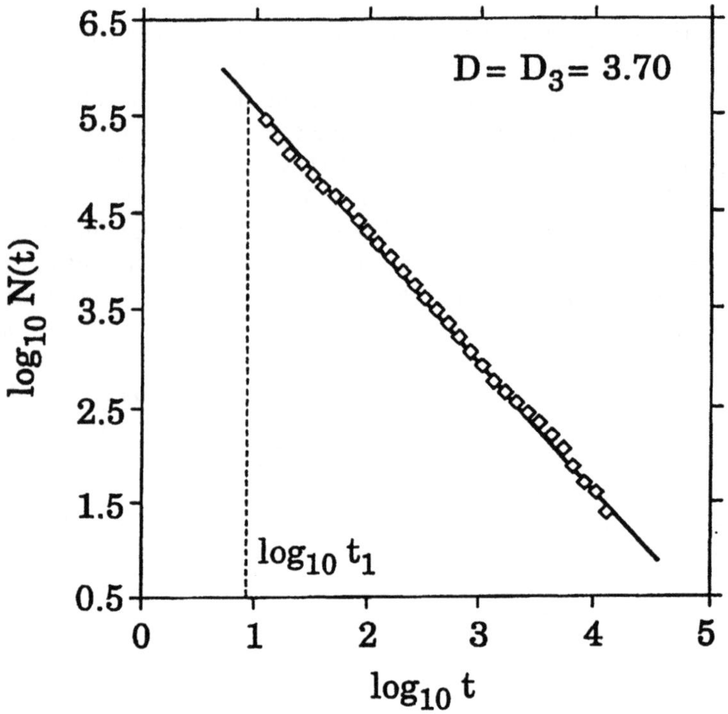

Fig. 5. $\log_{10} N(t)$ versus $\log_{10} t$ plot for $D = 3.70$

D_1 to D_2, the boundary between the two regions deforms slightly but otherwise remains essentially intact. In contrast, the area occupied by islands undergoes reduction from Fig. 7(a) to Fig. 7(b). The implication is that, on the average, particles exit the potential faster for $D_2 = 3.62$ than for $D_1 = 3.60$, thereby furnishing an explanation to the observed relationship $z_2 > z_1$. Note that if the experiments were to be performed for an infinitely long time the algebraic decay may yield an overall exponent which is the same for both parameter values.

We now turn to the second pair of D values, D_3 and D_4. For $D = D_4$, the decay exponent z attains a local minimum: $z_4 \approx 0.88$ (Fig. 6). The phase space plots for D_3 and D_4 show drastic differences [Figs. 8(a) and 8(b)]. In particular, we note that while the two regions, A and B, in Fig. 8(a) are still visibly separated by KAM surfaces, the boundary between the two regions in Fig. 8(b) appears to have been destroyed. This indicates that the chaotic component previously enclosed in region B is now accessible to the exploration of particles initialized in region A. In geometrical terms, the two chaotic regions lying on different sides of the KAM surfaces have been combined into a single connected chaotic set.

It is known that Cantori, immediately after the breakup of KAM surfaces,

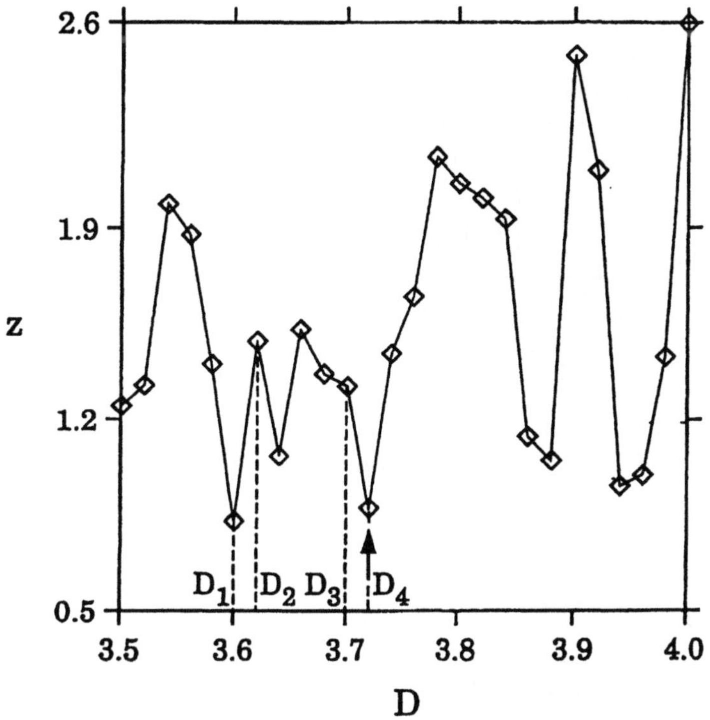

Fig. 6. Algebraic decay exponent z as a function of the parameter D for $3.5 \leq D \leq 4.0$

serve as effective barriers to particle transport [24]. The typical time for a particle to penetrate the Cantori thus constitutes a new time scale in the system we denote by t_c. The presence of this new time scale leads to observable consequences which we shall discuss below. For particles that exit the potential in time less than $t = t_c$, the corresponding trajectories lie entirely outside the newly created Cantori. This implies that the exponent measured over the time interval $t_1 < t < t_c$ should be roughly the same as that before the breakup of KAM surfaces. Particles that stay longer than $t = t_c$, however, are likely to penetrate the Cantori and explore the chaotic component previously enclosed within the KAM surfaces. For these particles their trajectories encounter additional KAM island chains before they finally exit. Reflected in the decay dynamics this corresponds to a slower escape process, thus a smaller decay exponent measured over the interval $t > t_c$. The numerical results for D_3 and D_4 are shown in Figs. 5 and 9, respectively. There are clearly two regions of scaling behavior in Fig. 9. The exponent measured over the first interval of scaling behavior $t < t_c$ is $z \approx 1.27$, while the exponent measured over the remaining interval $t > t_c$ is $z \approx 0.88$. This confirms the theoretical consideration given above. The crossover

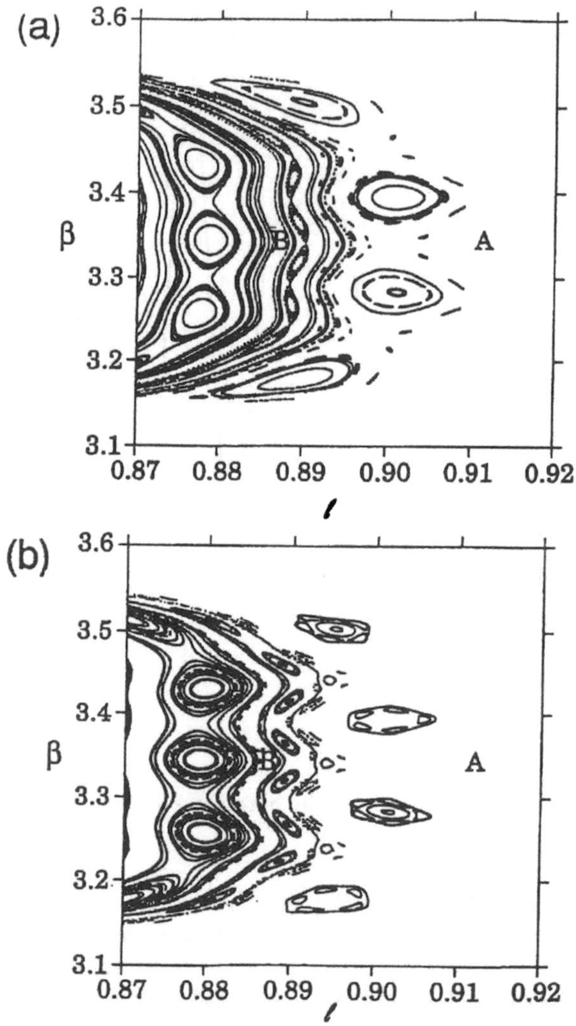

Fig. 7. Phase space plots for (a) $D_1 = 3.60$ and (b) $D_2 = 3.62$

time $t = t_c$ in Fig. 9 is roughly 1200 iterates. As indicated, this time represents the number of iterates for a typical trajectory to penetrate the Cantori. To demonstrate this point we plot the trajectories for two sets of judiciously chosen initial conditions. One set yields a trajectory which escapes the potential before $t = t_c$. The other leads to an orbit that stays longer than $t = t_c$. Figures 10(a) and 10(b) show the two respective trajectories. One sees that the conclusion alluded to is self-evident. Although we showed one stage of crossover in Fig. (9), the mechanism responsible for this crossover phenomenon suggests that, due to the continual nature of KAM surface breakup on all scales, one

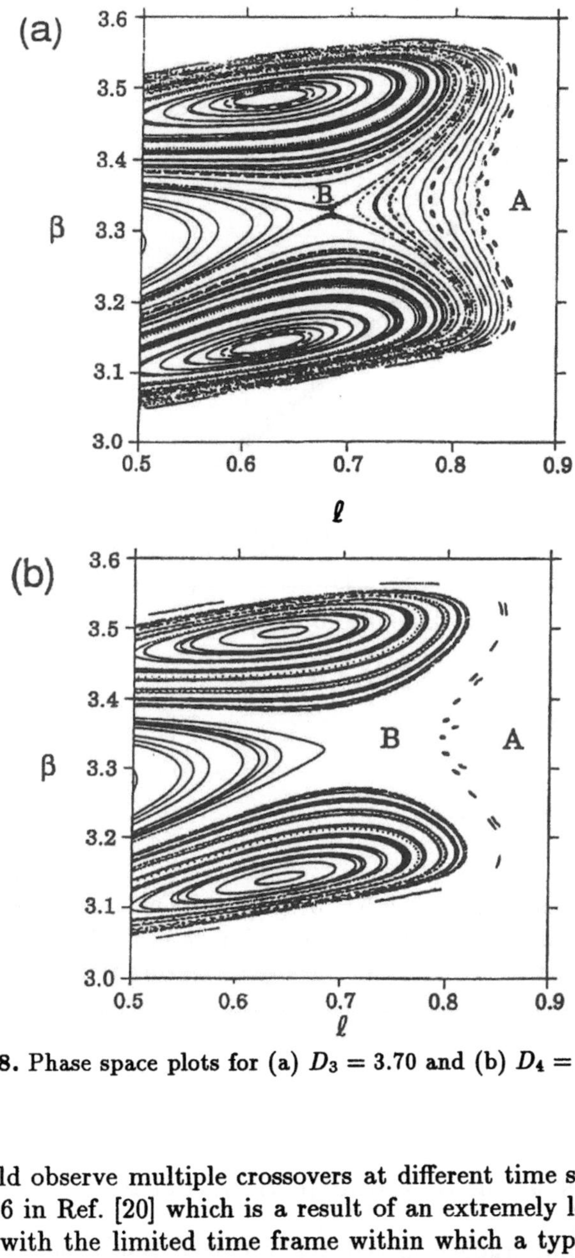

Fig. 8. Phase space plots for (a) $D_3 = 3.70$ and (b) $D_4 = 3.72$

should observe multiple crossovers at different time scales. This conforms with Fig. 6 in Ref. [20] which is a result of an extremely long numerical simulation. But with the limited time frame within which a typical experiment is carried out, only crossover associated with major changes in the phase space can be observed. Since a metamorphosis in the phase space exemplified by the destruction of KAM surfaces and subsequent exposure of a new layer of KAM islands may manifest itself through variations in the decay exponent, this relationship can also be used to interpret unexpected findings in physical systems. One of such examples is the study reported in Ref. [25] where we considered a theoretical

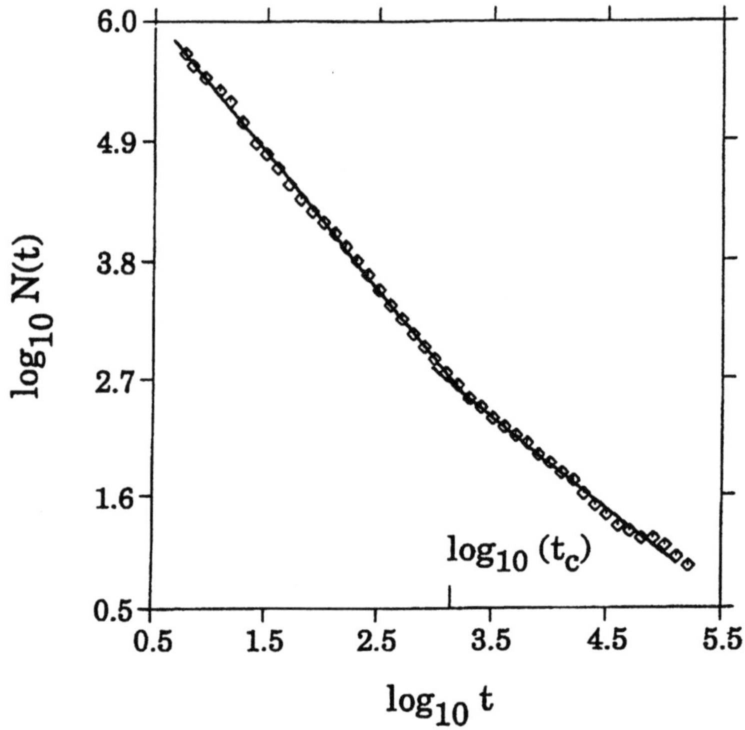

Fig. 9. $\log_{10} N(t)$ versus $\log_{10} t$ plot for $D_4 = 3.72$

model of microwave ionization of hydrogen Rydberg atoms. Our results indicate that, contrary to naive physical intuition, the ionization rate, analogous to the decay exponent z in our case, is not a monotonically increasing function of the field strength. In many instances the increment of the field strength actually leads to a decline of the ionization rate. The reasons behind the finding are precisely what we report here, namely, the complicated metamorphosis patterns in the phase space.

5 Quantum Manifestation of Chaotic Scattering

Quantum scattering is usually studied through the S-matrix. When the classical scattering is chaotic, the corresponding quantum S-matrix elements typically exhibit random fluctuations as a function of a system parameter such as the energy [13, 14, 5]. It is thus convenient to study the energy autocorrelation function of an S-matrix element [13], which is defined as

$$C_{ij}(\varepsilon) = < S^*_{ij}(E) S_{ij}(E+\varepsilon) >_E, \tag{12}$$

where ε is the energy difference (small) and $< >_E$ denotes an average over a band of energies centered at some value $E = E_0$ and of width ΔE, where ΔE is classically small ($\Delta E/E_0 \ll 1$) but "quantum mechanically large" ($\Delta E \gg \varepsilon$). More specifically, we define the average by $< \ldots >_E = \frac{1}{\Delta E}\int \ldots f[(E-E_0)/\Delta E]dE$, where in our numerical work we choose the following the weighting function, $f(x) = (1-|x/\sqrt{6}|)/\sqrt{6}$ for $|x| < \sqrt{6}$ and $f(x) = 0$ for $|x| \geq \sqrt{6}$.

To evaluate $C_{ij}(\varepsilon)$, we use the semiclassical theory in which the S-matrix elements are given by [26],

$$S_{ij}(E) = \sum [P^{(s)}_{j\to i}(E)]^{1/2} \exp[\frac{i}{\hbar}\Phi^{(s)}(E) - \frac{i\pi}{2}\nu^{(s)}], \tag{13}$$

where (s) denotes a classical path, $P^{(s)}_{j\to i}(E)$ is the classical transition probability from state j to state i along the path s, $\Phi^{(s)}(E)$ is the classical (reduced) action along the path (s), and $\nu^{(s)}$ is the Maslov index. The sum is over all the classical paths connecting state j and state i. The classical transition probability is given by [26] $P^{(s)}_{j\to i}(E) = (\frac{1}{2\pi})|\frac{\partial I_i}{\partial \theta_j}|_s^{-1}$, where (I_j, θ_j) and (I_i, θ_i) are the action-angle variables in the states j and i, respectively. For chaotic scattering, the number of classical paths connecting j and i increases exponentially with the length of the path. Hence, the sum in Eq. (13) is difficult to evaluate. Nonetheless, Eq. (13) is useful for the heuristic estimation of average quantities [13].

Substituting Eq. (13) into the definition of $C_{ij}(\varepsilon)$, approximating $\Phi^s(E+\varepsilon)$ as $\Phi^s(E) + \varepsilon\frac{\partial \Phi^s}{\partial E}$ and neglecting the contribution from cross terms in the double sum over the classical paths, one gets [13]

$$C_{ij}(\varepsilon) \sim \int dt < P_{ij}(E,t) >_E \exp(i\varepsilon t/\hbar), \tag{14}$$

where t is the delay time of the particle in the scattering region. In Eq. (14), $P_{ij}(E,t)dt$ is the classical probability that a $j \to i$ transition occurs while the delay time is in the interval $[t,\ t+dt]$. For hyperbolic chaotic scattering, $P_{ij}(E,t) \sim P(E,t) \sim \exp(-\gamma t)$. Substituting this exponential law into Eq. (14), one obtains [13]

$$C(\varepsilon) \sim \frac{1}{\varepsilon/\hbar + i\gamma}, \tag{15}$$

which is Lorentzian. For small ε we have $|C(\varepsilon)|_{\varepsilon\to 0} \approx constant + O(\varepsilon^2)$. For nonhyperbolic chaotic scattering, $P(E,t) \sim t^{-z}$ for *large* t. Substituting this algebraic law into Eq. (14), we can evaluate the integral asymptotically for small ε. We obtain [14]

$$C(\varepsilon) = C_0 + C_1(\frac{\varepsilon}{\hbar})^{z-1}, \tag{16}$$

where $C(\varepsilon) \equiv |C_{ij}(\varepsilon)|$, $C_0 \equiv C(0)$ and C_1 is a negative constant.

From Eq. (16), we observe that $\frac{dC(\varepsilon)}{d\varepsilon} \sim (\frac{\varepsilon}{\hbar})^{z-2}$. So in typical cases where $1 < z < 2$ ($z \approx 1.45$ in Karney's numerical experiment [20], $z \approx 1.34$ in Chirikov and Shepelyansky's numerical experiment [22], $z < 1.96$ in a theory based on a Markov-Tree model by Meiss and Ott [23]), we have $\frac{dC(\varepsilon)}{d\varepsilon}|_{\varepsilon\to 0} \longrightarrow \infty$. Hence,

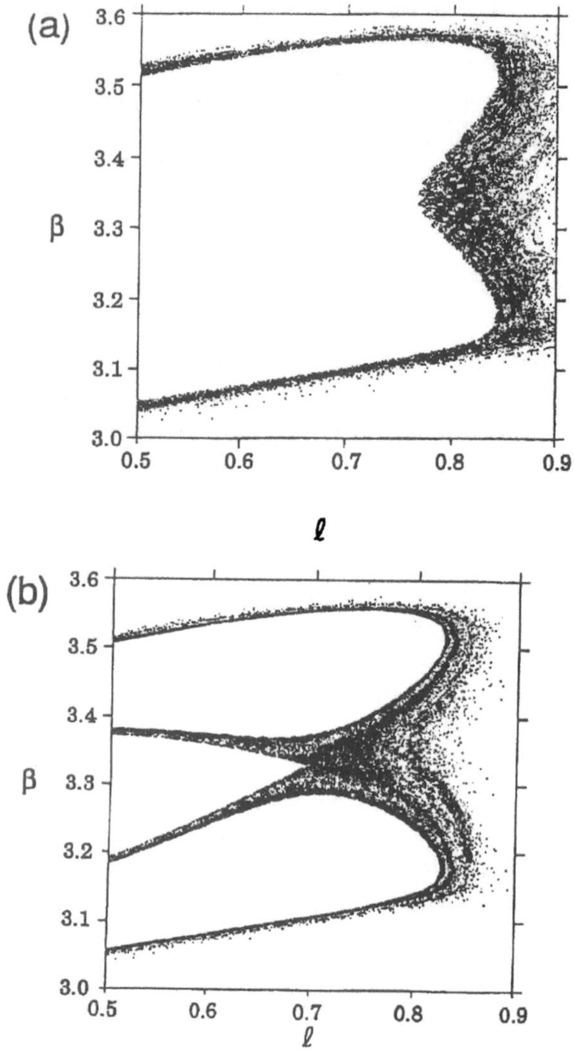

Fig. 10. Two particle trajectories corresponding to escape time (a) $t < t_c$ and (b) $t > t_c$ ($D = 3.72$)

the energy autocorrelation function exhibits a *cusp* near $\varepsilon = 0$. That is, $C(\varepsilon)$ decreases very rapidly from $\varepsilon = 0$ [14].

The above argument for the existence of Lorentzian behavior of $C(\varepsilon)$ in hyperbolic chaotic scattering and the cusp behavior of $C(\varepsilon)$ in nonhyperbolic chaotic scattering is nonrigorous since it involves the long time limit of the semi-classical approximation [14, 27]. Thus we have undertaken a series of numerical experiments to test Eqs. (15) and (16). We consider the Troll-Smilansky system with the Woods-Saxon potential at the following set of fixed parameter values:

$V_0 = 10$, $R_0 = 1$, $a = 0.1$ and $D = 4$, as in Sec. (3). When the particle energy is large, we observe that the phase space consists of KAM islands and chaotic regions, as shown in Fig. 11(a) for $E = 10$. In this case, the scattering is nonhyperbolic and the decay of particles from the scattering region is well-fit by an algebraic dependence over the time range tested with a decay exponent of $z \approx 1.62$. This exponent was calculated by initializing a large number of particles uniformly distributed in the region ($l \in [4.5, 6.5]$, $\beta \in [0, 2\pi]$) which apparently does not contain KAM islands. As the particle energy decreases from $E = 10$ to $E = E_{crit}$ ($1 < E_{crit} < 10$), the KAM surfaces are destroyed. At lower values of E, we find numerically that there are no KAM surfaces and the decay of particles from the scattering region is exponential. In this case, the scattering is hyperbolic, as shown in Fig. 11(b) (the stable manifolds of the chaotic invariant sets) for $E = 1$.

Quantum mechanically, the effect of any single scatterer is completely specified once the phase shifts $\sigma_l(E)$ are given [13]. Because of the discrete translational symmetry of the system in the y direction, the wave function satisfies $\psi^{(\alpha)}(x, y) = \exp(i\alpha y)\varphi^{(\alpha)}(x, y)$, where $\varphi^{(\alpha)}(x, y)$ is D-periodic in y. So we have $\psi^{(\alpha)}(x, y+nD) = \exp(in\alpha D)\psi^{(\alpha)}(x, y)$. Thus the y component of the momentum q_n is quantized $q_n = \alpha + 2n\pi/D$. Since by the conservation of energy $q_n^2 \leq k^2$, where k is the wavevector for the energy $[E = (\frac{\hbar^2}{2m})k^2]$, there exists a maximum integer N_B (maximum number of open channels) above which q_n is not allowed and corresponds to evanescent waves. To calculate the quantum S-matrix, we make use of the KKR method [28] adapted to the present problem [13].

To compare the quantum calculation with the semiclassical prediction of Eqs. (15) and (16), we must use sufficiently small values of Planck's constant $\hbar$, so that there are a large number of waves on the scale of the system length D. In the following, we shall compute results for a nonhyperbolic case, with nominal energy $E_0 = 10$, and a hyperbolic case with nominal energy $E_0 = 1$. For $E_0 = 10$, we choose $\frac{\hbar^2}{2m} = 10^{-3}$ so that the corresponding wavevector is $k_0 = 100$ and there are 127 open channels ($128 > 2D/\lambda > 127$). In the case of hyperbolic scattering where $E_0 = 1$, we choose $\frac{\hbar^2}{2m} = 10^{-4}$ (hence, we also have $k_0 = 100$) so the number of open channels (127) is the same as that in our nonhyperbolic scattering case. We then calculate the autocorrelation function $C(\varepsilon)$ using $\Delta E/E_0 \approx 0.1$. As a function of energy, an S-matrix element contains both a smooth part and a fluctuating part. To compare with the semiclassical theory, we subtract the smooth part and only keep the fluctuating part. Figure 12 shows the magnitude squared of the quantum autocorrelation functions $|C(\varepsilon)|^2$ for both the $E_0 = 10$ case (nonhyperbolic, plotted as diamonds) and the $E_0 = 1$ case (hyperbolic, plotted as triangles). For Fig. 12, we also perform an average with respect to a small block of matrix elements around a combination i and j ($i = 14, j = 25$). We observe that the $C(\varepsilon)$ curves thus obtained are essentially independent of i and j. Also shown in Fig. 12 are the semiclassical predictions from Eqs. (15) and (16); the lower thick solid curve corresponds to $E_0 = 10$, and the upper thick solid curve corresponds to $E_0 = 1$. To calculate the semiclassical correlation functions, we use the classical decay data and substitute them directly into Eq.

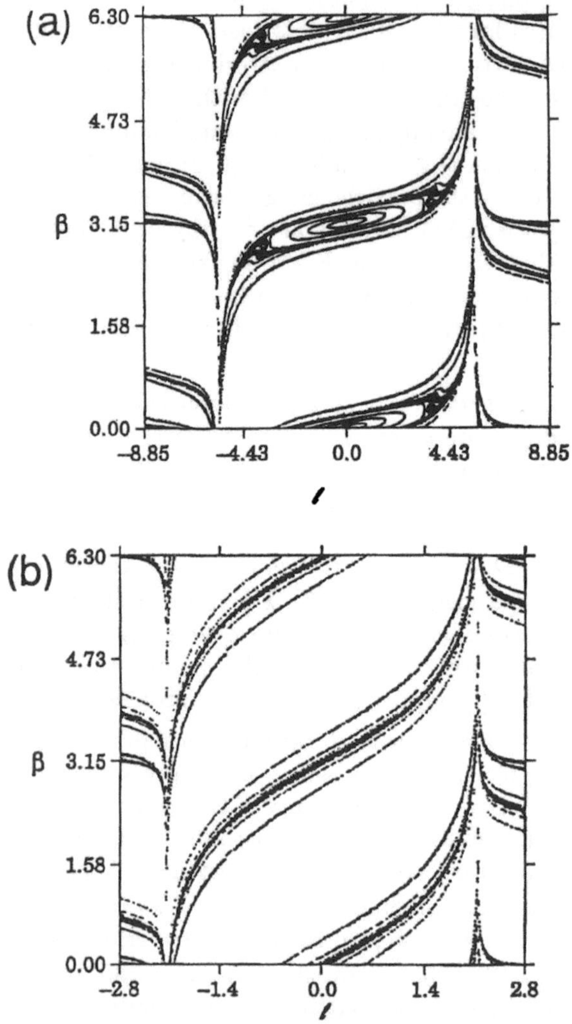

Fig. 11. (a) Nonhyperbolic chaotic scattering for $E = 10$ (chaotic invariant sets and KAM islands) and, (b) Hyperbolic chaotic scattering for $E = 1$ (chaotic invariant set)

(14). From Fig. 12, we see that for the hyperbolic case, at small ε the energy autocorrelation function deviates somewhat from the semiclassical Lorentzian prediction. This deviation may result from the failure of the assumptions used in deriving Eq. (14) from the semiclassical expression Eq. (13) for our finite $\hbar$ numerical calculation or from the failure of the validity of the long time limit of the semiclassical theory itself [14, 27]. For the nonhyperbolic case, Fig. 3 suggests that the quantum calculation agrees reasonably well with the semiclassical theory at small ε, and the energy autocorrelation function appears to follow the predicted cusp near $\varepsilon = 0$. Even though the hyperbolic $C(\varepsilon)$ function deviates from the predicted Lorentzian, its magnitude is still higher than that of the nonhyperbolic case. This means that as compared with the hyperbolic scattering case, the fine scale fluctuations of S-matrix elements with energy are greatly enhanced in the nonhyperbolic case.

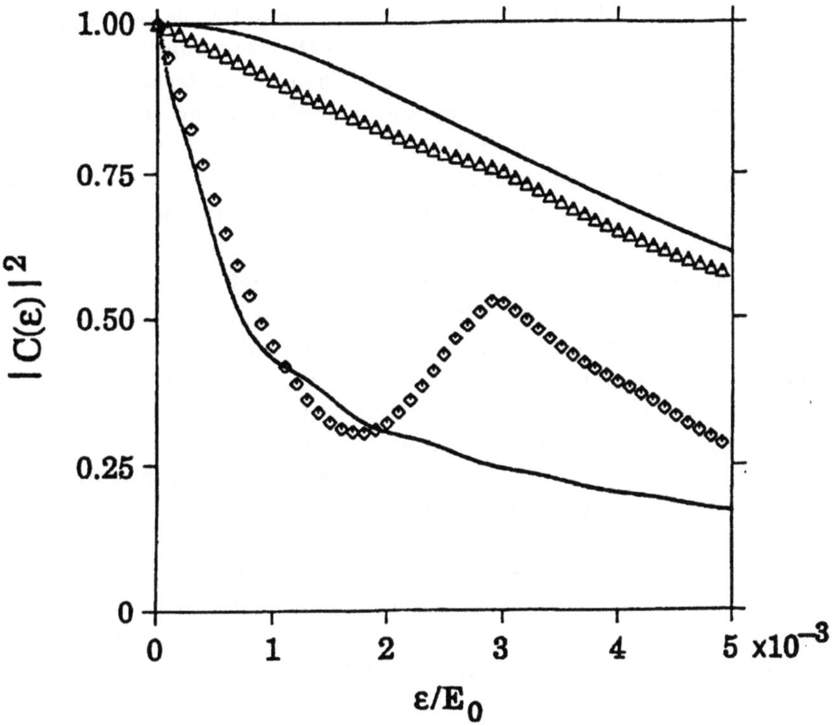

Fig. 12. Magnitude squared of S-matrix autocorrelation functions [diamonds for the nonhyperbolic case ($E_0 = 10$) and triangles for the hyperbolic case ($E_0 = 1$)] and the corresponding semiclassical predictions (the thick solid curves)

6 Conclusions and Discussions

We have presented new phenomena in both classical and quantum chaotic scattering. Our main results are as follows.

1. Crisis in classical hyperbolic chaotic scattering is characterized by heteroclinic tangencies of stable and unstable foliations of symmetric chaotic invariant sets. The physical manifestation is an increase of the fractal dimension

in the Cantor set of singularities in the scattering function. Consequently, chaotic scattering is enhanced due to the crisis. Crisis and the subsequent increase in the fractal dimension ***depend only on the occurrence of an infinite number of heteroclinic tangencies of the stable and unstable foliations.*** Since fractal stable and unstable foliations are a general feature of chaotic scattering systems, we expect crisis in chaotic scattering to be typical.

2. The algebraic decay exponent in classical nonhyperbolic chaotic scattering typically exhibits fluctuations as a system parameter changes when examined at finite times. This can be attributed to the continual process of breakup of KAM surfaces in the phase space. When the outermost KAM surfaces enclosing some central island undergo transformations from being absolute barriers to partial barriers (Cantori), the survival probability $P(t)$ displays two different regions of scaling behavior with different decay exponents. The time scale where this crossover takes place is found to coincide with the typical time for a particle to penetrate the newly created Cantori.

   It is a generally acknowledged notion that the algebraic decay in the presence of KAM surfaces should be a universal phenomenon with a single exponent. This notion is partly derived from the numerical works of Karney [20] and Chirikov and Shepelyansky [22]. In our opinion this may indeed be the case given the hierarchical island structure in the phase space. But to attain such a universal exponent one needs a prohibitively long time to perform the needed numerical calculations. In the case of actual experiments the required long period of observation time becomes even more unrealistic. On the other hand, the algebraic decay behavior itself is found to occur on relatively short time scales. The value of the decay exponent measured over the time interval accessible to numerical experiments is influenced by the presence of dominant island structures. In this regard the variations of the exponent can, in fact, be utilized to reflect the major qualitative changes in the system. For this reason we believe that the phenomena occurring on short time scales, such as these investigated in this paper, should be given appropriate emphasis and considerations.

3. For quantum manifestations of both hyperbolic and nonhyperbolic chaotic scattering, we have presented theoretical and numerical evidence that, in the semiclassical regime, the quantum energy autocorrelation function decreases more rapidly away from $\varepsilon = 0$ in the case of nonhyperbolic chaotic scattering. Thus the fine energy scale fluctuations of S-matrix elements are characteristically greatly enhanced in nonhyperbolic chaotic scattering as compared with the case of hyperbolic chaotic scattering.

Acknowledgments

I deeply thank my collaborators at various stages of the works reported in this paper. They are: Prof. C. Grebogi, Dr. R. Blümel, Prof. I. Kan, Prof. M. Ding and Prof. E. Ott. This work was partially supported by the Kansas Institute for Theoretical and Computational Science through the NSF/K*STAR program.

References

1. D.L. Rod, J. Diff. Eqs. **14**, 129(1973);
R.C. Churchill, G. Pecelli and D.L. Rod, J. Diff. Eqs. **17**, 329(1975);
M.C. Gutzwiller, Physica **D7**, 341(1983);
C. Jung, J. Phys. **A19**, 1345(1986);
B. Eckhardt, Europhys. Lett. **61**, 329(1988);
P. Gaspard and S.A. Rice, J. Chem. Phys. **90**, 2225, 2242, 2255(1989);
Z. Kovács and T. Tél, Phys. Rev. Lett. **64**, 1617(1990);
Y.C. Lai and C. Grebogi, Int. J. Bif. & Chaos **1**, 667(1991). For recent advances, see Chaos **3** (4), Focus Issus on Chaotic Scattering, eds. T. Tél and E. Ott (1993).
2. M. Hénon, Icarus **66**, 536(1986); Physica **D33** 132(1988);
P.T. Boyd and S.L.W. McMillan, Chaos **4**, 507(1993).
3. B. Eckhardt and H. Aref, Trans. Soc. Roy. Lond. **A326**, 655(1988);
J.B. Kadtke and E.A. Novikov, Chaos **3**, 543(1993);
C. Jung, E. Tél and E.M. Ziemniak, Chaos **3**, 555(1993);
T.H. Soloman, E.R. Weeks and H.L. Swinney, Physica **D76**, 70(1994);
E.M. Ziemniak, C. Jung and T. Tél, Physica **D76**, 123(1994).
4. D.W. Noid, S. Gray and S.A. Rice, J. Chem. Phys. **84**, 2649(1986).
5. R.A. Jalabert, H.U. Baranger and A.D. Stone, Phys. Rev. Lett. **65**, 2442(1990);
H.U. Baranger, R.A. Jalabert and A.D. Stone, Chaos **4**, 665(1993);
C.M. Marcus, R.M. Westervelt, P.F. Hopkins and A.C. Gossard, Chaos **4**, 643(1993);
W.A. Lin, J.B. Delos and R.V. Jensen, Chaos **4**, 655(1993).
6. S. Bleher, E. Ott and C. Grebogi, Phys. Rev. Lett. **63**, 919(1989);
S. Bleher, C. Grebogi and E. Ott, Physica **D46**, 87(1990).
7. M. Ding, C. Grebogi, E. Ott and J.A. Yorke, Phys. Rev. **A42**, 7025(1990).
8. M.J. Feigenbaum, J. Stat. Phys. **19**, 25(1978).
9. M. Ding, C. Grebogi, E. Ott and J.A. Yorke, Phys. Lett. **A153**, 21(1991).
10. Y.C. Lai, C. Grebogi, R. Blümel and I. Kan, Phys. Rev. Lett. **71**, 2212(1993);
Y.C. Lai and C. Grebogi, Phys. Rev. **E49**, 3761(1994).
11. Y.T. Lau, J.M. Finn and E. Ott, Phys. Rev. Lett. **66**, 978(1991).
12. Y.C. Lai, M. Ding, C. Grebogi and R. Blümel, Phys. Rev. **A46**, 4661(1992).
13. R. Blümel and U. Smilansky, Phys. Rev. Lett. **60**, 477(1988) and Physica **D36**, 111(1989).
14. Y.C. Lai, R. Blümel, E. Ott and C. Grebogi, Phys. Rev. Lett. **68**, 3491(1992).
15. G. Troll and U. Smilansky, Physica **D35**, 34(1989);
G. Troll, Physica **D50**, 276(1991); Chaos **4**, 459(1993).
16. C. Grebogi, E. Ott and J.A. Yorke, Phys. Rev Lett. **48**, 1507(1982) and Physica **D7**, 181(1983);
C. Grebogi, E. Ott, F. Romeiras and J.A. Yorke, Phys. Rev. **A36**, 5365(1987);
Y.C. Lai, C. Grebogi and J.A. Yorke, in *Applied Chaos*, pp. 441–455, EPRI Workshop on Applications of Chaos, eds. J.H. Kim and J. Stringer, John Wiley & Sons, 1992.
17. R.D. Woods and D.S. Saxon, Phys. Rev. **95**, 577(1954);
P. Ring and P. Schuck, *The Nuclear Many-Body Problem*, Springer-Verlag, New York, Heidelberg, Berlin, 1980.
18. C. Grebogi, S.W. McDonald, E. Ott and J.A. Yorke, Phys. Lett. **A99**, 415(1983);
S.W. McDonald, C. Grebogi, E. Ott and J.A. Yorke, Physica **D17**, 125(1985).

19. C. Grebogi, H.E. Nusse, E. Ott and J.A. Yorke, in *Springer Lecture Notes in Mathematics*, Vol. 1342 (Dynamical systems), Ed. J. C. Alexander (Springer-Verlag, New York), pp.220–250;
H.E. Nusse and J.A. Yorke, Commun. Math. Phys. **150**, 1(1992).
20. C.F.F. Karney, Physica **D8**, 360(1983).
21. J.D. Meiss, J.R. Cary, C. Grebogi, J.D. Crawford, A.N. Kaufman and H.D.I. Abarbanel, Physica **D6**, 375(1983);
22. B.V. Chirikov and D.L. Shepelyansky, Physica **D13**, 394(1984).
23. J.D. Meiss and E. Ott, Phys. Rev. Lett. **55**, 2741(1985); Physica **D20**, 387(1986).
24. R.S. MacKay, J.D. Meiss and I.C. Percival, Phys. Rev. Lett **52**, 697(1984).
25. Y.C. Lai, C. Grebogi, R. Blümel and M. Ding, Phys. Rev. **A45**, 8284(1992).
26. W.H. Miller, Adv. Chem. Phys. **30**, 77(1975).
27. Y.C. Lai, E. Ott and C. Grebogi, Phys. Lett. **A173**, 148(1993).
28. J. Koringa, Physica **13**, 392(1947);
W. Kohn and N. Rostoker, Phys. Rev. **94**, 1111(1954);
M.V. Berry, Ann. Phys. (New York) **131**, 163(1981).

From Fractals to Stochastic Differential Equations

Andrzej Lasota

[1] Institute of Mathematics, Silesian University,
PL–40–007 Katowice
[2] Institute of Mathematics, Polish Academy of Sciences,
PL–40–013 Katowice

Abstract: New sufficient conditions for asymptotic stability of Markov operators on metric spaces are presented. These criterion are applied to iterated function systems and stochastic differential equations with Poisson type perturbations.

Introduction

The purpose of this lecture is to present two sufficient conditions for asymptotic stability of Markov operators acting on a metric space. The first condition is based on double contraction principle which is a generalization of the classical Banach fixed point theorem. The second is related to the lower bound technique developed in [12].

We will apply our criterion to Markov operators generated by iterated function systems. This class of systems under the name of "systems with a complete connection" was already introduced by Onicescu and Mihoc [13] in 1935. Now iterated function systems are studied because of the close connection to fractals [1], [9], learning models [6], [7] and the theory of recursive estimation [3].

We will also discuss stochastic differential equations with Poisson type perturbations. Using double contraction principle we will show an effective criterion of asymptotic stability for such equations.

It should be underlined that our stability criterion are valid in a general class of metric spaces (so called Polish spaces) which are not necessarily locally compact. Thus these results are applicable to infinite dimensional systems.

The paper is organized as follows. Section 1 contains some notations and definitions concerning measures on metric spaces. In Section 2 we recall basic notions from the theory of Markov operators. Section 3 is devoted to the proof of double contraction principle and its corollaries for Markov operators. In Section 4 we discuss the relationship between iterated function systems and fractals and using double contraction principle we prove simultaneously convergence of measures and sets. In Section 5 we study a Poisson driven differential equation and we show its relation to some biological problems (cell cycle). Section 6 contains an outline of the proof of Lower bound theorem and Section 7 its application to learning systems.

1 Preliminaries

Let (X, ρ) be a metric space. Throughout this lecture we assume that X is complete and separable. In other words *X is a Polish space.* This assumption will not be repeated in the statements of theorems.

By $\mathcal{B}$ we denote the σ-algebra of Borel subsets of X and by $\mathcal{M}$ the family of all finite Borel measures (nonnegative, σ-additive) on X. By $\mathcal{M}_1$ we denote the subset of $\mathcal{M}$ such that $\mu(X) = 1$ for $\mu \in \mathcal{M}_1$. The elements of $\mathcal{M}_1$ will be called *distributions.* Further

$$\mathcal{M}_{sig} = \{\mu_1 - \mu_2 : \mu_1, \mu_2 \in \mathcal{M}\}$$

is the space of finite signed measures.

As usually by $B(X)$ we denote the space of all bounded Borel measurable functions $f : X \to R$ (R = real line) and by $C(X)$ the subspace of all bounded continuous functions. In both spaces the norm is

$$\|f\| = \sup_{x \in X} | f(x) | \quad .$$

For an $f \in B(X)$ and $\mu \in \mathcal{M}_{sig}$ we write

$$\langle f, \mu \rangle = \int_X f(x)\mu(dx) \quad .$$

The indicator function of a set A will be denoted by 1_A. Thus we have

$$\langle 1_A, \mu \rangle = \mu(A) \qquad \text{for } A \in \mathcal{B} \ .$$

In the space $\mathcal{M}_{sig}$ we introduce the *Fortet–Mourier norm* ([4], [14]) by setting

$$\|\mu\|_{\mathcal{F}} = \sup\{|\langle f, \mu \rangle| : f \in \mathcal{F}\} \ , \tag{1.1}$$

where

$$\mathcal{F} = \{f \in C(X) : \|f\| \le 1 \ \text{ and } \ |f(x) - f(y)| \le \rho(x, y) \ \text{ for } \ x, y \in X\} \ .$$

The space $\mathcal{M}_{sig}$ with the norm $\|.\|_{\mathcal{F}}$ is a normed vector space. In general, however, $(\mathcal{M}_{sig}, \|\cdot\|_{\mathcal{F}})$ is not a complete space. On the other hand the space $\mathcal{M}_1$ with the Fortet–Mourier metric

$$\|\mu_1 - \mu_2\|_{\mathcal{F}} \qquad \text{for} \qquad \mu_1, \mu_2 \in \mathcal{M}_1$$

is a complete metric space. A simple proof of this fact is given in [16].

In the space $\mathcal{M}_1$ we may introduce another distance called *Hutchinson metric* [5] by the formula

$$\|\mu_1 - \mu_2\|_{\mathcal{H}} = \sup\{| \langle f, \mu_1 - \mu_2 \rangle |: f \in \mathcal{H}\} \ , \tag{1.2}$$

where

$$\mathcal{H} = \{f \in C(X) : | f(x) - f(y) | \leq \rho(x,y) \text{ for } x,y \in X\} .$$

The metric (1.2) is always defined but for some $\mu_1, \mu_2 \in \mathcal{M}_1$ may be infinite. In the case when X is a bounded space, the Fortet–Mourier metric (1.1) and Hutchinson metric (1.2) are equivalent. More precisely

$$||\mu_1 - \mu_2||_{\mathcal{F}} \leq ||\mu_1 - \mu_2||_{\mathcal{H}} \leq c||\mu_1 - \mu_2||_{\mathcal{H}} \qquad \text{for} \qquad \mu_1, \mu_2 \in \mathcal{M}_1 ,$$

where

$$c = \text{diam}X = \sup\{\rho(x,y) : x,y \in X\} .$$

We say that a sequence $(\mu_n), \mu_n \in \mathcal{M}_1$, *converges weakly* to a measure $\mu \in \mathcal{M}_1$ if

$$\lim_{n\to\infty} \langle f, \mu_n \rangle = \langle f, \mu \rangle \qquad \text{for } f \in C(X) . \tag{1.3}$$

It should be underlined that in this lecture the weak convergence of measures is only defined for $\mu_n, \mu \in \mathcal{M}_1$. It can be proved that (1.3) is equivalent to

$$\lim_{n\to\infty} ||\mu_n - \mu||_{\mathcal{F}} = 0 .$$

If $x \in X$ and $A \subset X$, $A \neq \emptyset$ are given, then

$$\rho(x,A) = \inf\{\rho(x,y) : y \in A\}$$

denotes the distance from x to A. Further

$$N(A,r) = \{y \in X : \rho(y,A) \leq r\}$$

is the closed neighbourhood of the set A with radius r ($r > 0$). In particular $N(x,r)$ is the closed ball with the center x and radius r.

Given $\mu \in \mathcal{M}$ we define the *support* of μ by the formula

$$\text{supp}\mu = \{x : \mu(N(x,r)) > 0 \text{ for every } r > 0\} .$$

Evidently suppμ is a closed set and $\mu(A) = 0$ for $A \subset X\setminus$ suppμ. For a function $f : X \to R$ we write

$$\text{supp}f = \text{cl}\{x : f(x) \neq 0\} ,$$

where cl stands for the closure.

2 Markov Operators

A linear mapping $P : \mathcal{M}_{sig} \to \mathcal{M}_{sig}$ is called a *Markov operator* if $P(\mathcal{M}_1) \subset \mathcal{M}_1$. Thus, for every distribution μ the measure $P\mu$ is also a distribution.

A Markov operator P is called a *Feller operator*, if there is an operator $U : B(X) \to B(X)$, dual to P, such that

$$\langle Uf, \mu \rangle = \langle f, P\mu \rangle \qquad \text{for } f \in B(X), \ \ \mu \in \mathcal{M}_{sig} \tag{2.1}$$

and

$$Uf \in C(X) \qquad \text{for } f \in C(X) \ . \tag{2.2}$$

Setting $\mu = \delta_x$ in (2.1) we obtain

$$Uf(x) = \langle f, P\delta_x \rangle \qquad \text{for } f \in B(X), \ \ x \in X \ , \tag{2.3}$$

where $\delta_x \in \mathcal{M}_1$ is a point (Dirac) measure supported at x.

From formula (2.3) it follows immediately that U is a linear operator satisfying the following conditions

$$Uf \geq 0 \qquad \text{for } f \in B(X), \ \ f \geq 0 \ , \tag{2.4}$$

$$U1_X = 1_X \ . \tag{2.5}$$

Further using Lebesgue monotone convergence theorem applied to the integral $\langle f, P\delta_x \rangle$ we obtain the following implication

$$\left. \begin{array}{l} f_n \in B(X) \\ f_{n+1} \leq f_n \\ \lim_{n\to\infty} f_n(x) = 0 \end{array} \right\} \Longrightarrow \lim_{n\to\infty} Uf_n(x) = 0 \ . \tag{2.6}$$

Conditions (2.4)–(2.6) are quite important. They allow to reverse the role of P and U. Namely assume that a linear operator $U : B(X) \to B(X)$ satisfying conditions (2.4), (2.5) and (2.6) is given. Then we may define an operator $P : \mathcal{M}_{sig} \to \mathcal{M}_{sig}$ by setting

$$P\mu(A) = \langle U1_A, \mu \rangle \qquad \text{for } \mu \in \mathcal{M}_{sig}, \ A \in \mathcal{B} \ . \tag{2.7}$$

From the linearity of U it follows that $P\mu$ is a finitely additive set function. Further (2.6) implies that $P\mu$ satisfies the continuity condition

$$\left. \begin{array}{l} A_n \in \mathcal{B} \\ A_{n+1} \subset A_n \\ \bigcap_{n=1}^{\infty} A_n = \emptyset \end{array} \right\} \Longrightarrow \lim_{n\to\infty} P\mu(A_n) = 0 \ .$$

Thus $P\mu$ is countably additive. The verification of the remaining conditions appearing in the definition of a Markov operator is trivial. It is also easy to show by an approximation argument that P given by (2.7) satisfies (2.1). Finally, if U satisfies (2.2) then P is a Markov operator.

The possibility of reversing the role of P and U is convenient in applications, since in many cases it is much easier to find an explicit formula for U then for P. This will be visible in Section 4 devoted to iterated function systems.

Assume now that P and U are given. If $f : X \to R_+$ ($R_+ = [0, \infty)$) is a Borel measurable function, not necessarily bounded, we may admit that

$$Uf(x) = \lim_{n\to\infty} Uf_n(x) \ , \tag{2.8}$$

where (f_n), $f_n \in B(X)$, is an increasing sequence of functions converging pointwise to f. From the Lebesgue monotone convergence theorem it follows that Uf defined by (2.8) satisfies (2.1) and (2.3). In particular (2.3) shows that this extension does not depend of the particular choice of the sequence (f_n). Of course Uf may not be bounded or even finite valued function.

The operators P and U have an important probabilistic interpretation. Assume that we have given a probabilistic space $(\Omega, \mathcal{F}, \text{prob})$ and a family of random elements $x_{n,\mu} : \Omega \to X$ indexed by two parameters $n = 0, 1, \ldots$ and $\mu \in \mathcal{M}_1$. Further assume that

$$\text{prob}(x_{0,\mu} \in A) = \mu(A) \qquad \text{for } \mu \in \mathcal{M}_1 \ .$$

For every fixed μ the sequence $(x_{n,\mu})$ may be considered as a process starting from the initial point $x_{0,\mu}$ which has the distribution μ. From the statistical point of view this process is described by the sequence of distributions

$$\mu_{n,\mu}(A) = \text{prob}(x_{n,\mu} \in A) \ .$$

If there is a Markov operator P such that

$$\mu_{n+1,\mu} = P_{n,\mu} \qquad \text{for } \ n = 0, 1, \ldots, \quad \mu \in \mathcal{M}_1$$

we say that P is the *transition operator* for the process $(x_{n,\mu})$.

The theory of Markov processes assures that for every Markov process such operator always exists. By this reason P is called a Markov operator. However we will not use here any results from the theory of stochastic and in particular Markov operators. We have recalled the probabilistic interpretation of P only to support our intuition.

The same family $(x_{n,\mu})$ allows also to interpret the dual operator U. Namely from formula (2.3) it follows that $Uf(x)$ is the mathematical expectation of the value $f(x_{1,\mu})$ if $\mu = \delta_x$. In other words $Uf(x)$ is the expectation in time $n = 1$ of the parameter f defined on the phase space X assuming that the initial distribution ($n = 0$) was concentrated at the point x.

It is customary to omit the subscript μ in $x_{n,\mu}$ and $\mu_{n,\mu}$. Thus we have

$$\mu_{n+1} = P\mu_n, \qquad \mu_n(A) = \text{prob}(x_n \in A) \quad \text{for } \ n = 0, 1, \ldots; \quad A \in \mathcal{B} \ . \tag{2.9}$$

In our study of the asymptotic behaviour of the iterates (P^n) of a Markov operator P the notions of a stationary measure and asymptotic stability play an important role.

A measure μ is called *stationary* (or *invariant*) with respect to a Markov operator P if $P\mu = \mu$. A Markov operator is called *asymptotically stable if* there is a stationary distribution μ_* such that

$$\lim_{n\to\infty} \|P^n\mu - \mu_*\|_{\mathcal{F}} = 0 \qquad \text{for } \mu \in \mathcal{M}_1 \ . \tag{2.10}$$

A distribution μ_* satisfying (2.10) is unique. However in general condition (2.10) does not imply that μ_* is stationary. This implication holds for Feller operators.

3 Double Contraction Principle

Both norms $\|\cdot\|_{\mathcal{F}}$ and $\|\cdot\|_{\mathcal{H}}$ introduced in Section 1 have some disadvantages. The first one is not decreasing by application of many important asymptotically stable Markov operators. The second, Hutchinson norm, is infinite for some measures on unbounded metric spaces. However the simultaneous application of both norms may be quite efficient.

To make such application possible we prove an extension of the Banach contraction principle which we call *double contraction principle.*

Theorem 3.1. *Assume that $T : E \to E$ defined on a complete metric space (E, d) is nonexpansive, i.e.*

$$d(T(u), T(v)) \leq d(u, v) \qquad \text{for } u, v \in E \ . \tag{3.1}$$

Assume, moreover, that there is a subset $E_0 \subset E$ and a metric $d_0 : E_0 \times E_0 \to R_+$ such that the following conditions are satisfied:

(i) *E_0 is dense in E with respect to the metric d and invariant under T, i.e. $T(E_0) \subset E_0$.*
(ii) *d_0 is stronger then d, i.e.*

$$d(u, v) \leq d_0(u, v) \qquad \textit{for} \quad u, v \in E_0 \ . \tag{3.2}$$

(iii) *T restricted to E_0 is contractive with respect to d_0, i.e.*

$$d_0(T(u), T(v)) \leq \lambda d_0(u, v) \qquad \textit{for} \quad u, v, \in E_0 \ , \tag{3.3}$$

where $\lambda < 1$ is a constant.
Then T has in E a unique fixed point u_* and

$$\lim_{n\to\infty} d(T^n(u), u_*) = 0 \qquad \textit{for} \quad u \in E \ . \tag{3.4}$$

Proof. As in the proof of Banach contraction principle first we are going to show that for every $u \in E$ the sequence of iterates $(T^n(u))$ satisfies the Cauchy condition with respect to the metric d. Observe that from (3.1) and (3.2) the inequality

$$d(T^n(u), T^m(u)) \leq 2d(u, v) + d_0(T^n(v), T^m(u)) \tag{3.5}$$

follows which is valid for every $u \in E$, $v \in E_0$, and every pair of positive integers n, m. Given $u \in E$ and $\varepsilon > 0$ we may choose $v \in E_0$ such that $d(u, v) < \varepsilon/4$. Further using (3.3) it can be verified that there is an integer n_0 such that

$$d_0(T^m(v), T^n(v)) \leq \frac{\varepsilon}{2} \qquad \text{for } n, m \geq n_0 \ .$$

Thus inequality (3.5) implies that $(T^n(u))$ satisfies the Cauchy condition. Since (E, d) is a complete metric space the sequence $(T^n(u))$ is convergent to an element $u_* \in E$. The point u_* is evidently (continuity of T) a fixed point of T. It remains to prove that T has at most one fixed point. To show this assume that u_1 and u_2 are fixed points. Again using (3.1) and (3.2) we have

$$\begin{aligned} d(u_1, u_2) = &\ d(T^n(u_1), T^n(u_2)) \\ \leq &\ d(u_1, v_1) + d(u_2, v_2) + d_0(T^n(v_1), T^n(v_2)) \end{aligned} \tag{3.6}$$

for arbitrary integer $n \geq 0$ and $v_1, v_2 \in E_0$. According to (3.3) the last term in (3.6) converges to zero as $n \to \infty$. Therefore

$$d(u_1, u_2) \leq d(u_1, v_1) + d(u_2, v_2) \qquad \text{for } v_1, v_2 \in E_0 \ .$$

Since E_0 is dense in E, this shows that $d(u_1, u_2) = 0$ and completes the proof. □

Now we are going to apply Theorem 3.1 to Markov operators.

Theorem 3.2. *Let $P : \mathcal{M}_{sig} \to \mathcal{M}_{sig}$ be a Markov Feller operator and let U be its dual. Assume that there is a constant $\lambda < 1$ such that*

$$| Uf(x) - Uf(y) | \leq \lambda \rho(x, y) \qquad \text{for } x, y \in X \qquad \text{and } f \in \mathcal{H} \ . \tag{3.7}$$

Moreover, assume that for some $\mu_0 \in \mathcal{M}_1$

$$\|\mu_0 - P\mu_0\|_{\mathcal{H}} < \infty \ . \tag{3.8}$$

Then P is asymptotically stable.

Proof. First we are going to verify that P with respect to Fortet–Mourier norm is nonexpansive. For $f \in \mathcal{F}$ we have $|f| \leq 1$ which gives $|Uf| \leq 1$ because of the normalization property of U. This and condition (3.7) imply that $Uf \in \mathcal{F}$ for $f \in \mathcal{F}$. Consequently

$$\begin{aligned} \|P\mu_1 - P\mu_2\|_{\mathcal{F}} = &\ \sup_{f \in \mathcal{F}} | \langle f, P\mu_1 - P\mu_2 \rangle | \\ = &\ \sup_{f \in \mathcal{F}} | \langle Uf, \mu_1 - \mu_2 \rangle | \\ \leq &\ \sup_{f \in \mathcal{F}} | \langle f, \mu_1 - \mu_2 \rangle | = \|\mu_1 - \mu_2\|_{\mathcal{F}} \ . \end{aligned}$$

Analogous calculation shows that P is contractive with respect to Hutchinson norm. Namely, according to (3.7) we have $\lambda^{-1}Uf \in \mathcal{H}$ for $f \in \mathcal{H}$ (we may always assume $\lambda > 0$). Therefore

$$\begin{aligned}
\|P\mu_1 - P\mu_2\|_{\mathcal{H}} &= \sup_{f\in\mathcal{H}} \mid \langle f, P\mu_1 - P\mu_2\rangle \mid \\
&= \sup_{f\in\mathcal{H}} \mid \langle Uf, \mu_1 - \mu_2\rangle \mid \\
&= \lambda \sup_{f\in\mathcal{H}} \mid \langle \lambda^{-1}Uf, \mu_1 - \mu_2\rangle \mid \\
&\leq \lambda \sup_{f\in\mathcal{H}} \mid \langle f, \mu_1 - \mu_2\rangle \mid = \lambda\|\mu_1 - \mu_2\|_{\mathcal{H}} \ .
\end{aligned}$$

Now define

$$\mathcal{M}_0 = \{\mu \in \mathcal{M}_1 : \|\mu_0 - \mu\| < \infty\} \ .$$

The set $\mathcal{M}_0$ is invariant. In fact, if $\mu \in \mathcal{M}_0$ then

$$\begin{aligned}
\|\mu_0 - P\mu\|_{\mathcal{H}} &\leq \|\mu_0 - P\mu_0\|_{\mathcal{H}} + \|P\mu_0 - P\mu\|_{\mathcal{H}} \\
&\leq \|\mu_0 - P\mu_0\|_{\mathcal{H}} + \lambda\|\mu_0 - \mu\|_{\mathcal{H}} \ .
\end{aligned}$$

Since $\mu \in \mathcal{M}_0$, the right hand side is finite.

It remains to verify that $\mathcal{M}_0$ is dense in $\mathcal{M}_1$ with respect to Fortet–Mourier norm. Given $\mu \in \mathcal{M}_1$ fix a point $x_0 \in X$ and choose $r > 0$ such that

$$\mu(N_r) = \mu(N(x_0, r)) > 0 \ .$$

Define

$$\mu_r(A) = c_r\mu(A \cap N_r) + \mu_0(A \cap (X \setminus N_r)) \ ,$$

where $c_r = \mu_0(N_r)/\mu(N_r)$. Evidently $\mu_r \in \mathcal{M}_1$ and the support of $\mu_0 - \mu_r$ is contained in N_r. Thus $\|\mu_0 - \mu_r\|_{\mathcal{H}} \leq r$ which shows that $\mu \in \mathcal{M}_0$. Further an elementary calculation shows that

$$\|\mu - \mu_r\|_{\mathcal{F}} \leq |1 - c_r| + \mu_0(X \setminus N_r) \ .$$

Since $c_r \to 1$ and $\mu_0(X \setminus N_r) \to 0$ as $r \to \infty$, this completes the proof. □

In order to make condition (3.8) more effective we may calculate $\|\mu_0 - P\mu_0\|_{\mathcal{H}}$ for $\mu_0 = \delta_{x_0}$, where $x_0 \in X$ is an arbitrary point. For $f \in \mathcal{H}$ we have

$$|\langle f, \delta_{x_0}\rangle - \langle f, P\delta_{x_0}\rangle| = |f(x_0) - Uf(x_0)| \leq (U\varrho_{x_0})(x_0) \ ,$$

where $\varrho_{x_0}(x) = \varrho(x, x_0)$. Thus, condition (3.8) may be replaced by

$$U\varrho_{x_0}(x_0) < \infty \ . \tag{3.9}$$

Observe that now all conditions of Theorem 3.1 are expressed in terms of the dual operator U. We may summarize this result in the following

Corollary 3.3. *If P is a Feller operator and its dual satisfies* (3.7) *and* (3.9) *with some $\lambda < 1$ and $x_0 \in X$, then P is asymptotically stable.*

4 Iterated Function System (IFS)

An Iterated Function System (IFS) is given by a sequence of continuous transformations

$$S_i : X \to X, \qquad i = 1, \dots, N \ ,$$

and a probabilistic vector

$$p_i : X \to R, \qquad i = 1, \dots, N \ ,$$

where p_i are continuous function satisfying

$$p_i(x) \geq 0, \quad \sum_{i=1}^{N} p_i(x) = 1 \quad \text{for } x \in X \ .$$

Such a system is shortly denoted by

$$(S, p)_N = (S_1, \dots, S_N;\ p_1, \dots, p_N) \ .$$

The action of an IFS can be roughly described as follows. We choose an initial point x_0 and we randomly select from the set $\{1, \dots, N\}$ an integer in such a way that the probability of choosing k is $p_k(x_0)$. If a number k_0 is drawn, we define $x_1 = S_{k_0}(x_0)$. Having x_1 we select k_1 in such a way that the probability of choosing k is $p_k(x_1)$. Now we define $x_2 = S_{k_1}(x_1)$ and so on. This process can be easily formalized. Namely given $f \in B(X)$ we may calculate the mathematical expectation of $f(x_1)$ in the case when $x_0 = x$ is a constant. The probability of gaining the value $f(S_i(x))$ is $p_i(x)$ and the expectation of $f(x_1)$ is equal to

$$\sum_{i=1}^{N} p_i(x) f(S_i(x)) \ .$$

On the other hand as we know this expectation is equal to $Uf(x)$ where U is dual to the transition operator P. Thus we have

$$Uf(x) = \sum_{i=1}^{N} p_i(x) f(S_i(x)) \ . \tag{4.1}$$

Evidently U given by (4.1) satisfies conditions (2.2), (2.4), (2.5) and (2.6) of the dual operator and we may calculate P by formula (2.7). This gives

$$P\mu(A) = \langle 1_A, P\mu \rangle = \langle U1_A, \mu \rangle$$

$$\int_X \left\{ \sum_{i=1}^{N} p_i(x) 1_A(S_i(x)) \right\} \mu(dx)$$

or

$$P\mu(A) = \sum_{i=1}^{N} \int_{S_i^{-1}(A)} p_i(x)\mu(dx) \ . \tag{4.2}$$

We admit (4.2) as the formal definition of the *transition operator* for the iterated function system $(S,p)_N$. To simplify the language we will say that the iterated function system $(S,p)_N$ has an invariant density or is asymptotically stable if the Markov operator (4.2) has the corresponding property.

In the remaining part of this section we will mainly consider the case when the vector $(p_1,\dots,p_N)$ is constant. Further we assume that transformations S_k are Lipschitzean with constants L_k i.e.

$$\varrho(S_k(x),S_k(y)) \le L_k\varrho(x,y) \qquad \text{for} \quad x,y\in X;\ k=1,\dots,N\ . \tag{4.4}$$

In this case, using double contraction principle we obtain the following criterion.

Theorem 3.2. *If the iterated function system* $(S,p)_N$ *with constant* p_k *and Lipschitzean* S_k, $k=1,\dots,N$, *satisfies*

$$\sum_{i=1}^{N} p_i L_i < 1\ ,$$

then it is asymptotically stable.

Proof. Since p_i are constant, the dual operator has the form

$$Uf(x)=\sum_{i=1}^{N} p_i f(S_i(x))$$

and for every $f\in\mathcal{H}$ we obtain

$$\begin{aligned}|Uf(x)-Uf(y)| &\le \sum_{i=1}^{N} p_i|f(S_i(x))-f(S_i(y))|\\ &\le \Big(\sum_{i=1}^{N} p_iL_i\Big)\varrho(x,y)\ .\end{aligned}$$

Thus U satisfies condition (3.7) with

$$\lambda = \sum_{i=1}^{N} p_iL_i < 1\ .$$

Further

$$U\varrho_{x_0}(x_0) = \sum_{i=1}^{N} p_i\varrho(S_i(x_0),x_0)$$

is evidently a finite number. An application of Corollary 3.3 completes the proof.
□

The above theorem can be extended to iterated function systems with variable p_i under additional assumption that X is locally compact. We will discuss this problem in Section 6.

Iterated function systems with constant p_i give almost ideal tool for constructing fractals. This fact is well known and we are going to present it here briefly, underlying the difficulties related to the lack of compactness.

First we introduce an operator acting on subsets of X. Namely, given an iterated function system $(S,p)_N$, we define the Barnsley operator by the formula

$$F(A) = \text{ cl} \bigcup_{i=1}^{N} S_i(A) \qquad \text{for} \quad A \subset X \ . \tag{4.5}$$

The symbol cl denotes closure. It is not necessary when X is compact or even locally compact.

The relation between P and F is given by the following

Proposition 4.2. *Let $(S,p)_N$ be an iterated function system and let P and F denote corresponding Markov and Barnsley operators. Assume that $\inf p_k > 0$ for $k = 1, \ldots, N$. Then*

$$\text{supp}(P\mu) = F(\text{supp}\mu) \qquad \textit{for} \quad \mu \in M_1 \ . \tag{4.6}$$

The proof of similar proposition given in [9] can be extended without difficulties to our case.

From condition (4.6) it follows by an induction argument that

$$\text{supp}(P^n\mu) = F^n(\text{supp } \mu) \qquad \text{for} \quad \mu \in M_1 \tag{4.7}$$

and it is interesting to find under what condition this equality will be preserved in the limit as $n \to \infty$. To give the precise answer we need to introduce the notions of the Hausdorff distance and hyperbolic iterated function system.

Let $\mathcal{F}$ be the family of all closed bounded nonempty subsets of X. The *Hausdorff distance* between sets $A, B \in \mathcal{F}$ is given by

$$\text{dist } (A, B) = \inf \{r > 0 : \ A \subset N(B, r) \text{ and } B \subset N(A, r)\} \ ,$$

where $N(A, r)$ and $N(B, r)$ denote the neighbourhoods of the sets A, B with radius r. In other words dist(A, B) is the infimum of numbers r which have the following property: for every $x \in A$ there is $y \in B$ such that $\varrho(x, y) < r$ and vice versa for every $y \in B$ there is $x \in A$ such that the same inequality holds.

An iterated function system $(S,p)_N$ is called *hyperbolic* if p_i are constant and S_i are Lipschitzean with Lipschitz constants L_i which satisfy

$$L = \max_i L_i < 1 \ . \tag{4.8}$$

The main result giving the connection between operators F and P can be formulated as follows.

Theorem 4.3. *Let $(S,p)_N$ be a hyperbolic iterated function system and let A_* be the support of its invariant measure μ_*. Then*

$$\mathrm{dist}(F^n(A), A_*) = 0 \qquad \textit{for} \quad A \in \mathcal{F} \ . \tag{4.9}$$

Proof. Using the assumption (4.8) it is easy to verify that

$$\mathrm{dist}(F(A), F(B)) \leq L\ \mathrm{dist}(A,B) \qquad \text{for} \quad A, B \in \mathcal{F} \ .$$

By an induction argument this implies

$$\lim_{n\to\infty} \mathrm{dist}(F^n(A), F^n(B)) = 0 \qquad \text{for} \quad A, B \in \mathcal{F} \ . \tag{4.10}$$

Since $P\mu_* = \mu_*$, we have $F^n(A_*) = A_*$ according to (4.7). Substituting $B = A_*$ into (4.10) we obtain (4.9). □

From (4.9) it follows that A_* is the unique set in $\mathcal{F}$ satisfying $F(A_*) = A_*$. It is called the attractor of the iterated function system $(S,p)_N$.

According to Barnsley ([1] p. 80) fractals are simply the attractors of hyperbolic iterated function systems. It is interesting that using relatively simple tool – double contraction principle – it is possible to develop the theory of iterated function systems in Polish spaces.

5 Poisson Driven Stochastic Differential Equation

The iterated function systems discussed in preceding section were defined by a finite family of transformations. Analogous theory can be construct for infinite family. We are going to illustrate this possibility for a Poisson driven stochastic differential equation in a Banach space.

Consider differential equation of the form

$$du = a(u)dt + \sigma(u)dp \qquad \text{for} \quad t \geq 0 \tag{5.1}$$

with the initial condition

$$u(0) = x \ . \tag{5.2}$$

In (5.1) $u: R_+ \to X$ is an unknown function (stochastic process) with values in a Banach space $(X, \|\ \|)$. Further we assume that the coefficients $a, \sigma : X \to X$ are Lipschitzean and $p(t)$ is a Poisson process with intensivity λ. The initial value x in (5.2) is a random vector independent on $p(t)$. The problem of the uniqueness and the existence of solution for the initial value problem (5.1), (5.2) can be easily solved. To make it precise we will assume that every realization of $u(t)$ is right hand side continuous and satisfies (5.1) except of a countable number of points. The Poisson process $p(t)$ can be written in the form

$$p(t) = \sum_{t_n < t} \chi(t - t_n) \ ,$$

where χ is the Heaviside function

$$\chi(t) = \begin{cases} 0 & \text{for} \quad t < 0 \\ 1 & \text{for} \quad t \geq 0 \end{cases}$$

and (t_n), $n = 1, 2, \ldots$, is a sequence of random variables such that the increaments

$$\Delta t_1 = t_1 - t_0, \qquad \Delta t_2 = t_2 - t_1, \ldots \qquad (t_0 = 0)$$

are independent and have the same density distribution function $g(t) = \lambda e^{-\lambda t}$. Denoting by $v(t) = \pi(t, x)$ the solution of the unperturbed Cauchy problem

$$\dot{v} = a(v), \qquad v(0) = x \tag{5.3}$$

we have for the solution u of (5.1), (5.2) the explicit formula

$$u(t) = \pi(t - t_n, q(u(t_n))) \qquad \text{for} \quad t_n \leq t < t_{n+1} \ , \tag{5.4}$$

where $u(t_0) = u(0) = x$ and

$$q(z) = z + \sigma(z) \ .$$

For every realization of $p(t)$ formula (5.4) allows calculate the corresponding realization of $u(t)$.

In many applications we are mostly interested in the values of the solution u at the switching points t_n. Setting $x_n = u(t_n)$ and

$$S_t(x) = q(\pi(t, x)) \qquad \text{for} \quad t \geq 0, \ x \in X$$

we have according to (5.4)

$$x_{n+1} = S_{\Delta t_n}(x_n) \qquad \text{for} \quad n = 0, 1, \ldots \tag{5.5}$$

Equality (5.5) defines an iterated function system with infinite family of transformation $\{S_t\}$, $t \in R_+$. At each step the new transformation is selected according to density distribution function $\lambda e^{-\lambda t}$.

It is easy to find the transition operator for this system. As before we may start from the dual operator U. If $f \in B(X)$ is given, then the mathematical expectation of $f(x_1)$, provided $x_0 = x$ is constant, is equal to

$$\lambda \int_0^\infty f(S_t(x)) e^{-\lambda t} dt = \lambda \int_0^\infty f(q(\pi(t, x))) e^{-\lambda t} dt \ .$$

On the other hand the same expectation is given by $Uf(x)$. Therefore

$$Uf(x) = \lambda \int_0^\infty f(q(\pi(t, x))) e^{-\lambda t} dt \qquad \text{for} \quad x \in X, \ f \in C(X) \tag{5.6}$$

and

$$P\mu(A)=\int_X\left\{\int_0^\infty 1_A(q(\pi(t,x)))e^{-\lambda t}\right\}\mu(dx) \qquad \text{for} \quad A\in\mathcal{B},\ \mu\in\mathcal{M}_{sig}\ . \tag{5.7}$$

In order to formulate a simple criterion of stability we will assume that the solution $\pi(t,x)$ of unperturbed system satisfies

$$\|\pi(t,x)-\pi(t,y)\|\le e^{\alpha t}\|x-y\| \qquad \text{for} \quad x,y\in X\ , \tag{5.8}$$

where α is a real number not necessarily negative. Further we denote by L_q the Lipschitz constant of q, i.e.

$$\|q(x)-q(y)\|\le L_q\|x-y\| \qquad \text{for} \quad x,y\in X\ . \tag{5.9}$$

Having this we may formulate the following

Proposition 5.1. *Assume that X is a separable Banach space and the functions π and q satisfy conditions* (5.8), (5.9) *where the constants $\lambda>0,\ L_q\ge 0$ and $\alpha\in R$ satisfy inequality*

$$\frac{\alpha}{\lambda}+L_q<1\ . \tag{5.10}$$

Further assume that for some $x_0\in X$ the solution $\pi(t,x_0)$ of the unperturbed system is bounded, i.e.

$$\sup_{t\ge 0}\|\pi(t,x_0)\|<\infty\ . \tag{5.11}$$

Then the Markov operator P defined by (5.7) *is asymptotically stable.*

Proof. The space (X,ϱ) with metric $\varrho(x,y)=\|x-y\|$ is a Polish space. Thus, in order to complete the proof it suffices to verify conditions (3.7), (3.9) of Corollary 3.3. Using (5.6), (5.8) and (5.9) we have

$$|Uf(x)-Uf(y)|=\lambda\int_0^\infty|f(q(\pi(t,x)))-f(q(\pi(t,x)))|e^{-\lambda t}dt$$

$$\le\lambda L_q\|x-y\|\int_0^\infty e^{(\lambda-\alpha)t}dt=\frac{\lambda L_q}{\lambda-\alpha}\|x-y\| \qquad \text{for} \quad f\in\mathcal{H}\ .$$

Inequality (5.10) implies that $\lambda>\alpha$ and $\alpha L_q/(\lambda-\alpha)<1$. Condition (3.7) is satisfied. Further

$$U\varrho_0(x_0)=\lambda\int_0^\infty\|q(\pi(t,x_0))-x_0\|e^{-\lambda t}dt$$

$$=\|q(\pi(t,x_0)-x_0\|\le\|q(\pi(t,x_0\|+\|x_0\|$$

which according to (5.11) implies (3.9). The proof is completed. □

Equations similar to (5.4), (5.5) are discussed in the mathematical theory of the cell cycle [10], [11]. In this case $X = R^d$ and we assume [10] that each cell contains d–substances whose masses at time t are denoted by the vector

$$u(t) = (u^1(t), \dots, u^d(t)) \ .$$

The values $t_1, t_2, \dots$ denote the birth times and $x_n = u(t_n)$ represents the distribution of substances of cells just before mitosis in the nth generation. Thus it is naturally to assume that $q(x) = \frac{1}{2}x$ (or $\sigma(x) = -\frac{1}{2}x$), since after mitosis each daughter cell obtains exactly half of the components of the mother cell. In this case inequality (5.10) admits the form

$$\frac{\alpha}{\lambda} < \frac{1}{2} \tag{5.12}$$

which shows that even quite unstable system (5.3) governing the growth of the cell is stabilized be a relatively rapid cell proliferation. May be this is one of the reasons why all the living organisms are built of cells.

6 Lower Bound Theorem

We will present here a new sufficient condition for the asymptotic stability of Markov operators. In Section 7 it will be applied to iterated function systems with state dependent probability vector $(p_1(x), \dots, p_u(x))$. The proofs will be only outlined. In the case when X is locally compact (bounded closed subsets are compact) the complete proofs are given in [12]. For the Polish space X they will be published in [8].

First we will discuss the problem of the existence of an invariant distribution. For a Markov operator which is not a Feller one the invariant distribution may not exist, even if X is a compact space. Further the invariant distribution may not exist if X is not compact (e.g. real line) and the sequence of measures $(P^n\mu)$ "escapes" to infinity. Thus for the existence of a nontrivial invariant measure two conditions are important: 1) the Feller property, 2) a concentration condition which assures that at least a part of a measure $P^n\mu$ remains on a compact set when $n \to \infty$. This two conditions are easily seen in the following theorem.

Theorem 6.1. *Assume that P is a Feller operator and that there is a compact set $Y \subset X$ and a measure $\mu_0 \in \mathcal{M}_1$ such that*

$$\limsup_{n\to\infty} \frac{1}{n} \sum_{k=1}^{n} P^k\mu(Y) > 0 \ . \tag{6.1}$$

Then P has an invariant distribution.

Proof (outlined). Using (6.1) we may define an increasing sequence of integers (k_n) and an $\alpha > 0$ such that

$$\frac{1}{k_n} \sum_{i=0}^{k_n} P^i\mu_0(Y) \geq \alpha \qquad \text{for} \quad n = 1, 2, \dots \tag{6.2}$$

Then we define a functional $\varphi \in C^*(X)$ by the formula

$$\varphi(f) = \operatorname{Lim}\left(\frac{1}{k_n}\sum_{i=1}^{k_n}\langle f, P^i\mu_0\rangle\right) \tag{6.3}$$

where Lim denotes a Banach limit. This functional is evidently invariant, namely

$$\begin{aligned}\varphi(Uf) &= \operatorname{Lim}\left(\frac{1}{k_n}\sum_{i=1}^{k_n}\langle Uf, P^i\mu_0\rangle\right)\\ &= \operatorname{Lim}\left(\frac{1}{k_n}\sum_{i=1}^{k_n}\langle f, P^{i+1}\mu_0\rangle\right) = \varphi(f) \ .\end{aligned} \tag{6.4}$$

If X is compact, then by Riesz representation theorem there exists $\mu \in \mathcal{M}$ such that

$$\varphi(f) = \langle f, \mu\rangle \qquad \text{for} \quad f \in C(X) \ .$$

We have according to (6.4)

$$\langle f, P_\mu\rangle = \langle Uf, \mu\rangle = \varphi(Uf) = \varphi(f) = \langle f, \mu\rangle$$

for every $f \in C(X)$. This implies $P\mu = \mu$. Further from (6.1) and (6.2) it follows that $\varphi(1_X) \geq \alpha$ and consequently

$$\mu(X) = \langle 1_X, \mu\rangle = \varphi(1_X) \geq \alpha \ .$$

Evidently $\mu_* = \mu/\mu(X)$ is an invariant distribution.

When X is not compact, we may decompose φ into the sum

$$\varphi = \varphi_* + \varphi_0 \ ,$$

where $\varphi_*, \varphi_0 \in C^*(X)$ are nonnegative functionals and φ_0 is the maximal one given by a measure. Then again we may prove that φ_0 is invariant and nontrivial. □

The most important notion in this section is nonnexpansivness. We say that a Markov operator P is nonexpansive if

$$\|P\mu_1 - P\mu_2\|_{\mathcal{F}} \leq \|\mu_1 - \mu_2\|_{\mathcal{F}} \qquad \text{for} \quad \mu_1, \mu_2 \in \mathcal{M}_1 \ . \tag{6.5}$$

Observe that nonexpansivness was already used in Theorem 3.2 simultaneously with the contraction condition (3.3). Now condition (3.3) will be replaced by a concentration property.

Theorem 6.2. *Assume that P is a nonexpansive Markov operator. Further assume that for every $\varepsilon > 0$ there is a Borel set A with diam $A \leq \varepsilon$ and a number $\alpha > 0$ such that*

$$\liminf_{n\to\infty} P^u\mu(A) \geq \alpha \qquad \text{for} \quad \mu \in \mathcal{M}_1 \ . \tag{6.6}$$

Then P is asymptotically stable.

Proof (outlined). First we may show that P is a Feller operator. Namely settling

$$Uf(x) = \langle f, P\delta_x \rangle$$

we define U as our candidate for the dual operator. For every $f \in C(X)$ satisfying the Lipschitz condition

$$|f(x) - f(y)| \le k_f \varrho(x,y) \qquad \text{for} \quad x, y \in X$$

we have

$$\begin{aligned} |Uf(x) - Uf(y)| &= |\langle f, P\delta_x - P\delta_y \rangle| \le \|P\delta_x - P\delta_y\|_{\mathcal{F}} \\ &\le c_f \|\delta_x - \delta_y\| \le c_f \varrho(x,y) \ , \end{aligned}$$

where $c_f = \max(k_f, \sup |f|)$. Thus for Lipschitzean $f \in C(X)$ the function Uf is Lipschitzean. By an approximation argument (easy in compact spaces) it can be proved that Uf is continuous for every $f \in C(X)$ and that condition (2.1) holds. Thus P is a Feller operator. Now observe that condition (6.6) implies (6.1) and according to Theorem 6.1 the operator P has an invariant distribution.

Finally using (6.5) and (6.6) one can verify that for every $\mu_1, \mu_2 \in \mathcal{M}_1$ the sequence of norms

$$\|P^n \mu_1 - P^n \mu_2\|_{\mathcal{F}}$$

is decreasing and converges to zero. This and the existence of an invariant measure imply that P is asymptotically stable. □

7 Learning Systems

An iterated function system $(S,p)_N$ with state dependent probability vector

$$p(x) = (p_1(x), \ldots, p_N(x))$$

is quite often called a learning system. The system learns because in a new position x_n it uses a new strategy $p(x_n)$ for choosing the next stop.

The asymptotic behaviour of a learning system heavily depends on the properties of functions p_i. First of all they must be strictly positive. Consider, for example, the system $(S,p)_2$ acting on the space $X = [0,1]$ with the following transformations:

$$S_1(x) = 0, \ \ S_2(x) = 1; \quad p_1(x) = 1 - x, p_2(x) = x \qquad \text{for} \quad x \in [0,1] \ .$$

These assumptions imply that for $x_0 = 0$ the random variable x_1 satisfies $x_1 = S_1(x) = 0$ with probability one and further by induction $x_n = 0$ with probability one for every $n \ge 0$. Analogously if $x_0 = 1$ then also $x_n = 1$ with probability one for every $n \ge 0$. Thus in the first case $\mu_n(\{0\}) = 1$ and in the second $\mu_n(\{0\}) = 0$ for all n. This shows that the system is not asymptotically stable. Evidently it has not concentration property required by condition (6.6).

Thus we will often assume the condition

$$\inf_{x\in X} p_k(x) > 0 \qquad \text{for} \quad k = 1, \dots, N \ . \tag{7.1}$$

Further we should formulate an assumption which assures the nonexpansivness. This property also depends on p_k. We say that a probability vector $(p_1(x), \dots, p_N(x))$ satisfies the Dini condition if there is a continuous, nondecreasing, concave function $\omega : R_+ \to R_+$ such that

$$\sum_{k=1}^{N} |p_k(x) - p_k(y)| \le \omega(\varrho(x, y)) \qquad \text{for} \quad x, y \in X \tag{7.2}$$

and

$$\int_0^1 \frac{\omega(t)}{t} dt < \infty \ . \tag{7.3}$$

Recall that ω is said to be concave if

$$\omega(\lambda t + (1-\lambda)s) \ge \lambda\omega(t) + (1-\lambda)\omega(s) \qquad \text{for} \quad s, t \in R_+; \ \ 0 \le \lambda \le 1 \ .$$

Observe that the Dini condition is more restrictive than continuity (inequality (7.3) implies $\omega(0) = 0$) but less restrictive than the Lipschitz or Hölder condition.

Our next theorem belongs to Barnsley and coauthors [2]. It was not formulated there but used as a part of a proof.

Theorem 7.1. *Assume that the iterated function system $(S, p)_N$ satisfies inequality*

$$\sum_{k=1}^{N} p_k(x)\varrho(S_k(x), S_k(y)) \le r\varrho(x, y) \qquad \text{for} \quad x, y \in X \ , \tag{7.4}$$

where $r < 1$ is a nonnegative constant. Assume, moreover, that the probability vector $p(x)$ satisfies the Dini condition. Then there is a continuous increasing concave function $\varphi : R_+ \to R_+$ ($\varphi(0) = 0, \ \varphi(\infty) = \infty$) such that the transition operator P corresponding to $(S, p)_N$ is nonexpansive in the metric $\varphi(\varrho(x, y))$.

Proof. Write

$$\varphi(t) = t + \sum_{n=1}^{\infty} \omega(r^n t) \qquad \text{for } t \in R_+ \ . \tag{7.5}$$

From the Dini condition it follows that the series in (7.4) converges uniformly on compact subsets of R_+. Thus φ given by (7.5) is continuous, increasing, concave and such that $\varphi(0) = 0, \ \varphi(\infty) = \infty$. Moreover φ satisfies the inequality

$$\omega(t) + \varphi(rt) \le \varphi(t) \qquad \text{for} \quad t \in R_+ \ . \tag{7.6}$$

In the metric space $(X, \varphi \circ \varrho)$ the Fortet Mourier norm is given by the formula

$$||\mu_1 - \mu_2||_{\mathcal{F}_\varphi} = \sup\{|\langle f, \mu_1 - \mu_2\rangle| : \ f \in \mathcal{F}_\varphi\} \ ,$$

where $\mathcal{F}_\varphi$ is the set of continuous functions satisfying

$$|f(x)| \le 1, \quad |f(x) - f(y)| \le \varphi(\varrho(x,y)) \qquad \text{for} \quad x, y \in X \ . \tag{7.7}$$

Using the dual operator U we have

$$\|P\mu_1 - P\mu_2\|_{\mathcal{F}_\varphi} = \sup_{\mathcal{F}_\varphi} |\langle f, P\mu_1 - P\mu_2\rangle| = \sup_{\mathcal{F}_\varphi} |\langle Uf, \mu_1 - \mu_2\rangle| \ .$$

Thus in order to show that P is $\varphi \circ \varrho$ nonexpansive it suffices to verify that $Uf \in \mathcal{F}_\varphi$ for $f \in \mathcal{F}_\varphi$. Assume that f satisfies (7.7). From formula (4.1) which defines Uf we have

$$|Uf(x)| \le \sum_{i=1}^N p_i(x)|f(S_i(x))| \le \sum_{i=1}^N p_i(x) = 1 \ .$$

Thus it remains to verify that

$$|Uf(x) - Uf(y)| \le \varphi(\varrho(x,y)) \ . \tag{7.8}$$

We have

$$\begin{aligned} |Uf(x) - Uf(y)| &\le \sum_{i=1}^N |p_i(x)f(S_i(x)) - p_i(y)f(S_i(y))| \\ &\le \sum_{i=1}^N |p_i(x) - p_i(y)||f(S_i(y))| + \sum_{i=1}^N p_i(x)|f(S_i(x)) - f(S_i(y))| \\ &\le \sum_{i=1}^N |p_i(x) - p_i(y)| + \sum_{i=1}^N p_i(x)\varphi(\varrho(S_i(x)), S_i(y))) \ . \end{aligned}$$

Now using (7.2), and the concavity of φ we obtain

$$|Uf(x) - Uf(y)| \le \omega(\varrho(x,y)) + \varphi\left(\sum_{i=1}^N p_i(x)\varrho(S_i(x), S_i(y))\right) \ .$$

From this (7.4) and (7.6) it follows condition (7.8). The proof is completed. □

Now we are in a position to formulate a single condition for asymptotic stability of learning systems.

Theorem 7.2. *Let $(S,p)_N$ be an iterated function system satisfying the following conditions:*

1° The probability vector p satisfies (7.1) and the Dini condition.

2° The transformations S_i are Lipschitzean with Lipschitz constants L_i and

$$\sum_{k=1}^N p_i(x)L_i \le r < 1 \ . \tag{7.9}$$

3° *There is a number* $k \in \{1, \dots, N\}$ *such that* S_k *is compact, i.e.* $\mathrm{cl}\ S_k(X)$ *is a compact set.*

Then $(S, p)_N$ *is asymptotically stable*

Proof. As usually denote by P the transition operator of system $(S, p)_N$. Since inequality (7.9) is stronger than (7.4), we may apply Theorem 7.1 and construct a function φ such that P is nonexpansive with respect to metric $\varphi \circ \varrho$. In order to apply Theorem 6 we need only to verify the concentration property (6.6) with respect to $\varphi \circ \varrho$. Fix an $\varepsilon > 0$. From condition (7.9) it follows that there is an integer j such that $L_j < 1$. We have

$$\mathrm{diam}_\varrho S_j^n(\mathrm{cl}\ S_k(X)) \le L_j^n d \ ,$$

where $d = \mathrm{diam}_\varrho S_k(X)$ and the subscript ϱ in diam_ϱ denotes that the diameter is calculated with respect to metric ϱ. Since $L_j < 1$ and φ is continuous there is an integer $n_0 > 0$ such that

$$\mathrm{diam}_{\varphi\circ\varrho}(S_j^{n_0}(\mathrm{cl}\ S_k(X)) \le \varepsilon \ .$$

Now for an arbitrary $\mu \in \mathcal{M}_1$ we have

$$P^{n_0+1}\mu(A) \ge \overline{p}_k \overline{p}_j^{n_0} \mu(S_k^{-1}(S_j^{-n_0}(A))) \quad \text{for} \quad A \in \mathcal{B} \ , \tag{7.10}$$

where $\overline{p}_i = \inf_X p_i(x)$. Define

$$A_0 = S_j^{n_0}(\mathrm{cl}\ S_k(X)) \ .$$

The set A_0 is compact and

$$S_k^{-1}(S_j^{n_0}(A_0)) = X \ .$$

Thus substituting A_0 into (7.10) we obtain

$$P^{n_0+1}\mu(A_0) \ge \overline{p}_k \overline{p}_j^{n_0} \ .$$

The last condition holds for an arbitrary $\mu \in \mathcal{M}_1$. In particular, substituting the measure $P^{n-n_0-1}\mu$ instead of μ, we get

$$P^n\mu(A_0) = P^{n_0+1}\left(P^{n-n_0-1}\mu\right)(A) \ge \overline{p}_j \overline{p}_k^{n_0} \qquad \text{for} \quad n \ge n_0 + 1, \ \mu \in \mathcal{M}_1 \ .$$

Thus the concentration condition (6.6) of Theorem 6 is satisfied with $\alpha = p_j p_k^{n_0}$ and metric $\varphi \circ \varrho$. Consequently the operator P is asymptotically stable in the metric space $(X, \varphi \circ \varrho)$. However the metrics ϱ and $\varphi \circ \varrho$ define the same space of continuous functions $C(X)$ and the weak convergence of a sequence of measures in the space (X, ϱ) and $(X, \varphi \circ \varrho)$ is the same. This proves that P is also asymptotically stable in (X, ϱ). The proof is completed. □

When X is a locally compact space (bounded closed sets are compact) the assumptions of Theorem 7.2 can be released. In particular, inequality (7.1) can be replace by

$$p_k(x) > 0 \qquad \text{for} \quad x \in X, \ k = 1, \dots, N$$

and condition 3° can be completely omitted [12]. In the general case of a Polish space X new results are obtained by T. Szarek [15].

References

1. M.F. Barnsley, *Fractals Everywhere*, Academic Press, New York, 1988.
2. M.F. Barnsley, S.G. Demko, J.H. Elton and J.S. Geronimo, *Invariant measures arising from iterated function systems with place dependent probabilities*, Ann. Inst. Henri Poincaré **24**, 367–394(1988).
3. J.H. Elton and M. Piccioni, *Iterated function systems arising from recursive estimation problems*, Probab. Theory Relat. Fields, **91**, 103–114 (1992).
4. R. Fortet and B. Mourier, *Convergence de la répartition empirique vers la répartition théorétique*, Ann. Sci. École Norm. Sup. **70**, 267–285 (1953).
5. J. Hutchinson, *Fractals and self–similarity*, Indiana Univ. J. **30**, 713–747 (1981).
6. M. Iosifescu and R. Theodorescu, *Random Processes and Learning*, Springer––Verlag, New York, 1969.
7. S. Karlin, *Some random walks arising in learning models*, Pacific J. Math. **3**, 725–756 (1953).
8. A. Lasota, *A lower bound theorem for Markov operators on Polish spaces*, Ann. Polon. Math. (to be published).
9. A. Lasota and M.C. Mackey, *Chaos, Fractals and Noise–Stochastic Aspects of Dynamics*, Springer–Verlag, New York 1994.
10. A. Lasota and M.C. Mackey, *Why do cells divide?* (to be published).
11. A. Lasota, M.C. Mackey and J. Tyrcha, *The statistical dynamics of recurrent biological events*, J. Math. Biology, **30**, 775–800 (1992).
12. A. Lasota and J.A. Yorke, *Lower bound technique for Markov operators and iterated function systems*, Random and Computational Dynamics, **2**, 41–77 (1994).
13. O. Onicescu and G. Mihoc, *Sur les chaines de variables statistiques*, Bull. Sci. Math. **59**, 174–192 (1935).
14. S.T. Rachev, *Probability Metrics and the Stability of Stochastic Models*, John Wiley, New York, 1991.
15. T. Szarek, *Asymptotic stability of Markov operators acting on Polish spaces*, Bull. Pol. Acad. Sci., Mathematics (to be published).
16. T. Szarek, *Iterated function systems depending on previous transformation*, Univ. Iagellonicae Acta Math. (to be published).

References

[illegible]

Dissipative Structures and Weak Turbulence

Paul Manneville

LadHyX — Laboratoire d'Hydrodynamique
École Polytechnique, 91128 Palaiseau, France
and
Service de Physique de l'État Condensé
Centre d'Études de Saclay, 91191 Gif-sur-Yvette, France

Abstract: We present a brief overview of the current understanding of temporal and spatio-temporal chaos, both termed *weak turbulence* according to the context [1]. The process which allows one to reduce the primitive problem to a low-dimensional dynamical system is discussed. It turns out to be appropriate as long as *confinement effects* are sufficiently strong to freeze the space dependence of unstable modes, hence temporal chaos only. Otherwise modulated patterns arise, yielding genuine space-time chaos. The corresponding theory rests on *envelope equations* providing a useful framework for weak turbulence in a globally super-critical setting. *Spatio-temporal intermittency* analyzed next is the relevant scenario in the sub-critical case. Finally, the connection with hydrodynamic turbulence and the more general relevance of some of the ideas developed here are examined.

1 On the 'Nature' of Turbulence

Turbulence presents itself as a wildly fluctuating flow regime the understanding of which is important for many fields of basic and applied research, as its mixing and transfer properties are strongly enhanced with respect to their molecular counterparts. Following the suggestion of Landau [2]a, one can think of turbulence as resulting from a superposition of an infinite number of oscillation modes (Fig. 1, top), the randomness of the regime arising from the absence of knowledge of the initial phase of each mode. However, as stressed originally by Ruelle and Takens [2]b, this idea of a mere superposition is in some sense too linear. In fact, nonlinear interactions among a small number of modes generically yield chaotic behavior characterized by a decay of correlations that more appropriately describes a turbulent regime (Fig. 1, bottom) [3].

The main consequence of this remark is to place the problem of the transition to turbulence into the mathematical setting of the theory of dynamical systems, i.e., systems of ordinary differential equations (ODE) formally written as

$$\frac{d}{dt}\mathbf{X} = \mathbf{F}_r(\mathbf{X};t) \qquad \text{with} \quad \mathbf{X} = \{X_1, \ldots, X_n\}\,, \tag{1}$$

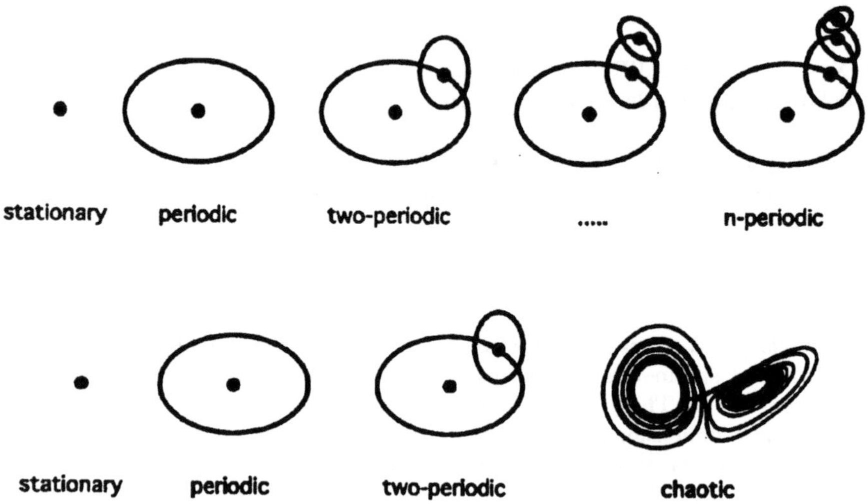

Fig. 1. Scenarios of transition to turbulence according to Landau (top) and Ruelle & Takens (bottom): oscillatory motions are suggested by ellipses, a miniature of the celebrated Lorenz attractor features chaos.

or their discrete time analog, namely iterations:

$$\mathbf{X}_{k+1} = \mathbf{G}_r\left(\mathbf{X}_k; k\right) , \tag{2}$$

where $\mathbf{X}$ denotes the instantaneous state of the system (a point in the appropriate *phase space*), r some set of *control parameters*, and $\mathbf{F}_r$ or $\mathbf{G}_r$ the evolution laws.

While degrees of freedom are discrete at the most microscopic scale, our current experience is with *continuous media* governed by partial differential equations (PDE), vis., $\mathbf{H}_r\left(\mathbf{X}; \partial_t, t; \partial_{\mathbf{x}}, \mathbf{x}\right) = 0$, e.g., Navier–Stokes equations for concrete realizations of incompressible fluid flows. However most results have been obtained in the framework of *low dimensional* systems (1) or (2) involving a small number of variables, mainly focusing on notion of *chaos* and *scenarios* of transition to turbulence. Some supplementary work is therefore needed to understand how this can be applied to physical (or more general [3]) systems which are *a priori* not explicitly written as low dimensional dynamical systems.

2 Coherence and Slaving

The reduction process from the primitive problem to a low–dimensional one is of great interest in itself since it is the manifestation of self–emerging space–time coherence within the system and non–linear processes associated to it. The relevant modes result from couplings between primitive variables (fields) *via* mechanisms acting over ***specific scales***. They can be be identified on the basis of conventional ***stability analysis***.

For example, natural convection in a layer of fluid at rest heated from below arises from an unstable density stratification with cold heavy fluid on top of hot light fluid: a small blob of fluid slightly hotter (and lighter) than its surroundings tends to rise in an ever colder (and denser) environment. A temperature fluctuation is then amplified provided that the imposed temperature difference ΔT between top and bottom exceeds some critical value ΔT_c so that thermal diffusion and viscosity, the stabilizing dissipative processes, become unable to withstand the gain of gravitational energy corresponding to a redistribution of matter. The result of this co–operative interaction of velocity and temperature fluctuations is an overturning of the layer in the form of stationary convection cells. Natural convection is the prototype of processes leading to the formation of so–called ***dissipative structures*** [4, 1].

This selection by an instability mechanism corresponds to a drastic reduction of the effective number of degrees of freedom since, out of (infinite dimensional) potential solutions of the governing PDEs, the system extracts an ***instability mode*** that can be written as $\mathbf{X}(\mathbf{x},t) = A(t)\overline{\mathbf{X}}(\mathbf{x})$ where $\overline{\mathbf{X}}(\mathbf{x})$ defines the (essentially frozen) spatial structure of the mode and $A(t)$ its ***amplitude*** slowly function of time. On general grounds, a whole basis of such eigen–modes is obtained, on which the dynamics can be projected yielding

$$\frac{dA_m}{dt} = L_r A_m - g_m\left(\{A_p; p = 1, \ldots\}\right), \qquad m = 1, \ldots, \tag{3}$$

where L_r is the linear stability operator ($L_r A_m = s_m A_m$ with $s_m = \sigma_m - i\omega_m$). The term g_m in (3) gathers all nonlinear contributions to the dynamics of mode m. In this way, the degrees of freedom of the problem are now the amplitudes A_m and the progress achieved lies in the fact that the coherence implied by the existence of the instability mechanism is properly taken into account.

The modes can be ordered by decreasing values of the real part σ_m of their growth rate s_m. Strongly ***stable*** modes (subscript "s") have $\sigma < 0$ with $|\sigma|$ large, while ***central*** ("c"), i.e., nearly neutral, modes have $|\sigma|$ small, either negative or positive. Situations of interest are those where the number n_c of central modes is sufficiently small and the stable modes are sufficiently strongly damped ($|\sigma_m|$ large for $m > n_c$) so that there is a wide gap in the spectrum between the stable and central parts. The amplitudes of the stable modes then quickly relax toward values imposed by the coupling to unstable modes. One says that they are ***enslaved*** to the central modes [5]. Owing to this property, one can eliminate them "adiabatically" and obtain an effective system with just the set of central modes coupled together. To illustrate this important process, let us consider a system of only two real modes X_c (amplitude A_c) and X_s (A_s) coupled through a dynamical system that formally reads:

$$\frac{dA_c}{dt} = \sigma_c A_c + g_c(A_c, A_s) ,$$

$$\frac{dA_s}{dt} = \sigma_s A_s + g_s(A_c, A_s) ,$$

with σ_c positive or negative but small (mode X_c nearly marginal) and σ_s negative and large (mode X_s strongly stable). If the amplitude A_c evolves slowly, as these assumptions imply, then the second equation can be solved for A_s, with A_c considered as a parameter. After a brief transient of duration $\mathcal{O}(1/|\sigma_s|)$, A_s settles to the root of $\sigma_s A_s + g_s(A_c, A_s) = 0$, which defines a manifold in the space (A_c, A_s). The value found for A_s, denoted as $\tilde{A}_s(A_c)$, is then inserted in the first equation which becomes

$$\frac{dA_c}{dt} = \sigma_c A_c + g_c\left(A_c, \tilde{A}_s(A_c)\right) = \sigma_c A_c + g_{\text{eff}}(A_c) ,$$

i.e., an effective equation for A_c that parametrizes the position of the system on the previously defined manifold. Here the degree of freedom A_s has been effectively eliminated and the number of degrees of freedom reduced to one (see Fig. 2) but, at least heuristically, this extends itself easily to cases with $n_c > 1$.

3 Bifurcations Towards Temporal Chaos

This possibility of a reduction is an essential prerequisite for interpreting the transition to turbulence in continuous media that are infinite–dimensional *a priori*. The bifurcations of the resulting system can then be studied using tools developed for low–dimensional deterministic, dissipative, autonomous dynamical systems (especially *Poincaré sections* which generalize the stroboscopic analysis of periodically forced systems and transform continuous–time systems into discrete–time ones with their dimensions decreased by one). In practice only the "$t \to \infty$" asymptotic regimes of such systems, represented by *attractors* in phase space, are of interest. The complexity of the associated dynamics is mainly a function of the dimension of the center manifold. Systems close enough to equilibrium relax towards time–independent states represented by *fixed points* in phase space. The transition between them can be accounted for by (essentially) one–dimensional real systems, the prototype of which reads

$$\frac{dA}{dt} = rA - a_2 A^2 - a_3 A^3 \ldots - H . \tag{4}$$

The study of (4) is elementary. The behavior of the bifurcated solution essentially depends the symmetries of the problem and on whether the instability saturates or not beyond threshold. For example, if the unperturbed system ($H = 0$) is invariant in the change "$A \mapsto -A$" the expansion contains no even powers of A

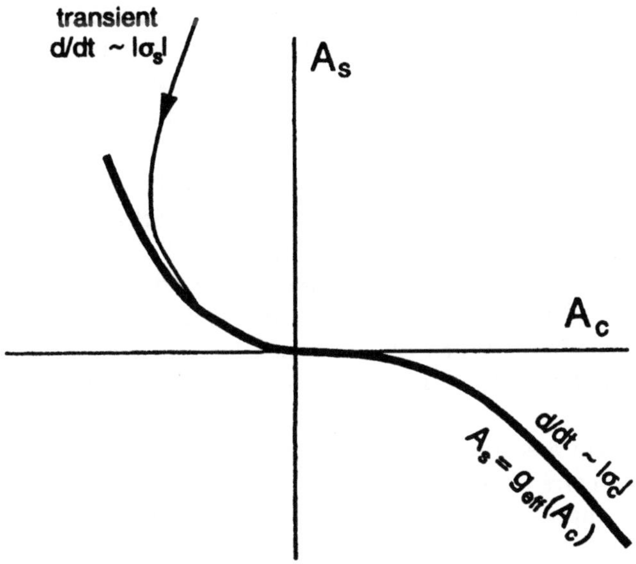

Fig. 2. Adiabatic elimination of stable modes yields an effective dynamics for central modes on a low-dimensional manifold.

and saturation is controlled by the sign of a_3. If a_3 is positive, the bifurcation is *super-critical* with A tending to $\pm\sqrt{r/a_3}$ for $t \to \infty$ when $r > 0$. When a_3 is negative, this non-trivial solution exists for $r < 0$ and is unstable, while the trivial solution is still stable. Adding a stabilizing five-order term $-a_5A^5$ with $a_5 > 0$ yields a discontinuous *sub-critical* bifurcation characterized by multi-stability and related hysteresis upon variation of the control parameter. The effect of a perturbation $H \neq 0$ and that of a symmetry breaking term $-a_2A^2$, both on the super-critical bifurcation, are easily guessed.

By driving a system further away from equilibrium, one can obtain less trivial regimes. In order of complexity, the next step is the occurrence of a periodic attractor represented by a *limit cycle* in phase space. The corresponding so-called *Hopf* bifurcation can happen only for $n_c \geq 2$. In two-dimensions it is best described by a model analogous to (4) but now with a complex amplitude A:

$$\frac{dA}{dt} = (r - i\omega)A - (g' + ig'')|A|^2A + \dots \tag{5}$$

Driving the system even further, one can reach multi-periodic regimes when other oscillatory modes enter play (corresponding attractors are then represented by *tori*). All these behaviors are characterized by the regular "recurrence" of states belonging to the corresponding attractors (cf. Fig. 1, top). But there is also the possibility of an aperiodic behavior called ***chaos*** [6] which turns out to be generic for systems with sufficiently many degrees of freedom: n being the effective dimension of the system, chaos can be observed in ***non–invertible iterations*** with $n \geq 1$, e.g., the dyadic map in Fig. 3a (top), in ***invertible iterations*** with $n \geq 2$, e.g., the Hénon map, and in ***differential systems*** $n \geq 3$, e.g., the Lorenz model [7] (cf. Fig. 1, bottom).

Chaos is characterized by the instability of individual trajectories on the attractor, i.e., the indefinite amplification of infinitesimal perturbations which makes two initially close trajectories diverge inexorably [6]. This property called ***sensitivity to initial conditions*** [3] is easily illustrated using the dyadic map: let two trajectories starting at X_0 and $X_0+\delta X_0$, then, after k iterations the distance between them is $\delta X_k = 2^k \delta X_0$. When $\delta X_k \sim 1$, i.e., for $k \sim \log_2(1/\delta X_0)$, the trajectories are so distant that their subsequent evolutions seem independent, see Fig. 3b.

Strange attractors [2]b associated with chaotic behavior most often display a *fractal transverse structure* able to manage with both stretching due to instability and contraction imposed by dissipation. As a trivial example we display in Fig. 3d the case of the *baker map* sketched in Fig. 3a (bottom) which, when dissipative, contracts more than it stretches.

The notion of chaos as accounted for by strange attractors is then associated with long term *unpredictability* in spite of short term *determinism* (see Fig. 3c), which makes them compatible with apparent *stochasticity* [6]. Recent work in this field has been devoted to practical applications including the reconstruction of the dynamics and the control of chaos [8].

4 Instability in Continuous Media

The assumptions made to achieve the elimination of enslaved modes require further discussion. Indeed, up until now the domain in which the instabilities develop has not been specified. On the one hand, a fully discrete spectrum is clearly relevant. On the other hand, stability analysis is usually developed in the context of an infinite medium with specific symmetries (translation, rotation, parity) which yields, not isolated modes, but branches of modes parametrized by continuous separation variables, e.g., wavevectors for modes now taken in the form $X(x,t) = A\exp(ikx)\exp(st)$. Different cases can be distinguished from the consideration of the dispersion relation $s = s(k;r)$ relating the (complex) growth rate s to the wavevector k of the modes and to the control parameter r. The classification rests on the character of the most unstable mode, i.e., the mode whose growth rate s has the largest real part σ. The marginal stability condition for this mode, $\sigma = 0$, is first reached at a critical value r_c of the control parameter (the instability threshold) to which corresponds a critical critical wavevector k_c and a possibly non–zero critical frequency ω_c.

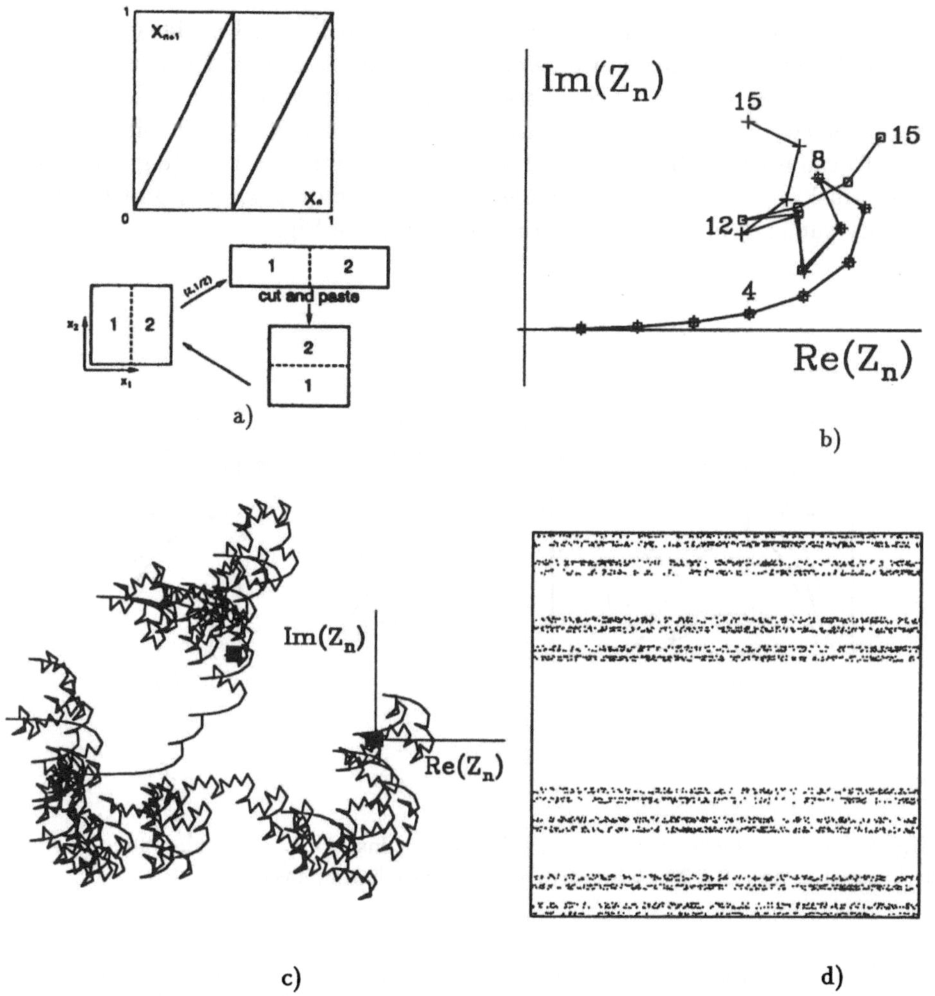

Fig. 3. (a) Top: Dyadic map $X \mapsto 2X \pmod 1$; Bottom: conservative Baker map with stretching by 2 and contraction by 1/2. (b) Divergence of neighboring trajectories represented as a walk in the complex plane through $Z_n = Z_{n-1} + \exp(2\pi i X_n)$. (c) Long term unpredictability of a trajectory using the same representation; start and stop points are marked by black squares. (d) Fractal structure typical of strange attractors, here for the dissipative baker map (contracting areas by a factor 2/3).

The instability is termed *cellular* when $k_c \neq 0$ and $\omega_c = 0$ (stationary mode, see Fig. 4a). This is the case of plain convection in ordinary fluids introduced earlier. The growth rate s is real. The critical branch displays a maximum as a function of k at $k \sim k_c$ negative below threshold ($r < r_c$) and positive above ($r > r_c$); a second, always stable, branch is represented, corresponding to a different physical mode. Dissipative waves correspond to $k_c \neq 0$ and $\omega_c \neq 0$ (Fig. 4b). They may happen in convecting binary mixtures of specific compositions. The real part of the (now complex) growth rate varies like in Fig. 4a as a function of k and r but, in addition, we have $\omega \neq 0$ so that a critical frequency emerges for $k = k_c$ and $r = r_c$. The phase velocity of the wave is $c = \omega/k$ and its group velocity $c_g = \partial\omega/\partial k$. The case of a uniform oscillatory instability ($k_c = 0$, $\omega_c \neq 0$) can arise in chemical systems of reaction–diffusion type (e.g., Belousov–Zhabotinsky reaction): the most unstable wavevector is $k_c = 0$ and the group velocity vanishes in a parity–invariant medium, see Fig. 4c.

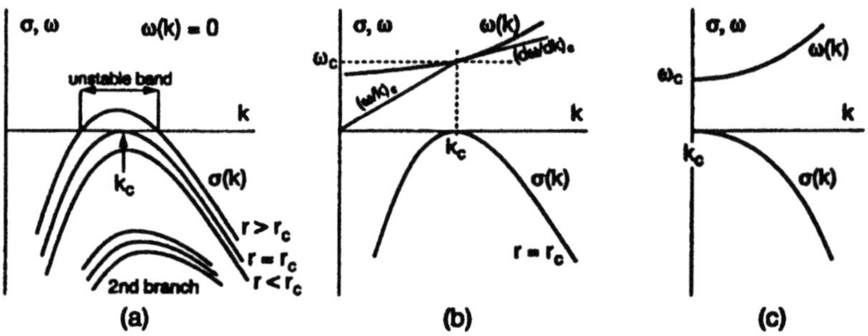

Fig. 4. Typical aspect of dispersion relations for instabilities in unbounded systems. a) Cellular instability. b) Dissipative wave. c) Oscillatory instability.

The role of lateral confinement effects is best understood when the instability mechanism generates a periodic structure with a typical size of the order of the critical wavelength $\lambda_c = 2\pi/k_c$. Considering a system of size ℓ, one can appreciate the importance of boundary effects from the value of the *aspect ratio* which measures ℓ in units of λ_c:

$$\Gamma = \ell/\lambda_c \ .$$

In all cases, the presence of lateral boundaries is expected to discretize the spectrum. However the infinite–system limit remains relevant when Γ is large, which yields a quasi–continuous spectrum. By contrast, when Γ is small (i.e., of order unity) end effects are predominant and one gets a fully discrete spectrum with well–distinct modes. Clearly, the temporal–chaos picture of the transition to turbulence emerging from the reduction process sketched in section 3 is only appropriate for the latter case (strongly *confined* systems). The opposite case of *extended* systems requires a different approach: owing to the branch structure of the linear stability problem in the unbounded limit, adiabatic elimination can be conceived as a two-step process: inter-branch coupling and intra-branch interaction.

For systems with a single relevant physical mode close enough to threshold, the result of the inter-branch elimination is a single effective equation for a real or complex amplitude A in the form (4) or (5), governing the *local* intensity of the mode. But, for an extended system, a whole band of wavevectors of width $\sim \sqrt{\delta r}$ is linearly unstable beyond threshold (see Fig. 4a) so that its state actually results from interferences among these wavevectors. These interferences can be interpreted as amplitude modulations that usually behave in a diffusive way at lowest order. This transforms the starting effective ODE into a PDE formally written as

$$\tau_0 \partial_t A = F_r(A) + \xi_0^2 \partial_{xx} A \tag{6}$$

and known under the generic name of *Ginzburg-Landau equation*. Here τ_0 and ξ_0 are characteristic time and length scales that can be derived from the linear analysis (intra-branch part of the calculation). The structure of F_r depends on the instability mode, stationary or oscillatory, and on the super- or sub-critical, i.e., continuous or discontinuous nature of the bifurcation.[1] Such equations can be derived rigorously *via* multi-scale expansions [9].

5 Envelopes and Space-Time Chaos

Large aspect ratio systems can be conceived as assemblies of individual cells for which lateral boundary effects are not able to maintain spatial coherence in the middle of the system. Modulated patterns or *textures* are therefore expected to be the rule rather than the exception. Modulations can have basically two origins. The most obvious one is extrinsic, i.e., imposed by the presence of lateral boundaries at large but finite distance; for example, $A(x = \pm\ell/2) = 0$. The other one is intrinsic, either because the uniform amplitude solution is unstable (secondary instability) or because the field A can support topological defects.

For stationary instabilities such as Rayleigh-Bénard convection, many cell arrangements are possible with defects such as the foci, dislocations or grain boundaries visible in Fig. 5. The occurrence of space-time chaos is mainly a problem of pattern selection since the time dependence is very slow. Its main features can be accounted for by a minimization of the potential from which the envelope equation then derives [11, 1]. By contrast, in the oscillatory case the variational structure is lost from the beginning and the time dependence is expected to be much wilder already for the one-dimensional case on which we concentrate our attention now.

Modulations around a uniform oscillatory state[2] are thus governed by spatial extension of (5), which after appropriate rescalings leads to the celebrated and

[1] In fact, A is always taken complex, even when dealing with a stationary cellular instability for which a real amplitude should be appropriate. This is done in order to account for translational invariance more easily. As a result, F_r has real coefficients in the stationary case whereas it is fully complex in the oscillatory case.

[2] (or a propagating wave in a frame moving at its own group velocity)

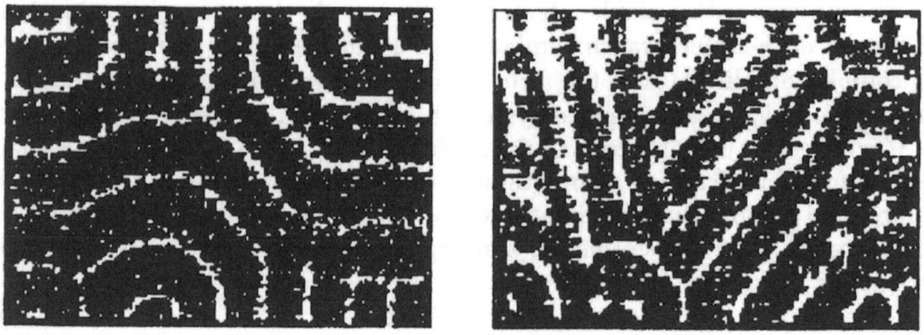

Fig. 5. Two examples of patterns observed in high Prandtl number convection (courtesy: P. Bergé and M. Dubois).

much studied cubic complex Ginzburg–Landau equation (CCGLE) explicitly rewritten here as:

$$\frac{\partial A}{\partial t} = A + (1 - i\alpha)\frac{\partial^2 A}{\partial x^2} - (1 + i\beta)|A|^2 A\,. \tag{7}$$

Forgetting about boundary–induced modulations we consider only intrinsic modulations generated by secondary instabilities. In this context, the dangerous modes are related to the translation invariance of the problem at the infinite–size limit. A neutral mode is associated to this property and, in some circumstances, perturbations close to it may become unstable depending on the value of α and β. This is the case for all uniform rotating waves ($A_q \exp[i(qx - \omega_q t)]$ with $A_q = 1 - q^2$ and $\omega_q = \beta - (\alpha + \beta)q^2$) when $1 - \alpha\beta < 0$, which is Newell's criterion for the Benjamin–Feir (BF) instability [12].

Local translations are directly interpreted in terms of the phase ϕ of the complex amplitude $A = |A| \exp(i\phi)$ and long–wavelength modulations can be studied through a perturbation expansion in $\partial_x \phi$ with the modulus $|A|$ being enslaved to it. Once truncated at the first significant order, the result of this *phase formalism* [13] is an equation called the Kuramoto–Sivashinsky equation (KSE) [14]

$$\partial_t \phi + \tfrac{1}{2}\lambda\,(\partial_x \phi)^2 = D\partial_{xx}\phi - K\partial_{xxxx}\phi\,,$$

where the coefficients λ, D and K depend on α and β. The most important one is the effective diffusion coefficient $D = 1 - \alpha\beta$, seen to change sign at the threshold of the BF instability. Solutions to the KSE for $D < 0$ are turbulent provided that the system is wide enough, hence accounting for a regime of *phase turbulence* [15] of the CCGLE, i.e., a regime where $|A|$ remain bounded away from zero so that the phase ϕ can be defined everywhere. However this is true only in the immediate vicinity of the BF threshold. If the system is driven further in the unstable phase domain, a "revolt" of the modulus $|A|$ occurs and the CCGLE enters a regime of *amplitude turbulence* where it varies wildly and sometimes reaches zero so that the phase can no longer be defined everywhere. Figure 6 displays the nature of the regime as a function of α and $1/\beta$ as determined by numerical experiments [17] : BF is the Benjamin–Feir line; line L_1 marks a smooth transition from amplitude turbulence to phase turbulence; in the domain above line L_2 and below line L_3 spatio–temporal intermittency is observed (see next section) involving defects and plane waves below the BF line [region (II)] or defects and phase turbulence above it ["bi–chaos" in region (III)].

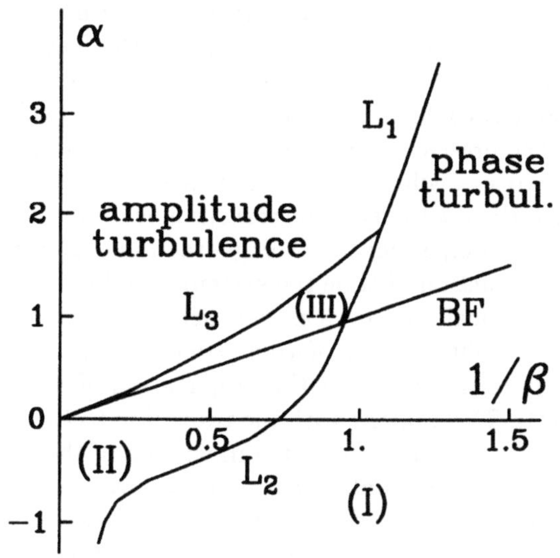

Fig. 6. Different regimes observed in the CCGLE as a function of $1/\beta$ and α.

Model equations extending (7) to higher dimensions or with different nonlinearities are of considerable help in understanding the transition to spatio–temporal chaos in realistic cases as they offer a framework that is far more simple than primitive equations and yet is sufficient to deal with universal features of pattern dynamics [18].

To conclude this section, let us notice that, up until now, we have dealt with a transition to turbulence that can be considered as globally super–critical in the following sense: 1) a primary instability is involved with a gentle saturation of the amplitude; 2) the relevant nonlinear dynamics is associated with smooth secondary instabilities mainly involving phase diffusion (with possibly additional features such as structural defects); 3) spatio–temporal chaos can be understood as resulting from a chaotic temporal dynamics of a generalized phase variable; turbulence is weak mainly because the gradients of the modulations remain "small" and can increase progressively. Questioning these three points will lead to a different scenario toward which we turn now.

6 Spatio–Temporal Intermittency

The main property associated with a *sub–critical* instability is the coexistence of different attractors in some control parameter range. For low–dimensional dynamical systems accounting for the strongly confined case, this generically leads to *crises* and *intermittency*. As stressed by Pomeau [19], in extended geometry attractor coexistence can be observed in physical space owing to the allowance for modulations leading to the formation of *walls* or *fronts*.

When one of the competing regimes is temporally chaotic and the other regular, the system can then be divided into *turbulent* and *laminar* domains. An example of a sub–critical instability with well–defined fronts between turbulent and laminar domains is given by spiral turbulence observed in the Taylor–Couette flow when the cylinders are counter–rotating [20]. According to Pomeau, the frontier between laminar and turbulent domains is expected to display small–scale fluctuations owing to the presence of local chaos, so that front propagation becomes a contamination process, which results in a new route to turbulence called *spatio–temporal intermittency* (STI).[3] The transition is monitored by measuring the "turbulent fraction," i.e., either the average fraction of space occupied by the turbulent state or the fraction of time the system remains in the chaotic state at a given place. In statistical physics, a similar behavior is displayed by a purely stochastic process called *directed percolation* (DP) that models problems such as the flow through a porous medium or the propagation of disease, forest fires, etc., see Fig. 7 (top) [21]. The analogy with the transition to turbulence *via* STI rests on the two–state (laminar/turbulent) decomposition of the system into an array of units and the definition of contamination rules between these units. Its interest stems from the fact that DP displays scaling behavior at threshold typical of critical phenomena and the subsequent identification of *universality classes*.

In order to go a little further in the study of Pomeau's conjecture, and especially in an attempt to understand how DP, a fully stochastic system, could be relevant in a deterministic context, intermediate models defined in terms of *coupled map lattices* (CML [22]) have been considered. Such systems generically read

$$X_j^{n+1} = \sum_{j' \in \mathcal{V}_j} W_{jj'} f(X_{j'}^n) ,$$

where j and n are discrete space and time indices, f a local map acting on a continuous degree of freedom X, and $W_{jj'}$ a coupling term between variables in the neighborhood $\mathcal{V}_j$ of a given site j, usually a simple diffusion [22, 23]. Contrary to Pomeau's initial guess that STI was in the universality class of DP, detailed numerical experiments of such models showed that, the STI transition

[3] The possibility of a specific transition when no stable bifurcated state exists in the neighborhood of the basic state was already noted by Landau [2]a.

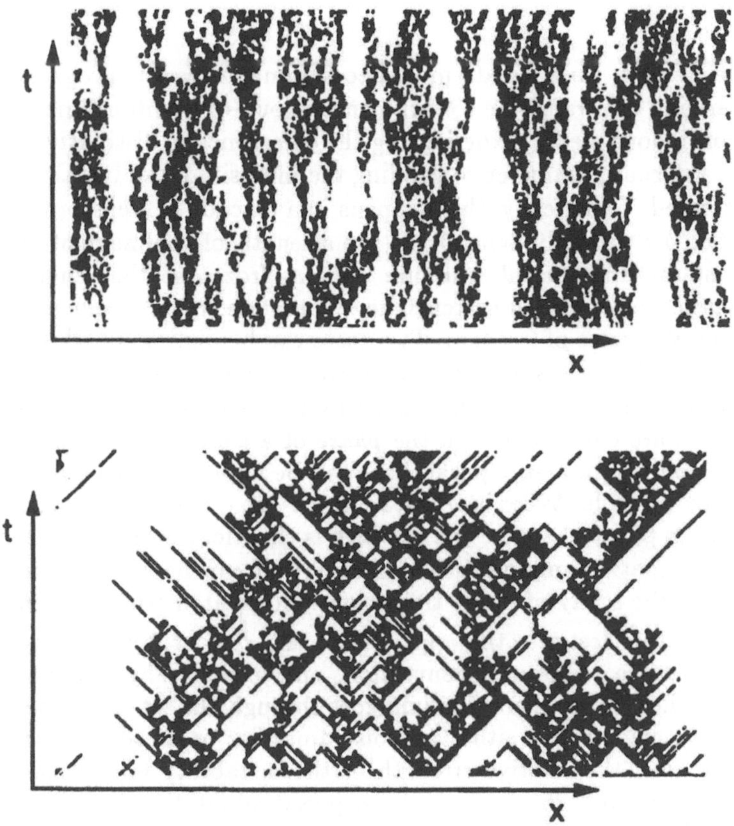

Fig. 7. Top: directed percolation. Bottom: spatio-temporal intermittency in a lattice of coupled maps. In both pictures time is running upwards, contaminated (= turbulent) sites are black and "healthy" (= laminar) sites are white. The local map is chosen so as to emphasize the role of the deterministic-but-chaotic dynamics contrasting with the fully stochastic behavior of DP.

could be smooth ("second order" in the phase transition terminology) but with exponents different from those of DP or, worse, that it could be discontinuous ("first order") depending on the dimension of space and on details of the models (i.e., the local dynamics, see Fig. 7, bottom), so that the universality concept was not tenable. What remains however is the introduction of methods and tools of statistical mechanics in an apparently disconnected new field of research [24].

This transition has been observed also in laboratory experiments, convection, "printer" instability, etc. [?]. The understanding gained in its study can also help us better describe other transitions involving turbulent spots, e.g., in Couette or boundary layer flows.

7 Conclusion

In this brief review, we have attempted to summarize recent progress achieved in understanding the transition to turbulence in out-of-equilibrium macroscopic systems (continuous media) experiencing the emergence of dissipative structures when some external parameter controlling the intensity of an instability mechanism is increased. Generically, this happens in a cascade of steps but the nature of this turbulence strongly depends on the strength of confinement effects. For confined systems, the spatial structure remains frozen so that chaos is essentially *temporal*. For extended systems, key-words are modulations and patterns involving many degrees of freedom. Two cases can be distinguished according to whether the instability is *super*-critical or *sub*-critical. When it is super-critical, a gentle saturation of the unstable mode takes place beyond threshold; "dangerous" modes are connected with the *phase* of a complex envelope, and most of the transition to spatio-temporal chaos can be understood as temporal chaos for this phase variable. When the instability is sub-critical, the coexistence of several states in physical space, one of which is chaotic, results in a transition to turbulence *via* spatio-temporal intermittency that can be analyzed using tools borrowed from the study of phase transitions and related critical phenomena.

Apparently this restricts the approach to the case of flows taking place in closed containers, e.g., natural convection, and excludes hydrodynamic flows called "open," i.e., flows where the fluid goes through the set-up, tube, channel, etc. The latter can behave either as noise amplifier or as self-excited oscillators depending on the relative strength of the instability mechanism and the downstream advection [26]. In fact, the required adaptation involves the same approach in terms of envelopes that are variants of the Ginzburg-Landau equations introduced before so that what we have learned remains of interest. However, further steps are obviously necessary before an understanding of developed turbulence is obtained, in particular with respect to the emergence of scaling in the "inertial range" where the energy cascades from large scales where it is injected by instability mechanisms to small scales where it is dissipated by viscous damping. While it would seem possible to understand this hierarchy of scales as resulting from the successive excitation of modes in the spirit of the Landau cascade, the stress has rather been put recently on the dynamics of persistent coherent structures [16].

Though the account given here might suggest that the advances thus far are mainly conceptual, their application to concrete situations of fluid dynamics is within reach. At any rate, use can certainly be made of the methods developed (modeling by envelope equations, coupled maps) and the vast catalog of behaviors already discovered in pattern dynamics (amplitude and phase turbulence, defect stability, spatio-temporal intermittency, etc.). Also, analog reasoning resting on the study of generic features of weak turbulence can be of great help for clarifying the respective relevance of temporal chaos and spatio-temporal chaos in specific cases.

References

1. For a more developed introduction, consult:
   P. Manneville, *Dissipative Structures and Weak Turbulence* (Academic Press, 1990).
2. Seminal articles are:
   a) L.D. Landau, Dokl. Akad. Nauk SSSR **44**(1944)339, reprinted in *Collected Papers of L.D. Landau*, D. ter Haar, Ed. (Pergamon Press, 1965);
   b) D. Ruelle & F. Takens, Comm. Math. Phys. **20**, 167 (1971).
3. D. Ruelle, Ann. N.Y. Acad. Sc. **316**, 408 (1979); and also "Where can one hope to profitably apply the ideas of chaos," Physics Today, July 1994.
4. G. Nicolis & I. Prigogine, *Self-organization in nonequilibrium systems, from dissipative structures to order through fluctuations* (Wiley, 1977).
5. For a physical viewpoint consult:
   H. Haken, *Synergetics* (Springer–Verlag, 1983); and for mathematical aspects: J. Guckenheimer & P. Holmes, *Nonlinear oscillations, dynamical systems and bifurcation of vector fields* (Springer–Verlag, 1983); or also:
   G. Iooss, "Reduction of the dynamics of a bifurcation problem using normal forms and symmetries," in *Instabilities and nonequilibrium structures*, E. Tirapegui & D. Villaroel, Eds. (Reidel, 1987).
6. For a introduction to chaos, consult:
   P. Bergé, Y. Pomeau & Ch. Vidal: *Order within Chaos* (Wiley, 1984); and for a more advanced review:
   H.G. Schuster: *Deterministic chaos* – 2nd edition (Physics Verlag, 1988); or else [1] chapters 4–7.
7. Two classical models are:
   M. Hénon, Comm. Math. Phys. **50**, 69 (1976); and:
   E. N. Lorenz, J. Atm. Sc. **20**, 130 (1963).
8. Problems related to the reconstruction of chaotic dynamics are also of great interest; consult:
   F. Takens, in *Dynamical Systems and Turbulence* D.A Rand, L.S. Young, eds. (Springer–Verlag, 1981);
   N.H. Packard *et al.*, Phys. Rev. Lett. **45**, 712(1980); and, for reviews:
   N.A. Gershenfeld, "An experimentalist's introduction to the observation of dynamical systems," in: Hao Bai-lin, ed., *Directions in Chaos* Vol. 2 (World Scientific, 1989); or
   N. A. Gershenfeld & A.S. Weigeng, Eds., *The Future of Time Series: Learning and Understanding* (Addison–Wesley, 1993); see in particular the lucid introduction by the editors of the book. A recent review of chaos control is given by:
   T. Shinbrot *et al.*, Nature **363**, 411 (1993).
9. Envelope equations have been introduced to describe modulated patterns in the context of convection by:
   A.C. Newell & J.A. Whitehead, J. Fluid Mech. **38**, 279(1969); and
   L.A. Segel, J. Fluid Mech. **38**, 203(1969). See also:
   M.C. Cross & A.C. Newell, Physica **10D**, 299(1984), and further:
   A.C. Newell, "The dynamics of patterns: a survey," in [10];
   A.C. Newell, T. Passot & J. Lega, Ann. Rev. Fluid Mech. **25**, 399 (1993).
10. J.E. Wesfreid *et al.*, Eds.: *Propagation in systems far-from-equilibrium* Springer Series in Synergetics vol. 41 (Springer, Heidelberg, 1988).
11. M.C. Cross, Phys. Rev. **A25**, 1065(1982).

12. A.C. Newell, Lectures in Applied Mathematics **15**, 157 (1974).
13. The general framework is introduced by:
H. Brand, "Phase dynamics — a review and a perspective," in [10].
14. Original introductions of the KSE are by:
G.I. Sivashinsky, Acta Astronautica **4**, 1177 (1977), and
Y. Kuramoto & T. Tsuzuki, Prog. Theor. Phys. **55**, 356(1976); see also:
Y. Kuramoto, Prog. Theor. Phys. **71**, 1182(1984), and *Chemical oscillations, waves and turbulence* (Springer–Verlag, 1984); for a review, consult also:
P. Manneville, "The Kuramoto–Sivashinsky equation: a progress report," in [10].
15. H. Chaté, P. Manneville, "Phase turbulence," in [16].
16. For an account of recent advances in turbulence research consult: *A tentative dictionary of turbulence*, O. Cardoso & P. Tabeling, Eds. (Plenum Press, 1995).
17. H. Chaté, in: *Spatiotemporal Patterns in Nonequilibrium Complex Systems*, P.E. Cladis & P. Palffy-Muhoray, Eds. (Addison-Wesley, 1995); also:
D.A. Egolf & H.S. Greenside, Phys. Rev. Lett. **74**, 1751 (1995).
18. Recent advances are reviewed by: M.C. Cross & P.C. Hohenberg, Rev. Mod. Phys. **65**, 851 (1993).
19. Y. Pomeau, in: *Spatio-temporal coherence and chaos in physical systems*, A.R. Bishop *et al.*, Eds. (North-Holland, 1986).
20. D. Coles, J. Fluid Mech. **21**, 285 (1965);
C.D. Andereck, S.S. Liu & H.L. Swinney, J. Fluid Mech. **164**, 155(1986);
J.J. Hegseth, C.D. Andereck, F. Hayot & Y. Pomeau, Phys. Rev. Lett. **62**, 257 (1989).
21. see, e.g., W. Kinzel, "Directed Percolation" in: G. Deutscher *et al.*, eds., *Percolation Structures and and Processes*, Annals of the Israel Phys Soc., **5** (1983).
22. K. Kaneko, Prog. Theor. Phys. **74**, 1033(1985).
23. H. Chaté & P. Manneville, Phys. Rev. Lett. **58**, 112(1987); Physica **32D**, 409(1988); Europhys. Lett. **6**, 591(1988); Physica **37D**, 33(1989); J.Stat.Phys. **56**, 357(1989).
24. Reviews with references by H. Chaté & P. Manneville can be found in P. Coullet & P. Huerre, Eds. *New trends in nonlinear dynamics and pattern forming phenomena* (Plenum, 1991), or in [16].
25. For experimental evidence of spatio-temporal intermittency in convection, consult:
P. Bergé, Nucl. Phys. B (Proc.Suppl.) **2**, 247(1987);
S. Ciliberto & P. Bigazzi, Phys. Rev. Lett. **60**, 286 (1988);
F. Daviaud, M. Dubois & P. Bergé, Europhys. Lett. **9**, 441(1989); for the "printer" instability:
M. Rabaud *et al.*, Phys. Rev. Lett. **64**, 184(1990); consult the review by
F. Daviaud, "Experiments in 1D turbulence," in [16].
26. See the review by P. Huerre & P. Monkewitz, Ann. Rev. Fluid Mech. **22**, 473(1990).

Entropy and Quantum Characteristic Exponents. Steps Towards a Quantum Pesin Theory

R. Vilela Mendes

Grupo de Física-Matemática
Complexo II - Universidade de Lisboa
Av. Gama Pinto, 2 - P1699 Lisboa Codex, Portugal

Abstract: Classical ergodic invariants like the Lyapunov exponent and the entropy, defined in a space of densities acted upon by the Perron-Frobenius operator, provide an intuitive framework to construct the corresponding notions in quantum mechanics.

Rigorous existence conditions obtained for quantum characteristic exponents and a definition of quantum entropy based on the Brin-Katok construction are steps towards a quantum ergodic theory, where a rigorous definition of quantum chaos may be established.

1 Characteristic Exponents

Classical mechanics deals with trajectories of dynamical systems whereas quantum mechanics deals with wave functions. This is a source of difficulty when transferring to quantum mechanics the notions of the ergodic theory of classical dynamical systems. The modulus squared of the wave function is a probability density in configuration space. Therefore we would be closer to the framework of quantum mechanics, and presumably in better conditions to develop a *quantum ergodic theory,* if classical concepts like the invariant measure, Lyapunov exponents and entropy were formulated in terms of densities rather than orbits.

Let (X, Σ, μ)be a measure space and $T : X \rightarrow X$ a measurable measure-preserving transformation, that is

(i) $T^{-1}(A) \in \Sigma$ for all $A \in \Sigma$

(ii) $\mu(T^{-1}(A)) = \mu(A)$ for all $A \in \Sigma$

In the context of the ergodic theory of classical dynamical systems X is the state space and T the operator defining time-evolution.

A *density* in X is a non-negative, normalized, integrable function, the space of densities being denoted by $\mathcal{D}$

$$\mathcal{D} = \{\rho \in L^1(X, \Sigma, \mu) : \rho \geq 0, ||\rho||_1 = 1\} \tag{1}$$

$\mathcal{D}$ is the space of functions that, by the Radon-Nikodym theorem, characterize the measures that are absolutely continuous with respect to μ. However, for some of the dynamical concepts to be discussed later, it will turn out that it is convenient to further restrict the space of *admissible densities*.

Time-evolution in the space of densities is described by the weakly continuous Perron-Frobenius operator, $P : L^1 \rightarrow L^1$

$$\int_A (P\rho)(x)d\mu(x) = \int_{T^{-1}(A)} \rho(x)d\mu(x) \tag{2}$$

$A \in \Sigma$, or by the Koopman operator, $U : L^1 \rightarrow L^1$

$$U\rho(x) = \rho(T(x)) \tag{3}$$

these two operators being adjoint to each other, in the sense

$$\int_X (P\rho)(x)g(x)d\mu(x) = \int_X \rho(x)Ug(x)d\mu(x) \tag{4}$$

with $\rho \in L^1$ and $g \in L^\infty$.

The basic notions of ergodic theory, usually defined in terms of the images or inverse images of sets of points in X under the action of T, may also be expressed in terms of the behavior of densities under the action of the Perron-Frobenius or the Koopman operator[1]. For example:

- μ is T-invariant if and only if $P1 = 1$.
- T is ergodic if and only if the sequence $\{P^n\rho\}$ is Cesaro convergent to 1 for all $\rho \in \mathcal{D}$.
- T is mixing if and only if $\{P^n\rho\}$ is weakly convergent to 1 for all $\rho \in \mathcal{D}$.

The mathematical framework needed to characterize the operators of time evolution and the notions of ergodicity and mixing in terms of densities is essentially the same that is needed to formulate the same notions in terms of orbits, namely a measure space and the duality of L^1 and L^∞. The situation is different when dealing with the ergodic invariants that quantify the notions of sensitive dependence and information loss. Consider for example the existence theorem for Lyapunov exponents[2][3]. Let $X = R^d$. Then, given an ergodic invariant measure μ, there is a sequence of numbers

$$\lambda_1 > \lambda_2 > ... > \lambda_\ell$$

$\ell \leq d$, and for μ-almost every point x, a decreasing sequence of vector spaces

$$R^d = E_1(x) \supset E_2(x) \supset ... \supset E_\ell(x) \supset E_{\ell+1}(x) = \{0\} \tag{5}$$

such that $DT_x E_r(x) = E_r(T(x))$ and

$$\lim_{n\rightarrow\infty} \frac{1}{n} \log ||DT_x^n v|| = \lambda_r \tag{6}$$

for $v \in E_r(x) \setminus E_{r+1}(x)$.

In Eq.(6) the action of the tangent map on the vector v measures the rate of separation of two orbits with initial conditions at x and at x plus an "infinitesimal" in the direction of v. When dealing with densities the notion that corresponds to this perturbation of the initial condition at x is that of a Gateaux derivative D_{Φ_x} along a (generalized) function Φ_x with support at x. Furthermore the proper definition and existence of the Lyapunov exponent in Eq.(6) requires the existence of a measure μ in the space of the initial conditions x. Likewise we will need a measure on the infinite-dimensional space of the generalized functions Φ_x. These two requirements, namely the possibility to define Gateaux derivatives along generalized functions with point support and the need for a well-defined σ-additive measure in an infinite-dimensional functional space, lead almost uniquely to the choice of mathematical framework. Namely we need to restrict the *admissible densities* to a nuclear space.

Densities are non-negative L^1-functions. Therefore to each $\rho \in \mathcal{D}$ we may associate a non-negative square root $\rho^{\frac{1}{2}}$ and $\rho \in L^1(\mu)$ implies $\rho^{\frac{1}{2}} \in L^2(\mu)$. We now construct a Gelfand triplet[4]

$$E^* \supset L^2(\mu) \supset E \tag{7}$$

E being a nuclear space and E^* its dual. For each non-negative $f \in E$, with $\|f\|_2 = 1$, $\rho = f^2$ is an admissible density. The generalized functions Φ_x needed to define the local perturbations of the densities will be in E^*. Furthermore, nuclearity of E is important to insure the existence of σ-additive measures on E^*. For definiteness let $E = \mathcal{S}(R^d, \mu)$, the space of functions of rapid decrease topologized by the family of semi-norms $\left\|x^\alpha D^\beta f\right\|_2$, and $E^* = \mathcal{S}'(R^d, \mu)$ its dual. For simplicity μ is assumed to be a T-invariant measure absolutely continuous with respect to Lebesgue measure. Because $\mathcal{S}$ is an algebra, $\rho^{\frac{1}{2}} \in \mathcal{S}$ implies $\rho \in \mathcal{S}$. We now consider a family of vector-valued functionals of a density $\rho \in \mathcal{S}$

$$\mathcal{F}_n(\rho) = \int d\mu(y)\; y\; P^n \rho(y) \qquad\qquad y \in X \tag{8}$$

and define the Lyapunov exponent as follows

$$\lambda_v = \lim_{n\to\infty} \frac{1}{n} \log \left\| -v^i D_{\partial_i \delta_x} \left(\int d\mu(y)\; y\; P^n \rho(y) \right) \right\| \tag{9}$$

v being a vector in R^d. Let X be a compact in R^d and $P^n\rho \in \mathcal{S}$. Then the Gateaux derivative of $\mathcal{F}_n(\rho)$ along $\partial_i \delta_x$ is well-defined[5] and a simple calculation leads to the same result as in Eq.(6). Existence of the limit in Eq.(9) is therefore insured in the same conditions as the limit in Eq.(6). However it is possible to prove directly the existence of the limit in Eq.(9). The proof follows essentially the same steps as used below for the quantum characteristic exponents and I leave it as an exercise for the reader.

Essential to the proofs is the existence of an appropriate measure in E^*, namely a measure for which $\cup_{x \in X} \{\partial_i \delta_x\}$ is not a null set. The following result shows how a class of such measures may be constructed.

Lemma 1.1

Let μ be a normalized measure in R^d and let $\{\Phi_x\}$, a family of elements of $\mathcal{S}'(\mu)$ indexed by the points of R^d, be integrable, that is

$$\int d\mu(x) \; < \Phi_x, \xi > < \infty \qquad\qquad \forall \xi \in \mathcal{S}(\mu) \tag{10}$$

Then the characteristic functional

$$C(\xi) = \int d\mu(x)\; e^{i<\Phi_x,\xi>} \qquad\qquad \xi \in \mathcal{S}(\mu) \tag{11}$$

defines a measure ν in $\mathcal{S}'$ for which the set $\cup_x \Phi_x$ is not a null set. Furthermore if μ is invariant for a mapping $T : R^d \to R^d$, ν is invariant for the corresponding Perron-Frobenius operator P_T if the transformation law of Φ_x is $P_T\Phi_x = \Phi_{Tx}$.

Proof:

(i) $C(0) = 1$

(ii) $\sum_{j,k} \alpha_j \alpha_k^* \; C(\xi_j - \xi_k) = \int d\mu(x) \; \left|\sum_j \alpha_j e^{i<\Phi_x,\xi>}\right|^2 \geq 0$

(iii) $C(\xi)$ is a continuous functional in $\mathcal{S}$. Follows from $\Phi_x \in \mathcal{S}'$ and (10).

Let $\mathcal{B}$ be the Borel σ-algebra generated by the cylinder sets $A_{\xi_1,\ldots,\xi_d,B}$ in $\mathcal{S}'$

$$A_{\xi_1,\ldots,\xi_d,B} = \{f \in \mathcal{S}' : (< f, \xi_1 >, \ldots, < f, \xi_d >) \in B\} \tag{12}$$

with B a Borel subset of R^d. From (i),(ii),(iii) and the Bochner-Minlos theorem[6] it now follows that there is in $(\mathcal{S}',\mathcal{B})$ a unique σ-additive measure ν such that

$$C(\xi) = \int_{\mathcal{S}'} d\nu(\chi)\; e^{i<\chi,\xi>} \tag{13}$$

To show that $\cup_x \Phi_x$ has positive measure consider an orthonormal basis $\{\varsigma_n\}$ of generalized eigenvectors in $\mathcal{S}'$[7]. Eq.(11) becomes

$$C(\xi) = \int d\mu(x)\; e^{i\sum_n \Phi_{x,n}\xi_n} \tag{14}$$

$\{\Phi_{x,n}\}$and $\{\xi_n\}$ being the coefficient sequences for the expansions of Φ_x and ξ, convergence of the sum being insured by the fact that $\xi \in \mathcal{S}$ and $\Phi_x \in \mathcal{S}'$. The measure ν in $\mathcal{S}'$ is then a measure in the space of tempered sequences and from the definition of the characteristic functional one sees that the corresponding measure has support on the $\{\Phi_{x,n}\}$ sequences. Informally

$$d\nu(\chi) = \int d\mu(x) \prod_n \delta(\chi_n - \Phi_{x,n})\; d\chi_n \tag{15}$$

Invariance of ν for P_T is an easy consequence of the T-invariance of μ and the transformation law for Φ_x. Notice that invariance of ν for P_T is equivalent to invariance of $C(\xi)$ for the Koopman operator U. □

Actually there is, for the classical case, a simpler way to construct measures in $\mathcal{S}'(\mu)$ with support in $\cup_x \Phi_x$. However the construction through characteristic functions has the merit of being generalizable to the quantum case.

So far I have been concerned with the formulation of the Lyapunov exponents in terms of densities. Notice however that by replacing in Eq.(9) the Gateaux derivative $D_{\partial_i \delta_x}$ by a Gateaux derivative D_Φ along other $\Phi \in E^*$, other ergodic invariants may be obtained. To prove the existence of such invariants relies on the construction of the appropriate measure, which in some cases may be obtained from lemma 1.1. For example for the higher-order Lyapunov exponents discussed by Farmer and Sidorovich[8] and Taylor[9], we would have

$$\Phi_x = \partial_{i_1} \cdots \partial_{i_n} \delta_x \tag{16}$$

Other choices of Φ may lead to other ergodic invariants[10].

An approach that would bring us closer to the formulation of quantum mechanics would be to consider the functional $\mathcal{F}_n(\rho)$ in Eqs.(8,9) as a functional of $\rho^{\frac{1}{2}}$ rather than of ρ, with the same Gateaux derivative. This means that the differential perturbation at x is performed on the square root of the density, which is like a "classical wave function". However, because of the regularity conditions on $\rho^{\frac{1}{2}} \in \mathcal{S}$ and the $\frac{1}{n}$ limit, the result is exactly the same.

We now turn to quantum mechanics. Let U^k (k continuous or discrete) be the unitary operator of quantum time evolution acting on an Hilbert space $\mathcal{H}$. $\mathcal{H}$ is taken to be the space of functions on the configuration manifold M and $\widetilde{Y}$ is a coordinate operator in M. The quantum version of the functional $\mathcal{F}_n(\rho)$ is

$$\mathcal{F}_n(\Psi) = \left(U^n\Psi, \widetilde{Y}U^n\Psi\right) \tag{17}$$

and taking the Gateaux derivative

$$v^i D_{\partial_i \delta_x} \left(U^n\Psi, \widetilde{Y}U^n\Psi\right) = 2v^i Re < U^n \partial_i \delta_x, \widetilde{Y}U^n\Psi > \tag{18}$$

Proper definition of the right-hand side of (18) requires the construction of a Gelfand triplet as in (7). It is convenient to obtain E as a countably normed space, projective limit of a sequence of Hilbert spaces $\mathcal{H}_p$

$$E^* \supset \ldots \supset \mathcal{H}_{-p} \supset \ldots \supset \mathcal{H}_{-1} \supset \mathcal{H} \supset \mathcal{H}_1 \supset \ldots \supset \mathcal{H}_p \supset \ldots \supset E$$

with ordered norms

$$\|\cdot\| \leq \|\cdot\|_1 \leq \cdots \leq \|\cdot\|_p \leq \cdots$$

The standard way to construct these spaces is to use a self-adjoint operator A to define the norms

$$\|f\|_p^2 = (A^p f, A^p f) \tag{19}$$

If A is bounded from below and there is a $k \in N$ for which A^{-k} is Hilbert-Schmidt, the projective limit space is a nuclear space. Notice that for a density $\rho \in \mathcal{S}$ the limit in (9) for the classical Lyapunov exponent turns out to be independent of ρ. To make the corresponding limit in the quantum characteristic exponent to be independent of Ψ, except for regularity conditions, we will need a norm for the functionals.

One now writes the *quantum characteristic exponent* as follows

$$\lambda_{v,x} = \lim_{n\to\infty} \frac{1}{n} \log \left\| U^{-n}\widetilde{Y}U^n v^i \partial_i \delta_x \right\|_{-p} \tag{20}$$

$\|.\|_{-p}$ being the norm in $\mathcal{H}_{-p}$

$$\|\cdot\|_{-p} = \sup_{\xi \in E, \|\xi\|_p \leq 1} |\langle \cdot, \xi \rangle| \tag{21}$$

p must be greater than or equal to the order of the functional $U^{-n}\widetilde{Y}U^n v^i \partial_i \delta_x$. Notice that U^n, being an element of the infinite-dimensional unitary group, has a natural action in E^*[6]. Let $\mathcal{B}$ be the Borel σ-algebra generated by the cylinder sets of E^* and ν a probability measure on $(E^*,\mathcal{B})$, invariant under the element U of the infinite-dimensional unitary group. Then, existence of the limit in Eq.(20) is a corollary of the following theorem, stated in a weaker form in[11]

Theorem 1.1

Let $\Phi \in \mathcal{H}_{-p}$, $\mathcal{H}_p$ invariant under $U^{-k}\widetilde{Y}U^n$ and ν a U-invariant probability measure on $(E^*,\mathcal{B})$ such that

$$G(\Phi) = \log \frac{\left\| \widetilde{Y}U\Phi \right\|_{-p}}{\left\| \widetilde{Y}\Phi \right\|_{-p}} \qquad \text{and} \qquad F_n(\Phi) = \frac{\left\| U^{-n}\widetilde{Y}U^n\Phi \right\|_{-p}}{\left\| U^{-n+1}\widetilde{Y}U^n\Phi \right\|_{-p}} \tag{22}$$

are $L^1(\nu)$ functions, with $F_n(\Phi)$ converging in $L^1(\nu)$. Then the limit

$$\lambda_\Phi = \lim_{n\to\infty} \frac{1}{n} \log \left\| U^{-n}\widetilde{Y}U^n\Phi \right\|_{-p}$$

exists ν-almost everywhere in $\mathcal{H}_{-p}$.

The proof[22] uses Birkhoff's theorem and one of its corollaries[24]. The conditions on the functions in (22) impose restrictions on the quantum evolution operator U. Much weaker conditions are sufficient if one uses lim sup in the definition of the characteristic exponent or a norm adapted to the dynamics[22].

The quantum characteristic exponent defined above may be explicitly computed in simple cases[13] and its physical meaning seems to be very close to the classical notion of Lyapunov exponent.

As follows from the preceding theorem, existence of a quantum characteristic exponent, analogous to the Lyapunov exponent, requires the existence of a U-invariant measure with support in the set $\{\Phi_x\} = \{\partial_i \delta_x\}$. When the quantum evolution is induced by a substitution operator as in one of the sectors of the four-dimensional Arnold's cat[13], the construction of the measure is the same as in lemma 1.1. In more general cases one may use a construction similar to the classical Bogolyubov-Krylov theorem. Let

$$C(\xi) = \int d\gamma(x)\; e^{i<\Phi_x,\xi>} \tag{23}$$

where $\gamma(x)$ is an arbitrary normalized measure in the configuration space M such that

$$\int d\gamma(x)\; <\Phi_x,\xi> < \infty \qquad \forall \xi \in E$$

Because $\left|C(U^k\xi)\right| \leq 1\ \forall k$, the sequence of partial sums

$$C_n(\xi) = \frac{1}{n}\sum_{k=0}^{n-1} C(U^k\xi)$$

has limit points and one defines

$$\overline{C}(\xi) = \lim_{n\to\infty} \sup \frac{1}{n} \sum_{k=0}^{n-1} C(U^k \xi) \tag{24}$$

$\overline{C}(\xi)$ is continuous and

$$\overline{C}(U\xi) = \overline{C}(\xi)$$

Hence $\overline{C}(\xi)$ is the characteristic function of an U-invariant measure in E^* with support in the set $\{\Phi_x\}$.

Before moving to a discussion of quantum entropy let me point out that other definitions of characteristic exponents in infinite-dimensional spaces have been proposed in the past by several authors[25][18][26][27][28]. They characterize several aspects of the dynamics of linear and non-linear systems. The definition discussed in this paper, and that was proposed for the first time in [13], attempts however to be as close as possible to the spirit of the classical definition of Lyapunov exponent. The essential point is to measure, in a suitable norm, the time evolution of a singular perturbation of the initial condition and, keeping this in mind, the definition (20) may even be further simplified[22].

2 Entropy

The notion of entropy is used in several different senses in mathematics and physics. In the ergodic theory of dynamical systems, the metric (or Kolmogoroff-Sinai) entropy is an asymptotic characterization of the number of different orbits. Usually the Kolmogoroff-Sinai (K-S) entropy is defined by means of the entropy of a state space partition, refined by the time evolution. Carrying over this construction to quantum mechanics, the role of the partition is played by a finite-dimensional Abelian subalgebra of a von Neuman algebra or an Abelian sublattice in the lattice of projections[14][15][16][17]. However, the refinement of the quantum partition defined in this way has a nature different from the classical case. In the classical case the assignment of a trajectory, at a certain time, to an element of the phase-space partition does not change the orbit. Therefore the rate of refinement of the partition does not depend on the measurement process, only on the diversity of orbit behavior. In the quantum mechanical construction, the projections change the dynamical state, hence the construction may depend as much on the reduction features of the quantum measurement process as on the intrinsic dynamical complexity. This point is discussed in more detail elsewhere[18]. Notice also that the definition of entropy proposed in a recent paper by Słomczyński and Życzkowski[19] also involves in an essential way both a dynamical system and a measurement instrument.

As I have already pointed out[18] there is, in the classical case, a definition of entropy, alternative to the one of K-S, which may be generalized to quantum mechanics without involving the measurement process in the intermediate steps of the quantum evolution. This is the Brin-Katok local entropy[20] which, for a compact metric space K is equivalent to the Kolmogoroff-Sinai entropy.

Let $T : K \to K$ be continuous and preserving the measure μ. Define

$$B_\epsilon(T,n,x) = \{y \mid d(T^k(x),T^k(y)) \leq \epsilon,\ 0 \leq k \leq n\} \tag{25}$$

$$h_\epsilon^+(T,x) = \lim\sup_{n\to\infty} \left(-\frac{1}{n}\log\mu(B_\epsilon(T,n,x))\right) \tag{26}$$

$$h_\epsilon^-(T,x) = \lim\inf_{n\to\infty} \left(-\frac{1}{n}\log\mu(B_\epsilon(T,n,x))\right) \tag{27}$$

Then

$$\lim_{\epsilon\to 0} h_\epsilon^+(T,x) = \lim_{\epsilon\to 0} h_\epsilon^-(T,x) = h(T,x) \tag{28}$$

for μ-almost every x and $h(T,x)$ is integrable

$$h_\mu(T) = \int_K h(T,x)d\mu(x) \tag{29}$$

$h_\mu(T)$ coinciding with the Kolmogoroff-Sinai entropy.

I will use the Brin-Katok form to generalize the entropy for the quantum case. As for the characteristic exponent, the first step will be to express the Brin-Katok formulas in terms of densities. Consider a compact $K \subset R^d$ and the Euclidean distance. Then

$$B_\epsilon(T,n,x) = \left\{y : \left|D_{(\delta_x-\delta_y)}\left(\int d\mu(z)\ z\ P^k\rho(z)\right)\right| \leq \epsilon, 0 \leq k \leq n\right\} \tag{30}$$

where linearity in ρ of the functional is used.

The measure to be considered is now a measure ν in $\mathcal{S}'$ with support on the $\{\delta_x\}$ set. It will be the measure induced from μ by the construction of lemma 1.1.

For the quantum case the ϵ-ball may be defined using a wave function

$$B_\epsilon(U,n,\delta_x,\Psi) = \left\{\delta_y : \left|Re < U^k(\delta_x-\delta_y), \widetilde{Y}U^k\Psi >\right| \leq \epsilon, 0 \leq k \leq n\right\} \tag{31}$$

or using norms as in Eq.(21)

$$B_\epsilon(U,n,\delta_x) = \left\{\delta_y : \left\|U^{-k}\widetilde{Y}U^k(\delta_x-\delta_y)\right\|_{-p} \leq \epsilon, 0 \leq k \leq n\right\} \tag{32}$$

and we may now define the quantum local entropy as

$$h(U,x) = -\lim_{\epsilon\to 0}\lim_{n\to\infty}\frac{1}{n}\log\nu(B_\epsilon(U,n,\delta_x)) \tag{33}$$

ν being a measure in $\mathcal{S}'$ invariant for the quantum evolution induced by the unitary operator U. Existence of the limit in Eq.(33) depends on the properties of the measure. I state a few conditions for the existence of the limit.

(i) There is an p such that $U^{-k}\widetilde{Y}U^k(\delta_x-\delta_y) \in \mathcal{H}_{-p}$

(ii) There is an U-invariant measure in $\mathcal{S}'$ such that

$$F_{\epsilon,n}(\delta_x) = \log \frac{\nu(B_\epsilon(U,n,\delta_x))}{\nu(B_\epsilon(U,n-1,U\delta_x))}$$

is an integrable sequence of functionals converging to an integrable functional $F_\epsilon(\delta_x)$. Writing

$$\frac{1}{n}\log\nu(B_\epsilon(U,n,\delta_x)) = \frac{1}{n}\sum_{j=0}^{n-1} F_{\epsilon,n-j}(U^j\delta_x) + \frac{1}{n}\log\nu(B_\epsilon(U,0,U^n\delta_x))$$

existence of $\lim_{n\to\infty}\frac{1}{n}\log\nu(B_\epsilon(U,n,\delta_x))$ follows from condition (ii) by the application of Birkhoff's theorem.

An intuitive derivation of the Pesin formula, relating the entropy and the sum over the positive Lyapunov exponents, follows from using the multipole expansion

$$\delta_y - \delta_x = \sum_{\{i_1\cdots i_n\}} \frac{1}{(\sum i_k)!}(y_1-x_1)^{i_1}(y_2-x_2)^{i_2}\cdots(y_n-x_n)^{i_n}\partial_1^{i_1}\partial_2^{i_2}\cdots\partial_n^{i_n}\delta_x$$

which requires an extension of the mathematical setting to ultradistribution spaces[21]. The conclusion is that a theory for the quantum entropy and a quantum analogue of the Pesin formula may be established in this framework. What is needed however is to relate the properties of the quantum evolution operators to the nature of the measures in $\mathcal{S}'$. For this I refer to work in progress[22].

3 Conclusions

From the rigorous results in Section 1 concerning the quantum characteristic exponents and the proposed definition of a quantum entropy in Section 2, I conclude that it seems possible to develop an ergodic theory for quantum mechanics that closely resembles the corresponding construction in classical mechanics. This might provide a rigorous framework to define quantum chaos. Furthermore, as explained elsewhere[13], this framework may also provide an explanation for the puzzling time reversibility in numerical simulations of quantum systems that are classically chaotic.

The Lyapunov exponent and the entropy play an important role in characterizing chaotic classical dynamical systems. However, except for the simplest systems, it is in general not possible to compute these numbers analytically. Therefore the existence of accurate numerical algorithms for the ergodic invariants plays an important role in the study of such systems. If the quantum characteristic exponents and the quantum entropy are to play a similar role in the study of quantum systems, developing adequate numerical algorithms will also be an important requirement[23].

References

1. A. Lasota and M.C. Mackey; *Probabilistic Properties of Deterministic Systems*, chapter 4, Cambridge Univ. Press, Cambridge 1985.
2. V.I. Oseledets; Trans. Mosc. Math. Soc. **19**, 197 (1969).
3. D. Ruelle; *Chaotic Evolution and Strange Attractors*, Cambridge Univ. Press, Cambridge 1989.
4. I.M. Gelfand and N.Ya. Vilenkin; *Generalized Functions*, vol. **4**, Academic Press, New York 1964.
5. J. Schwartz; *Nonlinear Functional Analysis*, Gordon and Breach, New York 1969.
6. T. Hida; *Brownian Motion*, Springer, Berlin 1980.
7. K. Maurin; *General Eigenfunction Expansions and Unitary Representations of Topological Groups*, Polish Scientific Pub., Warsaw 1968.
8. J. Doyne Farmer and J. Sidorovich; *Evolution, Learning and Cognition*, Y.C. Lee(Ed.), page 277, World Scientific, Singapore 1988.
9. T.J. Taylor, Nonlinearity **6**, 369 (1993).
10. A. Carreira, M.O. Hongler and R. Vilela Mendes; Phys. Lett. **A155**, 388 (1991).
11. R. Vilela Mendes; Phys. Lett. **A187**, 299 (1994).
12. J.F.C. Kingman; J.R. Statist. Soc. **B30**, 499 (1986); Ann. Prob. **1**, 883 (1973).
13. R. Vilela Mendes; Phys. Lett. **A171**, 253 (1992).
14. M.D. Srinivas; J. Math. Phys. **19**, 1952 (1978).
15. A. Connes, H. Narnhofer and W. Thirring; Commun. Math. Phys. **112**, 691 (1987).
16. H. Araki; in *Proc. 8th Int. Congr. on Mathematical Physics*, M. Mebkhout and R. Seneor (Eds.), page 354, World Scientific, Singapore 1987.
17. G. Lindblad; in *Quantum Probability and Applications III*, L. Accardi and W. von Waldenfelds (Eds.), page 183, Lecture Notes in Mathematics 1303, Springer, Berlin 1988.
18. R. Vilela Mendes; J. Phys. A: Math. Gen. **24**, 4349 (1991).
19. W. Słomczyński and K. Życzkowski; J. Math. Phys. **35**, 5674 (1994).
20. M. Brin and A. Katok; in *Geometric Dynamics*, J. Palis (Ed.), page 30, Lecture Notes in Mathematics 1007, Springer, Berlin 1983.
21. J. Sebastião e Silva; Math. Annalen **174**, 109 (1967).
22. R. Vilela Mendes and R. Coutinho; *A Quantum Pesin Theory*, in preparation.
23. R. Vilela Mendes and R. Coutinho; *Numerical Algorithms for Quantum Characteristic Exponents*, in preparation.
24. R. Mañé; *Ergodic Theory and Differentiable Dynamics*, Springer, Berlin 1987.
25. D. Ruelle; Ann. Math. **115**, 243 (1982).
26. F. Haake, H. Wiedemann and K. Życzkowski; Ann. Physik **1**, 531 (1992).
27. K. Życzkowski, H. Wiedeman and W. Słomczyński; Vistas in Astronomy **37**, 153 (1993).
28. W. Majewski and M. Kuna, J. Math. Phys. **34**, 5007 (1993).

Rigorous Numerics of Chaotic Dynamical Systems*

Marian Mrozek

Instytut Informatyki, Uniwersytet Jagielloński,
PL–30–059 Kraków, Poland
email–mrozek@ii.uj.edu.pl

1 Introduction

The importance of computers in the present day research on dynamical systems cannot be questioned. Todays physics or chemistry would not function without access to permanently faster and larger computers. However, from the mathematical point of view, the role of computers is still very limited in this process. Of course, mathematics provides numerical algorithms for solving differential equations which gover the respective dynamical systems. By means of these algorithms computers find discrete models of the equations. Standard numerical analysis also tells us how much the numerically computed solutions in the discrete model differ over a given compact time-space domain from the real solutions. This is usually supplemented by a theorem which says that the solutions of the discrete model limit to some real solutions with the complexity of the discrete model (size of grid discretizing the space) going to infinity. Classical numerical analysis essentially does not go beyond this point. This is maybe sufficient for an ingenieur interested in approximating one particular solution representing the physical object he wants to construct. However, when global understanding of dynamics is of interest such an approach is definitely not satisfactory for at least two reasons. First of all, it does not answer the question to what extend the global asymptotic properties of the constructed discrete model reflect the properties of the original equation. One can construct examples where global phenomena present in all discrete approximations disappear in the limit. Secondly, even assuming that the computations on real numbers are performed exactly, one can only study the behaviour of solutions of the discretized model over some finite time period, which always may differ from the asymptotic behaviour. Thus the physicist or chemist is left to decide on his own, by means of his intuition and experience, if the discrete model he uses appropriately reflects the dynamics of his interest. I do not want to say that physicists fail in this respect. On the contrary, I am pretty sure they are very good in their judgments. But I do not believe they are totally uninterested in making rigorous the

* Research supported by KBN, grant 0449/P3/94/06

claims they draw from the computer simulations of the discrete models about the asymptotic behaviour of the solutions of the equation. They only got used to the fact that the analytic methods of classical mathematics cannot tell much about global dynamics of concrete equations.

The purpose of this paper is to show that things start to change in this respect due to a proper combination of analytic and computer based methods. First of all let us mention that recently a new field emerges: the dynamics of numerics. The aim of this field is to give criteria under which one can claim that the qualitative properties of discrete models are preserved in limit. Whenever the criteria are satisfied, one knows that if only his discrete model is sufficiently close to the original equation then its asymptotic behaviour adequately reflects the behaviour of the equation. The interested reader is referred to [3, 4, 19] for information on the achievements of the new field.

The methods of dynamics of numerics constitute an enormous step forward but do not solve fully the problem. Namely, one no longer needs to worry about carrying over the qualitative properties of the discrete model to the original equations if only one knows that the discrete model is sufficiently close to the equation. However, the question whether the model is sufficiently close may be answered again by experience only.

An alternative approach is provided by the so called computer assisted proofs. Roughly speaking mathematical conditions fall into two categories: verifiable in a finite direct computation like an inequality among two rationals or nonverifiable in a finite direct computation like the existence of a zero of a polynomial. Since the conditions of interest in dynamics usually belong to the other class, in order to successfully perform a computer assisted proof in dynamics one needs a theorem which converts verifiable conditions into nonverifiable conditions, i.e. a theorem which has some verifiable conditions as assumptions and the nonverifiable conditions of interest as assertions.

Let us explain the feasibility of such an approach on a problem which actually is the easiest one among the problems solvable analytically but just for that reason it illustrates in the easiest way the ideas we want to present. Namely consider the problem of the existence of stationary points of a scalar differential equation. It reduces to finding zeros of the vector field. Since in general one can expect the zeros to be irrational numbers, it is unlikely to find them in a finite computation via verifying some clever guesses. The zeros may be found approximately by the Newton method for instance, but the method, even if the successive iterates behave like a convergent sequence, does not guarantee that a zero indeed exists. The standard analytic approach is to look for two points where the vector field has opposite signs and then to use the Darboux property to establish the existence of a zero. If determining analytically the signs of the vector field at a given point is hard then we can use a numerical procedure to determine the signs. This may be done in a finite computation and rigorously if rigorous error estimates are included. Thus a theorem which converts the verifiable conditions into the existence of a zero of a vector field may be the Darboux property.

Via the shooting methods the Darboux property may be fruitfully applied to computer assisted proofs of the existence of phenomena much harder to achieve analytically then just the existence of stationary points. A good example is the proof of the existence of parameter values of the Lorenz equations for which there exists a trajectory homoclinic to the origin. An analytic proof of this fact based on the shooting methods was proposed by Hastings and Troy [6] and later Hassard and Zhang [7] gave a computer assisted proof of this fact which provided very narrow bounds for the parameter admitting the homoclinic trajectory. Similar methods where used to show complicated behaviour of solutions of the Lorenz equations in [8]. Let us also mention two other interesting computer assisted proofs by Neumaier and Rage [17] and also by Rage, Neumaier and Schlier [18]. In the first paper they prove chaos in the standard map. In the other chaos is obtained in a Hamiltonian system govering a molecular model. In both cases the proof consists in verifying in a rigorous computation the existence of a transversal homoclinic point.

2 Multivalued Maps and Inheritable Properties

The approach we want to propose is based on finite multivalued dynamics and the concept of inheritable properties. Multivalued upper–or lower semicontinuous maps which admit only a finite set of values provide a natural bridge between the uncountable dynamics of differential equations and the finite dynamics treatable by computer. The point is that on one hand properties of such maps are often verifiable in a finite computation and on the other hand many such properties are inherited by single valued continuous selectors of such maps.

In order to be more concrete let us introduce some definitions. First of all observe that only a finite set of numbers $\hat{\mathbb{R}} \subset \mathbb{R}$ can be used in computer arithmetic. We will call the elements of $\hat{\mathbb{R}}$ *representable numbers*. Since for obvious reasons the arithmetic operations on representable numbers cannot be exact, some rounding tracking mechanism is needed. The most elegant approach is known as the interval arithmetic (comp. [1]). Assume A is a subset of $\mathbb{R}$. The set of *intervals over* A is defined as

$$\mathcal{I}(A) := \{[a,b] \mid a,b \in A, a \leq b\}$$

We treat the set of real numbers as a subset of $\mathcal{I}(\mathbb{R})$ via the following natural embedding

$$\mathbb{R} \ni x \rightarrow [x,x] \in \mathcal{I}(\mathbb{R})$$

Let $\diamond \in \{+,-,*,/\}$. We extend the arithmetic operations to $I, J \in \mathcal{I}(\mathbb{R})$ by

$$I \diamond J := \{a \diamond b \mid a \in I, b \in J\}$$

The intervals over the set of representable numbers will be called *representable intervals*. By the *representation* of a real number $x \in \mathbb{R}$ we mean any representable interval containing x. Similarly we define the *representation* of an interval $I \in \mathcal{I}(\mathbb{R})$. The *optimum representation* of I is given by

$$\updownarrow I := \bigcap \{J \in \mathcal{I}(\hat{\mathbb{R}}) \mid I \subset J\}$$

We define the adjusted arithmetic operations on representable intervals $I, J \in \mathcal{I}(\hat{\mathbb{R}})$ by

$$I \mathbin{\hat{\diamond}} J := \updownarrow (I \diamond J)$$

Note that the adjusted arithmetic operations on representable intervals are not associative.

Assume $Q : \mathbb{R}^p \to \mathbb{R}$ is a polynomial. By an *implementation* of Q we mean a function $\mathcal{Q} : \hat{\mathbb{R}}^p \to \mathcal{I}(\hat{\mathbb{R}})$ obtained by replacing every arithmetic operation in Q by the corresponding adjusted arithmetic operation in $\mathcal{I}(\hat{\mathbb{R}})$.

Note: A polynomial may have several implementations depending on the order of evaluation of the elementary arithmetic operations.

Proposition 2.1 *Assume $\mathcal{Q}$ is an implementation of a polynomial Q. Then*

$$\forall (x_1, x_2, \ldots x_p) \in \mathbb{R}^p \; Q(x_1, x_2, \ldots x_p) \in \mathcal{Q}(x_1, x_2, \ldots x_p)$$

The interval arithmetic allows us to find rigorously multivalued maps enclosing the exact maps of our interest. The procedure is as follows. First we define the collection of *elementary representable sets* as the family of all singletons $\{a\}$, where $a \in \hat{\mathbb{R}}$, and all open intervals (a, b), where $a, b \in \hat{\mathbb{R}}$ are such that b is an immediate successor of a in $\hat{\mathbb{R}}$. The Cartesian product of d elementary representable sets will be called an *elementary representable set.* A *representable set* is defined as the union of a collection of elementary representable sets. Let $A \subset \mathbb{R}^p$. A multivalued map $F : A \to 2^{\mathbb{R}^q}$ will be called *representable* if $A = \operatorname{dom} F$ and the values of F are representable and F restricted to an elementary representable set is constant. Now let $U \subset \mathbb{R}^p$ and let $f : U \to \mathbb{R}^q$ be an L-Lipschitz continuous single valued map. We will say that the multivalued map F is a *representation of f* if F is representable and f is a selector of $F_{|U}$, i.e. $U \subset \operatorname{dom} F$ and for every $x \in U$ we have $f(x) \in F(x)$. A multivalued representation of f may computed in the following way. First of all a finite collection $\mathcal{B}$ of representable sets covering U is chosen. Then a point x_B is selected in every representable set $B \in \mathcal{B}$ and an approximate value of the map f at the point is found. Now an appropriate error analysis (reducing to to the interval computations if f is a polynomial) allows one to find a representable set containing the exact value $f(x_B)$. Using the Lipschitz constant one finds a representable set $\mathcal{F}_B$ such that $f(B) \subset \mathcal{F}_B$. One then defines the following two maps

$$F^- : \bigcup \mathcal{B} \ni x \to \bigcap \{\mathcal{F}_B \mid x \in B \in \mathcal{B}\} \in \mathcal{P}(\mathbb{R}^q)$$

$$F^+ : \bigcup \mathcal{B} \ni x \to \bigcup \{\mathcal{F}_B \mid x \in B \in \mathcal{B}\} \in \mathcal{P}(\mathbb{R}^q)$$

It is easy to verify that both F^-, F^+ are representations of f and additionally F^- is lower semicontinuous and F^+ is upper semicontinuous. Moreover, such multivalued representations may be chosen arbitrarily close to f if only a sufficiently powerful computer is available.

The importance of representable multivalued maps lied in the fact, that most their properties may be verified in a finite, i.e. combinatorial computation. However, we need some mechanism to transfer such properties to single valued selectors. For that reason we introduce the following notion. Assume $\mathcal{A}$ is a collection of mv maps and $\varphi(F)$ is a property of such maps. We will say that φ is *inheritable* (comp. [14]) if the fact that $\varphi(F)$ holds for some multivalued map $F \in \mathcal{A}$ implies that it holds for every single-valued selector of F in $\mathcal{A}$. An inheritable property φ will be called *strongly inheritable* if for any single valued map $f \in \mathcal{A}$ such that $\varphi(f)$ and for any sequence $\{F_n\} \subset \mathcal{A}$ satisfying $F_n \quad \to f$ we have $\varphi(F_n)$ for n sufficiently large.

Here are some examples. Take $\mathcal{A} = \{F : \mathbb{R}^p \to 2^{\mathbb{R}}\}$ and consider the following three properties

$$\alpha(F) \Leftrightarrow \exists x \in \mathbb{R}^p \; F(x) > 0$$

$$\beta(F) \Leftrightarrow \exists x \in \mathbb{R}^p \; F(x) \geq 0$$

$$\gamma(F) \Leftrightarrow \exists x \in \mathbb{R}^p \; F(x) \ni 0$$

The first property is strongly inheritable. The second property is inheritable but not strongly inheritable. The last property is not inheritable.

Assume $\mathcal{A} = \{F : \mathbb{R}^p \to 2^{\mathbb{R}^p}\}$ and let B be the unit ball in $\mathbb{R}^p$. Consider the following three properties

$$\mu(F) \Leftrightarrow F(B) \subset \operatorname{int} B$$

$$\nu(F) \Leftrightarrow F(B) \subset B$$

$$\kappa(F) \Leftrightarrow \exists x \in B \; x \in F(x)$$

Again the first property is strongly inheritable, the second property is inheritable but not strongly inheritable and the last property is not inheritable.

We will explain the importance of the inheritable properties in the following sections.

3 A Geometric Criterion of Chaos

Let (X, X_0) be a pair of topological spaces. Let

$$\Gamma(X, X_0) := \{\, [\xi] \mid \xi : ([0,1], \{0,1\}) \to (X, X_0)\,\}$$

denote the family of homotopy classes of arcs in X with endpoints in X_0.

We will say that (A, A_0) is an I-pair if it is homotopically equivalent to the pair $([0,1], \{\,0,1\,\})$. It is easy to verify that if (A, A_0) is an I-pair then A is contractible and A_0 has exactly two arcwise connected components, which we will denote by A_0^-, A_0^+. Moreover, if $\xi_0, \xi_1 \in \Gamma(A, A_0)$ then $[\xi_0] = [\xi_1]$ iff $\xi_0(0), \xi_1(0)$

and $\xi_0(1), \xi_1(1)$ are in the same components of A_0. In particular $\Gamma(A, A_0)$ has exactly four elements $[\kappa_{i,j}]$ for $i, j \in \{0,1\}$, where $\kappa_{i,j} : ([0,1], \{0,1\}) \to (A, A_0)$ is an arc such that $\kappa_{i,j}(0) \in A_0^-$ and $\kappa_{i,j}(1) \in A_0^+$.

Let $J_0 := [-2,-1]$, $J_1 := [1,2]$, $J := J_0 \cup J_1$. For $i = 0, 1$ define the map

$$\mu_i : J_i \ni s \to 6s + (-1)^i 9 \in \mathbb{R}$$

and put $\mu := \mu_0 \cup \mu_1 : J \to \mathbb{R}$. Assume that $N = T \times J$, where T is a contractible space. Put $N^h := T \times \{-2,-1,1,2\}$, $N^v := \text{cl}(\text{bd}\, N \setminus N^h)$, $N_i := T \times J_i$, $N_i^h := T \times \text{bd}\, J_i$, $N_i^\pm := T \times \text{bd}^\pm J_i$, where $\text{bd}\, J_i$ stands for the doubleton consisting of the endpoints of J_i and $\text{bd}^\pm J_i$ denote the singletons consisting of the left and right endpoints of J_i respectively. For a point $c \in T$ and $i = 0, 1$ define a map $\mu_i^c : ([0,1], \{0,1\}) \to (X \setminus N^v, X \setminus N)$ by $\mu_i^c(t) := (c, \mu_i(t))$.

The following theorem is a variant of a result by Zgliczy/'nski (comp. [14, 21]).

Theorem 3.1 *Assume $f : X \to X$ is a homeomorphism and there exists an I-pair (A, A_0) such that*

(i) $A \cap N^v = \emptyset, A_0 \cap N = \emptyset$
(ii) f maps the triple (N_i, N_i^-, N_i^+) into the triple (A, A_0^-, A_0^+).
(iii) $i : (A, A_0) \to (X \setminus N^v, X \setminus N)$ maps $[\kappa_{0,1}]$ onto $[\mu_0^c] = [\mu_1^c] \in \Gamma(X \setminus N^v, X \setminus N)$

Then N is an isolating neighborhood for f and there exists a continuous surjection $\varphi : (\text{Inv}\, N, f) \to (\Sigma_2, \sigma)$. Moreover, for any periodic sequence $\alpha \in \Sigma_2$ there exists an $x \in N$ such that $\varphi(x) = \alpha$ and x is periodic with the same principal period as α.

Note that in the above theorem the only assumption concerning f is the inclusion condition on the image of N_i, N_i^-, N_i^+ under f and such a condition is obviously inheritable. Thus in order to apply the above theorem it is enough to check this condition for a multivalued representation of f which is sufficiently close to f.

4 A Few Examples

The following three theorems were obtained in computer assisted proofs based on the described geometric principle.

Theorem 4.1 *(see [21]) Consider the Hénon map $H_{ab} : \mathbb{R}^2 \to \mathbb{R}^2$ given by*

$$H_{ab}(x, y) := (1 + y - ax^2, bx).$$

Then there exists a compact set $N \subset \mathbb{R}^2$ such that for (a,b) sufficiently close to $(1.4, 0.3)$ there exists a continuous surjection $\rho : (\text{Inv}\, N, H_{ab}^7) \to (\Sigma_2, \sigma)$. Moreover, for any periodic sequence $\alpha \in \Sigma_2$ there exists an $x \in \text{Inv}\, N$ such that $\rho(x) = \alpha$ and x is periodic with the same principal period as α.

The following theorem follows from Theorem 3.1 and from the computations performed in [11, 12].

Theorem 4.2 *Consider the Lorenz equations*

$$\begin{aligned} \dot{x} &= s(y-x) \\ \dot{y} &= Rx - y - xz \\ \dot{z} &= xy - qz, \end{aligned} \tag{1}$$

and the plane

$$P := \{(x,y,z) \mid z = 53\}.$$

For all parameter values (s, R, q) sufficiently close to $(45, 54, 10)$, there exists a Poincaré section $N \subset P$ such that the Poincaré map g induced by (1) is Lipschitz and well defined. Furthermore, there exists a continuous surjection $\rho : (\operatorname{Inv} N, g) \to (\Sigma_2, \sigma)$. Moreover, for any periodic sequence $\alpha \in \Sigma_2$ there exists an $x \in \operatorname{Inv} N$ such that $\rho(x) = \alpha$ and x is periodic with the same principal period as α.

Theorem 4.3 *(see [22]). Consider the Rössler equations*

$$\begin{aligned} \dot{x} &= -(y+z) \\ \dot{y} &= x + by \\ \dot{z} &= b + z(x-a), \end{aligned} \tag{2}$$

and the plane

$$P := \{(x,y,z) \mid x = 0\}.$$

For all parameter values in a sufficiently small neighbourhood of $(a, b) = (5.7, 0.2)$, there exists a Poincaré section $N \subset P$ such that the second iterate of the Poincaré map g induced by (2) is Lipschitz and well defined. Furthermore, there exists a continuous surjection $\rho : (\operatorname{Inv} N, g^2) \to (\Sigma_2, \sigma)$. Moreover, for any periodic sequence $\alpha \in \Sigma_2$ there exists an $x \in \operatorname{Inv} N$ such that $\rho(x) = \alpha$ and x is periodic with respect to g^2 with the same principal period as α.

The computations necessary to establish the above results took respectively (approximately) 20sec, 90h, 44h CPU time of an IBM PC with a 486DX 50MHz processor.

5 Isolating Neighbourhoods

Although the computations necessary to establish the chaotic behaviour in a Poincaré map by means of Theorem 3.1 are many times faster then those based on the existence of a transversal homoclinic point criterion (tens instead of thousands of hours of CPU time –compare [18]), they still are quite expensive in case of a Poincaré map. A solution would be to look for a chaos criterion based on the behaviour of the vector field only in order to avoid the expensive computation of the Poincaré map. In the following sections we introduce some machinery, which shows that such an approach is indeed possible.

Assume X is a locally compact metric space and $f : X \to X$ is a homeomorphism. Then f generates a *discrete dynamical system*

$$\pi_f : X \times \mathbb{Z} \ni (x,n) \to f(x,n) := f^n(x) \in X$$

denoted often simply by f.

Let $N \subset X$ be a compact set. The *invariant part* of N is given by

$$\operatorname{Inv} N := \{x \in N \mid \forall n \in \mathbb{Z}\ f^n(x) \in N\}$$

The set N is called an *isolating neighborhood* if any of the following two equivalent conditions is satisfied

$$\forall x \in \operatorname{bd} N\ \exists n \in \mathbb{Z}\ f(x,n) \notin N$$

$$\operatorname{Inv} N \subset \operatorname{int} N.$$

If N is an isolating neighborhood then the set $S := \operatorname{Inv} N$ is called an *isolated invariant set* . Replacing in the above definitions $\mathbb{Z}$ by $\mathbb{R}$ one obtains analogous concepts for flows.

Definition 5.1 *The pair $P = (P_1, P_2)$ of compact subsets of N will be called an index pair of S in N iff the following three conditions are satisfied*

$$x \in P_i,\ f(x) \in N \ \Rightarrow\ f(x) \in P_i, \quad i = 1,2$$

$$x \in P_1,\ f(x) \notin N \ \Rightarrow\ x \in P_2$$

$$\operatorname{Inv} N \subset \operatorname{int}(P_1 \backslash P_2).$$

Theorem 5.2 *Every isolating neighbourhood admits at least one index pair P.*

It can be easily derived from the definition of the index pair that f induces a map of pairs

$$f_P : (P_1, P_2) \ni x \to f(x) \in (P_1 \cup f(P_2), P_2 \cup f(P_2))$$

and the inclusion

$$i_P : (P_1, P_2) \ni x \to x \in (P_1 \cup (P_2), P_2 \cup f(P_2)) \ .$$

induces an isomorphism in Alexander-Spanier cohomology.

Definition 5.3 *The endomorphism*

$$H^*(f_P) \circ H^*(i_P)^{-1} \text{ of } H^*(P),$$

where H^* denotes the Alexander-Spanier cohomology, will be called the index map associated with the index pair P and denoted by I_P.

6 The Conley Index

Let $\mathcal{E}$ be a category. We consider the *category of endomorphisms* of $\mathcal{E}$ denoted by $\mathrm{Endo}(\mathcal{E})$ and defined as follows. The objects of $\mathrm{Endo}(\mathcal{E})$ are pairs (A, a), where $A \in \mathcal{E}$ and $a \in \mathcal{E}(A, A)$ is a distinguished endomorphism of A. The set of morphisms from $(A, a) \in \mathcal{E}$ to $(B, b) \in \mathcal{E}$ is the subset of $\mathcal{E}(A, B)$ consisting of exactly those morphisms $\varphi \in \mathcal{E}(A, B)$ for which $b\varphi = \varphi a$. We write $\varphi : (A, a) \to (B, b)$ to denote that φ is a morphism from (A, a) to (B, b) in $\mathrm{Endo}(\mathcal{E})$. We define the category of automorphisms of $\mathcal{E}$ as the full subcategory of $\mathrm{Endo}(\mathcal{E})$ consisting of pairs $(A, a) \in \mathrm{Endo}(\mathcal{E})$ such that $a \in \mathcal{E}(A, A)$ is an automorphism, i.e., both an endomorphism and an isomorphism in $\mathcal{E}$. The category of automorphisms of $\mathcal{E}$ will be denoted by $\mathrm{Auto}(\mathcal{E})$. There is a functorial embedding

$$\mathcal{E} \ni A \to (A, \mathrm{id}_A \in \mathrm{Auto}(\mathcal{E}))\,,$$

$$\mathcal{E}(A, B) \ni \varphi \to \varphi \in \mathrm{Auto}(\mathcal{E})(A, B)\,,$$

hence we can consider the category $\mathcal{E}$ as a subcategory of $\mathrm{Auto}(\mathcal{E})$.

Definition 6.1 *We will say that $F : \mathcal{C} \to Auto(\mathcal{E})$ is* normal, *if for each $(A, a) \in \mathcal{C}$ the morphism $F(a)$ is equal to the automorphism distinguished in $F(A, a)$.*

Theorem 6.2 *(see [13]) Assume $\mathcal{E}$ admits inverse limits. Then the inverse limit functor which sends $(A, a) \in Endo(\mathcal{E})$ to the inverse limit of the sequence*

$$\ldots \xrightarrow{a} A \xrightarrow{a} A \xrightarrow{a} A$$

is a normal functor.

Let us denote this functor by L.

Theorem 6.3 *(see [13]) Assume $f : X \to X$ is a homeomorphism and N is an isolating neighborhood with respect to f. Then $L(H^*(P), I_P)$ does not depend on the particular index pair in N.*

We define the *Conley index* of the isolating neighborhood N as $\mathrm{Con}^*(N) := L(H^*(P), I_P)$, where P is an index pair in N. Let us emphasize that the Conley index has the form of a pair

$$\mathrm{Con}^*(N) = (CH^*(N), \chi^*(N))\ ,$$

where $CH^*(N)$ is a graded vector space and $\chi^*(N) : CH^*(N) \to CH^*(N)$ is a graded automorphism.

7 The Conley Index for Multivalued Maps

Assume $F : X \times \mathbb{Z} \to X$ is an usc compact-valued map. We will say that F is a *multivalued discrete dynamical system* (mvdds) on x if the following conditions are satisfied

$$F(x, 0) = x$$

$$y \in F(x, n) \Leftrightarrow x \in F(y, -n)$$

$$nm \geq 0 \Rightarrow F(F(x, n), m) = F(x, n + m)$$

One can easily observe that if X is compact then any usc mv map $F : X \to X$ induces a mvdds.

Assume $N \subset X$. The set $\operatorname{Inv}(N, F)$ is defined as the set of $x \in N$ such that there exists a function $\sigma : \mathbb{Z} \to N$ satisfying $\sigma(0) = x$ and $\sigma(n+1) \in F(\sigma(n))$.

The set N is said to be a pseudo isolating neighbourhood for F if

$$\forall x \in \operatorname{bd} N \; \exists n \in \mathbb{Z} \; F(x, n) \cap N = \emptyset$$

The set N is said to be an isolating neighbourhood for F if

$$B_{\operatorname{diam}_N F}(\operatorname{Inv} N) \subset \operatorname{int} N,$$

where $\operatorname{diam}_N F$ is the maximal diameter of the values of F in N.

Proposition 7.1 *(see [14]) Both the pseudo isolating neighborhood and the isolating neighborhood are inheritable properties. The isolating neighborhood is easily seen to be also strongly inheritable.*

Let N be an isolating neighbourhood for F. A pair $P = (P_1, P_2)$ of compact subsets $P_2 \subset P_1 \subset N$ is called an *index pair* if the following conditions are satisfied:

$$F(P_i) \cap N \subset P_i, \; i = 1, 2;$$

$$F(P_1 \backslash P_2) \subset N;$$

$$\operatorname{Inv} N \subset \operatorname{int}(P_1 \backslash P_2)$$

Theorem 7.2 *(see [9]) Every isolating neighborhood admits an index pair.*

Proposition 7.3 *If P is an index pair for N then the inclusion*

$$i_P : F(P) \subset T(P) := (P_1 \cup F(P_2), P_2 \cup F(P_2))$$

induces an isomorphisms in the Alexander-Spanier cohomology.

In order to define the index map in the multivalued case we need mv maps which induce maps in cohomology. For this purpose we restrict our attention to mvdds such that F is given by a morphism in the sense of Górniewicz & Granas (comp. [5]).

Let $F_{P,T(P)} : P \to 2^{T(P)}$ be the restriction of F to domain P and codomain $T(P)$. The endomorphism $I_P := H^*\left(F_{P,T(P)}\right) \circ H^*(i_P)^{-1}$ of $H^*(P)$ is called the *index map* associated with the index pair P.

Theorem 7.4 *(see [9]) Let N be an isolating neighbourhood. Then $(L(H^*(P), I_P)$ is independent of the choice of an index pair P in N.*

As in the discrete case we define the *Conley index* of the isolating neighborhood N as $\mathrm{Con}^*(N) := L(H^*(P), I_P)$, where P is an index pair in N.

Theorem 7.5 *(see [14]) The Conley index is a strongly inheritable property.*

We say that N is an *isolating block* if

$$B_{\mathrm{diam}_N F}(F^{*-1}(N) \cap N \cap F(N)) \subset \mathrm{int}\, N$$

Theorem 7.6 *(see [14]) Assume F is usc and N is an isolating block for F. Define*

$$N^- := N \backslash F^{-1}(\mathrm{int}\, N).$$

Then (N, N^-) is an index pair for F in N.

Theorem 7.7 *(see [14]) Assume F is lsc and N is an isolating block for F. Define*

$$N^- := N \backslash \mathrm{int}(N \cap F^{-1}(N)).$$

Then (N, N^-) is an index pair for F in N.

8 The Conley Index and Chaos

The following criterion is the earliest topological criterion for chaos.

Theorem 8.1 *(see [10]) Assume that $N = N_0 \cup N_1$ is an isolating neighbourhood under f where N_0 and N_1 are disjoint compact sets and for $k = 0, 1$*

$$\mathrm{Con}^n(N_k) = \begin{cases} (Q, \mathrm{Id}) & \textit{if } n\textit{=1} \\ 0 & \textit{otherwise.} \end{cases}$$

Then $N_{kl} := (N_k \cap f(N_k)) \cup (N_k \cap f(N_l)) \cup (N_l \cap f(N_l))$ for $k, l \in \{0, 1\}, k \neq l$, are isolating neighbourhoods and if additionally $\chi^*(N_{kl})$ is not conjugate to identity then there exists a $d \in N$ and a continuous surjection $\rho : (\mathrm{Inv}\, N, f) \to (\Sigma_2, \sigma)$.

The following theorem was conjectured in [10] and proved by Szymczak in [20].

Theorem 8.2 *Assume X is a compact ANR and $N = N_0 \cup N_1 \subset X$ is an isolating neighbourhood under f where N_0 and N_1 are disjoint compact sets and the Conley index of N_0 is as in the previous theorem, whereas $\mathrm{Con}(N) = 0$. Then there exists a $d \in N$ and a continuous surjection $\rho : (\mathrm{Inv}\, N, f) \to (\Sigma_2, \sigma)$. Moreover, for any periodic sequence $\alpha \in \Sigma_2$ there exists an $x \in \mathrm{Inv}\, N$ such that $\rho(x) = \alpha$ and x is periodic with respect to g^d with the same principal period as α.*

The above two theorems are formally weaker then Theorem 3.1 but their potential strength in case of a Poincaré map lies in the possibility of regaining the index information just from the vector field. We want to explain these ideas now.

If $\pi : X \times \mathbb{R} \to X$ is a flow and N is an isolating neighborhood with respect to π then the Conley index of $\pi_t : X \ni x \to \pi(x,t) \in X$ does not depend on t for $t > 0$. Thus the Conley index for flows is well defined. Numerical computations of such an index are much less expensive than the computations of the Conley index of a Poincaré map, because one needs to follow the flow only for a short period of time. If N admits a cross-section Ξ then there is a well defined Poincaré map $\Pi : U \to \Xi$, for some open $U \subset \Xi$. Unfortunately it turns out that it is impossible to regain the Conley index of the Poincaré map from the Conley index for flows.

Let S^1 denote the unit circle in $\mathbb{R}^2$ with a fixed orientation.

Assume N is an isolating neighborhood with respect to a flow π admitting a cross-section Ξ. For each $x \in N$ denote by $r(x)$ the time of first return of the trajectory of x to Ξ. W.l.o.g. we may assume that $r(x) = 1$ for $x \in N \cap \Xi$. Thus r defines a surjective continuous map $\Gamma : N \to S^1$.

Let $P = (P_1, P_2)$ be an index pair in N. We will say that the index pair P is regular if the map

$$\omega_P : P_1 \ni x \to \sup\{t > 0 | \pi(x,[0,t]) \subset P_1 \backslash P_2\} \in \mathbb{R} \tag{3}$$

is continuous. One can show that for any open neighborhood U of S there exists a regular index pair in N.

Let $\Gamma(P)$ denote the disjoint union $P_1 \cup S^1$ divided by the following equivalence relation

$$x \approx y \text{ iff } x = y \text{ or } (x \in P_1 \text{ and } y = \Gamma(x)) \tag{4}$$

Notice that S^1 may be considered as a subsapce of $\Gamma(P)$.

Theorem 8.3 *(see [15]) The cohomology $H^*(\Gamma(P), S^1)$ does not depend of the choice of a regular index pair, i.e. it is an invariant of S.*

We will call this cohomology the Conley index of S over S^1.

Assume $\Delta \subset S^1$ is a closed interval. The fixed orientation of S^1 induces an order on Δ.

We will say that (X,p) is a trivial semibundle over Δ if $p : X \to \Delta$ is a continuous map such that for some $a \in \Delta$ there exists a continuous surjection $h : p^{-1}(a) \times \Delta \to X$ satisfying the following two properties

(i) $p(h(x,t)) = t$ for any $t \in \Delta, x \in p^{-1}(a)$

(ii) $h(x,t) = h(y,t), s > t$ imply $h(x,s) = h(y,s)$

We will say that (X,p) is a semibundle over S^1 if for any $a \in S^1$ there exists a closed interval Δ such that $a \in \operatorname{int}\Delta$ and $(p^{-1}(\Delta), p|_{p^{-1}(\Delta)})$ is a trivial semibundle over Δ.

Theorem 8.4 *(see [15]) Assume (X,p) is a semibundle over S^1. Then there is a collection of maps $\{p_{s,t} : p^{-1}([t]) \to p^{-1}([s]),\ s,t \in \mathbb{R}, s \leq t\}$, unique up to a homotopy, such that*

$$p_{s+r,t+r} = p_{s,t} \tag{5}$$

$$p_{t,t} = \mathrm{id}_{p^{-1}(t)} \tag{6}$$

$$p_{r,s}p_{s,t} = p_{r,t} \tag{7}$$

The map $p_{t,t+1}$ will be called the monodromy map associated with the leave $p^{-1}([t])$.

Theorem 8.5 *(see [15]) Let P be an index pair in N. Then $\Gamma(P)$ is a semibundle over S^1. Let m be a monodromy map. Then $L(H^*(\Gamma(P), S^1), m)$ coincides with the discrete Conley index of the Poincaré map associated with S.*

The above theorem shows that the knowledge of the Conley index over S^1 allows one to reconstruct the Conley index of the Poincaré map. The numerical computation of the Conley index over S^1, like the computation of the classical Conley index in the flow sense, is much less expensive than the computation of the Conley index of a Poincaré map. Thus one can expect that the two presented Theorems characterizing chaos for discrete dynamical systems will become very useful in the rigorous numerical verification of chaos for flows admitting cross-section. Research in this direction is in progress.

References

1. O. Aberth, *Precise Numerical Analysis*, William C. Brown Publishers, Dubuque, Iowa, 1988.
2. C. Conley, On a generalization of the Morse index, in *Ordinary Differential Equations*, 1971 NRL–MRC Conference, ed. L. Weiss, Academic Press, New York, 27–33 (1972).
3. B.M. Garay, Discretization and some qualitative properties of ordinary differential equations about equilibria, *Acta Math. Univ. Comeniane*, **62**, 249–275(1993).
4. B.M. Garay, On structural stability of ordinary differential equations with respect to discretization methods, preprint.
5. L. Górniewicz, Topological Degree of Morphisms and its Applications to Differential Inclusions, *Raccolta di Seminari del Dipartimento di Matematica dell'Universita degli Studi della Calabria*, No. 5, 1983.

6. S.P. Hastings, W.C. Troy, A shooting approach to the Lorenz equations, *Bull. AMS*, **27**, 298–303(1992).
7. B. Hassard and J. Zhang, Existence of a homoclinic orbit of the Lorenz system by precise shooting, *SIAM J. Math. Anal.*, **25**, 179–196(1994).
8. B. Hassard, S.P. Hastings, W.C. Troy, J. Zhang, A computer proof that the Lorenz equations have "chaotic" solutions, *Appl. Math. Letter*, to appear.
9. T. Kaczyński and M. Mrozek, Conley index for discrete multivalued dynamical systems, *Topology & its Appl*, accepted.
10. K. Mischaikow and M. Mrozek, Isolating neighbourhoods and Chaos, accepted.
11. K. Mischaikow and M. Mrozek, Chaos in Lorenz equations: a computer assisted proof, *Bull. AMS*, in print.
12. K. Mischaikow and M. Mrozek, Chaos in Lorenz equations: a computer assisted proof, Part II: details, in preparation.
13. M. Mrozek, Leray Functor and the Cohomological Conley Index for Discrete Dynamical Systems, *Transactions of the American Mathematical Society*, **318**, 149–178(1990).
14. M. Mrozek, Topological invariants, multivalued maps and computer assisted proofs in dynamics, in preparation.
15. M. Mrozek, J. Reineck, R. Srzednicki, The Conley index over S^1, in preparation.
16. M. Mrozek and K.P. Rybakowski, Discretized ordinary differential equations and the Conley index, *J. of Dynamics and Differential Equations*, **4**, 57–63(1992).
17. A. Neumaier, Th. Rage, Rigorous chaos verification in discrete dynamical systems, *Physica D*, **67**, 327–346(1993).
18. Th. Rage, A. Neumaier, Ch. Schlier, Rigorous verification of chaos in a molecular model, *Physical Rev.* **E50**, 2682–2688(1994).
19. A.M. Stuart, Numerical analysis of dynamical systems, *Acta Numerica*, in print.
20. A. Szymczak, The Conley index and symbolic dynamics, preprint.
21. P. Zgliczyński, Fixed point index for iterations of maps, topological horseshoe and chaos, preprint.
22. P. Zgliczyński, in preparation.

The Effect of Symmetry Breaking on Random Walks and Brownian Motion

L.E. Reichl and P. Alpatov

Center for Studies in Statistical Mechanics and Complex Systems
The University of Texas at Austin
Austin, Texas 78712

Abstract: Many aspects of the dynamics of random walks and Brownian motion with detailed balance can be understood using techniques of classical and quantum chaos theory. Symmetry breaking coupling between degrees of freedom can lead to mutual repulsion of decay rates and in some cases to a change in the rate of approach to equilibrium.

The breaking of symmetry in conservative nonlinear dynamical systems can have a profound effect on their dynamics [1]. Symmetries give rise to constants of the motion and constants of the motion give structure to dynamical systems. In classical mechanical systems, constants of the motion define surfaces (called KAM surfaces after Kolmogorov, Arnold, and Moser) in phase space and reduce the effective dimension of the phase space available to a system. Symmetries may be broken (and constants of motion destroyed) when the various degrees of freedom interact and resonant. Resonance occurs if the natural frequencies (which generally depend on the energy) associated to those degrees of freedom become commensurate. The various degrees of freedom may be coupled either by interaction forces or by symmetry breaking boundaries. In nonlinear systems, when two or more resonances occur on one scale in the phase space, they generate families of resonances on all smaller scales. If the coupling between degrees of freedom is weak enough, resonances create isolated localized topological changes in the structure of phase space. In the regions dominated by the resonances, the KAM surfaces defined by the global constants of motion (in the absence of coupling) can be destroyed. If the coupling between degrees of freedom is strong enough, resonance regions can overlap and destroy any remaining KAM surfaces between them. The structure of phase space then becomes so complex in these regions that the motion of trajectories is chaotic and trajectories can diffuse through those regions of phase space which are dominated by the resonances.

In quantum mechanical systems, constants of motion are the source of the 'good' quantum numbers which determine the dynamics of the quantum system. 'Good' quantum numbers can also be destroyed by symmetry breaking forces or boundaries. When symmetry breaking occurs it is generally accompanied by a change in the statistical properties of the energy level spectrum and spreading of energy eigenstates in the Hilbert space of the original 'good' quantum numbers.

The fact that there may be a change in spectral statistics was first indicated by von Neumann and Wigner [2]. They showed that the breaking of a symmetry makes it much less likely to find close spacings between energy eigenvalues. Indeed, we now know that integrable quantum systems (systems with as many global constants of motion as degrees of freedom) will have nearest neighbor spacings that are random and have a distribution qualitatively similar to the Poisson distribution, $P_P(s) = \frac{1}{D}\exp\left(-\frac{s}{D}\right)$, where s is the spacing and D is the average spacing between levels. In systems with broken symmetry, energy level spacings repel and the distribution tends toward a Wigner distribution, $P_W(s) = \frac{\pi s}{2D^2}\exp\left(-\frac{\pi s^2}{4D^2}\right)$

We want to show that phenomena similar to the above can be observed in stochastic systems. We will begin with a random walk on a two dimensional finite square lattice in which each site of the lattice has four nearest neighbors (except at the boundaries) and the walker jumps only between nearest neighbor sites. We assume that the lattice has $N=(L+1)^2$ sites (L an even integer) which we label (n_1, n_2) where n_1 and n_2 are integers with values $-\frac{L}{2}\leq n_1 \leq +\frac{L}{2}$ and $-\frac{L}{2}\leq n_2 \leq +\frac{L}{2}$. The site labeled $\mathbf{n} = (n_1, n_2)$ lies at the spatial point $\mathbf{x} = x_1\hat{\mathbf{i}} + x_2\hat{\mathbf{j}}$, where $x_1 = n_1\Delta$, $x_2 = n_2\Delta$, $\hat{\mathbf{i}}$ and $\hat{\mathbf{j}}$ are unit vectors along the horizontal and vertical axes, respectively, and Δ is the spacing between lattice sites. The probability, $P(\mathbf{x},t)$, to find the walker at point $\mathbf{x}$ (site (n_1,n_2)) at time t is described by the master equation,

$$\frac{\partial P(\mathbf{x},t)}{\partial t} = \sum_{x_1'=-\frac{L\Delta}{2}}^{\frac{L\Delta}{2}} \sum_{x_2'=-\frac{L\Delta}{2}}^{\frac{L\Delta}{2}} P(\mathbf{x}',t)W(\mathbf{x}'|\mathbf{x}) \ , \tag{1}$$

where the transition matrix, $W(\mathbf{x}'|\mathbf{x}) = \omega(\mathbf{x}'|\mathbf{x}) - \delta_{\mathbf{x},\mathbf{x}'}\sum_{\mathbf{x}''}\omega(\mathbf{x}|\mathbf{x}'')$ and $\omega(\mathbf{x}'|\mathbf{x})$ is the transition rate from site $\mathbf{x}'$ to site $\mathbf{x}$.

Let $P_0(\mathbf{x}) = \lim_{t\to\infty} P(\mathbf{x},t)$ denote the long time probability distribution of the random walker. If the random walk satisfies detailed balance then $P_0(\mathbf{x})\omega(\mathbf{x}|\mathbf{x}') = P_0(\mathbf{x}')\omega(\mathbf{x}'|\mathbf{x})$, and we can symmetrize the transition matrix. To see this let us introduce a new function $\tilde{P}(\mathbf{x},t) = \frac{P(\mathbf{x},t)}{\sqrt{P_0(\mathbf{x})}}$. Then the master equation takes the form $\frac{\partial \tilde{P}(\mathbf{x},t)}{\partial t} = \sum_{\mathbf{x}'}\tilde{P}(\mathbf{x},t)V(\mathbf{x}'|\mathbf{x})$, where $V(\mathbf{x}'|\mathbf{x})$ is a symmetric matrix defined $V(\mathbf{x}'|\mathbf{x}) = \sqrt{\frac{P_0(\mathbf{x}')}{P_0(\mathbf{x})}}\omega(\mathbf{x}'|\mathbf{x}) - \delta_{\mathbf{x},\mathbf{x}'}\sum_{\mathbf{x}''}\omega(\mathbf{x}|\mathbf{x}'')$. Since $V(\mathbf{x}'|\mathbf{x})$ is symmetric, it has a complete orthonormal set of eigenvectors $|\phi_i>$ and real eigenvalues, λ_i ($i = 0,..,N-1$). If we let $<\mathbf{x}|\phi_i>$ denote the entry of the i–th eigenvector due to lattice site $\mathbf{x}$, the solution to the master equation takes the form

$$P(\mathbf{x},t) = \sum_{i=0}^{N-1}\sum_{\mathbf{x}'}\sqrt{\frac{P_0(\mathbf{x})}{P_0(\mathbf{x}')}}P(\mathbf{x}',0) < \mathbf{x}'|\phi_i > \mathrm{e}^{\lambda_i t} < \phi_i|\mathbf{x}> \ . \tag{2}$$

The eigenvalue, $\lambda_0 = 0$, and eigenvalues $\lambda_i < 0$ for $i = 1,...,N-1$.

Let us now consider some specific examples of random walks. We will choose the transition rates from a potential, $U(x_1, x_2)$, so that

$$\omega(x_1, x_2|x_1 \pm \Delta, x_2) = \frac{g}{2\Delta^2} \mp \frac{(U(x_1 \pm \Delta, x_2) - U(x_1, x_2))}{2\Delta^2} \quad \text{and}$$

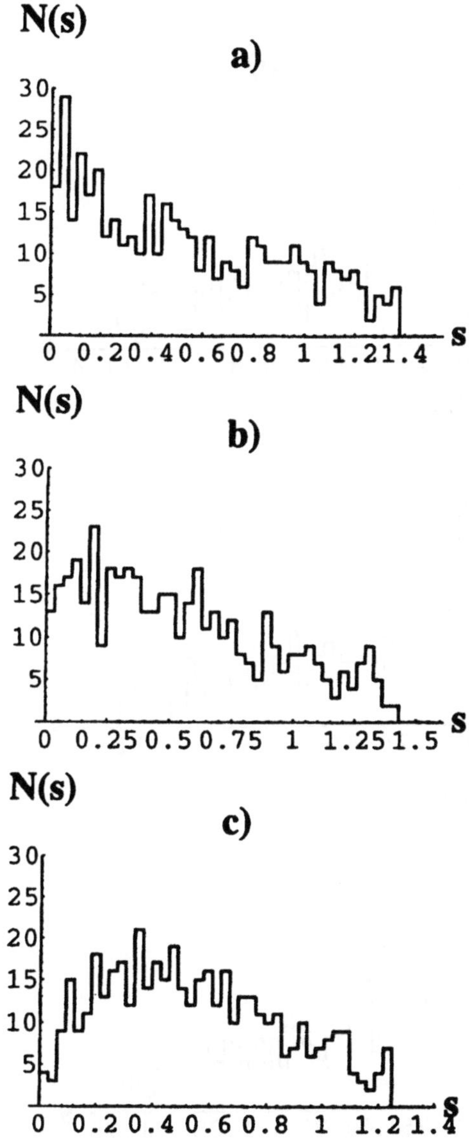

Fig. 1. Nearest neighbor spacing histograms of decay rates for a random walk with transition rates given by $U_1(x_1, x_2)$ and (a) $g = 0.2$, $\epsilon = 0.0$, (b) $g = 0.2$, $\epsilon = 0.1$, and (c) $g = 0.2$, $\epsilon = 1.0$.

$$\omega(x_1, x_2|x_1, x_2 \pm \Delta) = \frac{g}{2\Delta^2} \mp \frac{(U(x_1, x_2 \pm \Delta) - U(x_1, x_2))}{2\Delta^2} \tag{3}$$

We first choose the potential, $U(x_1, x_2) = U_1(x_1, x_2)$ where

$$U_1(x_1, x_2) = 2x_1^4 + \frac{3}{5}x_2^4 + \epsilon x_1 x_2 (x_1 - x_2)^2 \ . \tag{4}$$

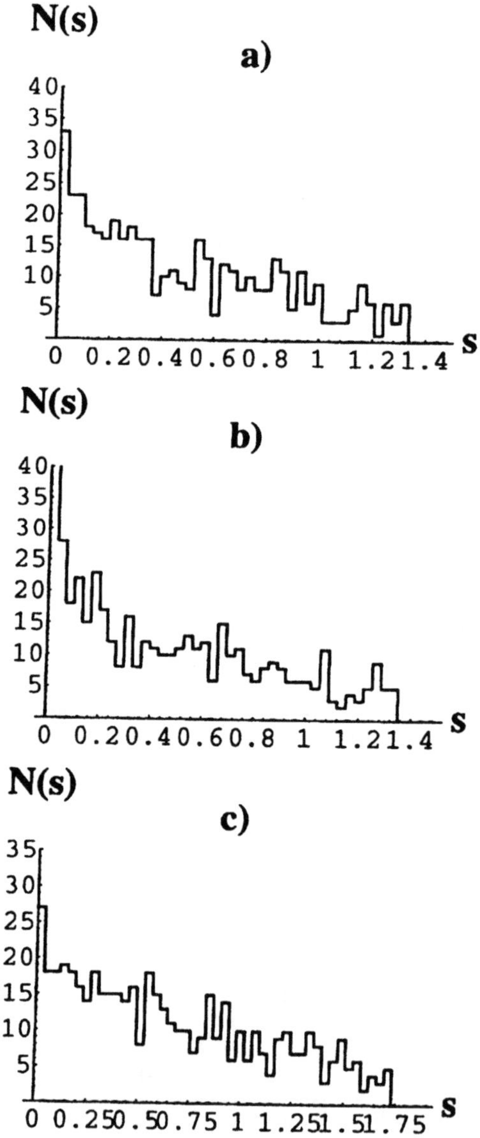

Fig. 2. Nearest neighbor spacing histograms of decay rates for a random walk with transition rates given by $U_2(x_1, x_2)$ and (a) $g = 0.2$, $\epsilon = 0.0$, (b) $g = 0.2$, $\epsilon = 0.1$, and (c) $g = 0.2$, $\epsilon = 0.5$.

When $\epsilon = 0$, x_1 and x_2 are independent and the problem decomposes into two independent one–dimensional problems. When $\epsilon \neq 0$, the two degrees of freedom are coupled and can resonate. In Fig. (1), we show the spectral spacing histograms for $g = 0.2$ and $\epsilon = 0.0, 0.1, 1.0$. Note the change from a Poisson–like distribution to a Wigner–like distribution as ϵ increases.
To support the point that this change is due to coupling and resonance between the degrees of freedom, we will repeat the calculation for a similarly shaped potential, $U(x_1, x_2) = U_2(x_1, x_2)$, where

$$U_2(x_1, x_2) = -\frac{1}{2}\epsilon(x_1 - x_2)^2 + \frac{13}{80}(x_1 - x_2)^4 + \frac{13}{80}(x_1 + x_2)^4 \ . \tag{5}$$

Note that there is no coupling of the degrees of freedom for any value of ϵ. The histogram of nearest neighbor spacings of the decay rates for this case is given in Fig. (2). The level spacing stays Poisson–like.

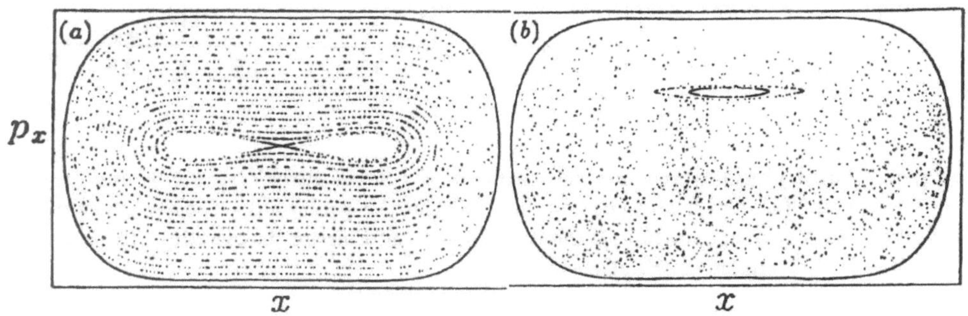

Fig. 3. Poincaré surface of section for classical trajectories for Hamiltonian, H, when (a) $\epsilon = 0$ and (b) $\epsilon = 0.1$ [5].

It is interesting to look at the limiting case of the walker on the lattice when the lattice spacing goes to zero ($\Delta \to 0$) and the transition rates approach infinity. In this limit, we can make a direct connection to conservative chaos theory. A Kramers–Moyal expansion of the master equation [3,4] leads to a Fokker–Planck equation with corrections of order Δ^2 and smaller,

$$\frac{\partial P(\mathbf{x}, t)}{\partial t} = \nabla_{\mathbf{x}} \cdot [P(\mathbf{x}, t) \nabla_{\mathbf{x}} U(\mathbf{x})] + \frac{g}{2} \nabla_{\mathbf{x}}^2 P(\mathbf{x}, t) + O(\Delta^2) \ . \tag{6}$$

If the small correction terms are neglected, this is exactly the Fokker–Planck equation used by Millonas and Reichl [5] to study Brownian motion in the two dimensional potential, $U(\mathbf{x}) = U_1(x_1, x_2)$. The Fokker–Planck equation can be transformed to a Schrödinger–like equation and a direct connection can be made to classical and quantum chaos theory (a similar connection was made by Alpatov and Reichl [6] for a periodically driven Brownian rotor). First make the transformation

$$P(x_1,x_2,t) = \exp\left(-\frac{U(x_1,x_2)}{g}\right)\Psi(x_1,x_2,t) \tag{7}$$

and substitute into the Fokker–Planck equation to obtain an equation for $\Psi(x_1,x_2,t)$. We find that $\Psi(x_1,x_2,t)$ satisfies the Schrödinger–like equation $-\frac{\partial\Psi}{\partial t} = \hat{H}\Psi$, where the self–adjoint operator, $\hat{H}$, is given by

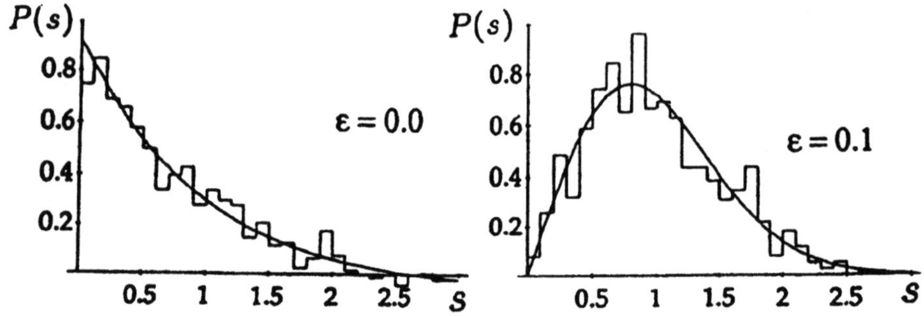

Fig. 4. Nearest neighbor spacing histograms of the decay rates of the Fokker–Planck equation when (a) $\epsilon = 0$ and (b) $\epsilon = 0.1$. $P(s)$ is the probability density of nearest neighbors with spacings, s. In (a), the solid line is the Poisson distribution. In (b), the solid line is the Wigner distribution [5].

$$\hat{H} = -\frac{g}{2}\nabla^2 + V(x,y) \ . \tag{8}$$

The operator, $\hat{H}$, may be thought of as a Hamiltonian operator with the diffusion coefficient, g, playing the role of Plancks' constant. Note that this transformation is the analog for the Fokker–Planck equation of the transformation to a symmetric matrix used for the random walk. When $U(x_1,x_2) = U_1(x_1,x_2)$, the potential, $V(x_1,x_2)$, is given by

$$\begin{aligned} V(x_1,x_2) = \frac{1}{2g}(\nabla U_1)^2 - \frac{1}{2}\nabla^2 U_1 = 32x_1^6 + \frac{72}{25}x_2^6 - g(12x_1^2 + \frac{18}{5}x_2^2) \\ +\epsilon\{\frac{1}{2}\epsilon x_1^6 + \frac{1}{2}\epsilon x_2^6 + (24-4\epsilon)x_1^5 x_2 \\ +(\frac{36}{5} - 4\epsilon)x_1 x_2^5 - (32 - \frac{31}{2}\epsilon)x_1^4 x_2^2 - (\frac{48}{5} - \epsilon\frac{31}{2})x_1^2 x_2^4 \\ +(\frac{52}{5} - 24\epsilon)x_1^3 x_2^3 + 2g(x_1^2 + x_2^2 - 3x_1 x_2)\} \ . \end{aligned} \tag{10}$$

The probability, $P(x_1, x_2, t)$ can be expanded in terms of eigenvalues, λ_i, and eigenfunctions, $\Psi_n(x_1, x_2)$, of the self–adjoint operator, $\hat{H}$,

$$P(x_1, x_2, t) = \Psi_0^2(x_1, x_2) + \sum_{n=1}^{\infty} a_n e^{-\lambda_n t}\, \Psi_0(x_1, x_2)\, \Psi_n(x_1, x_2) \ . \qquad (11)$$

Thus the spectrum of decay rates, λ_n, for the Fokker–Planck equation is determined by a Schrödinger–like equation. If the classical dynamics governed by the "Hamiltonian", $\hat{H}$, goes chaotic, then the Fokker–Planck equation will exhibit the 'manifestations of chaos' in a manner similar to what is seen in quantum systems.

The classical Hamiltonian for this system can be written

$$H = \frac{p_1^2}{2} + \frac{p_2^2}{2} + V(x_1, x_2) = \lambda \ , \qquad (12)$$

where $p_1 = i\sqrt{g}\frac{\partial}{\partial x_1}$ and $p_2 = i\sqrt{g}\frac{\partial}{\partial x_2}$. When $\epsilon = 0$, the two degrees of freedom are uncoupled and the system is integrable. When $\epsilon \neq 0$, the system is nonintegrable and exhibits a transition to global chaos. This can be seen in Fig. (3) where we show Poincaré surfaces of section for $\epsilon = 0$ and $\epsilon = 0.1$. For $\epsilon = 0$, the phase space has only KAM tori. For $\epsilon = 0.1$, the phase space is chaotic.

This transition to 'chaos' exhibits itself in the spectrum of decay rates of the Fokker–Planck equation. We show this in Fig. (4) where we give histograms of nearest neighbor spacings for decay rates, λ_n, for the Fokker–Planck equation. For $\epsilon = 0.0$ the nearest neighbor spacings have a distribution very close to that of a Poisson distribution, indicating in analogy to quantum systems that the stochastic system is integrable. For $\epsilon = 0.1$, the nearest neighbor spacings have a histogram which is very close to that of a Wigner distribution, indicating that symmetries of the system have been broken and the stochastic problem is non–integrable.

Acknowledgments

The authors wish to thank the Welch Foundation (Grant no.1052) for partial support of this work. They also wish to thank the San Diego Supercomputer Center and the University of Texas Computation Center for use of their computer facilities.

References

1. L.E. Reichl, *The Transition to Chaos in Conservative Classical Systems: Quantum Manifestations* (Springer–Verlag, Berlin, 1992).
2. J. von Neumann and E.P. Wigner, Phys. Z. **30**, 467 (1929).
3. C.W. Gardiner, *Handbook of Stochastic Methods* (Springer–Verlag, Berlin, 1983).
4. N.G. van Kampen, *Stochastic Processes in Physics and Chemistry* (North–Holland, New York, 1981).
5. Mark M. Millonas and L.E. Reichl, Phys. Rev. Lett. **68**, 3125 (1992).
6. P. Alpatov and L.E. Reichl, Phys. Rev. **E49**, 2630 (1994).

Quantum Dynamical Entropy

Gert Roepstorff

Institut für Theoretische Physik, RWTH Aachen
D–52056 Aachen, Germany

1 An Example

Given a two–by–two matrix $T \in SL(2, \mathbb{Z})$ with tr$T > 2$ acting on the 2–torus, it defines a discrete dynamical system with both, an expanding and a contracting direction, since the eigenvalues $\lambda_\pm$ of T are real and satisfy $0 < \lambda_- < 1 < \lambda_+$. This system known as "Arnold's cat map" has strong mixing properties. In particular, its KS entropy is $\log \lambda_+$.

The q–deformed torus, where $0 \le q < \pi$, is a *–algebra generated by two elements u, v satisfying

$$uv = e^{i2q} vu, \quad u^* u = v^* v = uu^* = vv^* = \mathbb{1}.$$

The algebra reduces to the classical case (function algebra on the torus) if $q = 0$. For $q > 0$, one may think of the defining relations as some discrete–time version of the Heisenberg canonical commutation relation (CCR) in quantum mechanics. By analogy with Weyl's treatment of the CCR, we may introduce Weyl operators

$$W(n) = \exp(-iqn_1 n_2) u^{n_1} v^{n_2}, \qquad n = \{n_1, n_2\} \in \mathbb{Z}^2.$$

Since we look upon $\mathbb{Z}^2$ as the dual object of the group $U(1) \times U(1)$ (i.e., the torus), there is an induced dual action of $T \in SL(2, \mathbb{Z})$ on $\mathbb{Z}^2$ that takes n into $T^* n$ with T^* the transpose of the matrix T. The induced map T_q that takes $W(n)$ into $W(T^* n)$ extends, by linearity, to a *–automorphism of the algebra for which there exists a preferred invariant state (ground or vacuum state):

$$\omega(W(n)) = \begin{cases} 1 & \text{if } n = \{0, 0\} \\ 0 & \text{otherwise} \end{cases}$$

It extends to a tracial state on the enveloping (type II_1) von Neumann algebra. Question: is the q–deformed cat map chaotic in a specified sense and what is the dynamical entropy $H(T_q, \omega)$ of this system? It is by now agreed that this system is in fact chaotic and its entropy coincides with the classical value, $\log \lambda_+$, irrespective of q.

The above example constitutes one of the simplest discrete quantum systems [1–4] and serves as a testing ground for those who invent new schemes for computing quantum entropy. At present, there is however little hope that we will ever be able to calculate, in closed form, the entropy of a realistic continuous–time quantum system using any scheme.

2 A Survey over Possible Approaches

If we look into the literature for help with regard to extensions of classical concepts to the quantum regime, the variety of possible approaches offered is bewildering. Here is a short list of the main attempts to define quantum dynamical entropy:

1. **Connes, Størmer** (1975)[5]. KS entropy for automorphisms of a type II_1 von Neumann algebra. A purely mathematical concept.
2. **Lindblad** (1979)[6]. KS entropy for a quantum stochastic process, i.e., for an iterated completely positive map. Applies to open systems.
3. **Connes, Narnhofer, Thirring** (1987)[7]. KS entropy for a (nuclear) C^*–dynamical system with respect to some invariant state. Applies to infinite–volume systems at finite density, hence is a concept designed to deal with quantum statistical systems where local algebras are of type III. We refer to it as the CNT approach.
4. **Sauvageot, Thouvenot** (1992)[8]. KS entropy for a C^*–dynamical system. Coincides with CNT for hyperfinite algebras and is simpler.
5. **Hudetz** (1991)[9]. *Topological* entropy for a C^*–dynamical system. Similar to CNT but does not need an invariant reference state.
6. **Alicki, Fannes** (1994)[10]. KS entropy. A simplified approach close to Lindblad's. There is a marked difference between the CNT approach and the AF approach for at least one example (noncommutative shift).

Apart from the above list there are many other attempts, often emphasizing the role of quantum measurement rather than that of the underlying algebraic structure. An almost complete list of references and an excellent discussion of the various approaches may be found in the article by Słomczyński and Życzkowski [11]. For a nice approach that utilizes the Brin–Katok metric entropy see [12] and the contribution of Vilela Mendes to this conference. Still another attempt [13] by the author borrows from the theory of quantum histories initiated by Omnès, Griffiths, Haag and others.

Two puzzling questions remain. Why are there so many options for finding the quantum analogue of the KS entropy? When do these definitions give the same result in spite of their different setup? Before trying to give an answer we need to recall the classical approach to the problem.

3 Review of the Classical KS Entropy

The classical definition of entropy starts from the formula

$$S(p) = -\sum_i p_i \log p_i, \qquad 0 < p_i < 1, \ \sum_i p_i = 1,$$

given a probability vector p, and proceeds in steps:

1. Define the entropy of a finite partition
$$\xi = \{A_1, \ldots, A_n\} \qquad (A_i \subset \Omega)$$
of the phase space Ω with respect to a probability measure m as $H(\xi) = S(p)$ where $p_i = m(A_i)$. Extend this idea to countable partitions if necessary.
2. Think now of a partition as an *experiment* and contemplate what is meant by the combined experiment $\xi \vee \eta$. The classical interpretation (intersection of sets) is the obvious solution.
3. Suppose T is a measure–preserving transformation, i.e., (Ω, m, T) is a discrete–time dynamical system. Since T acts naturally on partitions, define the entropy of T with respect ξ as
$$H(T, \xi) = \lim \frac{1}{n} H(T^{-n+1}\xi \vee \cdots \vee T^{-1}\xi \vee \xi).$$
Classically, the order of factors in the $\vee$–product does not matter. Convergence is monotone, and "lim" can in fact be replaced by "inf". Interpretation: $H(T, \xi)$ is the average information gain in an experiment when repeated in constant time intervals.
4. Define the entropy of T as the maximal information (per time step) obtained by performing all kinds of experiments:
$$H(T) = \sup_\xi H(T, \xi)$$
5. Prove that the entropy is proportional with time:
$$H(T^n) = nH(T).$$
This is a key property. It justifies defining the entropy of a flow $\phi_t : \Omega \to \Omega$ as the time–one–entropy: $H[\phi] = H(\phi_1)$, knowing that this is the same as $n^{-1}H(\phi_n)$, $(n \geq 2)$. Using the above property one proves that $H(\phi_t) = |t|H(\phi_1)$ for all $t \in \mathbb{R}$.

The only known technique to analytically compute $H(T)$ is through the existence of a *generating partition* ξ for which $H(T) = H(T, \xi)$. Recall that a countable partition is said to be generating for an (measure–preserving) transformation T if the partition $T^{-n+1}\xi \vee \cdots \vee T^{-1}\xi \vee \xi$ ultimately separates points in the phase space (neglecting a set of zero measure). Popular examples of positive entropy include:

- The dyadic map $Tx = 2x(\mathrm{mod} 1)$ of $[0,1]$ has entropy $\log 2$.
- The Gauss map $Tx = x^{-1}(\mathrm{mod} 1)$ of $[0,1]$ has entropy $\pi^2/(6\log 2)$: see the lecture by Micheal Keane, this conference.

A translation into the algebraic language is immediate. Namely, we replace:

phase space Ω	$\rightarrow$	*function algebra* $C(\Omega)$
subset	$\rightarrow$	*characteristic function*
intersection	$\rightarrow$	*product*
partition	$\rightarrow$	*decomposition of the unit*
discrete–time transf.	$\rightarrow$	*automorphism*
flow	$\rightarrow$	*semigroup of automorphisms.*
Hamilton equation	$\rightarrow$	*Liouville equation*

Quantum theory replaces $C(\Omega)$ by some algebra $\mathcal{A}$ of operators. Where lie the problems? One problem certainly is that there are many candidates for what should be the analogue of a partition. Another is that noncommutativity of operators translates into noncommutativity of $\vee$–products spoiling many classical proofs of existence and uniqueness. Let us first look at the options for "partition" to see the confusion.

1. A set of orthogonal projectors $\{P_i\}$ such that $\sum P_i = \mathbb{1}$. This is considered a bad choice since products of projectors are no longer projectors (unless they commute).
2. A set $\xi = \{X_1, X_2, \ldots, X_n\}$ of operators $X_i \in \mathcal{A}$ such that $\sum X_i^* X_i = \mathbb{1}$. Bad also: there are trivial partitions such as $\xi = \{c_1\mathbb{1}, c_2\mathbb{1}, \ldots, c_n\mathbb{1}\}$, $\sum |c_i|^2 = 1$, which correspond to measurements producing no information whatsoever.
3. A completely positive map $\phi : \mathcal{A} \rightarrow \mathcal{A}$ such that $\phi(\mathbb{1}) = \mathbb{1}$. Think of ϕ as given by
$$\phi(A) = \sum_i X_i^* A X_i \qquad (X_i \in \mathcal{A}).$$
Such map may result from iterating two basic physical operations (description within the Heisenberg picture): (1) time evolution via some unitary operator, (2) measurement of some observable with discrete spectrum. Therefore, each X_i is thought of as a product of unitaries and projectors. Notice: if we define the $\vee$–product in the natural way, i.e., $(\phi \vee \phi')(A) = \phi(\phi'(A))$, then
$$\phi \vee \phi' \neq \phi' \vee \phi$$
holds in general: noncommutativity of the $\vee$–product is unavoidable.
4. A finite–dimensional subalgebra $\mathcal{B} \subset \mathcal{A}$. Problem: the union $\mathcal{B} \vee \mathcal{B}'$ (smallest subalgebra containing $\mathcal{B}$ and $\mathcal{B}'$) need not be finite–dimensional.
5. A completely positive map $\gamma : \mathcal{B} \rightarrow \mathcal{A}$ where $\mathcal{B}$ is some finite–dimensional algebra (matrix algebra). Problem: the concept lacks physical intuition.
6. A completely positive map
$$\xi : \mathcal{A} \otimes M_n \rightarrow \mathcal{A} \quad , \quad (A_{ik}) \mapsto \sum_{i,k} X_i^* A_{ik} X_k \qquad (A_{ik} \in \mathcal{A},\ i,k = 1, \ldots, n)$$

(M_n = algebra of complex $n \times n$ matrices). Though this notion of a partition looks complicated, it is the best choice so far. For it describes a coupling of the system to a measuring device (having n degrees of freedom). See the discussion of the AF scheme below.

Mathematics aside, it is important to realize that the state of a quantum system changes not only due to the underlying dynamics but also *due to the measurement* performed. The latter change of state is absent in the classical regime. Consequently, there are two sources of entropy production in the quantum regime:

1. The dynamical law (we really wish to focus on).
2. The influence of the observer (we would like to eliminate).

I strongly believe that the entropy, in finite time, cannot be written as a sum of two terms that completely separate the two influences. I also think that the reason for seeing so many different versions of entropy in the literature is, in each case, an attempt, either unsuccessful or incomplete, made by the author, to eliminate the observer from the concept of entropy.

Assuming you are dissatisfied with all previous attempts, I shall offer a check list to go through if you are seriously thinking of of defining your own quantum dynamical entropy; a new one, of course – I call it $h(T)$ for distinction – :

- Whatever your concept of partition ξ, make sure that any trivial partition is assigned zero entropy relative to your choice of state, ω.
- Whatever your concept of product $\xi \vee \eta$, make sure that the entropy is *monotone* and *subadditive* in some sense. For instance,

$$h(\xi \vee \eta, \omega) \geq \max\{h(\xi, \omega'), h(\eta, \omega)\}$$

$$h(\xi \vee \eta, \omega) \leq h(\xi, \omega') + h(\eta, \omega)$$

  Here, ω is the initial state whereas ω' is the state after the first measurement, η, has been performed (no result recorded). Of course, $\omega' = \omega$ in the classical context.
- Whatever your definition of $h(T)$, make sure that the identity $T = id$ is assigned zero entropy. For if $h(id) > 0$, it means that you are calculating the (maximal) entropy production with respect to repeated measurements when there is no time evolution (trivial dynamics). More generally, check that $h(T) = 0$ if $T^n = id$ for some n.
- Even stronger, prove that $h(T^n) = nh(T)$ for all $n = 0, 1, 2, \ldots$. This property is crucial. It says that entropy is proportional with time allowing an extension of your concept to flows.

If, however, you decide to work with one of the existing constructions of entropy – let me call it $H(T)$ for simplicity – , I suggest doing the following large–time expansion,

$$H(T^n) = nh(T) + a_0 + a_1 n^{-1} + a_2 n^{-2} + \ldots,$$

assuming again it exists. Then $h(T^n) = nh(T)$ automatically, while the remaining coefficients a_i, though depending on T, measure the entropy production owing to a change of state induced by the observer. Hopefully, the leading term $nh(T)$ will come out always the same while the remaining coefficients a_i depend on the details of the construction.

4 The Setup of Alicki and Fannes

I shall briefly introduce you to the recent, very promising definition of $H(t)$. Let $(\mathcal{A}, T, \omega)$ be a discrete–time quantum dynamical system and $\mathcal{A}_0$ a *–subalgebra of "local observables". As for our purpose, $\mathcal{A}$ may be replaced by $\mathcal{A}_0$.

Finite partition: $\xi = \{X_1, X_2, \ldots, X_n\}$ with $X_i \in \mathcal{A}_0$ such that $\sum X_i^* X_i = \mathbb{1}$.

Product: $\xi \vee \eta = \{X_i Y_k\}$ if $\xi = \{X_i\}$, $\eta = \{Y_k\}$. The construction of entropy again proceeds in steps.

Step 1. To any partition ξ and state ω one associates a density matrix

$$\rho_\omega(\xi)_{ik} = \omega(X_k^* X_i)$$

It seems however best to think of a partition as a completely positive map

$$\xi : M_n \to \mathcal{A}\ , \quad a \mapsto A = \sum_{i,k} a_{ik} X_i^* X_k$$

so that $\omega(A) = \mathrm{tr}(\rho a)$. The map $\xi^* : \omega \mapsto \rho = \rho_\omega(\xi)$, introduced by Alicki and Fannes mapping states into density matrices, is simply the dual of ξ.

Step 2. With $S(\rho) = -\mathrm{tr}(\rho \log \rho)$ the von Neumann entropy, define the entropy of a partition (with respect ω) as

$$H(\xi, \omega) = S(\rho_\omega(\xi))$$

As desired, $\xi = \text{trivial} \Rightarrow \mathrm{rank}(\rho) = 1 \Rightarrow S(\rho) = 0$

Step 3. Define the entropy of T (with respect to ξ and ω) as

$$H(T, \xi, \omega) = \limsup_n n^{-1} H(\xi_n, \omega)$$

where $\xi_n = T(\xi_{n-1}) \vee \xi$ recursively ($\xi_0 = \xi$), and ω is invariant.

Step 4. Define the entropy of T (with respect to ω) as

$$H(T, \omega) = \sup_\xi H(T, \xi, \omega)$$

Certain remarks are in order.

(1) Adopting the Heisenberg picture, our preferred way to analyze a measurement, hence a partition, is to view it as a completely positive map between algebras:

$$\xi : M_n \otimes \mathcal{A} \to \mathcal{A}\ , \qquad (A_{ik}) \mapsto \sum_{ik} X_i^* A_{ik} X_k$$

($A_{ik} \in \mathcal{A}$, $i, k = 1, \ldots, n$) Reason: $M_n \otimes \mathcal{A}$ corresponds to the coupling of an instrument, M_n, to the quantum system $\mathcal{A}$. Obviously, the dual ξ^* maps states ω on the system $\mathcal{A}$ to states of the combined system $M_n \otimes \mathcal{A}$. The dual version of a measurement obviously relates to the Schrödinger picture. The density matrix ρ constructed above describes the state of the apparatus when decoupled from the system. The number $S(\rho)$ thus measures the entropy which is transferred from the system to the apparatus.

(2) Due to the lack of monotonicity and subadditivity properties, there is little (perhaps no) control of the limsup or sup operations (step 3 and 4 above). The entropy may come out infinite.

(3) One easily proves that $H(T^n) \geq nH(T)$. Hence, either $H(id) = 0$ or $H(id) = +\infty$. There exist examples for which $H(id) = +\infty$ (Alicki, Fannes: private communication). Assume now that the limit

$$h(T, \xi, \omega) = \lim n^{-1} H(T^n, \xi, \omega)$$

exists. We shall refer to this intermediate procedure (before taking the sup with respect to ξ) as *stretching the time scale*. As a result, we get the property $h(T^n) = nh(T)$ at the end simply by construction.

(4) The CNT entropy differs from the AF entropy at least for the so-called noncommutative shift which is the natural shift T performed on the lattice algebra

$$\cdots \otimes M_d \otimes M_d \otimes M_d \otimes \cdots$$

(a quantum version of the twosided Bernoulli shift). In more detail [14], for any extremal invariant state ω with entropy density $s(\omega)$,

$$H_{\mathrm{AF}}(T, \omega) = s(\omega) + \log d, \qquad H_{\mathrm{CNT}}(T, \omega) = s(\omega)$$

and $H(T^n, \omega) = nH(T, \omega)$ in either case (H. Narnhofer: private communication). It seems that there is no way to make the difference between CNT and AF vanish. We should be prepared for the fact that quantum systems have more invariants than classical systems and thus allow for a finer classification.

References

1. S. Weigert, Zeitschr. Phys. **B 80**, 3 (1990).
2. F. Benatti, H. Narnhofer, and G.L. Sewell, Lett. Math. Phys. **21**, 157 (1991).
3. J. Andries, M. Fannes, P. Tuyls, and R. Alicki, *The dynamical entropy of the quantum Arnold cat map*, preprint 1994.
4. S. Klimek and A. Leśniewski, *Quantized chaotic dynamics and non-commutative KS entropy*, Harvard preprint 1995.
5. A. Connes and E. Størmer, Acta Math. **134**, 289 (1975).
6. G. Lindblad, Commun. Math. Phys. **65**, 281 (1979).
7. A. Connes, H. Narnhofer, W. Thirring, Commun. Math. Phys. **112**, 691 (1987).
8. J-L. Sauvageot and J-P. Thouvenot, Commun. Math. Phys. **145**, 411 (1992).
9. T. Hudetz, in *Information Dynamics* (H. Almanspacher and H. Scheingraber, eds.), pp. 27–41, Plenum Press, New York 1991.

10. R. Alicki and M. Fannes, Lett. Math. Phys. **32**, 75 (1994).
11. W. Słomczyński and K. Życzkowski, J. Math. Phys. **35**, 5674 (1994).
12. R. Vilela Mendes, J. Phys. **A 24**, 4349 (1991).
13. G. Roepstorff, *Quantum Dynamical Entropy Based on Histories*, Preprint, RWTH Aachen 1995.
14. R. Alicki and H. Narnhofer, *Comparison of Dynamical Entropies for the Noncommutative Shifts*, preprint 1994.
15. M. Ohya and D. Petz: *Quantum Entropy and Its Use*, Springer, New York 1993.

Strange Attractors in Nonlinear Oscillators

Wanda Szemplińska–Stupnicka

Institute of Fundamental Technological Research
Polish Academy of Sciences,
PL–00–049 Warsaw, Poland

1 Introduction

In the lecture we consider nonlinear vibrating systems governed by second order nonautonomous differential equation:

$$\ddot{x} + f(x, \dot{x}) - F \cos \nu t = 0. \tag{1}$$

It is assumed that the system is dissipative and that at $F = 0$ it has a unique stable equilibrium state (equilibrium $x = \dot{x} = 0$ or a limit cycle). The function $f(x, \dot{x})$ is approximated by a finite Taylor series in the form:

$$f(x, \dot{x}) = \Omega^2 x + h\dot{x} + \alpha_2 x^2 + \alpha_3 x^3 - \mu(1 - x^2)\dot{x} + ... \tag{2}$$

where $\Omega^2, h, \alpha_2, \alpha_3, \mu$ are constant coefficients.

It is worth to note, that equation (1, 2) can be reduced to the continuous dynamical system of the third order

$$\frac{dy_i}{dt} = f_i(y_1, y_2, y_3), \qquad i = 1, 2, 3 \tag{3}$$

where $y_1 \equiv x, y_2 \equiv \dot{x}, y_3 \equiv t$.

Such a system has an extensive literature. A now classical approach to the study of the system behaviour, such as that presented in the popular books by [Bogoliubov & Mitropolsky, 1961], [Nayfeh, 1979], [Hayashi, 1985], [Stocker, 1951]. is the theoretical analysis based on approximate analytical methods with experimental verification employing electric circuits or electronic computers. In these studies the system is assumed to tend to steady-state oscillation when started with any initial conditions and steady-state solutions are often the main point of interest. Approximate analytical solutions describing various types of resonances and analysis of local stability of the solutions and their domains of attraction provided us with a great deal of knowledge about the system behaviour. The results show a variety of nonlinear phenomena such as: principal, sub, ultra and subultra harmonics resonances and jump phenomena associated with stability limits on resonance curves, which seem to leave no room for any irregular, random-like and unpredictable solutions in the deterministic systems.

Although chaotic motions in simple deterministic dynamic systems have attracted a great deal of attention in the last decades, results showing "strange attractors" in as classical a vibrating system as that governed by equations (1, 2), were a great surprise [Ueda, 1979, 1980, 1991], [Ueda & Akamatsu, 1981], [Guckenheimer & Holmes, 1983], [Thompson & Stewart, 1986].

It is pretty obvious that direct applications of the approximate theory of nonlinear vibrations to theoretical study of chaotic motion is impossible and so the return to qualitative topological studies seemed to be the only alternative. Nevertheless one might be tempted to seek a relation between the phenomena of nonlinear resonances determined by low order approximate solutions and irregular solutions obtained by computer simulation in order to see the chaotic zones against the background of the classical concepts of resonance curves, stability limits, and jump phenomena. Results on chaotic behaviour obtained by computer simulation allow us to make observations, which make such an idea an appealing one: one can readily notice that:

- chaotic motion appears as a transition zone between sub and subultra mT-periodic resonance and the principal T-periodic resonance,
- chaotic motion often borders and coexists with periodic motion, the motion which can be described by low order approximate solutions.

The idea of interpreting and studying regions of chaotic motion from the point of view of the approximate theory of nonlinear vibrations brings a great number of interesting questions:

- where are the chaotic motions zones located relative to the known phenomena of principal and subharmonic resonances?
- how does chaotic motion develop from periodic solutions if the periodic motion is described by a low order approximate solution and the approximate theory of nonlinear vibrations does not leave room for irregular solutions?
- whether and how routes to chaotic motion can be approximately described with the use of concepts and mathematical language of the approximate theory of nonlinear vibrations?
- can the low order approximate solution make foundations for approximate criteria of system parameter critical values—the parameters for which one might expect chaotic behaviour?

It is an attempt of this work to seek answers to some of the questions. A consistent use of mathematical tools and concepts commonly accepted in the approximate theory of nonlinear vibrations, and their adaptation to an analysis of routes to chaotic motion, is one of the main points of interest.

Our considerations are confined to two types of nonlinear forced oscillators: 1.) An unsymmetric oscillator with a unique equilibrium position ($x = \dot{x} = 0$ for $F = 0$), i.e., Duffing system with quadratic term governed by equation of motion as

$$\ddot{x} + h\dot{x} + \alpha_2 x^2 + \alpha_3 x^3 - F\cos\nu t = 0, \tag{4}$$

$h > 0$ and small, $\alpha_2, \alpha_3 > 0$.
2.) The van der Pol-Duffing system governed by equation of motion as

$$\ddot{x} - \mu(1 - x^2)\dot{x} + x^3 - F\cos\nu t = 0, \tag{5}$$

$\mu > 0$. The system has unique limit cycle for $F = 0$ [van der Pol, 1927].

The approximate theoretical analysis reveals that the systems may exhibit a variety of T-periodic principal and superharmonic resonances and mT-periodic subharmonic resonances, and particularly, the self-excited system (5) also exhibits almost-periodic response [Hayashi, 1986], [Quin *et al.*, 1989], [Belogortsev *et al.*, 1993].

In the years 1979-81 Ueda reported that both systems may also exhibit random-like, irregular response, now called chaotic motions. In the early publications chaotic motion in the system (1, 2) was characterized by "strange attractors" illustrated in Poincaré sections with sampling time equal to the period of excitation [Kawakami & Hayashi 1981], [Dmitriev *et al.*, 1983], [Dmitriev *et al.*, 1992], [Endo & Saito, 1990], [Ruelle, 1980].

2 The Unsymmetric System (equation 4)

Chaotic motions that we consider are characterized by strange attractor depicted in Fig. 1 and by positive values of the largest Lyapunov exponent $\sigma_{max} > 0$. The attractor was obtained by computer simulation for a particular set of the system parameters. To find out, where the chaotic zone is located relative to the principal and subharmonic resonance, we consider the approximate solution of principal resonance by assuming [Szemplińska-Stupnicka & Bajkowski, 1986]:

$$x(t) = x(t+T) = C_0 + C_1\cos(\nu t + \varphi_1), \qquad \nu \approx \omega_0, \tag{6}$$

where the unknown amplitude C_0, C_1 and the phase angle φ_1 are determined by the harmonic balance method. The amplitude $C_1 = C_1(\nu) \equiv A_1$ of the fundamental harmonic component is depicted in Fig. 2. The resonance curve possesses two types of unstable regions: (a) the branch of $C_1(\nu)$ between points with vertical tangent A and B is unstable in the classic sense so that at the stability limits "jump phenomena" are predicted in the approximate theory of nonlinear oscillations; (b) the portion of resonance curve in the neighbourhood of $\nu = 2$ is unstable in the sense, that stable T-periodic solution (6) does not exist in the region and is replaced by $2T$-periodic, 1 : 2 subharmonic resonance [Bogolubov & Mitropolsky, 1961], [Hayashi, 1985], [Szemplińska-Stupnicka, 1986].

At the stability limit of the type (b), the T-periodic solution bifurcates into $2T$-periodic solution, the new born solution can be approximated by three harmonic components:

$$x(t) = x(t+2T) = A_0 + A_1\cos(\nu t + \varphi_1) + A_{1/2}\cos(\frac{\nu t}{2} + \varphi_2). \tag{7}$$

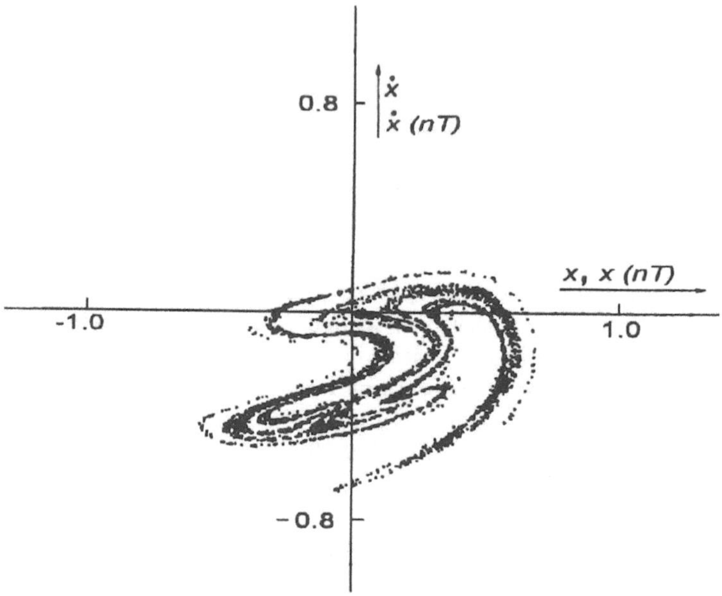

Fig. 1. System (1): $\nu = 1.0$, $\omega_0^2 = 0.38$, $\alpha_2 = 1.07$, $\alpha_3 = 1.0$, $F = 0.16$, $h = 0.05$, $\sigma_{max} > 0$.

The amplitudes A_0 and A_1 in eqs. (7) are not equal to C_0 and C_1 in eqs. (6), but the difference is small and for the sake of clarity we show only the new subharmonic term amplitude $A_{1/2}(\nu)$ and $A_1 = C_1$ in Fig. 2. The resonance curve $A_{1/2}(\nu)$ has two classic stability limits, i.e., two points with vertical tangent, but most part of the upper branch of the curve is unstable due to period-doubling instability.

In our consideration, we focus on the system behaviour at decreasing frequency (so we deal with the lower branch of the resonance curve $A_{1/2}(\nu)$) and note that our $2T$-periodic solution loses stability at point D (see Fig. 2). Therefore, according to the approximate theory, jump phenomenon is expected at the frequency, which corresponds to the point. At this frequency, amplitude $A_{1/2}$ should decay uniformly, and A_1 should grow to reach the value at the upper branch of the principal resonance. The phenomenon really occurs, if the value of the forcing parameter is sufficiently low. For the set of parameters considered, instead of transient and "jump phenomena," we observe a steady-state chaotic motion that occur in a finite band of the driving frequency. This chaotic motion zone plays a role of transition zone between the two types of of periodic resonances and evidently is related to the lost of stability of 1 : 2 subharmonic resonance.

Consequently, we look at the chaotic motion as response which develops from

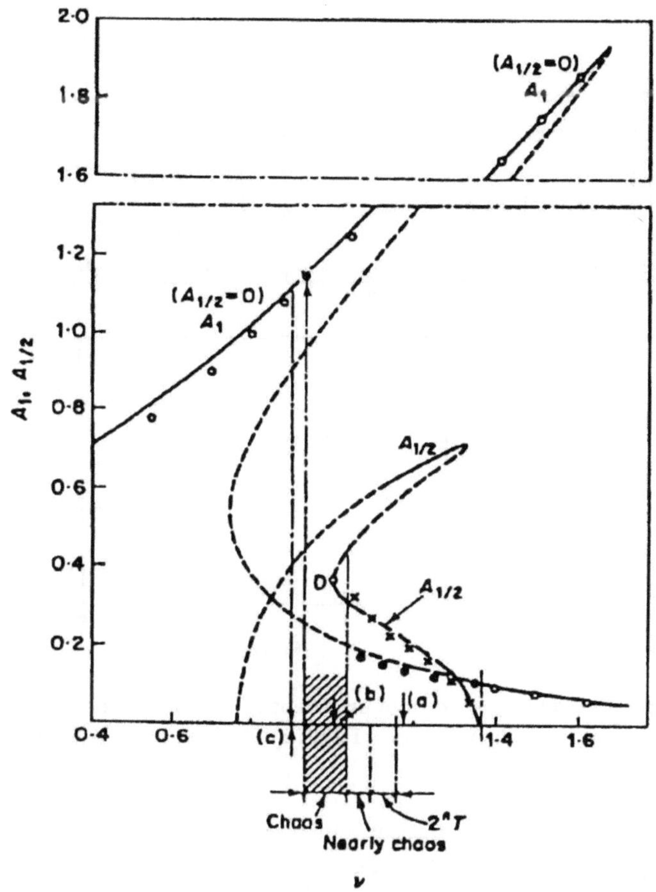

Fig. 2. System (1): resonance curves and chaotic motion zone. Solid lines—stable, dashed lines—unstable theoretical curves. Computer simulation results: •—A_1 at $A_{1/2} \neq 0$; o—A_1 at $A_{1/2} = 0$; ×—$A_{1/2}$.

1/2 subharmonic resonance solution (7). This is illustrated in Fig. 3, where the pre-chaotic frequency spectrum at $\nu = 1.22$, the chaotic continuous at $\nu = 1.0$ and post-chaotic T-periodic spectrum at $\nu = 0.9$ are displayed. The three values of the frequency are denoted in Fig. 2 by (a), (b) and (c). A rough look at the development of the frequency spectrum enables us to notice that the continuous segment of the frequency spectrum in Fig. 3(b) is that which spreads around the $\frac{1}{2}\nu$ subharmonic component: i.e., the harmonic component which approaches its stability limit and is due to decay. This brings an idea of seeking and insight into the nature of chaotic behaviour by observing the time history of the component of motion, which corresponds to the continuous frequency segment. To make such

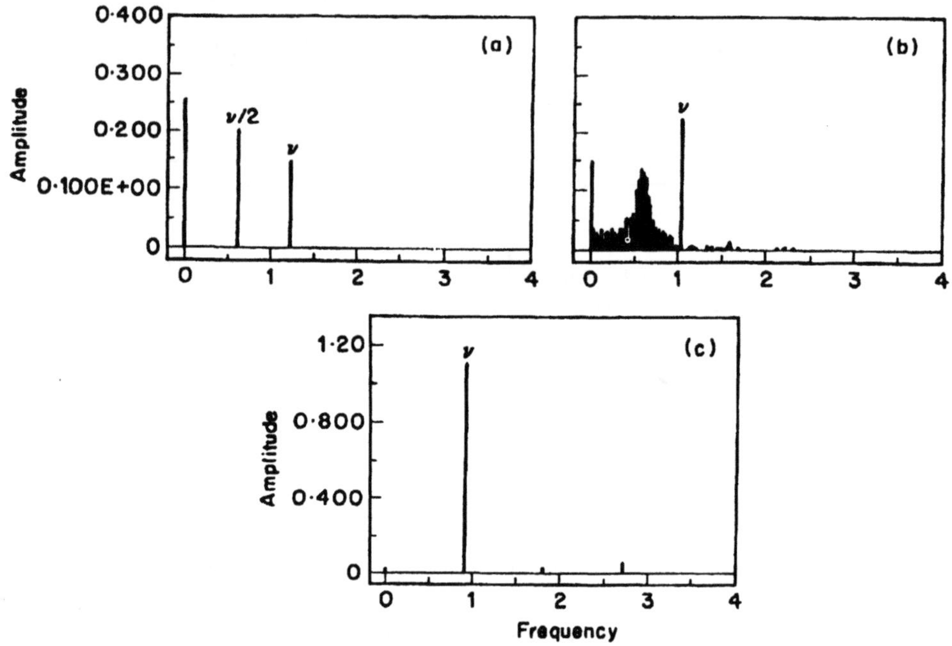

Fig. 3. System (1): frequency spectra of pre-chaotic, chaotic and post-chaotic responses at decreasing driving frequency. (a) $\nu = 1.22$; (b) $\nu = 1.0$; (c) $\nu = 0.90$.

an analysis possible, a numerical technique was developed to obtain the "filtered" response by reproducing a partial signal from the FFT data of the complete chaotic response. In the example considered we have only one continuous segment and the filtered response is obtained simply by cutting off the constant term and the harmonic components higher than $\nu = 0.9$. A sample of the complete chaotic response and the corresponding filtered response are presented in Fig. 4. Analysis of many samples revealed that the filtered response can be approximated by harmonic function of time with randomly modulated amplitude and frequency close to $\frac{1}{2}\nu$, but also varying with time. Moreover, we notice that the frequency is related to the amplitude. Eventually we approximate the filtered response as

$$\bar{x}(t) = x(t + 2T) = a(t)\cos[\bar{\omega}(a)t + \phi(t)], \tag{8}$$

where $a(t)$ and $\bar{\omega}(a)$ and $\phi(t)$ are random-like quantities fluctuating with time.

3 The van der Pol-Duffing System (equation 5)

Computer simulations results reported by Ueda [1981] revealed that chaotic motion occur in two different regions of the parameters F and ν (the parameter μ was fixed: $\mu = 0.2$).

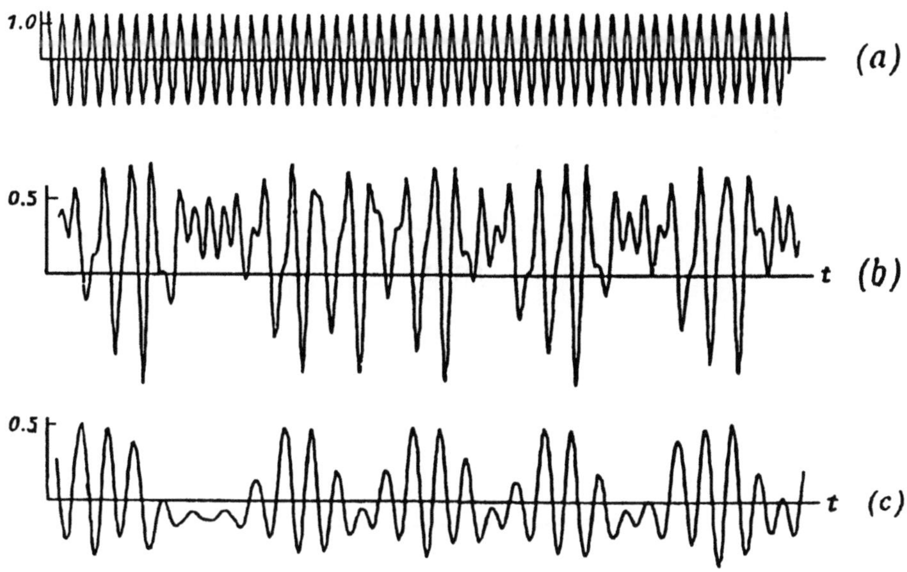

Fig. 4. Frequency spectrum and Poincaré map of the chaotic response for $F = 1.0$, $\nu = 0.94$, $\mu = 0.2$.

(a) at low value of F and ν
$F \approx 1.0$ and ν close to 1.0.
(b) at high value of F and ν
$F \approx 15$ and ν close to 4.0.

The corresponding strange attractors are depicted in Figs. 5 and 6.

(a)

First we focus on the chaotic motions which occur at the low values of F, ν and attempt to interpret it in the light of the approximate theory of nonlinear oscillations [Szemplińska-Stupnicka & Rudowski, 1994]. Preliminary observations indicate that the region of (F, ν) considered is close to the principal resonance (i.e., where the driving frequency is close to the frequency of the limit cycle of the autonomous system). Therefore we consider the first approximation solution for the principal resonance in the form

$$x(t) = C\cos(\nu t + \varphi) = x(t + 2T), \qquad T = \frac{2\pi}{\nu} \tag{9}$$

and determine the resonance curve $C = C(\nu)$ by the aid of an approximate method. For the harmonic solution all the commonly used techniques, such as

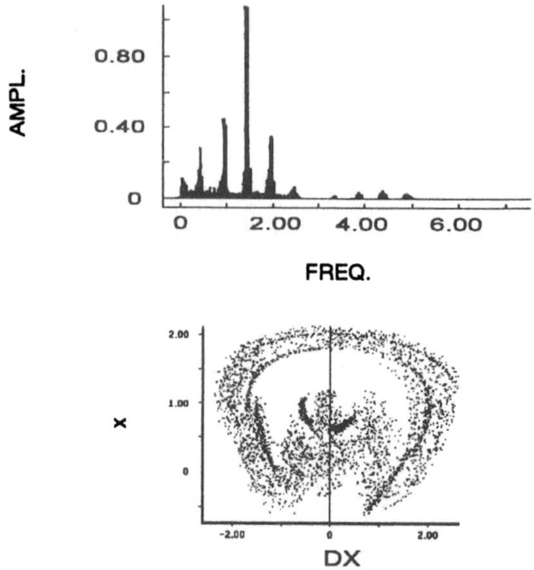

Fig. 5. System (1): $\nu = 1.0$; (a) forcing term; (b) chaotic response; (c) "filtered" response $\bar{x}(t)$.

averaging, asymptotic KBM method, multiple-scale method and harmonic balance method leads us to the same result [Szemplińska-Stupnicka, 1990]

$$C = \frac{F}{\sqrt{[\Omega^2(C) - \nu^2]^2 + \mu^2\nu^2(1 - \frac{C^2}{4})^2}}, \tag{10}$$

$$tg\varphi = \frac{\mu\nu(1 - \frac{1}{4}C^2)}{\Omega^2(C) - \nu^2}, \tag{11}$$

$$\Omega^2(C) = \Omega_0^2 + \frac{3}{4}C^2. \tag{12}$$

Note, that the autonomous system, i.e., the system (5) for $F = 0$, possesses the limit cycle, which also can (for the low value of the coefficient μ considered) be approximated by the harmonic function of time [Stocker, 1950], [Szemplińska-Stupnicka, 1990], [Hayashi 1985]:

$$x_{F=0}(t) = a \cos \omega_0 t, \tag{13}$$

where $a = 2$, and the unperturbed limit cycle frequency for $\Omega_0 = 0$:

$$\omega_0 = \sqrt{\frac{3}{4}a^2} = \sqrt{3}. \tag{14}$$

Computer simulations reveal that the true unperturbed limit cycle has the amplitude $a = 1.81$ and $\omega_0 = 1.625$. Therefore we assume that the first approximate solution gives satisfactory results. The same is concluded for the solution of the complete, nonautonomous system in the region of the principal resonance (eqs. 9–12). To examine local stability of the harmonic solution (10–12) we add a small disturbance to the steady-state solution:

$$\tilde{x} = x^0 + \delta x, \qquad x^0 \equiv C\cos(\nu t + \varphi) \tag{15}$$

and derive the linear variational equation with the result:

$$\ddot{\delta x} + \delta x G(t) + \dot{\delta x} R(t) = 0, \tag{16}$$

where

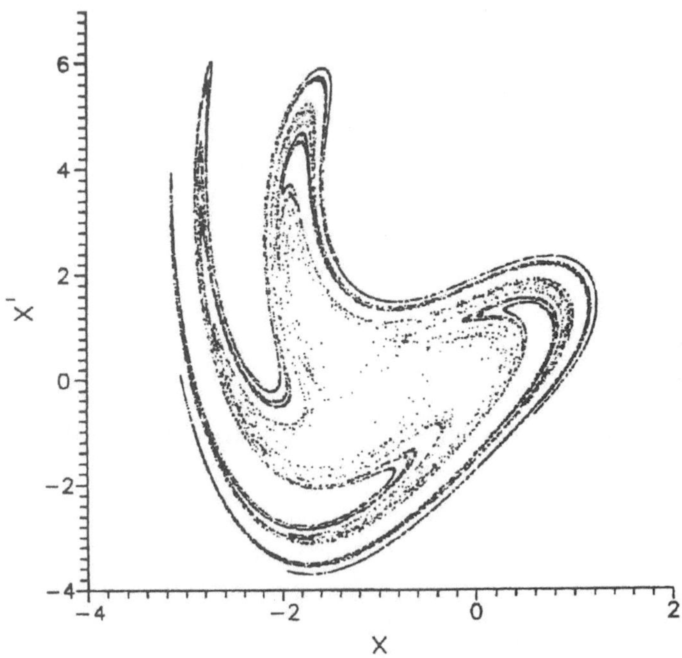

Fig. 6. Strange attractor, $F = 12.0$, $\mu = 0.2$, $\nu = 3.7$, $\sigma_{max} = 0.3$.

$$G(t) = \Omega_0^2 + \frac{3}{2}C^2 - \mu C^2 \nu \sin 2\nu t + \frac{3}{2}C^2 \cos 2\nu t,$$
$$R(t) = -\mu + \frac{1}{2}\mu C^2 + \frac{1}{2}\mu C^2 \cos 2\nu t.$$

To apply the Floquet theorem we first transform eq. (16) into the Hill's equation by introducing new variable η:

$$\delta x = \eta \exp\,[-\frac{1}{2}R(\tau)d\tau]. \tag{17}$$

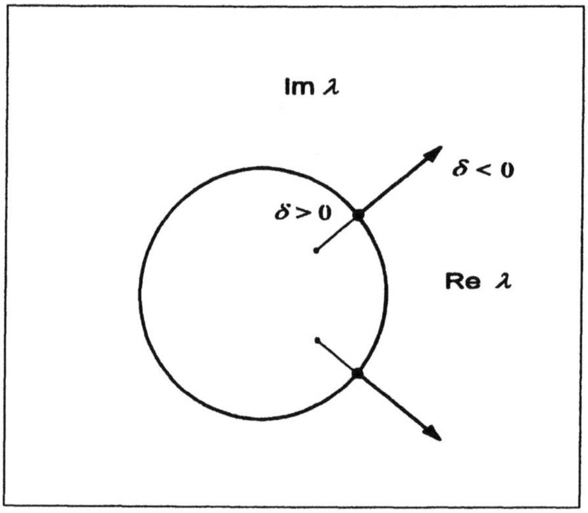

Fig. 7. Neimark bifurcation diagram.

This yields:

$$\ddot{\eta} + \eta[\lambda_0 + \lambda_{2C}\cos 2\nu t + \lambda_{2S}\sin 2\nu t + \lambda_{4C}\cos 4\nu t] = 0, \tag{18}$$

where

$$\begin{aligned} \lambda_0 &= \Omega_0^2 + \frac{3}{2}C^2 + \frac{\mu^2}{4}(C^2 - 1 - \frac{3}{8}C^4), \\ \lambda_{2C} &= \frac{3}{2}C^2 + \frac{\mu^2}{4}C^2(1 - \frac{1}{2}C^2), \\ \lambda_{2S} &= -\frac{1}{2}\mu C^2 \nu, \\ \lambda_{4C} &= -\frac{\mu^2}{32}C^4. \end{aligned} \tag{19}$$

The transformation (17) is reduced to:

$$\delta x = \eta \exp\,[-\delta t - \frac{\mu C^2}{8\nu}\sin 2\nu t], \tag{20}$$

$$\delta = \frac{\mu}{2}(\frac{C^2}{2} - 1). \tag{21}$$

Now the Floquet theorem tells us that we can seek a particular solution of eq. (18) in the form [Magnus & Winkler, 1966], [Bolotin, 1964], [Hayashi, 1985]:

$$\eta = e^{\varepsilon t}\phi(t), \tag{22}$$

where $\phi(t)$ is a periodic function of time and $\varepsilon_{1,2}$ are either imaginary $\varepsilon_{1,2} \equiv \mp i\bar{\varepsilon}$, or real, one of them being positive, $\varepsilon_1 \equiv \bar{\varepsilon}$, $\varepsilon_1 \equiv -\bar{\varepsilon}$ ($\bar{\varepsilon}$—real and positive).

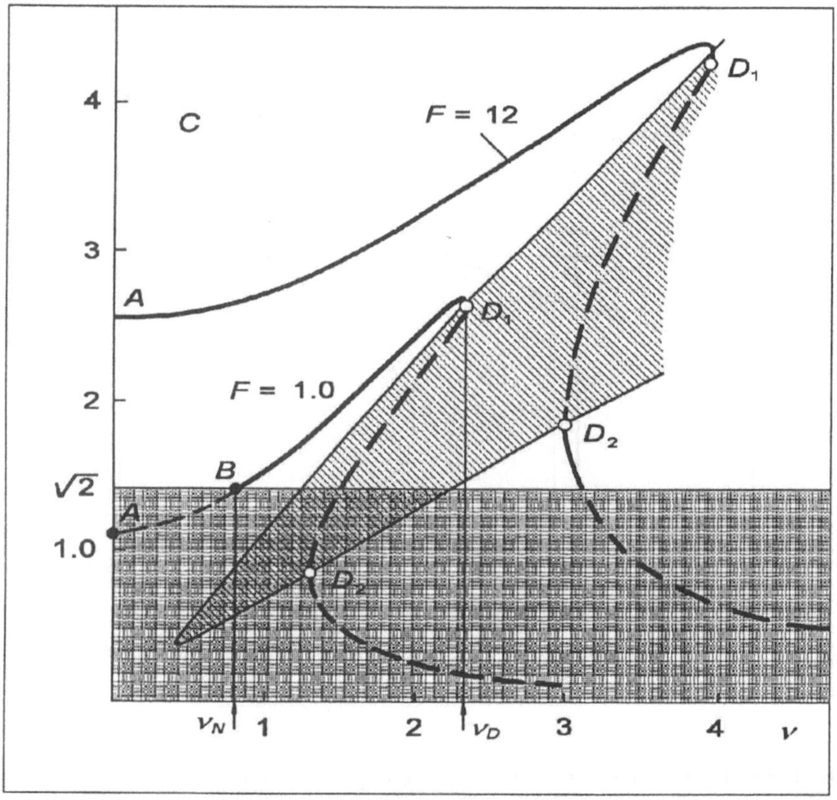

Fig. 8. Resonance curves and unstable regions for two values of the forcing parameter.

Equations (22) offers a good starting point for constructing an approximate solution for $\eta(t)$: if we develop $\phi(t)$ into a Fourier series, and assume the harmonic terms, which are consistent with the periodic solution considered, then the harmonic balance method would provide us with reasonable approximate estimates for the local stability for $\eta(t)$.

Then we can discuss stability of the variations $\delta x(t)$ and so the solution (9–12) by using the change of variables (20–21), which yields:

$$\delta x(t) = e^{(\varepsilon-\delta)t}\bar{\phi}(t), \tag{23}$$

where $\bar{\phi}(t)$ slightly differs from $\phi(t)$ (22) because of the time periodic term in the exponential term in eq. (20). We notice now, that stability of our harmonic solution (9) depends exclusively on the exponent coefficients $(\varepsilon - \delta)$ in eq. (23). Generally, if the real part of the coefficient $\varepsilon - \delta$ is negative, the variation $\delta x(t)$ and the solution (9) is stable, if it is positive—the solution considered is unstable.

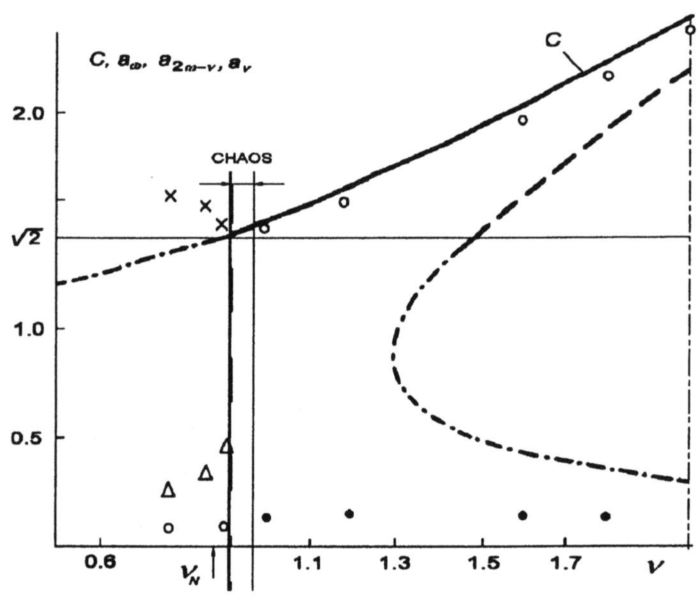

Fig. 9. Theoretical resonance curves and computer simulation results: $F = 1.0$. o, •, $a_\nu, a_{3\nu}$—amplitudes of the fundamental and third harmonic components; ×, a_ω—amplitude of the forced limit cycle; $\triangle$, $a_{2\omega-\nu}$—amplitude of the harmonic component with frequency $2\omega - \nu$.

Let us consider all possible cases:

I. If our system is within the region where the variation $\eta(t)$ is stable, that is, if $\varepsilon_{1,2} \equiv \mp i\bar{\varepsilon}$, the solution for $\delta x(t)$ is stable if $\delta > 0$. From eq. (21) we conclude the condition is satisfied if the amplitude C in eq. (3) is higher than $\sqrt{2}$. Finally we can write the condition:

$$\text{if} \quad \varepsilon_{1,2} \equiv \mp i\bar{\varepsilon}, \quad (\bar{\varepsilon}\text{—real and positive}) \quad \text{then} \tag{24}$$

the solution (6) is unstable if $C < \sqrt{2}$, and is stable if $C > \sqrt{2}$.
II. If our system is within the region where the variation $\eta(t)$ is unstable:

$$\text{if} \quad \varepsilon_{1,2} \equiv \mp i\bar{\varepsilon}, \quad (\bar{\varepsilon}\text{—real and positive}) \quad \text{then} \tag{25}$$

then the solution for $\delta x(t)$ can be stable if $\delta > 0$, and $\delta^2 > \varepsilon^2$, otherwise is unstable.

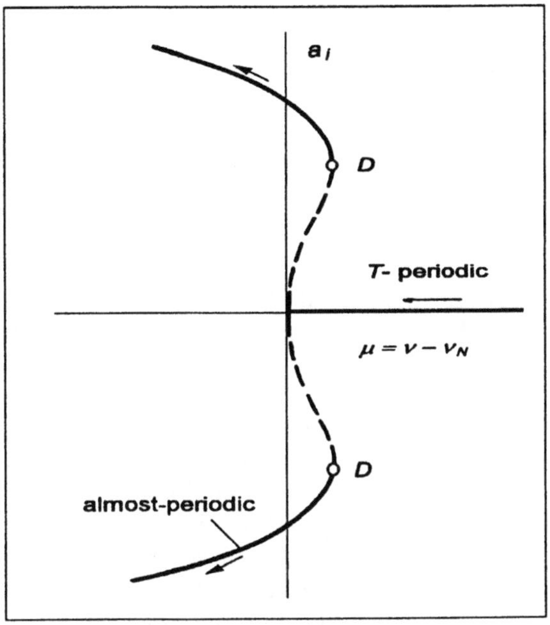

Fig. 10. Bifurcation diagram of the subcritical Neimark bifurcation $a_i, (i = 1, 2, 3\ldots)$ – amplitudes of the bifurcated almost–periodic solution.

Let us now discuss the form of instability which is defined by eq. (24), when applied to the T-periodic harmonic solution (9). In the case we can confine our attention to the first harmonic in the Fourier series of function $\phi(t)$ by setting:

$$\phi(t) = \phi(t+T) = b_1 \cos \nu t + b_2 \sin \nu t. \tag{26}$$

Therefore, the particular solution for $\delta x(t)$ within the unstable region defined by (24) can be written as:

$$\delta x(t) = e^{(-\delta \mp i\varepsilon)t} \phi(t), \tag{27}$$

and the general solution, on inserting eq. (26) yields

$$\delta x(t) = e^{-\delta t}[B_1 \cos[(\nu + \bar{\varepsilon})t + \Theta_1] + B_2 \cos[(\nu - \bar{\varepsilon})t + \Theta_2]], \tag{28}$$

where $B_1, B_2, \Theta_1, \Theta_2$ are constant.

Therefore, the form of instability defined by eqs. (24), (28) (at $\delta < 0$) results in a build-up of new harmonic components with the frequencies $\nu + \bar{\varepsilon}$, $\nu - \bar{\varepsilon}$, which are, in general case, incommensurate with the frequency ν of the periodic solution (9). The instability can be interpreted as Neimark instability and it gives rise to Neimark bifurcation, because the corresponding pair of the Floquet multipliers cross the unit circle as complex conjugates, see Fig. 7. The Floquet multipliers $\lambda_{1,2}$, which are defined by the relation

$$\delta x(t+T) = \lambda \delta x(t), \qquad T = \frac{2\pi}{\nu}, \tag{29}$$

are obtained as (eq. 25, 27).

$$\lambda_{1,2} = e^{(-\delta \mp i\bar{\varepsilon})T} \equiv \lambda_{Re} \mp \lambda_{Im}. \tag{30}$$

We notice that the multipliers lie within the unit circle $| \lambda |< 1$ and so $\delta x(t)$ is stable as long as $\delta > 0$. They reach the unit circle $| \lambda |= 1$ when $\delta = 0$ and cross the unit circle for $\delta < 0$ [Thompson & Stewart, 1986], [Szemplińska-Stupnicka & Rudowski, 1994].

Let us now look at the theoretical stability limit on the resonance curves $C = C(\nu)$ to find out where we can expect the Neimark bifurcation to occur. First, Fig. 8 depicts the resonance curves $C = C(\nu)$ and the two unstable regions, defined by the conditions (24) and (25). The former occurs at $C < \sqrt{2}$, the latter corresponds to the classic first order instability, which covers the branch of the resonance curve between two saddle-node stability limits (points with vertical tangents). Only the branch denoted by thick solid line appears to be stable in the sense of both conditions. The Neimark bifurcation is expected at the frequency, where the stable branch of the resonance curve crosses the critical boundary $C = \sqrt{2}$, at $\nu = \nu_N$. We notice immediately, that at low frequency this does not occur for $F = 12$, and, generally does not occur if $F > 2.12$. Therefore we choose the value of the forcing parameter $F = 1.0$ for further computer simulations analysis. The other crossing point, at higher frequency, corresponds to the branch of the resonance curve which is already unstable.

The computer simulation results are collected and illustrated in Fig. 9. We notice that, indeed, the system exhibits T-periodic oscillations close to the single harmonic function of time at the region of frequencies approximately between the two stability limits: and (see Fig. 8, Fig. 9 does not cover point D_1). The multi-frequency almost-periodic oscillations occur at the frequency $\nu < 0.9$, that is slightly below the approximate theoretical value of the Neimark bifurcation. Only three dominating harmonics of the almost-periodic oscillations are depicted in Fig. 9; a_ω—the amplitude of the forced limit cycle with frequency $\omega_F \approx 1.5$, $a_{2\omega-\nu}$—the amplitude of the harmonic component with the frequency $2\omega - \nu$ and the fundamental harmonic of frequency ν. It is striking that amplitude of the fundamental harmonic component is very low, approximately 0.05 of the forced limit cycle amplitude a_ω.

Between the two regular oscillations, within a narrow range of driving frequency $0.9 < \nu < 0.98$ the system exhibits chaotic motion (Fig. 5). It follows, that the narrow frequency strip of the chaotic motion is related to the transition from T-periodic to multi- frequency almost-periodic oscillations. The transition, however, does not occur in a smooth, continuous way, as it is supposed to be if the Neimark bifurcation was a supercritical one.

The transition resembles that observed in dissipative system with cubic and quadratic nonlinearity, where it was found to be related to cyclic fold (saddle-node) bifurcation of the secondary resonance, the bifurcation which is associated with jump phenomena to occur.

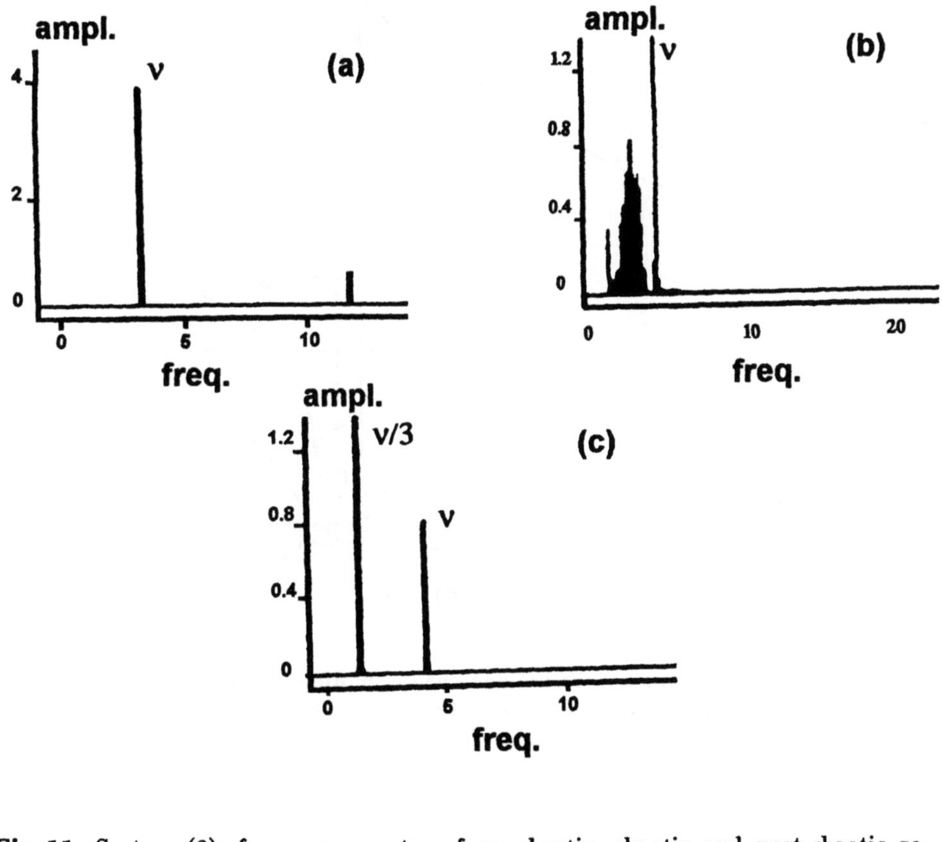

Fig. 11. System (2): frequency spectra of pre-chaotic, chaotic and post-chaotic responses at increasing frequency: (a) $\nu = 3.5$; (b) $\nu = 3.7$; (c) $\nu = 4.15$.

The observations bring us to a hypothesis that the occurrence of chaotic motion zone in the system considered close to $\nu = \nu_N$, in the region of driving frequency that separates the T-periodic from the multi-frequency almost-periodic oscillations can be interpreted as a subcritical Neimark bifurcation. As the amplitudes of the bifurcated, almost-periodic solution increase, the solution becomes stable when it undergoes the saddle-node bifurcation. This is illustrated in Fig.

10, where amplitudes $a_i, (i = 1,2,3)$ of the bifurcated almost-periodic solution versus the bifurcation parameter $\nu - \nu_N$ are plotted. Thus we deal here with both Neimark bifurcation and saddle-node bifurcation, the bifurcation, which is related to "jump phenomena".

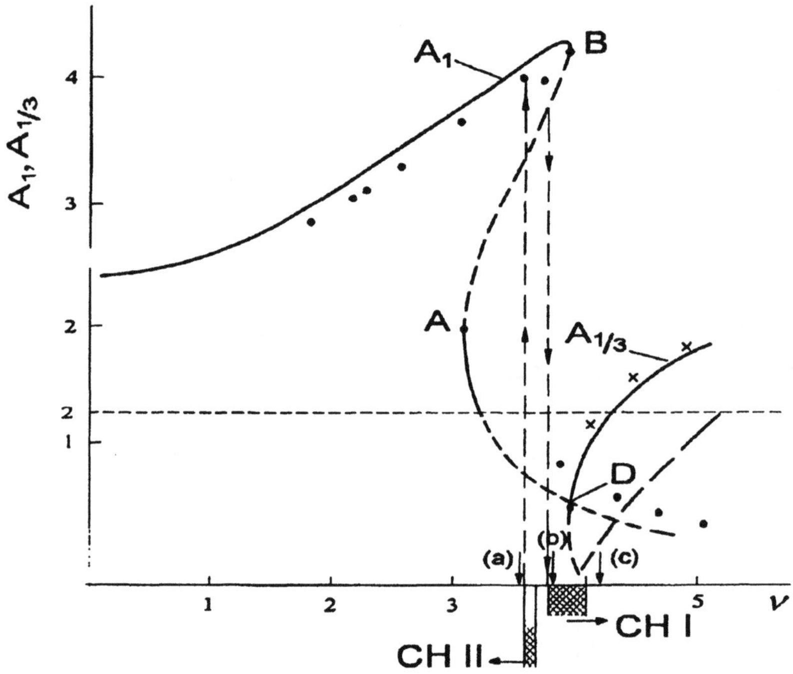

Fig. 12. System (2): Resonance curves and chaotic motion zones: solid lines—stable, dashed lines—unstable theoretical curves. Computer simulation results: dots—A_1; crosses—$A_{1/3}$. CH I—chaos at increasing frequency, CH II—chaos at decreasing frequency.

(b)

Then we turn to the chaotic motion which occurs at the high values of the parameters F, ν (see Fig. 6) and select the specific value $F = 12$ for detail study.

First we explore the frequency spectra of the response in the neighbourhood of the chaotic motion zone and from Fig. 11 we conclude that the chaotic motion occurs at the transition zone between the T-periodic principal resonance region and the $3T$-periodic subharmonic resonance.

Therefore we consider the 1/3 subharmonic resonance solution which, in the first approximation, is assumed as

$$x(t) = x(t+T) = A_{1/3}\cos(\frac{\nu}{3}t + \varphi_3) + A_1\cos(\nu t + \varphi_1), \tag{31}$$

where $A_{1/3}, A_1, \varphi_3, \varphi_1$ are constants to be determined by one of many approximate techniques available.

The amplitude curves depicted in Fig. 12 are calculated by a perturbation method, so that the amplitude A_1 in eqs. (31) is assumed to be equal to that cálculated for harmonic solution (9–10), i.e., $A_1 = C$ of Fig. 8. Then we focus on the classic instability limit, that one which occurs at the points of the curve, that has vertical tangents (point D in Fig. 12). Therefore we conclude that chaotic motion is related to the lost of stability of the $3T$-periodic subharmonic resonance.

References

Arrowsmith D.K. & Taha, K.J. [1983] *Meccanica* **18**, 195.

Belogortsev, A.B., Vavriv D.M. & Tretyakov, O.A. [1993] *Appl. Mech. Rev.* **46**, 372.

Bogoliubov, N.N. & Mitropolsky, Y.A. [1961] *Asymptotic Methods in the Theory of Nonlinear Oscillations* (Gordon and Breach, New York).

Bolotin, V.V. [1964] *Dynamic Stability of Elastic Systems* (Holden-Day, San Francisco).

Dimitriev, A.S., Kishov W.A & Spiro, A.G. [1983] *Radiotechnika i Elektrotechnika* **12**, 2430.

Dimitriev, A.S., Komlev, U.A. & Turaev, D.V. [1992] *Int. J. Bifurcation and Chaos* **2**, 93.

Endo, T. & Saito, T. [1990] *Transactions of the IEICE* **E73**, 763.

Guckenheimer, J. & Holmes, P. [1983] *Nonlinear Oscillations, Dynamical Systems, and Bifurcations of Vector Fields* (Springer-Verlag, New York).

Hayashi, C. [1985] *Nonlinear Oscillations in Physical Systems* (Princeton University Press, Princeton).

Kawakami, H. & Hayashi, C. [1981] *26 Int. Wiss. Coll. TH Ilmenau 1981.*

Magnus, W. & Winkler, S. [1966] *Hill's Equation* (Interscience, New York).

Nayfeh, A.H. & Mook, D.T. [1979] *Perturbation methods* (John Wiley and Sons, New York).

Parlitz, U. & Lauterborn, W. [1987] *Phys. Rev.* **A36**, 1428.

van der Pol, B. [1927] *Phil Mag.* **7**, 65.

Quin, G., Dong, D., Li, R. & Wen, X. [1989] *Phys. Lett.* **A141**, 412.

Ruelle, D. [1980], *The Mathematical Intelligencer*, 126.

Steel, W.H. & Kunick, A. [1987] *Int. J. Non-Linear Mech.* **22**, 349.

Stocker, J.J. [1950] *Nonlinear Vibrations in Mechanical and Electrical Systems* (Intersc. Publ. Inc., New York).

Szemplińska-Stupnicka, W. [1987] *J. Sound. and Vib.* **113**, 155.

Szemplińska-Stupnicka, W. [1990] *The Behavior of Nonlinear Vibrating Systems* (Kluwer Academic Publisher, Dordrecht) Szemplińska-Stupnicka, W. [1993] *ASME* **DE-56**, 421.

Szemplińska-Stupnicka, W. & Bajkowski, J. [1986] *Int. J. Non-Linear Mechanics* **21**, 375.

Szemplińska-Stupnicka, W. & Rudowski, J. [1994] Physics Letters A **192**, 201.

Szemplińska-Stupnicka, W. & Niezgodzki, P. [1990] *J. Sound. and Vib.* **141**, 181.

Thompson, J.M.T. & Stewart, H.B. [1986] *Nonlinear Dynamics and Chaos* (J. Wiley and Sons, New York, 1986).

Tomita, K. [1982] *Phys. Reports* **86**, 113.

Ueda, Y. [1979] *J. Statistical Physics* **20**, 181–196.

Ueda, Y. [1980] *Annals of the N.Y. Academy of Sciences* **357**–Nonlinear Dynamics, 422.

Ueda, Y. [1991] *Chaos, Solitons and Fractals* **1**, 199.

Ueda, Y. & Akamatsu, N. [1981] *IEEE Trans. of Circuits and Systems* **CAS-28**, 217.

Wave Packet Propagation, Nonlinear Dynamics, and Constructing Chaotic Eigenstates

Steven Tomsovic

Department of Physics
Washington State University
Pullman, WA 99164-2814

1 Introduction

Wave packets have by now a long history of contributing to our understanding of many physical phenomena and the relationship between quantum and classical mechanics. They forge one important, intuitive bridge between phase space trajectories and evolving wave functions. A recent excellent discussion of the semiclassical propagation theory and molecular applications is Heller's 1989 Les Houches session LII course entitled "*Wave packet dynamics and quantum chaology*" published in 1991 [1]. As that name suggests, wave packets are playing an important role in attempts to understand the Correspondence Principle for chaotic systems and is one primary motivation for the work presented in this course.

One may wonder whether the subject of wave packet dynamics could have possibly advanced in a significant way since as recently as 1991. Surprisingly enough the reply is that the semiclassical theory has taken a leap forward. For quite awhile, the semiclassical theory of wave packet propagation was synonymous with linearized wave packet dynamics (LWPD). The method was limited to short propagation times except in a few simple cases like free particle motion and the harmonic oscillator. In many applications, short time dynamics is sufficient to extract the essential physics such as the dissociation of some simple molecules. However, for other applications and especially for the study of chaotic motion and its manifestations in quantum mechanics, the inability to incorporate nonlinear dynamics in a time-dependent framework is a severe handicap. In the last four years, this impediment has been overcome and in these lectures, we shall describe the main concepts of work done in collaboration with Heller et al. [2] and Barnes et al. [3]. Applications involving nonlinear dynamical information are given here to the radial Coulomb problem [3] and the fully chaotic stadium billiard [2]. Space limitations prevent our first reviewing LWPD on which the nonlinear theory is intimately linked. We find it necessary to assume the reader is familiar with all the necessary background material as found in Ref. [1]. More details on the course matter are published in Ref. [2, 3].

2 Incorporating Nonlinear Dynamics

As long as the natural wavelength of a problem is sufficiently small and a wave packet is well localized in position and momentum, only the local behavior of the potential in the neighborhood of the wave packet's moving center is relevant to its evolution. The trajectory underlying the motion of the wave packet's center represents the local classical motion to within small linear deviations. All the neighboring trajectories needed to construct the wave packet are taken into account with a stability analysis of the central trajectory. This is how ordinary linear wave packet dynamics works – one reference trajectory represents a group of like trajectories that collectively fully support the construction of the evolving wave packet [4, 5]. As nonlinearities in the dynamics begin to emerge, the wave packet is no longer described by this single reference trajectory and its stability parameters. The errors grow first in the evolving wave function's tails.

We found that it was possible to incorporate nonlinear dynamics by making a series of small generalizations of LWPD without getting involved in methods which would rapidly collapse in chaotic systems in any case such as complicated higher-order expansions of the dynamics. The first extension comes from the recognition that it is possible to expand about any reference trajectory, not just the one associated with the wave packet centroid. One can say that the tails are as describable semiclassically as the central region of the state. If a reference trajectory is chosen in the tail of the wave packet, it will be the center and far tail regions where the approximation first fails under evolution. With strong nonlinearities present in the dynamics it is not possible to find any single trajectory which globally describes the evolving wave function, however parts of the wave function locally associated with the neighborhood of the reference trajectory will be well described for some time.

Taking the overlap of the evolving wave function at time t with some localized wave packet is an instrumental means of focusing or probing the information given by the local classical motion. It is thus helpful to consider correlation functions in generating the semiclassical theory. One of the most important in physical applications is the autocorrelation function, $C_\beta(t)$, which is measurable in experiments of the so-called pump-probe variety [6]-[12]. For a localized wave packet taken as the initial wave function, one has

$$C_\beta(t) = \langle\beta|\beta(t)\rangle = \int dx\ \psi_\beta^*(x,0)\psi_\beta(x,t), \tag{1}$$

where β stands for the parameters (center and variance of the wave packet) that uniquely specify the initial state $\psi_\beta(x,0)$. Clearly, in an autocorrelation function, contributing trajectories are those that nearly return. The problem of selecting an appropriate reference trajectory is system dependent since it must be handled differently for integrable versus chaotic systems; see discussions ahead.

The next generalization is crucial for a full description of nonlinear dynamics. A reference trajectory is only useful for a neighborhood of trajectories related by small perturbations. Generally at long times, a decomposition over multiple reference trajectories is required to capture the properties of all returning orbits.

Typically the orbits naturally separate into leaves in phase space. Each leaf supports a *local wave front* constructed as if it is part of an independent, linearized wave packet. The reference trajectory is, in principle, any trajectory well within the leaf ending in the region of the final state.

The coherent summation of these wave fronts gives the full semiclassical contributions to correlation functions. They therefore are written as sums over local solutions, one for each reference trajectory denoted by j. The autocorrelation function is

$$C_\beta(t) = \sum_j \langle\beta|\beta(t)\rangle_j . \tag{2}$$

The nonlinear dynamics now resides entirely in the summation and the particular choice of the set of reference trajectories.

3 Reference Trajectories to Wave Fronts

Following [3], we briefly outline the first-mentioned generalization by extending the original derivation of Ref. [4]. For our purposes, the one-degree-of-freedom problem is sufficient here. The Schrödinger equation (in units where $\hbar$ and m are unity) is

$$i\frac{\partial\psi(x,t)}{\partial t} = -\frac{1}{2}\frac{\partial^2\psi(x,t)}{\partial x^2} + V(x)\psi(x,t). \tag{3}$$

where $V(x)$ is the potential. Let us assume that we are interested in the local approximation associated with one particular reference trajectory which is specified by its collection of position and momentum values for all times which we denote $\{q_t, p_t\}$. A solution of Eq. (3) best suited to the neighborhood of such a trajectory starts by expanding the potential $V(x)$ about q_t up to the quadratic form

$$V(x) = V(q_t) + V'(q_t)(x - q_t) + \frac{1}{2}V''(q_t)(x - q_t)^2. \tag{4}$$

Since q_t is evolving, the expansion is continuously changing with time.

Suppose that at $t = 0$, the wave function $\psi_\beta(x,0)$ is a Gaussian. Equations (3, 4) govern the motion of a quantum harmonic oscillator with time varying frequency and center. Therefore, any initial state with a Gaussian or wave packet form will under evolution always retain a Gaussian form, albeit distorted. In that case, the solution may be expressed as

$$\psi_\beta(x,t) = A\exp\{i[\xi_t(x - q_t) + \alpha_t(x - q_t)^2 + \gamma_t]\}. \tag{5}$$

The variables ξ_t, α_t and γ_t are complex functions depending on time and the initial conditions β. The normalization factor A is time independent. In LWPD, the initial conditions of the reference trajectory are the position and momentum centroids of $\psi_\beta(x, t = 0)$. The usual results where $\xi_t = p_t$ is the momentum centroid and q_t is the position centroid as described by the classical trajectory (q_t, p_t) are then recovered. This constraint is relaxed here.

Substitution of the Gaussian wave function of Eq. (5) and the approximate potential Eq. (4) into Schrödinger's equation, and equating the coefficients of like powers of $(x - q_t)$ yields the ordinary differential equations

$$\begin{aligned}\dot{\alpha}_t &= -2\alpha_t^2 - \frac{1}{2}V''(q_t),\\ \dot{\gamma}_t &= \xi_t\dot{q}_t - \frac{1}{2}\xi_t^2 - V(q_t) + i\alpha_t,\\ \dot{\xi}_t &= 2\alpha_t\dot{q}_t - 2\alpha_t\xi_t - V'(q_t),\end{aligned} \qquad (6)$$

where q_t and p_t satisfy the classical Hamiltonian equations of motion

$$\begin{aligned}\dot{q}_t &= p_t,\\ \dot{p}_t &= -V'(q_t).\end{aligned} \qquad (7)$$

These equations are solved in Ref. [3].

A unit-normalized initial Gaussian wave packet of variance σ_β^2, centered at q^β and with average momentum of p^β, has the form

$$\psi_\beta(x,0) = (\pi\sigma_\beta^2)^{-1/4}\exp\left[ip^\beta(x-q^\beta) - \frac{(x-q^\beta)^2}{2\sigma_\beta^2}\right]. \qquad (8)$$

Comparing to the notation of Eq. (5), we obtain at $t = 0$

$$\begin{aligned}\alpha_0 &= i/2\sigma_\beta^2,\\ \gamma_0 &= (q_0 - q^\beta)[\alpha_0(q_0 - q^\beta) + p^\beta],\\ \xi_0 &= p^\beta + 2\alpha_0(q_0 - q^\beta),\end{aligned} \qquad (9)$$

and $A = (\pi\sigma_\beta^2)^{-1/4}$. Terms with factors of the form $(q_0 - q^\beta)$ or $(p_0 - p^\beta)$ result from our generalization.

Substituting the initial Gaussian wave packet $\psi_\beta(x,0)$ and the propagated wave packet $\psi_\beta(x,t)$, both given by Eq. (5), into $\langle\beta|\beta(t)\rangle_j$, an individual term of the autocorrelation function is

$$\begin{aligned}\langle\beta|\beta(t)\rangle_j = \Bigg\{&\left(\frac{2\alpha_0}{\alpha_t+\alpha_0}\right)^{1/2}\exp\Bigg[i(\gamma_t-\gamma_0^*) - \frac{i(\xi_t-\xi_0^*)^2}{4(\alpha_t+\alpha_0)}\\ &+\frac{i(\alpha_t\xi_0^* + \alpha_0\xi_t)(q_0-q_t) + i\alpha_t\alpha_0(q_0-q_t)^2)}{(\alpha_t+\alpha_0)}\Bigg]\Bigg\}_j,\end{aligned} \qquad (10)$$

where j labels the particular reference trajectory $\{q_t, p_t\}_j$, and α_t, γ_t, and ξ_t for the specific solutions of the radial Coulomb problem are given ahead. α_0, γ_0, and ξ_0 are given by Eq. (9), all evaluated for each particular reference trajectory. Although the square root is double valued, the appropiate branch is generally obtained by the requirement that the phase must evolve continuously with time.

For the special case that $q_t = q_0$, Eq. (10) simplifies considerably to the special form

$$\langle\beta|\beta(t)\rangle_j = \left\{\left(\frac{2\alpha_0}{\alpha_t+\alpha_0}\right)^{1/2} \exp\left[i(\gamma_t-\gamma_0^*) - \frac{i(\xi_t-\xi_0^*)^2}{4(\alpha_t+\alpha_0)}\right]\right\}_j . \tag{11}$$

To obtain the full autocorrelation function $C_\beta(t)$ of Eq. (2), a summation over reference trajectories j is made where the individual terms are given by Eq. (10) (or Eq. (11) if $q_t = q_0$). Only classical information enters into the evaluations of the parameters in Eq. (10).

4 The Radial Coulomb Problem

For the one-dimensional (radial, $J = 0$) Coulomb potential $V(q) = -1/q$ where $q > 0$, the classical equations of motion can be solved in parametric form, with

$$\begin{aligned} q_t &= \frac{1}{2E}(1-\cos\theta), \\ p_t &= (2E)^{1/2}\frac{\sin\theta}{(1-\cos\theta)}, \end{aligned} \tag{12}$$

where θ is implicitly determined by the time t

$$t = \frac{1}{(2E)^{3/2}}(\theta - \sin\theta) + t_0, \tag{13}$$

$E = 1/q_0 - p_0^2/2$ is the magnitude of the Kepler energy for the bound orbit starting at (q_0, p_0), and the constant t_0 is determined by what θ is at $t = 0$. In [3] the parameters $(\alpha_t, \xi_t, \gamma_t)$ are found to be

$$\begin{aligned} \alpha_t &= \frac{f_t}{2g_t} \\ \xi_t &= p_t + \frac{2\alpha_0(q_0 - q^\beta) + p^\beta - p_0}{g_t} \\ \gamma_t &= \gamma_0 + S + \frac{i\ln g_t}{2} - \frac{(\xi_0 - p_0)^2 g^{(h)}}{2g_t} \end{aligned} \tag{14}$$

where

$$\begin{aligned} g_t &= \frac{\alpha_0}{E}(2p_0q_t - 2q_0p_t - 3tp_0p_t) + \frac{1}{2q_0^2E}(2q_t - 3tp_t - q_0^2p_0p_t) \\ f_t &= \frac{1}{q_t^2E}\left[\alpha_0(3tp_0 + 2q_0 - q_t^2p_0p_t) + \frac{1}{2q_0^2}(3t + p_0q_0^2 - q_t^2p_t)\right] \end{aligned}$$

$$g^{(h)} = \frac{1}{2E}(2p_0 q_t - 2q_0 p_t - 3tp_0 p_t)$$
$$S = 3Et + 2(q_t p_t - q_0 p_0). \qquad (15)$$

The initial conditions have been chosen to be $g_0 = 1$ and $f_0 = 2\alpha_0 = i/\sigma_\beta^2$.

For simplicity, we choose the initial wave packet $\psi_\beta(x,0)$ [Eq. (8)] to be centered at the outer classical turning point of the Coulomb potential, so that $q^\beta = 1/E_\beta$ and $p^\beta = 0$. The initial variance of this Gaussian wave packet is σ_β^2. The Kepler period of the center of the Gaussian is $\tau_\beta = 2\pi/(2E_\beta)^{3/2}$.

To evaluate the autocorrelation function for the Coulomb potential, a suitable set of reference trajectories must be chosen. The initial decay of the wave packet for $t < \tau_\beta/2$ and the recurrences of the wave packet for $t > \tau_\beta/2$ must be handled separately; see [3] for details.

For times greater than half the Kepler period of the central trajectory, the propagated state soon has more than one contribution, as the leading edge of the propagated wave packet has traveled around the Coulomb center more times than the trailing edge. The classical version of this feature is illustrated in Fig. (1) where after close to 4 1/2 Kepler periods of motion, some trajectories are returning for their fifth time while the outer group only for their fourth. The ideal reference trajectories are shown boxed. In this case, they are given by the periodic points along the q axis in phase space at each fixed time t ($> \tau_\beta/2$). The periodic points which correspond to our choice of reference trajectories are always near the points of maximum *local* overlap between the initial classical density and the propagated density as illustrated by Fig. (2). These points are simple to obtain analytically, one has only to tune the energies of the periodic points such that they close at exactly time t.

Let the jth periodic point be denoted $(q_j, 0)$. Then $(q_0, p_0) = (q_t, p_t) = (q_j, 0)$. The periodic orbits will return to the starting classical turning points q_j which are given by

$$q_j = (\frac{t}{j\tau_\beta})^{2/3} q^\beta = (\frac{t'}{j})^{2/3} q^\beta, \qquad (16)$$

where $t' = t/\tau_\beta$ is the time in units of the Kepler period τ_β of the central trajectory and j is the number of times the periodic orbit has returned to the classical turning point during time t'. Such periodic orbits have an energy $E_j = 1/q_j$ and a period $\tau_j = 2\pi/(2E_j)^{3/2}$, and so $j\tau_j = t = t'\tau_\beta$. Therefore, the correlation function for $t' > 0.5$ is given by Eq. (11) where we calculate every time dependent quantity by setting $q_t = q_0 = q_j$ and $p_t = p_0 = p_j = 0$. Thus, the autocorrelation function (excluding the initial decay) for an initial Gaussian wave packet centered around the turning point $q^\beta = 1/E_\beta$ and a variance σ_β^2 is given by

$$C_\beta(t) = \sum_{j=1}^{\infty} \left(\frac{t'}{t' + \mu_\beta j^2} \right)^{1/2} \exp\left\{ i\frac{3\tau_\beta}{q^\beta}(j^2 t')^{1/3} + i\frac{2\alpha_0 \mu_\beta q^{[\beta]2} j^2 [(t'/j)^{2/3} - 1]^2}{(t' + \mu_\beta j^2)} \right\}, \qquad (17)$$

where $\mu_\beta = 3\tau_\beta/8\alpha_0 q^{[\beta]3}$ is a purely imaginary unitless quantity. Note that the major contributions in the sum come from the jth terms where j is close to t'.

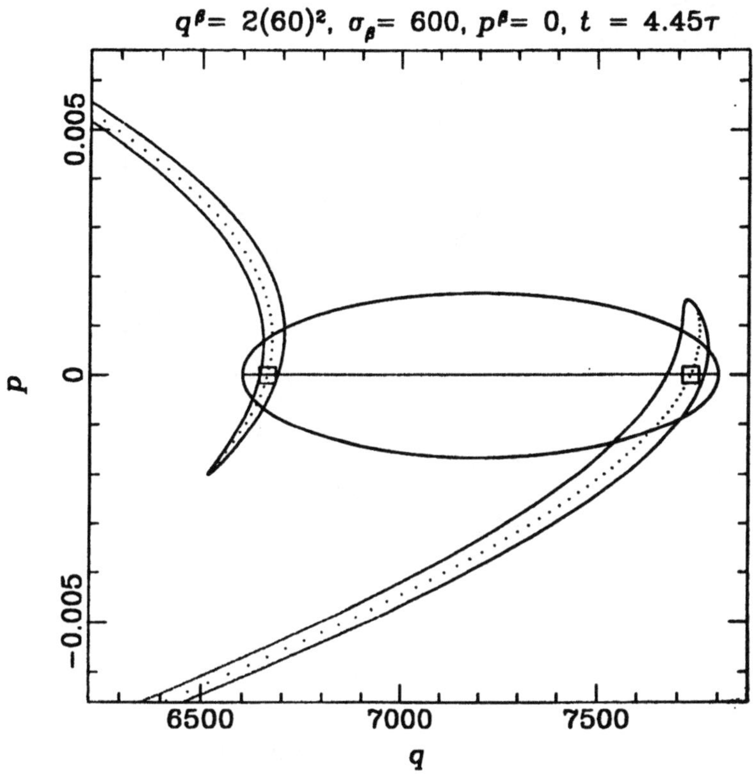

Fig. 1. An ellipse in phase space representing a one σ contour of the initial Gaussian with parameters $q^\beta, p^\beta, \sigma_\beta$ is propagated classically to a time $t = 4.45$. The inner branch is formed by trajectories on their fifth return, and the outer branch are those on their fourth return. Likewise, the p=0 line inside ellipse is propagated to give the final dotted line. The intersections correspond to the boxed initial and final points of the reference trajectories used to calculate the autocorrelation function. Reprinted from Ref. [3].

We now compare the autocorrelation function obtained with our semiclassical method and that obtained from an exact quantum calculation for the case where $q^\beta = 7200$ and $\sigma_\beta = 600$. This is very close to experimentally realizable wave packets [6]-[12]. In this case, the central orbit corresponds to a quantum number of $n = 60$. Figure (3) illustrates the excellent agreement between the real parts of the semiclassical and quantum autocorrelation functions for the initial decay and first recurrence of the Gaussian wave packet. For illustration the comparison between the semiclassical and quantum autocorrelation functions is extended to 10 Kepler periods of the central trajectory by plotting the absolute values of the autocorrelation function in Fig. (4).

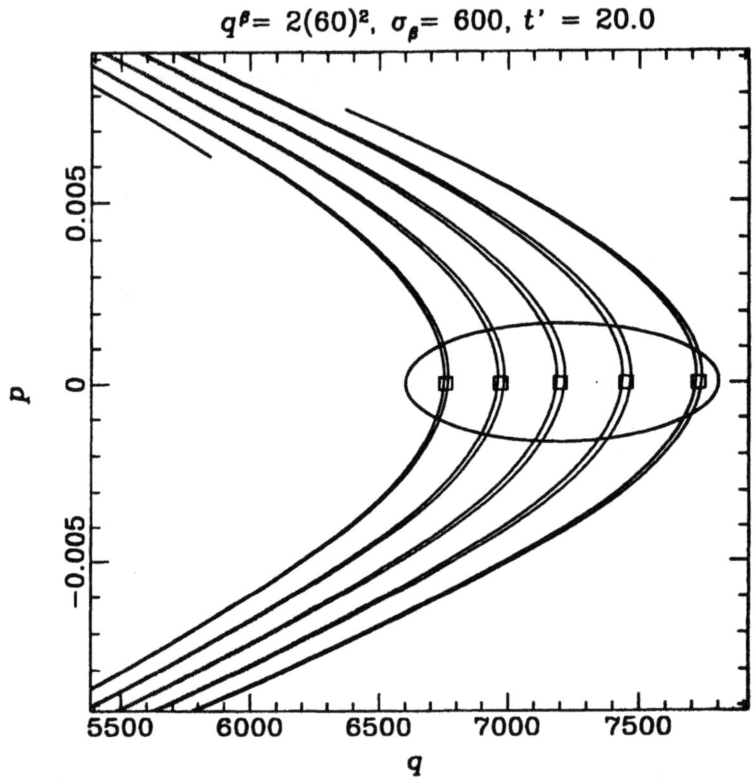

Fig. 2. Same as Fig. (1), but at $t' = 20.0$. The periodic points have traveled through the Coulomb singularity 18 to 22 times from right to left, respectively. Reprinted from Ref. [3].

It turns out for the Coulomb problem, the agreement continues just as well to as long a time one desires. An interesting feature of integrable systems (as this system is) is the appearance of revivals and fractional revivals [11]-[13]. After spreading out, the wave packet relocalizes into one or a small number of superposed localized wave packet states. The time scale for these phenomena are well within the scope of semiclassical theory. All the interference required by the quantum dynamics to relocalize is captured entirely with classical information; see [3, 14].

In addition to the autocorrelation function, the semiclassical approximation accurately reproduces the quantum spectrum. The Fourier transform of $C_\beta(t)$ gives the intensity weighted spectrum,

$$S(E) = \int \mathrm{d}t \exp(iEt) C_\beta(t), \tag{18}$$

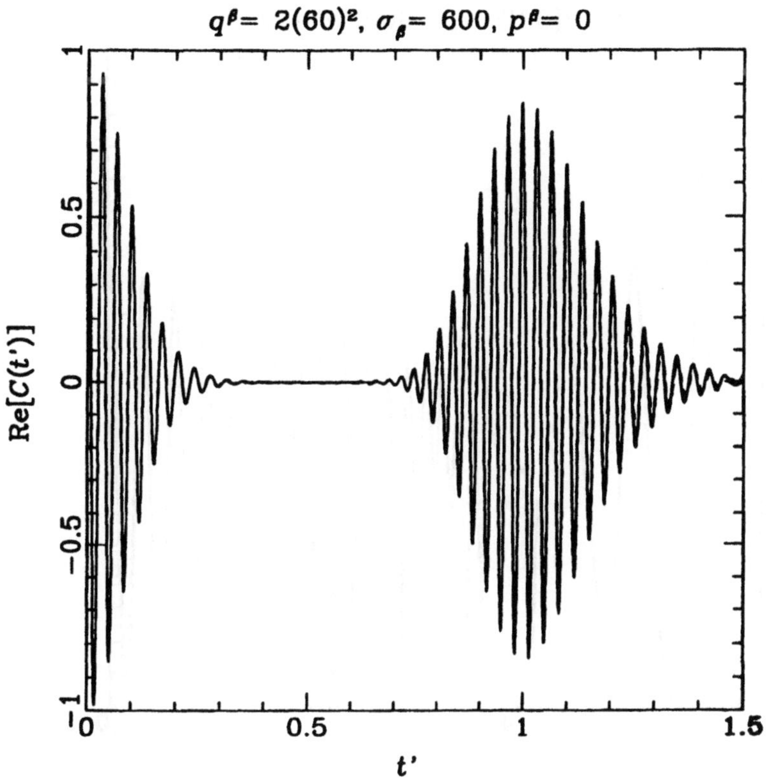

Fig. 3. A comparison of the real part of the semiclassical (solid) and the real part of quantum (solid) autocorrelation functions $C_\beta(t')$, where $q^\beta = 7200$ and $\sigma_\beta = 600$ for the initial decay and initial recurrence of the Gaussian wave packet. Reprinted from Ref. [3].

and Fig. (5) shows the comparison with the exact spectrum, given by

$$S(E) = \sum |\langle\beta|E_n\rangle|^2 \delta(E - E_n). \tag{19}$$

The semiclassical spectrum was computed with a finite time resolution of 50 Kepler periods. This remarkable reproduction of the familiar quantum Coulomb spectrum is obtained with classical reference trajectories that have a continuous energy range. No quantization conditions on the actions of the classical reference trajectories are imposed as they are with standard WKB quantization.

5 Time Dependent Green Functions

Though the introduction of multiple reference trajectories to capture nonlinear dynamical effects was introduced above in a purely intuitive way, Green function

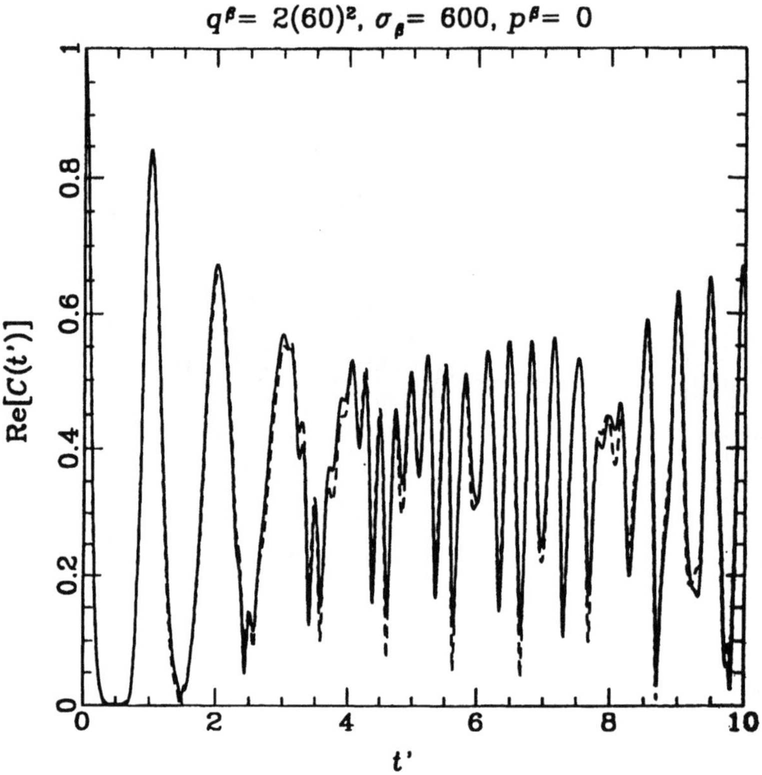

Fig. 4. A comparison of the absolute value of the semiclassical (solid) and quantum (dashed) autocorrelation functions $C_\beta(t')$, where $q^\beta = 7200$ and $\sigma_\beta = 600$ for the first 10 Kepler periods of the center of the Gaussian wave packet. Reprinted from Ref. [3].

techniques can be used to put their use on a more solid foundation. The success with the Coulomb problem is not a random accident.

The time dependent Green function of the Schrödinger equation for an autonomous Hamiltonian may be expressed

$$G(\mathbf{q}, \mathbf{q}'; t) = \langle \mathbf{q} | \mathrm{e}^{-i\hat{H}t/\hbar} | \mathbf{q}' \rangle \tag{20}$$

whose construction is formally given by Feynman's path integral [15]. The propagation of a wave function $\Psi_\alpha(\mathbf{q}) = \langle \mathbf{q} | \alpha \rangle$, can then be written as a configuration space integral

$$\Psi_\alpha(\mathbf{q}; t) = \int_{-\infty}^{\infty} d\mathbf{q}'\, G(\mathbf{q}, \mathbf{q}'; t) \Psi_\alpha(\mathbf{q}') \tag{21}$$

as can correlation functions

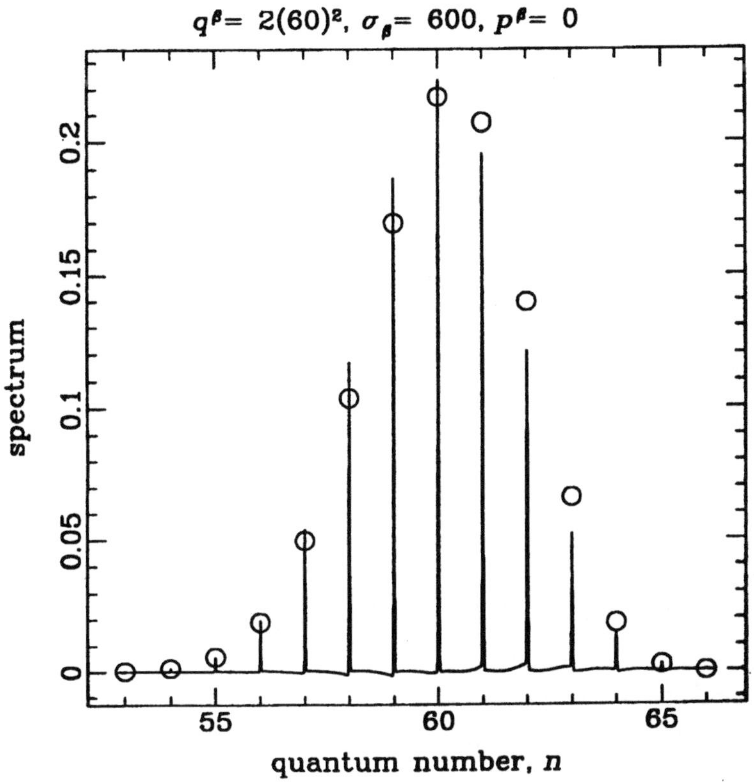

Fig. 5. A comparison between the semiclassical (solid line) and quantum (circles) spectra for the case of $q^\beta = 7200$ and $\sigma_\beta = 600$. Reprinted from Ref. [3].

$$C_{\beta\alpha}(t) = \int_{-\infty}^{\infty} d\mathbf{q}\Psi_\alpha(\mathbf{q};t)\Psi_\beta^*(\mathbf{q}) \tag{22}$$

Applying the method of stationary phase to the path integrals gives the semiclassical Green function

$$G(\mathbf{q},\mathbf{q}';t) \approx G_{sc}(\mathbf{q},\mathbf{q}';t) = \left(\frac{1}{2\pi i\hbar}\right)^{d/2} \sum_j \left|\mathrm{Det}\left(\frac{\partial^2 S_j(\mathbf{q},\mathbf{q}';t)}{\partial\mathbf{q}\partial\mathbf{q}'}\right)\right|^{1/2} \times$$
$$\exp\left(iS_j(\mathbf{q},\mathbf{q}';t)/\hbar - \frac{i\pi\nu_j}{2}\right) \tag{23}$$

In this expression, the sum over j is for all trajectories connecting $\mathbf{q}'$ to $\mathbf{q}$ in time t and d is the number of degrees of freedom. The determinant plays the role of the square root of a classical probability, and the phase is determined by the classical action $S_j(\mathbf{q},\mathbf{q}';t)$ and an index based on the properties of the conjugate

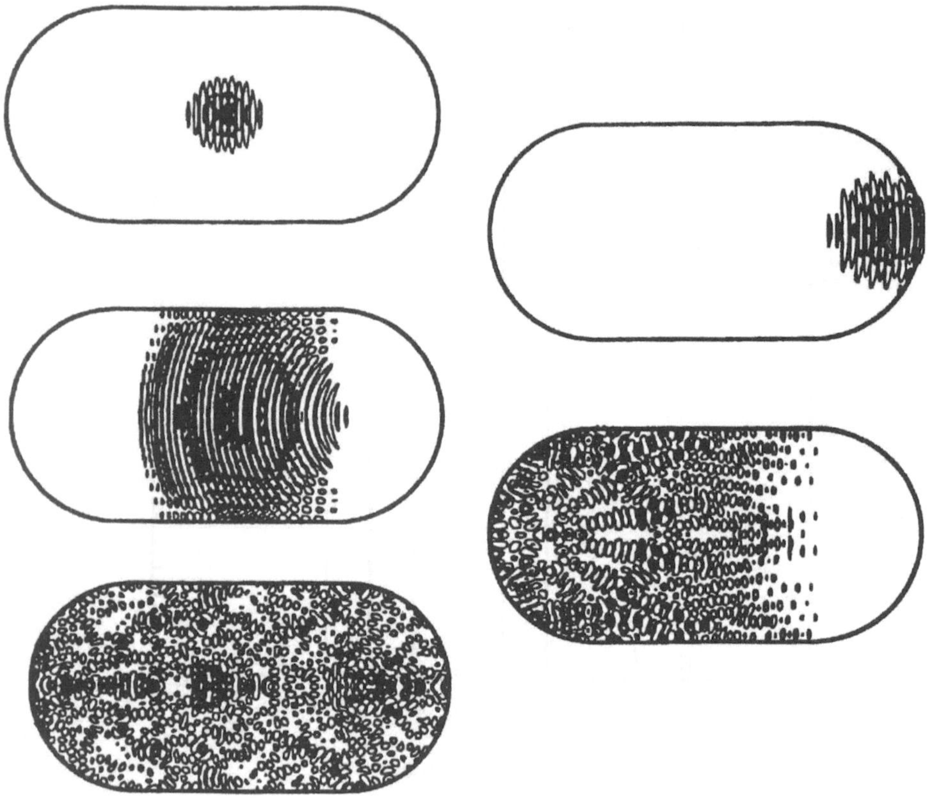

Fig. 6. The evolution of a wave packet in the stadium billiard. The time is scaled such that the classical particle traverses the horizontal length of the stadium in $t = 1$. Respectively pictured are $t = 0.0$, 0.4, 1.0, 2.0, and 6.0. Equally spaced contours are drawn of the real part of the wave function. Reprinted from Ref. [2].

points (like focal points), ν_j. $S_j(\mathbf{q}, \mathbf{q}'; t)$ is specified by the time integral of the Lagrangian $\mathcal{L}$

$$S_j(\mathbf{q}, \mathbf{q}'; t) = \int_0^t dt' \mathcal{L} = \int_0^t dt' \left\{\mathbf{p}(t') \cdot \dot{\mathbf{q}}(t') - H(\mathbf{p}(t'), \mathbf{q}(t'))\right\} \qquad (24)$$

along the jth classical path. H is the classical Hamiltonian which is presumed to be the classical limit of $\hat{H}$. Eq. (23) was originally written down by van Vleck in 1928 without the summation or index ν and was thus fundamentally limited to extremely short times at best. Gutzwiller used this extension to $G_{sc}(\mathbf{q}, \mathbf{q}'; t)$ for situations involving conjugate points and multiple paths in his derivation of a trace formula expressing the quantum density of states as a weighted sum phases determined by the properties of the classical periodic orbits [16].

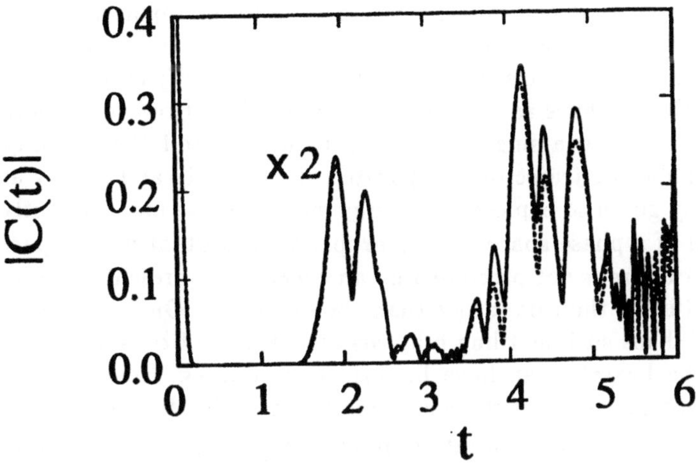

Fig. 7. The autocorrelation function for the wave packet evolution pictured in the previous figure. The first recurrence near $t = 2$ is the time of the shortest unstable periodic orbit (mid-stadium horizontal bounce orbit). The quantum result is denoted by a solid line and the semiclassical homoclinic orbit sum by the dashed line. Adapted from Ref. [2].

Semiclassical Green functions can be difficult objects to work with directly. One encounters problems such as singularities and divergences in numbers of paths. Chaos only exacerbates the situation. This is not terribly surprising in light of the mathematical subtleties of the path integral (even before applying stationary phase), but it must be confronted. Fortunately, many of the complications ease when the Green function is applied to propagating smooth states which is its basic purpose afterall. Typically, integration is possible over the majority of singularities and divergences. Though this may lead to inaccuracies, it simplifies computational methods. At this level, introducing Gaussian wave packet states is a matter of convenience. Note that all phase space trajectories appear in the integrations for $C_\beta(t)$. However, effectively only orbits whose initial and final points reside within a few q or p uncertainty widths can survive the integration. It is these orbits which reflect the quantum dynamics.

6 Heteroclinic Motion: Hallmark of Chaos

Chaos, without further specification, is a rather vague concept by virtue of its numerous interpretations across many different fields. To narrow it down, we shall restrict our attention to low-dimensional, conservative Hamiltonian systems possessing a compact energy surface in phase space. 'Chaotic' is *not* to be understood as synonymous with 'stochastic' since fairly simple, deterministic equations of motion are supposed existant. Perhaps the simplest intuitive picture of chaos is that of a single trajectory uniformly exploring the available phase space in a seemingly "random" way. The scale of exploration becomes finer as time increases. The selection of the particular trajectory is irrelevant as almost all of them behave as "randomly" as any other. However, it is important to consider more than a single trajectory to understand the full implications of chaos. Neighboring trajectories separate exponentially and locally are highly unstable. They go on to explore the energy surface quite independently. This is an expression of a phase space mixing property. Important phase space organization and structure is not precluded by an everywhere present local instability. In fact, structure is the rule rather than the exception. Only the most abstract paradigms of chaos such as tiling billiards on surfaces of constant negative curvature and the baker's map have homogeneous mixing. Relaxation is system specific and not captured by stochastic modeling of the dynamics.

For simplicity, imagine a situation in which the classical equations of motion are known exactly, precise propagation is possible, and initial and final conditions can be measured accurately, but there is a practical limit to the accuracy. The practical limit of uncertainty defines a local neighborhood or probability density of phase points. Predictions as to the future of the system depend on the evolution of the density and can only be made probabilistically. Instead of evolving the density directly, it is possible to organize the same information into a complete set of time correlation functions, $\mathcal{P}_{\alpha\beta}(t)$, defined as the probability of beginning in some neighborhood, ρ_α, of phase space and finishing in some other, ρ_β;[1] that is

$$\mathcal{P}_{\alpha\beta}(t) = (\rho_\beta, \rho_\alpha(t)) \tag{25}$$

How could we set out to predict a particular correlation function?

The initial ρ_α occupies a tiny fraction of the available phase space. The system is assumed smooth enough that the density remains simply connected as it propagates under the dynamics for all times, albeit incomprehensibly distorted. The density must preserve its measure by conservation of flux while its shape distorts, stretching and collapsing, at the rapid exponential rate that the individual trajectories are separating. It soon stretches across the phase space and must fold back upon itself in order to continue stretching. The phase space is being foliated by the density to an increasingly fine scale. The final ρ_β becomes laminated with fine layers of phase points contributing to $\mathcal{P}_{\alpha\beta}(t)$. At any fixed time, there are a finite number of layers. Further, all the trajectories within one layer are exceedingly similar; we will call them topologically identical. On the other hand, no two layers are anything alike whether neighboring or far apart. Within ρ_β, a linearization of the dynamics about any individual member trajectory of a layer captures the essence of the layer to a high degree of accuracy. The calculation of $\mathcal{P}_{\alpha\beta}(t)$ can be reduced to a sum of contributions each determined

[1] Admittedly this organization scheme is motivated by the fact that eventually we are interested in making connections to quantum mechanics.

by a single reference trajectory (labeled by γ) and its linearized dynamics,

$$\mathcal{P}_{\alpha\beta}(t) = \sum_{\gamma} (\rho_{\beta}, \rho_{\alpha}(t))_{\gamma} \tag{26}$$

There is a particularly apropos set of reference trajectories - they are the so-called heteroclinic trajectories (homoclinic if $\alpha = \beta$). There is one and only one per layer and they organize the complete dynamics. By the turn of the century, Henri Poincare had already discussed them as doubly asymptotic trajectories in his third volume on celestial mechanics [17]. In this case, they would be asymptotic to the central trajectory of ρ_α as $t \to -\infty$ and asymptotic to the central trajectory of ρ_β as $t \to \infty$.

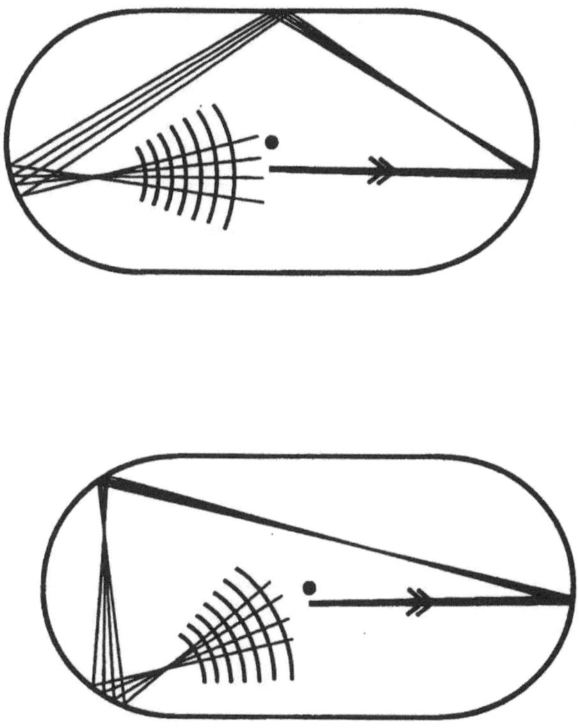

Fig. 8. Two collections of trajectories leaving with nearly identical initial conditions return by topologically distinct paths. Each group carries a wave amplitude and corresponds to a single homoclinic recurrence. Reprinted from Ref. [2].

The heteroclinic orbit decomposition of $\mathcal{P}_{\alpha\beta}(t)$ very neatly separates global and local dynamical information. Within a single heteroclinic contribution, the linearization of the local dynamics contains all the stability information, i. e. how fast the trajectories are exponentially separating, which directions are stretching, which are collapsing, etc. Global dynamical structure, on the other hand, is reflected in the properties of the set of heteroclinic orbits. For instance, if a portion of phase space is difficult to enter and return from, there will be a dearth of the shorter heteroclinic orbits passing through that region. The intricacies of the mixing property which is the essence of chaos is captured in every detail by the heteroclinic orbits.

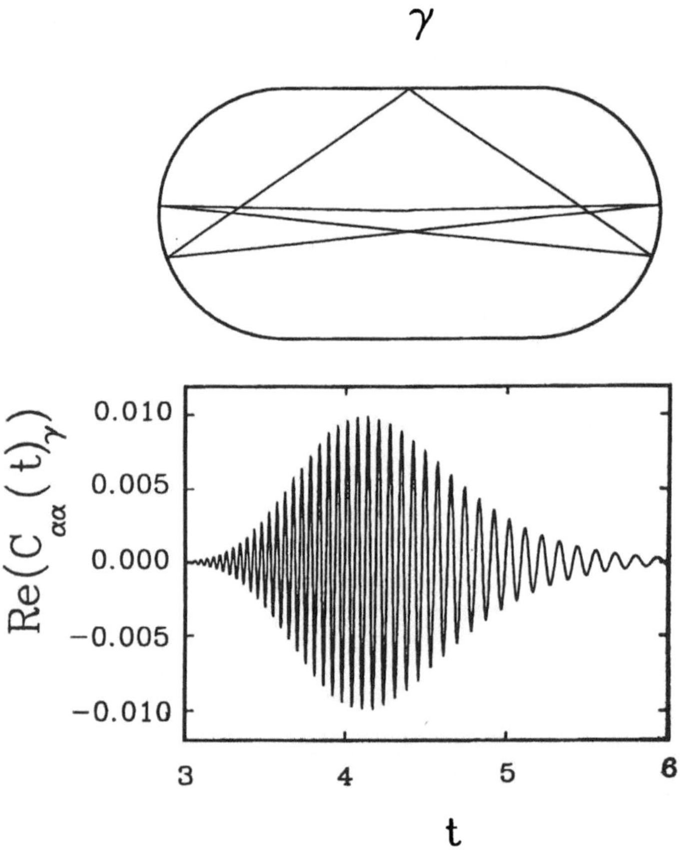

Fig. 9. An important primary homoclinic orbit is drawn above its contribution to $C_\beta(t)$. Nevertheless its magnitude is on a far smaller scale than that of $C_\beta(t)$. The $\alpha\alpha$ on the y-axis should be a β. Reprinted from Ref. [2].

It is important to point out the truly inefficient character of the heteroclinic orbit summation. At very long times correlation functions approach an ergodic average equal to the ratio of the uncertainty phase space volume to the available phase space volume. In contrast, the number of heteroclinic orbits continues to grow exponentially without bound (while their summed contributions approach a constant). No advantage is being taken of the loss of information due to the chaos. Exploring the information loss quickly leads to Markov partitioning. We stop short here and ahead only make a connection between heteroclinic orbits and quantum mechanics, though it would be extremely interesting to pursue the question of how information loss, in this context, occurs in quantum mechanics and how it differs from its classical mechanical counterpart.

Linearizing the dynamics For the semiclassical theory, linearizing the dynamics crudely means quadratically expanding the classical action. The reference trajectories are naturally the same heteroclinic orbits. Once the action is expanded, the correlation function is expressed as a Gaussian integral. All the details can be found in Ref. [1] and [2]. The final expressions involve positions, momenta and stability matrix elements, but are not terribly illuminating and are not given here. It does turn out that this approach gives exactly the same results as the quadratic expansion of the potential in the Schrödinger equation as discussed for the Coulomb problem assuming the same reference trajectories are considered.

7 The Stadium Billiard

The Bunimovich stadium billiard has been the object of at least ten experiments [18]. Interest in the stadium derives also from the following points – ergodicity of its dynamics has been proven [19], and it has a large positive (non-zero) Lyapunov exponent meaning that the motion is exceedingly unstable and represents a difficult challenge.

The classical stadium consists of a free point particle specularly reflecting off hard walls. The domain is bounded by two semicircles of radius one joined by straight edges of length two. A convenient property of the stadium is that the classical dynamics at any two energies are connected by simple scaling relations. The geometry of an orbit is independent of energy. Only the time to traverse a portion of the orbit changes, decreasing with increasing energy.

The quantum mechanics are governed by the time-dependent Schrödinger equation

$$2i\hbar\frac{\partial\Psi(\mathbf{q};t)}{\partial t} = -\nabla^2\Psi(\mathbf{q};t) \tag{27}$$

subject to the appropriate boundary conditions; in the calculations $\hbar = m = 1$. Numerical solutions pictured ahead are obtained by a modified fast Fourier transform method.

Time dependent behavior The d-degree-of-freedom version of Eq. (8) is immediate and allows one to probe a localized (h^d) volume around the phase point $(\mathbf{q}_\beta, \mathbf{p}_\beta)$ with uncertainty $\sigma_q\sigma_p = \hbar/2$ for each degree of freedom. Fig. (6) shows a wave packet starting at the stadium's center and moving toward the semicircle.

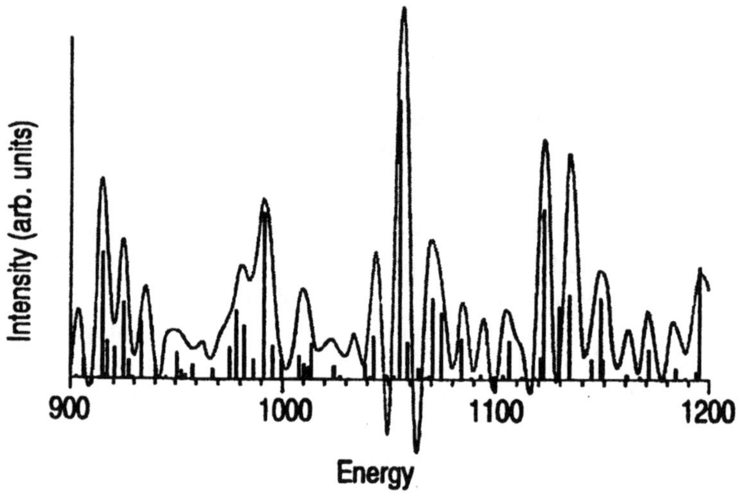

Fig. 10. Comparison of quantum and semiclassical spectra in the neighborhood of the 1100th eigenstate of the stadium. The smooth, broadened spectrum, $S_{t=8}(E)$, is semiclassical and the sharp spectrum, $S(E)$, is the quantum solution. The intensity was chosen arbitrarily and the energy scale is fixed by the parameters given in the text. Reprinted from Ref. [2].

As it heads toward the first collision with the boundary, it moves under free dynamics for which the linearized theory is exact. $C_\beta(t)$ rapidly decreases from one to essentially zero; see Fig. (7). Shortly after the first reflection, linearized wave packet dynamics has already become totally inadequate to describe the wave function. Not too much later, the wave function is completely delocalized and eventually it evolves into a random-looking pattern reminiscent of water waves on the surface of a pool. By this time, $C_\beta(t)$ mostly resembles noise. A similar look at the circular billiard, an integrable system, would show a wave that never attains such an unruly appearance.

Fig. (8) graphically illustrates how the semiclassical Green function of Eq. (23), locally approximated, works in conjunction with wave packets. The weights and phases of the rays create wavefronts centered on homoclinic trajectories.[2] The two homoclinic paths shown originate from nearly identical initial conditions, an expression of the high degree of instability in the stadium's dynamics. Why should we expect the independent superposition of these two wavefronts to be relevant since they come from imperceptible initial condition differences on a scale defined by $\hbar$? This kind of reasoning was at the heart of some of the more pessimistic objections to semiclassical chaotic dynamics. However, the more ap-

[2] Efficient algorithms for locating homoclinic solutions in the stadium can be developed that reduce the problem to a one parameter search.

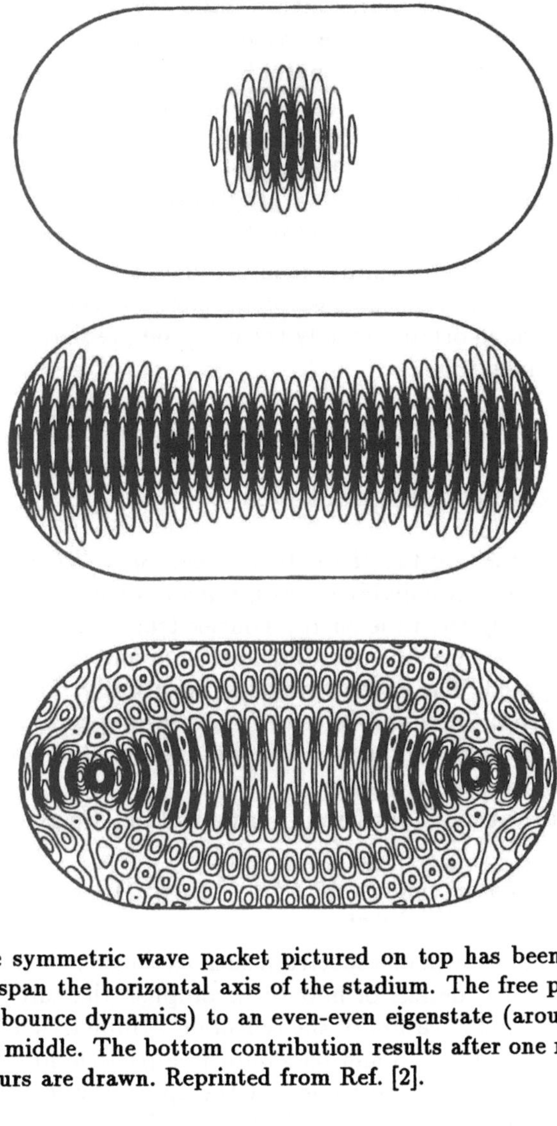

Fig. 11. The symmetric wave packet pictured on top has been chosen such that 15 wavelengths span the horizontal axis of the stadium. The free particle motion contribution (zero-bounce dynamics) to an even-even eigenstate (around the 320th state) is shown in the middle. The bottom contribution results after one reflection. Ten equally spaced contours are drawn. Reprinted from Ref. [2].

propriate question to ask is whether the two paths differ by at least a radian in the phases they accumulate. If so, it is proper to simply superpose their contributions by the usual rules of semiclassical approximations. Instead of expecting a breakdown time scale involving the logarithm of $\hbar$ [20], careful consideration of this point suggests algebraic time scales in $\hbar$ [2]. The distinction is very important because chaotic mixing only begins to reveal itself on a logarithmic time scale. If that were the end of the quantum-classical correspondence, quantum chaology would be an almost pointless endeavor. On the other hand, extraordinarily complex dynamics occur by algebraic times. Even if it is insufficient to resolve individual eigenproperties, there still must exist deep connections between chaotic motion and quantum mechanics.

Each homoclinic orbit adds a characteristic "echo" to $C_\beta(t)$; see Fig. (9). The high momentum component travels fastest along the orbit so the leading edge of the contribution has the shortest wavelength. Since the wave packet has a finite momentum uncertainty, wavelengths too far from the mean are exponentially damped. After summing over $\sim 30,000$ homoclinic contributions, the semiclassical $C_\beta(t)$ is obtained to times of $t \approx 6$. The surprising accuracy of the homoclinic sum is seen by comparing the dashed and solid lines in Fig. (7). Longer times require vastly more orbits (at $t \approx 8$, $O(10^6)$). For times $t > 3$, no single orbit is anywhere close to dominating the sum; it requires them all. One frustration with the newfound access to nonlinear semiclassical propagation is that so far we have been unable to sum enough orbits to study the eventual predicted breakdown of the approximation at long times.

8 Chaotic Spectra and Eigenstates

How close are we approaching the Heisenberg time, $t_H = h/D$ (D being the mean level spacing), at which individual eigenproperties can be resolved on average? We need only apply the time-energy Fourier transform of Eq. (18). The result is given in Fig. (10) where the quantum spectrum and its semiclassical approximation are compared. The spectra coincide beautifully. The resolution of the semiclassical spectrum to around twice the mean level spacing is constrained solely by the number of classical orbits we were able to calculate and not by any limitation of the semiclassical mechanics.

The eigenstates, to within an irrelevant normalization, are given by,

$$\Psi_n(\mathbf{q}) = \lim_{t\to\infty} \int_{-t}^{t} dt' \; e^{iE_n t'/\hbar} \phi_\alpha(\mathbf{q}; t') \tag{28}$$

where E_n is chosen at some peak in a spectrum such as the one given in Fig. (10). We leave out the technical details of how to accomplish the full propagation. The shortest time dynamical feature of the reflection symmetrized wave packet pictured in Fig. (11) is its forward and backward free particle motion up to its first collision with the wall. The projection of the spreading wave packet and its simple dynamics generate the drawing in the middle frame. The waves are then reflected back from the wall leading to the wave function in the lower frame. The limit of the continuing sum of these functions is the semiclassical eigenstate. Fig. (12) shows the semiclassical eigenstate resulting from carrying out the sum to times of the order of 10 bounce dynamics ($t \approx 6$). The semiclassical and quantum eigenstate are very similar to the eye. In fact, the overlap of the unit normalized pair is 0.96.

The semiclassical description of a wave that impinges on the joint between the semicircle and straight edge has a line of discontinuity. The joint therefore is responsible for producing diffraction. Beyond a short propagation time, the wavefunction is no longer localized away from the joint. Where is the evidence of diffraction in the eigenstate? Note also that the heteroclinic orbit summation

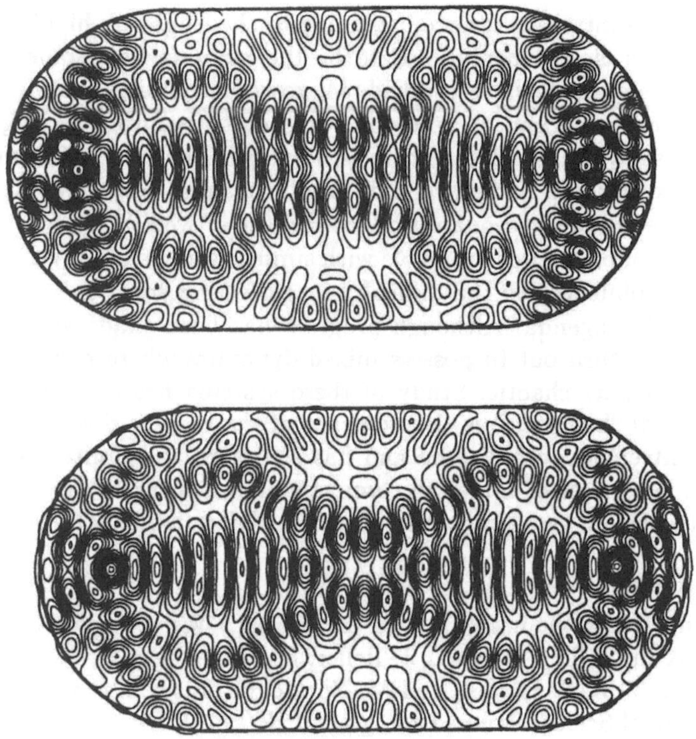

Fig. 12. Shown on top is the semiclassical state that includes all the dynamics to about 10 bounces. Below is the eigenstate it is approximating. The 13 contours are equally spaced and placed at exactly the same levels for both states (the states were normalized to unity first). Reprinted from Ref. [2].

is no better than a "poor man's" version of the method of stationary phase. We have not actually identified the trajectories which are the stationary phase solutions. What is their connection and why does this method apparently function so well? Would even better accuracy result with access to longer time semiclassical dynamics?

9 Conclusions

We have shown two examples, beginning in a time-dependent framework, where judicious application of semiclassical mechanics can lead to strikingly quantitative approximations. On the other hand, most research in the field has been conducted in the energy domain. For this reason and because of the nature of the many new experimental capabilities, there appears to be significant prospects for continued work on the development of time-dependent approaches.

Going back to the "Old Quantum Theory", only integrable systems were understood. Recent advances are beginning to shed light on highly chaotic systems and chaos is not turning out to be the nemesis that some of the more pessimistic assessments have implied. Microscopic detailed classical dynamics beyond the scale of $\hbar$ can be used to approximate quantum propagation, even if the converse is not possible. There is no reason that the stadium should be considered exceptional in this regard. Similar results were generated with the quantum baker's map and quantum kicked rotor as well [2].

The range of phenomena possible with simple systems of only 2 or 3 degrees-of-freedom combined with the difficulty in answering some the open questions makes for a full agenda. Although progress has been made with pure chaos, generic systems turn out to possess mixed dynamics where part of the motion is regular and part chaotic. Study of these systems has only just begun [21]. An ever present "nuisance" is the existence of the regular-chaotic motion interface. It is extremely complicated and has structure to infinitesimal scales. To unravel the mysteries of mixed dynamical systems, a far better understanding of multidimensional tunneling and diffraction theory is likely to be required.

Acknowledgments

I heartily thank the organizers of the Karpacz Winter School. I would also like to thank my collaborators E. J. Heller, M. Nauenberg, P. W. O'Connor, M. A. Sepúlveda, I. M. S. Barnes, and M. Nockleby. This course was supported by the US National Science Foundation grant PHY-9421153.

References

1. E. J. Heller in "*Chaos and Quantum Physics*", Eds. M.J. Giannoni, A. Voros, and J. Zinn-Justin, Les Houches Session LII, 1989 (Elsevier Science Publishers B.V., Amsterdam, 1991).
2. S. Tomsovic and E. J. Heller, *Phys. Rev. Lett.* **67**, 664 (1991); *Phys. Rev.* **E 47**, 282 (1993); *Phys. Rev. Lett.* **70**, 1405 (1993);
   P. W. O'Connor, S. Tomsovic and E.J. Heller, *Physica* **D 55**, 340 (1992);
   M.A. Sepúlveda, S. Tomsovic, and E.J. Heller, *Phys. Rev. Lett.*, **69**, 402 (1992);
   E. J. Heller and S. Tomsovic, *Physics Today*, **46**, 38 (1993).
3. I.M. Suárez Barnes, M. Nauenberg, M. Nockleby, and S. Tomsovic, *Phys. Rev. Lett.* **71**, 1961 (1993); *J. Phys. A: Math. Gen.* **27**, 3299 (1994).
4. E. J. Heller, *J. Chem. Phys.* **62**, 1544 (1975).
5. R. G. Littlejohn, *Physics Reports* **138**, 193 (1986).
6. J.A. Yeazell and C.R. Stroud Jr., *Phys. Rev. Lett.* **60**, 1494 (1988).
7. L.D. Noordam, A. ten Wolde, H.G. Muller, A. Lagendijk, H.B. van Linden van den Heuvell, *J. Phys.* **B***: At. Mol. Opt. Phys.* **21**, L533 (1988).
8. A. ten Wolde, L.D. Noordam, A. Lagendijk and H.B. van Linden van den Heuvell, *Phys. Rev. Lett.***61**, 2099 (1988); *Phys. Rev. Lett.* **40**, 485 (1989).
9. H.B. van Linden van den Heuvell, H.G. Muller and A. ten Wolde, *Electronic and Atomic Collisions* (1988) (Amsterdam: Elsevier)
10. A. ten Wolde, L.D. Noordam, H.G. Muller and H.B. van Linden van den Heuvell, *Lecture Notes in Physics* (1989) (Berlin: Springer)

11. J. A. Yeazell, M. Mallalieu and C. R. Stroud, *Phys. Rev. Lett.* **64**, 2007 (1990); *Phys. Rev.* **A 43**, 5153 (1991).
12. D. R. Meacher. P. E. Meyler, I. G. Hughes and P. Ewart, *J. Phys.* **B 24**, L63 (1991).
13. G. Alber, H. Ritsch and P. Zoller, *Phys. Rev.* **A 34**, 1058 (1986);
J. Parker and C.R. Stroud Jr., *Phys. Rev. Lett.* **56**, 716 (1986);
Z. Dačić Gaeta and C.R. Stroud Jr. *Phys. Rev.* **A 42**, 6308 (1990);
M. Nauenberg, *Phys. Rev.* **A 40**, 1133 (1989); *J. Phys.* **B***: At. Mol. Opt. Phys.* **23**, L385 (1990);
I.Sh. Averbukh and N.F. Perelman, *Physics Letters* **A 139**, 449 (1989); *Sov. Phys. JETP* **69**, 464 (1990); *Sov. Phys. USP* **34**, 572 (1991).
14. M. Mallalieu, *Semiclassical Dynamics of Rydberg Electronic Wave Packets*, Dissertation, University of Rochester, (1994).
15. R. P. Feynman and A. R. Hibbs, "*Quantum Mechanics and Path Integrals*", McGraw-Hill, New York, (1965);
L. S. Schulman, "*Techniques and Applications of Path Integration*", Wiley, New York, (1981).
16. M. C. Gutzwiller, *J. Math. Phys.* **12**, 343 (1971), and references therein.
17. H. Poincare, "*Methodes Nouvelles de la Mecanique Celeste*", English translation, ed. D. L. Goroff, AIP, Woodbury, (1993).
18. C. M. Marcus,
A. J. Rimberg, R. M. Westervelt, P. F. Hopkins, and A. C. Gossard, *Phys. Rev. Lett.* **69**, 506 (1992);
H. J. Stöckmann and J. Stein, *Phys. Rev. Lett.* **64**, 2215 (1990);
J. Stein and H. J. Stöckmann, *Phys. Rev. Lett.* **68**, 2867 (1992);
H. D. Graf, H. L. Harney, H. Lengeler, C. H. Lewenkopf, C. Rangacharyulu, A. Richter, P. Schardt and H. A. Weidenmüller, *Phys. Rev. Lett.* **69**, 1296 (1992);
R. Blumel, I. H. Davidson, W. P. Reinhardt, H. Lin, and M. Sharnoff, *Phys. Rev.* **A45**, 2641 (1992);
A. Kudrolli, S. Sridhar, A. Pandey, and R. Ramaswamy, *Phys. Rev.* **E49**, R11 (1994);
E.J. Heller, M. F. Crommie, C. P. Lutz, D. M. Eigler, *Nature* **369**, 464 (1994);
M.W. Keller, O. Millo, A. Mittal, D. E. Prober, and R. N. Sacks, *Physica*, **B194-6**, 1029 (1994);
T. H. Seligman, private communication;
A. M. Chang, H. U. Baranger, L. N. Pfeiffer, and K. W. West, *Phys. Rev. Lett.* **73**, 2111 (1994).
19. L. A. Bunimovich, *Funct. Anal. Appl.* **8**, 254 (1974); *Comm. Math. Phys.* **65**, 295 (1979).
20. M. V. Berry and N. L. Balazs, *J. Phys.* **A12**, 625 (1979);
M. V. Berry, N. L. Balazs, M. Tabor and A. Voros, *Ann. Phys. (NY)* **122**, 26 (1979).
21. O. Bohigas, S. Tomsovic and D. Ullmo, *Phys. Rep.* **223**, (2), 45 (1993).

Chaotic Dynamics of Weakly Nonlinear Systems

D.M. Vavriv

Radio Astronomy Institute of the Ukrainian Academy of Sciences
310002 Kharkov, Ukraine

Abstract: The progress made in recent years in the study of chaotic states of weakly nonlinear systems is reviewed. We concern with the class of chaotic states pertaining to physical systems with any degree of nonlinearity however small. The conditions for, and the mechanisms of, the transition to chaos are discussed for the weakly nonlinear oscillators and compared with that for the strongly nonlinear ones. Considerable attention is given to analytical methods of the chaos onset prediction. The dynamics of Duffing-type oscillators is considered to illustrate these results.

1 Introduction

It has recently been realized that the chaotic states are typical not only for strongly nonlinear systems, but they can arise in weakly nonlinear dissipative systems as well [1-6]. A wide range of typical situations was found where the chaotic states arise under the weakly nonlinear excitation conditions, including various types of oscillators with a quasi periodic forcing, multimode autonomous and non autonomous systems, distributed systems. The existence of the chaotic states in the quasilinear limit implies that the condition for the chaos to arise is not very stringent for many practical situations, and that the influence of the chaotic instabilities on the dynamics of real systems can be much more severe as compare with the recent thinking. The main results in this direction of research have been obtain for the systems described by the equations of motion of single or coupled oscillators

$$\ddot{x}_i + \omega_i^2 x_i = \varepsilon F_i\left(t, x_1, \dot{x}_1, x_2, \dot{x}_2 \ldots x_N, \dot{x}_N\right), \qquad i = 1, 2, \ldots, N$$

where x_i are the generalized coordinates of the oscillators ω_i, are their natural frequencies F_i, are the functions describing nonlinearity and dissipation, ε is the parameter of nonlinearity, being small for the weakly nonlinear systems, N is the number of interacting oscillators. It is well known that this mathematical model naturally arises when studying the dynamics of a variety of systems with a weak nonlinearity. The chaotic states of weakly nonlinear oscillators manifest themselves in the form of a weak chaotic modulation of the amplitude and phase of a periodic or quasiperiodic oscillation having the frequencies close to the natural frequencies of the oscillators. The characteristic time scale of the modulation is ω/ε, which is much more grater as compare with $1/\omega_i$. Due to it, the direct numerical study of these states by using the initial system of equations leads to a cumbersome computational procedure. The are also no analytical methods which can be used to detect the chaotic states in this case. Moreover, the weakly nonlinear systems are characterized by specific mechanisms of the transition to chaos as compare with strongly nonlinear systems. Due to these reasons, a demand arose for adequate methods of the weakly nonlinear systems study. A large experience thus far is accumulated in this direction, and the investigations of the chaotic dynamics of the weakly nonlinear systems constitute now a separate direction of the research in the field of chaos being important both for the basic study and applications. In this paper some general approaches to the study of chaotic phenomena in weakly nonlinear systems are described along with the results of the investigations of some practical systems. For better illustration of the obtained results, at first, we review in Section 2 the conditions of the formation of chaotic states in the strongly nonlinear oscillators, and then we compare this process with that going on in the quasilinear limit. When studying the chaotic dynamics of the weakly nonlinear oscillators, we introduce the concept of induced saddle states [7] which appears to be useful for the understanding of the mechanism of the chaos onset. Then we describe several analytical approaches to the chaos onset prediction by using the quasiperiodically forced Duffing-type oscillator as an example. In Section 3 characteristic features of the dynamics of parametrically forced oscillators are considered. The mechanisms of the transition to chaos are discussed along with the conditions of the arising of multistable states. These findings are applied to the stability analysis of parametrical amplifiers.

2 Conditions for Chaos in Strongly and Weakly Nonlinear Systems

In this part of the paper we discuss the main differences in mechanisms of the transition to chaos in strongly and weakly nonlinear oscillators. To illustrate these results, we start from the simplest model, as the quasiperiodically forced Duffing-type oscillator

$$\ddot{x} + \omega_0^2 x = \varepsilon \left[- \left(\delta_0 + \delta_1 x^2 \right) \dot{x} + \gamma x^3 + A_1 \cos \omega_1 t + A_2 \cos \omega_2 t \right] \quad . \qquad (2.1)$$

Here x is a generalized coordinate of the oscillator, $\delta_0 > 0$ and $\delta_1 > 0$ are the coefficients of linear and nonlinear damping, γ is the nonlinearity parameter, ω_0 is the natural frequency of the oscillator, A_1 and A_2 the amplitudes of parametric and external forcing with the incommensurate frequencies ω_1 and ω_2. The ε–factor is introduced in Eq.(2.1) in order to express in the explicit form the degree of nonlinearity of the oscillator.

Note that the oscillator (2.1) can be considered as the Duffing type oscillator due to the limitations imposed on the coefficients: $\delta_0 > 0$ and $\delta_1 > 0$. The case of $\delta_0 < 0$ and $\delta_1 > 0$ corresponds to the Van der Pol oscillator, and it is not considered here. The interested reader may refer to Refs. [8-10] for the later case.

A great deal of work has already been done to study chaotic solutions of this equation and its special cases. One of such cases is the harmonically forced Duffing oscillator, whose chaotic dynamics has been much investigated starting from the initial works in the field of chaos [11-13]. It was shown that the chaotic motion arises because of the formation of a homoclinic structure in the phase space of the system due to the transverse intersections between the stable and unstable manifolds of hyperbolic periodic orbits. The next important conclusion was made that a strong nonlinearity is required for the chaos onset in any single–degree–of–freedom dissipative oscillator with periodic excitation. The dynamics of the strongly nonlinear oscillator with quasiperiodic forcing was studied, for example, in Refs. [14,15]. Notice that beginning with the above-mentioned paper and up to now, the systems possessing a homoclinic loop (or loops) to a saddle point(s) in their Hamiltonian limit have been mainly considered. With respect to the equation under consideration such saddle points exist in the unperturbed oscillator with $\gamma > 0$ when the dissipation and amplitudes of external forcing are equal zero. We call this type of singular points as the original saddle points. The systems possessing an original saddle point are amenable to a theoretical treatment by using a global perturbation technique developed by Melnikov [1], and that is possibly the main reason why these systems have received much consideration (see Refs.[16,17]). This technique was generalized by Wiggins [18] to quasiperiodically forced oscillators. In the later case, the chaos origin was associated once again with the existence of the original saddle point, and chaotic states of the strongly nonlinear oscillators were considered as for the harmonically forced ones.

However, in the case of quasiperiodically forced oscillators, there is also another way of the chaos arising that is not related with the existence of homoclinic orbits in the unperturbed oscillators, and where the strong nonlinearity of the oscillators is not required. This possibility was predicted and studied independently by several authors in Refs.[1,2,4] and proved experimentally in Refs.[10,11]. The alternative way of the transition to chaos is associated with the occurrence of new saddle orbits in the phase space of the systems under the action of one of the external harmonic component. We call such orbits as the induced saddle orbits. With the availability of an additional incommensurate spectral component of low intensity in the external force, a normally hyperbolic invariant torus is formed in the phase space instead of the saddle orbit. The chaos results from the transverse intersection between the stable and unstable manifolds of the torus when the perturbation is increased. The formation of a homoclinic structure occurs here at much more lower values of amplitudes of the perturbation as compared to the case when this structure is formed on the base of an original saddle point. Now let us consider in more details these two mechanisms of the chaos onset, beginning from the case of the strongly nonlinear oscillator.

2.1 Strongly Nonlinear Oscillators

Let us rewrite Eq.(2.1) as a system of first-order equations:

$$\begin{aligned} \dot{x} &= y \\ \dot{y} &= -\omega_0^2 x + \varepsilon f(x, y, t) \ , \end{aligned} \tag{2.2}$$

where:

$$f(x,y,t) = -\left(\delta_0 + \delta_1 x^2\right) \dot{x} + \gamma x^3 + A_1 \cos(\omega_1 t) + \ A_2 \cos(\omega_2 t) \tag{2.3}$$

To obtain analytical conditions of the onset of chaos through Melnikov's method, we make use of the fact that at $\delta_0 = \delta_1 = A_1 = A_2 = 0$ the system under investigation is described by the following Hamiltonian [2]

$$H(x,y) = \frac{y^2}{2} + \frac{\omega_0^2 x^2}{2} - \frac{\varepsilon \gamma x^4}{4} \ . \tag{2.4}$$

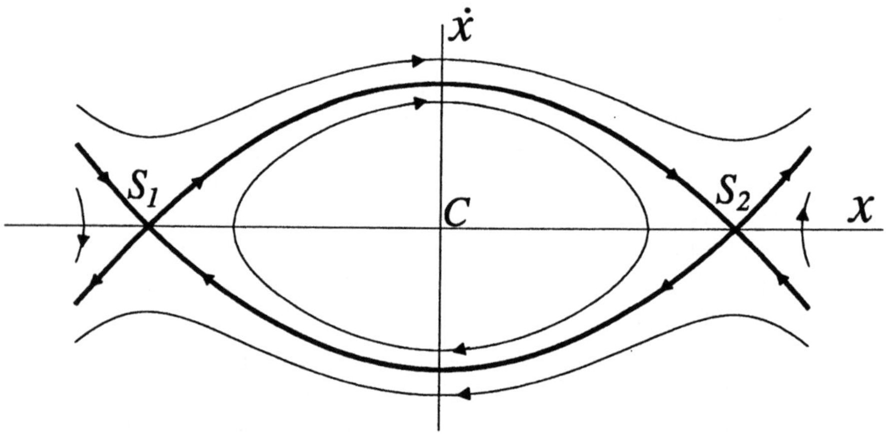

Fig. 1. The phase portrait for the Hamiltonian system (2.4), S_1 and S_2 are saddle points, C is center.

Provided that $\gamma > 0$, this system possesses an original saddle points in the phase space (x, y) with the coordinates

$$y = 0, \qquad x = \pm\omega_0 \Big/ \sqrt{\varepsilon\gamma} \tag{2.5}$$

and a heteroclinic orbit (separatrix loop) shown in Fig. 1. Note that a center-type singular point is located at $y = 0$, $x = 0$ in the phase space of the system (2.4).

The influence of the other terms of the function (2.3) not included in Eq.(2.4) is considered as perturbations. It permits to apply the standard Melnikov technique developed for similar systems (see Refs.[1-3] for details). We eventually come to the following condition of the homoclinic structure formation:

$$A_1\omega_1\mathrm{cosech}\left(\frac{\pi\omega_1}{\sqrt{2}\omega_0}\right)+A_2\omega_2\mathrm{cosech}\left(\frac{\pi\omega_2}{\sqrt{2}\omega_0}\right)\geq\left|\frac{2\omega_0^3}{3\pi\sqrt{\varepsilon\gamma}}\left(\delta_0+\frac{\delta_1\omega_0^2}{5\varepsilon\gamma}\right)\right| \qquad (2.6)$$

This criterion is considered as a necessary condition for the chaos to arise. It incorporates a variety of particular cases that have been studied up to now. For example, assuming $\delta_1 = 0$, we have from Eq.(2.6) the condition of a homoclinic structure arising obtained in Ref.[2] for the oscillator with the two-frequency excitation. Setting in addition $A_2 = 0$ in Eq. (2.6), we come to the result given in Ref.[1] for the harmonically forced oscillator.

It is evident from Eq.(2.6) that the chaotic oscillations can arise when the oscillator is excited only periodically. Under the quasiperiodic excitation, the threshold for chaos to arise is not changed essentially. Now let us consider the transition to the case of the quasilinear oscillator that corresponds to the limit: $\varepsilon \to 0$. According to the expression (2.5), the manifolds cease to intersect for any values of A_1 and A_2 if the ε–value is sufficiently small. Due to it, this mechanism of the homoclinic structure formation can not lead to the chaos onset in the weakly nonlinear systems.

However, this result is true only when we deal with the formation of homoclinic structures associated with the original saddle points (2.5). The situation is changed dramatically, if we take into account that additional saddle states can arise under the action of the external force. In the later case other homoclinic structures can arise on the base of homoclinic loops associated with this states due to the frequencies' interaction of the external forcing. It leads to the chaotic states formation in the weakly nonlinear limit. To predict the chaos onset analytically in the weakly nonlinear systems, one should at first detect the induced saddle states. In the case under consideration such states are saddle orbits arising under the action of one or each of the harmonic components of the external forcing. The direct application of the initial equations (2.1) or (2.2) do not allow detect analytically these orbits as well as to predict the chaos arising. One way around this problem is application of the averaging method [21], and it is demonstrated in the next section.

2.2 Induced Saddle States

Let us apply the method of averaging to the system (2.2), assuming that the right-hand side of Eq.(2.2) is small, that is, $\varepsilon|f(x,y,t)| << 1$. We are interesting in the case of the resonant excitation of the oscillator, when

$$\left|\omega_1-\omega_0\right| = O(\varepsilon\omega_0), \quad \text{and} \quad \left|\omega_2-\omega_0\right| = O(\varepsilon\omega_0) \qquad (2.7)$$

Then, by using the transformation

$$x = U\cos(\omega_1 t) + V\sin(\omega_1 t)$$
$$y = -U\omega_1\sin(\omega_1 t) + V\omega_1\cos(\omega_1 t) \tag{2.8}$$

and applying the standard averaging procedure, we come to the following system of averaged equations for the slowly varying time functions $U(\tau)$, $V(\tau)$:

$$\frac{dU}{d\tau} = -\left[\alpha_0 + \alpha_1\left(U^2+V^2\right)\right]U - \left[\Delta + \beta\left(U^2+V^2\right)\right]V - p_2\sin\Omega\tau$$

$$\frac{dV}{d\tau} = -\left[\alpha_0 + \alpha_1\left(U^2+V^2\right)\right]V + \left[\Delta + \beta\left(U^2+V^2\right)\right]U - p_1 + p_2\cos\Omega\tau \tag{2.9}$$

where dimensionless parameters are introduced: $\tau = \varepsilon\omega_1 t$, $\alpha_0 = \delta_0/(2\omega_1)$ $\alpha_1 = \delta_1/(8\omega_1)$, $p_1 = A_1/(2\omega_1^2)$, $p_2 = A_2/(2\omega_1^2)$, $\Delta = (\omega_1 - \omega_0)/(\varepsilon\omega_1)$, $\Omega = (\omega_1 - \omega_s)/(\varepsilon\omega_1)$, and $\beta = 3\gamma/(8\omega_1^2) > 0$.

For the case of the pure periodic excitation ($p_2 = 0$) this system is reduced to a second order autonomous system. On this basis we conclude again that the weakly nonlinear oscillators with harmonic excitation can not have any chaotic states. Such states can arise here only due to the interaction of the external frequencies provided that one of the frequencies induces a saddle orbit in the phase space of the system (2.9).

The transition from the original equation to the averaged ones allows to reduce the investigation of bifurcations of two-dimensional tori in the phase space of the original system (2.1) to the analysis of bifurcations of periodical orbits in the phase space of the system (2.9). It is important to note that the latter system does not contain ε as independent parameter. The ε changes lead only to the variation of the time scale of the excited oscillations, provided that values of the parameters Δ and Ω are kept constant. From this follows that, if the system (2.9) demonstrates chaotic behavior, than such behavior can arise for any degree of ε however small, i.e., in the quasilinear limit.

The averaged equations are convenient tool for the detection of the induced saddle orbits of the initial equation (2.1). In terms of the averaged equations these orbits are seen as saddle singular points. They can be induced by any of the external frequency components. We shall consider the case when the amplitude of one of the component, say p_2, is relatively small and the induced saddle states can arise mainly due to the second component. Assuming that $p_2 = 0$ and the dissipation is absent, we have instead of (2.9) the following Hamiltonian system:

$$\frac{dU}{d\tau} = -\left[\Delta + \beta\left(U^2+V^2\right)\right]V$$
$$\frac{dV}{d\tau} = \left[\Delta + \beta\left(U^2+V^2\right)\right]U - p_1 \tag{2.10}$$

with Hamiltonian energy given by

$$H(U,V) = \frac{\beta}{4}\left(U^2+V^2\right)^2 + \frac{\Delta}{2}\left[V^2+U^2\right] - p_1 U\ . \tag{2.11}$$

Unlike the previous case (see Eq.(2.4)), this Hamiltonian contains the amplitude of the external force p_1 as a parameter. As long as $p_1 = 0$, there is only a center-type singular point in the origin of the coordinate, thus, we come to the result following from the Hamiltonian (2.4). For $p_1 \neq 0$ and $-\Delta/\beta > 0$, the center is split into three singular points: two centers and a saddle. This situation is illustrated by the phase portrait of the system (2.11) in Fig.2. Note that the phase portrait is characterized by the existence of a double homoclinic orbit (separatrix) going out and in the saddle point with the coordinates $U = U_s$, $V = 0$. Here U_s is the minimum of the real roots of the following cubic equation,

$$U_s^3 + \frac{\Delta}{\beta} U_s - \frac{f}{4} = 0 \ , \tag{2.12}$$

where $f = 4p_1/\beta$.
The separatrix equation was found in Ref. [22] and can be written in the form:

$$U_l(\tau') = U_s + S(\tau') \left[2 \left(\delta - U_s^2 \right) + S(\tau') \right] \Big/ f \ , \tag{2.13}$$

where $S(\tau') = (q^2 - b^2)/(2(q \cosh \tau' - b))$; $\tau' = \beta\mu(\tau - \tau_0)$; $q = \pm\sqrt{-2fU_s}$; $b = -4(\delta - U_s^2)$; $\mu^2 = -2fU_s - b^2/4$, $\delta = -\Delta/\beta$, τ_0 is an initial moment of the "slow" time. The signs "+" and "−" correspond to the small and large loops of the separatrix, accordingly. The regions in the parameter space where the saddle states exist are reduced under the influence of dissipation. Let us consider this effect by introducing into the consideration the corresponding dissipative terms into the system (2.10). Then instead of it, we have:

$$\begin{aligned} \frac{dU}{d\tau} &= -\left[\alpha_0 + \alpha_1 \left(U^2 + V^2\right)\right] U - \left[\Delta + \beta \left(U^2 + V^2\right)\right] V \ , \\ \frac{dV}{d\tau} &= -\left[\alpha_0 + \alpha_1 \left(U^2 + V^2\right)\right] V - \left[\Delta + \beta \left(U^2 + V^2\right)\right] U - p_1 \ . \end{aligned} \tag{2.14}$$

By means of the stability analysis it can be shown that this system has a saddle point in its phase space when the following conditions are hold:

$$p_1^2 > p_{\mathrm{cr}}^2 \equiv \frac{8\alpha_0^3 \left(\alpha_1^2 + \beta^2\right)}{3\sqrt{3} \left(\beta - \sqrt{3}\alpha_1\right)^3} \tag{2.15}$$

$$\Delta < \Delta_{\mathrm{cr}} \equiv -\alpha_0 \frac{\alpha_1 + \sqrt{3}\beta}{\beta - \sqrt{3}\alpha_1} \tag{2.16}$$

$$\beta > \sqrt{3}\alpha_1 \tag{2.17}$$

When the value of the parameter of nonlinear dissipation α_1 is negligibly small, these conditions read

$$p_1^2 > p_{cr}^2 \equiv \frac{8\alpha_0^3}{3\sqrt{3}\beta} \tag{2.18a}$$

$$\Delta < \Delta_{cr} \equiv -\sqrt{3}\alpha_0 \tag{2.18b}$$

The above given conditions determine the regions in the parameter space where the induced saddle states arise and due to it, they can be considered as necessary conditions for the chaos onset in the weakly nonlinear. We complete these conditions in the following section by using several analytical approaches to the chaos onset study.

2.3 Conditions of Quasiperiodic Oscillations Destruction in the Weakly Nonlinear Oscillator

2.3.1 Melnikov's Criterion

We start from the application of Melnikov's method to the averaged equations (2.9) to find necessary conditions for the chaos onset with respect to the parameters of the second external spectral component. Assuming that the dissipation and the amplitude of this component are small, we write down the Melnikov function $\Delta_M(\tau_0)$ which determines the distance between the stable and unstable manifolds of the saddle point in the Poincare map of the averaged equations:

$$\Delta_M(\tau_0) = \int_{-\infty}^{\infty} \left[R_1(U_l, V_l, \tau) \cdot Q_0(U_l, V_l) - Q_1(U_l, V_l, \tau) \cdot R_0(U_l, V_l)\right] d\tau \ , \tag{2.19}$$

where

$$R_0 = -\left[\Delta + \beta\left(U_l^2 + V_l^2\right)\right] V_l \ ,$$

$$Q_0 = \left[\Delta + \beta\left(U_l^2 + V_l^2\right)\right] U_l - p_1 \ ,$$

$$R_1 = -\left[\alpha_0 + \alpha_1\left(U_l^2 + V_l^2\right)\right] U_l - p_2 \sin(\Omega\tau) \ ,$$

$$Q_1 = -\left[\alpha_0 + \alpha_1\left(U_l^2 + V_l^2\right)\right] U_l - p_2 \cos(\Omega\tau) \ .$$

The functions $U_l(\tau)$ and $V_l(\tau)$ are given by the relations (2.11)-(2.13). After performing the integration in Eq.(2.19), one can find from the condition of the manifolds intersections $\Delta_M(\tau_0) = 0$ the following necessary condition of the chaos onset [22, 23]:

$$\frac{p_2}{\alpha_0} \le \left| \frac{f \sinh(\pi\sigma)}{2\pi\mu^2\sigma^2 \exp(-\sigma \cdot \mathrm{sign} q \cdot \arccos(-b/q))} \left\{ 3\mu - 8\delta \arctan\frac{q+b}{2\mu} + \right.\right. \tag{2.20}$$

$$\left.\left. + \frac{\alpha_1}{\alpha_0}\left[7\delta\mu + \left(8\delta\left(U_s^2 - 2\delta\right) - \frac{3}{4}q^2\right) \arctan\frac{q+b}{2\mu}\right]\right\}\right| \ ,$$

where $\sigma = (2\Omega)/(\beta\mu)$. The equality sing corresponds to $p = p_{cr}$, where p_{cr} is the lowest threshold of the chaotic motion arising with respect to the amplitude of the external force. It was found in Refs. [22,23] that this condition taken simultaneously with (15)–(17) gives possibility to define quite accurately the regions of system control parameters where the chaotic state arise.

2.3.2 Current Lyapunov Exponents Technique

The current Lyapunov exponents technique is based on the investigation of the local properties of the motion on an attractor (see Appendix). The possibility of the constructive use of this technique to determine the analytical conditions for the chaos onset was pointed out in the papers [24,25]. Let us apply it to the system under consideration. We first write down the equation for the amplitude ρ and the phase Φ of a vector in the tangent space of the system (2.9)

$$\begin{aligned} \frac{d\rho}{d\tau} &= \left(\rho - \alpha_0 - 2\alpha_1 a^2 + a^2 \{\beta \sin[2(\Phi-\gamma)] - \alpha_1 \cos[2(\Phi-\gamma)]\}\right) \\ \frac{d\Phi}{d\tau} &= \left(\Delta + 2\beta a^2 + a^2 \{\beta \cos[2(\Phi-\gamma)] + \alpha_1 \cos[2(\Phi-\gamma)]\}\right) \end{aligned} \tag{2.21}$$

The expression for the largest Lyapunov exponent can be written in the following form

$$\mu_1 = -\alpha_0 - \alpha_1 a^2 \left\{-2 + \sqrt{1+\beta^2/\alpha_1^2}\sin[2(\Phi-\gamma+\Theta)]\right\} , \tag{2.22}$$

where $tan(2\Theta) = -\alpha_1 I\rho$. From this expression we immediately obtain that positive values of $\mu_1(\tau)$ can only be attained when

$$\beta > \sqrt{3}\alpha_1 \tag{2.23}$$

Hence, a necessary condition for the chaos onset is that the perimeter of nonlinearity β should exceed a certain threshold determined by the value of the nonlinear dissipation. Note that this condition coincides with the condition (2.17) of the saddle state arising under the action of a harmonic perturbation. According to the expressions (2.21) (2.23), this condition does not depend upon the form and parameters of the external action. The inequality which is the inverse of (2.23) is the sufficient condition for the stability of the oscillator.

A second condition, which is necessary for the onset of chaos, is found from Eq. (2.22) with respect to the amplitude $a(\tau)$. The function $\mu_1(\tau)$ can take positive values in finite time intervals only in the case when

$$\max\{a(\tau)\} > a_{\text{th}} \equiv \left[\frac{\alpha_0}{(\alpha_1^2+\beta^2)^{1\ 2} - 2\alpha_1}\right]^{1\ 2} \tag{2.24}$$

that is, the onset of chaotic oscillations is a threshold effect with respect to the amplitude of the forced oscillations. Condition (2.24) enables one to obtain an estimate for the smallest value of the amplitude of the spectral components of an external action, starting from which the development of stochastic instability is possible. In order to do this, we make use of the fact that the greatest amplitude of the excited oscillation of the harmonically forced oscillator does not exceed a value of $p_1 I \alpha_0$ and is identical to it when α_1. Let us assume that the amplitude of the second spectral component of the external action is relatively small which means that we may also adopt the above-mentioned value $p_1 I \alpha$ as an estimate of $\max\{a(\tau)\}$ in this case. Summing up what has been said, it is possible to write the following approximate condition for the onset of chaos

$$p_1 > p_{1\text{th}} \equiv \alpha_0 \left[\frac{\alpha_0}{(\alpha_1^2 + \beta^2)^{1\ 2} - 2\alpha_1} \right]^{1\ 2} \tag{2.25}$$

When the nonlinear dissipation is small, then instead of (2.25) we have

$$p_1 > p_{1\text{th}} \equiv \alpha_0 \left[\alpha_0 I \beta \right]^{1\ 2} \tag{2.26}$$

This condition is identical to relationship (2.18a) apart from a factor which is close to unity. Thus, the condition for the chaos onset (2.26) is equivalent to the onset of saddle states during a harmonic action on the oscillator.

When the inequalities (2.23), (2.24) are simultaneously satisfied, then to obtain positive values of $\mu_1(\tau)$ it is that a certain "phase" condition should be satisfied, which according to Eq. (2.22), is

$$\sin\left[2(\Phi - \gamma + \Theta)\right] > 0 \ ,$$

which corresponds to the orientation of the vector in the tangent space in directions where its stretching occurs. The results of the paper [25] indicate that this condition takes place due to a parametric instability of the solution of the linearized equations (2.21), and it is typical mechanism of the chaos onset in passive single–degree–of–freedom system with external excitation.

2.3.3 Mapping

Under some reasonable assumptions the distraction of quasiperiodic oscillations in weakly nonlinear oscillator can be reduced to the study of a discrete map [26,27]. To obtain the map for the equation (2.1), let us set the number of the spectral component of the external forcing equal to an infinity instead of the two ones in the following way

$$A_1 \cos \omega_1 t + A_2 \cos \omega_2 t \to \sum_{n=-\infty}^{n=\infty} A_n \cos(\omega_n t) \ . \tag{2.27}$$

where $\omega_n = \omega_1 + n\omega$, $\omega = \omega_2 - \omega_1$. This new expression contains two incommensurable frequencies and, due to it, holds main properties of the previous one.

When employ such substitution, we also use the well-known fact that the most significant influence on the behavior of the weakly nonlinear oscillator is exerted by the harmonic components of the external force with frequencies close to the natural frequency of the oscillator. One can expect that when $\omega > \varepsilon\omega_0$, then the substitution will be almost equivalent. Assuming that all the amplitudes A_2 are equal to same value A_0 and using the following representation for the Dirac δ–function,

$$\sum_{n=-\infty}^{n=\infty} \cos(n\Omega\tau) = T \sum_{k=-\infty}^{k=\infty} \delta(\tau - kT) \ ,$$

where $T = 2\pi/\Omega$, the system (2.10) is replaced by the following one[26]

$$\frac{dU}{d\tau} = -\left[\alpha_0 + \alpha_1 (U^2 + V^2)\right] U - \left[\Delta + \beta (U^2 + V^2)\right] V$$

$$\frac{dV}{d\tau} = -\left[\alpha_0 + \alpha_1 (U^2 + V^2)\right] V - + \left[\Delta + \beta (U^2 + V^2)\right] U + \qquad (2.28)$$

$$+ A_0 T \sum_{k=-\infty}^{k=\infty} \delta(\tau - kT)$$

We come to the system of equations with an infinite sequence of the δ–pulses. Now let us use the fact that during the time intervals between the pulses, equations (2.28) become autonomous. This allows one to write the solution of this system by connecting the solutions of the corresponding autonomous equations between neighboring δ–pulses. The result of that is the discrete mat which for $\alpha_1 = 0$ takes the form

$$\hbar z = z \exp\left[-\alpha_0 T + \left(i - \Delta T - E\mu\beta T|z|^2\right)\right] + A_0 T \qquad (2.29)$$

Here $z = V(kT+0) + iU(kT+0)$ and $\hbar z = V(kT+T+0) + iU(kT+T+0)$ are the complex variables representing the envelope of the oscillation at the time moments corresponding to two consequent, k-th and k+1-th, δ–pulses and $\mu = [1 - \exp(-2\alpha_0 T)]/(2\alpha_0 T)$.

Fixed and periodic points of the map (2.29) correspond to periodic orbits in the phase space of the averaged equations (2.10) and, therefore, to two-dimensional tori of the initial equation. Thus, the problem of studying of a quasiperiodic motion of the weakly nonlinear oscillator is essentially simplified.

Each fixed point Z of the map on the complex phase plane z satisfies the equation $\hbar z(Z) = Z$ which can be transformed to the following one

$$2 \exp(-\alpha_0 T) \left[\cosh(\alpha_0 T) - \cos(\Delta T + \mu\beta T I)\right] I = A_0 T^2 \qquad (2.30)$$

where $I = |Z|^2$ is the intensity of the forced oscillations. Analyzing the stability of the fixed points, we obtain the following equation for the boundaries of the instability regions

$$\cos(\Delta T + \mu\beta T I) + \mu\beta T I \sin(\Delta T + \mu\beta T I) = -\cosh(\alpha_0 T) \ . \qquad (2.31)$$

Here sign (+) corresponds to the tangent (or saddle-node) bifurcation when one of the characteristic multipliers of the fixed point becomes equal to 1. This curve surrounds the regions of the existing of an induced saddle fixed point. Sign (–) in Eq. 2.31 corresponds to the period- doubling bifurcation occurring when one of the characteristic multipliers becomes equal to –1. When the response curve of the oscillator (2.30) crosses over the latter region, the corresponding fixed point of the map undergoes the period-doubling bifurcation.

The boundaries of the chaos onset in a parameter plane are usually not far from that of period doubling. It allows to estimate the condition for chaos by using Eq. 2.31. Let us suppose that $\exp(-\alpha_0 T) << 1$, then from Eqs. (2.30),(2.31) one can obtain the following condition for the chaos onset

$$A_0^2 \beta T^2 e^{-\alpha_0 T} \alpha_0^{-1} >> 1 \tag{2.32}$$

The left-hand side of this condition achieves its maximal value when

$$\alpha_0 T = 2 \tag{2.33}$$

It means that the most favorable conditions for the arising of the chaotic motion appear in the case when the relaxation time of the system has the same order as the period T. If the condition (2.33) is hold, then from Eq. (2.32) we find the expression for the threshold value of the amplitude of the external force

$$A_{0\mathrm{thr}} \cong 1.35 \alpha_0^{\frac{3}{2}} \beta^{-1/2}$$

which coincides (with accuracy up to a numerical multiplier) with the same value found above (see Eq. (2.26)).

3 Chaotic States of Nonlinear Oscillators with Parametric Excitation

The phenomena of parametric generation, amplification of oscillations and frequency conversion are distinguishing features of the dynamics of a variety of physical systems. For the non autonomous single–degree–of–freedom systems which are investigated in this paper, these phenomena can be adequately described within the framework of a universal mathematical model, like the motion equation of the following nonlinear oscillator:

$$\ddot{x} + \left(\delta_0 + \delta_1 x^2\right) \dot{x} - \gamma x^3 + \omega_0^2 \left[1 - M \cos\left(\omega_p t\right)\right] x = A \cos\left(\omega_s t\right) \quad , \tag{3.1}$$

subjected to combined parametric and external forcing. Here x is a generalized coordinate of the oscillator, $\delta_0 > 0$ and $\delta_1 > 0$ are the coefficients of linear and nonlinear damping, γ is the nonlinearity parameter, ω_0 is the natural frequency of the oscillator, M and A are the amplitudes of parametric and external forcing with the incommensurate frequencies ω_p and ω_s. For example, in the case of optical and microwave parametric amplifiers, M and A are respectively proportional to the amplitude of a pumping oscillation and a signal wave to be amplified.

The unperturbed oscillator (3.1) possesses the same properties as that studied in the previous part of the paper (see (2.1)) with the same original saddle points and corresponding heteroclinic orbit (see Fig.1). The condition of the homoclinic structure formation on the base of this orbit can be found by using the Melnikov method. The final result reads [7]

$$A\omega_s \operatorname{cosech}\left(\frac{\pi\omega}{\sqrt{2}\omega_0}\right)+\frac{M\omega_0^2\omega_p^2}{\sqrt{2\varepsilon\gamma}}\operatorname{cosech}\left(\frac{\pi\omega}{\sqrt{2}\omega_0}\right)\geq\left|\frac{2\omega_0^3}{3\pi\sqrt{\varepsilon\gamma}}\left(\delta_0+\frac{\delta_1\omega_0^2}{5\varepsilon\gamma}\right)\right| \quad (3.2)$$

One can see from this equation that the chaotic oscillations can arise when there is only parametric excitation (A=0) or only external one (M=0). Under the combined excitation ($A \neq 0$ and $M \neq 0$), the threshold for chaos to arise is not changed essentially. It is easy to check that the formation of a homoclinic structure can take place only in the oscillator with a strong nonlinearity. In the weakly nonlinear limit ($\varepsilon \to 0$) this criterion does not predict the chaos onset. However, the chaotic states are also typical for the parametrically forced oscillator (3.1) under weakly nonlinear excitation conditions. The mechanism of the chaos arising is determined by induced saddle states, and by the interaction of the parametric and external resonances. In this part of the paper we describe general conditions of the transition to chaos and study the properties of chaotic and regular states of the system (3.1). To carry out this plan we divide the problem into two parts. The first one is devoted to the investigation of the induced saddle states in the oscillator. We consider the case when the amplitude of the external forcing p is relatively small and the induced saddle states can arise mainly due to the parametric excitation. Thus, this problem is reduced to the study of the nonlinear parametric resonance (see Section 3.1). This case is of prime interest for many practical situations, for example, for microwave and optical parametric amplifiers where amplitude of the pumping wave ($\propto m$) is much more greater than that of signal wave ($\propto p$). The opposite case was considered by Yagasaki [28]. The second part of the problem deals with the formation of homoclinic structures on the base of the induced saddle states, and with phenomena pertaining to the chaotic states arising (see Section 3.2).

3.1 Nonlinear Parametric Resonance

Hereinafter we shall study the condition of the chaotic states arising in the weakly nonlinear oscillators, interesting in the case of the main parametric resonance when

$$|\omega-\omega_0| = O(\varepsilon\omega_0), \quad \text{and} \quad |\omega_s-\omega_0| = O(\varepsilon\omega_0) \quad (3.3)$$

where $\omega \equiv \omega_p/2$. Then, by using the transformation

$$\begin{aligned} x &= U\cos\omega t + V\sin\omega t \\ y &= -U\omega\sin\omega t + V\omega\cos\omega t\ , \end{aligned} \quad (3.4)$$

and neglecting the terms $O(\varepsilon^2)$, we come to the following system of averaging equations for the slowly varying time functions $U(\tau)$, $V(\tau)$:

$$\frac{dU}{d\tau} = -\left[\alpha_0 + \alpha_1 \left(U^2 + V^2\right)\right] U - \left[\Delta - m + \beta \left(U^2 + V^2\right)\right] V - p \sin \Omega\tau$$
$$\frac{dV}{d\tau} = -\left[\alpha_0 + \alpha_1 \left(U^2 + V^2\right)\right] V + \left[\Delta - m + \beta \left(U^2 + V^2\right)\right] U - p \cos \Omega\tau \quad (3.5)$$

where dimensionless parameters are introduced: $\tau = \varepsilon\omega_p t/2$, $\alpha_0 = \delta_0/\omega_p$, $\alpha_1 = \delta_1/(4\omega_p)$, $m = M/4$, $p = 2A/\omega_p^2$, $\beta = 3\gamma/(2\omega_p^2) > 0$, $\Delta = (\omega_p - 2\omega_0)/(\varepsilon\omega_p)$ and $\Omega = (\omega_p - 2\omega_s)/(\varepsilon\omega_p)$.

Let us consider the conditions of the induced saddle states arising under the influence of the parametric modulation. Assuming that the dissipation and the external forcing are absent, we have instead of (3.5) the following Hamiltonian system:

$$\frac{dU}{d\tau} = -\left[\Delta - m + \beta \left(U^2 + V^2\right)\right] V$$
$$\frac{dV}{d\tau} = \left[\Delta + m + \beta \left(U^2 + V^2\right)\right] U \quad (3.6)$$

with Hamiltonian energy given by

$$H(U,V) = -\frac{\beta}{4}\left(U^2 + V^2\right)^2 - \frac{1}{2}\left[(\Delta - m) V^2 + (\Delta + m) U^2\right] \ . \quad (3.7)$$

As long as $m = 0$, there is only a center-type singular point in the origin of the coordinate. For $m \neq 0$ one can find from (3.6) that the center is split into three or five singular points depending upon the relation between m and Δ. The two possible situations are illustrated by the phase portraits of the system (3.6) in Fig. 2. Note that these phase portraits constitute Poincare maps of the initial system for the case under consideration: $\delta_0 = \delta_1 = A = 0$, and small values of the amplitude of the forced oscillations. When $|\Delta| < m$, there are three singular points: a saddle point in the origin of the coordinates ($U = 0$, $V = 0$) and two centers with coordinates $(U,V) = (0, \pm\sqrt{(m - \Delta)/\beta})$ which are denoted in Fig. 2a as C_1, C_2. In the second case, when $|\Delta| > m$, there are three centers C_1, C_2, C_3 and two saddle points with coordinates $U = \mp\sqrt{(-m - \Delta)/\beta}$, $V = 0$ in the phase space as shown in Fig. 2b. Because of these saddle points, the system possesses induced homoclinic or heteroclinic orbits (separatrix) indicated in Fig. 2. Owing to the induced centers, the system acquires a multistability property when $\alpha_0 \neq 0$ (see below).

Let us consider in more detail the structure of the separatrix loops for the two cases. For $|\Delta| < m$, there is a double symmetric saddle loop. The solution of the motion equation $U \equiv U_l(\tau)$ and $V_l \equiv V_l(\tau)$ on this separatrix was found in Ref.[29] and can be written in the form:

$$U_l^2(\tau) = \frac{4 b_l^2 b_2 \left(1 - \cosh\left[\alpha_1(\tau - \tau_0)\right]\right)}{\left((b_1 - b_2)\cosh\left[\alpha_1 (\tau - \tau_0)\right] - b_1 - b_2\right)^2} \ ,$$
$$V_l^2(\tau) = \frac{4 b_l^2 b_2 \left(1 + \cosh\left[\alpha_1(\tau - \tau_0)\right]\right)}{\left((b_1 - b_2)\cosh\left[\alpha_1 (\tau - \tau_0)\right] - b_1 - b_2\right)^2} \ , \quad (3.8)$$

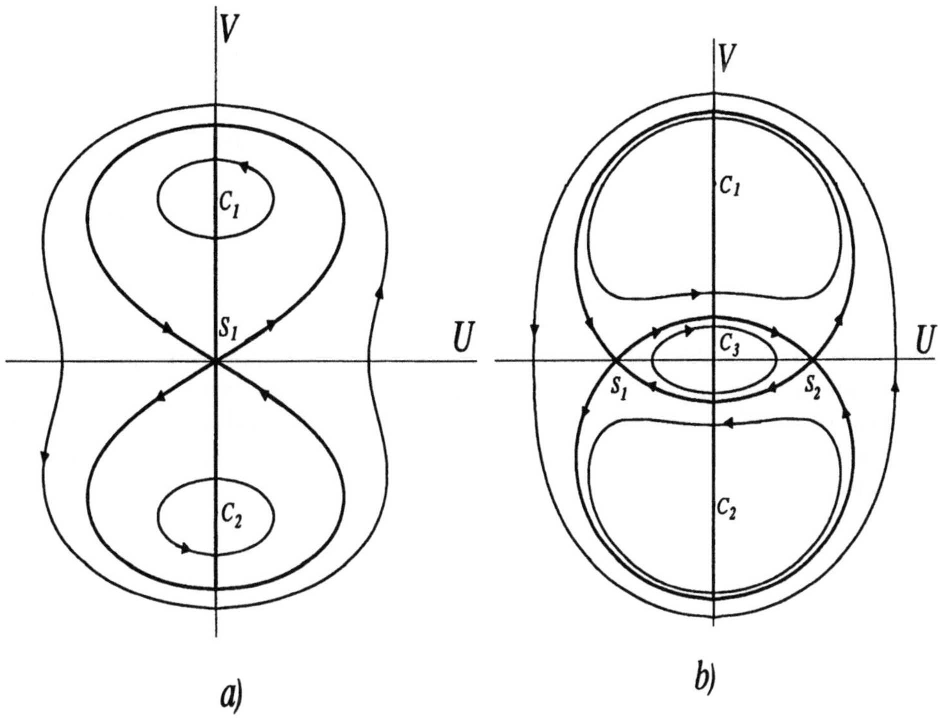

Fig. 2. Phase portraits of the Hamiltonian system (3.6) for: (a) $|\Delta| < m$ and (b) $|\Delta| > m$.

where $b_{1,2} = (\pm m - \Delta)/\beta$, $\alpha_1 = 2\beta\sqrt{-b_1 b_2}$, τ_0 is an initial moment of the "slow" time

In the second case shown in Fig. 2b, because of the two saddle points, double heteroclinic loops arise: the small and large ones. After intensive calculations one can find the following equation of these loops:

$$U_l(\tau) = -\frac{\sqrt{b_2}(b_1+b_2)\sinh[\alpha_2(\tau-\tau_0)]}{(b_1+b_2)\cosh[\alpha_2(\tau-\tau_0)] - q_\pm(b_1-b_2)}\,,$$
$$V_l(\tau) = -\frac{\sqrt{2(b_1-b_2)}b_2 q_\pm}{(b_1+b_2)\cosh[\alpha_2(\tau-\tau_0)] - q_\pm(b_1-b_2)}\,. \tag{3.9}$$

where $\alpha_2 = \beta\sqrt{2b_2(b_1-b_2)}$, $q_\pm = \pm\sqrt{(b_1+b_2)/(b_1-b_2)}$, q_+ and q_- correspond to the large and small loops of the separatrix, accordingly. The regions in the parameter space where the saddle states exist are reduced under the influence of dissipation. However, for any value of the linear ($\alpha_0 < \infty$) and nonlinear ($\alpha_1 < \infty$) dissipation these regions have finite dimensions. Let us show it by introducing into the consideration the corresponding dissipative terms into the system (3.6), thus we have

$$\frac{dU}{d\tau} = -\left[\alpha_0 + \alpha_1 \left(U^2 + V^2\right)\right] U - \left[\Delta - m + \beta \left(U^2 + V^2\right)\right] V \ ,$$
$$\frac{dV}{d\tau} = -\left[\alpha_0 + \alpha_1 \left(U^2 + V^2\right)\right] V + \left[\Delta + m + \beta \left(U^2 + V^2\right)\right] U \ . \tag{3.10}$$

This system yields the following equation for the equilibrium states.

$$\Delta = -\beta W^2 \pm \sqrt{m^2 - \left(\alpha_0 + \alpha_1 W^2\right)^2} \ , \tag{3.11}$$

where $W^2 = U^2 + V^2$ and W is the amplitude of oscillation. Note that $W = 0$ is also an equilibrium state. The amplitudes of stationary states versus Δ (the response curve of the oscillator) are shown in Fig. 3. The saddle states are marked by dashed curve. It is easy to check that the saddle states with W=0 exist in the following parameter region:

$$\alpha_0 < m \ , \tag{3.12}$$

$$|\Delta| < \sqrt{m^2 - \alpha_0^2} \ , \tag{3.13}$$

whereas such states with $W \neq 0$ exist when

$$\frac{\beta\alpha_0 - m\sqrt{\alpha_1^2 + \beta^2}}{\alpha_1} < \Delta < -\sqrt{m^2 - \alpha_0^2} \ , \tag{3.14}$$

provided that the amplitude of modulation m satisfy simultaneously the condition (3.12) and the following one

$$\beta\alpha_0 - m\sqrt{\alpha_1^2 + \beta^2} < -\alpha_1\sqrt{m^2 - \alpha_0^2} \ . \tag{3.15}$$

The above given conditions (3.12)-(3.15) determine the regions in the parameter space where Melnikov's method can be applied to the system (3.5). These conditions should be considered as additional ones to Melnikov's criterion of the chaos arising that will be obtained in the next sections. They determine the threshold of the chaos arising with respect to the amplitude of the parametric force and the resonant condition for the chaos to exist.

3.2 Chaotic States of the Oscillator with Induced Heteroclinic Orbits

In this section we illustrate the mechanism of the transition to chaos for the oscillator with the heteroclinic orbits associated with the pair of saddle points as shown in Fig. 2b. The second case was considered in Ref. [29]. We start from the application of Melnikov's method to the averaged equations (3.5) to find necessary conditions for the chaos onset. Performing the corresponding calculation

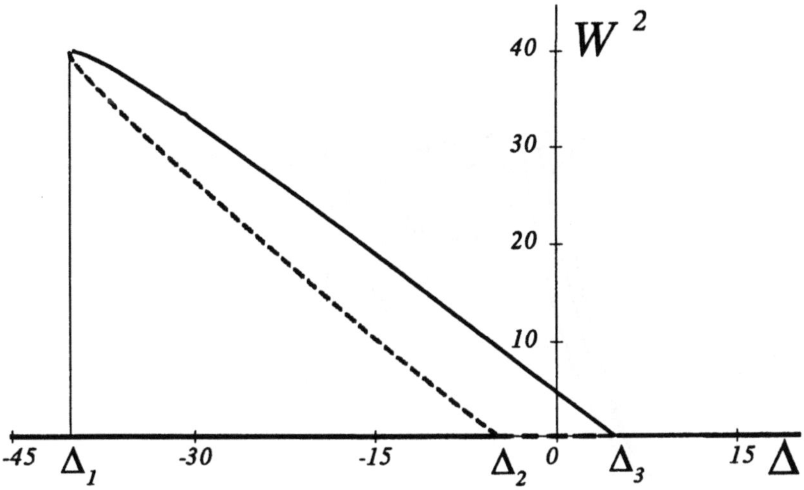

Fig. 3. Response curve of the system (15) for $\alpha_0 = 1$, $\alpha_1 = 0.1$, $\beta = 1$, $m = 5$. Solid curves correspond to stable states and dashed ones - saddle states. The boundaries of the regions with different types of the system behavior are $\Delta_1 = (\beta\alpha_0 - m\sqrt{\alpha_1^2 + \beta^2})/\alpha_1$, $\Delta_{2,3} = \mp\sqrt{m^2 - \alpha_0^2}$.

in the way similar to that described in the Section 2.3, we come to the following condition of the manifolds' intersections:

$$p \geq \left| \frac{2\sqrt{m/\beta}\sinh(\pi\sigma_2)}{\pi\Omega\left(\pm\sigma_2 \arccos\left(\mp\sqrt{-\frac{m}{\Delta}}\right)\right)} \times \right.$$
$$\times \left\{ \alpha_0 \left[r_2 \mp \Delta \arccos\left(\mp\sqrt{-\frac{m}{\Delta}}\right)\right] + \right. \tag{3.16}$$
$$\left. \left. - \frac{\alpha_1}{\beta}\left[3r_2\Delta \pm (2m\Delta - \Delta^2)\arccos\left(\mp\sqrt{-\frac{m}{\Delta}}\right)\right]\right\}\right|$$

Here $r_2 = \sqrt{-m(\Delta + m)}$, $\sigma_2 = \Omega/(2r_2)$.

A typical example of the bifurcation diagram in the parameter plane (p, Ω) obtained numerically along with Melnikov's criterion (curve 1) are given in Fig. 4. The best way to explain all possible behavior of the system is to start from the Hamiltonian case, i.e., Fig. 2b, and then put $\alpha_0 \neq 0$. We obtain three stable foci instead of three centers, and all trajectories will eventually come to one of them, depending on initial conditions. The attractors which arise in the vicinity of the centers C_1, C_2, C_3 we shall mark as A_1, A_2, A_3 accordingly. That is, the system is three-stable initially. Provided that the amplitude p is small enough, the foci turn to stable periodic orbits, and the three of them will exist simultaneously in the phase space. The attractor A_3 undergoes the symmetry breaking crises on curve 2, and to the left of this curve four attractors (A_1, A_2, A_3, A_3') coexist

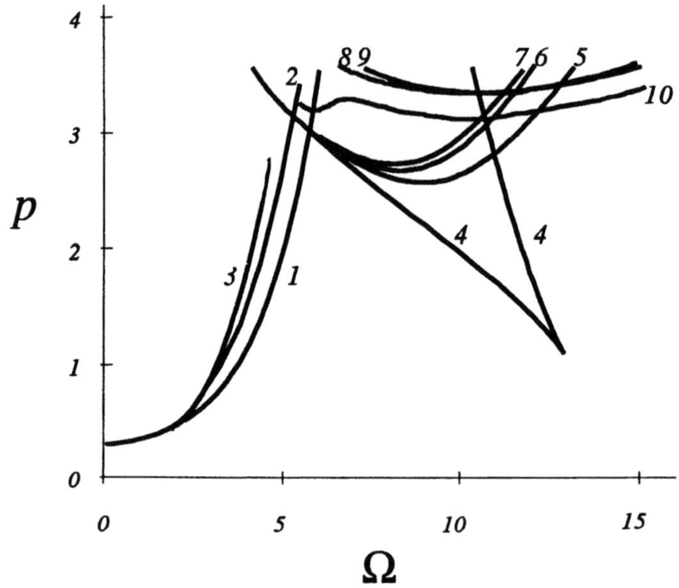

Fig. 4. Bifurcation diagram of the system (10) for and $\alpha_0 = 1$, $\alpha_1 = 0$, $\beta = 1$, $\Delta = -6$, $m = 5$. Curves 1-3 correspond to the attractors formed on the base of the small heteroclinic loop: (1) boundary of homoclinic structure existence according to Melnikov's criterion, (2) - symmetry breaking, (3) - first period doubling bifurcation. Curves 4-6 correspond to the attractors due to the saddle orbit induced by the external force: (4) - tangent bifurcation, (5) and (6) - first and second period doubling bifurcations, (7) - attractors crises. Curves 8, 9 correspond to the attractors formed on the base of the large heteroclinic loop, and represent boundaries of homoclinic structure existence found numerically and with Melnikov's criterion correspondingly. Curve (10) denotes the boundary of the strange attractors arising due to the intersection of stable manifold of the small loop and unstable manifold of the large loop.

in the phase space. At curve 3 both attractors A_3, A_3' undergo the first period doubling bifurcation, and then a pair of strange attractors appears through the period doubling cascade. In Fig. 5a the situation is given, when these two attractors merge and a unified strange attractor arises. It turns out that the strange attractors here are very sensitive to the variation of parameters and exist only in very narrow band adjacent to curve 3. Obviously, their basins of attraction are small and the majority of initial conditions lead to two stable attractors coexisting with them.

In this region of the parameter plane the strange attractors arise due to the manifolds' intersections of the small loop, and other roads to chaos were not observed. In this case Melnikov's criterion works well and good predictions can be developed on its basis.

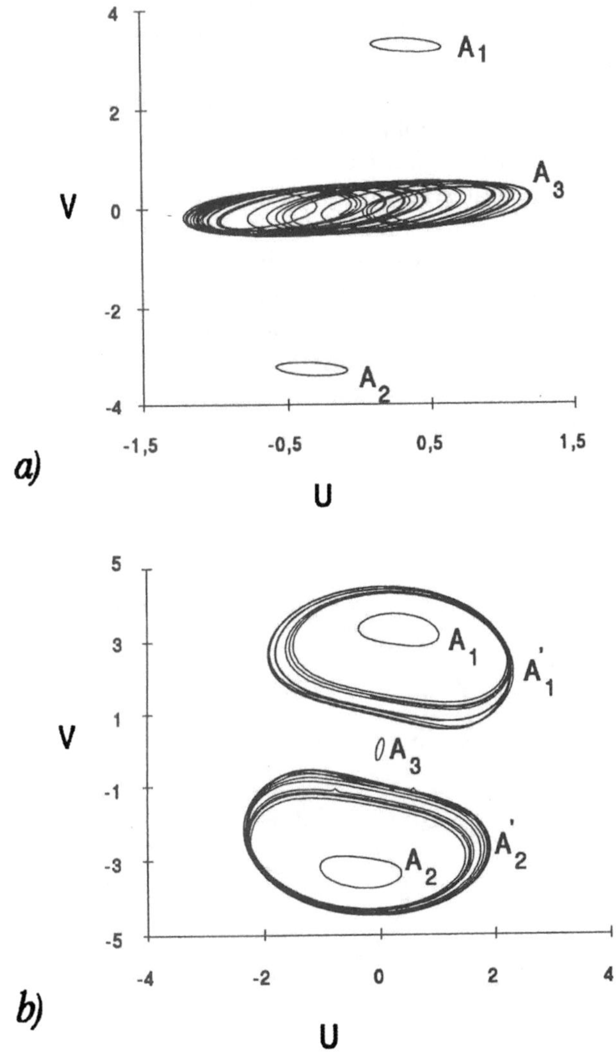

Fig. 5. All coexisting attractors of the system (10) at: (a) $\Omega = 3.95$, $p = 1.9$, (b) $\Omega = 9.2$, $p = 2.8$ for $\alpha_0 = 1$, $\alpha_1 = 0$, $\beta = 1$, $\Delta = -6$, $m = 5$.

Clearly our way of the Melnikov's technique application does not allow to predict the appearance of these strange attractors. However, it can be done by considering the parametric force as a perturbation, rather then the external force.(see, i.e., Ref. [28]).

The intersections of the manifolds of the large loop associated with the parametrically induced saddles $S_{1,2}$ (see Fig. 2b) took place at relatively large values of the external amplitude indicated by curve 8 in Fig. 5. This curve practically coincides with that found from Melnikov's criterion (curve 9). One can conclude that this criterion works here well again. The chaotic oscillations exist in a narrow layer adjacent to these lines, and they arise through the period doubling cascade. We have not detected here any significant influence of the saddle orbits arising after each doubling bifurcation on the system dynamics. Instead, another layer-like chaotic region has been found next to curve 10 with a peculiar kind of the homoclinic structure formation. In this region the chaos onset is due to the intersection of the manifolds of different loops - large and small ones, not some of them. This situation is illustrated in Fig. 6, where the crossings of the unstable manifold of the large loop with the stable manifold of the small loop take place. To our knowledge, such mechanism of the transition to chaos has never been observed. It is interesting to find other system exhibiting this road to case, as well as to develop an analytical tool to detect the presence of such homoclinic structure.

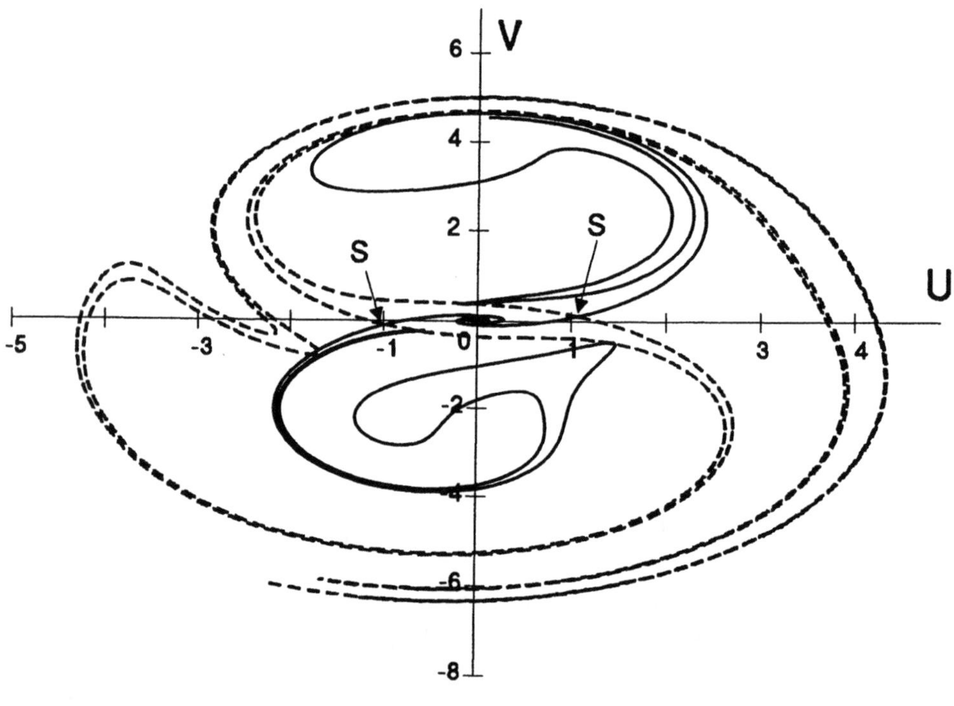

Fig. 6. Intersections of stable (dashed curve) manifold of the small loop and unstable (solid curve) manifold of the large loop in the Poincare section for $p = 3.2$, $\Omega = 12$, $\alpha_0 = 1$, $\alpha_1 = 0$, $\beta = 1$, $\Delta = -6$, $m = 5$.

4 Discussions

In this section, we have considered the conditions of the chaotic states arising in the weakly nonlinear Duffing-type oscillators subjected to the combined parametric and external forcing. It was shown that homoclinic or heteroclinic orbits induced by one of the external harmonic component play crucial role in the oscillator dynamics. Because of these orbits the formation of a homoclinic structure took place here in the weakly nonlinear limit under the action of the another incommensurate frequency, and the chaos onset is due to the distraction of two-dimensional torus. The generality of these results does not depend on the perturbations being considered, and they most probably hold for a variety of single–degree–of–freedom systems with quasiperiodic forcing. We have also detailed some of phenomena which accompany the chaos onset. Multistability of the oscillator conditioned by the formation of several attracting sets in the phase space is one of such phenomenon. This study along with the foregoing ones [22, 30] suggests that the multistability is a typical feature of the quasiperiodically forced weakly nonlinear oscillators. It has been shown in this paper that the splitting of a center-type singular point into several stable and unstable orbits under the action of the periodic force is primarily responsible for the multistability property of the oscillator, at least for relatively small values of the force amplitude. This study along with the previous results indicates that there are three typical roads of strange attractors arising depending upon the mechanism of the homoclinic structure formation. Let us cite them as they are seen in term of the averaged equations. The first road is through the intersection of the stable and unstable manifolds of some parametrically or externally induced homoclinic or heteroclinic loop. The second road described in Ref. [7] is through the manifolds' intersections of an additional saddle orbit arising after one of the period doubling bifurcation. The third one is through the intersection of the stable and unstable manifolds associated with different loops (see Fig. 7)

An important motivation for this work were results of experimental investigations of microwave parametric amplifiers which indicate that the amplifiers that stable under the action of only pumping oscillation or signal wave loose their stability when this oscillation and the signal wave are applied simultaneously [31]. The results of this paper give an explanation of such phenomenon and provide mathematical tools for its study and prediction. It is also obviously that our results are applicable, within certain limits, to other types of similar devices, say, Josephson- junction parametric amplifiers, optical amplifiers, etc. It is worth noting that the factors which are responsible for the chaos onset in the weakly nonlinear limit and, hence, for the low stability threshold of the parametric amplifiers are precisely the same as that providing low noise amplification from the point of view of the conventional theory of parametric devices. Indeed, according to the classical results of this theory, parametric amplifiers posses a low noise output level because they consist of a reactive circuit (anharmonic oscillator) and they utilize an ac power supply (pumping oscillation). In the mathematical model used the reactive type of nonlinearity is described by the term with parameter γ (see (3.1)). Proceeding from the obtained results, it is clear that exactly the combination of this type of nonlinearity and the parametric excitation leads to the possibility of the chaos onset in the weakly nonlinear limit when the external signal is applied. This is the main reason why the parametric amplifiers are extremely susceptible to the chaotic instabilities. These findings work also for Josephson-junction parametric amplifiers, SQUIDs and other types of similar devices [32-36].

Acknowledgments

The author wishes to thank O.A. Tretyakov, A.B. Belogortsev, I.Yu. Chernyshov, V.B. Ryabov, S.A. Sharapov, and Yu.A. Tsarin for useful discussions and collaborations. This work was partly supported by the International Science Foundation (grant # U33000).

References

1. J. Miles, Proc. Nat. Acad. Sci., Phys. Sci. USA, **81**, 3919–3923 (1984).
2. A.B. Belogortsev, D.M. Vavriv and O.A. Tretyakov, Sov. Phys. JETP, **65**, 737–740 (1987).
3. A.B. Belogortsev, D.M. Vavriv and O.A. Tretyakov, Sov. Phys. Tech. Phys. **33**, 174–179 (1988).
4. K. Yagasaki, M. Sakata, and K. Kimura, ASME J. of Appl. Mech. **57**, 209 (1990).
5. K. Yagasaki, ASME Journal of Appl. Mech., **59**, 161–167 (1992).
6. A.B. Belogortsev, D.M. Vavriv, and O.A. Tretyakov, Appl. Mech. Rev., **46**, 372–384 (1993).
7. D.M. Vavriv, V.B. Ryabov, and S.A. Sharapov, Phys. Rev. E. (pending).
8. D.M. Vavriv and I.Yu. Chernyshov, Radiotekhnika i Elektronika, **35**, 151–158 (1990).
9. A.B. Belogortsev, M. Poliashenko, O.A. Tretyakov, and D.M. Vavriv, Electronic Lett., **26**, 1354–1355.
10. A.B. Belogortsev, D.M. Vavriv and O.A. Tretyakov, Radiotekhnika i Elektronika, **35**, 1300–1307 (1990)
11. V.K. Melnikov, Trans. Moscow Math. Soc. **12**, 3–57 (1963).
12. A.D. Morosov, Diff. Equations **12**, 164–174 (1976).
13. P. Holmes, Philos. Trans. R. Soc. **A292**, 419–448 (1979).
14. T. Kapitaniak, J. Sound Vib., **121**, 259–265 (1988).
15. K. Yagasaki, SIAM J. Math. Anal., **23**, 1230–1254 (1992).
16. J.M. Guckenheimer and P. Holmes, Nonlinear oscillations, dynamical systems and bifurcations of vector fields (Springer, Berlin, 1983).
17. K. Ide and S. Wiggins, Physica **D34**, 169 (1989).
18. S. Wiggins, Phys. Lett. **A124**, 138–142 (1987).
19. A.B. Belogortsev, D.M. Vavriv, and B.A. Kalugin, Zh. Tekh. Fiz. **57**, 559 (1987). [Sov. Phys. Tech. Phys. **32**, 337 (1987)].
20. I.Yu. Chernyshov and D.M. Vavriv, in: Proc. Int. Conf. on Noise in physical systems, Ed. A. Ambrozy (Akademiai Kiado, Budapest, 1990) p.651.
21. N.N. Bogoliubov and Yu.A. Mitropolski, Asymptotic methods in the theory of nonlinear oscillations (Gordon & Breach, New York, 1961).
22. V.B. Ryabov and D.M. Vavriv, Phys. Lett. **A153**, 431–436 (1991).

23. D.M. Vavriv, V.B. Ryabov, and S.A. Sharapov, Radiotekhnika i Electronika, **38**, 464–472 (1993).
24. D.M. Vavriv and V.B. Ryabov, Dokl. Akad. Nauk Ukr SSR, **A2**, 50–55 (1990).
25. D.M. Vavriv and V.B. Ryabov, Comput. Maths & Math. Phys., **32**, 1259–1269 (1992).
26. A.B. Belogortsev, Nonlinearity, **5**, 889–897 (1992).
27. A.B. Belogortsev, Phys. Lett., **A161**, 352–356 (1992).
28. K. Yagasaki, ASME J. of Appl. Mech. **58**, 244–256 (1991).
29. D.M. Vavriv, V.B. Ryabov and I.Yu. Chernyshov, Sov. Phys. Tech. Phys., **36**, 1325–1331 (1991).
30. D.M. Vavriv and V.B. Ryabov, Pis'ma v Zh. Tekh. Fiz. **17**, 55–60 (1991).
31. I.Yu. Chernyshov and D.M. Vavriv, Phys. Lett., **A165**, 117–123 (1992).
32. D.M. Vavriv, O.A. Tretyakov, I.Yu. Chernyshov, Pis'ma v Zh. Tekh. Fiz., **14**, 903–908 (1991).
33. D.M. Vavriv and O.A. Tretyakov, Theory of Resonant Amplifiers with Distributed Interaction of O-type (Naukova Dumka Press, Kiev,1987).
34. D.M. Vavriv and A. Oksasoglu, Electronics Lett., **30**, 462–463 (1994).
35. A. Oksasoglu and D.M. Vavriv, IEEE Trans. Circuits Syst. I, **41**, 669–672 (1994).
36. S.A. Bulgakov, V.B. Ryabov, V.I. Shnyrkov, and D.M. Vavriv, J. Low Temp. Phys. **83** (5/6), 241–255 (1991).

Computer Simulation of Lévy α–Stable Variables and Processes

Aleksander Weron *
and
Rafał Weron

The Hugo Steinhaus Center for Stochastic Methods,
Technical University of Wrocław, 50–370 Wrocław, Poland.

Abstract: The aim of this paper is to demonstrate how the appropriate numerical, statistical and computer techniques can be successfully applied to the modeling of some physical systems. We propose to use a fast and accurate method of computer generation of Lévy α–stable random variates.

1 Introduction

The past few years have witnessed an explosive growth in interest in physical and engineering systems that could be studied using stochastic and chaotic methods, see Berliner (1992), Chatterjee and Yilmaz (1992), and Shao and Nikias (1993). "Stochastic" and "chaotic" refer to nature's two paths to unpredictability, or uncertainty. To scientists and engineers the surprise was that chaos (making a very small change in the universe can lead to a very large change at some later time) is unrelated to randomness. Things are unpredictable if you look at the individual events; however, one can say a lot about averaged–out quantities. This is where the stochastic stuff comes in. Stochastic processes are recognized to play an important role in a wide range of problems encountered in mathematics, physics and engineering. Recent developments show that in many practical applications leading to appropriate stochastic models a particular class of Lévy α–stable processes is involved. While the attempt at mathematical understanding of these processes leads to severe analytical difficulties, there exist very useful approximate numerical and statistical techniques (see Janicki and Weron (1994b)). Also non–Gaussian statistical methods in stochastic modeling are important when noises deviate from the ideal Gaussian model. Stable distributions are among the most important non–Gaussian models. They share defining characteristics with the Gaussian distribution, such as the stability property and central limit theorems, and include in fact the Gaussian distributions as a special case. To help the interested reader better understand the stable models and necessary

* Research of the first author was partially supported by the KBN Grant No.2 P302 065 04.

methodologies we discuss here tutorial examples of α–stable Ornstein–Uhlenbeck process and continuous–time random walks subject to α–stable displacement.

2 Computer Generation of Lévy α–Stable Distributions

The most common and convenient way to introduce Lévy α–stable random variables is to define their *characteristic function*

$$\log \phi(t) = \begin{cases} -\sigma^\alpha |t|^\alpha \{1 - i\beta sign(t) \tan \frac{\pi\alpha}{2}\} + i\mu t, & \alpha \neq 1, \\ -\sigma|t|\{1 + i\beta sign(t) \frac{2}{\pi} \log |t|\} + i\mu t, & \alpha = 1, \end{cases} \tag{2.1}$$

where $\alpha \in (0,2]$, $\beta \in [-1,1]$, $\sigma > 0$, $\mu \in R$.

Since (2.1) is characterized by four parameters we will denote α–stable distributions by $S_\alpha(\sigma, \beta, \mu)$ and write

$$X \sim S_\alpha(\sigma, \beta, \mu) \tag{2.2}$$

to indicate that X has the stable distribution $S_\alpha(\sigma, \beta, \mu)$. When $\sigma = 1$ and $\mu = 0$ the distribution is called *standard stable*.

The *location parameter* μ shifts the distribution to the left or right. The *scale parameter* σ compresses or extends the distribution about μ. Some authors use $\gamma = \sigma^\alpha$ instead of σ, but it no longer has the natural interpretation of the scale parameter. The *characteristic exponent* α lies in the range $(0,2]$ and determines the rate at which the tails of the distribution taper off. When $\alpha = 2$, a Gaussian distribution results, with mean μ and variance $2\sigma^2$. When $\alpha < 2$, the variance is infinite. When $\alpha > 1$, the mean of the distribution exists and is equal to μ. In general, the p–th moment of a stable random variable is finite iff $p < \alpha$. The fourth parameter, which determines the *skewness* of the distribution, is designated β and lies in the range $[-1,1]$. When β is positive, the distribution is skewed to the right. When it is negative, it is skewed to the left. When $\beta = 0$, the distribution is symmetrical. As α approaches 2, β loses its effect and the distribution approaches the symmetrical Gaussian distribution regardless of β.

The complexity of the problem of simulation of sequences of stable random variables results from the fact that there are no analytic expressions for the inverse F^{-1} of the distribution function. The only exceptions are the Gaussian $S_2(\sigma, 0, \mu) = N(\mu, 2\sigma^2)$, Cauchy $S_1(\sigma, 0, \mu)$ and Lévy $S_{1/2}(\sigma, 1, \mu), S_{1/2}(\sigma, -1, \mu)$ distributions, for which simple methods of simulation have been found.

Even recently new ways for simulating stable random variables are sought for in physical literature. For example, Mantegna (1994) proposes a "fast and accurate" algorithm for the symmetric case ($\beta = 0, \mu = 0$) based on Bergstrom series expansion. This algorithm may be fast but there are no proofs of its accuracy, except visual similarities of densities. Moreover, arguments used in the paper are not reliable since the author mixes the basic concepts of probability theory like density, distribution and stochastic process.

We propose to use a well known in mathematical literature method of computer generation of a symmetric α–stable random variable $X \sim S_\alpha(1,0,0)$. For $\alpha \in (0,2]$

– generate a random variable V uniformly distributed on $(-\frac{\pi}{2}, \frac{\pi}{2})$ and an independent exponential random variable W with mean 1;
– compute

$$X = \frac{\sin(\alpha V)}{(\cos(V))^{1/\alpha}} \times \left(\frac{\cos(V - \alpha V)}{W}\right)^{(1-\alpha)/\alpha}. \tag{2.3}$$

In a similar way to (2.3), reparametrizing the algorithm of Chambers et al. (1976) or using the theorem of Weron (1995a), we can construct a method of computer generation of a skewed random variable $X \sim S_\alpha(1, \beta, 0)$. For $\alpha \in (0, 2]$ and $\beta \in [-1, 1]$

– generate a random variable V uniformly distributed on $(-\frac{\pi}{2}, \frac{\pi}{2})$ and an independent exponential random variable W with mean 1;
– for $\alpha \neq 1$ compute

$$X = S_{\alpha,\beta} \times \frac{\sin(\alpha(V + B_{\alpha,\beta}))}{(\cos(V))^{1/\alpha}} \times \left(\frac{\cos(V - \alpha(V + B_{\alpha,\beta}))}{W}\right)^{(1-\alpha)/\alpha}, \tag{2.4}$$

where

$$B_{\alpha,\beta} = \frac{\arctan(\beta \tan \frac{\pi\alpha}{2})}{\alpha},$$

$$S_{\alpha,\beta} = \left[1 + \beta^2 \tan^2 \frac{\pi\alpha}{2}\right]^{1/(2\alpha)};$$

– for $\alpha = 1$ compute

$$X = \frac{2}{\pi}\left[(\frac{\pi}{2} + \beta V)\tan V - \beta \log\left(\frac{W \cos V}{\frac{\pi}{2} + \beta V}\right)\right]. \tag{2.5}$$

Formula (2.4) was initially presented by Janicki and Weron (1994a). However, there is a misprint in the form for $C_{\alpha,\beta}$ (the denominator is $1-|1-\alpha|$ instead of α, Formula (3.5.2), page 50), which corresponds to our $B_{\alpha,\beta}$, and a computationally more complicated form for $D_{\alpha,\beta}$ (our $S_{\alpha,\beta}$).

We have given formulas for simulation of standard stable random variables. Using the following property, which follows from the form of the characteristic function, we can simulate a stable random variable for all admissable values of the parameters α, σ, β and μ:

If $X \sim S_\alpha(1, \beta, 0)$ then

$$Y = \begin{cases} \sigma X + \mu, & \alpha \neq 1, \\ \sigma X + \frac{2}{\pi}\beta\sigma \log \sigma + \mu, & \alpha = 1, \end{cases}$$

is $S_\alpha(\sigma, \beta, \mu)$.

We regard the methods defined by (2.3), (2.4) and (2.5) as good techniques of computer generation of α–stable random variables, stochastic measures and processes of different kinds. For more details see Janicki and Weron (1994a) and Weron (1995a).

3 Simulation of Lévy α–Stable Stochastic Processes

Now we describe rather general technique of approximate computer simulation of univariate α–stable stochastic processes $\{X(t) : t \in [0,T]\}$ with independent increments, which is based on a construction of a discrete time process of the form $\{X^{\tau}_{t_i}\}_{i=0}^{I}$, defined by the formula

$$X^{\tau}_{t_i} = X^{\tau}_{t_{i-1}} + \mathcal{F}(t_{i-1}, X^{\tau}_{t_{i-1}}) + Y^{\tau}_i, \tag{3.1}$$

with a given X^{τ}_0, and where $Y^{\tau}_i{}'$s form a sequence of i.i.d. α–stable random variables.

In computer calculations each random variable $X^{\tau}_{t_i}$ defined by (3.1) is represented by its N independent realizations, i.e. a random sample $\{X^{\tau}_i(n)\}_{n=1}^{N}$. So, let us fix $N \in \mathcal{N}$ large enough. The algorithm consists in the following:

1. simulate a random sample $\{X^{\tau}_0(n)\}_{n=1}^{N}$ for X^{τ}_0;
2. for $i = 1, 2, ..., I$ simulate a random sample $\{Y^{\tau}_i(n)\}_{n=1}^{N}$ for α–stable random variable $Y^{\tau}_i \sim S_\alpha(\tau^{1/\alpha}, 0, 0)$, with appropriately chosen τ;
3. for $i = 1, 2, ..., I$, in accordance with (3.1), compute the random sample $X^{\tau}_i(n) = X^{\tau}_{i-1}(n) + \mathcal{F}(t_{i-1}, X^{\tau}_{i-1}(n)) + Y^{\tau}_i(n), \quad n = 1, 2, ..., N$;
4. construct kernel density estimators $f_i = f_i^{I,N} = f_i^{I,N}(x)$ of the densities of $X(t_i)$, using for example the optimal version of the Rosenblatt–Parzen method.

Observe that we have produced N finite time series of the form $\{X^{\tau}_i(n)\}_{i=0}^{I}$ for $n = 1, 2, ..., N$. We regard them as "good" approximations of the trajectories of the process $\{X(t);\ t \in [0,T]\}$.

In particular, the above described algorithm can be succesfully applied to the construction of approximate solutions to the following linear stochastic differential equation driven by an α–stable Lévy motion

$$X(t) = X_0 + \int_0^t (a(s) + b(s)X(s-))\, ds + \int_0^t c(s)\, dL_\alpha(s) \quad \text{for} \quad t \in [0, \infty), \tag{3.2}$$

with $X(0) = X_0$ a given α–stable or discrete random variable.
Let us notice that this linear stochastic equation is of independent interest because, as is easily seen, the general solution belongs to the class of α–stable processes. It may be expressed in the following form

$$X(t) = \Phi(t,0)X_0 + \int_0^t \Phi(t,s)\, a(s)\, ds + \int_0^t \Phi(t,s)\, c(s)\, dL_\alpha(s),$$

where $\Phi(t,s) = \exp\left\{\int_s^t b(u)\, du\right\}$.

This explains why outliers or heavy tails appear in the constructed approximate solutions $\{X^{\tau}_i(n)\}_{i=0}^{I}$, $n = 1, 2, ..., N$, to (3.2), which can be directly derived as a special case of (3.1). It is enough to define the set $\{t_i = i\tau, \quad i =$

$0, 1, ..., I\}$, $\tau = T/I$, describing a fixed mesh on the interval $[0, T]$, and the sequence of i.i.d. random variables $\Delta L^{\tau}_{\alpha,i}$ playing the role of the random α–stable measure of the interval $[t_{i-1}, t_i)$, i.e. an α–stable random variable defined by

$$\Delta L^{\tau}_{\alpha,i} = L_\alpha([t_{i-1}, t_i)) \sim S_\alpha(\tau^{1/\alpha}, 0, 0); \tag{3.3}$$

and to choose $X^{\tau}_0 = X_0 \sim S_\alpha(\sigma, 0, \mu)$, computing

$$X^{\tau}_{t_i} = X^{\tau}_{t_{i-1}} + (a(t_{i-1}) + b(t_{i-1})\, X^{\tau}_{t_{i-1}})\, \tau + c(t_{i-1})\Delta L^{\tau}_{\alpha,i}, \tag{3.4}$$

for $i = 1, 2, ..., I$.

An appropriate convergence result justifying the method can be found in Janicki, Michna and Weron (1994).

In order to obtain a graphical computer presentation of the discrete time stochastic process of the form (3.1), we propose the following approach:

1. fix a rectangle $[0, T] \times [c, d]$ that should include the trajectories of $\{X(t)\}$;
2. for each $n = 1, 2, ..., n_{max}$ (with fixed $n_{max} \ll N$) draw the line segments determined by the points $(t_{i-1}, X^{\tau}_{i-1}(n))$ and $(t_i, X^{\tau}_i(n))$ for $i = 1, 2, ..., I$, constructing n_{max} approximate trajectories of the process X;
3. fixing values of a parameter $p_j \in (0, 1)$, $j = 1, 2, ..., J$, it is possible do derive from each statistical sample $\{X^{\tau}_i(n)\}_{n=1}^{N}$ with fixed $i \in \{0, 1, ..., I\}$ estimators of corresponding quantiles $q^{i,j} = F_i^{-1}(p_j)$, where $F_i = F_i(x)$ denotes the unknown density distribution function of the random variable $X^{\tau}_{t_i}$ represented by the statistical sample $\{X^{\tau}_i(n)\}_{n=1}^{N}$. In this way we obtain approximation of the, so called, *quantile lines*, i.e., the curves $q_j = q_j(t)$ defined by the condition $P\{X(t) \geq q_j(t)\} = p_j$.

3.1 Construction of an α–Stable Ornstein–Uhlenbeck Process

To describe a motion of a particle in a fluid in a model with infinite variance (see, e.g., West and Seshadri (1982)) we have to consider a random strength. In the classical Langevin equation we obtain that this strength is Gaussian and this equation has a form

$$V(t) = V(0) - \lambda \int_0^t V(s)\, ds + \mu \int_0^t dB(s), \tag{3.5}$$

where V is a velocity of the particle and $-\lambda \int_0^t V(s)\, ds$ describes the strength of a resistance of an environment according to Stokes law and $\mu \int_0^t dB(s)$ is a random strength (μ and λ denoting some given constants).

Generalizing the model we can consider an α–stable random strength and we can rewrite the previous equation as follows

$$V(t) = V(0) - \lambda \int_0^t V(s)\, ds + \mu \int_0^t dL_\alpha(s), \tag{3.6}$$

where $\{\mathbf{L}\}_\alpha = \{L(t)\}$ is an α–stable Lévy motion.

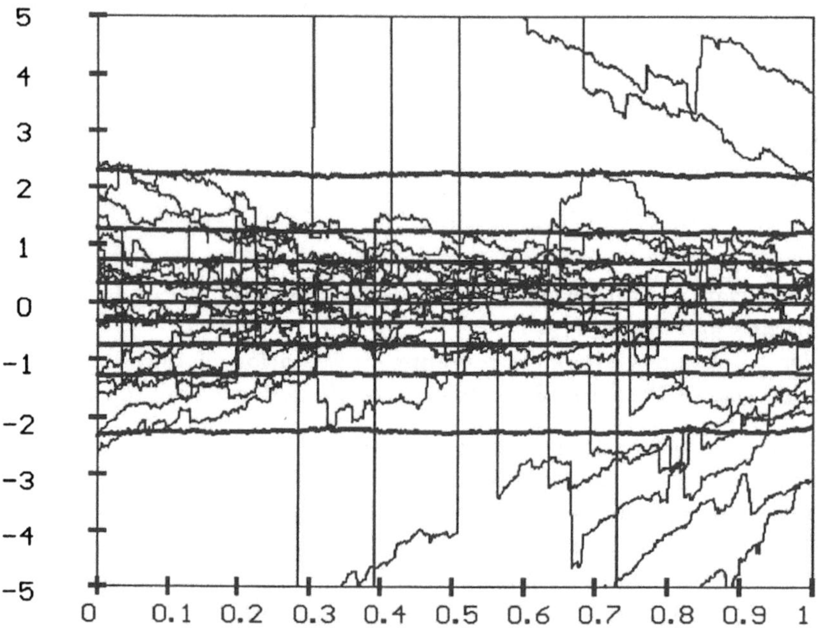

Fig. 3.1. *Solution to (3.6) in the case of* $\lambda = 2.0$ *and* $\mu = 2.0$.

Applying the result of Janicki, Michna and Weron (1994) with some obvious modifications we obtain the process $\mathbf{V}_n$ approximating the process $\mathbf{V}$. This leads to the following formula

$$V_n(t) = V(0)\exp(-\lambda t) + \mu \int_0^t \exp(-\lambda(t-s))\, dL_\alpha^{(n)}(s), \tag{3.7}$$

where $L_\alpha^{(n)}(t) = \sum_{j=1}^{[nt]} \frac{Y_j}{\phi(n)}$, and $\sum_{j=1}^{n} \frac{Y_j}{\phi(n)} \xrightarrow{\mathcal{L}} L_\alpha(1)$.

Fig. 3.1 – 3.2 contain graphical results of computer simulations of solutions to equation (3.6) for $\alpha = 1.3$. The problem was to find a value for λ assuring stationarity of the process $\{V(t) : t \geq 0\}$, provided $V(0) \sim S_{1.3}(1,0,0)$ and $\mu = 2.0$ were fixed. The figures present the results obtained for two different values of $\lambda = 2.0, 4.0$. It follows from the presented computer experiment that the proper value of λ should be chosen close to 2.0. This can be read from the shape of quantile lines, because horizontal quantile lines characterize stationary processes.

4 Tests for Infinite Variance

In some physical applications, arises the question whether the stable distribution is Gaussian or non–Gaussian. There exist simple and computationally convenient

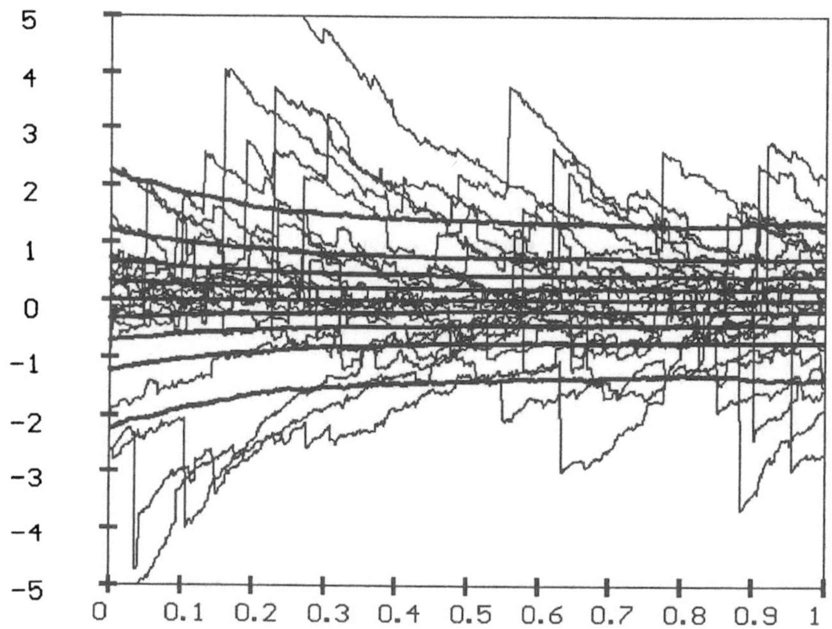

Fig. 3.2. *Solution to (3.6) in the case of* $\lambda = 4.0$ *and* $\mu = 2.0$.

methods that are used in practice. The handicap of all three mentioned here methods is that they rely on visual inspection.

4.1 Converging Variance Test

Given a sample $X_1, ..., X_n$ from a stable distribution, for each $1 \leq i \leq n$ form a statistics based on the first i observations

$$S_i^2 = \frac{1}{i-1} \sum_{k=1}^{i} (X_k - \overline{X}_i)^2 \tag{4.1}$$

where $\overline{X}_i = \frac{1}{i} \sum_{k=1}^{i} X_i$. Then, plot S_i^2 against i. If the population distribution has a finite variance, S_i^2 should converge to a finite value. Otherwise, S_i^2 should diverge. Distinguishing between these two cases is rather vague, especially when we have only one trajectory (one sample).

We have checked this method on four samples of size 4000 coming from $S_\alpha(1, 0, 0)$, for four different α's. Fig. 4.1 presents the results. For $\alpha = 2$, the distribution is Gaussian with mean $\mu = 0$ and a finite variance $\sigma^2 = 2$. As well known, in this case $ES_i^2 = \sigma^2 = 2$ and $\mathrm{Var} S_i^2 = \frac{2}{i-1}\sigma^4 = \frac{8}{i-1}$. When we took smaller α, the plot became more rugged and the values of S_i^2 were much larger

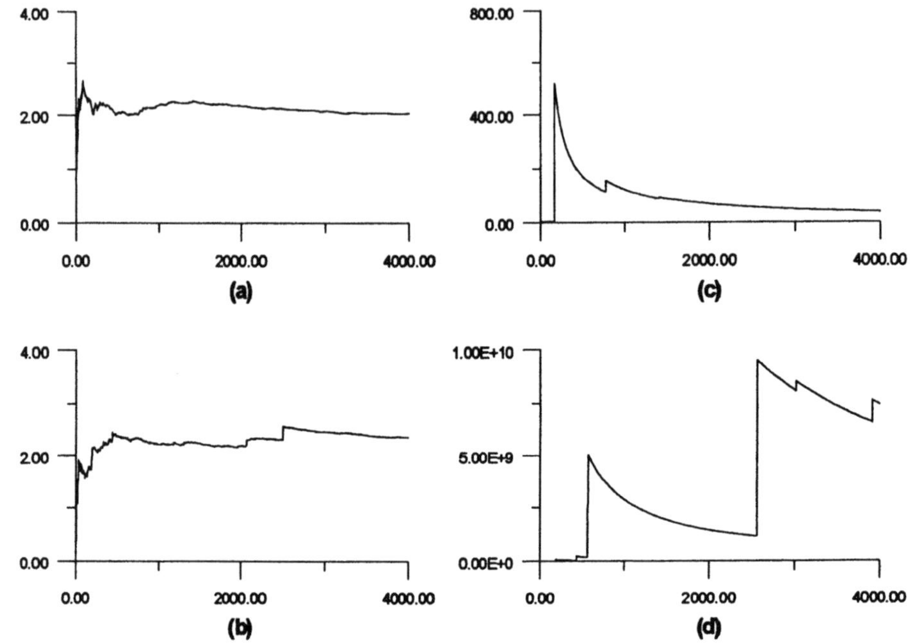

Fig. 4.1. *Plots of the statistics* S_i^2 *against* $i = 1, ..., 4000$ *for* $S_\alpha(1,0,0)$ *with (a)* $\alpha = 2.0$*; (b)* $\alpha = 1.95$*; (c)* $\alpha = 1.5$*; (d)* $\alpha = 0.5$.

(note the change of scale on plots (c) and (d)), which suggested infinite variance of the population distribution.

It is much easier to distinguish between finite and infinite variance when we have a few samples from each distribution. Fig. 4.2 shows five trajectories for every $\alpha = 2.0, 1.95, 1.5, 0.5$. In the latter case, the largest jumps in the trajectories differed by an order of magnitude and it was difficult to find two paths that would be visible on the same scale.

4.2 Q–Q Plot

Q–Q plots (or probability plots) are widely used since they provide quick estimates and a quick informal assessment about the fit of a distribution. Furthermore, when a distribution does not fit, the plot usually tells why this is the case, for instance is it because of a single outlying observation or because of a systematic departure from the assumptions.

If we want to test whether certain data $X_1, ..., X_n$ follow the distribution function F, we plot the data on F–probability paper. This is based on a transformation of the distribution function F (in this case Gaussian) into a straight line by applying F to the vertical scale. The graph of $y = F(x)$ against x is a straight line $y = x$.

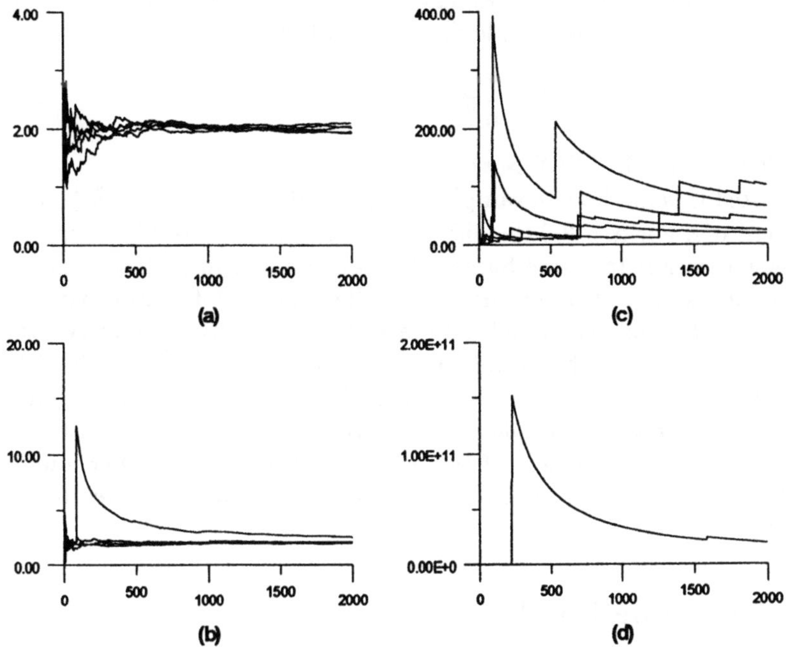

Fig. 4.2. *Plots of five trajectories of the statistics* S_i^2 *against* $i = 1, ..., 2000$ *for* $S_\alpha(1,0,0)$ *with (a)* $\alpha = 2.0$*; (b)* $\alpha = 1.95$*; (c)* $\alpha = 1.5$*; (d)* $\alpha = 0.5$.

The theoretical basis of plotting on probability paper (see Daniel (1976) or Embrechts et al. (1995)) is the fact that for F continuous the random variable $F(X_1)$ is uniform on $(0,1)$. Hence, for the ordered sample $X_{(1)} \le ... \le X_{(n)}$ we know that

$$\mathbb{E}[F(X_{(k)})] = \frac{k}{n+1}, \; k = 1, ..., n.$$

Consequently, we could plot $\frac{k}{n+1}$ against $F(X_{(k)})$. However, more common is to plot $X_{(k)}$ against $F^{-1}(\frac{k}{n+1})$.

If the plot is not approximately linear, the population from which the data seemed to be drawn was not F. In our case this means that the data probably came from a distribution with infinite variance, see Daniel (1976).

4.3 Log–Tail Test

This test examines the shape of the tails of the estimated distribution. Mandelbrot (1963) suggested that because a non–Gaussian stable variable satisfies the relation

$$\lim_{x \to \infty} x^\alpha P(X > x) = const, \tag{4.2}$$

then for large x a plot of $\log P(X > x)$ against $\log x$ should yield a straight line with slope $-\alpha$. We can use the sample distribution function

$$F_n(x) = \frac{1}{n}\#\{i : x_i \leq x\}, \tag{4.3}$$

to estimate $P(X \leq x) = 1 - P(X > x)$.

We have tested this method on samples of size 10000 coming from $S_\alpha(1,0,0)$, for four different α's. For $\alpha = 2.0$ the distribution is Gaussian and has exponential (not inverse power !) tails (see Samorodnitsky and Taqqu (1994), page 16). Therefore, the plot is not a straight line. For smaller α, the plot is approximately a straight line and the slope is a rough estimate of $-\alpha$. An obvious disadvantage of this method is the size of the sample needed. For small sizes ($n < 1000$) the plot becomes rugged and it is difficult to find the slope. We discourage from using this approach for small samples. It also gives only a poor estimate of α. Especially for $\alpha > 1.5$, as the tails of the distribution become smaller, our estimate is subject to a large error, since the plot significantly deviates from a straight line. See Fig. 4.3.

5 Continuous–Time Random Walk

Continuous–Time Random Walk (CTRW) is a stochastic process with random waiting times T_i between successive jumps of random length R_i. During the recent years it has been studied extensively and applied to turbulence, transport in disordered or fractal media, intermittent chaotic systems and relaxation phenomena. For references see Klafter et al. (1993). The common feature of these applications is that they exhibit anomalous diffusion, which is manifested through non-linear time dependence of a mean square distance R_t reached by a particle up to the moment t (with the initial condition $R_0 = 0$).

For simplicity we restrict ourselves to one-dimensional walks. The first instantaneous jump of random length R_1 is executed after a random waiting time T_1. Then the second instantaneous jump R_2 (i.e. jump of random length R_2) is executed after time T_2, etc. In general, the i–th jump R_i is dependent on its waiting time T_i, but the pairs (R_i, T_i) are independent for different i's. See Fig. 5.1. The special case when R_i is independent of T_i is called the *decoupled memory* CTRW as opposed to the *coupled* one with R_i dependent on T_i. If we define the random variable N_t as the number of jumps in the time interval $[0,t]$

$$N_t = \max\{k : \sum_{i=0}^{k} T_i \leq t\}, \tag{5.1}$$

then it is clear that the position R_t of the particle at the time t is equal to a random sum of N_t successive random jumps R_i, i.e.

$$R_t = \sum_{i=0}^{N_t} R_i. \tag{5.2}$$

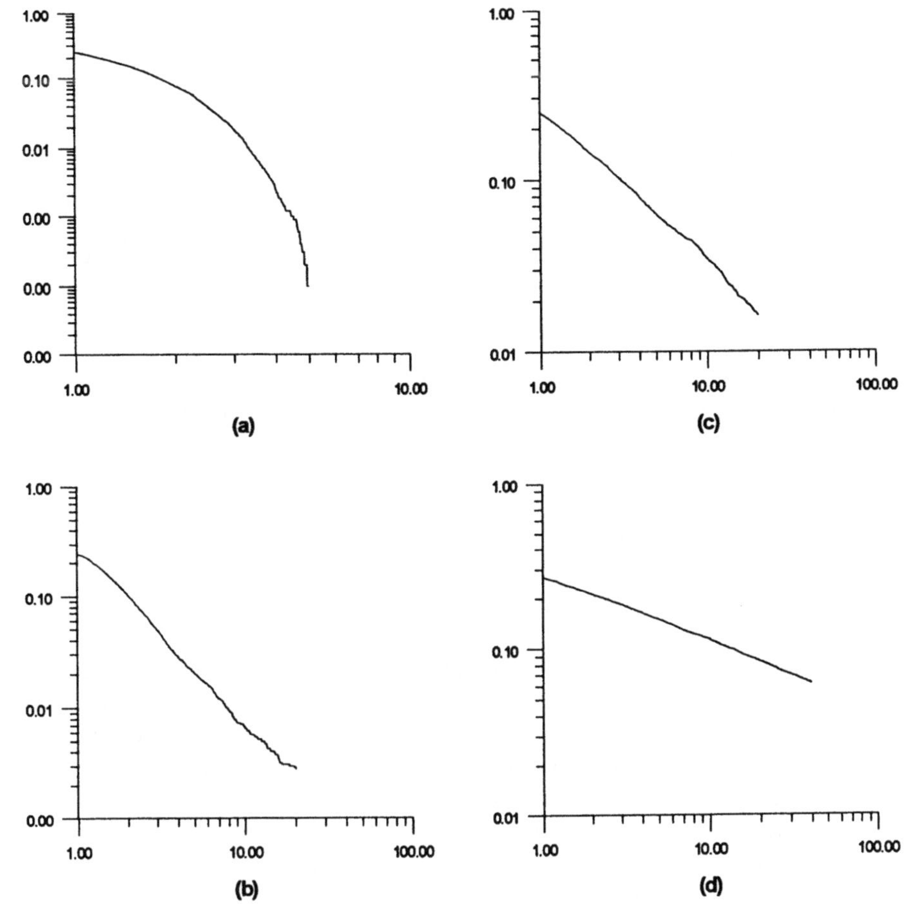

Fig. 4.3. *Log–Tail test: plots of* $\log(1 - F_n(x))$ *against* $\log x$ *for* $S_\alpha(1,0,0)$ *with (a)* $\alpha = 2.0$; *(b)* $\alpha = 1.5$; *(c)* $\alpha = 1.0$; *(d)* $\alpha = 0.5$.

Note, that (5.2) holds also for a random walk when waiting time intervals T_i are non-random and take a constant value.

Many authors were interested in finding the asymptotic distribution of the position R_t of a particle at time t. We concentrated our research on the following formula

$$\lim_{t\to\infty} \left(\frac{R_t}{(t/\tau)^{1/\alpha} c_2} < x \right) = S_\alpha(1,0,0)(x), \tag{5.3}$$

which is valid for $1 < \alpha < 2$ and $\tau = ET_i$. T_i has a density $g(t)$ such that $\lim_{t\to\infty} \frac{g(t)}{\alpha t^{-\alpha-1}} = 1$, and R_i satisfies $P(R_i = T_i) = P(R_i = -T_i) = \frac{1}{2}$. The

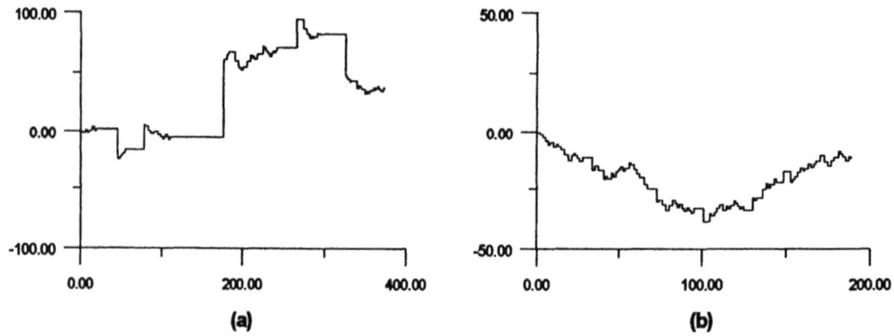

Fig. 5.1. *Two sample paths of a coupled memory CTRW: plots of the distance R_t against time with (a) $\alpha = 1.2$ and (b) $\alpha = 1.8$. There are 100 jumps in the trajectories on both plots. Note the scale difference on both axes.*

normalizing constant in (5.3) is given by $c_2 = c_\alpha^{-1/\alpha}$, where

$$c_\alpha = \frac{1-\alpha}{\Gamma(2-\alpha)\cos(\frac{\pi\alpha}{2})},$$

and Γ denotes the Euler Gamma function. This is a coupled memory CTRW. Fig. 5.1 presents two sample paths, with 100 jumps each, for two values of α. For $\alpha = 1.2$ the distribution of T_i, and thus of R_i, has longer tails then for $\alpha = 1.8$. This results in longer waiting times and jumps. A trajectory for $\alpha = 1.2$ would, in general, make twice less jumps in the same time then a trajectory for $\alpha = 1.8$.

In our simulations, we let T_i have a Pareto$(\alpha, 1)$ distribution, i.e. $g(t) = \alpha t^{-\alpha-1}$. A random variable X having this distribution can be easily constructed using the *inverse transform method*

- generate a random variable V distributed uniformly on $(0,1)$;
- compute $X = V^{-1/\alpha}$.

We have chosen three values of α: 1.2, 1.5 and 1.8 in the permissible range of $(1,2)$. For every α we ran 25 simulations with 2000 trajectories each. Then for every α and every stopping time $t = 500, 1000, 5000$ we used the regression method of Koutrouvelis (1980), which is the most reliable among the known statistical methods (see Weron (1995b)), to estimate α and σ (the distribution of R_t is symmetric). The results are summarized in Table 5.1.

The distribution of R_t is symmetric, i.e. $\beta = 0, \mu = 0$, thus we estimated only α and σ. For each parameter we give the mean, minimum, maximum and Mean Squarred Error i.e.

$$MSE_\theta = \frac{1}{n}\sum_{i=1}^{r}(\hat{\theta}_i - \theta)^2, \tag{5.4}$$

of the estimates $\hat{\alpha}_k$ and $\hat{\sigma}_k$ obtained in 25 simulations ($k = 1, ..., 25$).

α	t	$\overline{\alpha}$	α_{min}	α_{max}	MSE_α	$\overline{\sigma}$	σ_{min}	σ_{max}	MSE_σ
1.2	500	1.312	1.237	1.402	0.0139	1.629	1.530	1.739	0.3979
1.2	1000	1.292	1.220	1.352	0.0101	1.530	1.444	1.639	0.2836
1.2	5000	1.260	1.206	1.337	0.0047	1.364	1.321	1.412	0.1330
1.5	500	1.539	1.450	1.623	0.0033	1.081	1.020	1.131	0.0073
1.5	1000	1.519	1.436	1.601	0.0016	1.064	1.011	1.121	0.0048
1.5	5000	1.513	1.464	1.586	0.0013	1.037	0.952	1.082	0.0022
1.8	500	1.727	1.665	1.790	0.0064	0.807	0.781	0.835	0.0374
1.8	1000	1.735	1.642	1.796	0.0058	0.823	0.791	0.859	0.0318
1.8	5000	1.746	1.682	1.814	0.0040	0.840	0.819	0.871	0.0259

Table 5.1. *Parameter estimates using the regression method of* $\alpha = 1.2, 1.5, 1.8$ *and* σ *describing the stable distribution of* R_t *at stopping times* $t = 500, 1000, 5000$.

As we have stated before, the smaller α the larger are the waiting times and jumps, which is a consequence of longer tails of the distribution of T_i. This behavior can be observed in Table 5.1. The stopping time $t = 500$ for $\alpha = 1.8$ is more or less equivalent to the stopping time $t = 1000$ for $\alpha = 1.2$. This is the reason the convergence, to the population values of α and σ, for $\alpha = 1.8$ seems to be slower than for $\alpha = 1.2$. Moreover, for $\alpha = 1.2$ the true values are overestimated, for $\alpha = 1.8$ underestimated and for $\alpha = 1.5$ the estimators give almost a perfect match. This, in fact, confirms what other authors have recently observed for different tail thickness (they have only estimated the parameter α) of the distribution of T_i, see e.g. Zumofen et al. (1989). As α approaches 2 from below or 1 from above the convergence becomes very slow.

References

1. L.M. Berliner, *Statistics, probability and chaos*, Stat. Science **7**, 69–90 (1992).
2. S. Chatterjee and M. R. Yilmaz, *Chaos, fractals and statistics*, Stat. Science **7**, 49–68 (1992).
3. J. M. Chambers, C. L. Mallows and B. Stuck, *A method for simulating stable random variables*, J. Amer. Statist. Assoc. **71**, 340–344 (1976).
4. C. Daniel, *Applications of Statistics to Industrial Experimentation*, Wiley & Sons, New York, (1976).
5. L. Devroye, *A Course in Density Estimation*, Birkhäuser, Boston, (1987).
6. P. Embrechts, C. Kluppelberg and T. Mikosch, *Modelling Extremal Events in Insurance and Finance*, (1995), in preparation.
7. A. Janicki, Z. Michna and A. Weron,, *Approximation of stochastic differential equations driven by α–stable Lévy motion*, preprint, (1994).
8. A. Janicki and A. Weron,, *Simulation and Chaotic Behavior of α–Stable Stochastic Processes*, (1994), Marcel Dekker, New York.
9. A. Janicki and A. Weron, *Can one see α–stable variables and processes ?*, Stat. Science, **9**, 109–126 (1994).
10. M. Kanter, *Stable densities under change of scale and total variation inequalities*, Ann. Probab. **31**, 697–707 (1975).

11. J. Klafter, G. Zumofen and M.F. Shlesinger, *Fractal description of anomalous diffusion in dynamical systems*, Fractals, **1**, 389–404 (1993).
12. I.A. Koutrouvelis, *Regression-type estimation of the parameters of stable laws*, J. Amer. Statist. Assoc., **75**, 918–928 (1980).
13. A. Lasota and M. C. Mackey, *Chaos, Fractals, and Noise. Stochastic Aspects of Dynamics*, (1994), Springer, New York.
14. B. Mandelbrot, *The variation of certain speculative prices*, J. Business, **36**, 394–419 (1963).
15. B. Mandelbrot and J. W. Van Nes, *Fractional Brownian motions, fractional noises and applications*, SIAM Review, **10**, 422–437 (1968).
16. R.M. Mantegna, *Fast, accurate algorithm for numerical simulation of Lévy stable stochastic process*, Phys. Rev. **E49**, 4677–4683 (1994).
17. G. Samorodnitsky and M.S. Taqqu, *Stable Non-Gaussian Random Processes*, (1994) Chapman & Hall, New York.
18. M. Shao and C. L. Nikias, *Signal processing with fractional lower order moments: stable processes and their applications*, Proc. IEEE, **81**, 986–1010 (1993).
19. A. Weron, *Stable processes and measures: A survey*, in Probability Theory on Vector Spaces III (D. Szynal, A. Weron, eds.) 306–364, *Lecture Notes in Mathematics* 1080, (1984) Springer, New York.
20. R. Weron, *On the Chambers–Mallows–Stuck method for simulating skewed stable random variables*, Statist. Probab. Lett., (1995), submitted.
21. R. Weron, *Performance of the estimators of stable law parameters*, Appl. Math., (1995) submitted.
22. B.J. West and V. Seshadri, *Linear systems with Lévy fluctuations*, Physica, **A113**, 203–216 (1982).
23. G. Zumofen, A. Blumen, J. Klafter and M.F. Schlesinger, *Lévy walks for turbulence: A numerical study*, J. Stat. Phys., **54**, 1519–1528 (1989).

From Quantum Physics to Probability Theory and Back

J.C. Zambrini

Group of Mathematical Physics
University of Lisbon
1699 Lisboa Codex, Portugal

Table of Contents

1 Historical Introduction

One of the topics considered during the XXIV Karpacz Winter School on Theoretical Physics, in 1988, was Nelson's "Stochastic Mechanics," and this theory, along with some of its difficulties, was nicely summarized by E. Carlen in the proceedings volume of this school [1]. I was sitting in the audience, then, and gave later a little talk entitled "Euclidean Quantum Mechanics." Seven years after, I have been asked to give a survey of the approach which has slowly emerged from the last point of view sketched there.

In essence, the version of Euclidean Quantum Mechanics I will refer to here is due to E. Schrödinger (1931–32 [2]) but it has, of course been influenced by the major attempts, and failures, made since then to construct an intrinsically probabilistic version of quantum theory. A frequent reaction, among mathematical physicists, is to regard such a program as absurd or even self-contradictory. Indeed, it is known for ages that since most of the quantum observables of a physical system are noncommuting self-adjoint operators on an Hilbert space, the collection of all the observables of the system, in a given state, is not a set of random variables simultaneously defined on an unique probability space. For example, the uncertainty principle between position Q and momentum P can be interpreted as the inexistence of a joint measurement of the two observables. There is nothing wrong with that, however, since the experimental devices needed to perform those measurement exclude each other. When we describe classical phenomena in probabilistic terms (*i. e.* on a well defined probabilistic space (Ω, σ, P)[10]), and that two random variables can be measured during the same experiment, their joint probability is, in contrast, always defined. So a frequent conclusion is that probability theory does not help at all for the quantum theory of pure states (there are little controversy regarding the role of probability in quantum statistical mechanics). Although the above mentioned observations are undoubtedly true, there are serious doubts that the conclusion follows so easily.

At the end of the fourties, R. Feynman started a program of reinterpretation of basic quantum mechanics in terms which, if this was not for the irritating omnipresence of the factor $i = \sqrt{-1}$, would be regarded as fundamentally probabilistics [3]. Although everybody knows, today, that mathematically speaking it does not make sense (at least from this probabilistic point of view, otherwise, it does (cf. [4])) this may be, nowadays, the most popular approach, among theoretical physicists, to the quantization of classical dynamical systems, even in those exciting areas where what we mean exactly by "quantization" is not entirely obvious (cf., for example, [5]).

We believe that there is still much more to extract from Feynman's heuristic that what is generally taught, especially by some mathematical physicists. I will be satisfied if I am able at least to convey this feeling during my lectures.

We shall therefore start from what Feynman did, and review the main mathematical difficulties of his key results, as well as some theoretical ones, not as familiar as the first ones.

Then we will describe well defined probabilistic counterpart of Feynman's results. For the simplest case of the Hamiltonian of a single, unit mass, particle in a scalar potential V, *i. e.*

$$H = -\frac{\hbar^2}{2}\Delta + V \tag{1.1}$$

(but the method is in no way limited to this case) they involve well defined probability measures of a special class of time symmetric diffusion processes, whose construction was suggested by Schrödinger.

We shall, in those terms, construct a consistent (Euclidean) version of Feynman's theory of "transition elements" (cf. Chap. 7 of [3(2)]). Since our probability measures are well defined, this theory can also be regarded as a kind a "stochastic calculus of variations." This brings us back to Nelson's approach since this is there that Yasue suggested first to regard the probabilistic laws of motion as Euler-Lagrange equations for a certain stochastic action functional, with two fixed probability densities [6].

Although his initial idea, together with many works motivated by them (including some of mine before 1986) were plagued with severe mathematical difficulties, it was quite influential. Some of those intuitions (but not all) were progressively given a mathematical status (cf. E. Nelson, P. E. Meyer etc. and references in [7]) but it became progressively clear that Stochastic Mechanics itself had a probabilistic structure far too distinct from Feynman's path integrals approach to allow to go any further in this direction. As a matter of facts, the above mentioned mathematical puzzles were precisely at the origin of my early interest for Schrödinger's Euclidean Quantum Mechanics (cf. the Introduction of [8]) which happens to be, now, the mathematically consistent arena where all this makes sense.

It should be clearly stated, in order to avoid any misunderstanding, that Euclidean Quantum Mechanics is only an analogy of Quantum Physics. However, although it is still a rather incomplete one, it is a very strong analogy. The key reason for this is that probability enters into this Euclidean analogy exactly in the same way it enters into regular Quantum Physics. This was, besides, exactly the point of Schrödinger's original idea. In 1931–32, however, probability theory as we know it, *i. e.* axiomatized by A. N. Kolmogorov a short time after, did not even exist! Today we have at our disposal a huge arsenal of probabilistic tools of which, quite extraordinarily, a tiny proportion has been used in Quantum Physics. If we succeed, in the future, to construct indeed a probabilistic analogue of Quantum Physics which is reasonably complete, *i. e.*, in particular, not limited to one picture of the dynamics and such that, using some improved version of Wick rotation, we can recover all the predictions of Quantum theory, we will have achieved much more than a cumbersome translation of a familiar theory: its complete embedding into a framework which allows us to ask (and, hopefully, to answer!) questions that regular Quantum theory prevents us to consider. We will be bound, then, to discover new physical features, in real time. In addition, such an embedding would constitute, if achieved, a useful auxiliary in the semiclassical regime, notably, for the quantization of nonintegrable systems.

It is this rather conservative strategy that we call "Euclidean Quantum Mechanics" and that we are going, now, to summarize.

2 Feynman's Transition Elements

In §7.2 of their famous treatise on path integral [3(2)] Feynman and Hibbs introduce a "very general relation between transition elements" which, according to them, "might well serve as an alternative starting point for the definition of quantum mechanics." A few pages before, they had defined the transition element of "any" functional of the paths, $F[\omega(.)]$, defined on $\Omega = \{\omega(.) \in C(]0,\tau[;\mathbf{R}^3)\}$ by the integral

$$\langle\psi|F|\theta\rangle_S = \iiint_\Omega \theta(x)\, F[\omega(.)]\, e^{\frac{i}{\hbar}S[\omega(.)]}\overline{\psi}(y)\, \mathcal{D}\omega\, dx\, dy \qquad (2.1)$$

where the celebrated path integral representation of the integral kernel of the Schrödinger equation associated with the Hamiltonian H has been introduced, namely

$$\left(e^{-\frac{i}{\hbar}H\tau}\right)(x,y)=\int_{\Omega_x^y} e^{\frac{i}{\hbar}S[\omega(.)]}\mathcal{D}\omega, \tag{2.2}$$

and the bar $\bar{\ }$ denotes the complex conjugation.

On the left hand side of (2.2) we have a well defined object for a given self-adjoint Hamiltonian operator H, for example of the type (1.1) with V in the Kato class [9]. On the right hand side, one defines a formal sum over all the paths ω starting from the point $x \in \mathbf{R}^3$ at time 0 and ending in y at the later time τ.

The symbol "$\mathcal{D}\omega$" means, literally, the infinite product $\prod_{0\leq s\leq\tau} d\omega(s)$ and, therefore, should not be taken too seriously. Now the crucial point of Feynman's approach is, of course, the presence of the weight factor $\exp\left(\frac{i}{\hbar}S[\omega(.)]\right)$ in the formal expression (2.2). Indeed, $S[\omega(.)]=\int_0^\tau L(\dot\omega(s),\omega(s))\,ds$ is the action functional of the classical system to be quantized (for H like (1.1) the Lagrangian L is $L(\dot\omega,\omega)=\frac{1}{2}|\dot\omega|^2-V(\omega)$) so that the relation (2.2) provides a recipe to describe quantum dynamics (*i. e.* the one-parameter group of unitary operator $U_\tau=e^{-\frac{i}{\hbar}H\tau}$) only in terms of a classical data, the action functional of the system. In consequence, Feynman's method is a new method of quantization, not just a translation of what was already known before 1948.

As observed by the authors (p. 165 of [3(2)]) the physical intuition behind transition elements is weak. So they suggest a classical analogy in terms of Brownian motion! It is worth noticing here that this intuition has been essential all along Feynman's strategy, although quite rapidly forgotten by those who used his ideas. Since, in Quantum Mechanics, one starts from a state θ and arrive at some final state ψ, the integral (2.1) "requires integration over the variables x and y of the initial and final states." Since the classical analogy the authors have in mind is the one of a Brownian Bridge (cf. [24]) *i. e.* a Standard Brownian motion starting from x and conditioned to end at the point y at time τ, they add that this integration over x and y is unnecessary in the classical context. More precisely, in this case, both θ and ψ are Dirac's distributions.

At this point, they do a slight mistake. As we shall see (or, more precisely, as Schrödinger did see in 1931–32!), there is a large class of perfectly classical problems for which this integration is not only possible but even necessary.

Feynman and Hibbs go on and observe that any physical property of interest regarding the motion on the time interval $[0,\tau]$ corresponds to the calculation of some transition element like (2.1) for a certain F. They pick, as an example the "acceleration experienced by the particle at some particular time" (p. 166), which would correspond to a special choice of functional F in (2.1) and then to a particular weighted average.

This Chapter on "Transition elements," the central one regarding the mathematical structure of Feynman's path integral approach, uses intensively the above mentioned "analogy" with the theory of Brownian motion but, of course, the authors stress that this is only an analogy since "the weighting function in quantum mechanics is a complex function." As a result of this unfortunate feature, when $F = 1$, a transition element (one say, then, "transition amplitude between the states θ and ψ") is, generally, a complex function of the time τ. Only expressions bilinear in ψ and $\overline{\psi}$ are interpreted as probabilities. Most of the particularities of quantum theory with respect to the classical theory of Markov processes [10] have their origin in the nature of this transition amplitude. Indeed, suppose that $\psi = e^{-\frac{i}{\hbar}\tau H}\theta$. Then the transition amplitude between the states θ and ψ could also be written as

$$\iint \theta(x)\,\overline{\theta}(x)\,K(x,\tau,y)\,\frac{\overline{\psi}(y)}{\overline{\theta}(x)}\,dy\,dx \tag{2.3}$$

where we have denoted by $K(x,\tau,y)$ the integral kernel defined by (2.2). Clearly, the product $\theta(x)\,\overline{\theta}(x)$ could be regarded as density of an initial probability $\rho_0(x)\,dx = |\theta(x)|^2 dx$ and the complex function

$$K(x,\tau,y)\,\frac{\overline{\psi}(y)}{\overline{\theta}(x)} \equiv K(x,\tau,y)\,\frac{\overline{\psi}(y)}{\overline{\psi}(x,0)} \tag{2.4}$$

which propagate forward ρ_0, as the quantum analogue of the (forward) transition probability density with respect to dy of a Markov process Z_τ^F with initial probability $\rho_0(x)\,dx$. In this case, by definition [10] the result of the integral over x,

$$\int_{\mathbf{R}^3} |\theta(x)|^2 (dx)\,K(x,\tau,y)\,\frac{\overline{\psi}(y)}{\overline{\theta}(x)}\,dy$$

should coincide with the probability, for the underlying stochastic process Z^F to be in dy at time τ. Now, since $\psi = e^{-\frac{i}{\hbar}\tau H}\theta$, we obtain $P(Z_\tau^F \in dy) = |\psi(y)|^2 dy$, in accordance with Born interpretation of the final state. An important remark is in order here; since quantum theory is perfectly time reversible (as long as the forces are not time dependent) we could as well have chosen to start from the final probability density $\rho_\tau(y)\,dy = |\psi(y)|^2 dy$ and a quantum backward transition probability

$$\frac{\theta(x)}{\psi(y)}K(x,\tau,y)\,dx \equiv \frac{\psi(x,0)}{\psi(y)}K(x,\tau,y)\,dx \tag{2.5}$$

Such concept, although not as familiar to the physicists as the above mentioned forward one is, nevertheless, perfectly respectable in probability theory, *i. e.* when the existence of the underlying Markov process has been shown.

When the Hamiltonian is of the simple form (1.1), this Markov process should be a diffusion process. Suppose, for one minute, that such a Markovian diffusion does exist. Then it would be simple, using the properties of the complex integral kernel K, to compute formally (of course!) the two infinitesimal coefficients characterizing this diffusion. Denoting by $p_F(x,\tau,y)$ the expression (2.4), we would obtain respectively, via Taylor expansion of $\overline{\psi}$ around x at the origin of time

$$\begin{aligned} B_F(x,0) &= \lim_{\Delta\tau\downarrow 0} \frac{1}{\Delta\tau} \int (x-y)\, p_F(x,\tau,y)\, dy \\ &= i\hbar \frac{\nabla\overline{\psi}}{\overline{\psi}}(x) \end{aligned} \tag{2.6}$$

and

$$\begin{aligned} C_F(x,0) &= \lim_{\Delta\tau\downarrow 0} \frac{1}{\Delta\tau} \int \{(x-y)\otimes(x-y)\}\, p_F(x,\tau,y)\, dy \\ &= i\hbar\, \mathbf{1} \qquad (\mathbf{1} = 3\times 3 \text{ identity matrix}) \end{aligned} \tag{2.7}$$

From what is known of regular quantum mechanics, this is what we expect. Indeed, according to Born's interpretation of the wave function, this diffusion should have an initial probability density

$$\rho_0(x)\, dx = |\theta(x)|^2 dx \equiv |\psi(x,\tau=0)|^2 dx \tag{2.8}$$

The average of the "drift" B_F with respect to this probability measure provides the quantum expectation at time 0 ($\langle .|.\rangle_2$ denotes the L^2 scalar product, conjugate linear in the first vector)

$$\langle\psi|P\psi\rangle_2 \qquad \text{for } P = -i\hbar\nabla \tag{2.9}$$

so that what we did was just to find a random variable underlying the momentum operator on (a domain of) $L^2(\mathbf{R}^3)$ [11].

We could as well have described the infinitesimal characteristics of the same quantum diffusion in terms of its backward quantum transition probability p_F^*, as defined by (2.5). Its diffusion matrix C_F^* would have coincide with (2.7) but the drift at the final time τ would have been

$$B_F^*(y,\tau) = -i\hbar \frac{\nabla\psi}{\psi}(y) \tag{2.10}$$

so that the quantum expectation of B_F^* at time τ would have defined the same momentum observable as before. Unfortunately, as shown long ago by Cameron [12], the complex probability measure μ_F associated with the above mentioned Feynman's diffusion Z^F does not exist. This is definitely not, from a mathematical point of view, an elementary matter. Actually, until around 15 years after Feynman's publication [3(1)], various claims of existence of μ_F were made, even by very famous mathematicians.

As summarized by L. S. Schulman "the sum over paths for Brownian motion is a theory having substance, while its path integral image exists mainly in the eye of the beholder" [13]. In particular, all the above mentioned local characteristics indexed by F (for Feynman) do not characterize, after all, any well defined diffusion process.

Nevertheless Feynman and Hibbs go on with their Brownian analogy and, as we said before, present what they describe as this "very general relation between transition elements" adding, to make clear how central it is in their approach, that "this relation might well serve as an alternative starting point for the definition of quantum mechanics" (p. 170 of [3(2)]).

In order to stress the fact that the relation in question is independent of the form of the initial or final states θ, ψ of the system, the authors introduce the short notation $\langle F\rangle_S$ for the transition elements like $\langle\psi|F|\theta\rangle_S$ defined by (2.1). Then their result reads: ((7–30) p. 172)

$$\big\langle \delta F[\omega](\delta\omega)\big\rangle_S = -\frac{i}{\hbar}\big\langle F\delta S[\omega](\delta\omega)\big\rangle_S \tag{2.11}$$

where $\delta F[\omega](\delta\omega)$ is the variational derivative of the functional involved in (2.1) and $\delta S[\omega](\delta\omega)$ the one of the classical action functional of the system appearing in the weight of the "measure." Before to see what can be done with the integration by parts formula (2.11), let us make a few remarks regarding the general interpretation of Feynman path integral.

If the final state ψ is a Dirac distribution, we know that the path space associated with the underlying path integral is not Ω_x^y like for the integral kernel (2.2) itself but

$$\Omega^{y,\tau} = \big\{\omega(.) \in C([0,\tau];\mathbf{R}^3) \text{ such that } \omega(\tau) = y\big\}$$

This means that the underlying path integral provides a representation of the solution

$$\left(e^{-\frac{i}{\hbar}\tau H}\theta\right)(y) = \int_{\Omega^{y,\tau}} \theta(\omega(0))\, e^{\frac{i}{\hbar}\int_0^\tau L(\dot\omega,\omega)\,ds}\mathcal{D}\omega \tag{2.12}$$

of the Cauchy problem for the Schrödinger equation in $L^2(\mathbf{R}^3)$.

$$\begin{cases} i\hbar\dfrac{\partial\psi}{\partial\tau} = H\psi \\ \psi(y,0) = \theta(y) \end{cases} \tag{2.13}$$

The right hand side of (2.12) defines a functional on $\Omega^{y,\tau}$, which can also be regarded as a function of the space-time point (y,τ) itself. Then it solves the Schrödinger equation (2.13).

If, on the other hand, we consider the case where it is the initial state θ which is a Dirac distribution, then the relevant path integral in the transition element (2.1) becomes

$$\int e^{\frac{i}{\hbar}\int_0^\tau L(\dot\omega,\omega)\,ds}\,\overline{\psi}(\omega(\tau))\,\mathcal{D}\omega$$

After a formal time reversal of the paths ω (observing that our simple Lagrangian L, as well as the whole structure of classical mechanics, is invariant under this operation) this can be regarded as

$$\int e^{-\frac{i}{\hbar}\int_\tau^0 L(\dot\omega,\omega)\,ds}\,\overline{\psi}(\omega(0))\,\mathcal{D}\omega$$

In other words

$$\left(e^{\frac{i}{\hbar}H\tau}\overline{\psi}\right)(y) = \int_{\Omega_{y,\tau}} \overline{\psi}(\omega(0))\, e^{-\frac{i}{\hbar}\int_\tau^0 L(\dot{\omega},\omega)\,ds}\mathcal{D}\omega \tag{2.14}$$

is the path integral representation of the solution of

$$\begin{cases} -i\hbar\dfrac{\partial\overline{\psi}}{\partial\tau} = H\overline{\psi} \\ \overline{\psi}(y,0) = \overline{\psi}(y) \end{cases} \tag{2.15}$$

i. e. of the problem complex conjugate to (2.13). One may regard the relevant path space $\Omega_{y,\tau}$ as formal time reversed of the one of (2.12) *i. e.*, for $\tau < 0$,

$$\Omega_{y,\tau} = \{\omega(.) \in C([\tau,0];\mathbf{R}^3) \text{ such that } \omega(\tau) = y\} \tag{2.16}$$

In this sense, (2.14) is a path integral involving the whole future of the paths, starting from $\omega(\tau) = y$. This will be crucial afterwards.

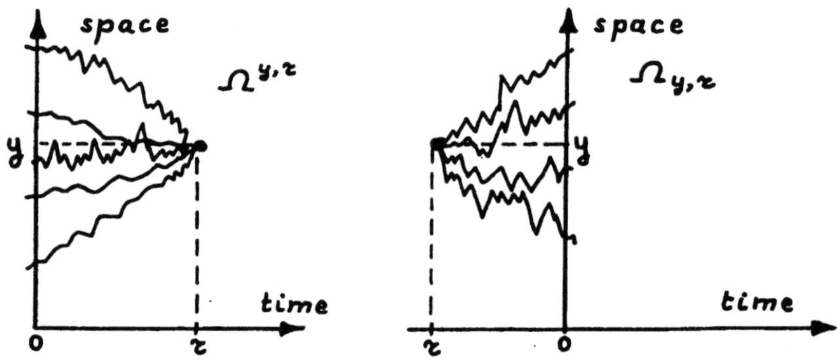

Fig. 1.

Notice that, in those pictures, the direction of the time is, actually, unchanged. What is changed is the relative position of the condition $\omega(\tau) = y$, respectively in the future or in the past of the time interval of interest. Also notice that the forward infinitesimal coefficients B_F and C_F of Feynman's "diffusion" depends only on the solution of Eq. (2.15) with final boundary condition, and the backward ones B_F^* and C_F^* only on the solution of the Cauchy problem (2.13). Afterwards it will be often more convenient to consider arbitrary time interval $[s,u]$ for the construction, or time intervals symmetric around 0, for example $[-T/2, T/2]$.

If Feynman's diffusion was well defined, we would be entitled to consider $\Omega^{y,\tau}$ and $\Omega_{y,\tau}$ as the path spaces associated, respectively, with conditional expectations in the future and in the past. Indeed, let us consider any bounded regular function g of the missing process, say Z^F. According to what has been said before, is expectation at time $\tau > 0$ is

$$\int \psi(y)\,\overline{\psi}(y)\,g(y)\,dy \equiv E[g(Z^F_\tau)]$$

$$= \iint \theta(x)\,K(x,\tau,y)\,\overline{\psi}(y)\,g(y)\,dx\,dy$$

$$= \int \theta(x)\,\overline{\theta}(x) \int g(y)\,K(x,\tau,y)\,\frac{\overline{\psi}(y)}{\theta(x)}\,dy\,dx$$

Now if Z^F was a well defined stochastic process the integral over y would be (cf. (2.4)) the "conditional expectation" of g at time τ, under the condition $Z^F_0 = x$, denoted $E^F[g(Z^F_\tau)\,|\,Z^F_0 = x]$. Since the extra integral (over x) denotes the (absolute) expectation with respect to Born initial probability $\rho_0(x)\,dx$, we would have just verified that

$$E\left[g(Z^F_\tau)\right] = E\Big[E^F\left[g(Z^F_\tau)\,|\,Z^F_0 = x\right]\Big] \tag{2.17}$$

a property which can be used as definition of the probabilistic concept of conditional expectation [10].

Alas, there is no such thing as a conditional expectation E^F in quantum physics, because there is no transition probability, at least in the usual sense of probability theory (since the integral kernel K is not positive, and not even real, we lack the positivity needed for the probabilistic construction). As a matter of fact, the definition of some substitutes of this fundamental concept is a traditional subject of controversies in conferences on the foundations of quantum mechanics [14].

Notice however that, from the physical point of view, there is nothing wrong with the idea of conditioning. When computing expectations for quantum particles we should, indeed, be able to do our statistics over paths starting all from the same point in space, for example, if the preparation of the state is compatible with this restriction. In Feynman path integral approach, conditioning has a physical meaning, even if it does not have a mathematical one. This is another frustration for those who believe that the probabilistic content of quantum mechanics should, in a way, play a central role in this theory.

Of course, the calculation of the expectation at the initial time 0 of a function of the position would have force us to use Feynman's "backward transition probability" (2.5) with a condition at the future time τ (if τ is still positive).

This discussion prepare us to understand that Feynman integration by parts formula (2.11) can be understood, if we wish, in terms of "conditional expectation" (in the above mentioned allegorical sense) for a condition $\omega(\tau) = y$ in the past or in the future of the time interval of interest.

Now let us look at the main results following, according to Feynman, from the relation (2.11).

The first application (p. 174 of [3(2)]) is the one of the above mentioned, unit mass, single particle in a scalar potential V, *i. e.* with classical action

$$S[\omega] = \int_0^T \left\{ \frac{1}{2} |\dot{\omega}(s)|^2 - V(\omega(s)) \right\} ds \tag{2.18}$$

and one chooses the simplest nontrivial functional F, namely $F = 1$, for Eq. (2.11). The left hand side is, of course, 0 and a formal use of the familiar calculus of variations for smooth functionals like S provides a relation valid for an arbitrary variation $\delta\omega(s)$ of the path. From this the authors conclude that

$$\left\langle \frac{d^2}{ds^2}\omega(s) \right\rangle_S = -\langle \nabla V(\omega(s)) \rangle_S \tag{2.19}$$

This equation (7.42 in [3(2)]) is interpreted as the "quantum-mechanical analogue of Newton's law," involving the average defined before in terms of the action S. As a matter of fact, Feynman will refer mostly to a time discretized version of the relation (2.19) for reasons which are going to be immediately obvious.

Indeed, for the same physical system, pick $F[\omega] = \omega(\tau)$. Then, after some formal manipulations, the consequence of the integration by parts formula is

$$\left\langle \omega_l(\tau) \frac{(\omega(\tau) - \omega(\tau - \Delta\tau))\, m}{\Delta\tau} \right\rangle_S - \left\langle \frac{(\omega(\tau + \Delta\tau) - \omega(\tau))\, m}{\Delta\tau} \omega_l(\tau) \right\rangle_S = i\hbar\delta_{lm} \tag{2.20}$$

$l, m = 1, 2, 3$, where we have denoted by l, m the components of the paths in $\mathbf{R}^3$.

Feynman observes that, once translated in the more conventional language of operators in Hilbert space Eq. (2.19) is consistent with the Heisenberg equation of motion and Eq. (2.20) with the commutation relations between position Q and momentum P, namely

$$Q_l P_m - P_m Q_l = i\hbar\, \delta_{lm} \tag{2.21}$$

Eq. (2.20) shows clearly why we are in trouble here if we try to take the continuum limit $\Delta\tau \downarrow 0$. If typical paths $\tau \mapsto \omega(\tau)$ were smooth, their instantaneous velocity at time τ would be defined and the limits of the forward and backward increments (w. r. to the time τ) would, of course, coincide. After a reorganization of the relation (2.20) Feynman shows, in fact, that

$$\left\langle \frac{[\omega(\tau + \Delta\tau) - \omega(\tau)]^2}{\Delta\tau} \right\rangle_S = i\hbar\, \mathbf{1} \tag{2.22}$$

and concludes, in a way which is at the origin of the excitement of the probabilistics reading his first papers and consistent with (2.7), that the typical paths of a quantum mechanical particle are nondifferentiable" (p. 177, [3(2)]). Of course, the content of (2.22) is the same as the one of (2.7). The divergence of this instantaneous velocity is the simplest example of those which have plagued quantum field theory from its beginning, the "ultraviolet" divergence [15]. It is also the prototype of those which should be avoidable if a proper probabilistic framework was available for all this, since it would provide for free the regularization needed.

As observed by Feynman, in the simplest problems, the transition elements can sometimes be evaluated directly. He gives the example of a one dimensional (unit mass) free particle going from x to y during the time interval $I = [0,T]$ (p. 179–180 of [3(2)]). The transition element of the position at time τ is shown to satisfy, on the basis of Eq. (2.19) for a potential $V = 0$,

$$\langle\omega(\tau)\rangle_S = \left[x + \frac{\tau}{T}(y-x)\right] \langle 1\rangle \tag{2.23}$$

where $\langle 1\rangle$ denotes a normalization factor. Now, up to this irrelevant factor, the r. h. s. of Eq. (2.23) coincide with the underlying classical trajectory $\tau \mapsto q(\tau)$ compatible with the two given boundary conditions, *i. e.*

$$\langle\omega(\tau)\rangle_S = q(\tau) \quad \text{such that} \quad \ddot{q}(\tau) = 0, \quad q(0) = x, \quad q(T) = y \tag{2.24}$$

This feature is, in fact, valid for any quadratic action S.

For the transition element $\langle\omega(s)\,\omega(\tau)\rangle_S$ of the same free particle, the authors obtain ((7–64 of [3(2)])

$$\langle\omega(s)\,\omega(\tau)\rangle_S = \begin{cases} q(s)\,q(\tau) + i\,\hbar\,\dfrac{s(T-\tau)}{T} & s < \tau \\ q(s)\,q(\tau) + i\,\hbar\,\dfrac{\tau(T-s)}{T} & \tau < s \end{cases} \tag{2.25}$$

and stress the purely quantum mechanical interpretation of the second term in the r. h. s. of Eq. (2.25).

More generally, a fundamental application by Feynman of his approach lies in his interpretation of the method of "characteristic functional."

Assuming the basic action S to be quadratic let us consider the transition element

$$\left\langle e^{\frac{i}{\hbar}\int_s^t e(\tau)\,\omega(\tau)\,d\tau}\right\rangle_S \tag{2.26}$$

Comparing with Eq. (2.1) this means that we are now dealing with a new action

$$S'[\omega(.)] = S[\omega(.)] + \int_s^t e(\tau)\,\omega(\tau)\,d\tau \tag{2.27}$$

which is also quadratic and is regarded as a perturbation of the first one. Then Feynman shows that, up to the usual multiplicative factor,

$$\left\langle \exp\left(\frac{i}{\hbar}\int_s^t e(\tau)\,\omega(\tau)\,d\tau\right)\right\rangle_S = \exp\left(\frac{i}{\hbar}(S'[q] - S[q])\right) \tag{2.28}$$

where the r. h. s. is now computed over classical paths $q(.)$. As an illustration, he takes up a case already considered in his Ph. D. thesis, the forced harmonic oscillator, *i. e.*, for k a constant,

$$S[\omega(.)] = \int \left\{ \frac{1}{2} |\dot{\omega}(\tau)|^2 - \frac{k^2}{2} \omega^2(\tau) \right\} d\tau \tag{2.29}$$

and obtain, when $\omega(s) = x$, $\omega(t) = y$

$$\begin{aligned} \left\langle \exp \frac{i}{\hbar} \int_s^t e(\tau)\omega(\tau)\, d\tau \right\rangle_S &= \\ = \exp\Bigg\{ \left[\frac{i}{\hbar} \frac{k}{2 \sin k(t-s)} \right] \times \\ \Bigg[\frac{2y}{k} \int_s^t e(\tau) \sin k(t_1 - s)\, dt_1 + \frac{2x}{k} \int_s^t e(\tau) \sin k(t - t_1)\, dt_1 + \\ \frac{2}{k^2} \int_s^t \int_s^{t_1} e(t_2) \sin k(t_2 - s) \sin k(t - t_1) e(t_1)\, dt_1\, dt_2 \Bigg] \Bigg\} \end{aligned} \tag{2.30}$$

As well known, the main use of characteristic functionals is their relation with the moments of the underlying random variable. For example Feynman shows that

$$\langle \omega(\tau) \rangle_S = \left. \frac{\delta S'}{\delta e} \right|_{e=0} \langle 1 \rangle \tag{2.31}$$

and higher order moments, like the simple example (2.25), follow from higher order differentiation of Eq. (2.28).

We shall conclude this summary of Feynman's transition elements theory by a warning regarding the relation, in this path integral approach, between quantum observables in Hilbert space and associated "random variables." We have already seen such an example: the drifts forward and backward (2.6) and (2.10) associated with the momentum observable P of Eq. (2.9). In general, one may get into difficulties when trying to find the natural "random variable" corresponding to a given quantum observable. This is what happens, notably, for the Hamiltonian observable H of (1.1). After an heuristic reasoning, Feynman find the two following, time discretized (in order to avoid ultraviolet divergences) candidates (7.121 of [3(2)]).

$$H_1 = \frac{1}{2} \left(\frac{\omega(t_{k+1}) - \omega(t_k)}{t_{k+1} - t_k} \right)^2 + \frac{\hbar}{2i(t_{k+1} - t_k)} + V(\omega(t_{k+1})) \tag{2.32}$$

$$H_2 = \frac{1}{2} \left(\frac{\omega(t_{k+1}) - \omega(t_k)}{t_{k+1} - t_k} \right) \left(\frac{\omega(t_k) - \omega(t_{k-1})}{t_k - t_{k-1}} \right) + V(\omega(t_k)) \tag{2.33}$$

both of those candidates would, of course, define the same Hamiltonian observable in the quantum Hilbert space. Notice that the average value $\langle . \rangle_S$ of (2.32) is finite when $\Delta\tau = t_{k+1} - t_k \to 0$, in contrast with the average of the first term alone. This is in using perturbative arguments that Feynman justifies the presence of the extra (nonclassical) term in Eq. (2.32) along a line that he will illustrate masterfully in field theory.

At last, let us stress that Feynman regards the lack of continuum limit as a nuisance of his approach, to be removed: "At present, it requires an unnatural and cumbersome subdivision of the time interval to make the meaning of the equations clear" ([3(1)] p. 384).

3 Schrödinger's Euclidean Quantum Mechanics

Schrödinger's starting idea [2], which lies at the foundations of Euclidean Quantum Mechanics (EQM, in short) seem to have been inspired by the following observation of A. S. Eddington (in Gifford lectures, Cambridge 1928, p. 216, quoted in [2]) regarding the physical interpretation of quantum mechanics: "The whole interpretation is very obscure, but it seems to depend on whether you are considering the probability after you know what has happened or the probability for the purpose of prediction. The $\psi\overline{\psi}$ is obtained by introducing two symmetrical systems of ψ waves traveling in opposite directions in time; one of these must presumably correspond to probable inference from what is known (or is stated) to have been the condition at a later time."

Schrödinger considers the following problem of classical statistical mechanics where, as he stresses himself, the only difference with respect to traditional statistical theories lies in the way to ask the question:

The position of a Brownian particle in a domain $D \subset \mathbf{R}^3$ is observed at time 0 and found to be distributed according to $\rho_0(x)\,dx = \chi_*(x)\,dx$. From the classical theory, its probability density at a later time t should coincide with the solution $\eta^*_{\chi^*}(x,t)$ of the free heat equation

$$\frac{\partial\eta^*}{\partial t} = \frac{1}{2}\Delta\eta^* \tag{3.1}$$

with the relevant (positive) initial condition

$$\eta^*_{\chi^*}(x,0) = \chi^*(x)$$

and some additional constraints on the boundary ∂D of the domain D. This is familiar to physicists since A. Einstein [17].

Now suppose that at the later time T the probability density is observed and found to be $\rho_T(x) \neq \eta^*_{\chi^*}(x,T)$. This unlikely but possible situation suggests to Schrödinger the following problem: What is the most probable evolution of the probability density $\rho(x,t)$ on $[0,T]$ compatible with the second observation as well as with the first one?

Using a classical argument of statistical mechanics (extremization of entropy), Schrödinger shows that $\rho(x,t)$ belongs to a class formed by all the products

$$\rho(x,t) = \eta(x,t)\,\eta^*(x,t), \qquad 0 < t < T \tag{3.2}$$

where η^* is a non-negative solution of the heat equation (3.1), $t > 0$, such that, now, only the boundary condition on ∂D is preserved, and $\eta(x,t)$ a non-negative solution of the adjoint (or backward) equation

$$-\frac{\partial \eta}{\partial t} = \frac{1}{2}\Delta\eta, \qquad t < T \tag{3.3}$$

with the same boundary condition on ∂D as before. Does this product class contain a unique $\rho(x,t)$ such that $\rho(x,0) = \rho_0(x)$ and $\rho(x,T) = \rho_T(x)$, our given boundary probability densities?

Schrödinger shows that to answer to this question amounts to prove the existence and uniqueness of the solution of the following nonlinear system of equations for $\eta^*(x,0) \equiv \eta_0^*(x)$ and $\eta(x,T) \equiv \eta_T(x)$:

$$\begin{cases} \eta_0^*(x) \displaystyle\int_{\mathbf{R}^3} h(x,T,z)\,\eta_T(z)\,dz = \rho_0(x) \\ \eta_T(z) \displaystyle\int_{\mathbf{R}^3} \eta_0^*(x)\,h(x,T,z)\,dx = \rho_T(z) \end{cases} \tag{3.4}$$

where $h(x,T,z)$ denotes the integral kernel associated with the heat equation (3.1), *i. e.*

$$h(x,T,z) = \text{kernel}\,(e^{1/2T\Delta})(x,z), \qquad x, z \in \mathbf{R}^3 \tag{3.5}$$

since $H = -\frac{1}{2}\Delta$ is the (free) Hamiltonian.

Schrödinger observes [2(299)] that he did not succeed to prove the existence and uniqueness of the solution $\{\eta_0^*, \eta_T\}$ of the system (3.4), a kind of boundary value problem which "seems to be new even in mathematics," but that he is convinced that it holds true.

This problem will be considered successively by S. Bernstein, R. Fortet and solved quite generally by A. Beurling almost 30 years after [18]. As long as the Hamiltonian H is such that the integral kernel h is bounded and strictly positive, the system (3.4) (referred to afterwards as the "Schrödinger system"), for ρ_0 and ρ_T two given strictly positive probability densities, has indeed one and only one strictly positive (not necessarily integrable) solution $\{\eta_0^*, \eta_T\}$.

S. Bernstein was the first probabilistic to think about Schrödinger's problem. He reinterpreted this one as the construction of a stochastic process from the data of the joint probability density m for its starting and ending position on $[0,T]$ and Eq. (3.4) as a consistency relation for the marginals of this joint probability. Clearly, the usual Markovian transition probability concept should be replaced here by the one taking into account a position in the past and another in the future ("Reciprocal property").

Bernstein paper's [18(1)] contains a lot of fascinating observations in the perspective of the above mentioned "probabilistic" strategies for quantum physics, going from Feynman to Yasue. We shall only mention 3 of them:

1) The fundamental symmetry under time reversal of classical mechanics is lost, usually, by regular Markov processes.
2) In order to restore this symmetry, "reciprocal chains" $Z_i \equiv Z(t_i)$ such that the probability law of Z_i is determined only when an anterior random variable Z_{i-h} and a posterior one Z_{i+k} are known should be introduced, so that the additional knowledge of Z_j, $j < i-h$ and $j > i+k$ does not change the probability law of Z_i.
3) Such reciprocal chains (or processes) correspond in probability theory, to mechanical trajectories from the variational point of view but, in contrast to those, the reciprocal chains cannot be, in general, extended arbitrarily outside their initial time interval of definition.

The nature of the stochastic process underlying Schrödinger's problem has, actually, not been investigated before Jamison's works in the mid seventies [19]. His construction, however, was not directly appropriate to a typical quantum physical problem, where the only data is an Hamiltonian H like (1.1), for example, and not any given process like in [19(2)]. What we are going to summarize, and to compare systematically with Feynman's path integral approach is the construction needed for the Hamiltonian (1.1), as initiated in [8].

Let $h(s,x,t,y) \equiv h(x,t-s,y)$ be the integral kernel of (3.5), but for any V in the Kato class [21]. We are, then, in the conditions of validity of Beurling Theorem. For $s \le t \le u$ in a given time interval I, x,y,z in $\mathbf{R}^3$, define the "Bernstein transition density"

$$p(s,x,t,y,u,z) = \frac{h(s,x,t,y)\,h(t,y,u,z)}{h(s,x,u,z)} \tag{3.6}$$

This plays the role of the 3-point analogue of Markovian transition probability whose study was suggested by Bernstein. Notice that, by the semigroup property, for any Borelian B of $\mathbf{R}^3$

$$B \mapsto \int_B p(s,x,t,dy,u,z) \equiv P(s,x,t,B,u,z)$$

is indeed a probability measure.

Jamison has axiomatized in [19(1)] the other properties needed for the Bernstein transition density and proved the following:

Given a Bernstein transition $P(s,x,t,B,u,z)$ and m a probability measure on $\mathbf{R}^3 \times \mathbf{R}^3$, there is a unique process Z_t such that if $\mathcal{P}_t$ denotes the associated past filtration [10] up to time t and $\mathcal{F}_t$ the future one,

a) $E[f(Z_t)|\mathcal{P}_S \cup \mathcal{F}_u] = E[f(Z_t)|Z_s, Z_u] \quad \forall s < t < u$ in I, for any f bounded measurable and where $E[\ldots|a]$ denotes the conditional expectation, given a. This is the mathematical version of Bernstein second observation. The given measure m will be the joint probability for the process to be in two location at different times. In analogy with the Markovian case, the Bernstein transition is a conditional probability:

b) $P(s, Z_s, t, B, u, Z_u) = P(Z_t \in B | Z_s, Z_u)$ for $s \leq t \leq u$ in I (3.7)

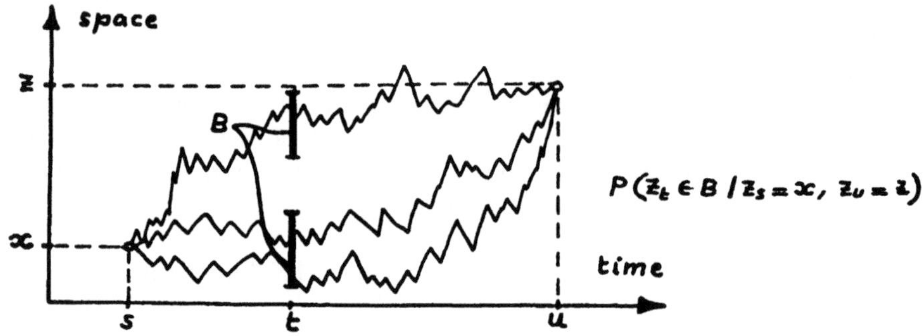

Fig. 2.

At last, the joint probability m and Bernstein transition P are used to produce the finite dimensional distributions on $[s, u]$:

c) $P(Z_s \in B_s, Z_{t_1} \in B_1, \ldots, Z_{t_n} \in B_n, Z_u \in B_u) =$

$$\int_{B_s \times B_1 \times \ldots \times B_n \times B_u} \cdots \int dm(x, z) \prod_{i=0}^{n-1} P(t_i, x_i, t_{i+1}, dx_{i+1}, u, z) \tag{3.8}$$

Of course if, following Schrödinger, we impose only the boundary probabilities on, say, $I = [s, u]$, $P_s(dx) = \rho_s(x)\, dx$ and $P_u(dz) = \rho_u(z)\, dz$, we are going to obtain a whole family of such processes (let's call them "Bernstein processes" [16][1]) since many joint probabilities m will have those 2 marginals. Only one, $m = M$, will be associated with a Markovian process, the one characterized precisely by Schrödinger's system (3.4):

$$\eta_s^*(x)\, h(x, u - s, z)\, \eta_u(z)\, dx\, dz = dm(x, z) \tag{3.9}$$

We claim that (3.9) is the probabilistic counterpart of the formal measure used by Feynman in Eq. (2.1), *i. e.*, in the notations used for (2.3)

$$\theta(x)\, K(x, u - s, z)\, \overline{\psi}(z)\, dx\, dz. \qquad (3.9)_F$$

Let us prove this.

[1] They have also been used sometimes under the name of "Schrödinger processes."

First, the substitution of the joint probability M (3.9) in (3.8) provides, in particular, (cf. (3.2))

$$P(z_t \in B) = \int_B \eta^*(y,t)\,\eta(y,t)\,dy \qquad \forall t \in I \tag{3.10}$$

where, by construction, η^* and η are two positive solutions on $[s,u]$ of

$$\begin{cases} -\hbar\,\dfrac{\partial \eta^*}{\partial t} = H\eta^* \\ \eta^*(y,s) = \eta^*_s(y) \end{cases} \quad \text{and} \quad \begin{cases} +\hbar\,\dfrac{\partial \eta}{\partial t} = H\eta \\ \eta(y,u) = \eta_u(y) \end{cases} \tag{3.11}$$

For an Hamiltonian H bounded below we can analytically continue the parameter t of Eqs. (3.10) and (3.11): $t \to i\tau$. Then Eqs. (3.10) and (3.11) become the usual (Born) probabilistic interpretation of the wave function in terms of a solution ψ of the Schrödinger equation (2.13) and its complex conjugate (2.15) (for appropriate boundary conditions). But, of course, in this case, the above mentioned construction of Bernstein measures will stop making sense. As a matter of fact, we will be back to Feynman's approach.

For obvious reason, we will refer to the relation (3.10) as the "Euclidean Born interpretation" [47]. It is the corner stone of Euclidean Quantum Mechanics and is, in principle, independent of the form of the Hamiltonian H.

The pair of heat equations (3.11) also shows clearly the fundamental role played, henceforth, by time reversal. This relation between η and η^* is the probabilistic counterpart of the L^2 complex conjugacy.

Notice that such observation has, of course, been made before EQM, [22] but not in relation with the construction of specific Markovian probability measures like Bernstein's ones.

Since our Bernstein measures are well defined, all the formal results of Feynman (Chap. 2) are going to make sense, after the analytical continuation in the time parameter defined by Eqs. (3.11).

As for transition elements, we can regard the stochastic process Z_t as a forward Markovian one. Then its transition probability density is, as suggested by (2.4),

$$q(s,x,t,y)\,dy = h(x,t-s,y)\,\frac{\eta(y,t)}{\eta(s,x)}\,dy \tag{3.12}$$

for $s \le t \le u$ in I, x, y in $\mathbf{R}^3$, and the finite dimensional distributions (3.8) reduce to the usual Markovian ones for an initial probability

$$\begin{aligned} P(z_s \in dx) &= P_s(dx) \\ &= \eta^*_s(x)\,\eta(x,s)\,dx \end{aligned} \tag{3.13}$$

in accordance with the Euclidean Born interpretation (3.10). In other words,

$$\begin{aligned} &P(Z_s \in B_s\,, Z_{t_1} \in B_1, \dots, Z_{t_n} \in B_n, Z_u \in B_u) = \\ &\int_{B_s} \int_{B_1} \cdots \int_{B_n} P_s(dx)\,Q(s,x,t,dx_1)\dots Q(t_n,x_n,u,B_u) \end{aligned} \tag{3.14}$$

where we have denoted $Q(s,x,t,B) = \int_B q(s,x,t,y)\,dy$.

But we can as well consider Z_t, $t \in [s,u]$, as a backward Markovian one, for the final probability

$$\begin{aligned} P(z_u \in dz) &= P_u(dz) \\ &= \eta^*(z,u)\,\eta_u(z)\,dz \end{aligned} \tag{3.15}$$

and the (backward) transition probability suggested by (2.5), namely

$$q^*(s,x,t,y)\,dx = \frac{\eta^*(s,x)}{\eta^*(t,y)}\,h(x,t-s,y)\,dx \tag{3.16}$$

so that the finite distributions (3.8) could be rewritten only in term of $P_u(dz)$ and

$$Q^*(s,B,t,y) = \int_B q^*(s,x,t,y)\,dx \tag{3.17}$$

It is now quite easy to guess (and to verify) that the well defined Bernstein processes associated with the Hamiltonian (1.1) are diffusion processes whose infinitesimal coefficients are just the one of Feynman, after analytical continuation in the time, namely, for any t in $[s,u]$,

$$B(y,t) = \hbar\,\frac{\nabla\eta}{\eta}(y,t) \tag{3.18}$$

$$C(y,t) = C^*(y,t) = \hbar\,\mathbf{1} \tag{3.19}$$

as suggested by (2.6) and (2.7), and

$$B_*(y,t) = -\hbar\,\frac{\nabla\eta^*}{\eta^*}(y,t) \tag{3.20}$$

as Feynman's formula (2.10) hints.

Notice that the forward and backward drifts B and B_* depends only on the respective positive solutions of the 2 heat equations (3.11) involved in the Euclidean Born interpretation (3.10). In consequence, no mixture will ever appear, in Euclidean Quantum Mechanics, between the information carried by the two filtrations $\mathcal{P}_t$ and $\mathcal{F}_t$ associated with the forward and backward descriptions of the processes. The backward one is the image of the forward one under time reversal (as long as the chosen boundary conditions reflect this symmetry, of course). Before going on with Feynman's analogy, one should emphasize that most of the Bernstein processes which are the rigorous counterpart of Feynman's formal diffusions will have little to do with quantum dynamics. Indeed the above mentioned construction works under quite general conditions and, in particular, no restriction on the boundary probabilities $P_s(dx)$, $P_u(dz)$ (except their positive character, needed to assure the uniqueness of the measure by Beurling's Theorem). The strong relation with quantum dynamic on $[s,u]$ will appear only for those P_s, P_u associated with a quantum evolution. A list of explicit, one dimensional, examples can be found in [23].

Now let us come to the famous Feynman's integration by parts formula (2.11). What we need is a well defined version of this formula for a class as large as possible of functionals $F[.]$ of Bernstein diffusions. Now consider a typical path space like the one underlying the path integral (2.14). Since we are going to need translation $\omega \mapsto \omega + \varphi$ in path space, like the ones introduced by Feynman to deal with Gaussian integrals (§3.5 in [3(2)]), it is sufficient to consider, for example

$$\Omega_0 = \{\omega \in C([0,T/2];\mathbf{R}^3);\ \omega(0) = 0\} \tag{3.21}$$

where we have taken advantage of the time symmetry of our construction by considering a time interval of definition of the process symmetric around the origin of time: $I = [-T/2, T/2]$.

Typical functional F of Brownian motion, like the ones handled by Feynman in (2.11) are highly irregular mathematical objects. For example, each of our Bernstein processes is solution of a certain stochastic differential equation (in the sense of K. Itô [24]) whose coefficients are given by (3.18–3.19). As such, it is a functional, generally quite complicated, of a Wiener process (or Brownian motion) W, not even continuous in the uniform norm topology. In addition, there are strong restrictions on the directions $\delta\omega$ allowed in variational derivatives $\delta F[\omega](\delta\omega)$. The key condition has been discovered by Cameron and Martin [25]: in its simplest version, *i. e.* when $s \mapsto \varphi(s)$ is deterministic, it says that $\varphi \in \Omega_0$ should be derivable and such that $\int_0^{T/2} |\dot{\varphi}(s)|^2 ds < \infty$. This defines the scalar product $(\varphi_1|\varphi_2)$ of an Hilbert space $\mathcal{H}$, the "Cameron-Martin" subspace of Ω_0. Consider the Wiener measure with parameter $\hbar$, denoted $\mathcal{M}_W^\hbar$ (we need this $\hbar$ because of Feynman's formula (2.22)). Then $(\Omega_0, \mathcal{H}, \mathcal{M}_W^\hbar)$ is called the "classical Wiener space" [20].

Thanks to Cameron-Martin restriction on φ, the measure induced by $\omega \mapsto \omega + \varphi$ is absolutely continuous with respect to the Wiener measure $\mathcal{M}_W^\hbar$ ($\varphi \in \mathcal{H}$ is necessary and sufficient for that) and there is a (nonnegative) density such that any expectation with respect to the new measure can be written in term of the original one. Notice, in contrast, that measures associated to processes with different diffusion coefficients are not absolutely continuous.

Consider a square integrable (*i. e.* $\in L^2(\Omega_0, \mathcal{M}_W^\hbar)$) functional F of the Wiener process, denoted $F[W]$. Let $U(s)$, $s \in [0, T/2]$ be any stochastic process such that $\int_0^{\cdot} U(s)\,ds$ exists and, instead of φ, denote now the "variation" by

$$\delta W(.) = \int_0^{\cdot} U(s)\,ds \tag{3.22}$$

The condition $E[\exp \lambda^2/2 \int_0^{T/2} U^2(s)\,ds] < \infty$, for a certain $\lambda > 0$, is a generalizations of the above mentioned Cameron-Martin one, sufficient to guarantee the absolute continuity of $\mathcal{M}_{W+\delta W}^\hbar$ with respect to $\mathcal{M}_W^\hbar$ when φ becomes random. Then we are allowed to define the variational derivative of $F[W]$ in direction δW by

$$\nabla F[W](\delta W) = \lim_{\epsilon \to 0} \left\{ \frac{F[W(.) + \epsilon\delta W(.)] - F[W(.)]}{\epsilon} \right\} \tag{3.23}$$

Feynman would denote this by $\delta F[W](\delta W)$.

Using the Cameron-Martin-Girsanov Theorem [37(2)] and the definition of $\mathbb{\nabla} F$, one verifies that the following integration by parts formula is valid

$$E[\mathbb{\nabla} F[W](\delta W)] = E\left[F[W]\int_0^{T/2}\frac{d}{ds}\delta W(s)\,dW(s)\right] \tag{3.24}$$

These are the prolegomena to Malliavin stochastic calculus of variations [26], as reinterpreted by Bismut [27], an elaborated mathematical machinery which provided the well defined tools needed to make sense of Feynman's functional calculus (as well as Yasue's one in [6], for example). Notice that since $\mathbb{\nabla} F$ can be regarded as a linear functional on the Hilbert space $\mathcal{H}$ the l. h. s. of (3.24) can also be written as $E[(\mathbb{\nabla} F|\delta Z)_1]$.

The integral in the right hand side of (3.24) is an Itô's integral [24], called the "divergence of δW." The relation (3.24) can be considerably generalized. In particular, we shall need its extension to measures of Bernstein diffusions absolutely continuous with respect to the Wiener measure $\mathcal{M}_W^{\hbar}$, with Radon-Nikodym density $\rho[Z(.)]$.

Given the form (3.12) of the forward transition probability of a Bernstein diffusion Z_t, and assuming that the potential V satisfies the hypothesis of Feynman-Kac formula [28], it is easy to show that [29]

$$\rho[Z] = \exp -\frac{1}{\hbar}\int_0^{T/2} V(Z(t))\,dt\,\frac{\eta(Z(T/2),T/2)}{\eta(Z(0),0)} \tag{3.25}$$

Using this, one proves that the well defined version of Feynman's integration by parts formula (2.11) is

$$E[(\mathbb{\nabla} F[Z]|\delta Z)_1] = \frac{1}{\hbar}E\Bigg[F[Z]\Bigg\{\int_0^{T/2}\frac{d}{ds}\delta Z(s)\,dZ(s) + \int_0^{T/2}\delta Z(s)\,\nabla V(Z(s))\,ds - $$

$$+\,\delta Z(T/2)\frac{\nabla\eta}{\eta}(Z(T/2)) + \delta Z(0)\frac{\nabla\eta}{\eta}(Z(0))\Bigg\}\Bigg] \tag{3.26}$$

The demonstration, as well as the regularity conditions of the functionals F and the variations δZ, can be found in [29].

Notice that the relation (3.26) "chooses a sense of time." Indeed, although Bernstein diffusions are time-symmetric, we have used their forward transition probability (3.12) to obtain (3.26) and the description of diffusions in terms of transition probability involves the past-future direction [10].

However, the time symmetry of our framework is restored by observing that we could as well derive an integration by parts formula with respect to the other filtration, in other words using the density ρ_* of the Bernstein process with respect to the backward Wiener measure $\mathcal{M}_{W^*}^{\hbar}$ on $C([-T/2,0];\mathbf{R}^3)$, namely

$$\rho_*[Z] = \frac{\eta^*(Z(-T/2),-T/2)}{\eta^*(Z(0),0)}\exp -\frac{1}{\hbar}\int_{-T/2}^{0} V(Z(t))\,dt \tag{3.27}$$

and the backward transition probability (3.16) (cf. [29]).

In order to extract useful information form formula (3.26) we need to introduce the concept of forward and backward derivatives of a regular function f of a Bernstein diffusion. In relation with quantum mechanics, they are due to Nelson [30] but were familiar to probabilistics before. By definition, the forward one is the limit conditional expectation

$$\begin{aligned} Df(Z(t),t) &= \lim_{\Delta t \downarrow 0} E\left[\frac{f(Z(t+\Delta t), t+\Delta t) - f(Z(t),t)}{\Delta t} \middle| Z(t) \right] \\ &= \left(\frac{\partial}{\partial t} + \hbar \frac{\nabla \eta}{\eta} \nabla + \frac{\hbar}{2} \Delta \right) f(Z(t),t) \end{aligned} \tag{3.28}$$

The second line results from the first one by Itô's Theorem [24] and the use of the definitions (3.18) and (3.19). Notice the last, purely quantum, additional term which would disappear if the path $t \mapsto Z(t)$ was smooth. Also observe that the conditioning $Z(t)$ is in the past of the time interval considered before taking the limit. For example, if we know that $Z(t) = y$, the formula (3.28) when $f(y,t) = y$ provides

$$DZ(t) = \hbar \frac{\nabla \eta}{\eta}(Z(t),t) \tag{3.29}$$

showing that the forward drift (3.18) results, indeed, from the best possible forward estimation of the velocity of a Bernstein diffusion given its position at an early time.

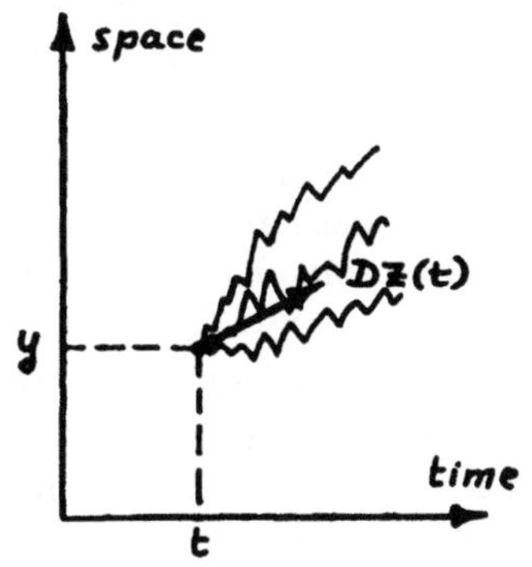

Fig. 3.

Again, Eq. (3.28) cannot be time symmetric since it involves the forward transition probability. The analogue definition for a conditioning in the past of the time interval of interest provides

$$D_* f(Z(t),t) = \left(\frac{\partial}{\partial t} - \hbar \frac{\nabla \eta^*}{\eta^*} \nabla - \frac{\hbar}{2} \Delta \right) f(Z(t),t) \tag{3.30}$$

involving the backward drift (3.20). The minus sign in front of the purely quantum term is due to the use of the decreasing filtration $\mathcal{F}_t$ in Itô's calculus. We take advantage, in Eqs. (3.28) and (3.30), of the existence of the Bernstein measures to take the continuum limit $\Delta t \downarrow 0$, in contrast with what Feynman can do with formal derivatives appearing, for example, in (2.20). We also observe that these limit random variables (3.28) and (3.30) are, in general, distincts. This is hardly a surprise since $t \mapsto Z(t)$ carries all the irregularities of the Brownian trajectories.

Since the Markov property kills off the past (or future) information relative to t, we could replace the conditioning $Z(t)$ in the definition (3.28) by $\mathcal{P}_t$.

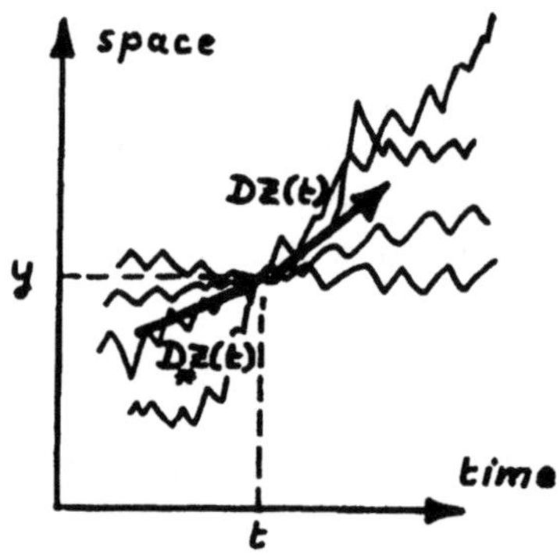

Fig. 4.

For Bernstein but not Markov diffusions such forward (and backward) derivatives would still make sense so that part of what we are going to obtain is of a larger generality than it seems (cf. [31]).

It would be a mistake to believe that those forward and backward derivatives are conceptually new in elementary quantum physics. Feynman's relation (2.20) shows clearly that, if we could take some continuum limit of the forward and backward increments $\frac{\omega(\tau+\Delta\tau)-\omega(\tau)}{\Delta\tau}$ and $\frac{\omega(\tau)-\omega(\tau-\Delta\tau)}{\Delta\tau}$ (and our inability to do this is merely due to mathematical difficulties) the two limit "random variables" would be distincts, otherwise the Heisenberg principle would be violated. And, to go back much early in the past, it is fascinating to notice that, in his discussion on the physical content of quantum kinematics [32], the argument used by W. Heisenberg to rule out the concept of trajectories for quantum particles is, precisely, that any position "is associated with two velocities."

Now we are ready to consider the Euclidean analogues of Feynman's formula (2.11) (cf. (2.18) and subsequent).

For $F[Z] = 1$, the left hand side of Eq. (3.26) is zero. The Itô's integral in the r. h. s. can be replaced by a usual (Riemann's) integral where $dZ(s)$ is replaced by $DZ(s)\,ds$. Since $\delta Z(s)$ is of bounded variation and such that $\delta Z(t) = 0$, for any regular f, $E\int_t^{T/2} \frac{d}{ds}\delta Z(s)\, f(Z(s),s)\,ds = -E\int_t^{T/2} \delta Z(s)\, Df(Z(s),s)\,ds +$

$E[\delta Z(T/2)\, f(Z(T/2),T/2)]$. So we obtain, if E denotes here the conditional expectation $E_{y,t}$,

$$0 = E\left[\int_t^{T/2} \delta Z(s)\left\{-DDZ(s) + \nabla V(Z(s))\right\} ds\right] +$$

$$+ E\left[\delta Z(T/2)\left\{DZ(T/2) - \hbar\frac{\nabla\eta}{\eta}(Z(T/2))\right\}\right]$$

and from the validity of this relation for a large class of variation δZ we deduce that

$$\begin{cases} DDZ(s) = \nabla V(Z(s)) & s \text{ in } [t,T/2] \\ Z(t) = y, \quad DZ(T/2) = \hbar\dfrac{\nabla\eta}{\eta}(Z(T/2)) \end{cases} \tag{3.31}$$

This is the probabilistic counterpart of Feynman's equation (2.19). The change of sign of the forces, with respect to Feynman, is expected in any Euclidean framework since $t \mapsto i\tau$. On the other hand, the Euler-Lagrange equation (3.31) holds without expectation (almost surely) and, therefore, is much stronger than Feynman's result (2.19). As in classical mechanics, of course, we do not need a variational argument to derive the equation of motion (3.31). Let us mention another way, quite rich in structures, both physical and mathematical. Actually, it will provide us with a direct probabilistic interpretation of Eq. (3.31).

Let us define, for η the underlying regular positive solution of Eq. (3.11),

$$A(y,t) = -\hbar \log \eta(y,t). \tag{3.32}$$

This is the Euclidean version of the nonlinear change of variable made originally by E. Schrödinger, in 1926 [33], in order to relate the classical theory of Hamilton-Jacobi (left hand side of (3.32)) with the solution of a linear equation bound to a certain success: his wave equation (2.13). Here, the Hamilton-Jacobi equation reads

$$\begin{cases} -\dfrac{\partial A}{\partial t} + \dfrac{1}{2}|\nabla A|^2 - \dfrac{\hbar}{2}\Delta A - V = 0, \quad t \le T/2 \\ A_{T/2}(y) = -\hbar \log \eta_{T/2}(y) \end{cases} \tag{3.33}$$

where the $\hbar\Delta A$ reflects the presence of the underlying Bernstein diffusion and, therefore, represents the quantum correction to classical mechanics.

This equation is quite famous in the theory of optimal stochastic control for continuous time Markov processes. In this context, it is called "Hamilton-Jacobi-Bellman equation" and Flemings refers to (3.32) as the "Logarithmic transformation" [34].

Eqs. (3.33) is associated with the solution of a problem of optimal control, namely the minimization of the functional

$$S = E_{y,t}\int_t^{T/2} L(B,X,s)\, ds - E_{y,t}\, A_{T/2}(X(T/2)) \tag{3.34}$$

whose Lagrangian is, here,

$$L(\dot{q}, q, s) = \frac{1}{2}|\dot{q}|^2 + V(q) \tag{3.35}$$

as expected for the simple physical system under consideration. It is assumed, for Eq. (3.34), that X is solution of a forward stochastic differential equation in the sense of Itô [24]:

$$dX(s) = B(s)\, ds + \sqrt{\hbar}\, dW(s) \tag{3.36}$$

whose unknown drift B is a so-called "admissible control process" which may, a priori, event depend on the past $\mathcal{P}_S$ (or, equivalently, is progressively measurable with respect to $\mathcal{P}_S$) and should satisfy integrability conditions of the form

$$E \int_t^{T/2} |B(s)|^m\, ds < \infty, \qquad m = 1, 2, \ldots$$

In Eq. (3.36), $W(s)$ denotes, as before, a $\mathcal{P}_S$-Wiener process. Then it is shown that the optimal control, solution of the minimization problem for (3.34) is, in fact, Markovian and precisely of the form

$$\begin{aligned} B(X, S) &= -\nabla A(X, S) \\ &= \hbar\, \frac{\nabla \eta}{\eta}(X, S) \end{aligned} \tag{3.37}$$

where the definition (3.32) has been used. In other words, the forward Bernstein diffusions Z constructed before can be entirely characterized in this way. The relation between this approach of optimal stochastic control and Schrödinger's Euclidean Quantum Mechanics has been established in 1985 (cf. [8]). Given the general nature of some of the methods developed since then (cf. [34]) it is expected that this line of thought will be even more influential, for EQM, in the future. We shall limit ourselves, here, to a few remarks:

1) The heuristic account given here refers to "classical," or smooth, solutions of the nonlinear Eq. (3.33). This is much too restrictive in general, and especially for quantum physics. Spectacular progress have been made about this, in recent years. The theory of "viscosity solutions" [35(1–2)] is a natural framework to study the Hamilton-Jacobi-Bellman equation and provides, under weak restrictions, existence and uniqueness of its solution for appropriate final condition (and at the boundary of the state space). Since (3.37) holds, this method should provide existence and uniqueness of Bernstein diffusions, under very weak conditions. Notice that those results should also be valid in infinite dimensions, *i. e.* for quantum field theory [35(3)].
2) In classical mechanics, the fundamental relation between the action S, regarded as a function of the initial point (y, t) of the trajectories over $[t, T/2]$, and the Lagrangian L is (cf. [36], p. 442)

$$\frac{d}{dt} S(q(t), t) = -L(\dot{q}(t), q(t), t) \tag{3.38}$$

The change of sign with respect to Goldstein is due to the fact that he considers instead the action as a function of the final point and therefore a "path space" of the form $\Omega^{y,t}$.

A key ingredient of the above mentioned stochastic generalization relies on the fact that the Hamilton-Jacobi equation (3.33) can also be read as the following identity on the (forward) Bernstein diffusion

$$DA(Z(t),t) = -L(DZ(t), Z(t), t) \qquad \text{(a. s)} \tag{3.39}$$

i. e. the stochastic version of the classical equation (3.38). Using Dynkin formula [37(1)], we identify the minimal value of the action S on the Bernstein diffusion with $A(y,t)$. Since Eq. (3.33) is the stochastic version of the classical Hamilton-Jacobi, it is therefore natural to define the energy random variable by

$$\begin{aligned}\mathcal{E}(y,t) &= -\frac{\partial A}{\partial t}(y,t) \\ &= -\frac{1}{2}B^2(y,t) - \frac{\hbar}{2}\nabla B(y,t) + V(y)\end{aligned} \tag{3.40}$$

where (3.37) has been used. We observe with satisfaction that, up to the analytical continuation in the time, it is consistent with Feynman's time discretized candidate (2.32) for the Hamiltonian random variable. His other candidate (2.33) has the same expectation as $\mathcal{E}$.

We conclude this remark by observing that, like in classical mechanics, the gradient of the stochastic Hamilton-Jacobi equation (3.33) coincides with the equation of motion (3.31).

3) All what as been said with respect to the forward filtration $\mathcal{P}_t$ has a backward counterpart, following from the backward version of integration by parts formula (3.26). So the following backward equation of motion holds on $[-T/2, t]$,

$$\begin{cases} D_* D_* Z(s) = \nabla V(Z(s)) & -T/2 \le s \le t \\ D_* Z(-T/2) = -\hbar \dfrac{\nabla \eta^*}{\eta^*}(Z(-T/2)), \quad Z(t) = y & \end{cases} \tag{3.41}$$

clearly relevant for the path space $\Omega^{y,t}$. The nonlinear change of variable

$$A^*(y,t) = -\hbar \log \eta^*(y,t) \tag{3.42}$$

for the positive solution of the other heat equation (3.11) provides an Hamilton-Jacobi equation adjoint to (3.33) with respect to the time, namely

$$\begin{cases} \dfrac{\partial A^*}{\partial t} + \dfrac{1}{2}|\nabla A^*|^2 - \dfrac{\hbar}{2}\Delta A^* - V = 0, \quad t \ge -T/2 \\ A^*_{-T/2}(y) = -\hbar \log \eta^*_{-T/2}(y) \end{cases} \tag{3.43}$$

The action functional associated with this equation is

$$S^* = E^{y,t} \int_{-T/2}^{t} L(D_* Z(s), Z(s), s)\, ds - E^{y,t} A^*_{-T/2}(Z(-T/2)) \qquad (3.44)$$

which corresponds to the minimal value $A^*(y,t)$ of a problem of stochastic optimal control, reached on the Markovian Bernstein diffusion $Z(s)$ solving Itô's $\mathcal{F}_t$-stochastic differential equation, with drift $D_* Z(s) = B_*(Z(s), s)$ as in (3.20).

Notice that, in the stochastic action functional (3.44), the conditioning $Z(t) = y$ is this time in the future of the relevant time interval so that, regarded as a function of y, this is the regularization of Goldstein classical action (cf. 2)), using the drift $D_* Z(s)$. It is worth observing that $S^* = S^*[Z(.)]$ is time reversed of the functional $S[Z(.)]$ defined by (3.34). Of course, we have

$$B_*(y,t) = \nabla A^*(y,t) = -\hbar \frac{\nabla \eta^*}{\eta^*}(y,t) \qquad (3.45)$$

and the backward energy corresponding to (3.40) is

$$\mathcal{E}_*(y,t) = \left(-\frac{1}{2} B_*^2 + \frac{\hbar}{2} \nabla B_* + V\right)(y,t) \qquad (3.46)$$

Notice the change of sign in (3.46) with respect to the forward version (3.40), due to the use of backward Itô's calculus (equivalently $B \mapsto -B_*$ under time reversal).

Let us consider, now, the Euclidean analogue of Feynman's second application of his integration by parts formula (2.1). Choosing, like him, $F[Z] = Z(t)$ in the formula (3.26) and a special direction $\delta Z(s)$ (cf. [29]) compatible with the conditions of validity of this formula, one shows that

$$E\big[Z^l(t)\, D_* Z^m(t) - D Z^m(t)\, Z^l(t)\big] = \hbar \delta^{lm} \qquad (3.47)$$

which is the probabilistic version of Feynman's heuristic one (2.20), where we have, once again, taken advantage of the existence of the continuum limit. The necessity of the existence of two "momenta random variables" at a given position $Z(t)$ is now quite obvious: it is a kinematical requirement of the quantum theory of particles, along the line suggested by Heisenberg and Feynman.

Of course, Feynman's relation (2.22) becomes just the definition of the diffusion coefficient of $Z(t)$:

$$\lim_{\Delta t \downarrow 0} E_t \left[\frac{(Z(t) - Z(t-\Delta t))^2}{\Delta t}\right] = \hbar \mathbf{1} \qquad (3.48)$$

This statement is invariant under time reversal.

In order to understand (2.28), it is better to do first some comments on the Euclidean path integrals and their semiclassical limit.

According to Eqs. (3.34) and (3.37) the past condition functional, evaluated on the minimizing Bernstein diffusion Z is

$$E_{y,t}\int_t^{T/2} L(DZ(s),Z(s))\,ds - E_{y,t}\,A_{T/2}(Z(-T/2))$$

But, from Eq. (3.39) and Dynkin formula, this is just $A(y,t)$, the solution of (3.33). Using (3.32) we obtain the following path integral representation of the positive solution η of the heat equation (3.11):

$$\eta(y,t) = \exp\left\{-\frac{1}{\hbar}E_{y,t}\int_t^{T/2} L(DZ(s),Z(s))\,ds + E_{y,t}\log\eta_{T/2}(Z(T/2))\right\} \tag{3.49}$$

and quite symmetrically for the solution of the adjoint heat equation,

$$\eta^*(y,t) = \exp\left\{-\frac{1}{\hbar}E^{y,t}\int_{-T/2}^{t} L(D_*Z(s),Z(s))\,ds + E^{y,t}\log\eta^*_{-T/2}(Z(-T/2))\right\} \tag{3.50}$$

The representations (3.49) and (3.50) are the Euclidean versions of Feynman's path integrals (2.14) and (2.12). Notice that although Feynman's formulas are, roughly, average of (complex) exponentials, we are dealing here with exponentials of expectations. On the other hand, both representations involve nothing else (this is essential) than the classical Lagrangian and boundary conditions. And, in the Euclidean formulation, there is nothing formal in the continuum limit, which is built into the path integrals from the beginning.

Let us consider the semiclassical limit of our path integrals, as a test of their validity.

Since, in contrast with Feynman, we have at our disposal well defined probability measures, it makes sense to consider expansion of the Bernstein diffusions $Z(t)\equiv Z^{\hbar}(t)$ in term of the Planck constant $\hbar$, as usually in the semiclassical limit. Functionals of $Z^{\hbar}(t)$ can be expanded in the same way (cf., for example, [37(2)]).

In one dimension, for simplicity, if

$$Z^{\hbar}(t) = q(t) + \sqrt{\hbar}\,Z^{sc}(t) + o\left(\sqrt{\hbar}\right) \tag{3.51}$$

where $q(t)$ denotes the relevant smooth solution of the classical (Euclidean) equation of motion and $Z^{sc}(t)$ a Gaussian diffusion, those $\hbar$-expansions of the path integrals (3.49) and (3.50) reduce respectively to

$$\eta(y,t) =$$

$$= \frac{1}{\sqrt{2\pi|F(t)|}}\exp -\frac{1}{\hbar}\left\{\int_t^{T/2} L(\dot q_{y,t}(s),q_{y,t}(s))\,ds + A^0(q_{y,t}(T/2)) + o(\hbar)\right\} \tag{3.52}$$

and

$$\eta^*(y,t) =$$

$$= \frac{1}{\sqrt{2\pi|F_*(t)|}} \exp -\frac{1}{\hbar}\left\{\int_{-T/2}^{t} L(\dot{q}^{y,t}(s), q^{y,t}(s))\, ds + A_*^0(q^{y,t}(-T/2)) + o(\hbar)\right\} \tag{3.53}$$

where $q_{y,t}(.)$ and $q^{y,t}(.)$ denote respectively the relevant smooth solutions of the Euclidean classical equations of motion, *i. e.* the classical limit of Eq. (3.31) and (3.41). The location of the suffix (y,t) indicates that those classical trajectories solve the same equations (since $D = D_* = \frac{d}{dt}$ at the classical limit) but with different boundary conditions. An asymptotic representation like (3.53) for the solution of the heat equation (3.11) is quite familiar ([38]). So, in this respect although, as a quantization method, Schrödinger's Euclidean Quantum Mechanics is distinct from the usual ones, the basic path integrals coincide with was we already know at the semiclassical limit. However, we deal with two path integrals and this provides us with new structures which are specific to this approach. In the representations (3.52) and (3.53), F and F_* are two linearly independent solutions of the Jacobi equation for the considered physical system. Let us summarize where they come from (cf. [40–41]).

For our simple Lagrangian (3.35) the Jacobi equation reduces to

$$(C^{-1}q)(t) \equiv \left(-\frac{d^2}{dt^2} + \nabla^2 V(q(t))\right) q(t) = 0, \qquad t \in [-T/2, T/2] \tag{3.54}$$

where C denotes the covariance operator. Using C^{-1}, the second variation of the classical action functional is often written as the norm of a Sobolev space of paths

$$\| q \|_1^2 = \int_{-T/2}^{T/2} q(\tau)\,(C^{-1}q)(\tau)\, d\tau.$$

The role of the boundary conditions, for Eq. (3.54), *i. e.* of the domain $\mathcal{D}(C^{-1})$, will be more subtle, here, than in the conventional Euclidean approaches (cf. [39] for example). Let $\begin{pmatrix} F_{T/2} \\ \dot{F}_{T/2} \end{pmatrix}$ be an arbitrary pair of final boundary conditions for Eq. (3.54), such that $|F_{T/2}| + |\dot{F}_{T/2}| > 0$. With this, the relevant $\mathcal{D}(C^{-1})$ for the ("future") time interval $[t, T/2]$ can be defined as the Sturm-Liouville ones:

$$\begin{aligned} \mathcal{D}_t(C^{-1}) = \{ & q(.) \text{ in } C^2([t,T/2]) \text{ such that} \\ & q(t) = y \qquad \text{and } F_{T/2}\dot{q}(T/2) - \dot{F}_{T/2}q(T/2) = \delta \} \end{aligned} \tag{3.55}$$

where δ is a constant.

Symmetrically, for $[-T/2, t]$, we shall use

$$\begin{aligned} \mathcal{D}^t(C^{-1}) = \{ & q(.) \text{ in } C^2([-T/2,t]) \text{ such that} \\ & \dot{F}^*_{-T/2}q(-T/2) - F^*_{-T/2}\dot{q}(-T/2) = \delta^* \qquad \text{and } q(t) = y \} \end{aligned} \tag{3.56}$$

where $\begin{pmatrix} F^*_{-T/2} \\ \dot{F}^*_{-T/2} \end{pmatrix}$ are initial boundary conditions for (3.54) not both zero but, otherwise, arbitrary and δ^* is a constant. Clearly the constants δ, δ^* are associated respectively to the boundary terms A^0 and A^0_* of (3.52) and (3.53). The two domains (3.55) and (3.56) introduce an arrow of time, due to the "conditioning $q(t) = y$." They correspond to what is needed for the representations (3.52) and (3.53) respectively. But we know that any Bernstein diffusion on $[-T/2, T/2]$ has, for any time in between, a probability density of the product form $\eta\eta^*(y,t)$. Using this for the path integrals (3.52), (3.53) one shows that ([40–41]) the underlying Bernstein diffusion is, as expected, Gaussian with expectation m and covariance c given respectively by

$$\begin{cases} m(t) = W^{-1}(F(t)\delta^* + F^*(t)\delta) \\ c(t) = W^{-1}(F(t)\,F^*(t)) \end{cases} \tag{3.57}$$

where $W = W(F, F^*) = \dot{F}^* F - F^* \dot{F}$ is the Wronskian of F and F^*. The expectation $m(t)$ is a solution of the Jacobi equation (3.54) compatible, now, with the boundary conditions at $-T/2$ and $T/2$ used in the two above mentioned domains $\mathcal{D}_t$, $\mathcal{D}^t$. When those two manifolds are themselves invariant under time reversal, F and F_* are time reversed of each other and the Gaussian measure determined by (3.57) comes, indeed, from quantum mechanics.

All the probability measures regarded as relevant to quantum mechanics by conventional approaches are included in the present class. With the many others (the relation between solutions of the Schrödinger equation and Bernstein

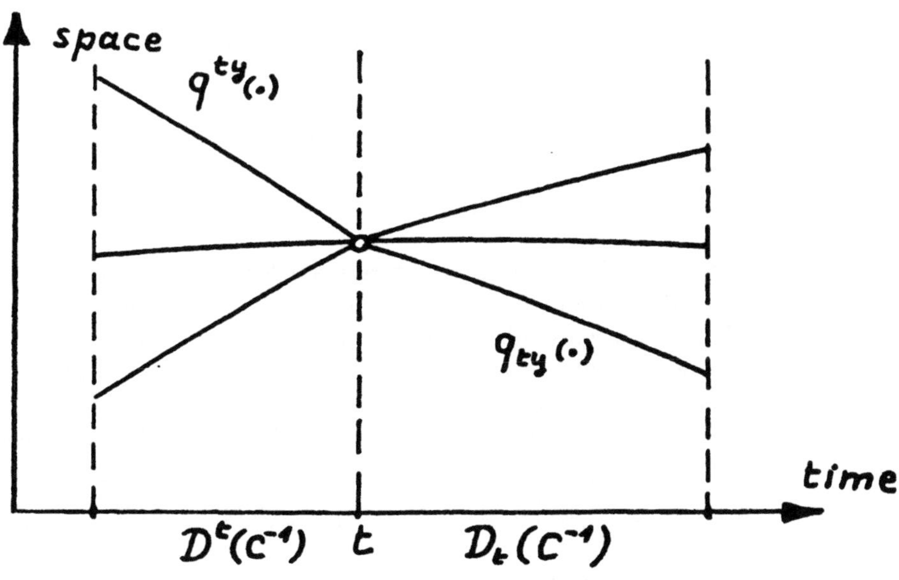

Fig. 5.

measures is one-to-one) we see emerge new mathematical structures dues intrinsically to the time symmetry of Schrödinger's Euclidean Quantum Mechanics. As a part of his Ph. D. thesis, A. Brandão has studied the symplectic structure of Gaussian Bernstein measures whose presence is manifested by the role of the Wronskian $W(F, F^*)$ [42].

With those results, all the Gaussian strategies of Feynman have their probabilistic analogue. For example, let us consider, like him, (2.28) the case of a Gaussian Bridge ($Z(s) = x, Z(t) = y$). After analytical continuation in the time parameter, the left hand side becomes a well defined probabilistic object:

$$E\left[\exp -\frac{1}{\hbar}\int_s^t e(\tau)\, Z(\tau)\, d\tau\right] \tag{3.58}$$

Now it is a familiar result of probability theory [43] that this functional of the Gaussian process Z can be computed explicitly and gives

$$\exp\left(-\frac{1}{\hbar}\left\{\int_s^t e(\tau)\, m(\tau)\, d\tau - \int_s^t \int_s^{t_2} e(\tau_1)\, K(\tau_1, t_2)\, e(t_2)\, d\tau_1\, dt_2\right\}\right) \tag{3.59}$$

where $m(\tau)$ is the expectation and $K(\tau_1, \tau_2)$ the correlation of the process $Z(\tau)$. Now consider the classical (Euclidean) action S' associated with the perturbation by the source $e(\tau)$ of the (quadratic) action S, as in (2.27). Then it is easy (but tedious) to see that those classical actions are related by

$$\begin{aligned} S'(s,x,t,y) = S(s,x,t,y) &+ \int_s^t e(\tau)\, m_x^y(\tau)\, d\tau \\ &- \int_s^t \int_s^{t_2} e(\tau_1)\, K_s^t(\tau_1, t_2)\, e(t_2)\, d\tau_1\, dt_2 \end{aligned} \tag{3.60}$$

where

$$\begin{cases} m_x^y(\tau) = W^{-1}(xF^t(\tau) + yF_s^*(\tau)) \\ K_s^t(\tau_1, t_2) = W^{-1}(F_s^*(\tau_1)F^t(t_2)), \qquad \tau_1 < t_2. \end{cases}$$

Here F_s^*, F^t both solve the Jacobi equation relevant for the quadratic action S, with past boundary conditions $F_s^*(s) = 0$, $\dot{F}_s^*(s) = 1$ and future ones $F_t^*(t) = 0$, $\dot{F}_t^*(t) = -1$. This is the particular case of (3.57) where $\delta^* = x$, $\delta = y$ (since the expectation $m(\tau)$ should be such, here, that $m(s) = x$, $m(t) = y$), and W is the Wronskian of F^t, F_s^*, namely the case of the "Bernstein Bridge" $Z_{sx}^{ty}(\tau)$ associated with S (cf. [40]). The comparison of (3.60) with (3.59) shows that

$$E\left[\exp -\frac{1}{\hbar}\int_s^t e(\tau)\, Z_{sx}^{ty}(\tau)\, d\tau\right] = \exp -\frac{1}{\hbar}(S'(q) - S(q)) \tag{3.61}$$

as predicted by Feynman in (2.28).

From this, the Euclidean version of correlations for quadratic cases follows easily. For example, for the free case of (2.24), (2.25) one obtains

$$E[Z(s)\,Z(t)] = \begin{cases} q(s)\,q(\tau) + \hbar\dfrac{s(T-\tau)}{T}, & s < \tau \\ q(s)\,q(\tau) + \hbar\dfrac{\tau(T-s)}{T}, & \tau < s \end{cases}$$

where $q(.)$ is defined as in real time, since the particle is free.

Much more could be said about the semiclassical case but we hope to have been able to convince the reader that Feynman's transition elements theory is entirely preserved in this Euclidean framework. Now let us describe a "quantum version" of a fundamental Theorem of Lagrangian mechanics, that Feynman did not find in his path integral approach.

4 The Theorem of Noether

The Theorem of Noether is, after the Least Action Principle, the most important Theorem of Lagrangian mechanics. It relates the invariance of the action functional $S[q(.)]$ under particular transformations of space and time to the existence of constants of motion for the Euler-Lagrange equations resulting from the Least Action Principle. One of the reasons of its importance is that it holds for dynamical systems with arbitrary number of degrees od freedom, *i. e.* also in classical field theories [36]. More precisely

If $S[q(.)]$ is invariant under a one parameter (α) family of transformations,

$$\begin{cases} q^\alpha = q + \alpha X(q,t) \\ \tau^\alpha = \tau + \alpha T(q,t) \end{cases} \tag{4.1}$$

where X and T are the infinitesimal generators of the transformation, *i. e.* for any given s, u,

$$\int_s^u L(\dot{q}(t), q(t), t)\,dt =$$
$$= \int_{\tau_s^\alpha}^{\tau_u^\alpha} \left\{ L(\dot{x}(\tau^\alpha), x(\tau^\alpha), \tau^\alpha) + \alpha \frac{d}{d\tau^\alpha}\phi(x(\tau^\alpha), \tau^\alpha) \right\} d\tau^\alpha + o(\alpha) \tag{4.2}$$

where $x(\tau^\alpha) \equiv q^\alpha(t)$ is a family of space-time changed trajectories according to (4.1) and ϕ is a given "divergence term," then, along each extremal of the action $S[q(.)]$,

$$\left[\left(\frac{\partial L}{\partial \dot{q}}\right) X - \left(L - \frac{\partial L}{\partial \dot{q}}\dot{q}\right) T + \phi \right] (q(t), t) = \text{constant} \tag{4.3}$$

One identifies the factor in front of X as the momentum at time t, $p(t) = \frac{\partial L}{\partial \dot{q}}$ and the one in front of T as minus the energy $-H(t) = (L - \frac{\partial L}{\partial \dot{q}}\dot{q})$.

For example if

1) $X = 0$, $T = 1$, $\phi = 0$ the action functional is invariant under time translation. This is the case when L is not explicitly dependent on the time. Then Noether's Theorem says that the energy H is conserved.
2) If $S[q(.)]$ is invariant under space translation $X = 1, T = 0, \phi = 0$, we obtain the conservation of momentum $\frac{d}{dt}p(t) = 0$

If one does not know Euclidean Quantum Mechanics, the task to generalize Noether's Theorem to some diffusion process seems hopeless. Among the problems jumping to the mind, one would ask: what is the time derivative of a diffusion, what are the action, the momentum, the energy, what is a constant along extremals of the action?

But notice that, at this point, only the last question is an open problem for us. For a given filtration, say the usual one $(\mathcal{P}_t)$, and our simple Lagrangian (3.35), the momentum should be the drift of the diffusion, $B(z,t) = \hbar\nabla\log\eta(z,t)$ and the energy should be the random variable $\mathcal{E}(z,t)$ defined in (3.40). As a matter of fact, it is easy to realize that, for a given observable A in the quantum mechanical Hilbert space, the formal recipe to produce random variables in a given state (here this means a given pair η, η^*) are

$$a(z,t) = \left(\frac{A\eta}{\eta}\right)(z,t) \qquad \text{or} \qquad a^*(z,t) = \left(\frac{A\eta^*}{\eta^*}\right)(z,t) \tag{4.4}$$

depending on the filtration considered. This observation is considerably elaborated in [44].

In order to understand what is, in this framework, a constant of motion, let us describe the relation between the generator D of a Bernstein diffusion and the quantum mechanical Hamiltonian H. For f regular enough, this is [45]

$$Df(z,t) = \frac{1}{\eta(z,t)}\left(\frac{\partial}{\partial t} - H\right)(f\eta)(z,t) \tag{4.5}$$

and a similar formula holds with respect to the decreasing filtration. This follows from Itô's calculus and the explicit form of the coefficients of a Bernstein diffusion. In the Hilbert space formulation of Euclidean Quantum Mechanics, it is clear that an observable A is a constant of motion if, on a properly defined dense domain (cf. [44]),

$$\begin{aligned}\frac{dA}{dt} &= [A,H] + \tfrac{\partial A}{\partial t} \\ &= 0\end{aligned} \tag{4.6}$$

since this is the quantum mechanical definition, up to analytical continuation in time. Now consider the (forward) random variable a canonically associated with such a constant observable A by (4.4). Then an easy calculation shows that, on a Bernstein diffusion $Z(t)$,

$$Da(Z(t),t) = 0 \tag{4.7}$$

This is an important feature of this approach. It means that constant observables correspond to $\mathcal{P}_t$-martingale, since Eq. (4.7) is nothing else than the definition of this fundamental concept of the modern theory of stochastic process (cf. [46] for example). In the same way, one shows that if a^* is the other random variable associated with A, according to (4.4), $D_* a^*(Z(t), t) = 0$ so that a^* is an $\mathcal{F}_t$-martingale [47].

For example, when L is as in (3.35), in particular with time independent forces, one verifies algebraically that $D\mathcal{E}(Z(t), t) = 0$, *i. e.* the energy is a $\mathcal{P}_t$-martingale, as well as $D_* \mathcal{E}_*(Z(t), t) = 0$.

Notice that since $\mathcal{E}$is consistent with Feynman's discretized version (2.32), this is an example where the existence of well defined measures (and the continuum limit) enables us to discover qualitative properties which have escaped to Feynman.

Let us consider the following regularized version of the classical invariance condition (4.2), relevant for the (forward) action functional used in Chapt. 3:

$$
\begin{aligned}
&E_s \int_s^{T/2} L(DZ(t), Z(t), t)\, dt = \\
&= E_{\tau_s^\alpha} \int_{\tau_s^\alpha}^{\tau_{T/2}^\alpha} \{L(DX(\tau^\alpha), X(\tau^\alpha), \tau^\alpha) + \alpha D\phi(X(\tau^\alpha), \tau^\alpha)\}\, d\tau^\alpha + o(\alpha)
\end{aligned}
\tag{4.8}
$$

where $X(\tau_t^\alpha) \equiv Z_t^\alpha$ and this diffusion solves

$$
\begin{cases}
dZ_t^\alpha = \hbar \nabla \log \eta^\alpha(Z_t^\alpha, \tau_t^\alpha)\, \varphi^\alpha\, dt + \hbar^{1/2} \sqrt{\varphi^\alpha}\, dW_t \\
d\tau_t^\alpha = \varphi^\alpha(Z_t^\alpha, \tau_t^\alpha)\, dt
\end{cases}
\tag{4.9}
$$

in one space dimension for simplicity. Formally, when $\hbar = 0$, we are back to the classical invariance condition (4.2), by construction. When $\alpha = 0$ we impose that $\varphi^0 = 1$, $\eta^0 = \eta$ so that $(Z_t^0, \tau_t^0) = (Z_t, t)$, where Z_t is the Bernstein diffusion critical point of the left hand side action in (4.8), as defined in Chapt. 3. We also suppose that φ^α is continuous, positive, and such that $\frac{d\tau^\alpha}{dt} = \varphi^\alpha$ defines a random time change for Z_t^α (cf. [48], for example).

For regular coefficients, an asymptotic expansion in α, of the form

$$
\begin{cases}
Z_t^\alpha = Z_t + \alpha X(Z_t, t) + o(\alpha) \\
\tau_t^\alpha = t + \alpha T(t) + o(\alpha)
\end{cases}
\tag{4.10}
$$

makes sense, for some "directions" X, T (not independent). Also we may define η^α as η_{χ_α}, where χ_α is a positive analytical vector in $\mathcal{D}(H)$ (cf. [44]) *i. e.* a family of initial conditions for the relevant heat equation of (3.11). The invariance condition can be rewritten exclusively in term of the parameter t, as an identity between two actions, $J^0[Z, t] = J^\alpha[Z^\alpha, t^\alpha]$. Therefore the associated variation of the action is zero and one shows that, on the critical Bernstein diffusion Z_t, this implies that

$$
D(B(Z_t, t)\, X(Z_t, t) + \mathcal{E}(Z_t, t)\, T(t) + \phi(Z_t, t)) = 0 \tag{4.11}
$$

The proof can be found in [49].

Notice that the theorem does not say where do come from the "proper" directions X, T, ϕ for the variation of the action functional. One shows that they are precisely the infinitesimal generators of the symmetry group of the underlying heat equation [49].

The simplest examples are the one considered in the above classical case:

1) When $L(\dot{q}, q)$ is time independent, $X = 0$, $T = 1$ and $\phi = 0$ are admissible. This invariance under time translation provides our regularized version of conservation of energy: $D\mathcal{E}(Z(t), t) = 0$.
2) If $L(\dot{q}, q) = 1/2|\dot{q}|^2$ (the free case), $X = 1$, $T = 0$ and $\phi = 0$ are admissible. This invariance under space translation implies that $DB(Z_t, t) = 0$, which is certainly true since this is the Euler-Lagrange equation (3.31) in this particular case.

So, from now on, we know that Schrödinger's Euclidean Quantum Mechanics enjoys all the symmetries needed to turn, in particular, this framework into a serious version of "Stochastic Mechanics." But we shall not use this terminology since it has been employed at least twice, for quite distinct approaches [30-50].

As suggested in the introduction, Nelson's Stochastic Mechanics shares some common features with the theory inspired by Schrödinger. As a matter of fact, it can be regarded as a special realization of this last approach, in the following sense:

The probabilistic interpretation of the Schrödinger equation (2.13) involves the product $\overline{\psi}\psi(y, \tau)\, dy$. Since quantum probabilities are more in heat equations than in the Schrödinger ones, is it not possible to use te Euclidean approach of Chapt. 3, with two adjoint heat equations whose solutions η, η^* are such that

$$\eta\eta^*(y, t)\, dy = \overline{\psi}\psi(y, t)\, dy \equiv \rho(y, t)\, dy \qquad \forall t? \tag{4.12}$$

Curiously enough, the answer to this question is positive. Schrödinger's Euclidean strategy can be applied and produce well defined diffusion processes $X(t)$: These are Nelson's diffusions.

However, there is a catch here: the Hamiltonian H needed for the two adjoint heat equations (3.11) satisfying the constraint (4.12) is of the form (1.1) but for a potential which is not the physical one V given in the Schrödinger equation. Instead, the modified potential becomes [8]

$$\mathcal{V}(y, t) = \hbar^2 \frac{\Delta \rho^{1/2}}{\rho^{1/2}}(y, t) - V(y) \tag{4.13}$$

The first term is quite familiar to those interested by the foundations of quantum mechanics: this is the "quantum potential" of D. Bohm (cf. [51]). From the physical point of view, it has been argued about the pertinence of a version of quantum theory using (4.13). A priori surprising quantum features, like the non-locality become obvious, mathematically if not physically, in the non probabilistic frameworks dealing with such quantum potentials [52]. Even better, such approaches have been sometimes instrumental in the discovery of new spectacular quantum effects [58]. Still, it remains that, from our dynamical point of view, a probabilistic theory involving (4.13) becomes rather complicated: even the case of the one dimensional free particle ($V = 0$) is seriously nontrivial since the potential $\mathcal{V}$ depends now on the initial conditions. As the constraint (4.12) is imposed, the first moment of Nelson's diffusion X_t will coincide with the prediction of regular quantum physics. But all the moments of order > 1 will be without mathematical relations with quantum mechanical ones. So the probabilistic structure of this stochastic mechanics will be full of puzzling surprises. For example, no path integral representation of the wave function ψ using exclusively the classical Lagrangian L is known in this context. Various candidates have been suggested, after Yasue's original one (cf. [50(1)], [8]) but all of them contain "non classical" additive terms, whose origin can be traced back to the potential $\mathcal{V}$. As a result, a probabilistic counterpart of Feynman's transition element theory, providing all the kinematical and dynamical properties of the particle at once is very hard to imagine in stochastic mechanics.

Here is a typical difficulty one faces in the attempt to develop this framework. In [50(1)], the present author is credited with having found a strong (*i. e.* without expectation) probabilistic counterpart of the conservation of energy. This claim has been withdrawn since then: actually nobody knows if such a strong probabilistic counterpart of constant of motion exists or not in Nelson's theory. It follows, in particular, that no rigorous Theorem of Noether has been proved in stochastic mechanics (Yasue's one, in [6], contains unsolved mysteries). Some of the additional difficulties regarding the basic relations between variational principles and dynamics in this context have been described by Carlen in [1].

Let us stress that Nelson's diffusions are, however, well defined mathematical objects and may also be regarded, if needed, as products of Schrödinger Euclidean strategy (cf. [53], for example). Nevertheless it would be a mistake to consider Nelson's diffusions, as I did for several years (maybe not alone) as some kind of rigorous version of Feynman's "processes." Nelson processes are still source of active research, both in physics and in mathematics (cf. [54] for a sample, as well as [23]). And without the experience gained in stochastic mechanics, of course, the present Euclidean strategy would not have been developed so easily.

Conclusions and Prospects

Schrödinger's Euclidean Quantum Mechanics is a general strategy, independent on the form of the Hamiltonian or the dimensionality of the underlying classical system. A few encouraging results in ∞ dimension are already known ([55]) but they are, as yet, more of a mathematical than physical nature. It will be interesting for example to look at the infinite dimensional symplectic structure for Gaussian Bernstein measures and to compare with Segal's approach [42–56].

An important issue has not be addressed here, the one of the physical interpretation of Schrödinger's basic idea. Well, it is not at all trivial and this is what is so exciting about it!

At the end of [2], Schrödinger was wondering if one should understand Eddington's observation (Chapt. 3) as suggesting to take quantum probabilities at two different times as boundary conditions for quantum dynamics. The first, emotional, reaction to such a proposal is that it is absurd or violates basic physical principles like causality. As a matter of fact, the discussion of the asymmetrical role of initial and final boundary conditions has been a recurrent theme since the beginning of quantum mechanics. Such discussion is, most of the time, associated with deep questions on the theory of quantum measurement and often with the question of time symmetry. In any cases, it seems that, in recent years, such approaches have reappeared with various motivations (cf. [57]).

When Schrödinger observed, long ago [2], that it was hard for him to believe that his Euclidean analogy was accidental, he had very little to argue in favour of his feeling. No doubt that he would be considerably more confident today.

The history of our understanding of the relations between probability theory and quantum physics is an history full of sound and fury, full of inspired guesses and seemingly fatal mistakes. It may be, however, that it has a meaning. My own hunch is that there is much more than merely intriguing mathematics in this area. If it is the case, Schrödinger's Euclidean Quantum Mechanics may show us the way out of this quantum chaos.

It is a pleasure to thank all the members of the organizing committee of this 31st Winter School for their cordiality and their remarkable organisation.

References

1. E. Carlen, "Progress and Problems in Stochastic Mechanics," *in* XXIV Karpacz Winter School: "Stochastic methods in Mathematics and Physics," Edit. R. Gielerak and W. Karwowski, World Scientific Publis. (1989).
2. E. Schrödinger, *Ann. Inst. H. Poincaré* **2**, 269 (1932).
3. R. Feynman, *Rev. Mod. Phys.* **20**, 367 (1948);
   R. Feynman, A. R. Hibbs, "Quantum Mechanics and Path Integrals," McGraw-Hill, N. Y. (1965).
4. E. Nelson, *J. Math. Phys.* **5**, 332 (1964);
   S. Albeverio, R. Høegh-Krohn, "Mathematical Theory of Feynman Path Integrals," Lect. Notes in Mathematics no. 523, Springer-Verlag (1976).
5. M. C. Gutzwiller, "Chaos in Classical and Quantum Mechanics," Springer-Verlag (1990).
6. K. Yasue, *J. of Functional Analysis* **41**, 327 (1981).
7. Ph. Blanchard, Ph. Combe, W. Zheng, "Mathematical and Physical Aspects of Stochastic Mechanics," Lect. Notes in Physics, 281, Springer-Verlag (1987).

8. J. C. Zambrini, *J. of Mathematical Physics* **27(9)**, 2307 (1986).
9. M. Reed, B. Simon, "Methods of Modern Mathematical Physics," II, Academic Press (N. Y.) (1975).
10. A. D. Wentzell, "A course in the theory of stochastic processes," McGraw-Hill (1981).
11. M. Schechter, "Operator methods in Quantum Mechanics," North Holland (1981).
12. R. H. Cameron, *J. of Math. and Physics* **39**, 126 (1960).
13. L. S. Schulman, "Techniques and applications of path integration," John Wiley & Sons, N. Y. (1981).
14. "Quantum theory and measurement," J. A. Wheeler and W. H. Zurek Ed., Princeton University Press (1983).
15. P. Ramond, "Field Theory: a modern primer," Benjamin Publ., Reading, Mass. (1981).
16. J. C. Zambrini, "New Probabilistic approaches to the classical heat equation" in Springer Lect. Notes in Math. no. 1325, A. Truman and I. M. Davies, Edit., Springer (1988).
17. A. Einstein, "Investigations on the theory of the Brownian Movement," Edit. by R. Fürth, Dover N. Y. (1956).
18. 1) S. Bernstein, "Sur les liaisons entre les grandeurs aléatoires," Verh. der Intern. Mathematikerkongr. Zürich (1932) Band 1.
    2) R. Fortet, "Résolution d'un systeme d'équations de M. Schrödinger," *J. Math. Pures et Appl. IX*, p. 83 (1940).
    3) A. Beurling, *Annals of Mathematics* Vol. 72 No. 1 p. 189 (1960).
19. 1) B. Jamison, *Z. Wahrscheinlichkeitstheorie vern. Geb.* **30**, 65 (1974).
    2) B. Jamison, *Z. Wahrscheinlichkeitstheorie vern. Geb.* **32**, 323 (1975).
20. N. Wiener, "Differentiable Space," *J. Math. Phys.* **2**, 131 (1923).
21. T. Kato, *Isr. J. Math.* **13**, 135 (1973).
22. K. Osterwalder and R. Schrader, *Commun. Math. Phys.* **31**, (1973); **42**, (1975) 281.
23. A. B. Cruzeiro and J. C. Zambrini, "Euclidean Quantum Mechanics. An Outline" *in* A. I. Cardoso *et al* (eds.), "Stochastic Analysis and Applications in Physics," Kluwer Academics, Boston, 59 (1994).
24. N. Ikeda and S. Watanabe, "Stochastic differential equations and diffusion processes," 2th ed., North-Holland, Amsterdam (1989).
25. R. H. Cameron and W. T. Martin, *Ann. of Math.* **45**, 386 (1944).
26. P. Malliavin, "Stochastic calculus of variations and hypoelliptic operators" *in* Proc. of Intern. Symp. Stochastic Differential Equations," Kyoto (1976) (K. Ito Edit.) Kinokuniya-Wiley, Tokyo (1978).
27. J. M. Bismut, "Large deviations and the Malliavin calculus," Progress in Math., Birkhäuser, Boston (1984).
28. M. Kac, "On some connections between probability theory and differential and integral equations" *in* Proc. of 2th Berkeley Symp. on Prob. and Statistics, J. Neyman Ed., Univ. of California Press, Berkeley (1951).
29. 1) A. B. Cruzeiro, J. C. Zambrini, *Journ. of Funct. Analysis*, Vol. 96, No. 1, 62 (1991).
    2) A. B. Cruzeiro, J. C. Zambrini, "Feynman's Functional Calculus and Stochastic Calculus of Variations" *in* "Stochastic Analysis and Applications" edit. by A. B. Cruzeiro and J. C. Zambrini, Progress in Probability, Vol. 26, 83 Birkhäuser, Boston (1991).

30. E. Nelson, *Physical Review* **150**, 1079 (1966).
31. M. Thieullen, *Prob. Th. Rel. Fields* **97**, 231 (1993).
32. W. Heisenberg, *Zeitschrift für Physik* **43**, 172 (1927) reprinted in "Quantum theory and measurement" Ed. by J. A. Wheeler, W. H. Zurek, Princeton Series in Physics, P. U. P., Princeton (1983).
33. E. Schrödinger, *Annalen der Physik* (4), **79**, (1926).
34. W. H. Fleming, H. M. Soner, "Controlled Markov Processes and viscosity solutions," Springer-Verlag, N. Y. (1993).
35. 1) P. L. Lions, Comm. PDE 8, Part I, 1101–1134, Part II, 1229–1276 (1983).
2) M. G. Crandall and P. L. Lions, *Trans. A. M. S.* 277, 1 (1984).
3) P. L. Lions, Act. Math. 161, 243 (1988).
36. H. Goldstein, "Classical Mechanics," Addison-Wesley (1980).
37. 1) E. B. Dynkin, "Markov Processes," Springer-Verlag Berlin (1965).
2) M. I. Freidlin and A. D. Wentzell, "Random Perturbations of Dynamical Systems," Springer-Verlag, N. Y. (1983).
38. V. P. Maslov, "Non-standard Characteristics in Asymptotical Problems," Proc. of Intern. Congress of Mathematicians, 1983, Warszawa, North-Holland (1984) Vol. 1 p. 139.
39. B. Simon, "Functional Integration and Quantum Physics," Acad. Press, N. Y. (1979).
40. T. Kolsrud, J. C. Zambrini, "J. Math. Physics, 33, 4, 1301 (1992).
41. J. C. Zambrini, "Time reversal and Gaussian measures in Quantum Physics," *in* "Stochastic and Quantum Chaos," Ed. Z. Haba *et al,* Kluwer Acad., Boston 195, (1995).
42. A. Brandão, Ph. D. Thesis, in preparation.
43. R. Z. Has'minskii, "Stochastic stability of differential equations," Sijthoff & Noordhoff, Rockville U. S. A. (1980).
44. S. Albeverio, K. Yasue, J. C. Zambrini, Ann. Inst. Henri Poincaré, Phys. Theor., **49**, no. 3, 259 (1989).
45. T. Kolsrud, J. C. Zambrini, "The mathematical framework of Euclidean Quantum Mechanics. An Outline," *in* "Stochastic Analysis and Applications," A. B. Cruzeiro and J. C. Zambrini, Progress in Probability Vol. 26, Birkhäuser, Boston (1991).
46. J. L. Dobb, "Stochastic Processes," Wiley, N. Y. (1953).
47. J. C. Zambrini, "An alternative starting point for Euclidean field theory: Euclidean Quantum Mechanics" IXth International Congress of Mathematical Physics, Swansea (U. K.), Edit. B. Simon, A. Truman, I. M. Davies, Adam Hilger, New York, 260 (1989).
48. I. I. Gihman and A. V. Skorohod, "The theory of stochastic processes," Vol. I, II, III, Springer-Verlag (1979).
49. M. Thieullen and J. C. Zambrini: "The Theorem of Noether in Euclidean Quantum Mechanics," Talk given at 5th European Symposium on Stochastic Analysis, Bonn, 4–8/10/94.
M. Thieullen and J. C. Zambrini: "The Theorem of Noether in Schrodinger's Euclidean Quantum Mechanics", to be published.
M. Thieullen and J. C. Zambrini: " A Stochastic Noether Theorem", to be publised.

50. 1) E. Nelson, "Quantum Fluctuations," Princeton Univ. Press, Princeton N. J. (1985).
2) J. M. Bismut, "Mécanique Aléatoire," Lect. Notes in Math. no. 866, Springer-Verlag, Berlin (1981).
51. D. Bohm, Phys. Rev. **85**, 166, 180 (1952); D. Bohm, J. P. Vigier, Phys. Rev. **96**, 208 (1954).
52. J. S. Bell, "Speakable and Unspeakable in Quantum Mechanics," Cambridge Univ. Press, Cambridge, (1987).
53. R. Carmona *in* Taniguchi Symp. PMMP, K. Itô and N. Ikeda Edit., Academic press, Boston (1987).
54. a) P. Garbaczewski and J. P. Vigier, Phys. Rev. **A46**, 4634 (1992); P. Garbaczewski and P. Blanchard, Phys. Rev. **E49**, No. 5, 3815 (1994); P. Garbaczewski, J. R. Klauder, R. Olkiewicz: "The Schödinger Problem, Lévy Processes and all that noise in Relativistic Quantum Mechanics," to appear.
b) T. Mikami, Comm. Math. Phys. **135**, 19 (1990); T. Misawa, J. Math. Phys. **34**, 775 (1993); A. Wakolbinger, J. Math. Phys. **30**(12), 2943 (1989); B. Djehiche, Potential Analysis **2**, 349 (1993); J. of Math. Phys. **33**(9), 3050 (1992).
55. 1) A. B. Cruzeiro, Z. Haba, J. C. Zambrini, "Bernstein diffusions and Euclidean Quantum Field Theory" *in* Monte Verità Meeting (Switzerland), Ed. E. Bolthausen and F. Russo, Birkhäuser, to appear.
2) A. B. Cruzeiro and J. C. Zambrini, "Malliavin Calculus and Euclidean Quantum Mechanics II" to appear in J. of Funct. Analysis.
56. J. C. Baez, I. E. Segal, Z. Zhou, "Introduction to algebraic and constructive Quantum Field Theory" Princeton Univ. Press (1992).
57. Y. Aharonov, B. Reznik: "On a time symmetric formulation of Quantum Mechanics" (Preprint).
58. Y. Aharonov, D. Bohm, Phys. Rev. **115**, No. 3, 485 (1959).

SEMINARS

Stochastic Approach to Many Bosons Physics

*Roman Gielerak and Robert Olkiewicz**

Institute of Theoretical Physics, University of Wroclaw
PL 50–204 Wrocław, Poland.

Abstract: The modular structure of the free Bose gas is completely described by a Gaussian periodic OS-positive generalized Markov process with values in $S'(\mathbb{R}^d)$. Methods of Euclidean quantum field theory are applied to study some Gibbsian perturbations of the corresponding free Bose gas stochastic thermal structures leading to constructions of some modular structures which describe interacting Bosons in the thermal equilibrium. In particular it is shown that certain perturbations of the free Bose gas in the critical regime of couplings preserve nonergodicity and this phenomenon seems to be connected to the preservation of the Bose-Einstein condensate in the interacting system.

1 Introduction

Although a variety of basic existence results for realistic (i.e. interacting Bose matter systems being in the thermal equilibrium have been obtained rigorously [1, 2, 3] the basic questions of many-body physics, like: the problem of the Bose-Einstein condensate (see [4]) for the present status of this problem in Mathematical Physics), superconductivity (see however the recent progress coming from [5]), modular structure for the interacting systems (see e.g. recent recent monograph by Sakai [6] for the presents status of this problem) etc. are still out of reach from the mathematical point of view.

In the present contribution we construct a class of (model) systems of interacting Bosons in which the above mentioned basic questions seems to be mathematically tractable using the methods of Constructive Quantum Field Theory (see e.g. [7]). The models that can be analysed by the methods invented below correspond (roughly) to the following perturbations of the free Bose gas Hamiltonian:

$$H_{int}\text{``}=\text{''}\lambda \int dx\, dy\, G(a_x + a_x^+)V(x,y)F(a_y + a_y^+)\,,$$

where a_x, a_x^+ are Bose anihilation (resp. creation) operators and G, F are some functions that are selected by thermodynamic stability requirement.

* Supported by KBN grant No. 2P 302 057 07.

2 Stochastic Description of the Free Bose Gas

A stochastic process ξ_τ indexed by the circle K_β of radius β and taking values in some (the real part of) conuclear space $\mathcal{J}_r'$ will be called a thermal process iff:

1. the process ξ_τ is reflection invariant on K_β i.e. for any $\tau \in K_\beta$; $f \in \mathcal{J}_r$: $\langle \xi_\tau, f \rangle = \langle \xi_{r\tau}, f \rangle$ in law, where r denotes reflection at $\tau = 0$.
2. the process ξ_τ is (locally) shift invariant, i.e. for any τ, $\tau' \in K_\beta$, such that $\tau + \tau' \in K_\beta$, any $f \in \mathcal{J}_r$: $\langle \xi_{\tau+\tau'}, f \rangle = \langle \xi_\tau, f \rangle$ in law.
3. the process ξ_τ is OS-positive on the circle K_β which means that for any integer n; $\tau_1, \dots, \tau_n \in [0, \frac{\beta}{2})$, $f_1, \dots, f_n \in \mathcal{J}_r$ and $c_1, \dots, c_n \in \mathbb{C}$:

$$\sum_{\alpha,\beta=1}^{n} \bar{c}_\alpha c_\beta E \langle \xi_{r\tau_\alpha}, f \rangle \langle \xi_{\tau_\beta}, f \rangle = 0 \,, \tag{2.1}$$

4. the map $\mathcal{J}_r \ni f \longrightarrow \langle \xi_\tau, f \rangle$ is continuous in probability for any $\tau \in K_\beta$ and the map $K_\beta \in \tau \longrightarrow \langle \xi_\tau, f \rangle$ for a fixed $f \in \mathcal{J}_r$ is stochastically continuous.

Let us assume that in the space $\mathcal{J}$ a symplectic form $\sigma(f,g) \equiv \mathrm{Im}\langle f|g \rangle$ is given, where the inner product $\langle \cdot | \cdot \rangle$ has the property that $\mathrm{Im}\langle f|g \rangle = 0$ for f, g real. The corresponding to $(\mathcal{J}, \sigma)$ Weyl algebra will be denoted by $\mathcal{W}(\mathcal{J}, \sigma)$ and its generating elements by $W(f)$. The abelian subalgebra $\mathcal{A}(\mathcal{J}, \sigma)$ of $\mathcal{W}(\mathcal{J}, \sigma)$ generated by $W(f)$ with f real, will play the basic role in the following. Following strictly the general constructions presented in [8], certain W^* KMS structure can be associated (essentially in a unique way) with any given thermal process.

Theorem 2.1 *Let $(\xi_\tau)_{\tau \in K_\beta}$ be a thermal process indexed by $\mathcal{J}_r$. Then, there exists a system $\{\mathcal{H}^\xi, \Omega^\xi; \pi^\xi, \alpha_t^\xi; \mathcal{M}^\xi\}$ where $\mathcal{H}^\xi$ is a Hilbert space; $\Omega^\xi \in \mathcal{H}^\xi$; π^ξ is a $*$-representation of $\mathcal{A}(\mathcal{J}, \sigma)$ in $B(\mathcal{H}^\xi)$, α_t^ξ is a *authomorphism group generated by some unitary group U_t^ξ with the property that $U_t^\xi \Omega^\xi = \Omega^\xi$; $\mathcal{M}^\xi$ is a W^*-algebra defined as*

$$\mathcal{M}^\xi = \{\alpha_{t_1}^\xi(\pi^\xi(W(f_1))) \dots \alpha_{t_n}^\xi(\pi^\xi(W(f_n))) \mid t_i \in \mathbb{R};\ f_i \in \mathcal{J}_r\}$$

such that the system $\mathbb{C}^\xi \equiv \{\mathcal{H}^\xi, \Omega^\xi; \pi^\xi, \alpha_t^\xi; \mathcal{M}^\xi\}$ forms a W^* KMS structure i.e. the vector Ω^ξ is cyclic and separating for $\mathcal{M}^\xi$ and the group α_t^ξ coincides with the canonical Tomita-Takesaki modular group associated with the pair $(\mathcal{M}^\xi, \Omega^\xi)$.

It follows from the results in [8] that the W^* KMS structure $\mathbb{C}^\xi$ is uniquely (up to unitary equivalence) determined by the following condition:

There exists a $*$-representation π^ξ of $\mathcal{A}(\mathcal{J}_r, \sigma)$ such that the multitime Euclidean Green functions of $\mathbb{C}^\xi$ restricted to the abelian subalgebra $\pi^\xi(\mathcal{A})$ of

$\mathcal{M}^\xi$ are given by

$$
\begin{aligned}
{}^E G^\xi_{f_1,\dots,f_n}(\tau_1,\dots,\tau_n) &\equiv \left\langle \Omega^\xi \, ; \, \alpha^\xi_{i\tau_1}(\pi^\xi(W(f_1)))\dots\alpha^\xi_{i\tau_n}(\pi^\xi(W(f_n)))\Omega^\xi \right\rangle \\
&= E\prod_{i=1}^{m} e^{i\langle \xi_{\tau_i}, f_i\rangle} \\
&= \int_{K_\beta \to \mathcal{J}'_r} d\mu^\xi(\phi) \prod_{i=1}^{n} e^{i\langle \phi(\tau_i,\cdot), f_i\rangle}
\end{aligned}
\tag{2.2}
$$

for any $-\frac{\beta}{2} \le \tau_1 \le \tau_2 \le \dots \le \tau_n \le \frac{\beta}{2}$; $f_1,\dots,f_n \in \mathcal{J}_r$; where in the last formulae we have denoted by $d\mu^\xi$ the corresponding to ξ_τ "path space measure" on the space of maps $K_\beta \longrightarrow J'_r$.

In particular taking $J \equiv S(\mathbb{R}^d)$ with the canonical symplecting form

$$\sigma(f,g) = 2\,\mathrm{Im}\int \overline{f}(x)g(x)\,dx$$

and defining

$$\widetilde{\alpha}^0_t(W(f)) \equiv W(e^{it(\Delta-\mu)}f) \tag{2.3}$$

and the free thermal state ω^β_0 on $\mathcal{W}(S(\mathbb{R}^d),\sigma)$ by

$$\omega^\beta_0(W(f)) = \exp -\frac{1}{4}C^\beta_0(f,f)\,, \tag{2.4}$$

where:

$$C^\beta_0(f,f) = \int dp\,|\widehat{f}(p)|^2\frac{1+ze^{-\beta p^2}}{1-ze^{-\beta p^2}}\,, \tag{2.5}$$

with $z = e^{-\beta\mu} \in (0,1)$, we have defined pre-KMS structure of the free Bose gas in the noncritical regime. The corresponding pre-KMS structure of the critical free Bose gas with an overcritical density ρ_0 defined on $\mathcal{W}(S(\mathbb{R}^d),\sigma)$ for $d \ge 3$ is determined by the thermal state:

$$
\begin{aligned}
\omega^{cr}_\beta &= \exp -\frac{1}{4}C^\beta_{cr}(f,f)\,,\\
C^\beta_{cr}(f,f) &= c_\sigma\cdot\rho_0|\widehat{f}(0)|^2 + \int dp\,|\widehat{f}(p)|^2\frac{1+ze^{-\beta p^2}}{1-ze^{-\beta p^2}}\,,
\end{aligned}
\tag{2.6}
$$

where $c_\sigma > 0$ is some numerical constant (see e.g. in [1]). The corresponding pre-dynamics is given by:

$$\widetilde{\alpha}^{cr}_t(W(f)) \equiv W(e^{it\Delta}f)\,. \tag{2.7}$$

It is well known that the application of the GNS construction to ($\mathcal{W}$, $\widetilde{\alpha}^0_\tau$, ω^β_0) (respectively to ($\mathcal{W}$, $\widetilde{\alpha}^{cr}_t$, ω^β_{cr}) leads to a W^* KMS structure(s) $\mathcal{C}^\omega_0 \equiv (\mathcal{H}^{\omega_0}, \Omega^{\omega_0}, \alpha^{\omega_0}_t; \pi^\omega_0(\mathcal{W}(S(\mathbb{R}^d), \sigma))'')$ (respectively $\mathcal{C}^{cr}_0 \equiv (\mathcal{H}^{\omega_0}_{cr}, \Omega^{\omega_0}_{cr}, \alpha^{cr_0}_t; \pi^\omega_{cr}(\mathcal{W}(\mathcal{J},\sigma))'')$). Computing the corresponding multitime Green functions restricted to the abelian subalgebra $\mathcal{A}$ of $\mathcal{W}(S(\mathbb{R}^d),\sigma)$ we obtain easily:

1. for $\mathfrak{A}_0^{\omega}$:

$$G^0_{f_1,\ldots,f_n}(\tau_1,\ldots,\tau_n) \equiv \omega_0^{\beta}(\tilde{\alpha}^0_{i\tau_1}(W(f_1))\ldots\tilde{\alpha}^0_{i\tau_n}(W(f_n))) = \prod_{i,j=1}^{n} \exp -\frac{1}{2} S_0^{\beta}(\tau_i - \tau_j;\, f_i \otimes f_j) \tag{2.8}$$

where

$$S_0^{\beta}(\tau_i - \tau_j\,;f_i \otimes f_j) \equiv \int dp\, \overline{f}_i(p) f_j(p) \widehat{S}_0^{\beta}(\tau_i - \tau_j, p); \tag{2.9}$$

$$\widehat{S}_0^{\beta}(|\tau|, p) = \frac{z^{\frac{|\tau|}{\beta}} e^{-|\tau| p^2} + z^{1-\frac{|\tau|}{\beta}} e^{-(\beta-\tau)p^2}}{1 - z e^{-\beta p^2}} \tag{2.10}$$

$$-\frac{\beta}{2} \le \tau_1 \le \tau_2 \le \ldots \le \tau_n \le \frac{\beta}{2};$$

2. for $\mathfrak{A}_0^{cr}$:

$$G^{cr}_{f_1,\ldots,f_n}(\tau_1,\ldots,\tau_n) = \prod_{i,j=1}^{n} \exp -\frac{1}{2} S_0^{cr}(|\tau_i - \tau_j|, f_i \otimes f_j), \tag{2.11}$$

$$S_0^{cr}(\tau, f_i \otimes f_j) \equiv \int dp\, \overline{f}_i(p) \widehat{S}_0^{cr}(\tau; p) f_j(p) \tag{2.12}$$

$$\widehat{S}_0^{cr}(\tau; p) \equiv c_{\sigma}\delta(p) + \frac{e^{-\tau p^2} + e^{-(\beta-\tau)p^2}}{1 - e^{-\beta p^2}}, \tag{2.13}$$

$$-\frac{\beta}{2} \le \tau_1 \le \ldots \le \tau_n \le .\frac{\beta}{2}$$

The important observation is that the thermal kernels S_0^{β} and S_0^{cr} are both OS-positive and stochastically positive on K_{β} and this leads to the following result.

Proposition 2.2 *1. For any $0 < z < 1$, $\beta > 0$ there exists an unique up to the stochastic equivalence Gaussian centered thermal process $(\xi^0_{\tau})_{\tau \in K_{\beta}}$ taking values in $S'_r(\mathbb{R}^d)$ and such that*

$$G^0_{f_1,\ldots,f_n}(\tau_1,\ldots,\tau_n) \equiv E \prod_{i=1}^{n} e^{i\langle \xi_{\tau_i}, f_i \rangle}, \tag{2.14}$$

for $f_1,\ldots,f_n \in S_r(\mathbb{R}^d)$, $-\frac{\beta}{2} \le \tau_1 \le \ldots \le \tau_n \le \frac{\beta}{2}$. The process (ξ^0_{τ}) can be extended to the space $L^2_r(\mathbb{R}^d)$ by continuity. The process is Euclidean invariant and ergodic under the action of the group of translations of $\mathbb{R}^d$.

2. *For any $\beta > 0$ $d \geq 3$ there exists (a unique up to the stochastic equivalence) Gaussian centered thermal process $(\xi_\tau^{cr})_{\tau \in K_\beta}$ taking values in $S'(\mathbb{R}^d)$ and such that*

$$G^{cr}_{f_1,\ldots,f_n}(\tau_1,\ldots,\tau_n) \equiv E\prod_{i=1}^{n} e^{i\langle \xi^{cr}_{\tau_i}, f_i\rangle}, \tag{2.15}$$

for $f_1,\ldots,f_n \in S_r(\mathbb{R}^d)$, $-\frac{\beta}{2} \leq \tau_1 \leq \ldots \leq \tau_n \leq \frac{\beta}{2}$. The process $(\xi_\tau^{cr})_{\tau \in K_\beta}$ being Euclidean invariant is however not ergodic under the action of translations of $\mathbb{R}^d$. The Dynkin entrance boundary of the process $(\xi_\tau^{cr})_{\tau \in K_\beta}$ can be parametrized by $(\rho,\alpha) \in [0,\infty) \times [0, 2\pi)$ and the corresponding ergodic components $\xi_\tau^{(\rho,\alpha)})_{\tau \in K_\beta}$ are again Gaussian thermal processes taking values in $S'_r(\mathbb{R}^d)$ with means:

$$E\left\langle \xi_\tau^{(\rho,\alpha)}, f\right\rangle \equiv c_\sigma^{\frac{1}{2}}\rho^{\frac{1}{2}}\cos\alpha \int f(x)\,dx \tag{2.16}$$

and variances

$$E\left(\left\langle \xi_\tau^{(\rho,\alpha)}, f\right\rangle, \left\langle \xi_{\tau'}^{(\rho,\alpha)}, g\right\rangle\right) \equiv S_0^{(\rho,\alpha)}(\tau-\tau', f\otimes g) \tag{2.17}$$

$$= \int dp\, \widehat{f}(p)\widehat{g}(p)\frac{e^{-|\tau-\tau'|p^2} + e^{-(\beta-|\tau-\tau'|p^2)}}{1-e^{-\beta p^2}}.$$

The resultant measure $d_0 r(\rho,\alpha)$ of the process $(\xi_\tau^{cr})_{\tau\in K_\beta}$ on the Dynkin entrance boundary can be read from the formula:

$$Ee^{i\int_0^\beta d\tau \langle \xi_\tau^{cr}, f\rangle g(\tau)} = \int_0^\infty d\rho \int_0^{2\pi} \frac{d\alpha}{2\pi}\frac{e^{-\frac{\rho}{\rho_0}}}{\rho_0} e^{i\int_0^\beta d\tau\langle \xi_\tau^{(\rho,\alpha)}, f\rangle g(\tau)} \tag{2.18}$$

for $g \in C^\infty(K_\beta)$.

The corresponding to the free thermal processes $(\xi_\tau^0)_{\tau\in K_\beta}$, $(\xi_\tau^{cr})_{\tau\in K_\beta}$, $(\xi_\tau^{(\rho,\alpha)})_{\tau\in K_\beta}$ path space measures on the space $S'_r(K_\beta \times \mathbb{R}^d)$ are denoted by μ_0^β, μ_0^{cr}, $\mu_0^{(\rho,\alpha)}$ respectively. Their characteristic functionals are given by

$$\int_{S'_r(K_\beta\times\mathbb{R}^d)} \mu_0^\Delta(d\Psi)e^{i\int_0^\beta d\tau(\Psi(\tau,\cdot), f)g(\tau)}$$
$$= \exp -\frac{1}{2}\int_0^\beta\int_0^\beta S^\Delta(|\tau-\tau'|, f\otimes f)g(\tau)g(\tau')\,d\tau\,d\tau', \tag{2.19}$$

where Δ corresponds to β or cr or (ρ,α) respectively and the corresponding kernels S^Δ are given by (2.9), (2.13) and (2.17) respectively.

The interesting aspect is that the constructed Gaussian stochastic structures completely describe the corresponding W^*-modular structures $\mathbb{C}_0^\omega$, $\mathbb{C}_0^{cr}$ and $\mathbb{C}_0^{(\rho,\alpha)}$, up to an unitary equivalence, where $\mathbb{C}^{(\rho,\alpha)}$ is the W^* KMS structure corresponding to $\omega_0^{(\rho,\alpha)}$. The following theorem has been proven in [9] (for noncritical gas) and in [10] (for the remaining cases).

Theorem 2.3 *The W^* KMS structures $\mathbb{C}(\xi_t^0)$, $\mathbb{C}(\xi_\tau^{cr})$ and $\mathbb{C}(\xi_\tau^{(\rho,\alpha)})$ obtained via theorem 2.1 are unitary equivalent to the canonical ones $\mathbb{C}_0^\omega$, $\mathbb{C}_0^{cr}$ and $\mathbb{C}_0^{(\rho,\alpha)}$, respectively.*

Remark
A similar result can be formulated in the case $d < 3$, for ξ_τ^{cr} by restricting the suitable indexing space.

3 Gibbsian Perturbations

From the Minlos theorem it follows that the path space measures $d\mu_0^\Delta$ (for $\Delta \in \{\beta, cr, (\rho.\alpha)\}$) can be realized as Gaussian measures an the space $S'(K_\beta \times \mathbb{R}^d)$. It is very easy to see that the corresponding covariances have dimension independent δ singularity which tells us that the typical realisations of ϕ are very singular and therefore in order to introduce nonlinear perturbations certain regularisation is necessary. In order not to destroy the crucial OS-positivity we regularize only over the space variables. For this purpose let $(\chi_\epsilon)_{\epsilon>0}$ be any mollifier taken from $C_c^\infty(\mathbb{R}^d)$ and then we define $\phi_\epsilon(\tau, x) \equiv (\phi * \chi_\epsilon)(\tau, x)$. Let $(V_\Lambda)_\Lambda$ be a family of functions, indexed by bounded subsets Λ of $\mathbb{R}^d$ and such that

(i) $\int_0^\beta d\tau \int_\Lambda V_\Lambda(\phi_\epsilon)(\tau, x) dx \in \bigcap_{p\geq 1} L_p(d\mu_0^\Delta)$,

(ii) $\mu_0^\Delta(\exp \lambda \int_0^\beta d\tau \int_\Lambda V_\Lambda(\phi_\epsilon)(\tau, x)dx) \leq \exp O(1)|\Lambda|$, for any $\lambda \in K \subset \mathbb{R}$, $\beta > 0$ and bounded $\Lambda \subset \mathbb{R}^d$. The constant $O(1)$ depends only on details of λV_Λ and β.

The perturbed (by $(V_\Lambda)_\Lambda$) free Bose gas measure $d\mu_0^\Delta$ shall be denoted by $d\mu_\Lambda^\lambda$ and is defined as:

$$\begin{aligned} d\mu_\Lambda^\lambda(\phi) &= (Z_\Lambda^\lambda)^{-1} \exp\left(\lambda \int_0^\beta d\tau \int_\Lambda V(\phi_\epsilon)(\tau, x)\, dx\right) d\mu_0^\Delta(\phi)\,, \\ Z_\Lambda^\lambda &\equiv \mu_0^\Delta\left(\exp \lambda \int_0^\beta d\tau \int_\Lambda V(\phi_\epsilon)(\tau, x)\, dx\right). \end{aligned} \tag{3.1}$$

Of particular interest are the limits (suitably defined) $\lim_\Lambda \mu_\Lambda^\lambda$. The proper meaning of this operation is based on the notion of Dobrushin-Lanford-Ruelle equations. For this goal we define a thermal Gibbs measure μ on $S'(K_\beta \times \mathbb{R}^d)$ and corresponding to Gibbsian perturbation of $d\mu_0^\Delta$ by V_Λ as any cylindric probability measure such that for any bounded $\Lambda \subset \mathbb{R}^d$, any bounded F

$$E_{\mu_\Lambda^\lambda}\{F \mid \Sigma(\Lambda^c)\} \equiv E_\mu\{F \mid \Sigma(\Lambda^c)\} \quad (\ \mu\text{-a.e.}) \tag{3.2}$$

where $E_\mu\{\cdot \mid \sum(\Lambda^c)\}$ stands for the conditional expectation values of the measure μ taken with respect to (μ-completed) σ-algebras $\Sigma(\Lambda^c)$ defined as those generated by random elements of the form $\langle\phi, f\rangle$ with f supported in $\Lambda^c \equiv \mathbb{R}^d \setminus \Lambda$. A Gibbs measure μ is called to be locally S_0^Δ-regular iff for any f supported in bounded Λ there exists a constant $C^\mu(\Lambda)$ such that:

$$\mu(\langle\phi, f\rangle^2) \leq C^\mu(\Lambda) S_0^\Delta(f, f). \tag{3.3}$$

The set of all locally regular Gibbs measures will be denoted by ${}^\Delta G_{(r)}^\lambda(\beta)$ where the superscript Δ will be kept only for $\Delta = cr$ case.

Theorem 3.1 *Any $\mu \in {}^\Delta G^\lambda(\beta)$ (provided it exists) defines thermal random process ξ_τ^μ and therefore to any $\mu \in {}^\Delta G^\lambda(\beta)$ there is associated an unique up to the unitary equivalence W^* KMS structure $\mathfrak{C}(\mu)$ such that the corresponding Euclidean multitime Green functions Γ^μ, restricted to the abelian sector $\pi(\mathcal{A})$ are given by the following functional integrals:*

$$\Gamma^\mu_{(f_1,\dots,f_n)}(\tau_1, \dots, \tau_n) \equiv \int\limits_{S'(K_\beta\times\mathbb{R}^d)} d\mu(\phi) \prod_{i=1}^n e^{i\langle\phi(\tau,\cdot), f_i\rangle} \tag{3.4}$$

for $f_i = \overline{f}_i$, $\quad -\frac{\beta}{2} \leq \tau_1 \leq \tau_2 \leq \dots \leq \tau_n$.

If additionally $\mu \in {}^\Delta G_r^\lambda(\beta)$ for $\Delta \in \{0, (\rho, \alpha)\}$ then the corresponding thermal process ξ_τ^μ can be extended to $L^2(\mathbb{R}^d)$ by continuity.

The crucial point of the proof is that the perturbations are local in time and therefore do not destroy OS-positivity on the circle K_β. Heuristically the corresponding thermal process ξ_τ^μ can be seen as "$\xi_\tau^\mu(\phi) = \phi(\tau, \cdot)$".

Several particular perturbations have been and are being studied by us. We list a few examples below.

Example: Gentle perturbations of the noncritical free Bose gas [9]

$$V_\Lambda(\phi_\epsilon)(\tau, x) = z \int dr(\alpha) \; : e^{i\alpha\phi_\epsilon(\tau,x)} : , \tag{3.5}$$

$z \in \mathbb{R}$; dr is a complex measure with the compact support an $\mathbb{R}$ and such that $dr(\alpha) = \overline{dr(-\alpha)}$.

$$\begin{aligned} : e^{i\alpha\phi_\epsilon(\tau,x)} : &= \exp \frac{\alpha^2}{2} S_\epsilon^\beta(0, 0) \; : e^{i\alpha\phi_\epsilon(\tau,x)} : , \\ S_\epsilon^\beta(0, 0) &\equiv (\chi_\epsilon \otimes \chi_\epsilon) * S_\epsilon^\beta)(0, 0). \end{aligned} \tag{3.6}$$

Theorem 3.2 *1. For z sufficiently small there exists an unique limit (as a weak limit) $\lim_\Lambda \mu_\Lambda^z \equiv \mu_\epsilon^z \in G_r(z)$. The limiting thermal field μ_ϵ^z is translationally invariant and ergodic under translations.*

2. If the measure dr in (3.5) is real and additionally fulfills $dr(-\alpha) = dr(\alpha)$ then for any $z > 0$ there exists an unique limit $\lim_\Lambda \mu_\Lambda \equiv \mu_\infty^z \in G_r(z)$. Moreover, for any regular value of $z > 0$ the set of translationally invariant, locally regular Gibbs measures G_r^T consists exactly of μ_∞^z.

The part 1 of this theorem is proven by the application of Kirkwood-Salsburg analysis and part 2 is proven by an application of the Fröhlich-Park type (see e.g. [9, 10, 11] for details and references).

Example: Polynomial perturbations Let

$$V_\Lambda(\phi_\epsilon)(x) = \lambda \mathcal{P}(\phi_\epsilon)(\tau, x) \int_\Lambda dy\, W(x-y)\mathcal{P}(\phi_\epsilon)(\tau, y)\,, \tag{3.7}$$

where $\lambda \in (-\infty, 0)$, $\mathcal{P}$ is a polynomial bounded from below and $W \in L^1(\mathbb{R}^d)$ or $W(x, y) = \delta(x-y)$.

The adoption of the (high temperature) cluster expansion (see e.g. [7]) leads to the proof of the following result

Theorem 3.3 *There exists $\lambda_0 > 0$ (depending on β and W) such that for any $-\lambda_0 < \lambda$ an unique thermodynamic limit $\mu_\infty^\lambda = \lim_\Lambda \mu_\Lambda$ exists and belongs to $G_r^\lambda(\beta)$. The limiting thermal field μ_∞^λ is Euclidean invariant and ergodic.*

The work on the following conjectures is now in progress:

Conjecture 3.4 *There exists λ_0 (depending on β and W) such that for any $-\lambda_0 < \lambda \leq 0 : G_r^\lambda(\beta) \equiv \{\mu_\infty^\lambda\}$.*

Conjecture 3.5 *If $d > 1$ then for $\mathcal{P}(\phi_\epsilon)(\tau, x) \equiv (\phi_\epsilon^2(\tau, x) - \Sigma)$ there exists λ^{cr} depending on β, Σ and W such that for all $\lambda < \lambda^{cr}$ the set $\partial G^\lambda(\beta)$ consists of at least two elements $\mu_\infty^{\lambda,+}$ and $\mu_\infty^{\lambda,-}$ such that $\mu_\infty^{\lambda,\pm}(\phi(\tau, x)) \neq 0$ and $\mu_\infty^{\lambda,+} \perp \mu_\infty^{\lambda,-}$.*

Perhaps the most interesting part of our work concerns the perturbations of the critical free Bose gas measure μ_0^{cr}. We have analysed the even gentle perturbations (given as in Th 3.2 2.) of the nonergodic Gaussian measure $d\mu_0^{cr}$.

Example: Perturbations of the free critical Bose gas

Theorem 3.6 *1. For any $z > 0$ there exists a unique (weak) limit:*

$$\lim_\Lambda \mu_\Lambda(d\phi) \equiv \frac{1}{Z_\Lambda} e^{U_\Lambda(\phi_\epsilon)} \mu_0^{cr}(d\phi) \equiv \mu_\infty^z(d\phi)$$

which is the thermal random field i.e. $\mu_\infty^z \in {}^{cr}G^z(\beta)$.

2. For small $0 < z < z_0$ the limiting thermal field μ_∞^z is nonergodic under translations. There exists a Borel probability measure $d\nu_{ren}$ on $K_{2\pi} \times [0, 2\pi]$ and ergodic thermal fields $\mu_\infty^{z,(\rho,\alpha)} \in {}^{cr}G^\lambda(\beta)$ such that

$$\mu_\infty^z \equiv \int_0^{2\pi} \int_0^\infty d\nu_{ren}(\rho, \alpha) \mu_\infty^{z,(\rho,\alpha)}\,, \tag{3.8}$$

where $d\upsilon_{ren}(\rho,\alpha)$ is not concentrated on a single point of $K_{2\pi}\times[0,\infty)$. The ergodic thermal fields $\mu_\infty^{z,(\rho,\alpha)}$ are defined as (unique) limits

$$\lim_{\Lambda}\frac{1}{Z_\Lambda^{\alpha,\rho}}e^{U_\Lambda(\phi_\epsilon)}d\mu_0^{(\rho,\alpha)}(d\phi),$$

which exists for $|z|$ sufficiently small.

Sketch of the proof (for details we refer to our forthcoming paper [10]):

The unique $\lim_\Lambda \mu_\Lambda(d\phi)\equiv\mu_\infty^z(d\phi)$ is constructed by applications of the Fröhlich-Park correlation inequalities [9, 10, 11]. For $|\Lambda|<\infty$ we have the following identity (see 2.18):

$$\mu_\Lambda(F)\equiv\int_0^{2\pi}\frac{d\alpha}{2\pi}\int_0^\infty d\rho\frac{e^{-\frac{\rho}{\rho_0}}}{\rho_0}\langle F\rangle_\Lambda^{\rho,\alpha}\frac{Z_\Lambda^{(\alpha,\rho)}}{Z_\Lambda}, \tag{3.9}$$

where: $\langle F\rangle^{\rho,\alpha}\equiv\mu_\Lambda^{\rho,\alpha}(F)$.

Although the correlation inequalities of the Fröhlich-Park fail for the measure $\mu_\Lambda^{\rho,\alpha}$ the Kirkwood-Salsburg analysis can be applied again to construct a unique ergodic $\mu_\infty^{\rho,\alpha}=\lim_\Lambda\mu_\Lambda^{\rho,\alpha}$ for small $|z|$. The most difficult point concerns the existence of

$$\lim_\Lambda \upsilon_0(d\alpha\otimes d\rho)\cdot\frac{Z_\Lambda^{(\alpha,\phi)}}{Z_\Lambda}\equiv \upsilon_{ren}(d\alpha\otimes d\rho)$$

in the sense of measures. For the existence of the measure $\upsilon_{ren}(d\alpha\otimes d\rho)$ and for the equivalence:

$$\mu_\infty(f)\equiv\int_0^{2\pi}\int_0^\infty \upsilon_{ren}(d\alpha\otimes d\rho)\,\langle F\rangle_\infty^{\rho,\alpha} \tag{3.10}$$

we refer to our forthcoming paper [10]. Finally we have to check that the measure $d\nu_{ren}$ is not supported on a single point (α_0,ρ_0). Assuming contrary we have:

$$0=\langle\phi(\tau,x)\rangle_\infty^z=\langle\phi(\tau,x)\rangle_\infty^{\rho_0,\alpha_0}. \tag{3.11}$$

However from the integration by parts formula it follows:

$$\begin{aligned}\langle\phi(\tau,x)\rangle_\infty^{\rho_0,\alpha_0}&=C_\sigma^{\frac12}\rho_0^{\frac12}\cos\alpha_0\\&+z\int dr(\alpha)\,\alpha\left\langle e^{i\alpha\phi_\epsilon(0,0)}\right\rangle_\infty^{(\alpha_0,\rho_0)}\int_0^\beta d\tau\int S_\epsilon(\tau,x)\,dx,\end{aligned} \tag{3.12}$$

which can be shown to be different from zero for sufficiently small zf.

References

1. O. Bratelli and O.W. Robinson, *Operator Algebra and QSM II*, Springer, New York, 1981, and refs therein.
2. Y.M. Park, J. Stat. Phys. **40**, 259 (1985).
3. D. Brydges, P. Federbush, Comm. Math. Phys. **53**, 19 (1977), and refs therein.
4. M. van der Berg, J.T. Lewis, Helv. Phys. Acta **59**, 1271 (1988) and refs therein.
5. G. Benfatto, G. Gallavotti, A. Procacci, B. Scoppola, Comm. Math. Phys. **160**, 93 (1994);
   J. Feldman, J. Mangan, V. Rivasseau, E. Trubowitz, Helv. Phys. Acta, **65**, 679 (1992).
6. S. Sakai, *Operator Algebras in Dynamical Systems: the theory of unbounded derivations in C^* Algebras*, Cambridge Univ. Press, Cambridge, 1991.
7. J. Glimm, A. Jaffe, *Quantum Physics: A Functional Integral Point of view*, Springer, 1981.
8. R. Gielerak, L. Jakóbczyk, R. Olkiewicz JMP **35**, 3726 (1994);
   R. Gielerak, L. Jakóbczyk, R. Olkiewicz JMP **35**, 6291 (1994).
9. R. Gielerak, R. Olkiewicz, BiBoS preprint, 1994, to appear in Journ. Stat. Phys. 1995.
10. R. Gielerak, R. Olkiewicz, papers in preparation.
11. J. Fröhlich, Y.M. Park, Comm. Math. Phys. **57**, 235 (1978);
    J. Fröhlich, Y.M. Park, Journ. Stat Phys. **23**, 701 (1980).

Ionization of Rydberg Atoms in a Low Frequency Field: Modelling by Maps of Transition to Chaotic Behavior

B. Kaulakys and G. Vilutis

Institute of Theoretical Physics and Astronomy,
2600 Vilnius, Lithuania

Abstract: We investigate a microwave ionization of highly excited atom in a low frequency field and show that such a process may be studied on the bases of map for the electron energy change during the period of the electron motion between two subsequent passages at aphelion. Simple approximate criterion results to the threshold field for transition to chaotic behavior very close to the numerical results. We show that transition from adiabatic to chaotic ionization mechanism takes place when the field frequency to the electron's Kepler frequency ration approximately equals 0.1.

1 Introduction

Highly excited atom in a monochromatic field is one of the simplest real strongly driven by an external driving field nonlinear system with stochastic behavior. That is why recently a large amount of effort, both theoretical and experimental, has been devoted to the studies of the Rydberg atoms in strong microwave fields (for review see [1–3] and references therein). The observation of excitation and ionization rates provides evidence for stochastic behavior of weakly bound electron: the ionization of Rydberg atoms exhibits a threshold dependence on the field amplitude and, at least for the low relative frequency of the field, appears as a diffusion like process. The classical dynamics of the excited electron may be described by the map, called the "Kepler map", rather than by differential equations of motion [4,5], which greatly facilitates the numerical and analytical investigation of stochasticity and ionization process. The map, mostly represented for the number of absorbed photons, is widely used for the analysis of chaotic processes for relatively high frequencies of the field [1–5]. In the very low frequency region the ionization is direct: the electron escapes over the potential barrier [5–6]. However, the threshold field for the direct low frequency (adiabatic) ionization is considerably higher than that for the transition to chaotic behavior in the medium and high frequency field. On the other hand, the derivation of the mapping equations of motion [4,5] is based on the classical perturbation theory. Therefore, the question of the validity of these maps for description of the ionization process in the low frequency relatively strong field arises. In addition, the transition from adiabatic to chaotic ionization mechanism is of great interest, especially as the classical ionization theory is proper for the low frequency ionization.

It is the purpose of this work to investigate the fitness of the classical maps [4,5] for the low–frequency ionization of Rydberg atoms and on this basis to analyse a transition from adiabatic to chaotic motion in the strongly driven by the slow external field system.

2 Equations of Motion and Dynamics

The direct way of coupling the electromagnetic field to the electron Hamiltonian is through the $\mathbf{A} \cdot \mathbf{P}$ interaction, where $\mathbf{A}$ is the vector potential of the field and $\mathbf{P}$ is the generalized momentum of the electron. Thus the Hamiltonian of the hydrogen atom in a linearly polarized field $F\cos(\omega t + \vartheta)$, where F and ω are the field strength amplitude and frequency, in atomic units has the form

$$H = \frac{1}{2}\left(\mathbf{P} - \frac{\mathbf{F}}{\omega}\sin(\omega t + \vartheta)\right)^2 - \frac{1}{r} \; . \tag{1}$$

The electron energy change due to the interaction with the external field follows from the Hamiltonian equations of motion [7]

$$\dot{E} = -\dot{\mathbf{r}} \cdot \mathbf{F}\cos(\omega t + \vartheta) \; . \tag{2}$$

Measuring the time of the field action in the field periods one can introduce the scale transformation [8] when the scaled field strength and energy are $F_s = F/\omega^{4/3}$, and $E_s = E/\omega^{2/3}$. However, it is convenient to introduce the positive scaled energy $\varepsilon = -2E_s$ and relative field strength $F_0 = Fn_0^4 = F_s/\varepsilon_0^2$, with n_0 being the initial principal quantum number.

Note, that eq. (2) is exact if $\dot{\mathbf{r}}$ is obtained from the equations of motion including the influence of the electromagnetic field. Using the parametric equations of motion we can calculate the change of the electron's energy in the classical perturbation theory approximation. We restrict our subsequent consideration to the one–dimensional model, which corresponds to the states with low orbital quantum numbers $l \ll n$. The integration of eq. (2) for the motion between two subsequent passages at the aphelion (where $\dot{x} = 0$ and there is no energy change) results to the map (see [4,5] for details)

$$\begin{cases} \varepsilon_{j+1} = \varepsilon_j - 4\pi F_0 \varepsilon_0^2 \varepsilon_{j+1}^{-1} \mathbf{J}'_{s_{j+1}}(s_{j+1}) \sin\vartheta_j \\ \vartheta_{j+1} = \vartheta_j + 2\pi\varepsilon_{j+1}^{-3/2} + G(\varepsilon_{j+1}, \vartheta_j) \end{cases} \tag{3}$$

where $s = \varepsilon^{-3/2} = \omega/(-2E)^{3/2}$ is the relative frequency of the field, $\mathbf{J}'_s(z)$ is the derivative of the Anger function and function G may be obtained from the requirement of area–preserving [4]. In the low frequency $s \ll 1$ limit $\mathbf{J}'_s(s) \simeq s/2$ and we have the map

$$\begin{cases} \varepsilon_{j+1} = \varepsilon_j - 2\pi F_0 \left(\varepsilon_0^2/\varepsilon_{j+1}^{5/2}\right) \sin\vartheta_j \\ \vartheta_{j+1} = \vartheta_j + 2\pi/\varepsilon_{j+1}^{3/2} + 5\pi F_0 \left(\varepsilon_0^2/\varepsilon_{j+1}^{7/2}\right) \cos\vartheta_j \ . \end{cases} \tag{4}$$

The results of the numerical analysis of maps (3) and (4) in the low frequency, $s \leq 1$, limit are presented in Fig. 1 and 2.

We see that the threshold ionization field approaches the static field ionization threshold $F_0^{st} \approx 0.13$ when $s_0 \to 0$ ($\varepsilon \to \infty$). Thus, the maps (3) and (4) are valid in the low frequency limit when the strength of the driving electric field is of the order of the Coulomb field.

For the low frequencies, $2\pi s = 2\pi/\varepsilon^{3/2} \ll 1$, the change of the angle ϑ after one step of iteration is small and we can transform the system of equations (4) to the differential equation

$$\frac{d(\cos\vartheta)}{d\varepsilon} = \frac{\varepsilon}{\varepsilon_0^2 F_0} + \frac{5\cos\vartheta}{2\varepsilon} \ . \tag{5}$$

The analytical solution of eq. (5) with the initial conditions $\varepsilon = \varepsilon_0$ when $\vartheta = \vartheta_0$ is

$$\cos\vartheta = z^5 \cos\vartheta_0 - (2/F_0) z^4 (1-z), \quad z = \sqrt{\varepsilon/\varepsilon_0} \ . \tag{6}$$

For the relatively low values of F_0 there is a motion in all the interval $0 \div 2\pi$ of the angle ϑ. However for $F_0 = 2z^4/5$ the increase of ϑ at $\vartheta \simeq \pi$ changes to the decrease and results to the fast decrease of ε and ionization process. The minimal value of F_0 for such a motion corresponds to $\vartheta_0 = 0$ and may be defined as a maximal value of F_0 resulting to the motion in the interval $0 \div 2\pi$, i. e., the maximum of the expression

$$F_0 = 2z^4(1-z)/(1+z^5) \ . \tag{7}$$

Such a maximum is at $z = z_0$, where z_0 is a solution of equation $z^5 + 5z - 4 = 0$, i. e., $z_0 = 0.75193$ and results to the $F_0^0 = 0.1279$ which is only 1% lower the adiabatic ionization value $F_0^{st} = 2^{10}/(3\pi)^4 \simeq 0.1298$.

The approximate criterion for transition to chaotic behavior is [5]

$$K = \max|\delta\vartheta_{j+1}/\delta\vartheta_j - 1| \geq 1 \tag{8}$$

and according to eqs. (4) yields to the expression for the threshold field strength

$$F_0^c = \varepsilon^5/6\pi^2\varepsilon_0^2 = z_c^{10}/6\pi^2 s_0^2 \tag{9}$$

where z_c is the solution of eq. (7) with $F_0 = F_0^c$. Note that $z_c^{10} \simeq 1 - 10F_0^c + 30(F_0^c)^2 \mp \ldots$ if $F_0^c \leq 0.1$. For $0.09 \leq s_0 \leq 0.5$ expression (9) gives the ionization threshold field very close to the numerical results (see Fig. 2).

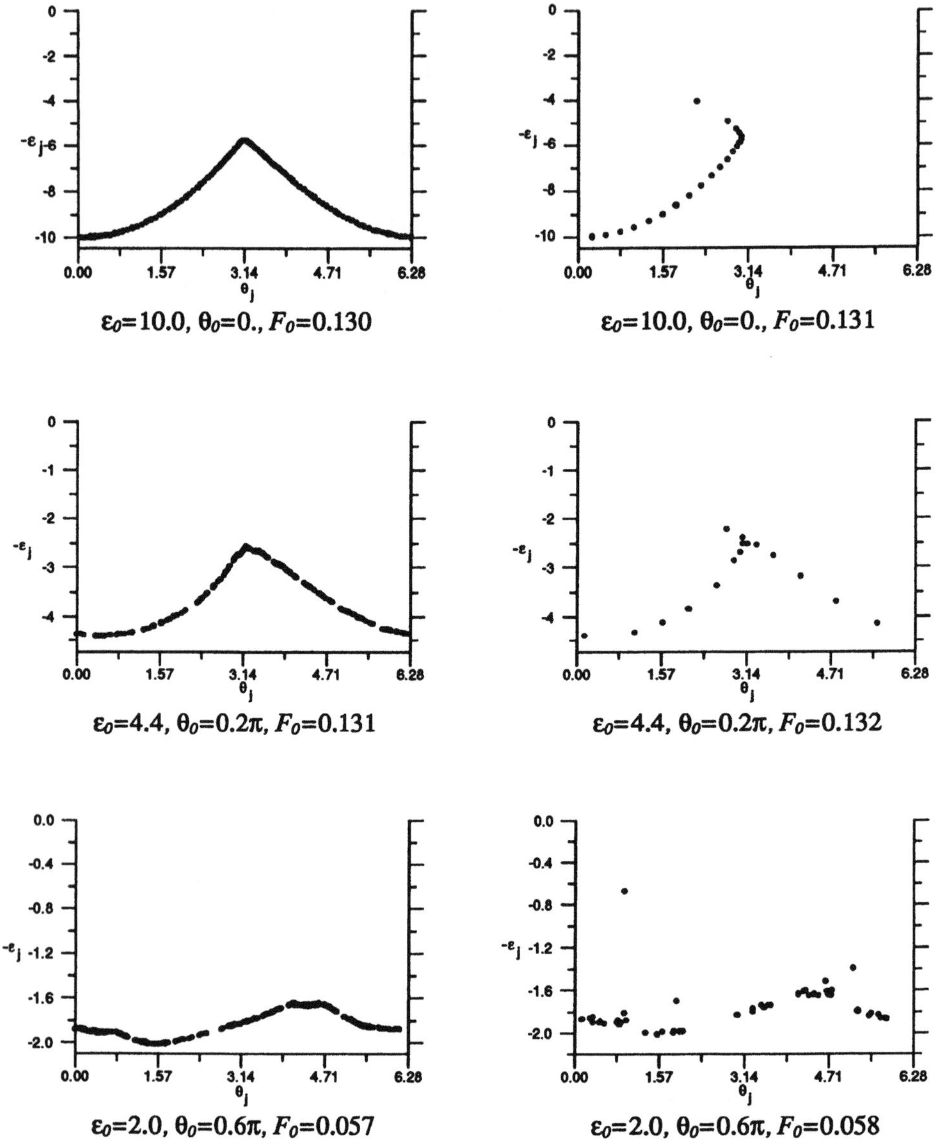

Fig. 1. Trajectories of the map (3) for the different initial conditions ϵ_0, θ_0 and relative field strength F_0. The pictures in the left-hand side corresponds to the regular quasiperiodic motion, while in the right-hand side represent the ionization process for a little bit stronger field. At $\epsilon_0 \cong 4.3$, i.e. at $s_0 \cong 0.11$ the transition from the adiabatic to the chaotic ionization mechanism takes place.

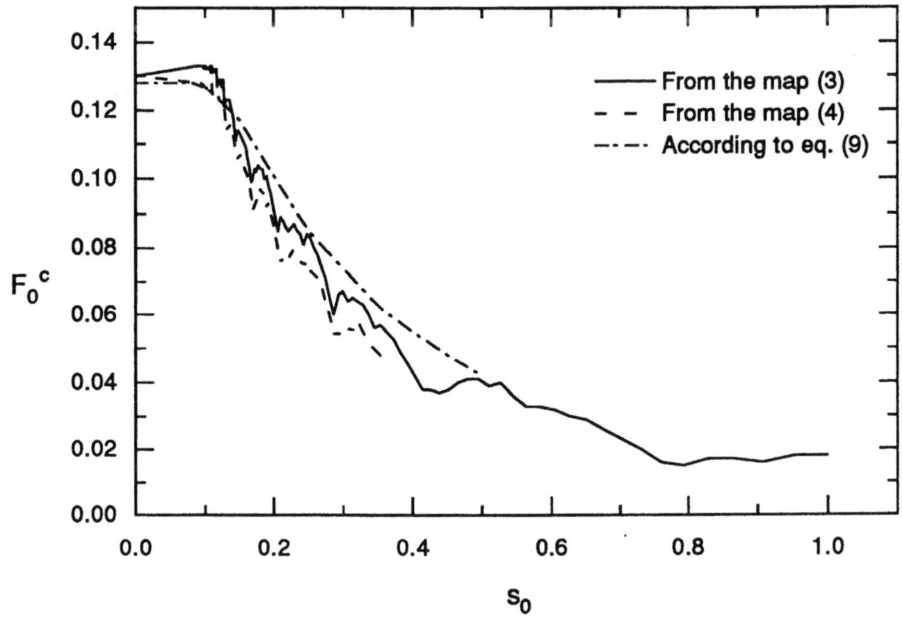

Fig. 2. The relative threshold field strengths for the ionization outset from the numerical analysis of the maps (3) and (4) and according to the approximate criterion (8)–(9).

3 Conclusion

From the above analysis we can conclude that the map at aphelion (3) is suitable for the investigation of transition to chaotic behavior and ionization of Rydberg atoms also in the low frequency field, even for adiabatic ionization, when the strength of the external field is comparable with the Coulomb field. For such a purpose there is no need to use the map for the two halves of intrinsic period [9]. Moreover, the approximate criterion (8) for transition to chaotic behavior yields to the threshold field strength very close to the numerical results if we take into account the increase of the electron's energy by the influence of the electromagnetic field. The transition from adiabatic to chaotic ionization occurs at the relative field frequency $s_0 \simeq 0.1$.

Acknowledgments

The research described in this publication was made possible in part by Grant No. LAA000 from the International Science Foundation.

References

1. G. Casati, I. Guarneri and D.L. Shepelyanski, IEEE J. Quant. Electron., **24**, 1420 (1988).
2. R.V. Jensen, S.M. Suskind and M.M. Sanders, Phys. Reports, **201**, 1 (1991).
3. P.M. Koch in *Chaos and Quantum Chaos*, Lecture Notes in Physics Vol. 411, ed. W. Dieter Heiss (Springer–Verlag, 1992), p. 167–224.
4. V. Gontis and B. Kaulakys, deposited in VINITI as No. 5087–V86, (1986) and Lit. fiz. sb. (Sov. Phys.–Collect.), **27**, 368 (111) (1987).
5. V. Gontis and B. Kaulakys, J. Phys. B: At. Mol. Phys., **20**, 5051 (1987).
6. R.V. Jensen, Phys. Rev. , **A30**, 386 (1989).
7. L.D. Landau and E.M. Lifshitz, *Classical Field Theory* (New York: Pergamon, 1975).
8. B. Kaulakys *et al.*, Phys. Lett. **A159**, 261 (1991); Lith. Phys. J., **33**, 354 (290) (1993).
9. B. Kaulakys, J. Phys. B: At. Mol. Opt. Phys., **24**, 571 (1991).

Periodic Perturbations of Chaotic Dynamics

Andrzej L. Kawczyński

Institute of Physical Chemistry,
Polish Academy of Sciences,
01224 Warsaw, Poland

Abstract: Small periodic perturbations of a chemical model exhibiting infinite number of period adding bifurcations are investigated. At some periods of the perturbations chaotic trajectories become periodic.

1 Introduction

Whereas periodic perturbations of oscillators have been studied for a long time [1] an influence of such perturbations on deterministic systems exhibiting chaotic behavior is an open problem. In this paper periodical perturbations of two model systems exhibiting chaotic behavior are studied. A first model consists of three ordinary differential equations which describe an open chemical system and a second one is 1D return map with a cusp shape.

2 Chemical Model and Results

The unperturbed model describes an open chemical system with coupled enzymatic reactions [2].

$$V \longrightarrow U \tag{1}$$

$$P \longrightarrow U \tag{2}$$

$$P \longrightarrow V \tag{3}$$

The block scheme describes two coupled enzymatic (catalytic) reactions (1) and (2) allosterically inhibited by an excess of their own substrate and the common product U. The irreversible reaction (3) which transforms P to V creates the desired coupling inside the scheme.

It is assumed that total concentrations of all enzymes are much lower than concentrations of substrates and product and as fast variables can be eliminated from kinetic equations using the Tikhonov theorem [3]. The dynamical behavior of the system is then described by three kinetic equations [2]. In dimensionless form they are given by:

$$\frac{dv}{dt} = A_1 - A_2 v - \frac{v}{(1+v+A_3v^2)(1+u)} \tag{4.a}$$

$$\frac{du}{dt} = \epsilon_2\left(\frac{v}{(1+v+A_3v^2)(1+u)} + \right.$$
$$\left. + B\left(\frac{p}{(1+p+B_3p^2)(1+Ku)}\right) + Dp - \frac{Cu}{L+u}\right) \tag{4.b}$$

$$\frac{dp}{dt} = \epsilon_3\left(B_1 - B_2 p - B\left(\frac{p}{(1+p+B_3p^2)(1+Ku)}\right)\right) \tag{4.c}$$

where v, p, u are dimensionless concentrations of V, P, U, and t is dimensionless time, and the remaining symbols denote parameters. ϵ_3 is treated as the bifurcation parameter, whereas others are fixed.

The period adding bifurcations [2] occur for ϵ_3 belonging to the interval $[1.28, 1.36]$. Trajectories obtained by numerical integrations can be roughly characterized by sequences of small (S) and large (L) loops or short and long ones respectively. At $\epsilon_3 = 1.28$ the system approaches the periodic trajectory with the sequence $SSLSL$ for all initial conditions whereas at $\epsilon_3 = 1.36$ the periodic trajectory with the sequence $SSLSLSL$ is the sole attractor. Decreasing ϵ_3 from 1.36 to 1.28 an infinite sequence of attracting, periodic trajectories $SSLSLSL(SSLSL)_1, SSLSLSL(SSLSL)_2, \ldots, SSLSLSL(SSLSL)_n$ with n going to infinity appears. Subintervals of ϵ_3 on which subsequent orbits exist are separated by windows on which trajectories are chaotic.

The behavior of the unperturbed system the can be characterized by the Poincaré section at the plane $u = 1.4343$, looking only at these cases when trajectories cross this plane with u(t) decreasing. The Poincaré section consists of five bunches of points which looks like 1D intervals. 1D return maps, constructed from points belonging to one of these bunches, have a shape of cusp. Therefore the sequence of the period adding bifurcations can be explained as shifting one of the cusp branch to tangency with the diagonal [4], [5]. It is worth to notice that the chaotic behavior appears in the system without warning. It means that very close to the windows no complications in periodic behavior is observed.

The system (4) is perturbed periodically by assuming that

$$\epsilon_3(t) = \epsilon_{03} + A\sin\left(\frac{2\pi t}{T}\right) \tag{5}$$

where A and T are the amplitude and the period of perturbation. Choosing a value of ϵ_{03} one can perturb a given periodic or chaotic trajectory. Values of A determine a number of bifurcations from a trajectory defined by ϵ_{03} to neighbouring orbits. Numerical calculations are performed for $\epsilon_{03} = 1.35$ which corresponds to chaotic orbit and $A = 0.01$ at which the perturbed system is switched between $SSLSLSL$ and $SSLSLSL(SSLSL)_1$ periodic trajectories. Various values of T from small ones (10,100) to large ones (40000,50000) are chosen.

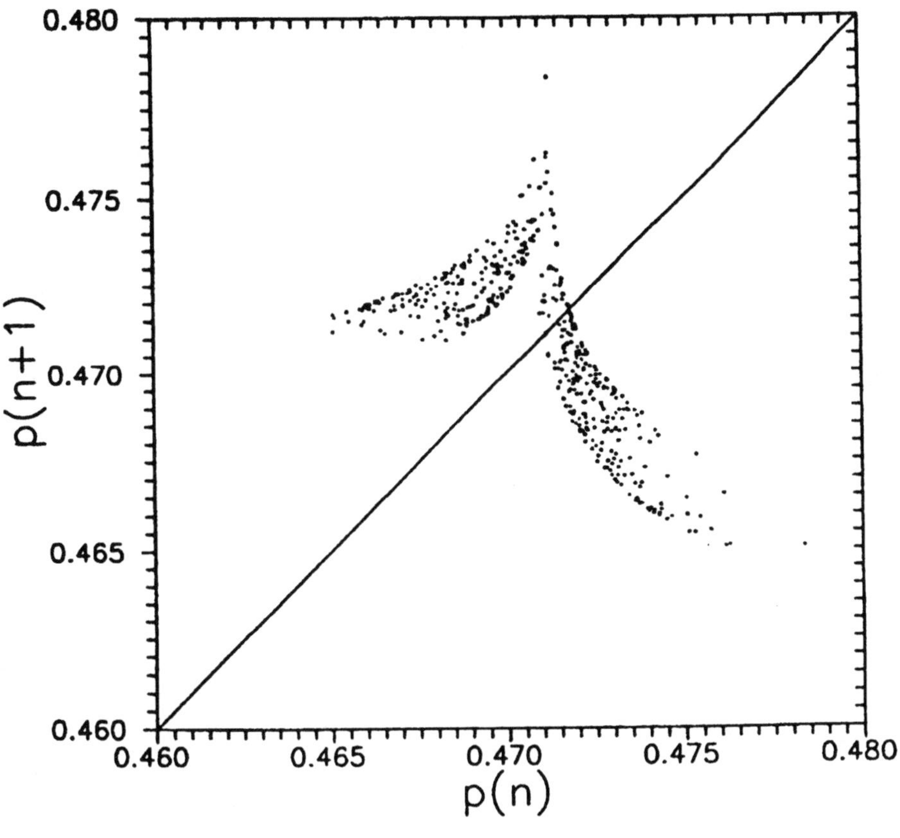

Fig. 1. The return map for the perturbed system (4),(5) with $\epsilon_{03} = 1.35, A = 0.01$ and $T = 20000$. The values of the remaining parameters are: $A_1 = 0.08928606601$, $A_2 = 0.01486767767$, $A_3 = 4$, $B = 0.04$, $B_1 = 0.000701754$, $B_2 = 0.000140351$, $B_3 = 4$, $C = 0.122$, $D = 0,001$, $K = 10$, $L = 0.74$ and $\epsilon_2 = 0.2$.

In order to characterize the behavior of the perturbed system the Poincaré section is constructed at the same plane as in the unperturbed model and next return maps are constructed from one of the bunches obtained. For the perturbed system the Poincaré section consists of five bunches but they do not look like apparently 1D intervals. Some distribution of points along the intervals for the unperturbed system is observed. Therefore return maps are not single valued functions but have some distributions too. An example of the return map is shown in Fig.1. This is a picture which one could expect in perturbed system. However, for some values of T an order in the maps appears (see Fig.2). The fifth iterate of the map shown in Fig.2 is close to a saddle-node bifurcation. Small changes of T induce periodicity of orbits. The periodic orbit has the sequence $SSLSL(SSLSLSL)_2$. This sequence does not appear in the period adding bifurcations for the unperturbed system.

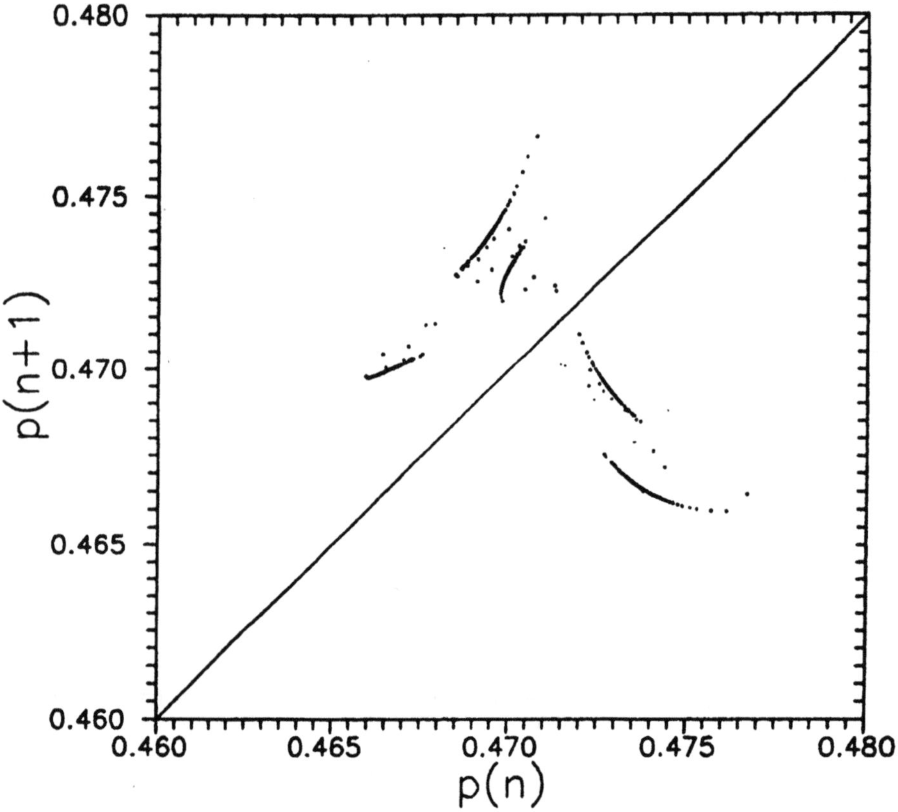

Fig. 2. The return map for the perturbed system (4),(5) with $\epsilon_{03} = 1.35, A = 0.01$ and $T = 39900$. All remaining parameters are the same as in Fig. 1.

3 Cusp Shape Map and Results

Let us consider the map which consists of two hyperbolic curves [4], [5]

$$f(x) = \begin{cases} f_l(x) = 1 - \gamma^2 \left(\dfrac{1}{\eta - \beta} - \dfrac{1}{\eta - x} \right) & \text{if } x \le \beta \ ; \\ f_r(x) = 1 - \dfrac{1-\varphi}{1-\beta}\left(1 - \dfrac{\beta - \varphi}{x - \varphi}\right) & \text{if } x \ge \beta \end{cases} \qquad (6)$$

The map (6) transforms the interval [0,1] onto itself for some ranges of the parameters. It is easy to check that $f_l(\beta) = f_r(\beta) = 1$ so that the parameter β determines the position of the intersection of above hyperbolic curves. The parameter γ determines the position of the branch f_l which moves up or down with changes of γ. It can be show that the family (depending on the parameter γ) of maps exhibits the same period adding bifurcations as discussed above [4], [5].

Let us fix the values of $\eta = 1$ and $\varphi = 0.73$ and change periodically the position of maximum and the position of the left branch of the map according to:

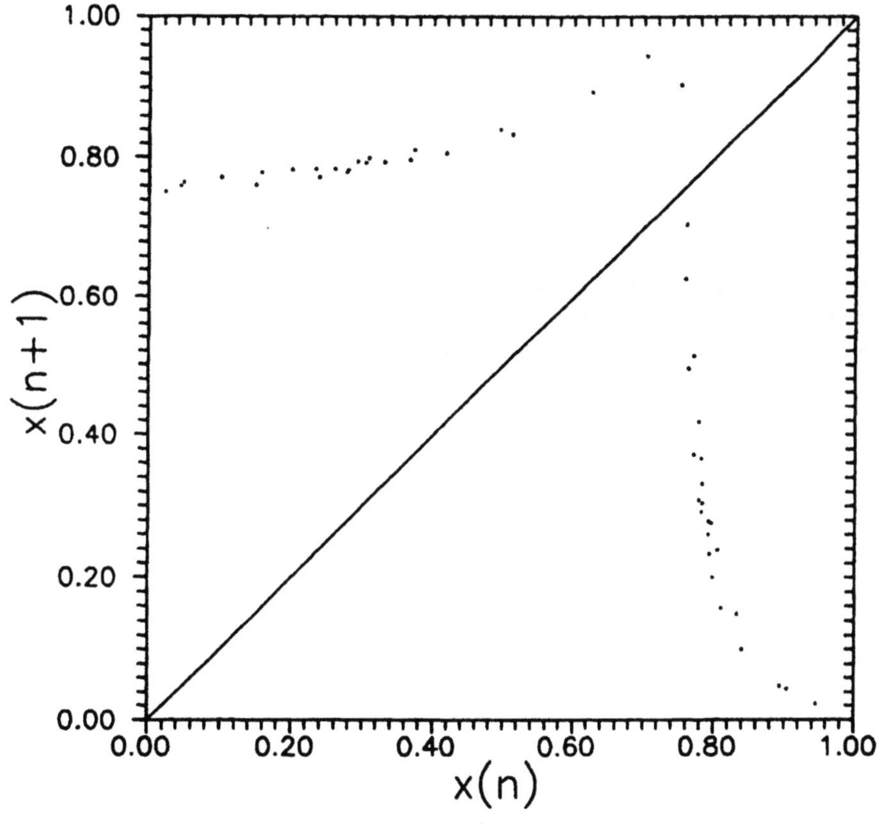

Fig. 3. The return map (6),(7) with $\beta_0 = 0.75, \gamma_0 = 0.286, A_\beta = A_\gamma = 0.003$ and $T = 50$. The orbit has the period equal to 50.

$$\beta = \beta_0 + A_\beta \sin\left(\frac{2\pi i}{T}\right) \tag{7.a}$$

$$\gamma = \gamma_0 + A_\gamma \cos\left(\frac{2\pi i}{T}\right) \tag{7.b}$$

where i numerates iterations of the map (6) and $A_\beta = A_\gamma = 0.003$, $\beta_0 = 0.75$, and $\gamma_0 = 0.286$. At these values of the parameters the unperturbed map has the chaotic orbit which is between the period 3 and period 4 attracting orbits. Iterations of an initial point for small T give the picture which is similar to Fig.2. However, for same values of T the periodic orbit is obtained (see Fig.3).

4 Conclusions

Two examples presented in this paper show that periodic perturbations can induce periodicity in systems exhibiting chaotic behavior. This way should be distinguished from the so called controlling chaos method [6].

References

1. N. Minorsky, *Nonlinear Oscillations*, D. van Nostrand Comp. Inc., Princeton, (1962).
2. F. Aicardi and A.L. Kawczyński, Complex Systems, **6**, 95 (1992).
3. A.N. Tikhonov, Mat. Sbor., **31**, 575 (1952) (in Russian).
4. A.L. Kawczyński and M. Misiurewicz, Z. Phys. Chem., –Leipzig **271**, 1037 (1990).
5. M. Misiurewicz and A.L. Kawczyński, Comm. Math. Phys., **131**, 605 (1990).
E. Ott, C. Grebogi and J.A. Yorke, Phys. Rev. Lett., **64**, 1196 (1990).

p-adic Stochastics with Applications to the Einstein–Podolsky–Rosen Paradox

*Andrew Khrennikov**

Institut für Mathematatik, Ruhr–Universität Bochum,
D–44780 Bochum, Germany

Abstract: In the last years much attention has been paid to p–adic quantum models (especially, in string theory). As usual, new physical models generate new mathematical methods. In our case there appears a new type of stochastics, the p–adic stochastics, which arises within the so called p–adic quantum physics. We apply this stochastics to propose a justification of the Einstein–Podolsky–Rosen theory of hidden variables, which was in contradiction with Bell's type inequality. Our main result is: if we consider the p–adic probability distribution of hidden variables, then no problems appears with the Bell inequality.

1 Introduction

The theory of hidden variables created on the basis of the Einstein–Podolsky (EPR) [1] "realistic philosophy" is very attractive. But there are difficulties ..., one of the main problems of EPR hidden variable description is the contradiction with quantum mechanics predictions in the connections with Bell's inequality [2] and its generalizations, see, for example, reviews [3]. Experimental tests on the basis of Bell–type inequalities show that it is impossible to apply hidden variables formalism to a lot of physical models.

We wish to apply to Bell's inequality paradox a mathematical apparatus developed inside of new physical (at the moment, only theoretical) models, the models of the p–adic quantum physics, see books [4], [5] and pioneer papers [6], about the p–adic numbers field Q_p see [7] and the section 1 of this article. The p–adic physical models is an attempt to describe the reality with the aid of another number field Q_p which has many properties very different from real or complex case.

Quantum mechanics, where wave functions assume values in Q_p, was one of numerous the p–adic models. It was developed in authors papers [8]. The main problem of this theory is a probability interpretation of these wave functions. A new mathematical theory, the p–adic valued theory of probability was proposed

* This research was realized on the basis of Alexander von Humboldt-Stiftung

in [9, [5] to resolve this problem. As usual we consider a probability as a limit of relative frequencies ν_n but with respect to another metric on the field of rational numbers Q. We play our game with the following evident fact. The only physical numbers are rational numbers. We can get in any experiment only finite fractions and not real, complex (or p–adic numbers). Then we can study these rational data with the aid of different mathematical methods. The p–adics helps us to find some additional information about these rational numbers which we cannot find on the basis of real numbers. In particular, there exist such random sequences, where ν_n oscillate between 0 and 1 with respect to the usual real metrics, but stabilize with respect to one of the p–adic metrics [9], [5]; you can find in these papers also the results of computer simulation. From the usual point of view such sequences are chaotic, it would be impossible to define real probabilities. But there is a well defined probability distribution from the p–adic point of view.

The p–adic probability has many properties of the standard real probability [9], [5]. But one property is very and very different. In usual case all probabilities belong to the segment [0,1] of R and this fact is evident, as relative frequencies $0 \leq \nu_n \leq 1$, then the limit is also between 0 and 1. But we can represent every rational number as the limit of ν_n with respect to the p–adic metric [9], [5]. And, in particular, every negative rational number can be presented as the limit of relative frequencies. This fact is the cornerstone of our further considerations. The negative probability distributions are the ordinary object in the p–adic theory of probability. We have in the same statistical connection with relative frequencies and negative probabilities. It was impossible in the usual theory of probability.

2 Few Words About the p–adic Numbers

The field of real numbers R is constructed as a completion of the field of rational numbers Q with respect to the metric $\rho(x, y) = |x - y|$, where $|\cdot|$ is the usual absolute value, norm. Fields of the p–adic numbers Q_p are constructed in the same way. There is an infinite sequence of the p–adic number fields, there is its own field for every prime number $p = 2, 3, 5, \ldots$. The p–adic norm $|\cdot|_p$ is defined in the following way. At the first, we define it for natural numbers. Every natural number n can be represented as the product of prime numbers: $n = 2^{r_2} 3^{r_3} \cdots p^{r_p} \cdots$. Then $|n|_p = p^{-r_p}$, by the definition $|0|_p = 0$, $|-n|_p = |n|_p$, and $|n/m|_p = |n|_p/|m|_p$. The completion of Q with respect to the metric $\rho_p(x, y) = |x - y|_p$ is a locally compact field Q_p.

And it is an intrinsic fact of the theory of numbers [7] that there is the only possibility to introduce a norm on Q to use a real one $|\cdot|$ or one of the p–adic valuations.

3 Frequency Approach to the p–adic Theory of Probability

The basis of the modern theory of probability is a measure theoretical axiomatic of A.N. Kolmogorov, 1933, [10]. A probability is a normalized measure μ assumed its values in the segment $[0,1] \subset R$. But why? To formulate his axioms, A.N. Kolmogorov had used properties of frequency probability defined by R. von Mises [11].

A basic object of Mises' theory is a *collective*. Let $\mathcal{S}$ be a random experiment and $T = \{\alpha_1, ..., \alpha_m\}$ be a set of all possible realizations of $\mathcal{S}$ (in the simplest case $T = \{0, 1\}$). An infinite sequence of realizations of $\mathcal{S}$

$$x = (x_1, ..., x_n, ...), x_i \in T \ , \tag{1}$$

is said to be a collective if for every label $\alpha \in T$ there exists a limit of relative frequencies $\nu_n(\alpha) = k(\alpha)/n : \mathbf{P}(\alpha) = \lim_{n\to\infty} \nu_n(\alpha)$. This limit is said to be a probability of α. In particular, why probability belongs to the segment [0,1] of the real line in the Kolmogorov axiomatic? No problem, $0 \leq \nu_n(\alpha) \leq 1$ and this why $0 \leq \mathbf{P} \leq 1$.

Now a sequence (1) is said to be the p–adic collective [9], [5] if a limit of relative frequencies exists with respect to the p–adic metric for every label $\alpha \in A$. This limit $\mathbf{P}_p(\alpha) = \lim_{n\to\infty} \nu_n(\alpha)$ is called the p–adic probability. In [9], [5] the p–adic collectives were considered from the theoretical and computer points of view. A series of computer statistical experiments was proposed. Relative frequencies oscillated between 0 and 1 with respect to real metric but stabilize very quickly with respect to one of the p–adic metrics (p–plays a role of parameter of the model).

4 No Bell's Inequality for the p–adic Probability Hidden Variables Distributions

At first let us consider a standard hidden variable approach to Bell's inequality. Let Λ be a set of hidden variables λ distributed with a probability (usual) density $\rho(\lambda)$. Then from the mathematical point of view all problems are contained in the integral

$$< AB >= \int_\Lambda A(\lambda)B(\lambda)d\mu(\lambda) \ , \tag{2}$$

where $A(\lambda)$ and $B(\lambda)$ are the random variables with values +1 or -1 and $d\mu(\lambda) = \rho(\lambda)d\lambda$. The estimate of such type mean values is based on two properties of probability: $\rho(\lambda)$ is nonnegative and $\mu(\Lambda) = 1$.

Now let us suppose that hidden variables λ are distributed with the p–adic probability law. Thus μ is Q_p–valued measure on Λ, $\mu(\Lambda) = 1$. And $A(\lambda)$ and $B(\lambda)$ are random variables with respect to μ assumed the same values +1 or -1. Then we cannot apply a usual estimate to (2) and Bell's inequality is not more valid for hidden variables distribution.

That is why, if we suppose that a physical model has the p–adic hidden variable distributions (in this case relative frequencies for hidden variables stabilize in the p–adic metrics) then there is no more contradiction between quantum mechanical experiments and hidden variables description.

References

1. A. Eistein, B. Podolsky, N. Rosen, Phys. Rev. **47**, 777–780 (1935).
2. J.S. Bell, Physics **1**, 195–200 (1965).
3. J.F. Clauser, A. Shimony, Rep. Progr. Phys. **41**, 1881–1927 (1978).
4. V.S. Vladimirov, I.V. Volovich, E.I. Zelenov, p–adic numbers in mathematical physics. World Sc. Publ., Singapore, 1993.
5. A.Yu. Khrennikov, p–adic valued distributions in mathematical physic. Kluwer Academic Publ., Dordrecht–London, 1994.
6. I.V. Volovich, Classical Quantum Grav. **4**, L83–L87 (1987).
   P.G.O. Freund and M. Olson, Phys. Lett. **B199** , 186–190 (1987).
7. W. Schikhof, Ultrametric Calculus. Cambridge Studies in Adv. Math. **4**. Cambridge U.P. Cambridge, 1984.
8. A.Yu. Khrennikov, Russian Math. Surveys **45**, 87–125 (1990).
   A.Yu. Khrennikov, J. Math. Phys. **32**, 932–937 (1991).
9. A.Yu. Khrennikov, Sov. Phys. Dokl. **37** (1992).
   A.Yu. Khrennikov, Probability interpretation of the p–adic valued quantum theories, Symposium on the Foundations of Modern Physics, Helsinki, June 1994, Abstracts, 134.
10. A.N. Kolmogoroff Grundbegriffe der Wahrscheinlichkeitsrechnung. Berlin, 1933. English translation by N. Morrison, New–York (1950).
11. R. von Mises: The Mathematical Theory of Probability and Statistics. Academic, London, 1964.

Quantum Chaos: Double Resonance Model and Its Physical Applications

Andrey R. Kolovsky

L.V. Kirensky Institute of Physics
660036 Krasnoyarsk, Russia

1 Introduction

One dimensional hamiltonian systems with a time-periodic perturbation are very popular models for studying of Classical Chaos. The number of the systems studied counts hundreds, but two of them – the kicked rotor model

$$H = \frac{I^2}{2} + \cos\theta \sum_n \delta(t - Tn) , \tag{1}$$

and so-called Double Resonance Model (DRM)

$$H = \frac{I^2}{2} + \cos\theta \cos(\omega t) , \quad 0 < \theta \le 2\pi \tag{2}$$

– have attracted a special attention. This is explained by the fact that these models are the simplest members of two large groups of the dynamical systems. The kicked rotor (1) represents the class of the systems with an infinite number of the nonlinear resonances and unbounded chaotic region of the phase space. On the contrary, there are only two nonlinear resonances in the system (2) (the minimally possible number for chaos arising) and this model represents the class of the dynamical systems with bounded chaotic component. In some sense these models are "two poles on the globe" of all one-dimensional time-periodic systems.

It is no wonder that models (1) and (2) became popular in studying of Quantum Chaos. Moreover, the kicked rotor seems to be the first system, which was numerically studied in connection with this problem [1], and now one considers the system (1) as a paradigm of quantum "chaotic" system. The study of DRM began later [2] and the results on its quantum dynamics are less known. In what follows we shall show that the quantum manifestations of chaos in DRM essentially differ from those in the kicked rotor model. Another aim of this contribution is to demonstrate that DRM is a very physical model and it has a number of applications among the real objects of the microworld.

2 Double Resonance Model

We recall some results on the classical dynamics of DRM. Let us rewrite the potential energy in the Hamiltonian (2) as a sum of two terms: $\cos\theta\cos(\omega t) = (1/2)[\cos(\theta-\omega t)+\cos(\theta+\omega t)]$. Then each term causes a nonlinear resonance. The term $\cos(\theta-\omega t)$ originates the nonlinear resonance at $I=\omega$, the term $\cos(\theta+\omega t)$ originates the resonance at $I=-\omega$ [see Fig. 1(a)]. The mutual influence of the nonlinear resonances is characterized by Chirikov overlap parameter $K=\sqrt{2}/\omega$, which is the ratio of the width of the resonances to the distance between them. When we decrease ω, the system has a transition to the chaotic dynamics [see Fig. 2(a)]. We note that in this case the system (2) has no other integrals of the motion besides the quasienergy

$$H_{eff} = \omega J + I^2 + \cos\theta\cos\vartheta \ . \tag{3}$$

(In Eq. (3) $\vartheta=\omega t$ is the phase of the field and J is the variable conjugated to ϑ).

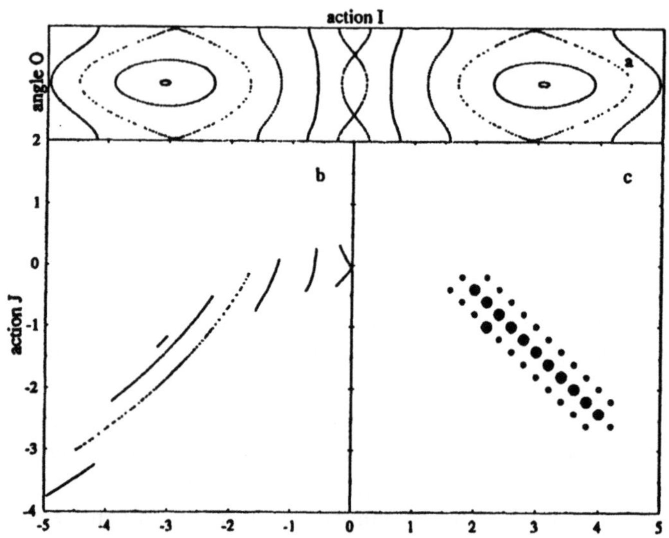

Fig. 1. The results of the numerical analysis of the system (3) for $\omega=3$ and $\hbar=0.2$: (a) Poincare cross section by (I,θ)-plane; (b) Poincare cross section by (I,J)-plane; (c) typical quasienergy function in (I,J)-representation, where we use big dot if $|c_{n,m}|>0.1$ and small dot if $0.01<|c_{n,m}|<0.1$. For the cases (b) and (c) only half of the plane is shown. The figure is adopted from [3].

We proceed to the quantum case ($I\rightarrow\hat{I}=i\hbar\partial/\partial\theta$). If we use free rotation states $\phi_n(\theta)=(2\pi)^{1/2}\exp(in\theta)$ as a reference basis, then the system (2) is just the set of the energy levels $E_n=\hbar^2n^2/2$ and time-periodic perturbation causes

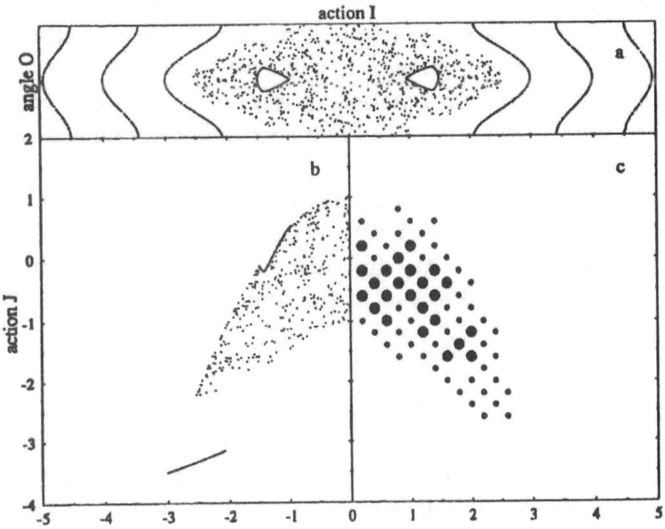

Fig. 2. The same as in the previous figure but $\omega = 1$.

transitions between the neighboring levels. The main questions, which should be answered in the quantum case, are the following: What is the dynamics of the transitions between the energy levels? What is the behavior of an observable? Which structure have the system quasienergy functions? And, of course, what is a difference between the cases $\omega > \sqrt{2}$ (regular regime) and $\omega < \sqrt{2}$ (chaotic regime)? We start with the question on the structure of the quasienergy states.

By definition the quasienergy functions are eigenfunctions of the unitary operator of the system evolution over period of the external field. Alternatively, the quasienergy functions can be defined as eigenfunctions of the effective Hamiltonian (3). We shall use the following expansion for the quasienergy states

$$\psi_l(\theta, \vartheta) = \exp\left(-\frac{i\epsilon_l t}{\hbar}\right) \sum_{n,m} c_{n,m}^{(l)} \exp(in\theta) \exp(im\vartheta) \ , \vartheta = \omega t \ , \qquad (4)$$

where ϵ_l denotes the quasienergy. Figure 1(c) and Fig. 2(c) show in a symbolic way the values of the expansion coefficients $c_{n,m}^{(l)}$ of typical quasienergy function for $\omega = 3$ (regular regime) and $\omega = 1$ (chaotic regime). The structural difference between these two cases is seen. Loosely speaking, this change in the structure of the quasienergy states makes basis for all other quantum manifestations of the transition to chaos in DRM.

Let us turn to the behavior of an observable $A(t) = \langle \psi(\theta, t) | \hat{A} | \psi(\theta, t) \rangle$, where $\hat{A}$ is an arbitrary operator. Substituting Eq. (4) into the latter formula, we obtain that independently of the value of ω (i.e., independently of the type of the classical dynamics) the behavior of $A(t)$ are quasiperiodic processes. However, there is a fundamental difference in the "complexity" of these quasiperiodic processes. In the case of the regular regime the spectrum of $A(t)$ usually consists of a few lines and the autocorrelation function of the process practically does not decay. In the case of the chaotic regime the spectrum is close to the spectrum of a random process, the correlation function of $A(t)$ decays and its recurrence time rapidly tends to infinity when $\hbar \to 0$. (The corresponding figures are given in the next section.)

The third question asked in the beginning was about the dynamics of the level populations. The detailed study of this question can be found in Ref. [4]. Here we mention only one result – the fluctuation of the level population in the chaotic regime was shown to obey one of the universal statistics for random variable, namely, χ_2^2-statistic.

3 Physical Applications of DRM

The simplest quantum object, which is described by DRM, is a polar molecule affected by the microwave field. The Hamiltonian of this system has the form

$$\hat{H} = B\hat{\mathbf{L}}^2 - (\mathbf{d}, \mathbf{E}\cos\omega t) \ . \tag{5}$$

In Eq. (5) B denotes the rotational constant of the molecule, $\mathbf{d}$ is the vector of its permanent dipole momentum, $\hat{\mathbf{L}}$ is the operator of the angular momentum, E and ω are the amplitude and the frequency of the linear polarized microwave field. Since the projection of the angular momentum on the direction of the field polarization is an integral of the motion, the system (5) reduces to the model (2), where the parameters of the molecule are related to the dimensionless parameters of the model (labeled by primes) by $\omega' = \hbar\omega(BEd)^{-1/2}$, $\hbar' = (B/Ed)^{1/2}$. We note that to speak about Quantum Chaos one need the scaled Planck constant $\hbar'$ to be less than unit. This requires a pretty large amplitude of the field. For example, the value $\hbar' = 0.2$ corresponds to $E = 3$ KV/cm for CsI molecule.

Another example is a molecule affected by modulated infrared field $E(t) = E\cos(\omega t)\cos(\Omega t)$ (the modulation can be done by using two lasers with a slightly different frequencies). If the carrier frequency Ω is chosen resonant to some vibrational transition, then the modulation frequency ω can be shown to be relevant to the dimensionless frequency ω' in DRM. Measuring the induced polarization of the molecule $P(t) = p_c(t)\cos(\Omega t) + p_s(t)\sin(\Omega t)$ one can easily distinguish between chaotic ($\omega \neq 0$) and regular ($\omega = 0$) regimes [see Fig. 3].

4 Extended Double Resonance Model

We shall get an interesting extension of DRM if we remove the periodic boundary condition. In the other words, we shall consider the system with the Hamiltonian

$$H = \frac{p^2}{2} + \cos x \cos(\omega t) \ , \ -\infty < x < \infty \ , \tag{6}$$

(a particle in a standing wave). The phase space of the system (6) consists of the bounded chaotic component, which has the form of a stripe along x-axis,

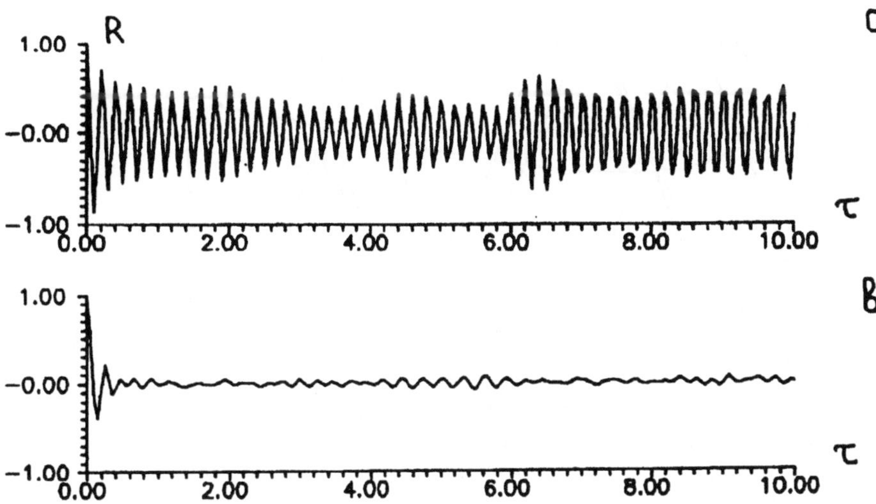

Fig. 3. The autocorrelation function $R(\tau) = \overline{p_s(t+\tau)p_s(t)}$ of the out-of-phase induced polarization in a molecule affected by a laser field: (a) regular regime, and (b) chaotic regime. The value of $\hbar' = 1/16$. The figure is addopted from [5].

and the regular ballistic trajectories above and below the chaotic component. If we choose the initial position of a classical particle in the chaotic region, then its motion should resemble the Brownian motion. In fact, it have been found in [6] that for some particular values of ω (for example, for $\omega = 0.3$ when there are no stability island inside the chaotic component) the uncertainty in the particle position follows to the diffusion law $\langle \Delta x^2(t) \rangle = D_{cl}t$. The question we are focused on in this section is the question about "diffusion" in the quantum case.

To answer this question we have to analyze the quasienergy spectrum of the quantum counterpart of the system (6) ($p \to \hat{p} = -i\hbar'\partial/\partial x$). Because of the translational symmetry the spectrum forms the bands $\epsilon_l(k)$, where k is the wave vector ($-1/2 < k \leq 1/2$) and l is a band index. Figure 5 shows the quasienergy spectrum of EDRM for $\omega = 0.5$ and $\hbar = 0.2$. It is seen that the spectrum structure is typical for a nonintegrable system and contains many avoided crossings.

The detailed analysis of the spectrum can be done by using the diabatic representation. In the diabatic representation all avoided crossings with the gaps less than some given number are substituted by the real crossings. Then the bands can be sorted into two groups – group of the relative narrow bands and group of the wide bands. The narrow bands are formed by the states associated with the chaotic component of EDRM. The states associated with the ballistic trajectories give rise to the wide bands with the dispersion law $\epsilon_l(k) \approx [\hbar^2(l+k)^2/2]_{\rm mod}\omega \approx [\hbar^2 lk + \hbar^2 l^2/2]_{\rm mod}\omega$, where $|l| > l_{max} = V/2\pi\hbar$ and by V we denote the phase volume of the chaotic component per one space

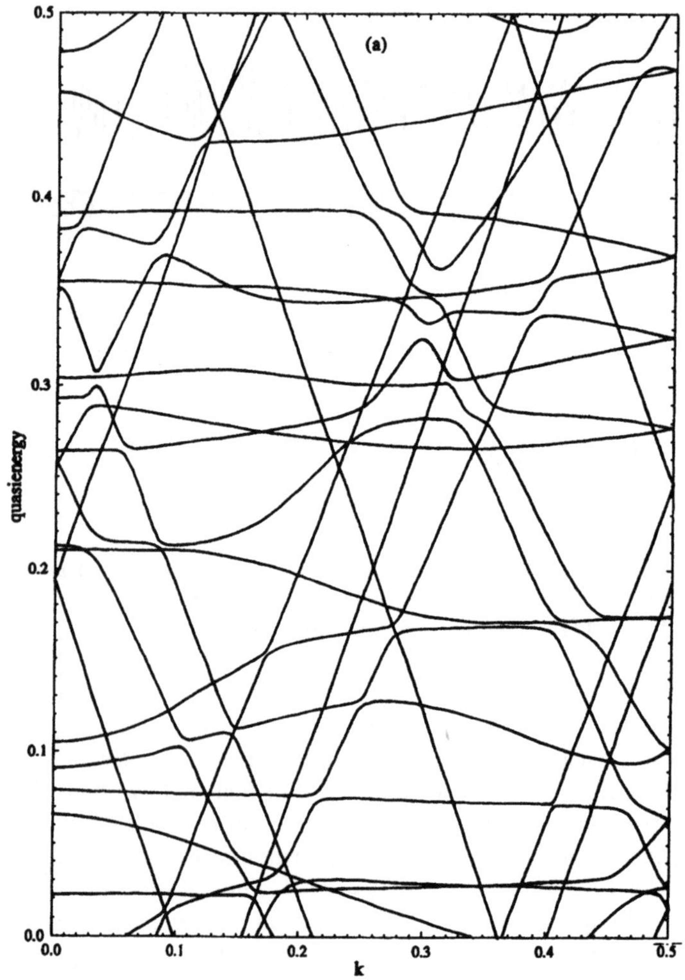

Fig. 4. (a) Quasienergy spectrum of EDRM for $\omega = 0.5$, $\hbar = 0.2$, and $0 \leq k \leq 1/2$ (the quasienergies are plotted in the units of $\hbar$).

period of the standing wave. The avoided crossings between the regular and chaotic bands removed in Fig. 5(b) is a kind of the resonant effect, which has no a classical analogous. For those particular values of k, where the avoided crossings take place, the quasienergy eigenfunction (in the adiabatic representation) is a superposition of the chaotic and ballistic states.

Having the quasienergy spectrum obtained, we can predict the dynamics of the system. We shall do this first in the diabatic picture. In this case the evolution

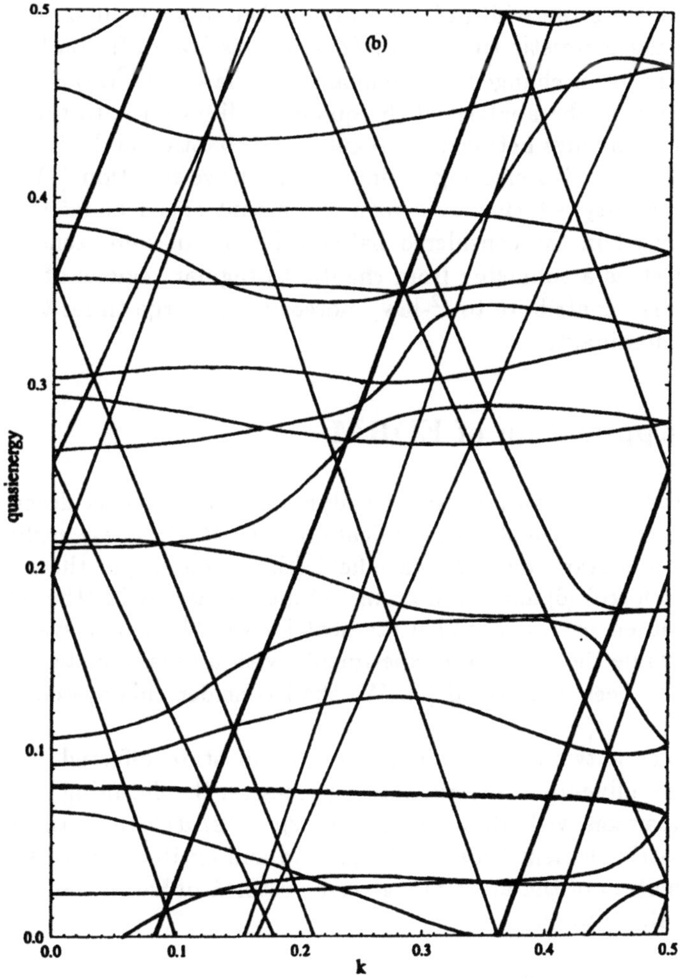

Fig. 5. (b) Adiabatic representation of the quasienergy spectrum. Only 6 regular bands with $l = \pm 8$, ± 9, ± 10 are shown. The figure is addopted from [6].

of $\psi(x,t)$ is defined by the following equation

$$\psi(x,t) = \sum_{l<l_{max}} \int_{-1/2}^{1/2} b_l(k)\phi_l^{(k)}(x,t)\exp\{i[kx - \epsilon_l(k)t/\hbar]\}\mathrm{d}k \ . \qquad (7)$$

In Eq. (7) the sum over l is restricted by the chaotic bands and we denote by $\phi_l^{(k)}(x,t)$ the quasienergy eigenfunctions, which are periodic both on x and t by definition. Let us denote the characteristic width of the chaotic bands by $\Delta\epsilon$. One can consider two time limits: $t < \hbar/\Delta\epsilon$, and $t \gg \hbar/\Delta\epsilon$. For the second time limit the integral over k in Eq. (7) can be estimated according to the stationary phase method. It gives the width of the wave packet to grow linear with time. Therefore, after characteristic time $t^* \sim \hbar/\Delta\epsilon$ the classical diffusion law for the coordinate uncertainty is changed by the quantum t^2-law: $\langle \Delta x^2(t) \rangle \sim t^2 , t \gg t^*$.

The second step in the analysis of the quantum diffusion is to take into account the avoided crossings between the regular and chaotic bands. It is possible to show that every avoided crossings brings to the wave function (7) a narrow peak of the probability, which moves with a constant speed equal to one-half of the particle speed in the considered ballistic band. (In some sense one can consider this effect as a tunneling from chaotic to regular regions of the phase space.) These peaks contribute to t^2-law, moreover, for large times this contribution becomes dominant.

5 Physical Application of EDRM

There are at least two quantum systems, dynamics of which is described by EDRM (more precisely, by a slight modification of EDRM). These system are a charged particle in a standing microwave field (the scattering of the particle is caused by Lorenz force) [6] and an atom in a standing laser field (the scattering of the atom is caused by the so-called gradient force). The latter system looks more attractive while the successful experiments with atoms in a standing laser wave have already been performed [7]. We shall consider this system in more details.

Let us consider a two-level atom in the ground state affected by a laser field. To avoid the influence of the relaxation processes, the detuning $\Delta\Omega$ should be large in comparison with Raby frequency $\Omega_R = 2Ed/\hbar$, d is the value of the transition matrix element between the ground and excited states. Providing $\Delta\Omega \gg \Omega_R$ the motion of the atom along wave vector k of the standing wave is governed by the Hamiltonian

$$\hat{H} = \frac{\hat{p}^2}{2m} + U\cos^2(kx) \ , U = \frac{\hbar\Omega_R^2}{\Delta\Omega} \ . \tag{8}$$

All we have to do now is to modulate the amplitude of the field: $E \to E(t) = E\cos(\omega t)$. Then the dynamics of the system considered becomes similar to that of EDRM. We conclude this section by the estimate for Ytterbium atom. Let Raby frequency to be $\sim 10^9\ Hz$ and the detuning $\Delta\Omega \sim 10^{10}\ Hz$. Then modulation frequency ω should be about $10^6\ Hz$ and scaled Planck constant $\hbar' = (k^2\hbar^2/mU)^{1/2}$ has the value $\hbar \approx 0.05$.

6 Conclusion

We have analyzed the dynamics of DRM and EDRM. The main conclusion from this analysis might sound surprising – there are no analogies with quantum kicked rotor at all. Thus the chaotic systems with bounded chaotic component is really different class of the dynamical system and they require a different approach in the quantum case. We note that the same conclusion have been independently done in Ref. [4].

We also would like to note an interesting chain of the transformations on the way from classics to quanta in DRM. In fact, in the classical approach any system trajectory is random in the chaotic regime. However, when we average over an ensemble of the trajectories, the dynamics of an arbitrary classical average becomes a pure regular process [5]. On the last stage, when we substitute the classical average by the quantum-mechanical average, its dynamics appears to be a quasirandom process. This "quasirandomness" is a sign of the underlaying classical chaos and it is absent when the classical dynamics is regular.

References

1. G. Casati, B.V. Chirikov, F.M. Izrailev and J. Ford, Lecture Notes in Physics **90**, 334 (Springer, Berlin, 1979).
2. G.P. Berman, G.M. Zaslavsky and A.R. Kolovsky, Sov. Phys. JETP **54**, 272 (1981).
3. A.R. Kolovsky, Phys. Rev. A **48**, 3072 (1993).
4. N. Moiseyev, H.J. Korsch and B. Mirbach, Z. Phys. D **29**, 125 (1994).
5. A.R. Kolovsky, Phys. Lett. **A157**, 474 (1991).
6. A.R. Kolovsky, S. Miyazaki and R. Graham, Phys. Rev. **E49**, 70 (1994).
7. C. Solomon, J. Dalibard, A. Aspect, H. Metcalf and C. Cohen–Tannoudji, Phys. Rev. Lett. **56**, 827 (1986); E.L. Moore, Phys. Rev. Lett. **73**, 2974 (1994).

Asymptotic Behavior of Generalized Lévy Walks

Marcin Kotulski

Hugo Steinhaus Center for Stochastic Methods
Technical University of Wrocław,
PL–50–370 Wrocław, Poland

Abstract: We propose a generalization of Lévy walk in one dimension allowing for an arbitrary bias and asymmetry of jumps. An asymptotic distribution (propagator) of distance $R(t)$ reached up to time t by a particle initially at the origin is found to be a possibly asymmetric Lévy-stable law $s_{\alpha,\beta}(r)$ or a positive law $h_\lambda(r)$. A probabilistic approach in terms of random variables R_i and T_i is applied.

A continuous-time random walk (CTRW) introduced in ref. [1] is a walk with random waiting-times T_i between successive random jumps R_i. The special case of CTRW, called Lévy walk [2], proved to be a useful model of various aspects of anomalous transport, e.g. in turbulence [2] or intermittent chaotic systems [3]. The signature of the anomalous diffusion is a non-Gaussian asymptotic distribution (propagator, diffusion front) of distance $R(t)$ reached up to time t by a particle initially at the origin.

Our aim is to find the asymptotic distribution of $R(t)$ for Lévy walks (the problem partially solved in refs. [4], [5]) and also for generalized Lévy walks in one dimension, which we propose in this paper in order to include a bias and a jump tails asymmetry into consideration. The limiting laws obtained are Lévy-stable law $s_{\alpha,\beta}(r)$ and a positive law $h_\lambda(r)$ which is a distribution of a random variable $1/Y^\lambda$, where Y has a positive Lévy-stable law $s_{\lambda,1}(r)$, $0 < \lambda < 1$.

In this paper we rely on ref. [6] where, for finding an asymptotic distribution of an arbitrary CTRW, a straightforward probabilistic approach in terms of random variables R_i and T_i is applied, instead of the usually used Tauberian theorems for the Fourier-Laplace transform.

We now begin with a detailed definition of a generalized Lévy walk. Let us consider a continuous-time random walk described by means of two sequences of random variables R_i and T_i, jump distances and waiting-time intervals, respectively, with a particle initially at the origin. For simplicity we restrict our attention to one-dimensional walks. The first instantaneous jump of random length R_1 takes place after a random waiting-time T_1, then the second instantaneous jump R_2 after time T_2, etc. In general, the i-th jump R_i may depend on its waiting-time T_i in an arbitrary way, but the pair (R_i, T_i) is independent of the preceding and succeeding pairs of jumps and its waiting-times (R_k, T_k). The special case when R_i is independent of T_i is called a decoupled memory CTRW as opposed to the coupled one with R_i depending on T_i. It is evident that the model is entirely determined by $\psi(r,t)$, a two-dimensional joint probability density of the pair (R_i, T_i), the same for each $i \geq 1$. The marginal densities $g(t)$ of T_i and $f(r)$ of R_i are

$$g(t) = \int_{-\infty}^{\infty} \psi(r,t)dr, \qquad f(r) = \int_{0}^{\infty} \psi(r,t)dt \ . \tag{1}$$

We define a random variable N_t as the number of jumps in the time interval $[0,t]$. Then, the position $R(t)$ of a walking particle at time t is equal to a random sum of N_t successive random jumps R_i, that is

$$R(t) = \sum_{i=0}^{N_t} R_i \quad \text{where} \quad N_t = \max\{k : \sum_{i=0}^{k} T_i \leq t\}, \quad R_0 = 0, \quad T_0 = 0. \tag{2}$$

Lévy walk [2] is a CTRW with the specific form of coupling between jump R_i and its waiting-time T_i, namely

$$\psi(r,t) = cg(t)\delta(|r| - t^{\nu}) \quad \text{or} \quad R_i = \pm T_i^{\nu} \ , \quad \text{where } \nu > 0, \ c \text{ are constants.} \tag{3}$$

In this paper we propose a *generalized* Lévy walk assuming

$$R_i = Y_i \ T_i^{\nu} + m \ , \quad \text{where } Y_i = \begin{cases} +1 & \text{with probab. } p \\ -1 & \text{with probab. } q = 1 - p \ , \end{cases} \tag{4}$$

$\nu \geq 0, 0 \leq p \leq 1$ and m are constants. Variables Y_i are random signs independent of T_i. For $\nu = 0$ Eq. (4) reduces to the decoupled case. Equation (4) allows us to take into account jumps R_i which are asymmetric (for $p \neq \frac{1}{2}$) with possible bias governed by a constant m. Typical densities $g(t)$ of T_i and $f(r)$ of R_i for a generalized Lévy walk are shown in Fig. 1.

Our aim is to find the limiting distribution of position $R(t)$ for generalized Lévy walks which, as we will see, depends on the tails of T_i and R_i. In order to find a relation between these tails we assume that $m = 0$ and denote by λ the tail exponent of the waiting-time T_i, that is

$$\Pr(T_i > x) \sim ax^{-\lambda} \ , \quad \text{as} \quad x \to \infty \ , \tag{5}$$

where a is a constant which determines a magnitude of the tail of the waiting-time distribution. Simple transformations yields for large x

$$\Pr(R_i > x) = p \Pr(T_i^{\nu} > x) = p \Pr(T_i > x^{1/\nu}) \sim pa(x^{1/\nu})^{-\lambda} = pax^{-\lambda/\nu}$$
$$\Pr(R_i < -x) = q \Pr(T_i^{\nu} > x) = q \Pr(T_i > x^{1/\nu}) \sim qax^{-\lambda/\nu} \ .$$

Hence, denoting by ρ the tail exponent of R_i, one obtains

$$\rho = \lambda/\nu \ , \quad \text{where} \quad \Pr(R_i < -x) + \Pr(R_i > x) \sim ax^{-\rho} \ , \tag{6}$$

as $x \to \infty$. The same relation is valid for any value of m.

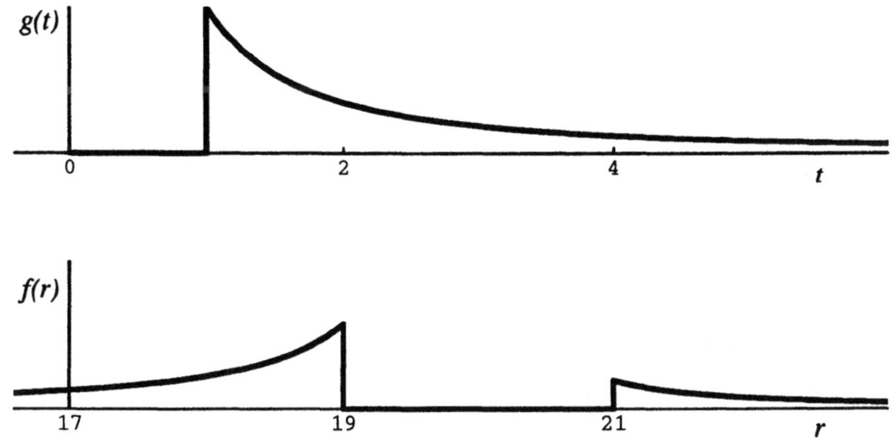

Fig. 1. Grahps of densities $g(t)$ of T_i and $f(r)$ of R_i for a generalized Lévy walk corresponding to $\nu = 1.3$, $p = 0.25$, $m = 20$ in Eq. (4).

We now briefly summarize some necessary probabilistic facts. The Central Limit Theorem for *non-random sum* of jumps R_i which are defined by Eqs. (4)-(6) reads

$$\Pr\left(\frac{\sum_{i=1}^{n} R_i - n\mu}{n^{1/\alpha} c_2} \leq r\right) \to S_{\alpha,\beta}(r) \equiv \int_{-\infty}^{r} s_{\alpha,\beta}(x)dx \ , \tag{7}$$

where $\alpha = \min(\rho, 2)$, $\beta = 2p - 1$, $\mu = \begin{cases} \langle R_i \rangle & \text{if } 1 < \alpha \leq 2 \\ 0 & \text{if } 0 < \alpha < 1 \end{cases}$

and $s_{\alpha,\beta}(r)$ is the standard Lévy-stable density defined by Eq.(9) below. The normalizing constant c_2 in Eq. (7) is $c_2 = (\text{Var}R_i/2)^{1/2}$ for $\alpha = 2$, or

$$c_2 = (a/C_\alpha)^{1/\alpha} \ , \quad \text{where} \quad C_\alpha = \frac{1-\alpha}{\Gamma(2-\alpha)\cos(\pi\alpha/2)} \quad \text{for } 0 < \alpha < 2 \ , \ \alpha \neq 1 \ . \tag{8}$$

Let us remark that for $0 < \alpha < 1$ the constant m defined by Eq. (4) does not affect Eq. (7), while for $1 < \alpha \leq 2$ the constant m governs the mean μ of the jump R_i. Also, parameter β satisfies $\frac{1+\beta}{2} = p$, $\frac{1-\beta}{2} = q$.

The standard Lévy-stable density $s_{\alpha,\beta}(r)$ with parameters α, β is defined by its characteristic function [7], [8]

$$\int_{-\infty}^{\infty} e^{iur} s_{\alpha,\beta}(r)dr = \exp\{-|u|^\alpha(1 - i\beta \text{sign}(u)\tan(\pi\alpha/2))\} \ , \tag{9}$$

where $0 < \alpha \leq 2$, $\alpha \neq 1$ and $-1 \leq \beta \leq 1$. Note that density $s_{2,0}(r)$ is Gaussian with mean 0 and variance 2 and $s_{2,0}(r) = s_{2,\beta}(r)$ for any β. The graphs of $s_{\alpha,\beta}(r)$ for $\alpha = 0.8$ are given in Fig. 2 and for $\alpha = 1.25$ in Fig. 3. We observe long tails which can be expressed [7], [8] by

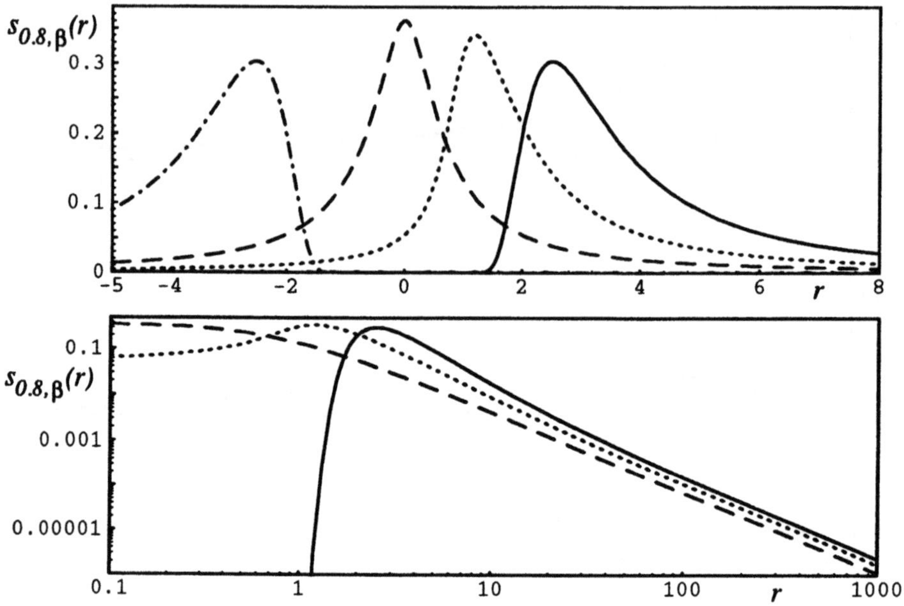

Fig. 2. Linear and logarithmic graphs of Lévy-stable densities $s_{0.8,\beta}(r)$ corresponding to the values $\beta = 1$ (full line), $\beta = 0.5$ (dotted line), $\beta = 0$ (dashed line) and $\beta = -1$ (dashed-dotted line).

$$s_{\alpha,\beta}(\pm r) \sim \alpha C_\alpha \frac{1 \pm \beta}{2} \, r^{-\alpha-1} + D_{\alpha,\beta} \, r^{-2\alpha-1} + \ldots , \quad \text{as } r \to \infty , \tag{10}$$

except for density $s_{\alpha,1}(r) = s_{\alpha,-1}(-r)$ for which the left tail is exponential, i.e.

$$s_{\alpha,1}(r) \sim A_\alpha |r|^{\frac{\alpha-2}{2-2\alpha}} \exp\left(-B_\alpha |r|^{\frac{\alpha}{\alpha-1}}\right), \quad \text{as} \begin{cases} r \to -\infty & \text{if } 1 < \alpha < 2 \\ r \downarrow 0 & \text{if } 0 < \alpha < 1 , \end{cases} \tag{11}$$

where A_α, B_α, $D_{\alpha,\beta}$ are constants and C_α is defined by Eq. (8).

We now turn to the study of asymptotic behavior of $R(t)$ for generalized Lévy walks defined by Eqs. (4)-(6). Here we make use of Eq. (2), which states that $R(t)$ is a *random sum* of jumps R_i.

In the case $\lambda > 1$ the mean waiting-time $\tau = \langle T_i \rangle$ is finite, and

$$\lim_{t\to\infty} N_t/t = 1/\tau \qquad \text{with probability 1} , \tag{12}$$

thus the number N_t of jumps up to moment t approximately equals t/τ. In order to obtain the limiting distribution of $R(t)$ we replace $\sum_{i=1}^{n} R_i$ with $\sum_{i=0}^{N_t} R_i = R(t)$ and n with t/τ in formula (7). As a result one finds that for $\mu = 0$ the limiting distribution is unchanged [9]-[11], while for $\mu \neq 0$ the limiting distribution may change [6]. Hence, we must distinguish two subcases.

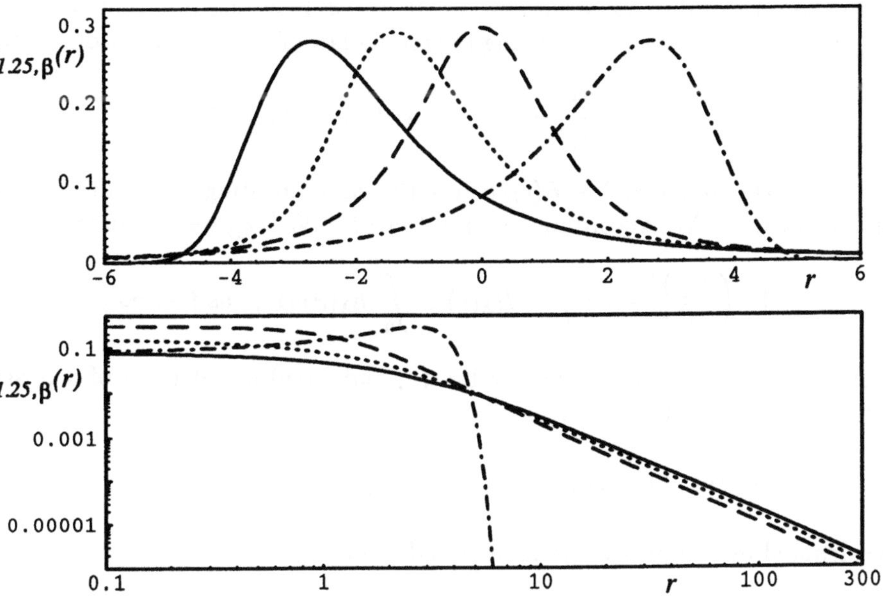

Fig. 3. Linear and logarithmic graphs of Lévy-stable densities $s_{1.25,\beta}(r)$ corresponding to the values $\beta = 1$ (full line), $\beta = 0.5$ (dotted line), $\beta = 0$ (dashed line) and $\beta = -1$ (dashed-dotted line).

For $1 < \lambda$, $\mu = 0$ or $\alpha < \lambda$, $1 < \alpha < 2$, $\mu \neq 0$ we have as $t \to \infty$

$$\Pr\left(\frac{R(t) - (t/\tau)\mu}{(t/\tau)^{1/\alpha} c_2} \leq r\right) \to S_{\alpha,\beta}(r) \equiv \int_{-\infty}^{r} s_{\alpha,\beta}(x)dx \ , \tag{13}$$

where c_2 is defined by Eq. (8) or, equivalently, denoting by $p_t(r)$ the density of position $R(t)$ we may express Eq. (13) for large t as

$$p_t(r + (t/\tau)\mu) \approx \frac{1}{(t/\tau)^{1/\alpha} c_2}\ s_{\alpha,\beta}\left(\frac{r}{(t/\tau)^{1/\alpha} c_2}\right) . \tag{14}$$

Combining Eq. (10) with Eq. (14), one obtaines

$$p_t(r + (t/\tau)\mu) \propto \begin{cases} \dfrac{1}{t^{1/\alpha}}\left(\dfrac{r}{t^{1/\alpha}}\right)^{-\alpha-1} = \dfrac{t}{r^{\alpha+1}} & \text{for large } \dfrac{r}{t^{1/\alpha}} \\ \dfrac{1}{t^{1/\alpha}} & \text{for small } \dfrac{r}{t^{1/\alpha}} \ . \end{cases} \tag{15}$$

For $\alpha > \lambda > 1$ and $\mu \neq 0$ one arrives at an unexpected conclusion [6]. The limiting distribution is determined only by the waiting-time tail exponent λ in Eq. (5), namely as $t \to \infty$

$$\Pr\left(\frac{R(t)/\mu - (t/\tau)}{t^{1/\lambda} c_3} \leq r\right) \to S_{\lambda,-1}(r)\ , \quad \text{where} \quad c_3 = (a/C_\lambda)^{1/\lambda} \tau^{-1-1/\lambda} \ . \tag{16}$$

In the case of dispersive motion, i.e. $0 < \lambda < 1$, we assume that $\mu \neq 0$, i.e. $\langle R_i \rangle \neq 0$, which entails $\alpha > 1$. Using the strong law of large numbers one obtains

$$\lim_{t\to\infty} \frac{R(t)}{N_t} = \mu \qquad \text{with probability 1.} \tag{17}$$

Hence the random variable $R(t)/\mu$ has the same limiting distribution as the random variable N_t. According to a limit law for N_t [12], we conclude that

$$\Pr\left(\frac{R(t)/\mu}{t^\lambda c_6} \le r\right) \to H_\lambda(r) \equiv \int_0^r h_\lambda(x)dx \ , \quad \text{as } t \to \infty \ , \tag{18}$$

where $c_6 = C_\lambda/a$, C_λ is defined by Eq. (8) and $h_\lambda(r)$ is a density of a positive distribution $H_\lambda(r)$ defined by

$$H_\lambda(r) = \Pr\left(1/Y^\lambda \le r\right) = 1 - S_{\lambda,1}\left(1/r^{1/\lambda}\right), \qquad r > 0 \ . \tag{19}$$

We denoted by Y a positive random variable having the Lévy-stable law $S_{\lambda,1}(r)$.

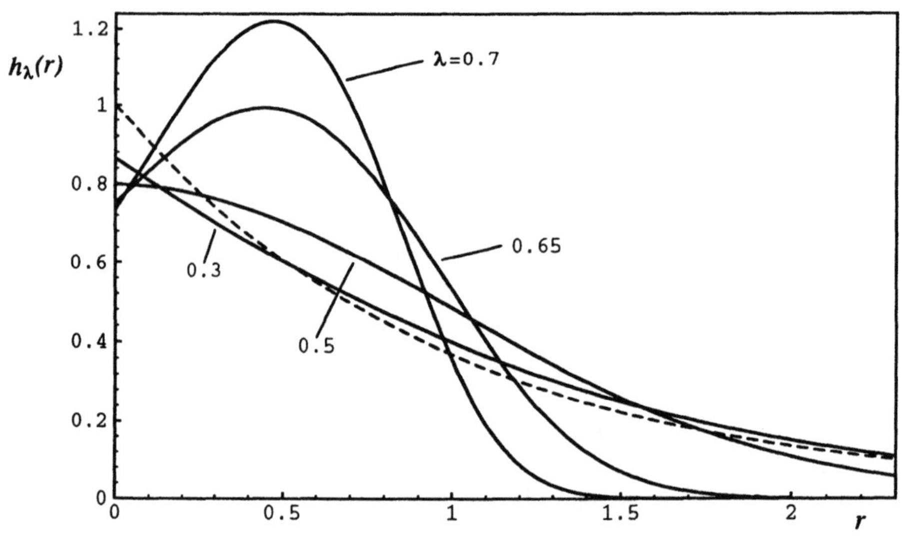

Fig. 4. Graphs of densities $h_\lambda(r)$ corresponding to the values $\lambda = 0.3$, 0.5, 0.65, 0.7 . The dashed line shows the density e^{-r} which is very close to $h_{0.1}(r)$, cf. Eq. (20).

An equivalent form of Eq. (18) has been given in refs. [13], [14] for the decoupled case, i.e. for $\nu = 0$ in Eq. (4). The graphs of densities $h_\lambda(r)$ are given in Fig. 4. Note that $h_\lambda(r)$ are discontinuous at $r = 0$ and they are non-zero only at points $r > 0$. Surprisingly [8],

$$h_\lambda(r) \to e^{-r} \qquad \text{as} \quad \lambda \downarrow 0 \ , \qquad \text{for} \quad r > 0 \ . \tag{20}$$

It is worth pointing out [8] that for $1/2 \leq \lambda < 1$ the density $h_\lambda(r)$ is simply the cut-off Lévy-stable density $s_{1/\lambda,-1}(r)$, so that it is restricted to the positive half-line only, namely

$$h_\lambda(r) = 1/\lambda \; s_{1/\lambda,-1}(r/c_8)/c_8 \,, \qquad \text{for } 1/2 \leq \lambda < 1 \,, \quad r > 0 \,. \qquad (21)$$

where the constant $c_8 = \cos(\pi\lambda/2)(-\cos(\pi/(2\lambda)))^\lambda$ is the scale shortening. Note that the cut-off normalization term $1/\lambda$ in Eq. (21) results from the relation $\lambda = 1 - S_{1/\lambda,-1}(0/c_8)$. An interesting feature of formula (18) is that, despite bias μ, there is no shift proportional to time t, as opposed to Eqs. (13) and (16). It can be attributed to the infinite mean waiting-time τ, i.e. to the lack of characteristic time scale. The remaining case $0 < \lambda < 1$, $\mu = 0$ will be investigated elsewhere.

We can conclude that proposed generalized Lévy walk is an extension of Lévy walk allowing for an arbitrary bias and tail asymmetry of jump density. The derived asymptotic distributions (propagators) of position $R(t)$ are the following: possibly asymmetric law $s_{\alpha,\beta}(r)$ in Eqs. (13), (16) and positive law $h_\lambda(r)$ in Eq. (18). The results are in agreement with previous studies [4], [5] which dealt with particular cases of Lévy walks. The probabilistic approach presented here appears to be a convenient way of deriving asymptotic behavior of random walks. It also provides analytical form for normalizing constants c_2, c_3, c_6 (prefactors) in Eqs. (13), (16) and (18) which are useful in computer simulations.

Acknowledgments

I would like to thank Prof. K. Weron for help with the preparation of this paper. This work was partially supported by KBN Grant no. 2 P302 065 04.

References

1. E. W. Montroll and G. H. Weiss, J. Math. Phys. **6**, 167 (1965).
2. J. Klafter, A. Blumen and M. F. Shlesinger, Phys. Rev. A **35**, 3081 (1987).
3. G. Zumofen, J. Klafter and A. Blumen, Phys. Rev. E **47**, 2183 (1993).
4. G. Zumofen, J. Klafter and A. Blumen, Chem. Phys. **146**, 433 (1990).
5. R. N. Mantegna, J. Stat. Phys. **70**, 721 (1993).
6. M. Kotulski, to appear in J. Stat. Phys. (1995).
7. A. Janicki and A. Weron, *Simulation and Chaotic Behavior of α-Stable Stochastic Processes*, (Marcel Dekker, New York, 1994).
8. V. M. Zolotarev, *One-dimensional Stable Distributions*, (AMS, 1986).
9. H. Wittenberg, Z. Wahrschein. verw. Geb. **3**, 7 (1964).
10. M. Csorgo and Z. Rychlik, Canadian J. Statis. **9**, 101 (1981), Theorem 3.
11. D. J. Aldous, Math. Proc. Cambridge Phil. Soc. **83**, 117 (1978), Theorem 7.
12. W. Feller, *An introduction to Probability Theory and its Applications*, Vol. 2, (Wiley, New York, 1966).
13. J. K. E. Tunaley, J. Stat. Phys. **12**, 1 (1975); **11**, 397 (1974).
14. H. Weissman, G. H. Weiss and S. Havlin, J. Stat. Phys. **57**, 301 (1989).

Stochastic Moore Loop Space

R. Léandre

Dpt de Mathématiques. Institut Elie Cartan
Faculté des Sciences. Université de Nancy I
54000. Vandoeuvre les Nancy. France
and
Mathematical Institute
Warwick University
Coventry CV4 7 Al. England

Introduction

Let M be a manifold of finite dimension. There is a product over the cohomology groups which is the wedge product. Generally, there is no others operations. But if the manifold is an H-space (See [M-M], [H]), there is a coproduct which is the pullback of the multiplication operation over the manifold, and the cohomology groups are endowed with richer structures.

Let us consider the based loop space of the manifold, that is the space of smooth applications γ from the circle over the manifold with the boundary conditions $\gamma_0 = \gamma_1 = x$. There is no product over this infinite dimensional manifold. For this reason, topologist consider another model of the based loop space, which is an H-space, that means it is endowed with a nice product. Namely if we consider a based loop in time s and another based loop in time t, we can stick them together in order to get a loop in time $s+t$. We consider the family of loop in all possible times. The cohomology of this extended loop space, called the Moore loop space, is equal to the cohomology of the simple loop space, and is an Hopf algebra: this means that its cohomology is endowed with a product an a coproduct, which are compatible in some sense.

The purpose of this paper is to try to clarify the situation for serie of forms over the Moore loop space. Our aim is that this serie of forms converges in some sense. For this, we introduce a measure over the Moore loop space, and we use a family of Brownian bridges. Bridges in all times are involved: small time estimates as well as long time estimates are useful.

The wedge product over the cohomology groups of the stochastic Moore loop space $L_M(M)$ is the pull-back of the diagonal map in $L_M(M) \times L_M(M)$. The diagonal is a set of infinite codimension in $L_M(M) \times L_M(M)$. Therefore the technics of the quasi-sure analysis ([Ge], [A.M$_1$]. See [Ug] for a preliminary work) are not sufficient to restrict over the diagonal functionals over the product, because these technics work over finite codimensional manifolds of the Wiener space. We used the definition of $A\hat{\otimes}B$ of [Gro], [Ru], and [Sch] to define the smaller tensor product of two Banach spaces. Therefore, we can define over this tensor product a wedge product which is continuous.

It is not the same for the coproduct, which is the pullback at the level of forms or of the cohomology of the product $L_M(M) \times L_M(M) \to L_M(M)$. If we take a loop in time t, the set of loop which are in a fixed time $s < t$ in x is a finite codimensional manifold. We can use technics of quasi-sure analysis to pullback forms and define a coproduct: long time and short time estimates of heat kernels are involved (See [L_1], [K_2], [Wa] for surveys.). But a form over this submanifold is not a product of a form over the based loop space between 0 and s and the loop space between s and t. This implies that we have to change the definition of the tensor product involved in order to define the coproduct.

If we consider Chen forms, or more precisely their algebraic counterparts, we have the same tensor product, and therefore an analytic Hopf algebra. This provides an algebraic treatment of the dilatation procedure of [Pa] in order to define dilatation over the stochastic Moore loop space, by using Adams operations. The technics of the quasi-sure analysis are inefficient: let us recall namely that the set of periodic paths is an infinite codimensional subspace of the Wiener space. Moreover long time estimates are involved, and we don't have the same criteria of convergences than in [J.L_1].

On the other hand, we consider an approximation of the exterior derivative over the Moore Brownian bridge: namely the Moore Brownian bridge appears more and less locally as a fibration in standard brownian bridges. The exterior derivative we consider works only in the fiber and not in the time basis.

Let us give now a brief historical review of the analysis over infinite dimensional spaces:

•) [Ku] has constructed diffusion processes over infinite dimensional manifolds.

•) A theory of infinite dimensional manifold was developed in the beginning of the seventies. Let us remark that [Ra] has studied the exterior derivative over finite codimensional forms over such manifolds.

•) The Russian school ([D.Fo]) has too performed a theory of infinite codimensional manifolds. In all these theories, the structure of the manifold is of the main importance.

•) It is not the same for the quasi-sure analysis: the curved manifold is given by an equation over the Wiener space ([Ge], [A.M_1]). [K_1] studied the case of forms.

•) The analysis over loop groups is purely global too ([A.M_2], [Gr_2], [Gr_3], [Sh_3]). The main tool are the integration by parts over loop groups which use the symmetries of the group ([A.H-K]).

The present analysis is more related to [J.L_1]: an L^p theory of forms is performed, in order that the stochastic Chen forms which satisfy criteria of convergences in the manner of [Co] are belonging to all the L^p. There is no differential operation in [J.L_1], because the lack of integration by parts. They are performed in [L_2]: the main tool are the Bismut integration by parts with geometrical meaning of [Bi] and short time asymptotics of densities (See [L_1], [K_2] and [Wa] for surveys). A simpler proof is given in [L.N]. This allows to define an Ornstein-Uhlenbeck operator over the free loop space invariant by rotation in [L_3]. In [L_3], a link between the Hochschild coboundary and the stochastic exterior derivative is performed, but no satisfying functional analysis is done in such direction.

It is not the same for [J.L$_2$], [L.R], [L$_4$]: non scalar operators over the free loop space are studied, their adjoint are computed, and after using a limit procedure, their possible index is computed. In particular, it is possible in [J.L$_2$] and in [L.R] to work with the adjoint of a regularised version of the exterior derivative over the free loop space. The price to pay is the following: we don't have a complex, but only an operator which is homotopically equivalent to the exterior derivative.

[L$_5$] shows that the definition of the stochastic exterior derivative over the free loop space is strongly related to the definition of anticipative integrals. [L$_6$] shows that the Adams-Chen theorem is still true for the stochastic loops space of an homogeneous manifold. The Hochschild cohomology is equal to the stochastic cohomology of loop spaces ([Ad$_1$], [Ch]). He used the proof of [G.J.P], with a stochastic meaning, and the ideas of [Gr$_2$] in order to get Sobolev logarithmic inequalities over a loop group: the free loop space of a Lie group is a fibration over the Lie group with fiber the based loop group, which uses deeply the quasi-invariance formulas of [A.H-K] (See [Sh$_1$] and [Ar] for the case of an homogeneous manifold).

We work over C by considering complexified objects if necessary. Multilinear expressions are written as standard products.

We thank D. Elworthy for his warm hospitality during our sojourn at the University of Warwick where this work was done, F. Patras for stimulating discussions as well as Y. Felix and D. Elworthy.

1 Moore Brownian Bridge

Let M be a compact Riemannian manifold. Let (Ω, F) the space of continuous applications from $[0,\infty[$ into M such that $\gamma_s = \gamma_0 = x$ for some $s \geq t$. F is the σ-algebra spanned by the cylindrical functionals. Let $dP_{t,x}$ the law of the Brownian bridge in time t over the canonical space Ω_t of the Brownian bridge in time t starting from x and coming back in time t in x. Let $g_p(t)$ a probability density over R^+, $g_p(t) > 0$. We endowe the Moore Brownian bridge with the probability measure $dP_{t,x} \otimes g_p(t)dt$ modulo a set of measure 0.

The tangent space of a Brownian loop $\gamma_{s,t}$ in time t is constructed as follows: we consider the parallel transport $\tau_{s,t}$ from $\gamma_{0,t} = x$ to $\gamma_{s,t}$, $s \leq t$, for the pinned brownian motion. We consider the space of continuous sections $X_{s,t}$ over $\gamma_{s,t}$ of the shape $\tau_{s,t}H_{s,t}$ and we suppose that the path $H_{s,t}$ in $T_x(M)$ has a finite variation, and we denote its derivative by $H'_{s,t}$. We choose as Hilbert structure ([Dr.R])

$$||X||_t^2 = \int_0^t ||H'_{s,t}||^2 ds \tag{1.1}$$

Moreover $X_{0,t} = X_{t,t} = 0$.

The Moore Brownian bridge appears as a fibration in Brownian bridges in time t. The basis is constituted by the time of the considered Brownian bridge and the fiber by the standard Brownian bridge in time t. The model we choose for forms over the Moore Brownian bridge is only involved with the fiber: a form σ_n is given by a family of kernels $\sigma_{n,t}(s_1,..,s_n)$ such that:

$$\sigma_{n,t}(X_t^1,..,X_t^n) = \int_{[0,t]^n} \sigma_{n,t}(s_1,..,s_n) d/dsH^1_{s_1,t}..d/dsH^n_{s_n,t}\, ds_1...ds_n \tag{1.2}$$

Moreover in time t, we work over the based loop space. This implies that the kernels satisfy the following averaging condition:

$$\int_0^t \sigma_t(s_1,..,s_{i-1},s_i,s_{i+1},..,s_n)ds_i = 0 \tag{1.3}$$

for all i. We consider the kernels $\sigma_t(s_1,..,s_n;t_1,..,t_m)$ of the covariant derivatives $\nabla^m\sigma_t(s_1,..,s_n)$ (See [L5], [L6] for the case when the based loop space depends on a parameter).

Let us denote two family of Sobolev norms as follows: for all $s_1,..,s_n;t_1,..,t_m$ belonging to a connected component of the complementary of the diagonals in $[0,t]^n \times [0,t]^m$, we have the following Nualart–Pardoux inequality:

$$\begin{aligned} &\|\sigma_t(s_1,..,s_n;t_1,..,t_m) - \sigma_t(s'_1,..,s'_n;t'_1,..,t'_m)\|_{L^p} \\ &\leq C(t,p;n,m)\left(\sum\sqrt{|t'_i - t_i|} + \sum\sqrt{|s'_i - s_i|}\right) \end{aligned} \tag{1.4}$$

We suppose in such case that $|s_i - s'_i|$ is bounded by 1 as well as $|t_i - t'_i|$. We say that σ_t satisfy the Nualart–Pardoux conditions.

The second family of Sobolev norms is given by

$$Sup\,\|\sigma_t(s_1,..,s_n;t_1,..,t_m)\|_{L^p} = C'(t,p;n,m) \tag{1.5}$$

The $W(p,m)$ Sobolev norm in the sense of Nualart–Pardoux of an n-form over the Moore Brownian bridge is given by:

$$\|\sigma_n\|_{W(p,m)} = \frac{2^{np}}{(n-p)!n!}\left\{\int_0^\infty (C(t,p;n,m)^p + C'(t,p;n,m)^p)g_p(t)dt\right\}^{1/p} \tag{1.6}$$

The $W(p,m)$ norm of a serie $\sigma = \sigma_n$ of n-forms is given by

$$\|\sigma\|_{W(p,m)} = \sum \|\sigma_n\|_{W(p,m)} \tag{1.7}$$

$W(p,m)$ is a Banach space. The space of smooth forms in the Nualart–Pardoux sense is $W_\infty = \cap\, W(p,m)$ endowed with the family of norms (1.7).

Let us introduce the partial exterior derivative in the fiber. It is given by

$$d_p\sigma_t(X_{1,t},..,X_{n+1,t} = \sum(-1)^{i-1} < d\sigma_t(X_{1,t},..,\hat{X}_{i,t},..,X_{n+1,t}), X_{i,t} >$$
$$+\sum_{i<j}(-1)^{i+j}\sigma_t([X_{i,t},X_{j,t}],X_{1,t},..,\hat{X}_{i,t},..,\hat{X}_{j,t},..,X_{n+1,t}) \tag{1.8}$$

where $\hat{}$ denotes the omission operator.

THEOREM I.1.: The partial exterior derivative d_p is continuous from W_∞ into W_∞ if $(t+1)^r g_p \leq C g_{p'}$ for all $p' > p$, $r > 0$.

PROOF:Let us choose two deterministic paths H and $\tilde{H}$. Let us denote by X and $\tilde{X}$ the corresponding vector fields over γ. Let us recall that:

$$[X,\tilde{X}]_{s,t} = \tau_{s,t}\int_0^s \tau_{u,t}^{-1}R(d\gamma_{u,t},\tilde{X}_{u,t})\tau_{u,t}H_{s,t} + antisymmetry \tag{1.9}$$

We find two parts in the partial exterior derivative.

The first one arises from the derivatives $\sigma_t(s_1,..,s_n;t)$ of the kernels and is continuous from $W(p';m+1)$ into $W(p,;m+1)$ for $p' > p$. We used to show this property that there are at most $n+1$ contributions which arise in the derivatives of σ and the fact that $\frac{2^{np}}{(n-p)!}n^k < C\frac{2^{np'}}{(n-p')!}$ for all k if $p' > p$ and the fact that $g_p < Cg_{p'}$.

The second part arises from the derivative of the parallel transport and leads as in [L5], [L6] to anticipatives Stratonovitch integrals. But we work over the based loop space instead of working over the path space as in [L5], lemma A.2. We have to improve this lemma.

Let us introduce a piecewise constant process $u_t(s)$, which is smooth for traditional Sobolev spaces over the Brownian bridge in time t (No regularity assumption is needed.) We take the classical Sobolev norm $\tilde{W}(p,m)$, this means the L^p norm of the Hilbert-Schmidt norm of its derivatives.

We put $U_t(s) = \int_0^s u_t(u)du$ for $s \leq t$. In order to recognize in $\int_0^t <\tau_{s,t}u_t(s), d\gamma_{s,t}>$ a curved Skorohod integral, we have to consider two counterterms:

•) The first one is $\int_0^t u_t(s;s)ds$.

•) The second one is $1/2\int_0^t < \tau_{s,t}S_{\tau_{s,t}U_t(s)}, \delta\gamma_{s,t} >$ where S denotes the Ricci tensor and δ the Ito differential. We integrate by parts in $U_t(s)$ and in $\int_0^{\cdot} \tau_{s,t}S_{\tau_{s,t}}.\delta\gamma_{s,t}$ by using the Ito formula; this counterterm is only apparently an anticipative Ito integral.

This allows to get the following lemma which extends the lemma A.2 of [L5]:

LEMMA I.2.: Let $u_t(s,s';u,\tilde{s})$ a random variable with values in T_x, which satisfies the regularity assumption (1.4), (1.5), s, s' and $\tilde{s}$ included. Let $I(s,s',\tilde{s})$ the following anticipative Stratonovitch integral:

$$I(s,s',\tilde{s}) = \int_s^{s'} < \tau_{u,t}u_t(s,s';u,\tilde{s}), d\gamma_{u,t} > \tag{1.10}$$

$I(s,s',\tilde{s})$ satisfies the regularity assumption (1.4), (1.5), s, s' and $\tilde{s}$ included. The two types of Nualart–Pardoux Sobolev norms of this anticipative Stratonovitch integral can be estimated in terms of the original Nualart–Pardoux Sobolev norms.

PROOF: We integrate by parts in order to compute explicitly $E[(I(s,s',\tilde{s}))^p]$ for some integer p. If $\int_s^{s'}$ satisfies the averaging condition $\int_s^{s'} u_t(u)du = 0$, its derivatives two. It remains to take the derivative of $\tau_{s,t}$ which is given by the Bismut formula:

$$\nabla_X \tau_{s,t} = \tau_{s,t} \int_0^s \tau_{u,t}^{-1} R(d\gamma_{u,t}, X_{u,t}) \tau_{u,t} \tag{1.11}$$

By integrating by parts, it is only apparently an anticipative Stratonovitch integral.

When we get by derivation of the parallel transport an integral of the type $\int_s^{s'} < \tau_{u,t} H(u), \delta\gamma_u >$, where $\int_s^{s'} H(u)du \neq 0$, we substract the quantity $\int_s^{s'} \frac{H(u)}{s'-s} du$ over the time interval $[s,s']$, in order to get a tangent vector over the based loop space. We begin by studying $E[(I(s,s',\tilde{s})^p]$ for a discrete approximation of the anticipative Stratonovitch integral by Riemann sum of $1/(t_{i+1}-t_i) \int_{t_i}^{t_{i+1}} u_t(s)ds$ for a suitable sudivision t_i of $[0,t]$. We find a finite sum of integrals over $[s,s']^k$ of polynomial expressions in the derivatives of u_t with possible contraction over the diagonals, and some universal expressions in the derivatives of the parallel transport and in the Ricci tensor, which checks the Nualart–Pardoux conditions. It is possible to divide this integral by a power k' of (s'-s), but we have always $k' + p/2 \leq k$. We deduce that the discrete approximation of the integral converges in all the L^p to $I(s,s',\tilde{s})$, this from the regularity assumption over the kernels due to (1.4) and (1.5), and that $\| I(s,s';\tilde{s}) \|_{L^p}$ is bounded by the Nualart- Pardoux Sobolev norms on u, c and C' multiplied by a certain power of t, because we integrate over an unbounded interval.

In order to check the regularity assumption in s, s', $\tilde{s}$, we suppose in order to simplify that $s < \tilde{s}_1 < .... < \tilde{s}_n < s'$ and that $v < \tilde{v}_1 < .... < \tilde{v}_n < v'$. We split the integral between s and s' into smaller integrals over the intervals defined by the contiguous times of the subdivision, as it was already done in the proof of the lemma A.2 of [L5]. We consider then the integral of the difference of $u_t(s,s,;\tilde{s})$ and $u_t(v,v';\tilde{v})$ over $[\tilde{s}_i,\tilde{s}_{i+1}] \cap [\tilde{v}_i,\tilde{v}_{i+1}]$, we substract the necessary counterterms in order to get an anticipative integral over the based loop space, and we get the expectation of an integer power of it as before. The main remark is that we remains in the same connected components of the parameter set outside the diagonals in the integrals which are got. If all the $|\tilde{s}_i - \tilde{v}_i|$ are smaller than 1, and if this property is still true for the time s, v, s' and v', we have the Nualart–Pardoux property (1.4), and the Nualart–Pardoux constants are estimated by the Nualart–Pardoux constant C of the first type of u, but we have to take care that the intervals of integration are not bounded and we have to multiply by a power of t the constant which are got, in comparison of the lemma A.2 of [L5] when the time was bounded . If we integrate outside the intersection, the distance between the extremities of the considered interval of integration is smaller than the sum of the distance between $\tilde{s}_i$ and $\tilde{v}_i$ and the distance between $\tilde{s}_{i+1}$ and $\tilde{u}_{i+1}$. The Nualart–Pardoux condition (1.4) is still checked; the Nualart–Pardoux constants are estimated by the a mixture between the Nualart–Pardoux constants of the two types of u, without to multiply by a power of t, because we integrate over bounded intervals.

The Nualart–Pardoux constants of the second type of I are estimated by a mixture of the Nualart–Pardoux constants of the two types of I, but we have in the estimates to introduce a suitable power of t because we can integrate over some big intervals.

For the derivatives of I, we deduce the result of the previous one by taking derivative under the sign $\int$, and by using the fact:

$$\nabla_X d\gamma_{s,t} = \tau_{s,t} H'_{s,t} ds \tag{1.12}$$

and by using the Bismut formula (1.11).

In order to conclude, let us precise the effect of the operation $H'(u) \to H'(u) - 1/(s'-s)\int_s^{s'} H'(v)dv$ over I in order to get a tangent vector over the based loop space. It has only the consequence to substract to the initial anticipative Stratonovitch integral the non anticipative Stratonovitch integral $\int_s^{s'} < \tau_{s,t}C, d\gamma_{s,t} >$.

◇

This lemma allows to deduce the theorem: namely the kernel of the second term appear as at most n^r anticipative integrals of the kernels of $\sigma_{n,t}$ which are involved with $n + C$ times. Since

$$\frac{n^r 2^{np}}{(n-p)!} \leq C \frac{2^{np'}}{(n-p')!} \tag{1.13}$$

for $r > 0$ and for $p' > p$, and since by hypothesis $(t+1)^r g_p(t) < C g_{p'}(t)$, we deduce the result.

◇

DEFINITION I.3.: $\mathrm{Ker} d_p / \mathrm{Im} d_p$ is called the partial cohomology group associated to the partial exterior derivative.

REMARK: Let us remark that the Moore Brownian bridge is constituted by a family of Brownian bridges indexed by R^+; the full exterior derivative is involved with a derivation in time t, and should give the same cohomology group than the based loop space as it is the case for the case of smooth loop.

2 Algebra of Forms of the Moore Brownian Bridge

We suppose in this section that for $p < p'$, $g_p(t) < C g_{p'}(t)$.

THEOREM II.1 : let σ belonging to W_∞, σ' belonging to W_∞. Then $\sigma \wedge \sigma'$ belongs to W_∞.

PROOF: Let $\sigma = \sum \sigma_n$ and $\tilde{\sigma} = \sum \tilde{\sigma}_n$. Then

$$\sigma \wedge \tilde{\sigma} = \sum_m \sum_{l+l'=m} \sigma_l \wedge \tilde{\sigma}_{l'} \tag{2.1}$$

$\sigma_{l,t} \wedge \tilde{\sigma}_{l',t}$ has a kernel which is modulo signs due to the antisymmetry the sum of $\frac{(l+l')!}{l!l'!}$ terms of the type $\sigma_t(s_1,..,s_l) \otimes \tilde{\sigma}_t(s'_1,..,s'_l)$. Moreover

$$C(t,p;l+l',m)(\sigma_l \wedge \tilde{\sigma}_{l'}) \leq$$

$$\leq C\frac{(l+l')!}{l!l'!}\left\{\sum_{j\leq m} C(t,2p;l,j)(\sigma_l) + \sum_{j\leq m} C'(t,2p;l,j)(\sigma_l)\right\} \tag{2.2}$$

$$\left\{\sum_{j\leq m} C(t,2p;l',j)(\tilde{\sigma}_{l'}) + \sum_{j\leq m} C'(t,2p;l',j)(\tilde{\sigma}_{l'})\right\}$$

and

$$C'(t,p;l+l',m)(\sigma_l \wedge \tilde{\sigma}_{l'}) \leq$$

$$\leq C\frac{(l+l')!}{l!l'!}\left(\sum_{j\leq m} C'(t,2p;l,j)(\sigma_l)\right)\left(\sum_{j\leq m} C'(t,2p;l,j)(\tilde{\sigma}_{l'})\right) \tag{2.3}$$

We use the fact if $l+l'=m$

$$\frac{2^{mp}}{(m-p)!m!} \leq \frac{2^{l2p}}{(l-2p)!}\frac{2^{l'2p}}{(l'-2p)!}\frac{1}{l!l'!} \tag{2.4}$$

Moreover

$$(A+B)^p \leq C(A^p + B^p) \tag{2.5}$$

We use the Cauchy-Schwartz inequality and the hypothesis inequality $g_p \leq Cg_{2p}$. We get:

$$\|\sigma_l \wedge \tilde{\sigma}_{l'}\|_{W(p,m)} \leq C\left(\sum_{j\leq m} \|\sigma_l\|_{W(2p,m)}\right)\left(\sum_{j\leq m} \|\tilde{\sigma}_{l'}\|_{W(2p,m)}\right) \tag{2.6}$$

Therefore

$$\|\sigma \wedge \tilde{\sigma}\|_{W(p,m)} \leq \sum_{l,l'} \|\sigma_l \wedge \tilde{\sigma}_{l'}\|_{W(p,m)} \leq$$

$$\leq C\left(\sum_{l,j\leq m} \|\sigma_l\|_{W(2p,j)}\right)\left(\sum_{l,j\leq m} \|\tilde{\sigma}_l\|_{W(2p,j)}\right) \tag{2.7}$$

This shows the theorem.
◊ In order to get a nicer structure, we put:

$$\|\sigma\|_{W'(p,r)} = \sum_{j\leq r} \|\sigma\|_{W(p,j)} \tag{2.8}$$

We have then the Cauchy-Schwartz inequality:

$$\|\sigma \wedge \tilde{\sigma}\|_{W'(p,r)} \leq C\|\sigma\|_{W'(2p,r)}\|\tilde{\sigma}\|_{W'(2p,r)} \tag{2.9}$$

Let B and B' be two Banach spaces. Let us recall how the smallest tensor product $B\hat{\otimes}B'$ is defined ([Gro], [Ru], [Sch]):

$$\|\sum b_i \otimes b'_i\|_{\hat{\otimes}} = Inf \sum \|c_i\|_B \|c'_i\|_{B'} \tag{2.10}$$

The infimum is taken over the sum of finite sum of elementary products equal to the original finite sum. This defines a norm for finite sum of products in $B\hat{\otimes}B'$. We complete this Hilbert space. There is a characterization of $B\hat{\otimes}B'$ ([Gro]): the elements of $B\hat{\otimes}B'$ are series $\sum \lambda_i b_i \otimes b'_i$ where $\|b_i\|_B = \|b'_i\|_{B'} = 1$ and $\sum |\lambda_i| < \infty$. Moreover, ([Gro], p342), there is a bicontinuous isomorphism between the space of applications which are n multilinear and continuous from ΠB_i into B and the space of the continuous linear applications from $\hat{\otimes} B_i$ into B.

Let us call $\cap W'(r,m)$ A_∞. $A_\infty\hat{\otimes}A_\infty$ means we take $\cap W'(r,m)\hat{\otimes}W'(r,m)$. We have an analogous definition for $A_\infty\hat{\otimes}A_\infty\hat{\otimes}A_\infty$. The wedge product is a continuous linear application from $A_\infty\hat{\otimes}A_\infty$ into A_∞, which is continuous. Moreover, A_∞ is a graded algebra (See [M.M]). We have the following commutative diagram:

$$\begin{array}{ccc} A_\infty\hat{\otimes}A_\infty\hat{\otimes}A_\infty & \rightarrow & A_\infty\hat{\otimes}A_\infty \\ \downarrow & & \downarrow \\ A_\infty\hat{\otimes}A_\infty & \rightarrow & A_\infty \end{array} \tag{2.11}$$

The first vertical map is $\wedge \otimes I$ and the last vertical map is $\wedge$. The first horizontal map is $I \otimes \wedge$ and the second horizontal map is $\wedge$.

This diagram reflects the associativity of the wedge product. The multiplication by 1 is the unity η and satisfies the following commutative diagram (see [M.M], p213):

$$\begin{array}{ccc} C \otimes A_\infty & \rightarrow & A_\infty\hat{\otimes}A_\infty \\ \downarrow & & \downarrow \\ A_\infty & \rightarrow & A_\infty \\ \uparrow & & \uparrow \\ A_\infty \otimes C & \rightarrow & A_\infty\hat{\otimes}A_\infty \end{array} \tag{2.12}$$

The vertical maps of the left are isomorphism; the vertical maps of the right are the wedge product. The first horizontal map is $\eta \otimes I$, the second one I, and the third one $I \otimes \eta$.

Moreover the twisting morphism T: $A_\infty\hat{\otimes}A_\infty \rightarrow A_\infty\hat{\otimes}A_\infty$ is continuous:

$$\sigma_l \otimes \tilde{\sigma}_{l'} \xrightarrow{T} (-1)^{ll'} \tilde{\sigma}_{l'} \otimes \sigma_l \tag{2.13}$$

and the graded algebra is commutative in the sense (See [M.M]) that:

$$\begin{array}{ccc} A_\infty\hat{\otimes}A_\infty & \rightarrow & A_\infty \\ \downarrow & & \downarrow \\ A_\infty\hat{\otimes}A_\infty & \rightarrow & A_\infty \end{array} \tag{2.14}$$

The first vertical map is T, the second one is I, and the horizontal maps are the wedge product.

3 Coproduct of Forms over the Moore Brownian Bridge

We will define another tensor product $A_\infty \otimes A_\infty$.

Let us consider a $n \times \tilde{n}$ form over $\Omega_t \times \tilde{\Omega}_{\tilde{t}}$ $\sigma_{t,\tilde{t}}(s_1, .., s_n; \tilde{s}_1, .., \tilde{s}_{\tilde{n}})$. It is a partial form over each element of the product and not globally a form in the both components. We can antisymmetrize in order to get globally a form over the product. We get by this way a sum of at most $2^{n+\tilde{n}}$ tensor products. We assimilate this element to this antisymmetrized counterpart. Let us consider the kernels of its covariant derivatives $\sigma_{t,\tilde{t}}(s_1, .., s_n; \tilde{s}_1, .., \tilde{s}_{\tilde{n}}; t_1, .., t_j; \tilde{t}_1, .., \tilde{t}_{\tilde{j}})$. We choose the Nualart–Pardoux Sobolev norm $\sum_{j \le m}(C_n(t;p,j) + C'_n(t;p,j))$, $\tilde{s}_1, .., \tilde{s}_{\tilde{n}}, ..\tilde{t}_1, .., \tilde{t}_{\tilde{j}}$, $\tilde{j} \le m$ fixed. After we take the supremum over these times of these Nualart–Pardoux Sobolev norms. We get an expression called $\sum_{j \le m} D_n(t;p,j) + D'_n(t;p,j)$, $\tilde{t}$ being fixed. We perform the same manipulations for $\tilde{\Omega}_{\tilde{t}}$, $\tilde{s}_i$, and $\tilde{t}_i$ and we get an expression called $\sum_{j \le m}(\tilde{D}_{\tilde{n}}(\tilde{t};p,j) + \tilde{D}'_{\tilde{n}}(\tilde{t};p,j))$. This depends on t and $\tilde{t}$.

Let us introduce the function $K_{n,\tilde{n}}(t,\tilde{t};p,m)$ defined as follows:

$$K_{n,\tilde{n}}(t,\tilde{n};p,m) = \sum_{j \le m}(D_n(t;p,j) + D'_n(t;p,j)) + \sum_{j \le m}(\tilde{D}'_{\tilde{n}}(\tilde{t};p,j) + \tilde{D}_{\tilde{n}}(\tilde{t};p,j)) \tag{3.1}$$

The Nualart–Pardoux Sobolev norm of $\sigma_{n\tilde{n}}$ is

$$\|\sigma_{n,\tilde{n}}\|_{\otimes,p,m} =$$

$$= \frac{2^{np}}{(n-p)!}\frac{2^{\tilde{n}p}}{(\tilde{n}-p)!}\frac{1}{n!\tilde{n}!}\left\{\int_0^\infty\int_0^\infty K_{n,\tilde{n}}(t,\tilde{t};p,m)^p g_p(t) g_p(\tilde{t}) dt d\tilde{t}\right\}^{1/p} \tag{3.2}$$

If $\sigma = \sum_{n,\tilde{n}} \sigma_{n,\tilde{n}}$, we take:

$$\|\sigma\|_{\otimes,p,m} = \sum_{n,\tilde{n}} \|\sigma_{n,\tilde{n}}\|_{\otimes,p,m} \tag{3.3}$$

If we take a product $\sigma_{n,\tilde{n},t,\tilde{t}} = \sigma_{n,t} \otimes \sigma_{\tilde{n},\tilde{t}}$, we have clearly:

$$\begin{aligned} K_{n,\tilde{n}}(t,\tilde{t},p,m)^p \le\ & C(p)\left(\sum_{j\le m} C(t,p;n,j)^p + \sum_{j \le m} C'(t,p;n,j)^p\right) \\ \le\ & \left(\sum_{j \le m} \tilde{C}(\tilde{t},p;\tilde{n},j)^p + \sum_{j\le m} \tilde{C}'(\tilde{t},p;\tilde{n},j)^p\right) \end{aligned} \tag{3.4}$$

since $(A+B)^p \le C(A^p + B^p)$.

This implies if σ belongs to A_∞, $\tilde{\sigma}$ belongs to A_∞

$$\|\sigma \otimes \tilde{\sigma}\|_{\otimes,m,p} \le C\|\sigma\|_{W'(m,p)}\|\tilde{\sigma}\|_{W'(m,p)} \tag{3.5}$$

We deduce that $A_\infty \otimes A_\infty$ contains $A_\infty \hat{\otimes} A_\infty$ and that the injection from $A_\infty \hat{\otimes} A_\infty$ into $A_\infty \otimes A_\infty$ is continuous. We have the same property for tensor products of a bigger length.

Let $\sigma_{n,t}$ be a form over Ω_t. It is given by kernels $\sigma_{n,t}(s_1, .., s_n)$. Let us choose a times s smaller than t, We stick together a loop in time s and a loop between s and t. We get a finite codimensional manifold of Ω_t and by pullback a form over the product of Ω_s and of Ω_{t-s} called $\{\sigma_{n,t} \,|\, \gamma_{s,t} = x\}$. Later, we will write γ_u for $\gamma_{u,t}$. A form after the time s is endowed with the parallel transport $\tau_{u,s}$ between γ_s and γ_u. On the other hand $\tau_{u,s} = \tau_u \tau_s^{-1}$. Therefore the kernel of $\{\sigma_{n,t} \,|\, \gamma_s = x\}$ is given as follows: We choose a subset P of $(1, .., n)$. They are 2^n. Over this subset, we keep the form in $u \leq s$. Outside this subset, we choose a form in $u > s$. Therefore

$$\{\sigma_{n,t} \,|\, \gamma_s = x\}(s_1, .., s_n) = \sum \tilde{\sigma}_{n,t,P}(s_1, .., s_n) =$$

$$= \sum_{B \subseteq (1,..,n)} \prod_{j \in P} 1_{s_j \leq s} \prod_{j' \in P^c} 1_{s_{j'} \geq s} \left\{ \sigma_{n,t}(s_1, .., s_n) \prod_{j' \in P^c} \tau_s^{-1} \,|\, \gamma_s = x \right\} \tag{3.6}$$

This last product means that we multiply each term of the cotensor which correspond to a time $s_{j'}$ belonging to the complementary of P by τ_s^{-1}, and that we restrict this functional over the finite codimensional manifold given by $\gamma_s = x$.

Moreover in (3.6), the kernels of $\{\sigma_{n,t} \,|\, \gamma_s = x\}$ does not check the condition of average 0. If s_j is a time before s, we should have

$$\int_0^s \sigma_{n,t,P}(s_1, .., s_n) ds_j = 0 \tag{3.7}$$

If s'_j is a time after s,

$$\int_s^t \sigma_{n,t,P}(s_1, .., s_n) ds'_j = 0 \tag{3.8}$$

We have to remove suitable orthogonal projections in order to deduce $\sigma_{n,t,P}$ from $\tilde{\sigma}_{n,t,P}$. We put:

$$\sigma_{n,t,P}(s_1, .., s_n) = \sum_{J \subseteq P; J' \subseteq P^c} (-1)^{|J|}(-1)^{|J'|} \int_{j \in J; s_j \leq s} \int_{j' \in J'; s_{j'} \geq s} \tilde{\sigma}_{n,t,P}(s_1, .., s_n) \prod_{j \in J} ds_j \prod_{j' \in J'} ds_{j'} \tag{3.9}$$

Let us consider the application

$$\sigma_{n,t} \to \sum_P \sigma_{n,t,P} \tag{3.10}$$

We choose s with the law $g_p(s)ds$ and $t-s=v$ with the law $g_p(v)dv$. We get an application Δ. We have to take care that we antisymmetrize the kernel at the beginning. So we have to consider $P=\{1,2,..,i\}$, $P^c=\{i+1,..,n\}$, because we have $\sigma_{n,t}(s_{\sigma(1)},..,s_{\sigma(n)})=(-1)^{sign(\sigma)}\sigma_{n,t}(s_1,..,s_n)$ for a permutation σ, this due to the antisymmetry of the form.

THEOREM III.1.: Let us suppose that for all r, for $p'\geq p$, we have $(t+1)^r g_p(t)\leq Cg_{p'}(t)$ and that $g_{p'}(t+t')\leq Cg_p(t)g_p(t')$ for $p'\gg p$. Then Δ is a continuous application from A_∞ into $A_\infty\otimes A_\infty$.

PROOF: Let F be a functional over Ω_t. We would like in a first step to study $\{F\mid\gamma_s=x\}$ its restriction to the loop which are in x in time s. For that, we introduce the measure μ_s:

$$f\to E_t[Ff(\gamma_s)] \tag{3.11}$$

for $s<t$.

We have four difficult cases:

a) $t\to 0$ and $s\to 0$.

b) t bounded. $s\to 0$ or $s\to t$.

c)$t\to\infty$. $s\to 0$ or $s\to t$.

d)$t\to\infty$. s bounded by lower value and $|s-t|>C>0$.

We consider the classical Sobolev norms of F:

$$||F||_{m,p}=\sum_{r\leq m}E\left[\left(\int|F(s_1,..,s_r)|^2ds_1..ds_r\right)^{p/2}\right]^{1/p} \tag{3.12}$$

where $F(s_1,..,s_r)$ denote the derivatives of F.

a) Let us study the first case. We suppose that $s<t/2$, the other case being symmetric. We perform the estimate in small time of the density in x of the measure μ_s. We embed the manifold into a linear manifold, we use near x the system of exponential coordinates, and modulo vanishing terms which arise from the exponential inequality, we have only to estimate the density of the measure ν_s:

$$f\to E_t\left[Ff\left(\frac{\gamma_s-s}{\sqrt{s}}\right)\right] \tag{3.13}$$

(see [L1], [K2], [Wa]). Namely the density in 0 of the second measure is equal to $s^{-d/2}$the density of μ_s in x.

Let us introduce the vector $X_0(big)$ equal to $s^{-1/2}\tau_u u\tau_s^{-1}X_0(\gamma_s)$ if $u\leq s$ and equal to $\tau_u(2s-u)s^{-1/2}\tau_s^{-1}X_0(\gamma_s)$ if u belongs to $[s,2s]$ and 0 outside. We can integrate by part in

$$(d^r f)(\frac{\gamma_s-x}{\sqrt{s}})X_1(\frac{\gamma_s-x}{\sqrt{s}})..X_r(\frac{\gamma_s-x}{\sqrt{s}})=d^r(f(\frac{\gamma_s-x}{\sqrt{s}}))X_1(big)..X_r(big) \tag{3.14}$$

if the X_i are constant vector fields in the linear space. In the first part, we consider the derivative of the function f which depends only of $(\frac{\gamma_s-x}{\sqrt{s}})$, and in the second part of the previous formula, we consider the derivative of the Brownian bridge functional $f(\frac{\gamma_s-s}{\sqrt{s}})$. d^rF means that we consider $d(d^{r-1}FX_1..X_{r-1})X_r$.

It is not a tensorial operations. The vector fields $X_i(big)$ are bounded in L^p as well as for their derivatives for the classical Sobolev norms, essentially because:

$$\int_0^s \left(\frac{1}{\sqrt{s}}\right)^2 du < C \tag{3.15}$$

We deduce from this that

$$\nu_s[d^r f X_1..X_r] \leq C\|f\|_\infty \tag{3.16}$$

and that C is bounded by the classical Sobolev norms of F. Therefore the density of μ_s in x is bounded by $Cs^{-d/2}$, C being estimated by the classical Sobolev norms of F. But the density of the law of γ_s is $\frac{p_s(x,y)p_{t-s}(y,x)}{p_t(x,x)}$, and in $\gamma_s = x$, it has a behaviour in $Cs^{-d/2}$ because $s < t/2$. We deduce that

$$|E_t[F \mid \gamma_s = x]| \leq \|F\|_{r,p} \tag{3.17}$$

for some r and some p independent from s and t small.

b) Let us study the second case. If $s \to 0$, we do as before, because $\frac{p_s(x,x)p_{t-s}(x,x)}{p_t(x,x)}$ behaves as $Cs^{-d/2}$, and because we consider the derivative of the parallel transport over a bounded interval. If $s \to t$, we proceed as before, but by starting from s. $X_0(big) = \frac{1}{\sqrt{t-s}}\tau_u(u+t-2s)\tau_s^{-1}X_0(\gamma_s)$ if $u \in [2s-t, s]$ and $X_0(big) = \frac{1}{\sqrt{t-s}}\tau_u(-u+t)\tau_s^{-1}X_0(\gamma_s)$ if $u \in [s,t]$ for a given vector field over M X_0 constant over the linear manifold after using a system of exponential chart near γ_t. $X_0(big)$ is equal to 0 outside these time intervals. We consider this time the auxiliary measure

$$f \to E_t\left[Ff\left(\frac{\gamma_s - x}{\sqrt{t-s}}\right)\right] \tag{3.18}$$

in order to conclude as in the case a).

c)When $t \to \infty$ and $s \to 0$, the computations are as in the first case because we consider vector fields which are not equals to 0 only on the bounded interval $[0, 2s]$; but in the divergence of these vector fields, we have the Ricci term contribution which appears, and therefore no bounded stochastic integrals appear. We have to multiply the control constant by a suitable power of t. If $s \to t$, we do the same transformation than in (3.18). The major difficulty is that we consider the parallel transport for big t, as well as non bounded stochastic integrals, which have derivatives bounded in L^p bounded by a power of t.

Moreover $p_t(x,x) \to 1$ when $t \to \infty$. We deduce the following bound:

$$|E_t[F \mid \gamma_s = x]| \leq Ct^\alpha \|F\|_{r,p} \tag{3.19}$$

for some power α.

d)When $t \to \infty$, s bounded by lower value and such that the distance between s and t is bounded by lower value, we don't have to do the rescaling (3.13) or (3.18) in order to conclude that

$$|E_t[F \mid \gamma_s = x]| \leq Ct^\alpha \|F\|_{r,p} \tag{3.20}$$

for some power α.

In order to write that we consider classical Sobolev norms in time t, we denote by $\|.\|_{t,r,p}$ these Sobolev norms. In conclusion, in all the cases

$$|E_t[F \mid \gamma_s = x]| \leq C(t+1)^{\alpha}\|F\|_{t,r,p} \tag{3.21}$$

for some power α. (3.21) is due to a mixture between the technics of the quasi-sure analysis and short time and rough long time asymptotics of densities of diffusion.

Let F be a smooth functional in the classical sense over Ω_t. Let us restrict it over the finite codimensional manifold $\gamma_s = x$. Let $F(s_1, .., s_n)$ its kernels. The kernels of its restriction is given by similar considerations than (3.6) and (3.9), but we have to take care that over $[s,t]$, we consider the vector $\tau_{u,s}H_u = \tau_u\tau_s^{-1}H_u$ with $H_0 = H_s = H_t = 0$. Therefore, we have to take the derivatives of τ_s which satisfy the Nualart–Pardoux conditions over $[0,s]$. In order to be precise, let us suppose that $d^r\{F \mid \gamma_s = x\}$ has a kernel $k(s_1, .., s_r)$, where we take the vector $\tau_{u,s}H_u$ over $[s,t]$ and that $k(s_1, .., s_r)$ is the restriction over $\gamma_s = x$ of a functional over Ω_t whose the traditional derivatives over Ω_t satisfy the Nualart–Pardoux conditions for $s_i \leq s$ and for $s_i' \in [s,t]$: we suppose that these Nualart–Pardoux constants can be estimated in terms of the Nualart–Pardoux norms of F multiplied by a certain power of t. Then

$$\begin{aligned} d^{r+1}\{F \mid \gamma_s x\} &= \\ &= \left\{k(s_1, .., s_r; u)1_{u\leq s} + k(s_1, .., s_r; u)1_{u>s}\tau_s^{-1}\tau(\gamma_s, x) \mid \gamma_s = x\right\} \end{aligned} \tag{3.22}$$

$\tau(\gamma_s, x)$ is the parallel transport along the unique geodesic joining x to γ_s which can be supposed close modulo a smooth cutoff, which depends only from γ_s; it satisfies therefore the Nualart–Pardoux conditions for $v \leq s$. We deduce by induction that $d^{r+1}\{F \mid \gamma_s = x\}$ satisfies the Nualart–Pardoux conditions over $[0,s]$ and over $[s,t]$ and that their Nualart–Pardoux norms can be estimated by a power of t and the Nualart–Pardoux Sobolev norms of F in (1.4) and in (1.5). Moreover the kernels in this way of $d^{r+1}\{F \mid \gamma_s = x\}$ do not satisfy the averaging conditions (3.7) and (3.8). We conclude by applying the procedure (3.9) in order to project the kernels over the kernels of the based loop space and by applying the lemma A.2 of [L5].

Therefore the kernels of $\{F \mid \gamma_s = x\}$ are satisfying the Nualart–Pardoux conditions over $\Omega_s \times \Omega_{t-s}$ and their Nualart–Pardoux conditions can be estimated by a power of t and the Nualart–Pardoux norms of F.

This works too for $\sigma_{n,t,P}$, where we include the tensor times of the kernels of $\sigma_{n,t,P}$ in the Nualart–Pardoux constants of the derivatives; let us recall that we have two types of tensor times, if this time belongs to $[0,s]$ or to $[s,t]$. In the expression of $\sigma_n \to \sum_P \sigma_{n,t,P}$ there are at most n terms, after using the identification given in the beginning of this chapter; it remains only the term in the sum in $P = \{1, .., i\}$. Therefore for given p and m, there exists $m' \geq m$ and

$p' \geq p$ such that:

$$\|\sum_P \sigma_{n,t,P}\|_{\otimes,m,p} \leq$$
$$\leq \; C^n \|\sigma_t\|_{m,p'}(t+1)^\alpha \left(\sum_{l+l'=n} \frac{2^{lp}}{(l-p)!l!}\frac{2^{l'p}}{(l'-p)!l'!}\right)\frac{(n-p')!n!}{2^{np'}} \tag{3.23}$$

By the Stirling formula

$$n_1!n_2! \geq C^{n_1+n_2}(n_1+n_2)! \tag{3.24}$$

We deduce that if $p' \gg p$

$$C^n \left(\sum_{l+l'=n} \frac{2^{lp}}{(l-p)!l!}\frac{2^{l'p}}{(l'-p)!l'!}\right)\frac{(n-p')!n!}{2^{np'}} < C \tag{3.25}$$

Therefore for $p' \gg p$

$$\|\sum_P \sigma_{n,t,P}\|_{\otimes,m,p} \leq C\|\sigma_{n,t}\|_{m'.p'}(t+1)^\alpha \tag{3.26}$$

where in the left hand side of (3.26) we take the Nualart–Pardoux Sobolev norms in time t of $\sigma_{n,t}$ (We dont integrate in time t the constant C and C' of (1.4) and (1.5), but we consider the weight which depends on the degree of the form as in (1.6)).

Let $\sigma_{t,n,P}$ a part of the pullback form over $\Omega_t \times \Omega_{\tilde{t}}$. This part is a l form in Ω_t and a l' form in $\Omega_{\tilde{t}}$. Let us come back to the notations of (3.2). We get for some $q \gg p$:

$$\begin{aligned}
&\left\{\int_0^\infty\int_0^\infty K_{l,l'}(t,t';p,m)^p g_p(t)g_p(t')dtdt'\right\}^{1/p} \\
&\leq C\left\{\int_0^\infty\int_0^\infty (t+t'+1)^\alpha\left\{\sum_{j\leq m'} C(t+t',q;l+l',j)\right.\right. \\
&\left.\left.+\textstyle\sum_{j\leq m'} C'(t+t',q;l+l',j)\right\}^p g_p(t)g_p(t')dtdt'\right\}^{1/p} \\
&\leq C\left\{\int_0^\infty\int_0^\infty (t+t'+1)^\alpha\left\{\sum_{j\leq m'} C(t+t',q;l+l',j)\right.\right. \\
&\left.\left.+\textstyle\sum_{j\leq m'} C'(t+t',q;l+l',j)\right\}^p g_{p''}(t+t')dtdt'\right\}^{1/p} \\
&\leq C\left\{\int_0^\infty (\sum_{j\leq m'} C(t,q',l+l',j) + \sum_{j\leq m'} C'(t,q';l+l',j))g_{q'}(t)dt\right\}^{1/q'}
\end{aligned} \tag{3.27}$$

for $q' \gg p" \gg p$, this from the hypothesis of the theorem III.1. From (3.26), we deduce our theorem.
◊

REMARK: Δ is continuous from A_∞ into $A_\infty \otimes A_\infty$. But if we take classical Sobolev spaces, Δ is continuous from A'_∞ into $A'_\infty \otimes A'_\infty$, where we take the classical Sobolev norms.

THEOREM III.2: Under the hypothesis of the theorem III.1, the diagram

$$\begin{array}{ccc} A_\infty & \rightarrow & A_\infty \otimes A_\infty \\ \downarrow & & \downarrow \\ A_\infty \otimes A_\infty & \rightarrow & A_\infty \otimes A_\infty \otimes A_\infty \end{array} \qquad (3.28)$$

is commutative. The first vertical map is Δ, the second one is $\Delta \otimes I$. The first horizontal map is Δ and the second one is $I \otimes \Delta$. The maps are continuous.

PROOF: Let us define first what we mean by $A_\infty \otimes A_\infty \otimes A_\infty$. Instead of working over $\Omega_t \times \Omega_{t'}$, we work over $\Omega_t \times \Omega_{t'} \times \Omega_{t"}$. We have 3 partial Nualart–Pardoux Sobolev norms instead of 2. We integrate in the 3 times t, t' and $t"$ with the probability law $g_p(t)g_p(t')g_p(t")dtdt'dt"$. For the L^p norm of a form in each coordinates $\sigma_{n,n',n"}$, we introduce the weight

$$\frac{2^{(n+n'+n")p}}{(n-p)!(n'-p)!(n"-p)!n!n'!n"!} \qquad (3.29)$$

The considerations of the theorem III.1 show us that $I \otimes \Delta$ and $\Delta \otimes I$ are continuous from $A_\infty \otimes A_\infty$ into $A_\infty \otimes A_\infty \otimes A_\infty$. Moreover the diagram (3.28) is clearly commutative. By the vertical map Δ, we divide $[0,t]$ in $[0,s]$ and $[s,t]$ and by the horizontal map, we divide $[s,t]$ in $[s,u]$ and $[u,t]$. By the horizontal map Δ, we divide $[0,t]$ in $[0,u]$ and $[u,t]$ and by the vertical map, we divide $[0,u]$ in $[0,s]$ and $[s,u]$.
◊

4 Analytic Cobar Construction and Stochastic Chen Forms

Let $\Omega_.(M)$ the space of forms of degree large than 1. Let $C \otimes \Omega_.(M)^{\otimes n}$ endowed with the Hilbert norm of a product of Hilbert spaces. Over $\Omega_.(M)$, we consider namely the Hilbert structure:

$$\|\omega\|_{2,k} = \|(dd^* + d^*d + 2)^k \omega\|_{L^2} \qquad (4.1)$$

where d is the exterior derivative and d^* its adjoint. Let $\tilde{\omega} = \sum \tilde{\omega}_n$ a serie of element of $C \otimes \Omega_.(M)^{\otimes n}$. We consider the Hilbert structure

$$\phi_{k,z}(\tilde{\omega})^2 = \sum \frac{z^n}{n!}\|\tilde{\omega}\|^2_{2,k} \qquad (4.1)$$

Let $A'_{k,z}$ the corresponding Hilbert space.

We have two algebraic operations:

•) The ϕ shuffle product (See [G.J.P]):

$$(\omega_1 \otimes .. \otimes \omega_n).(\omega_{(n+1)} \otimes .. \otimes \omega_{(n+m)}) = \sum_s sign\, \omega_{s(1)}..\omega_{s(i)} \otimes .. \otimes \omega_{s(n+m)} \quad (4.2)$$

We take the sum over the bijection from $\{1, .., n+m\}$ into $\{1, .., n+m\}$ which are increasing over $\{1, .., n\}$ and over $\{n+1, .., n+m\}$. The sign arises from the anticommutation relation over forms; let us recall that the degree of ω in this formalism is substracted from one unit. There are $\frac{(n+m)!}{n!m!}$ such bijections.

•) The Δ coproduct:

$$\Delta(\omega_1 \otimes ... \otimes \omega_n) = \sum_i (\omega_1 \otimes .. \otimes \omega_i) \bigotimes (\omega_{i+1} \otimes .. \otimes \omega_n) \quad (4.3)$$

$A'_{k,z}$ is an Hilbert space. $A'_{k,z} \otimes A'_{k,z}$ is still an Hilbert space as tensor product of Hilbert spaces. $A'_\infty = \cap A'_{k,z}$ and by definition $A'_\infty \otimes A'_\infty = \cap A'_{k,z} \otimes A'_{k,z}$.

THEOREM IV.1.: The shuffle product ϕ is a continuous application from $A'_\infty \otimes A'_\infty$ into A'_∞. The Δ coproduct is a continuous application from A'_∞ into $A'_\infty \otimes A'_\infty$.

PROOF: Let ω_i an orthonormal basis of eigenvectors of $dd^* + d^*d + 2$ associated to the eigenvalue λ_i. If $J = (i_1, .., i_n)$, let us write $\omega_J = \omega_{i_1} \otimes .. \otimes \omega_{i_n}$. We have

$$\tilde{\omega} = \sum_J \tilde{\alpha}_J \omega_J \quad (4.4)$$

We have

$$||\tilde{\omega}||^2_{z,k} = \sum_j \frac{z^{|J|}}{|J|!} \lambda_J^{2k} |\tilde{\alpha}_J|^2 \quad (4.5)$$

where $\lambda_J^{2k} = (\lambda_{i_1}..\lambda_{i_n})^{2k}$.

We perform the same computation for $\tilde{\omega}'$. We have

$$\tilde{\omega}' = \sum \tilde{\alpha}'_J \omega_J \quad (4.6)$$

We deduce that

$$\tilde{\omega}.\tilde{\omega}' = \sum \beta_J \omega_J \quad (4.7)$$

where

$$\beta_J = \sum sign\, \tilde{\alpha}_{J'} \tilde{\alpha}'_{J"} \quad (4.8)$$

where we sum over all the way such the union of the multi indices J' and J' is equal to J modulo a shuffle of J and J'. If we fix $|J'| = l$ and $|J"| = n - l$, there are $\frac{n!}{l!(n-l)!}$ terms in this sum. Therefore since $\frac{n!}{l!(n-l)!} \le 2^n$, we have:

$$|\beta_J|^2 \le C^{|J|} \sum_{J', J"} |\tilde{\alpha}_{J'}|^2 |\tilde{\alpha}'_{J"}|^2 \quad (4.9)$$

where we take the same sum than in (4.8).

Moreover we have clearly,

$$\frac{1}{n!} \leq \frac{1}{l!}\frac{1}{(n-l)!}C^l C^{n-l} \tag{4.10}$$

Moreover if we get J by a shuffle of J' and $J"$, we have:

$$\lambda_J^{2k} = \lambda_{J'}^{2k}\lambda_{J"}^{2k} \tag{4.11}$$

We deduce the inequality for $z' \gg z$

$$||\tilde{\omega}.\tilde{\omega}'||^2_{z,k} \leq C||\tilde{\omega}||^2_{z',k}||\tilde{\omega}'||^2_{z',k} \tag{4.12}$$

Namely when we take the expression

$$\sum \frac{\lambda_J^{2k}}{n!}|\beta_J|^2 \tag{4.13}$$

the expression $|\tilde{\alpha}_{J'}|^2|\tilde{\alpha}_{J"}|^2$ appears at most C^n times where $n = |J'| + |J"|$. We increase z' in consequence in (4.12). This shows the continuity of the ϕ shuffle product.

Let us study the continuity of the Δ coproduct. Let us consider

$$\tilde{\omega} = \sum \alpha_J \omega_J \tag{4.14}$$

Then

$$\Delta\tilde{\omega} = \sum_J \alpha_J \sum_{J',J"} \omega_{J'} \bigotimes \omega_{J"} \tag{4.15}$$

The second sum is over the way of cutting the multiindex J in two segments J' and $J"$ (See (4.3)) . We deduce that

$$||\Delta\tilde{\omega}||^2_{z,k} \leq \sum \lambda_J^{2k}|\alpha_J|^2 \sum_{j\leq|J|} \frac{z^j}{j!}\frac{z^{|J|-j}}{(|J|-j)!} \tag{4.16}$$

By the Stirling formula

$$\frac{C^n}{j!(n-j)!} \leq \frac{C_1^n}{n!} \tag{4.17}$$

Therefore

$$||\Delta\tilde{\omega}||^2_{z,k} \leq ||\tilde{\omega}||^2_{z',k} \tag{4.18}$$

for $z' \gg z$.

This shows that the Δ coproduct is continuous.

◊ THEOREM IV.2: A'_∞ is an analytic Hopf algebra.

PROOF: The continuity of the operations is checked. Moreover the counit ϵ applies $1 \otimes \omega_1 \otimes .. \otimes \omega_n$ to 0 if n is not equal to 0 and to 1 in the other case, and is clearly continuous. The twisting operator T is clearly continuous (let us recall that the degree of $\omega_1 \otimes .. \otimes \omega_n$ is equal to $\sum(deg(\omega_i) - 1)$. We have all the commutative diagram which were got before. Moreover ϵ satisfies the following commutative diagram:

$$\begin{array}{ccc} A'_\infty \otimes A'_\infty & \rightarrow & C \otimes A'_\infty \\ \uparrow & & \downarrow \\ A'_\infty & \rightarrow & A'_\infty \\ \downarrow & & \uparrow \\ A'_\infty \otimes A'_\infty & \rightarrow & A'_\infty \otimes C \end{array} \tag{4.19}$$

The vertical map of the left are the coproducts. The vertical maps of the right are isomorphisms. The first horizontal map is $\epsilon \otimes I$. The second one is I and the third one is $I \otimes \epsilon$.

Moreover, $\Delta \circ \phi$ is equal to the iteration of the following maps:

-) $\Delta \otimes \Delta$ in a first step which applies $A'_\infty \otimes A'_\infty$ over $A'_\infty \otimes A'_\infty \otimes A'_\infty \otimes A'_\infty$.
-) Secondly $I \otimes T \otimes I$ which applies $A'_\infty \otimes A'_\infty \otimes A'_\infty \otimes A'_\infty$ over himself.
-) $\phi \otimes \phi$ which applies this last tensor product over $A'_\infty \otimes A'_\infty$.

This property works classically for finite sum and is extended by continuity. This shows that the product and the coproduct are compatible , therefore finishes the proof of the theorem.

◇

Let us define the Adams operations. Let A and B two continuous applications from A'_∞ into A'_∞. $A * B$ is defined from A'_∞ into A'_∞ by the iteration of the following maps:

$$A'_\infty \xrightarrow{\Delta} A'_\infty \xrightarrow{A \otimes B} A'_\infty \otimes A'_\infty \xrightarrow{\phi} A'_\infty \tag{4.20}$$

This means that $A * B = \phi \circ (A \otimes B) \circ \Delta$. If A and B are continuous,$A * B$ are continuous.

DEFINITION IV.3 (Adams operations): there are denoted by Ψ^k and defined by $\Psi^k = I * I * I .. * I$ k times.

THEOREM IV.4: The Adams operations are continuous.

Let us choose the probability law on the time of the brownian bridge $g_p(t) = Cexp[-C(p)t^\alpha]$ where $C(p)$ increases with p and tends to ∞ when $p \rightarrow \infty$. We get:

$$sup\, t^n exp[-C(p)t^\alpha] < C^n n!^{1/\alpha} \tag{4.21}$$

Let us introduce the Hilbert structure

$$\phi_{\beta,k,z}(\tilde{\omega})^2 = \sum \frac{z^n}{n!^\beta} \|\tilde{\omega}_n\|^2_{2,k} \tag{4.22}$$

for $\beta < 1$. We get an Hilbert space called $A'_{\beta,k,z}$ and this intersection of these Hilbert spaces is called $A'_{\beta,\infty}$.

Let σ the application iterated Chen integrals. To $\omega_1 \otimes .. \otimes \omega_n$, we associates:

$$\sigma_t(\omega_1 \otimes .. \otimes \omega_n) = \int_{0<s_1<..<s_n<t} \omega_1(d\gamma_{s_1,t},.) \wedge .. \wedge \omega_n(d\gamma_{s_n,t},.) \tag{4.23}$$

for a loop $\gamma_{s,t}$ in time t.

THEOREM IV.5.: σ is continuous from $A'_{\beta,\infty}$ into A_∞ for some suitable β if $\alpha > 2$.

PROOF: We repeat the proof of the theorem II.1 of [L5]. We put $n' = \sum deg(\omega_i) - 1)$. We have:

$$(C(t,p;n',m)+C'(t,p;n',m))\sigma(\omega_1\otimes..\otimes\omega_n) \leq n!^{-1/2}\prod \|\omega_i\|_{K,\infty}(t^{n+C}+1)\alpha(n) \tag{4.24}$$

$\|.\|_{K,\infty}$ denotes the C^K uniform norm over the forms. K is a suitable integer which depends only on m. Moreover

$$\alpha(n) = n'!C^n n^{n'} \tag{4.25}$$

$n'!$ is due to the antisymmetry of the form, and $n^{n'}$ is due to the number of way to put n' times between n times (See [L5] proof of the theorem II.1) We have to divide $\alpha(n)$ by the weight $n'!(n'-p)!$, this due to (1.6) ($2^{n'p}$ is bounded by C^n). For this, we have to study after using the Stirling formula the maximum in K of

$$exp[(log n)K - K Log K] \tag{4.26}$$

It has a maximum in C^n. We deduce from (4.21) that

$$\|\sigma(\omega_1\otimes..\otimes\omega_n)\|_{W(p,m)} \leq \frac{C^n}{\sqrt{n!}} n!^{1/\alpha}\prod\|\omega_i\|_{K,\infty} \tag{4.27}$$

for K big.

Let us put:

$$\tilde{\omega} = \sum \alpha_J\omega_J \tag{4.28}$$

We get

$$\|\tilde{\omega}\|_{W(p,m)} \leq \sum|\alpha_J|\frac{C^{|J|}}{\sqrt{|J|!}}|J|!^{1/\alpha}\prod_{j\in J}\|\omega_j\|_{K,\infty} \tag{4.29}$$

By the Garding inequality ([Gi])

$$\|\omega_i\|_{K,\infty} \leq \lambda_i^{K'} \tag{4.30}$$

forK' big.

Therefore by the Cauchy-Schwartz inequality:

$$\begin{aligned}\|\tilde{\omega}\|_{W(p,m)} &\leq \sum|\alpha_J|\frac{C^{|J|}}{\sqrt{|J|!}}|J|!^{1/\alpha}\lambda_J^{K'} \leq \\ &\leq \left(\sum|\alpha_J|^2\frac{C^{|J|}}{|J|!^{\beta}}\lambda_J^{K''}\right)^{1/2}\left(\sum\lambda_J^{-\tilde{K}}\right)^{1/2}\end{aligned} \tag{4.31}$$

for a suitable β smaller than 1, K" big and $\tilde{K}$ big. on the other hand

$$\sum\lambda_{i_1}^{-\tilde{K}}...\lambda_{i_n}^{-\tilde{K}} = \left(\sum\lambda_i^{-\tilde{K}}\right)^n \tag{4.32}$$

and for $\tilde{K}$ big, $(\sum \lambda_i^{-\tilde{K}}) < 1$. Therefore the last sum in (4.30) is finite. This shows that

$$\|\sigma(\tilde{\omega})\|_{W(p,m)} \leq C\phi_{\beta,K",z}(\tilde{\omega}) \tag{4.33}$$

for a suitable β and the theorem follows.

◇

The theorem of the part II and of the part III remains valid for $A_{\beta,\infty}$. In particular, $A_{\beta,\infty}$ is an analytic Hopf algebra.

Moreover, let us recall that the Hochschild coboundary is defined as follows:

$$\begin{aligned} d(\omega_1 \otimes .. \otimes \omega_n) &= \sum sign\, \omega_1 \otimes .. \otimes \omega_i \otimes d\omega_i \otimes \omega_{i+1} .. \otimes \omega_n \\ &+ \sum sign\, \omega_1 \otimes .. \omega_{i-1} \otimes \omega_i \wedge \omega_{i+1} \otimes \quad \omega_{i+2} .. \otimes \omega_n \end{aligned} \tag{4.34}$$

For the precise definition of the signs, the reader can see [G.J.P] or [L_5]. We have:

THEOREM IV.6.: The Hochschild coboundary is continuous from $A'_{\beta,\infty}$ into $A'_{\beta,\infty}$.

PROOF: it is the same that in [L_5], theorem II.4.

References

[Ad₁] Adams J.F.: On the cobar construction. Proc. Nat. Ac. Sci. U.S.A. 42 (1956), 346-373.

[Ad₂] Adams J.F.: Infinite loop space. Princeton University Press. (1978);
[A. M₁] Airault H. Malliavin P.: Quasi sure analysis. Publication Universite Paris VI.

[A.M₂] Airault H. Malliavin P.: Integration on loop groups. Publication Universite Paris VI.

[A. H-K] Albeverio S. Hoegh-Krohn R.: The energy representation of Sobolev Lie groups. Compositio Math. 36 (1978), 37-52.

[Ar] Arnaudon M.: Semi-martingales dans les espaces homogenes. Ann. I.H.P. Proba. Stat. 29.2 (1993), 269-289.

[A.T] Atiyah M.F. Tall D.O.: Group representation, λ-rings and the J homomorphism. Topology 8 (1969), 253-297.

[Bi] Bismut J.M.: Large deviations and the Malliavin Calculus. Progress in Math. 45. Birkhauser (1984).

[Ch] Chen K.T.: Iterated path integrals of differentials forms and loop space homology. Ann. Math. 97 (1973), 213-237.

[Co] Connes A.: Entire cyclic cohomology of Banach algebras and characters of Θ-summable Fredholm modules. K.Theory 1 (1988) ,519-548.

[D. Fo] Daletskii Y. Fomin V.I.: Measures and differential equations in infinite dimensional spaces. Math. and Applications. Nauka (1988).

[D.S] Daletskii Y. Shnaiderman Y.: Diffusions and quasi-invariant measures on infinite dimensional Lie groups. Funct. Anal. and Applic. 3 (1969), 156-158.

[Dr. R] Driver B. Rockner M.: Construction of diffusions on path and loop spaces of compact Riemannian manifolds. C.R.A.S. 315 Serie I. (1992), 603-608.

[Fa. M] Fang S. Malliavin P.: Stochastic analysis on the path space of a Riemannian manifold. I. J.F.A. 118 (1993), 249-273.

[Ge] Getzler E.: Dirichlet form on a loop space. Bul. Science. Math. 2. 113 (1989), 157-174.

[G.J.P] Getzler E. Jones J.D.S. Petrack S.: Differential forms on loop spaces and the cyclic bar complex. Topology 30 (1991), 339-373.

[Gi] Gilkey P.: Invariance theory, the heat equation and the Atiyah-Singer theorem. Math.lect.series.11. Publish and Perish.

[Gr_1] Gross L.: Potential theory on Hilbert spaces. J.F.A. 1 (1967), 123-181.

[Gr_2] Gross L.: Logarithmic Sobolev inequalities on a loop group. J.F.A. 102 (1991), 268-312.

[Gr_3] Gross L.: Uniqueness of ground states for Schrödinger operators over loop groups. J.F.A. 112 (1993), 373-441.

[Gro] Grothendieck A.: La theorie de Fredholm. Bull. Soc. Math. France 84 (1956), 319-384.

[H] Hopf H.: Ueber die Topologie der gruppen-Mannigfaltigkeiten und ihren Verallgemeinerungen. Ann. Math 42 (1941), 22-52.

[$J.L_1$] Jones J.D.S. Leandre R.: L^p- Chen forms on loop spaces. In "Stochastic Analysis". M.T. Barlow N. Bingham edit. Cambridge University Press (1991), 104-162.

[$J.L_2$] Jones J.D.S. Leandre R.: A stochastic approach to the Dirac operator over the free loop space. In preparation.

[Ku] Kuo H.H.: Diffusion and Brownian motion on infinite dimensional manifolds. Trans. Amer. Math. Soc. 159 (1972), 432-451.

[K_1] Kusuoka S.: De Rham cohomology of Wiener-Riemannian manifolds. Preprint.

[K_2] Kusuoka S.: More recent theory of Malliavin Calculus. Sugaku 5. 2 (1992), 155-173.

[L_1] Léandre R.: Applications quantitatives et qualitatives du calcul de Malliavin. Col. Franco-Japonais. M. Metivier S.Watanabe edit. L.N.M. 1322. Springer (1988), 109-133. English translation: Geometry of Random motion. R. Durrett M. Pinsky edit. Contemporary Math. 73 (1988), 173-197.

[L_2] Léandre R.: Integration by parts formula and rotationally invariant Sobolev Calculus on free loop spaces. In " Infinite dimensional geometry in Physics". XXVIII Winter school of theoretical Physics. R. Gielerak A.Borowiec edit. Journal of Geometry and Physics 11 (1993), 517-528.

[L_3] Léandre R. : Invariant Sobolev Calculus on the free loop space. Preprint.

[L_4] Léandre R. : Brownian motion over a Kahler manifold and elliptic genera of level N. To be publish in "Stochastic Analysis and applications to physic." L. Streit edit.

[L_5] Léandre R. : Cohomologie de Bismut–Nualart–Pardoux et cohomologie de Hochschild entiere. Preprint.

[L_6] Léandre R.: Brownian cohomology of an homogeneous manifold. Preprint.

[L. N] Léandre R. Norris J.R.: Integration by parts and Cameron-Martin formulas for the free path space of a compact Riemannian manifold. Preprint.

[L. R] Léandre R. Roan S.S.: A stochastic approach to the Euler-Poincaré number of the loop space of a developable orbifold. To be publish in Journal of Geometry and Physics.

[Lo] Loday J.L.: Operations sur l'homologie cyclique des algebres commutatives. Invent. Math. 96 (1989), 205-230.

[M.M] Milnor J.W. Moore J.C.: On the structure of Hopf algebras. Ann. Math. 81 (1965), 211-264.

[N. P] Nualart D. Pardoux E.: Stochastics Calculus with anticipatings integrands. P.T.R.F. 78 (1988), 535-581.

[Pa] Patras F.: Thesis. Université Paris VII (1992).

[Ra] Ramer R.: On the de Rham complex of finite codimensional forms on infinite dimensional manifolds. Thesis. Warwick University (1974).

[Ru] Ruston A.: Direct products of Banach spaces and linear functional equations. Pro. London. Math.Socie. 1 (1951), 327-384.

[Sch] Schatten R.: A theory of cross spaces. Princeton University Press (1950).

[Sh_1] Shigekawa I.: Transformations of Brownian motion on a Riemannian symmetric space. Z.W 65 (1984), 493-522

[Sh_2] Shigekawa I. : De Rham-Hodge-Kodaira's decomposition on an abstract Wiener space. J. Math. Kyoto. Uni. 26 (1986), 191-202.

[Sh_3] Shigekawa I.: Differential Calculus on a based loop group. Preprint.

[Ug] Uglanov Y.: Surface integrals and differntial equations on an infinite dimensional space. Sov. Math. Dok. 20 (1979),917-920.

[Wa] Watanabe S.: Stochastic analysis and its applications. Sugaku 5. 1 (1992), 51-71.

Relativistic Chaos in Time–Driven Linear and Nonlinear Oscillators

Hai–Woong Lee

Department of Physics
Korea Advanced Institute of Science and Technology
Taejon, 305-701, S. Korea

Abstract: The relativistic dynamics of time–driven oscillators is studied by analyzing the resonance structure in the relativistic phase space. For the case of the harmonic oscillator and "hard" oscillators, a new series of resonances that are absent in the nonrelativistic description may appear when relativistic effects are taken into account. The overlap between these resonances can be considered as a signal for the onset of chaotic relativistic motion.

Relativistic chaos refers to chaos exhibited by a system moving with relativistic velocities. In this work we study the chaotic dynamics of periodically driven relativistic oscillators subject to a potential $V(q)$. The Hamiltonian for such a system can be written as

$$H = \sqrt{p^2c^2 + m^2c^4} + V(q) + qF_0 \cos \omega t \tag{1}$$

where F_0 and ω are, respectively, the amplitude and frequency of the driving force. Hamilton's equations of motion then take the form

$$\frac{dq}{dt} = \frac{p}{\sqrt{m^2 + p^2/c^2}} \tag{2}$$

$$\frac{dp}{dt} = -\frac{\partial V(q)}{\partial q} - F_0 \cos \omega t \tag{3}$$

It should be noted that Eq. (2) is in general nonlinear in p, and thus there exists a possibility for chaotic motion regardless of the form of the potential $V(q)$ [1, 2]. In view of the fact that the right hand side of Eq. (2) is simply linear in p in the nonrelativistic limit, it is obvious that the theoretical analysis of chaos, which has mainly been concerned with nonrelativistic systems, should be modified when the motion being considered becomes relativistic.

For our discussion of relativistic chaos, we have found the standard description [3, 4, 5] based upon nonlinear resonances and their overlap most useful. The location of resonances can be readily determined using the technique of action–angle variables. Restricting our consideration to a potential of the form

$$V(q) = A|q|^{\alpha}, \alpha > 0 \tag{4}$$

for simplicity of presentation, we first obtain an expression for the action variable I, in the limit $F_0 \to 0$,

$$I = I(E) = \frac{1}{2\pi}\oint p dq = \frac{2}{\pi c}\int_0^b \sqrt{[E - V(q)]^2 - m^2c^4}dq, \tag{5}$$

where b is the amplitude of oscillation defined as

$$E - mc^2 = V(b). \tag{6}$$

The period–n resonance occurs at energy at which

$$n = \frac{\omega}{\Omega(E)} = \omega\frac{\partial I(E)}{\partial E}. \tag{7}$$

For a given potential $V(q)$, one can compute the resonance energy E_n at which the period–n resonance is formed from Eqs. (5) and (7). A simple Chirikov overlap criterion[4] or its refined version such as the renormalization technique[5] can then be applied to estimate the critical force amplitude at which the onset of chaos occurs.

Let us first consider the simplest case of a driven relativistic harmonic oscillator for which $V(q) = kq^2/2$. In this case, Hamilton's equations of motion become

$$\frac{dq}{dt} = \frac{p}{\sqrt{m^2 + p^2/c^2}} \tag{8}$$

$$\frac{dp}{dt} = -kq - F_0 \cos\omega t. \tag{9}$$

In the nonrelativistic limit, the right hand side of Eq. (8) is simply p/m. Hamilton's equations are then linear in both q and p, and as a result no chaos can be associated with the driven nonrelativistic harmonic oscillator. When the motion becomes relativistic, however, Eq. (8) is nonlinear in p, and chaos can be exhibited even by the linear oscillator. The action variable in this case can be obtained by substituting $V(q) = kq^2/2$ into Eq. (5). We obtain

$$I = I(E) = \frac{4a}{3\pi c}\left[E\mathcal{E}(\frac{b}{a}) - mc^2K(\frac{b}{a})\right] \tag{10}$$

where $K(x)$ and $\mathcal{E}(x)$ are, respectively, the complete elliptic integrals of first and second kinds, and $a = \sqrt{2(E + mc^2)/k}$. Substituting Eq. (10) into Eq. (7), one can then determine the resonance energies E_n. In our earlier work[1] we reported the result of our computation for the Poincaré map with the parameters $m = 1, k = 1, \omega = 2$ and $c = 300$. At $F_0 = 500$ the map showed clearly that odd–period resonances are formed at energies determined by Eqs. (10) and (7). As the force amplitude was increased from 500 to 1500, however, the overlap between neighboring resonances was seen to occur and the chaotic sea occupies a large portion of the phase space.

For a nonlinear oscillator for which the potential is given by Eq. (4) with $\alpha \neq 2$, there may also occur a significant change in the resonance structure and consequently in the chaotic behavior of the oscillator when relativistic effects are considered. In the nonrelativistic description of resonances for a nonlinear oscillator, the order in which resonances occur in energy space is different depending on whether the oscillator is "hard" ($\alpha > 2$) or "soft" ($\alpha < 2$). For the case of a hard oscillator, the frequency of oscillatory motion is an increasing function of energy, and thus a lower–period resonance appears at a higher energy. For a soft oscillator, on the other hand, the frequency of oscillatory motion is a decreasing function of energy and a higher–period resonance appears at a higher energy. In contrast to nonrelativistic resonances, the order in which resonances appear in the relativistic phase space is the same whether the oscillator is hard or soft. This can be best seen by considering the ultrarelativistic limit in which the speed of the motion remains close to c for most of the times. A further increase in energy from an ultrarelativistic energy would not cause any appreciable change in the speed of the oscillator, and yet the amplitude of oscillation will increase. Thus, the frequency of oscillatory motion should be a decreasing function of energy in the relativistic case, and relativistic resonances always appear in the order of increasing period as one moves along the energy axis from $E = 0$ to $E \to \infty$.

For a soft oscillator, the orders in which nonrelativistic and relativistic resonances appear are the same. This means that relativistic considerations should have an effect of simply shifting the location of each nonrelativistic resonance. Relativistic effects show up more dramatically for a hard oscillator. Nonrelativistic resonances that appear in the order of decreasing period will be superimposed with relativistic resonances in the order of increasing period. Suppose, for example, the energy at which each period-n nonrelativistic resonance is formed is sufficiently low that relativistic effects are negligible. As one moves along the energy axis, one will first encounter a series of nonrelativistic resonances in the order of decreasing period all the way down to the period-1 resonance. As one moves forward a higher and relativistic energy region, one should then encounter a new series of resonances beginning with a period-1 resonance and then a period-3 resonance and so on. This new set of resonances is clearly absent in the nonrelativistic description and can be considered to be induced by relativity. Just as in the harmonic oscillator, a new set of resonances can be induced for the case of a hard oscillator when relativistic effects are considered.

As an illustration of the formation of a relativity–induced resonance, we show in Fig. 1 the Poincaré map for the Duffing oscillator with the potential given by

$$V(q) = -\frac{1}{4}q^2 + \frac{1}{8}q^4 \tag{11}$$

for the case when $m = 1, \omega = \pi/2$ and $c = 5.5$. There is a period-1 resonance centered at $q \simeq -3.5, p = 0$, corresponding to $E - mc^2 \simeq 15.6$. This resonance is also present in the nonrelativistic Poincaré map (not shown) obtained by taking $c \to \infty$. Fig. 1 shows another period-1 resonance centered at $q \simeq 4.7, p = 0$, corresponding to $E - mc^2 \simeq 54.2$. This resonance is totally absent in the nonrelativistic description, and thus represents a new lowest–order resonance induced by relativity. We should mention here that, at energy at which this resonance occurs, the Duffing potential of Eq. (11) can be considered to be approximately of the form $A|q|^\alpha$ with $\alpha = 4$.

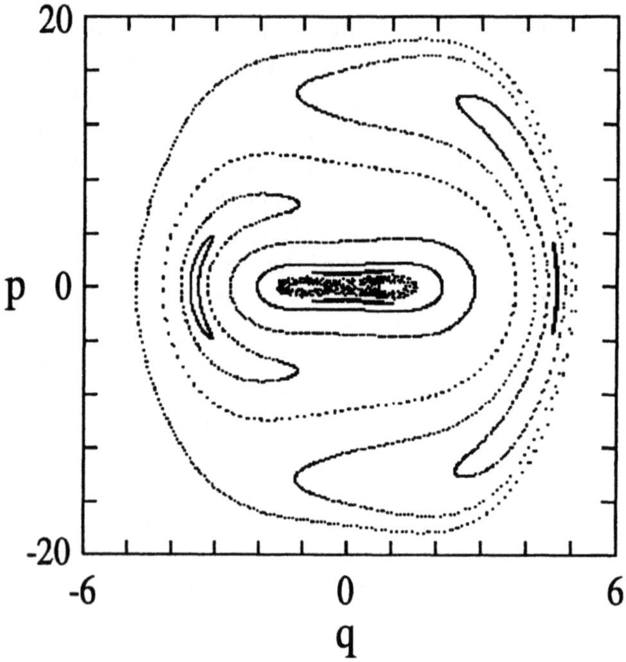

Fig. 1. Phase space map for the driven Duffing oscillator with the oscillator mass $m = 1$, the driving frequency $\omega = \pi/2$ and the amplitude of the driving force $F_0 = 0.1$ in a unit system in which $c = 5.5$.

Finally, we emphasize that the significance of the appearance of a new set of resonances in the relativistic phase space lies in the fact that the overlap between these relativistic resonances signals the onset of chaotic relativistic motion. The relativistic resonances play a central role in the theoretical analysis of chaotic motion exhibited by relativistic linear or nonlinear oscillators.

References

1. J. H. Kim and H. W. Lee, Phys. Rev. E (Feb. 1995).
2. I. Archip, B. Constantin and P. Sufitchi, Buletinul Institutului Politehnic Din Iasi, Sectia I, **37**, 147 (1991).
3. W. A. Lin and L. E. Reichl, Physica **19D**, 145 (1986).
4. B. V. Chirikov, Phys. Rep. **52**, 263 (1979).
5. D. F. Escande and F. Doveil, J. Stat. Phys. **26**, 257 (1981).

Applications of Quantum Characteristic Exponents

W.A. Majewski

Institute of Theoretical Physics and Astrophysics
Gdańsk University, PL–80–952 Gdańsk, Poland

Abstract: We show that using the recently introduced quantum characteristic exponents one can give a simple and natural characterization of stability of quantum dynamical maps.

1 Introduction

One of the fundamental problems of quantum dynamical system theory is the following question: Is there anything within quantum systems to compare with the chaotic properties of classical dynamical systems? To find an (affirmative) answer to this question we propose to use as a basic criterion for a study of stability of quantum systems the recently introduced quantum characteristic exponents [MK1], [MK2], [ENTS]. This method bears an analogy with the technique developed in classical system theory. However, such an approach to a characterization of stability of dynamical variables is done in purely quantum terms, i.e. we use the C^*–algebraic description of quantum dynamical systems and the Heisenberg form of equations of motion. The choice of the Heisenberg formula for equations of motion is dictated by its harmonious relation with the very rich algebraic structure of the matrix formulation of quantum mechanics. In particular, this approach gives a possibility to get nonlinear operator equations. To be more specific, let H be a hamiltonian of a physical system such that $H = H(A, B)$ is a function of several noncommuting dynamical variables. Then, the function

$$A \mapsto e^{iH(A,B)t} A e^{-iH(A,B)t} \tag{1}$$

does not have to be the linear one. We shall use quantum characteristic exponents to examine a stability of the time evolution given by (1). Consequently, a positivity of quantum characteristic exponents, λ^q, for such nonlinear maps will be taken as a measure of the lack of predictability due to an incomplete specification of the initial values of quantum dynamical variables. Therefore, we shall consider the positivity of λ^q as a basic signature of unstable dynamics. However, we do not identify this signature of chaos with the very chaos of dynamical system since to get a complete description of chaotic behaviour of dynamical maps some additional geometrical assumptions on the space of dynamical variables are necessary.

The central goal of this paper is to review examples of dynamical systems with explicitly computed two types of quantum characteristic exponents. We start with preliminaries (section II) and our examples of stable and unstable dynamical systems are presented in section III.

2 Definitions and Notation

In this section we introduce some necessary concepts. To fix the notation let us recall basic definitions of classical dynamical system theory. A classical dynamical system is defined as a pair $((\Omega, \Sigma, P), (\tau))$ where (Ω, Σ, P) is a probabilistic space, Ω a locally compact, Hausdorff space and $\{\tau : \Omega \to \Omega\}$ is a measure preserving transformation. Thus, the (discrete) evolution of the system is given by the iteration of τ. It is well known that if Ω admits a differential calculus then Lyapunov exponents can be defined as the following limit

$$\lambda^{\mathrm{cl}} = \lim_{n\to\infty} \frac{1}{n} \log |D_x \tau^n(y)|, \tag{2}$$

where $D_x\tau^n(y)$ denote the directional derivatives of τ composed with itself n times at a point x in a direction y (cf. [ER]). One can show that a positive Lyapunov exponent measures the average rate of growth of separation of orbits which at time zero differ by a small vector. This property can be taken as a basic motivation for a definition of deterministic chaos in terms of Lyapunov exponents.

However, its application to physical problems is limited since Nature, at least on a microscopic level, is described by quantum laws. To get a proper definition (cf. [MK1], [MK2]) let us replace the probabilistic structure (Ω, Σ, P) in the above definition of dynamical system by a non-commutative probability space $(\mathcal{A}, \phi)$ where $\mathcal{A}$ is a C^*–algebra, ϕ is a state on $\mathcal{A}$. Then, in a such defined non-commutative dynamical system τ maps $\mathcal{A}$ into $\mathcal{A}$, i.e. the quantum dynamical system $(\mathcal{A}, \tau, \phi)$ was defined. Consequently, the natural quantization of probabilistic structure leads to the Heisenberg picture.

As the first definition of quantum characteristic exponent we present (cf. [MK1], [MK2]):

$$\lambda^q(\tau; x, y) = \lim_{n\to\infty} \frac{1}{n} \log \|(D_x\tau^n)(y)\| \quad (\equiv \lambda^q) \tag{3}$$

where we used the same notation as in (2), i.e. $D_x\tau^n(y)$ denote the directional derivatives of τ composed with itself n times at a point x in a direction y. However, let us note that in (3) x and y are elements of C^*–algebra. Clearly, (3) can be generalized for continuous dynamical systems in an obvious way. This limit, if it exists, will be called *the quantum characteristic exponent.*

The noncommutative generalization (3) of characteristic exponent is not the only one posibility. Namely, within the Connes's noncommutative geometry the required differential structure can be introduced by derivations δ of the algebra $\mathcal{A}$ associated with a physical system. Here, by derivation δ we understand a linear map from a $*$-subalgebra $\mathcal{D}(\delta)$, the domain of δ, into $\mathcal{A}$ with the properties that

1. $\delta(A)^* = \delta(A^*), \qquad A \in \mathcal{D}(\delta)$
2. $\delta(AB) = \delta(A)B + A\delta(B), \qquad A, B \in \mathcal{D}(\delta)$
3. δ is the generator of a one-parameter group of automorphisms σ_s of $\mathcal{A}$.

Then (see [ENTS]) we can define

$$\Lambda^q = \lim_{t\to\infty} \frac{1}{t} \log ||\delta\tau_t(x)|| \tag{3a}$$

where $x \in \mathcal{D}(\delta)$ and τ_t stands for (continuous) dynamics of the system.
Remarks:

1. Definition (3a) is essentially designed for continuous dynamical systems. This is a result of the method of introducing the differential structure for $(\mathcal{A}, \tau, \varphi)$.
2. The presented two examples of definitions of quantum Lyapunov exponents do not exhausted this subject (cf. [V], also see the lecture of Vilela-Mendes in this proceedings).

In this review we shall restrict ourself to λ^q and Λ^q exponents. As we shall consider mostly λ^q let us give an account of its properties. To present a convincing argument that λ^q is well defined we give an example of sufficient conditions implying the existence of the limit in (3). Let us assume:

(i) τ is a completely, in general nonlinear, positive map (in fact, this is the most important assumption for the next theorem),
(ii) $||\tau^l(0)|| \le C_1$ for all $l \in N$ and some positive C_1, and

$$\Theta_\tau = \{x \neq 0 : ||\tau^l(x)|| \le C_2||x|| + ||\tau^l(0)||\} \neq \emptyset \tag{4}$$

for some positive C_2 and all $l \in N$.
(iii)

$$||D_x\tau^k(y)|| > C^k(x,y)$$

for some positive $C(x,y)$ and all large $k \in N$.

Under the above assumptions one can prove:
Theorem (cf. [MK2]): Let $\tau : \mathcal{A} \to \mathcal{A}$ be a map such that the assumptions (i), (ii), (iii) are satisfied. The following limit

$$\lim_{n\to\infty} \frac{1}{n} \log ||D_x\tau^n(y)|| \tag{5}$$

exists for $x \in \Theta_\tau$.
Remarks:

1. Examples of nonlinear completely positive maps as well as the structure theorem describing this class of maps can be found in [A], [ACh].

2. The just given theorem clearly shows that the notion of quantum characteristic exponent λ^q is well defined.

Having the well–established notion of quantum counterpart of Lyapunov exponents λ^q let us list its basic properties.

(i) $\lambda^q(x,y) = \lambda^q(x,ay)$ for $a \in \mathbf{R} \setminus \{0\}$.

(ii) Note that since the map $y \to D_x\tau^n(y)$ is linear one , it is natural to set

$$\lambda^q(x,0) = -\infty.$$

(iii) Let $\lambda^q(x,y) > \lambda^q(x,z) > -\infty$ and, additionally, let τ satisfy assumption (iii) of the theorem in the direction $y+z$. Then:

$$\lambda^q(x, y+az) \le \lambda^q(x,y)$$

where $a \in \mathbf{R}$.

(iv) The function $y \to \lambda^q(x,y)$ as the limit of continuous functions (in y) is, in general, the Baire function of type I. In particular, the set $\{y | y \to \lambda^q(x,y)$ is a continuous function$\}$ is dense in $\mathcal{A}$.

(v) Let $\mathcal{A}$ be the C^*–algebra generated by a fixed self–adjoint operator and identity on some Hilbert space, i.e. $\mathcal{A}$ is an abelian C^*–algebra. Consequently, such a dynamical system $(\mathcal{A}, \tau, \phi)$ should be considered as a classical one. Then, if some mild technical conditions are met (cf. [MK3]) the definition (3) leads to the largest classical characteristic exponent.

Let us note that the above listed properties of λ^q are reminiscent of the basic ones for characteristic exponents (cf. [AC], [P]). Therefore, *we conclude that λ^q is well defined quantum couterpart of characteristic exponent.*

To end the presentation of characteristic exponents let us turn to the second exponent Λ^q. Our first observation is that the definition of Λ^q has the very similar functional form to that of λ^q. Accordingly, one can expect that similar properties to that of λ^q hold for Λ^q. However, we have neither a version of Theorem for Λ^q nor the property (v). Nevertheless, it can be argued on the basis of (noncommutative) differential geometry (see [ENTS]) that this notion is also well defined.

3 Examples

Now we are in position to give examples of dynamical systems with analytically calculated exponents.

(1) A dynamical system composed of N–level quantum system with dynamics given by a linear dynamical semigroup S_t of contractions. The assumption of local nonexpansiveness is always satisfied but the condition of variability is satisfied for some directions. In general, $\lambda^q(x,\delta x) \in (-\infty, 0)$. As an illustration let us consider $S_t x = e^{-\lambda t}x$, where $t \ge 0$ and $\lambda > 0$. Then $\lambda^q(x, \delta x) = -\lambda$. This means the stability of the considered dynamics.

(2) A dynamical system with dynamics given by Hartree – type evolution equation (cf. [AM]):

$$\frac{d}{dt}\rho_t = -i[H(\rho_t), \rho_t] \tag{6}$$

$$\rho_t|_{t=0} = \rho_o,$$

where $H(\rho) = [Tr(\rho Q)]Q$ with $Q = Q^*$, and ϱ is a density matrix on a finite dimensional Hilbert space $\mathcal{H}$. Let us recall that this type of equation can be obtained in the mean field limit for interacting quantum systems. The explicit solution of (6) is given by

$$\Phi_t(\rho_o) = \exp\left[-iTr(Q\rho_o)Qt\right]\rho_o \exp\left[iTr(Q\rho_o)Qt\right]. \tag{7}$$

Thus taking the discrete time in (7) and putting $\tau = \Phi_1$ in (3) we get:

$$\lim_{n\to\infty} \frac{1}{n} \log \| -i[Tr(\delta\rho Q)Qn, \rho] + \delta\rho \| = 0. \tag{8}$$

In (8) $[\cdot,\cdot]$ denotes the commutator. Clearly, assumptions of our theorem are satisfied. It follows that Hartree–type evolution (7) is the stable one.

(3) As the next example let us consider the following map

$$\Phi(\rho) = \rho^2 \tag{9}$$

for $\rho^* = \rho$ from unit ball in $B(\mathcal{H})$ where $\mathcal{H}$ is a finite dimensional Hilbert space. It is easy to observe (cf. Euler theorem) that

$$\| D_\rho \Phi^n(\rho) \| = 2^n \| \rho \|^{2^n} \tag{10}$$

for $\rho \neq \mathbf{1}$ and $\rho \notin$ Projectors($\mathcal{H}$). To prove (10), one should use the C^*–property of the norm $\|\cdot\|$, i.e. $\|\rho^*\rho\| = \|\rho\|^2$. Clearly, the assumptions of the theorem are satisfied. Consequently, (10) implies:

$$\lambda^q(\rho,\rho) = \lim_{n\to\infty} \frac{1}{n}(2^n \log\|\rho\| + n \log 2).$$

As a conclusion, this mathematical example shows that the quantum exponent λ^q can also be positive (put $\rho : \|\rho\| = 1$). Thus the quadratic map, in some region, can be unstable.

Now we shall present examples where the Hilbert space $\mathcal{H}$ of a system *is not assumed to be of finite dimension.*

(4) Let us consider a quantum two–level system interacting with a single mode of the electromagnetic field.(cf. [FE]). The Hamiltonian for a such system can be taken as

$$H = \frac{1}{2}\hbar\omega_0\sigma_z + \hbar\omega(a^+a + \frac{1}{2}) + \hbar\lambda_0\sigma_x(a + a^+) \tag{11}$$

where σ_x, σ_z are Pauli matrices and a^+ and a stand for bose creation and anihilation operators. The energy separation of the two–levels is equal to $\hbar\omega_0$ while the frequency of the radiation mode is ω. The time evolution of the system is given by the following operator equation

$$\dot{A} = \frac{i}{\hbar}[H, A] \tag{12}$$

for an observable A, where the dot denotes a time derivative. Let us remark that the Heisenberg equation (12) gives for this model first order coupled nonlinear operator equations (cf. [FE]). Although the Hamiltonian H depends, in general, on the time t, one can solve (12) and subsequently compute the quantum exponents λ^q (see [KM1]). In particular one has:

$$\lambda^q(\sigma_i, \sigma_j) = 0 \tag{13}$$

for $i, j = 1, 2$, where $\sigma_1 = \sigma_x$ and $\sigma_2 = \sigma_z$. Consequently, this quantum optic model exhibits stable evolution.

(5) Non–commutative Anosov system [ENTS]. Let $(\mathcal{A}, \phi, \tau)$ be a (concrete) von Neumann dynamical system, i.e. $\mathcal{A}$ is a von Neumann algebra acting on a Hilbert space $\mathcal{H}$, ϕ a faithful normal state on $\mathcal{A}$ defined by a cyclic and separating vector $\Omega \in \mathcal{H}$, and τ represents the action of $\mathbf{R}$ on $\mathcal{A}$. In other words τ is a continuos homomorphism of $\mathbf{R}$ into the group $Aut\mathcal{A}$ of C^*–algebra automorphisms of $\mathcal{A}$ with the topology of (weak) pointwise convergence. We assume additionally that the state ϕ is τ–invariant, i.e. $\phi \cdot \tau = \phi$. The dynamical system $(\mathcal{A}, \phi, \tau)$ admits an *integrable Anosov structure* if there exists a collection

$$\{\sigma_i | i = 1, ..., k; k+1, ..., n\}$$

of *horocyclic* actions of $\mathbf{R}$ on $\mathcal{A}$ such that

- ϕ is σ_i invariant.
- $\tau(t)\sigma_j(s)\tau(-t) = \sigma_j(e^{-\lambda_j t}s)$ where $t, s \in \mathbf{R}$, $\lambda_j \in \mathbf{R}$ and

$$\lambda_1 \leq ... \leq \lambda_k < 0 < \lambda_{k+1} \leq ... \leq \lambda_n.$$

Let δ_j be the derivation generating the horocyclic action σ_j, i.e. a linear map from a $*$–subalgebra $\mathcal{D}(\delta_j)$, the domain of δ_j, into $\mathcal{A}$ with the properties described prior to Definition (3a).

Then, it was proved (see [ENTS])

Theorem: The quantum characteristic exponent Λ^q (cf. (3a)) for the non–commutative Anosov system exists, i.e.

$$\lim_{t\to\infty} \frac{1}{t} \log ||\delta_j \tau(t)(A)|| = \lambda_j$$

where $A \in \mathcal{D}(\delta_j)$ is such that $\delta_j(A) \neq 0$.

As an example of concrete dynamical system with positive quantum Lyapunov exponent Λ^q one can consider (see [ENTS]) an arbitrary relativistic quantum field on the Rindler wedge of Minkowski space with dynamics given by the modular automorphism group. Hence, the noncommutative Anosov system is an example of quantum dynamical system with very unstable evolution.

(6) This example is the quantum counterpart of a parametrically kicked nonlinear oscillator. Its hamiltonian H is of the form

$$H = \frac{\chi}{2}(a^*)^2 a^2 + i\hbar\frac{\kappa}{2}[(a^*)^2 - a^2] \cdot \sum_{n=-\infty}^{+\infty} \delta(t - nt_0) \tag{14}$$

where χ and κ are constants characterizing the system, a^+ and a stand for bose creation and anihilation operators, and t_0 is the period of free evolution (i.e. the evolution described by $H_{NL} = \frac{\chi}{2}(a^*)^2 a^2$). Let us remark that H_{NL} gives the nonlinear part of the evolution. Obviously, $\delta(\cdot)$ stands now for the Dirac distribution. Then, the Heisenberg equation of motion,

$$\dot{A} = \frac{i}{\hbar}[H, A] \tag{15}$$

can be solved in terms of self–adjoint operators Φ and Π, where Φ and Π are defined by: $a = \Phi + i\Pi$. The solution of the Heisenberg equation for the considered model can be written as:

$$\begin{pmatrix} \Phi(t_n^+) \\ \Pi(t_n^+) \end{pmatrix} = e^{\frac{-i\mu}{2}} \begin{pmatrix} e^r \cos\mu B_0 & e^r \sin\mu B_0 \\ -e^{-r} \sin\mu B_0 & e^{-r} \cos\mu B_0 \end{pmatrix} \begin{pmatrix} \Phi(t_{n-1}^+) \\ \Pi(t_{n-1}^+) \end{pmatrix} \tag{16}$$

where $\mu = \chi t_0$, r the effective constant for kicks, and $B_0 = a^* a - \frac{1}{2} = (\Phi^2 + \Pi^2 - 1)$.

Let us remark that (16) gives a nonlinear evolution since the matrix on the right hand side of (16) is a nonlinear function of operators Φ and Π. We want to study the stability properties of the evolution (16). We restrict ourself to an examination of the Lyapunov instabilities of the quadrature components of the electric field during the time evolution given by the formula (16). To do this let us define the quadrature operators:

$$\Phi^\epsilon = \frac{1}{2}[e^{i\epsilon} a + e^{-i\epsilon} a^*] \tag{17}$$

and

$$\Pi^\epsilon = \frac{1}{2i}[e^{i\epsilon} a - e^{-i\epsilon} a^*] \tag{18}$$

Let us note that the operators Φ^ϵ and Π^ϵ are related to the amplitude components of electric field. At this point in order to avoid any confusion some explanation is necessary. Namely, the hamiltonian (14) leads to *the nonlinear evolution map* (16). However, from physical point of view, an examination of the stability properties of dynamical map does not allow any change of anihilation (so also creation) operators nor any modification of the fixed hamiltonian. This is exactly the reason that to find *physical examples with positive quantum exponents* λ^q we are examining the quadrature operators. In other words, the quadrature operators are examples of dynamical variables in our model which can be varried without any change of a $(a^\dagger)$ and the hamiltonian.

To compute quantum exponents for quadrature operators in the considered model we should find an explicit formula for the norm of $||D_\epsilon \tau^n(y)||$, where in the example considered, by τ we denote the time evolution of the system between two succesive kicks and the effect of the first one. y denotes an element of the set $\{\Phi, \Pi\}$. Clearly, τ^n stands for τ composed with itself n times. D_ϵ is the derivative with respect to the phase angle ϵ which describes a "rotation" in (Φ, Π) variables and which is used in the definition of quadrature operators. But, since a^+ and a are unbounded operators, in order to compute $||D_\epsilon \tau^n(\cdot)||$ we should introduce $(\Phi_\delta, \Pi_\delta)$, the cut–off in the (Φ, Π) variables (cf. [KM2]). Then, replacing the original dynamics by its well defined approximation and changing the variables (Φ, Π) to $(\tilde{\Phi}, \tilde{\Pi})$ one can find such set of parameters (χ, κ, t_0) that $\lambda^q(\tilde{\Pi}) > 0$. In other words, for large enough r (r is the effective constant for kicks) and some values of $\mu = t_0\kappa$, the quantum characteristic exponent λ^q for quantum variable $\tilde{\Pi}_\delta$ is strictly positive. To comment on this result let us note that in polyparametric cases physics as well as mathematics allow numerous combinations of stability in certain directions and irregularity in others. Therefore, one can expect chaotic chaotic features of the evolution of our model for some values of χ, τ, κ and regularity for others and we get a confirmation of such behaviour ($\lambda^q(\tilde{\Phi}_\delta)$ can be negative for some values of parameters κ, χ, and r). Moreover, let us recall that the different approach to the considered problem gives the similar result (see [M]). We conclude that also this model exhibits the hyperbolic type of instabilities.

(7) In the last example we consider the squeezed light in a nonlinear medium having the second–order susceptibility ($\chi^{(2)}$). An analysis of such system leads to the following equations (see [Y], section 11):

$$\frac{d}{dz}a(z) = ka^\dagger(z), \tag{19}$$

$$\frac{d}{dz}a^\dagger(z) = k^* a(z), \tag{20}$$

where $a(z)$ is the annihilation operator, k is the coupling constant and k^* stands for the complex conjugation of k. The dependence of a on z is a result of the one dimensional propagation of the electric field along the z-axis. Obviously, the equations (19), (20) lead to

$$\frac{d^2}{dz^2}a(z) = |k|^2 a(z). \tag{21}$$

It is easy to observe that the same equation can be derived from the Hamiltonian

$$H = \hbar\omega a^\dagger a + i\hbar\frac{\kappa}{2}(a^{\dagger 2} - a^2) \tag{22}$$

where $\omega = 0$. Such a Hamiltonian would appear if we take into account an oscillatory time dependence of a, and then eliminate the free evolution by the interaction picture. The solution to equations (19), (20) is

$$a(z) = \cosh(|k|z)a(0) + \frac{k}{|k|}\sinh(|k|z)a^\dagger(0). \tag{23}$$

It is clear that we can apply our definition of Lyapunov exponent λ^q to analysis of stability of the propagating squeezed light. Taking the quadrature operators in the form

$$P_\alpha(z) = \frac{1}{2}\Big([e^{|k|z}\sin\alpha + e^{-|k|z}\cos\alpha]a(0)$$

$$+[-e^{|k|z}\sin\alpha + e^{-|k|z}\cos\alpha]\, a^\dagger(t)\Big) \tag{24}$$

we get

$$\lambda^q = \lim_{n\to\infty}\lim_{z\to\infty}\frac{1}{z}\ln\left\|\frac{dQ_\alpha^n(z)}{d\alpha}\right\| = |k| \geq 0 \tag{25}$$

where we put for simplicity $\frac{k}{|k|} = -1$, Q_α^n indicates the cut off in Q_α, and

$$P_\alpha(z) = \frac{1}{2}\left(e^{i\alpha}\,a(z) + e^{-i\alpha}\,a^\dagger(z)\right)$$

$$Q_\alpha(z) = \frac{1}{2i}\left(e^{i\alpha}\,a(z) - e^{-i\alpha}\,a^\dagger(z)\right)$$

Hence, this is another quantum optic model with very unstable dynamics. It is remarkable property of this model that the classical Lyapunov exponent calculated for the trajectory

$$t \mapsto \langle w|Q_\alpha(t)|w\rangle \tag{26}$$

has the same value as its quantum counterpart; in (26) w denotes a standard coherent state $a(0)|w\rangle = w|w\rangle$.

4 Conclusions

All presented examples clearly show the considerable utility of the quantum characteristic exponents for a characterisation of stability properties of various quantum dynamical maps. We have found examples of physical systems with positive characteristic exponents, i.e. with very unstable evolution. However, it should be stressed that the positivity of exponents is a necessery condition for a quantum chaos but not sufficient one. In other words, the genuine quantum chaos demands positive exponents as well as some geometric properties of dynamical system, e.g. boundedness of trajectories, a repeatibility. The presented examples do not possess such geometrical properties and therefore they constitute examples of quantum systems with very unstable dynamics.

Acknowledgments

The support from BW 5400-5-0089-4 is gratefuly acknowledged.

References

[A] W. Arverson, in *Operator Algebras and Mathematical Physics*, Contemporary Mathematics, **62**, 283–343 (1987) (American Mathematical Society, Providence, RI, 1987).

[AC] L. Arnold, H. Crauel, Random dynamical systems, in *Lyapunov exponents*, Lecture Notes in Mathematics 1486, Springer–Verlag (1991).

[ACh] T. Ando and M.D. Choi, in *Aspects of Positivity in Functional Analysis,* edited by Nagel et al. (North–Holland, Amsterdam, 1986), pp. 3–13.

[AM] R. Alicki, W.A. Majewski, *On a general characterization of nonlinear quantum dynamical maps*, Phys. Lett. **A148**, 69 (1990).

[CKM] M. Czachor, M. Kuna, W.A. Majewski, *Lyapunov instability of squeezing,* preprint, Gdansk University (1994).

[ENTS] G.G. Emch, H. Narnhofer, W. Thirring and G.L. Sewell, *Anosov actions on noncommutative algebras*, J. Math. Phys. **35**, 5582 (1994).

[ER] J.P. Eckmann, D. Ruelle, Rev. Mod. Physics **57**, 617 (1985).

[FE] R.F. Fox, J. Eidson, Phys. Rev. **34A**, 482(1986).

[KM1] M. Kuna, W.A. Majewski, *On quantum chaos and quantum characteristic exponents*, Rep. Math. Phys. **33**, 111 (1993).

[KM2] M. Kuna, W.A. Majewski, *An example of quantum dynamical system with the hyperbolic instabilities*, preprint, Gdansk University (1994).

[M] G.J. Milburn, *Coherence and chaos in a quantum optic system*, Phys. Rev. **A41**, 6567 (1990).

[MK1] W.A. Majewski, M. Kuna, *On the definition of characteristic exponents*, in Proceedings of the Workshop on Phase Transition: Mathematics, Physics,.., Ed. R. Kotecky, World Scientific (1993).

[MK2] W.A. Majewski, M. Kuna, *On quantum characteristic exponents*, J. Math. Phys. **34**, 5007 (1993).

[MK3] W.A. Majewski, M. Kuna, *Non–commutative systems with chaos*, Quantum Probability and Related Topics, vol. 9, in print.

[P] Ya.B. Pesin, Characteristic Lyapunov exponents and smooth ergodic theory, in *Dynamical Systems, Advanced series in nonlinear dynamics, Vol 1.*, World Scientific (1991).

[V] R. Vilela Mendes, *On the existence of quantum characteristic exponents*, Phys. Lett. **A187**, 299 (1994).

[Y] B. Yurke, *Squeezed Light*, Rochester University (1989).

Asymptotic Properties of the Fokker-Planck Equation

Ryszard Rudnicki

Institute of Mathematics, Polish Academy of Sciences
PL–40–013 Katowice, Poland
and
Institute of Mathematics, Silesian University,
PL–40–007 Katowice, Poland

1 Introduction

In this note we present some recent results concerning the asymptotic behaviour of the solutions of the following equation

$$\frac{\partial u}{\partial t} = \sum_{i,j=1}^{n} \frac{\partial^2 (a_{ij}(x)u)}{\partial x_i \partial x_j} - \sum_{i=1}^{n} \frac{\partial (b_i(x)u)}{\partial x_i} \qquad (x,t) \in \mathbf{R}^n \times [0,\infty) \ . \tag{1}$$

Equation (??) generates a semigroup $\{P_t\}_{t\geq 0}$ of Markov operators on $L^1(\mathbf{R}^n)$ given by the formula

$$(P_t u_0)(x) = u(x,t), \qquad u_0(x) = u(x,0) \ .$$

We recall that P_t is a Markov operator if P_t is linear and $P_t(D) \subset D$, where D is the set of all densities, i.e. $D = \{ f \in L^1(\mathbf{R}^n) : f \geq 0, \int f = 1 \}$.

We investigate the following properties of the semigroup $\{P_t\}_{t\geq 0}$:

(A) *Asymptotic stability.* Semigroup $\{P_t\}_{t\geq 0}$ is called asymptotically stable if there exists a density f_* such that $P_t f_* = f_*$ for $t \geq 0$ and

$$\lim_{t\to\infty} \|P_t f - f_*\| = 0 \qquad \text{for } f \in D \ .$$

(M) *Completely mixing.* Semigroup $\{P_t\}_{t\geq 0}$ is called completely mixing if for any two densities f and g

$$\lim_{t\to\infty} \|P_t f - P_t g\| = 0 \ . \tag{2}$$

This notion is connected with the probabilistic interpretation of equation (??). Namely, the solutions of equation (??) are densities of the distributions of some diffusion process. Assume that this process describes a movement of particles. Then condition (??) means that particles are mixed in such a way that after a long time their distribution does not depend on the initial distribution. If there exists an invariant density f_* ($P_t f_* = f_*$), then completely mixing is equivalent to asymptotic stability. However, the semigroup $\{P_t\}_{t\geq 0}$ can be completely mixing, but it can have no invariant density. For example, the heat equation $\frac{\partial u}{\partial t} = \Delta u$ generates the semigroup which is completely mixing and has no invariant density.

(S) *Sweeping.* Semigroup $\{P_t\}_{t\geq 0}$ is called sweeping if for every bounded measurable set A and for every $f \in D$

$$\lim_{t\to\infty} \int_A P_t f(x)\, dx = 0 \ .$$

We show that the semigroup $\{P_t\}_{t\geq 0}$ satisfies the Foguel alternative, namely, it is asymptotically stable or sweeping.

(L) *Limit distribution.* Let $S = \{x \in \mathbf{R}^n : |x| = 1\}$ and A be a measurable subset of S. Denote by $K(A)$ the cone spanned by A:

$$K(A) = \{x \in \mathbf{R}^n : x = \lambda y \ , \ y \in A,\ \lambda > 0\} \ .$$

Consider a diffusion process corresponding to (??). Then the function

$$p_A(t) = \int_{K(A)} P_t f(x)\, dx \ , \qquad f \in D \ ,$$

describes the mass of particles which are in the cone $K(A)$. If the semigroup $\{P_t\}_{t\geq 0}$ is completely mixing then the asymptotic behaviour of $p_A(t)$ does not depend on f. It is interesting when there exists the limit $p_A = \lim_{t\to\infty} p_A(t)$. If $\{P_t\}_{t\geq 0}$ is sweeping then nearly all particles are in a neighbourhood of ∞ for large t and p_A measures the *sectorial limit distribution* of particles.

2 Foguel Alternative

From now on we assume that $a_{ij} \in C^3(\mathbf{R}^n)$, $b_i \in C^2(\mathbf{R}^n)$ for all i, j, where $C^k(\mathbf{R}^n)$ is the space of k–times differentiable bounded functions whose partial derivatives of order $\leq k$ are continuous and bounded. We also assume that $\sum_{i,j=1}^n a_{ij}(x)\lambda_i\lambda_j \geq \alpha|\lambda|^2$ for some $\alpha > 0$ and for every $\lambda \in \mathbf{R}^n$.

Theorem 1. *Let $\{P_t\}_{t\geq 0}$ be a Markov semigroup generated by equation* (??). *If there exists an invariant density, then this semigroup is asymptotically stable. If there is no invariant density, then this semigroup is sweeping.*

Remark 1. In [?] a similar result is proved under an additional assumption that there exists a positive locally integrable function f_* such that $P_{t_0} f_* \leq f_*$ for some $t_0 > 0$.

Proof. According to Remark 1 it remains to prove that there exists a positive locally integrable function f_* such that $P_1 f_* \leq f_*$. The proof is based on the abstract theory of Markov processes (see [?] for details). Let $Pf = P_1 f$. Then P is an integral operator, i.e. there exists a Borel measurable function $k : \mathbf{R}^n \times \mathbf{R}^n \to \mathbf{R}$ such that $Pf(x) = \int k(x, y) f(y)\, dy$. The function $k(x, y)$ is strictly positive and continuous with respect to (x, y). Since the operator P has a strictly positive kernel, it is dissipative or conservative. If P is dissipative, then for every $f \in L^1(\mathbf{R}^n)$, $f > 0$, we have $f_* = \sum_{n=0}^{\infty} P^n f < \infty$ and $Pf_* \le f_*$. If P is conservative, then P is a Harris operator and, consequently, there exists a measurable function f_* such that $0 < f_* < \infty$ and $Pf_* = f_*$. It remains to check that f_* is locally integrable. Suppose, on the contrary, that there is a bounded set G such that $\int_G f_*\,dx = \infty$. Since $k(x,y)$ is a continuous strictly positive function, there exists $\varepsilon > 0$ such that $k(x,y) \ge \varepsilon$ for $x \in G$ and $y \in G$. This implies that

$$f_*(x) \ge Pf_*(x) = \int k(x,y) f_*(y)\,dy \ge \varepsilon \int_G f_*(y)\,dy = \infty$$

for $x \in G$, which is impossible.

3 Completely Mixing

Risken in his book [?] investigated the relative entropy of two solutions of equation (??). The relative entropy can be written down in the following way

$$H(t) = \int P_t f(x) \ln \frac{P_t f(x)}{P_t g(x)}\,dx\ , \qquad f, g \in D\ .$$

He claimed that $\lim_{t\to\infty} H(t) = 0$. But the result seems to be not true. Namely, if $H(t) \to 0$ for all $f \in D$ and $g \in D$ then the semigroup $\{P_t\}_{t\ge 0}$ is completely mixing (this theorem is a simple consequence of some general result on relative entropy [?]). Consider the conjugated semigroup $\{P_t^*\}_{t\ge 0}$ defined on the space $L^\infty(\mathbf{R}^n)$ given by

$$\int (P_t^* g) f\,dx = \int g(P_t f)\,dx\ .$$

Completely mixing implies that all fixed points of the semigroup $\{P_t^*\}_{t\ge 0}$ are constant functions. Consider one–dimensional equation

$$\frac{\partial u}{\partial t} = \frac{\partial^2}{\partial x^2}(a(x)u) - \frac{\partial}{\partial x}(b(x)u)\ . \tag{3}$$

Then all fixed points of the semigroup $\{P_t^*\}_{t\ge 0}$ are the solutions of the stationary equation

$$a(x)\frac{d^2u}{dx^2} + b(x)\frac{du}{dx} = 0\ . \tag{4}$$

The function u given by the following formula

$$u(x) = \int_{-\infty}^{x} \exp\left(-\int_0^y \frac{b(r)}{a(r)}\,dr\right) dy$$

is a solution of (??). This means that if u is a bounded function, then u is a fixed point of the semigroup $\{P_t^*\}$. For $a(x) = 1$ and $b(x) = \frac{2x}{1+x^2}$ the function u is bounded. Consequently, the relative entropy does not converge to 0 for some densities f and g.

Moreover, according to the above reasoning the necessary condition for completely mixing of (??) is the following

$$\int_{-\infty}^{\infty} \exp\left(-\int_0^y \frac{b(r)}{a(r)}\,dr\right)\,dy = \infty\ . \tag{5}$$

Conversely, from (??) it follows that the semigroup generated by (??) is completely mixing. Namely:

Theorem 2. *Semigroup $\{P_t\}_{t\geq 0}$ generated by equation (??) is completely mixing if and only if condition (??) holds.*

The proof of Theorem ?? is given in the paper [?]. It is worth pointing out that the assumption that b is bounded cannot be replaced with the assumption that b grows linearly.

Condition (??) has an interesting probabilistic interpretation. Namely, let X_t be a diffusion process corresponding to equation (??). Then (??) holds if and only if X_t has one of the following properties:
(a) $\lim_{t\to\infty} X_t = \infty$ with probability 1,
(b) $\lim_{t\to\infty} X_t = -\infty$ with probability 1,
(c) $\liminf_{t\to\infty} X_t = -\infty$ and $\limsup_{t\to\infty} X_t = \infty$ with probability 1.

Open problem. I do not know if Theorem ?? can be generalized to n–dimensional Fokker–Planck equation. I think that the semigroup $\{P_t\}$ generated by equation (??) is completely mixing if and only if all bounded solutions of the elliptic equation

$$\sum_{i,j=1}^{n} a_{ij}(x)\frac{\partial^2 u}{\partial x_i \partial x_j} + \sum_{i=1}^{n} b_i(x)\frac{\partial u}{\partial x_i} = 0$$

are constant. In particular, Brzeźniak and Szafirski [?] proved that the double–stochastic equation

$$\frac{\partial u}{\partial t} = \sum_{i,j=1}^{n} \frac{\partial}{\partial x_i}\left(a_{ij}(x)\frac{\partial u}{\partial x_j}\right)$$

generates a completely mixing semigroup.

4 Limit Distribution

The problem (L) in n–dimensional space is difficult. Some partial results can be obtained under additional assumption that all functions a_{ij} and b_i are periodic with the same periods (we recall that a function $f : \mathbf{R}^n \to \mathbf{R}$ is periodic if there exist independent vectors $v_1, \ldots, v_n$ such that $f(x + v_i) = f(x)$ for each $x \in \mathbf{R}^n$ and $i = 1, \ldots, n$).

In one–dimensional space we can considered the function $p_+(t) = \int_c^\infty u(x,t)\,dx$ which describes the mass of particles in the interval (c, ∞). The following theorem provides a criterion for the existence of the limit $\lim_{t\to\infty} p_+(t)$.

Theorem 3. *Let* $B(x) = \int_0^x \frac{b(y)}{a^2(y)}\,dy$ *be a bounded function and*

$$g(y) = \left(\int_0^y \frac{e^{B(x)}}{a(x)}\,dx\right) \Big/ \left(\int_0^y e^{-B(x)}\,dx\right) .$$

Assume that there exist limits $\lim_{y\to\infty} g(y) = \beta^2$ *and* $\lim_{y\to-\infty} g(y) = \gamma^2$, $\beta > 0$ *and* $\gamma > 0$. *Then*

$$\lim_{t\to\infty} p_+(t) = \frac{\beta}{\beta+\gamma} .$$

The proof of Theorem ?? is given in [?]. In the same paper we constructed an equation such that the following condition holds

$$\limsup_{t\to\infty} \frac{1}{t}\int_0^t p(s)\,ds = 1 \qquad \text{and} \qquad \liminf_{t\to\infty} \frac{1}{t}\int_0^t p(s)\,ds = 0 . \tag{6}$$

In this example $a(x) = 1$ and $b(x) \to 0$ as $|x| \to \infty$. The probabilistic interpretation of condition (??) is interesting. Consider a diffusion process corresponding to (??) with these coefficients. Then condition (??) means that particles synchronously oscillate between $+\infty$ and $-\infty$.

References

1. Z. Brzeźniak and B. Szafirski, *Asymptotic behaviour of L^1 norm of solutions to parabolic equations,* Bull. Pol. Ac.: Math., **39**, 1–10 (1991).
2. S. R. Foguel, *The Ergodic Theory of Markov Processes,* Van Nostrand Reinhold Comp., New York, 1969.
3. A. Lasota and M. C. Mackey, *Chaos, Fractals and Noise. Stochastic Aspects of Dynamics,* II edition, Springer Applied Mathematical Sciences **97**, New York, 1994.
4. K. Łoskot, R. Rudnicki, *Relative entropy and stability of stochastic semigroups,* Ann. Pol. Math., **53**, 139–145 (1991).
5. H. Risken, *The Fokker–Planck equation,* Springer Series in Synergetics, **18**, (1984).
6. R. Rudnicki, *Asymptotical stability in L^1 of parabolic equations,* J. Diff. Equations, **102**, 391–401 (1993).
7. R. Rudnicki, *Strangely sweeping one–dimensional diffusion,* Ann. Pol. Math., **58**, 37–45 (1993).

Spacetime Distortion as a Reason for Quantum Stochasticity

Yuri A. Rylov

Institute for Problems in Mechanics
Moscow 117526, Russia.

Abstract: One develops a theory which relates to the conventional quantum mechanics in the same way, as the statistical physics relates to the axiomatic thermodynamics. Quantum effects are shown to be a result of some spacetime deformation (distortion) which converts one–dimensional world lines of free pointlike particles into three–dimensional world tubes. The world tubes are random, and their thickness is determined by the Planck constant describing the distortion magnitude. A statistical description of the world tubes in the associated Minkowski spacetime is equivalent to the quantum description (Schrödinger equation). Application of the statistical description to the dynamic system described by the Pauli equation shows that the electron spin has a classical analog, which is a collective property of the statistical ensemble of electrons (not a property of a single electron).

A quantum system is a stochastic system in the sense that experiments with a single quantum system are irreproducible in general. As far as the quantum constant $\hbar$ is an universal constant, and quantum phenomena are universal, it should expect that the reason of quantum stochasticity is also universal and connected with the spacetime properties. We are trying to interpret quantum effects and quantum stochasticity as a result of some spacetime deformation (distortion). It is impossible in frames of the conventional Riemannian spacetime model, and one needs to look at the spacetime from a more general point of view which can be formulated as follows.

All properties of the spacetime are described exhaustively by the interval $S(P,Q)$ between any two points P and Q of the spacetime, or by the world function $\sigma = S^2/2$. Manifold is a derivative concept which can be described in terms of the world function. Mathematical possibility of such a construction was founded [1]. A crucial point of this construction is a new geometrical object: tube of nth order ($n = 0, 1, \ldots$) defined by n spacetime points. The geometry constructed on the basis of world function and tubes will be referred to as T–geometry, (a geometry of tubes), or the metric approach to geometry (MAG).

Constructing T–geometry is rather simple. Relations of the Euclidean geometry written in terms of the world function are claimed as relations of T–geometry. For instance, vector $\mathbf{P_0Q}$ is defined as an ordered set of two point P_0 and Q.

P_0 is the origin, and Q is the end of the vector $\mathbf{P_0Q}$. The length $|\mathbf{P_0Q}|$ of the vector $\mathbf{P_0Q}$ is defined by the relation

$$|\mathbf{P_0Q}| = S(P_0, Q) = [2\sigma(P_0, Q)]^{1/2} \ . \tag{1}$$

In T–geometry the scalar product $(\mathbf{P_0P}.\mathbf{P_0Q})$ of two vectors $\mathbf{P_0P}$ and $\mathbf{P_0Q}$, having a common origin P_0, and the angle ϑ between them are defined through the world function as follows

$$2\,|\mathbf{P_0P}||\mathbf{P_0Q}|\cos\vartheta = 2(\mathbf{P_0P}.\mathbf{P_0Q}) = |\mathbf{P_0P}|^2 + |\mathbf{P_0Q}|^2 - |\mathbf{PQ}|^2 \tag{2}$$

where the module $|\mathbf{PQ}|$ of the vector $\mathbf{PQ}$ is defined by Eq.(1). In the Euclidean space Eq.(2) is the cosine theorem.

By definition two vectors $\mathbf{P_0P}$ and $\mathbf{P_0Q}$ are parallel ($\uparrow\uparrow$), antiparallel ($\uparrow\downarrow$), or collinear ($||$), if the relations $\cos\vartheta = 1, -1, \pm 1$ respectively take place. The tube $\mathcal{T}_{P_0P}$ (of the first order) determined by two points P_0, P is defined as a set of such points Q, that $\mathbf{P_0Q} \parallel \mathbf{P_0P}$. According to Eq.(2) it means

$$\mathcal{T}_{P_0P}: \qquad (\mathbf{P_0P}.\mathbf{P_0Q})^2 - |\mathbf{P_0P}|^2 |\mathbf{P_0Q}|^2 = 0 \ . \tag{3}$$

One equation (3) in the n–dimensional space describes in general a tube as a $(n-1)$–dimensional surface. However in the n–dimensional Euclidean space the tube degenerates into one–dimensional straight line, because there only one unit vector $\mathbf{P_0Q}$ parallel to $\mathbf{P_0P}$ exists. Nondegeneracy of timelike tubes in the spacetime will be referred to as a spacetime distortion.

Application of MAG to the Minkowski spacetime M permits to answer such a very simple question why world lines of free real particles are timelike. The conventional geometry and physics cannot answer this simple question. The answer of MAG is as follows. Timelike tubes and spacelike ones of the Minkowski spacetime M have different shape. The timelike geodesics of M are tubes of M, whereas spacelike geodesics of M are not tubes of M. Spacelike tubes of M are two three–dimensional planes tangent to a set of light cones which have their vertices on the spacelike geodesic.

The spacetime distortion can be considered as a result of some deformation of the Minkowski spacetime M. Indeed at MAG any deformation of the spacetime leads to a variation of its world function. Vice versa, any variation of the world function means some deformation of the spacetime. In particular, there are such spacetime deformations which convert the Minkowski spacetime M into non–Riemannian space and cannot be expressed in terms of a manifold, i.e. in terms of conventional Riemannian geometry. Distortion is one of such spacetime deformations.

A uniform distortion is defined as follows [2]:

$$\sigma_M \rightarrow \sigma_D = \sigma_M + D(\sigma_M) \tag{4}$$

where σ_M is the world function of the Minkowski spacetime, σ_D is the world function of the distorted spacetime (DST). D is a distortion function. The real spacetime is DST. A distortion transforms timelike straight lines of the Minkowski spacetime into timelike three–dimensional tubes. The uniform distortion does not violate the spacetime isotropy and uniformity.

The world tube of a particle of the mass m is defined as a broken timelike tube.

$$\mathcal{T}_{br} = \bigcup_i \mathcal{T}_{[P_i,P_{i+1}]} , \qquad S_D(P_i, P_{i+1}) = \mu = m/b , \qquad i = 0, \pm 1, \pm 2, \dots \quad (5)$$

where $\mathcal{T}_{[P_i,P_{i+1}]}$ is a segment of the tube $\mathcal{T}_{P_i,P_{i+1}}$ between the points P_i, P_{i+1}, b is an universal constant of dimensionality $[b] = g/\text{cm}$. According to the first Newton law for a free particle the momenta $p = \mathbf{P}_i\mathbf{P}_{i+1}$ of different links are parallel, i.e. $\mathbf{P}_i\mathbf{P}_{i+1} \uparrow\downarrow \mathbf{P}_i\mathbf{P}_{i-1}, \quad i = 0, \pm 1, \pm 2 \dots$ Together with Eq.(5) it means

$$(\mathbf{P}_i\mathbf{P}_{i+1}.\mathbf{P}_i\mathbf{P}_{i-1}) = - \mid \mathbf{P}_i\mathbf{P}_{i+1} \mid\mid \mathbf{P}_i\mathbf{P}_{i-1} \mid = -\mu^2 , \qquad i = 0, \pm 1, \pm 2, \dots \quad (6)$$

In the Minkowski spacetime M the relations (5),(6) considered as equations for P_{i+1} with P_{i-1}, P_i fixed have a sole solution. It means that any link $\mathbf{P}_{i-1}\mathbf{P}_i$ of the world line (5) determines single–valuedly the next link $\mathbf{P}_i\mathbf{P}_{i+1}$. In DST the same equations (5), (6) have many solutions, in general. At fixed link $\mathbf{P}_{i-1}\mathbf{P}_i$ a position of the next link $\mathbf{P}_i\mathbf{P}_{i+1}$ is not determined single–valuedly, in general, because in DST there are many vectors $\mathbf{P}_i\mathbf{P}_{i+1}$ of the length μ antiparallel to $\mathbf{P}_i\mathbf{P}_{i-1}$. It is a reason of the stochasticity (uncertainty) of world tubes of free particles in the deterministic DST.

Thus, if the spacetime is distorted, world tubes of free pointlike particles are random [3]. Statistical description of stochastic world tubes of free non–relativistic particles is equivalent to the Schrödinger equation, provided at large σ_M the distortion function has the asymptotical form $D(\sigma_M) = \hbar/(2bc)$, where b is the universal constant, $\hbar$ is the Planck constant, and c is the speed of the light. Thus, at this approach the Planck constant describes the spacetime distortion, i.e. properties of the spacetime. Stochasticity of a world tube of a free particle in the **deterministic** DST is connected with an existence of many unit vectors $\mathbf{P}_0\mathbf{Q}$ parallel to a given vector $\mathbf{P}_0\mathbf{P}$.

Now let us define what does the term "statistical description" mean. Let $\mathcal{S}$ be a stochastic physical system. It means that experiments with a single system $\mathcal{S}$ are irreproducible. There are no dynamic equations for a single stochastic system. In order to investigate the stochastic system $\mathcal{S}$, let us consider a set $\mathcal{E}$ (statistical ensemble) of many like independent stochastic systems $\mathcal{S}$. Experiments with the statistical ensemble $\mathcal{E}$ are reproducible, and $\mathcal{E}$ can be considered as a deterministic dynamic system, associated with the stochastic system $\mathcal{S}$. There exist dynamic equations for $\mathcal{E}$. By definition the statistical ensemble $\mathcal{E}$ of stochastic systems $\mathcal{S}$ carries out a statistical description of the stochastic systems $\mathcal{S}$. In other words, the statistical description is a way of reduction of a stochastic system $\mathcal{S}$ to a deterministic system $\mathcal{E}$.

The main concept of the statistical description is a deterministic dynamic system called the statistical ensemble (not the probability density, or probability amplitude). Such a statistical description will be referred to as the statistical ensemble technique (SET). It is the most general way of the statistical description. The classical statistical physics and the quantum mechanics are special cases of SET, when the statistical ensemble is described respectively in terms of a probability density, or a probability amplitude.

Finally, any statistical description substitutes "a dynamics of stochastic systems" by a dynamics of the statistical ensemble. SET carries out a connection between the stochastic system and the statistical ensemble directly and independently of the statistical ensemble interpretation, i.e. it is of no importance what rules are used for calculation of average values.

The statistical physics and quantum mechanics associate a stochastic system with a statistical ensemble by means of intermediate concepts (respectively probability density and probability amplitude) which are connected with an interpretation of the statistical ensemble, i.e. with the rules of calculation of average values. These rules are different for classical and quantum stochastic systems and have nothing to do with the dynamics. A usage of these intermediate concepts leads to a conceptual distinction between the dynamics of classical systems and the dynamics of quantum systems. As a result many people consider that all that is not quantum is classical and vice versa. In reality, an existence of such a conceptual boundary is not necessary, and not all that is not quantum is necessarily classical.

In application to stochastic world tubes the SET means a construction of a statistical ensemble $\mathcal{E}$ consisting of stochastic world tubes. In the Minkowski spacetime M the statistical ensemble $\mathcal{E}$ of free classical deterministic particles of the mass m is described by the action

$$\mathcal{A}[\mathbf{x}] = -\int mc\sqrt{1 - c^{-2}(d\mathbf{x}/dt)^2}dt d\boldsymbol{\xi} \ , \tag{7}$$

$$\mathbf{x} = \mathbf{x}(t, \boldsymbol{\xi}), \qquad \mathbf{x} = \{x^1, x^2, x^3\} \ , \qquad \boldsymbol{\xi} = \{\xi^1, \xi^2, \xi^3\} \tag{8}$$

where $t, \mathbf{x}$ are coordinates of a particle in M, and $\boldsymbol{\xi}$ are Lagrangian coordinates labeling the ensemble particles. Using Eulerian coordinates $t, \mathbf{x}$ as independent variables instead of $t, \boldsymbol{\xi}$, the action (7) reduces to the form

$$\mathcal{A}[\rho, \mathbf{j}, \xi_0, \boldsymbol{\xi}] = \int \{-\sqrt{m^2c^2j^ij_i} - j^i\partial_i\xi_0 - j^ig^\alpha(\boldsymbol{\xi})\partial_i\xi_\alpha\}dtd\mathbf{x} \tag{9}$$

where $j^i = \{\rho, \mathbf{j}\}$ is 4–flux of particles, $g^\alpha(\boldsymbol{\xi})$, $\alpha = 1, 2, 3$ are arbitrary functions of their arguments, ξ_0 is the action variable (temporal Lagrangian coordinate), and $j_i = g_{ik}j^k$, g_{ik}=diag$\{c^2, -1, -1, -1\}$. Indeed, the Jacobians

$$j^i = \frac{\partial(x^i, \xi_1, \xi_2, \xi_3)}{\partial(x^0, x^1, x^2, x^3)} \ , \qquad i = 0, 1, 2, 3 \tag{10}$$

satisfy identically dynamic equations $\delta\mathcal{A}/\delta\xi_i = 0$, $i = 0, 1, 2, 3$. Adding relations (10) to the action (9) as additional constraints, one does not change the variational problem. Then eliminating j^i from Eq.(9) by means of Eq.(10) and reversing relations $\boldsymbol{\xi} = \boldsymbol{\xi}(t, \mathbf{x})$, one obtains Eq.(7) with $\mathbf{x} = \mathbf{x}(t, \boldsymbol{\xi})$. Let us note that the variational problems (7) and (9) are equivalent provided m =const. If m depends on the ensemble state, the equivalence of Eqs. (7) and (9) violates.

Let us imagine that the mass m in Eq.(9) depends on the ensemble state derivatives $\partial_i\rho, \partial_i\mathbf{j}$. Then the action (9) will describe an ensemble of stochastic particles, and the action (9) cannot reduce to Eq.(9). In particular, replacing m^2 in Eq.(9) by

$$m_q^2 = m^2 + \left(\frac{\hbar}{2c}\nabla\log\rho\right)^2 , \qquad m = \text{const} \tag{11}$$

setting $g^\alpha(\boldsymbol{\xi}) = 0, \quad \alpha = 1,2,3$ and using non–relativistic approximation $|\, \mathbf{j}/c \,| \ll \rho$, the action (9) reduces to the form

$$\mathcal{A}_q[\rho,\mathbf{j},\varphi] = \int \left\{\frac{\mathbf{j}^2}{2m\rho} - \rho\partial_t\varphi - \mathbf{j}\nabla\varphi - \frac{\hbar^2}{8m}\rho(\nabla\log\rho)^2\right\} dtd\mathbf{x} , \tag{12}$$

$$\varphi = \xi_0 - mc^2 t \ .$$

Substitution

$$\rho = \psi^*\psi , \qquad \mathbf{j} = -\frac{i\hbar}{2m}(\psi^*\nabla\psi - \nabla\psi^*\psi) , \qquad \psi = \sqrt{\rho}\exp\left(i\varphi/\hbar\right) \tag{13}$$

transforms this action into the action for the Schrödinger equation

$$\mathbf{A}_S[\psi,\psi] = \int \left[\frac{i\hbar}{2}(\psi^*\frac{\partial\psi}{\partial t} - \frac{\partial\psi^*}{\partial t}\psi) - \frac{\hbar^2}{2m}\nabla\psi^*\nabla\psi\right] d^4x \tag{14}$$

Thus the dependence (11) of the mass on the ensemble state ρ leads in the non–relativistic approximation to the Schrödinger equation.

Let us define the mass in the action (9) by means of the relation

$$m = b < \mu_{oM}(P) > , \tag{15}$$

where the mean oriented mass $< \mu_{oM} >$ is the projection of the momentum $p = \mathbf{P}_i\mathbf{P}_{i+1}$ onto the direction of the mean velocity $< u(P) >=< \mathbf{P}_i\mathbf{P} \mid \mathbf{P}_i\mathbf{P} \mid_M^{-1} >$ $|< \mathbf{P}_i\mathbf{P} \mid \mathbf{P}_i\mathbf{P} \mid_M^{-1} >|_M^{-1}$, b is the universal constant. Index "M" means that the corresponding scalar product and the module are calculated by means of the world function σ_M, i.e. in the Minkowski space associated with DST by Eq.(4), whereas segments $\mathcal{T}_{[P_iP_{I+1}]}$ of the broken tube are calculated by means of the world function σ_D of DST. Angular brackets mean averaging over the ensemble. Vectors $\mathbf{P}_i\mathbf{P}_{i+1}$ and $\mathbf{P}_i\mathbf{P} \mid \mathbf{P}_i\mathbf{P} \mid_M^{-1}, \quad P \in \mathcal{T}_{[P_iP_{I+1}]}$ are parallel in DST, but they are not parallel in the associated Minkowski space M. If the distortion vanishes $D(\sigma_M) = 0$, the mass μ_{oM} does not depend on the point P, and m=const. If $D(\sigma_M) \neq 0$, the momentum $\mathbf{P}_i\mathbf{P}_{i+1}$ and the 4-velocity $\mathbf{P}_i\mathbf{P} \mid \mathbf{P}_i\mathbf{P} \mid_M^{-1}$ are not parallel in M. Then the mass $< \mu_{oM} >= p_k < u^k(P) >$ depends on the ensemble state.

In the coordinate system, where spatial components of the 4–momentum vanish

$$p_i = \{\mu_M, 0\} \ , \quad < u^i >= \{(1-\beta^2)^{-1/2}, \beta(1-\beta^2)^{-1/2}\}, \quad \mu_M =\mid \mathbf{P}_i\mathbf{P}_{i+1} \mid_M \tag{16}$$

one has

$$< \mu_{oM} >= p_i < u^i >= \mu_M(1-\beta^2)^{-1/2} \simeq \mu_M(1+\beta^2/2) \ , \qquad \beta \ll 1 \ . \tag{17}$$

Here βc is the mean 3–velocity in the coordinate system, where $\mathbf{p} = 0$. ϑ=arctanhβ is the angle between vectors p_i and $< u^i >$ in M.

If the mean 3–velocity βc is determined by the relation

$$\beta c = -\frac{\hbar}{2b\mu_M}\nabla \log \rho \tag{18}$$

then the mass m defined by relations (15),(17),(18) coincides with m_q and in the non–relativistic approximation the action (9) for the ensemble of stochastic world tubes takes the form (12). In this case the statistical description of stochastic world tubes coincides with the quantum mechanic description. What form must have the distortion function $D(\sigma_M)$, in order that the relation (18) takes place? The answer is such [3]

$$\sigma_D = \sigma_M + \sigma_0/2 \ , \qquad D(\sigma_M) = \sigma_0/2 \ , \qquad \text{for} \qquad \sigma_M > \mu_e = m_e/b \ , \tag{19}$$

where $\sigma_0 = \hbar/bc$ =const and m_e is the mass of the lightest massive particle, i.e. m_e is the electron mass.

This result is obtained as follows. Let us use notations

$$\begin{gathered} \mu_D = [2\sigma_D(P_0, P_1)]^{1/2} \ , \qquad \mu_M = [2\sigma_M(P_0, P_1)]^{1/2} \\ \mu_M = g(\mu_D), \qquad \mu_D = g^{-1}(\mu_M) = [\mu_M^2 + 2D(\mu_M^2/2)]^{1/2} \ , \end{gathered} \tag{20}$$

where D is the distortion function defined by Eq.(4). The radius $R_M(\mu_M)$ of the tube segment $\mathcal{T}_{[P_0P_1]}$ measured at the point P in M can be presented in the form

$$R_M(\mu_D) = [g^2(\mu_D)/4 - g^2(\mu_D/2)]^{1/2} \ . \tag{21}$$

The point P is located at the middle of the intercept P_0P_1. The mean 3–velocity at the point P is determined by the contributions of adjacent world tubes. Any tube at the point P' contributes the 3–velocity $\beta c = R_M/\mu_M$ in the direction $\mathbf{P'P}$ with the weight $\rho(P')$. Averaging over all directions, one obtains the following result

$$\beta c = c\vartheta = -\frac{2}{3}\frac{R_M^2}{\mu_M}\frac{\nabla \rho}{\rho} \ . \tag{22}$$

Comparison of Eqs. (18) and (22) leads to a functional equation $R_M^2(\mu_D) = 3\sigma_0/4$ for the function g which has the solution $g^2(\mu_D) = \mu_D^2 - \sigma_0$. It corresponds to the result (19).

The quantity $(\sigma_0)^{1/2}$ is of the order of the tube radius and can be treated as an elementary length. Its value can be estimated on the base of the supposition that $\mu_M^2 > 0$ for any $\mu_D^2 \geq \mu_e^2 = m_e^2/b^2$. Setting $\mu_D = \mu_e$ and using that $\mu_D^2 < \sigma_0 = \hbar/bc$, one obtains $b < m_e^2 c/\hbar \simeq 10^{-16.5} g/cm$, $(\sigma_0)^{1/2} > \hbar/m_e c \simeq 10^{-10.5} cm$. If it appears that the distortion can explain quantum effects in other cases, but not only for free particles, one does not need the quantum principles, and the spacetime distortion can be treated as a reason of quantum effects.

Let us apply SET to the dynamic system $\mathcal{S}_P$ described by the Pauli equation for the non-relativistic electron. The action has the form

$$\mathbf{A}_P[\psi,\psi] = \int \left[\frac{i\hbar}{2}(\psi^* \frac{\partial\psi}{\partial t} - \frac{\partial\psi^*}{\partial t}\psi) - \frac{\hbar^2}{2m}\nabla\psi^*\nabla\psi\right] d^4x \tag{23}$$

$$\psi = \binom{\psi_1}{\psi_2}\ , \qquad \psi^* = (\psi_1^*, \psi_2^*)$$

with $j^0 = \rho$ and $\mathbf{j}$ defined by the equation

$$j^0 = \rho = \psi^*\psi\ , \qquad \mathbf{j} = -\frac{i\hbar}{2m}(\psi^*\nabla\psi - \nabla\psi^*\psi)\ .$$

By means of the transformation

$$\psi = \binom{\psi_1}{\psi_2} = \rho^{1/2}\exp(i\varphi)\boldsymbol{\sigma\eta}\binom{\chi_1}{\chi_2}\ , \qquad \chi = \binom{\chi_1}{\chi_2} = \text{ const }\ ,$$

$$\chi^*\chi = 1, \qquad \mathbf{b} = \chi^*\boldsymbol{\sigma}\chi\ , \qquad \boldsymbol{\eta} = \frac{\boldsymbol{\xi}}{|\boldsymbol{\xi}|} \tag{25}$$

the action for $\mathcal{S}_P$ reduces to the form

$$\mathcal{A}_P[\rho,\mathbf{j},\varphi,\boldsymbol{\xi}] = \mathcal{A}_{Pc}[\rho,\mathbf{j},\varphi,\boldsymbol{\xi}] + \mathcal{A}_{Pq}[\rho,\mathbf{j},\varphi,\boldsymbol{\xi}] \tag{26}$$

$$\mathcal{A}_{Pc}[\rho,\mathbf{j},\varphi,\boldsymbol{\xi}] = \int (\frac{m}{2\rho}\mathbf{j}^2 - \hbar j^i\partial_i\varphi - \hbar j^i\varepsilon_{\alpha\beta\gamma}\eta_\alpha\partial_i\eta_\beta b_\gamma)d^4x, \qquad j^0 \equiv \rho, \tag{27}$$

$$\mathcal{A}_{Pq}[\rho,\mathbf{j},\varphi,\boldsymbol{\xi}] = \int \frac{\hbar^2}{8m}\rho[4(\varepsilon_{\alpha\beta\gamma}\eta_\alpha\nabla\eta_\beta b_\gamma)^2 - 4(\partial_\alpha\eta_\beta)(\partial_\alpha\eta_\beta) - \frac{(\nabla\rho)^2}{\rho^2}]d^4x \tag{28}$$

where $\varepsilon_{\alpha\beta\gamma}$ is the Levi–Civita pseudo–tensor, $\mathbf{b}$ is a constant unit vector, $\boldsymbol{\sigma}$ are the Pauli matrices

$$\mathbf{s} = 2\boldsymbol{\eta}(\mathbf{b}\boldsymbol{\eta}) - \mathbf{b}\ , \qquad \boldsymbol{\eta} = \boldsymbol{\xi}/\,|\,\boldsymbol{\xi}\,| = (\mathbf{b}+\mathbf{s})[2(1+\mathbf{bs})]^{-1/2} \tag{29}$$

$$\mathbf{s} = \psi^*\boldsymbol{\sigma}\psi/(\psi^*\psi)\ ,$$

Eq.(27) describes a classical part of the action (26) which can reduce to the form

$$\mathcal{A}_{cl}[\mathbf{x}] = \int \frac{m}{2}\left(\frac{d\mathbf{x}}{dt}\right)^2 dt d\boldsymbol{\xi}\ , \tag{30}$$

The action (30) describes a statistical ensemble $\mathcal{E}_{Pc}$ of independent classical pointlike particles. It means that the action (26) describes the statistical ensemble of quantum particles. Indeed, supposing that the wave function ψ and the action $\mathcal{A}_P$ describe a single quantum electron, one cannot explain, why in the quasi–uniform limit

$$\mid \frac{\hbar\rho}{m\mid \mathbf{j}\mid}\nabla \log u \mid \ll 1, \qquad u=\rho,\eta_1,\eta_2,\eta_3 \ , \tag{31}$$

when one can neglect $\mathcal{A}_{Pq}$ as compared with $\mathcal{A}_{Pc}$, the action $\mathcal{A}_P = \mathcal{A}_{Pc}$ describes a classical statistical ensemble, but not a single classical particle. The statement that the dynamic system $\mathcal{S}_P$ is a statistical ensemble is the crucial point, where statistics penetrates into quantum dynamics.

The action for the ensemble of spinless particles described by the Schrödinger equation has the form (12). The classical part $\mathcal{A}_{Sc}$ of the spinless particle action (12) distinguishes from Eq.(27) by the term depending on the spin variables $\boldsymbol{\eta}$ and responsible for vorticity of the vector field $\mathbf{v} = \mathbf{j}/\rho$.

$$\nabla \times \mathbf{v} = \nabla \times \frac{\mathbf{j}}{\rho} = \frac{\hbar}{m}\varepsilon_{\alpha\beta\gamma}\nabla\eta_\alpha \times \nabla\eta_\beta b_\gamma \tag{32}$$

Both $\mathcal{A}_{Pc}$ and $\mathcal{A}_{Sc}$ can reduce to the form (30). It means that both classical ensembles $\mathcal{A}_{Pc}$ and $\mathcal{A}_{Sc}$ consist of the same pointlike particles. At the same time the actions $\mathcal{A}_{Pc}$ and $\mathcal{A}_{Sc}$ distinguish by the term containing the spin variable $\boldsymbol{\eta}$. Then one concludes that the electron spin has a classical analog which can be interpreted as a vorticity of the electron ensemble flow. The electron spin is a collective (statistical) property of the electron ensemble (not of a single electron).

Important remark. Electron spin as a statistical property cannot be understood from the standpoint of quantum axiomatics, because the conventional quantum dynamics does not contain any mention about statistical description. (For the same reason the temperature as a statistical property cannot be understood from the standpoint of axiomatic thermodynamics. It can be understood only from the standpoint of statistical physics.)

References

1. Yu.A. Rylov, J. Math. Phys. **31**, 2876 (1990).
2. Yu.A. Rylov, J. Math. Phys. **33**, 4220 (1992).
3. Yu.A. Rylov, J. Math. Phys. **32**, 2092 (1991).

Divergences of the Semiclassical S–matrix Beyond Hyperbolic Systems

Krzysztof Stefański

Institute of Astronomy, Nicholas Copernicus University
PL–87–100 Toruń, Poland

Abstract: It is shown that the absolute convergence of the semiclassical S–matrix formula in the case of irregular inelastic scattering depends only on the fractal dimension of the corresponding singular set of the scattering function, and the critical fractal dimension is derived. Ways of handling divergences are discussed.

Computing probabilities for collision–induced vibrational and/or rotational transitions in molecules belongs to most important problems in molecular physics and theoretical chemistry. Such a probability for transition between molecular states described by strings of quantum numbers $\mathbf{n}_i$ and $\mathbf{n}_f$ can be obtained from an appropriate element of the scattering matrix, or S–matrix (cf. ref [1])

$$P_{\mathbf{n}_f\mathbf{n}_i} = \left|S_{\mathbf{n}_f\mathbf{n}_i}\right|^2 . \tag{1}$$

If, as quite often happens (cf. refs [1-6], computations are limited to a system consisting of an atom and a diatomic molecule in a linear configuration, the internal state of the molecule is described by a single vibrational quantum number v.

Although at present one can relatively easily obtain S–matrix for such systems using fully quantum methods (see e.g. ref. [7]), a semiclassical approximation due to Miller [1] may be still of essential use if one is interested in obtaining a transition probability for a particular transition without necessity of computing all irrelevant S–matrix elements. Miller's formula may be also useful for "explaining" unusual properties of transition probabilities obtained by quantum methods. An element of the semiclassical S–matrix (ref. [1]) for a pair of molecular vibrational states, represented by quantized virational actions $J_i^Q = (v_i + 1/2)\hbar$ and $J_f^Q = (v_f + 1/2)\hbar$ is given as a sum

$$S_{J_f^Q J_i^Q} = \sum_{k=1}^{N} A_k \exp(i\tilde{\phi}_k) , \tag{2}$$

of contributions from N scattering trajectories fulfilling the condition

$$J_f(\varphi_i^k; E, J_i^Q) = J_f^Q , \tag{3}$$

where $J_f(\varphi_i; E, J_i^Q)$ is an excitation profile, which is a special type of the scattering function for inelastic scattering, for an assumed total energy E and the quantized initial vibrational action J_i^Q. Here the vibrational molecular angle φ_i, which is a canonical conjugate of the vibrational action J_i plays a role of the variable scattering parameter.

The phases $\tilde{\phi}_k$ in Eq.(2) are obtained from an action variable computed along the corresponding scattering trajectories. Although they are of crucial importance for efficiency of the formula in particular applications, for the problem to be considered here one can neglect their particular form and focus ones attention on the preexponential factors given by the formula

$$A_k = \left[2\pi\left|\frac{\partial J_f}{\partial \varphi_i}(\varphi_i^k)\right|\right]^{-\frac{1}{2}} . \tag{4}$$

If the number N of contributing trajectories is finite the only difficulty with applying the formula can follow from the trajectories corresponding to small vicinities of extrema of the excitation profile, and the problem can be removed if the primitive stationary phase approximation used in derivation of Eq.(2) is replaced by the uniform approximation (ref. [8]).

An essentially different problem may arise if the number of contributing trajectories is infinite as it happens in the case of the so called irregular scattering discovered by Rankin and Miller [2], and investigated for the same class of systems, i.e. inelastic and/or reactive scattering of an atom and a diatom in linear configuration, by a number of groups [3-6]. Recent results [4,6] of these investigations have shown that the excitation profile in the irregular interval $\Delta\varphi_{irr}$ of the initial vibrational angle consists of smooth (often almost parabolic) curves called icicles [4] hanging–down from the complementary fractal set of singular points at $J_f = J^{Max}$, corresponding to a marginal bonding of the triatomic system.

In such a case the semiclassical S–matrix formula is represented by an infinite series whose absolute convergence will be the main topic of our consideration. The absolute convergence of the infinite N version of the sum (1) is equivalent to the convergence of the series of the preexponential factors

$$S^{abs} = \sum_{k=1}^{\infty} A_k . \tag{5}$$

Since the preexponential factor for an individual contribution is given by Eq. (4), the problem of convergence of the series (5) can be reduced to the problem of convergence of the series

$$\bar{S}^{abs} = \sum_{k=1}^{\infty} \sqrt{\delta\varphi_k} , \tag{6}$$

where $\delta\varphi_k$ is the width of a k-th smooth subinterval, if only for certain $0 < \gamma \leq \Gamma < \infty$ the condition

$$\gamma\sqrt{\delta\varphi_{k'}} < A_{k'} < \Gamma\sqrt{\delta\varphi_{k'}} \tag{7}$$

holds for a typical subset of the smooth subintervals of the excitation profile in the irregular region. The available data from computer experiments show no evidence of falseness of the above assumption (cf. refs. [4-6]).

To formulate a criterion for the convergence of the series (6) one can notice that it is a special case for $D = 1/2$ of a more general series

$$\Sigma(D) \equiv \sum_{k}^{\infty} (\delta\varphi_k)^D . \tag{8}$$

For the latter family, due to its monotonicity in D and the obvious conditions

$$\Sigma(1) = \Delta\varphi_{irr} < \infty \quad \text{and} \quad \Sigma(0) = \infty , \tag{9}$$

a unique value $0 < D_C < 1$ of D always exists such that

$$\Sigma(D) < \infty \text{ for } D > D_C \text{ and} \tag{10}$$

$$\Sigma(D) = \infty \text{ for } D < D_C . \tag{11}$$

Therefore, the convergence criterion for the series (6) reads as

$$D_C < \frac{1}{2} . \tag{12}$$

Since strong evidences exist (Ref.[9]) that the convergence dimension D_C is identical with the Hausdorff dimension D_H of the singular set in φ, the convergence condition should finally read

$$D_H < \frac{1}{2} . \tag{13}$$

This criterion is essentially identical with the absolute convergence criterion recently obtained by Jensen [10] for simple models of one and two–dimensional elastic scattering. Despite the identity of the final results, the approach applied here differs from that of Jensen. It seems that the complexity of the smooth scattering systems, to which we refer, requires a sort of phenomenological approach presented above.

The problem of absolute divergence of a semiclassical series is dealt with for some time in bounded hyperbolic systems in the context of the Gutzwiller's trace formula giving a formal semiclassical approximation to the density of eigenenergies (see e.g. Berry [11]). Thus, looking from a very general point of view, the convergence criterion for the semiclassical S–matrix elements may look as bringing little new. However a few reasons allow one to argue for the opposite point of view.

First of all, little of the machinery used in hyperbolic systems can be directly applied in smooth scattering systems and this means that more specific tricks are necessary to obtain the convergence criterion explicitly. What's more, the approach presented above shows that the convergence criterion itself is absolutely independent of details of dynamics, and the latter decide only whether for a particular system the criterion is fulfilled or not. Second, the scattering convergence criterion shows that the transient chaoticity of the classical scattering need not interfere with applicability of the relevant semiclassical formula if the corresponding fractal singular set is sufficiently sparse (i.e. low–dimensional). And finally the methods of dealing with the cases, when the S–matrix formula is divergent, may and should be treated in a different way than the divergent trace formula for the density of states.

Contrary to the semiclassical density of eigenstates, the semiclassical S–matrix is a sort of semiproduct which is used in computing a physically interesting variable – the transition probability. This may be a very helpful circumstance. It might be virtually impossible to apply in smooth scattering systems such countermeasures for divergence as systematic resummation of the series or Lorentzian broadening of the energy resolution resulting in exponential quenching of small individual contributions [11]. On the other hand, one can use much simpler tricks thanks to the fact that ultimately the transition probabilities and not the S–matrix elements are to be approximated. Consequently, the series can be truncated at N contributions coming from the broadest icicles but upon computing the corresponding transition probability one should take into account the correction coming from the remaining terms, which can be written (omitting all indices) as

$$P \simeq \Big|\sum_{k=1}^{N} A_k \exp(i\tilde{\phi}_k)\Big|^2 + \sum_{k=N+1}^{\infty} \big(A_k\big)^2 \simeq \Big|\sum_{k=1}^{N} A_k \exp(i\tilde{\phi}_k)\Big|^2 + \bar{\gamma}^2 \,\Delta\varphi^{(N)} \tag{14}$$

Here $\bar{\gamma}$ denotes an average ratio between the inverse of the local derivative $|dJ/d\varphi|$ and the square root of the corresponding width of the smooth subinterval $\delta\varphi_k$. $\Delta\varphi^{(N)}$ denotes the total width of the irregular subintervals comprising all icicles except for those contributing with the N scattering trajectories to the truncated S–matrix element.

This seems to be a practically applicable prescription for dealing with divergence of the semiclassical S–matrix in smooth scattering systems but the price must be paid for it as ever. Such a transition probability should be treated as an approximation to an energy averaged quantum transition probability and not to an energy shell transition probability. In this respect a partial incompatibility of the quantum and classical description, faced in bounded hyperbolic systems, seems to extend on sufficiently strong classical irregularities in scattering. An extensive discussion of different aspects of this problem and other problems mentioned before can be found in ref. [9].

References

1. W.H. Miller, Adv. Chem. Phys. **25**, 69 (1974).
2. C.C. Rankin and W.H. Miller, J. Chem. Phys. **55**, 3150 (1971).
3. L. Gottdiener, Mol. Phys. **29**, 1585 (1975).
4. D.A. Noid, S.K. Gray and S.A. Rice, J. Chem. Phys. **84**, 2649 (1986).
5. R.T. Skodje, J. Chem. Phys. **90**, 6193 (1989).
6. K. Someda, R. Ramaswamy and H. Nakamura, J. Chem. Phys. **98**, 1156 (1993).

7. J.M. Bowman (ed.) *Advances in Molecular Vibrations and Collision Dynamics: Quantum Reactive Scattering*, JAI Press, Greenwich 1993.
8. S. Levit and U. Smilansky, Ann. Phys. **108**, 165 (1977).
9. K. Stefański, K. Someda and H. Nakamura, *Divergencies of the Classical S-Matrix Formula in Irregular Scattering* – to be published.
10. J.H. Jensen, Phys. Rev. Lett. **73**, 244 (1994).
11. M.V. Berry, in: *Lectures at Les Houches 1989, Session LII, Chaos and Quantum Physics* (eds. M.-J. Giannoni, A. Voros, J. Zinn–Justin), North Holland, Amsterdam 1991, p. 251.

Disturbance Propagation in Coupled Map Lattices

Alessandro Torcini

Theoretische Physik, Bergische Universität–Gesamthochschule Wuppertal,
D–42097 Wuppertal, Germany

Abstract: Propagation velocities of disturbances are analyzed in chaotic spatially extended systems with local and long–range couplings. For nearest–neighbours interactions two distinct speed selection mechanisms for propagating fronts are found. One mechanism can be interpreted within a linear–stability analysis, while the origin of the second one is fully non–linear. For long–range couplings, decaying as powers of the distance, the spreading–rate of disturbances increase exponentially in time.

1 Introduction

Propagation of fronts in spatially extended systems occurs in several fields of science ranging from fluid dynamics to liquid crystals, from the study of epidemics to chemical reactions [1, 2]. Several mathematical models, reproducing the propagation of fronts into stable (or metastable) steady states, have been widely studied [3, 4]. In the present contribution we will focus our attention on the analysis of disturbances spreading not into unstable steady states but in chaotic systems, when local and non–local spatial couplings are considered. Simple one–dimensional models that turn out to give quite useful results for such studies are represented by the so–called Coupled Map Lattices (CML) models [5]. In the case of nearest–neighbours coupling the CML model can be written as

$$x_i^{n+1} = f(\tilde{x}_i^n) \tag{1}$$

$$\tilde{x}_i^n = x_i^n + \frac{\varepsilon}{2}\left(x_{i+1}^n + x_{i-1}^n - 2x_i^n\right) \tag{2}$$

where ε is the diffusive coupling constant and the indices i and n label the spatial and time discrete variables, respectively.

The function $f(x)$ appearing in (1) is usually assumed to be a chaotic map of the interval, or of the plane, into itself. Within the same class of models a long range interaction can be mimicked replacing the expression (2) for $\tilde{x}_i^n$ with [6]

$$\tilde{x}_i^n = (1-2\varepsilon_s)x_i^n + \frac{1}{2}\left[\sum_{k=0}^{\mathrm{Log}_2(L)-1} \varepsilon_k (x_{i+2^k}^n + x_{i-2^k}^n)\right] \tag{3}$$

where $\varepsilon_s = \sum_{k=1}^{(L-1)/2} \varepsilon_k$ and in the particular case of a coupling decaying with an α power–law

$$\varepsilon_0 = c_1 \;; \qquad \varepsilon_k = \frac{c_2}{(2^k)^\alpha} \qquad \text{for} \qquad k \geq 1 \tag{4}$$

In order to obtain a more efficient scheme for computer simulations, the coupling in eq. (3) has been restricted to sites whose distances from the reference point are a power of 2. We have verified that this limitation does not influence the main aspects of the dynamics of these systems, which instead are strongly affected by the decay–law of the interactions. For both the reported types of coupling, periodic boundary conditions are assumed along a chain of lenght L ($x_i^n = x_{i\pm L}^n$).

In the next sections we will consider the propagation of disturbances arising once a single site (or a finite part) of a chaotic chain is initially perturbed. In the local–coupling case, two distinct regimes have been found to characterize the dynamics of front propagation [7]. Namely, in the first regime the velocity of the propagating front v_{L} can be recovered through a linear–stability analysis, while in the second situation the value of the velocity is determinated by the full nonlinear evolution of the system. For some particular class of 1-d maps, a transition between the two regimes has been observed when a parameter of the map is continuously varied [7]. When a long–range interaction is considered even the definition of a velocity can be quite non–obvious [2]. However, we will show that for the model described by eqs. (3) and (4) the spreading–rate of disturbances can be straightforwardly related to the maximal Lyapunov exponent.

2 Local Coupling

In this section we will limit our analysis to CML with nearest–neighbours coupling. For such models the propagation of fronts, representing interfaces between stationary (stable or unstable) states and a fully chaotic region, has been previously studied in Ref. [8]. We will instead consider disturbances moving along a chaotic chain.

A direct measurement of the front propagation speed v_{F} can be obtained considering two configurations of the lattice $\{x_i^n\}$ and $\{y_i^n\}$ which initially differ only in one site ($|x_0^0 - y_0^0| = \delta_0$) and coincide elsewhere. Starting from such initial states N successive iterations are performed, then we will determine the spatial interval $[-R_-(N), R_+(N)]$, where

$$R_+(N) = max\{i : |x_i^n - y_i^n| \geq \theta\}$$

$$\text{and} \tag{5}$$

$$R_-(N) = min\{i : |x_i^n - y_i^n| \geq \theta\} \;,$$

with $n \leq N$ and θ indicating a preassigned threshold. The front velocities are then defined as

$$v_{\mathrm{F}}^{\pm} = \lim_{N\to\infty} \frac{R_{\pm}(N)}{N} \tag{6}$$

For symmetric couplings, as considered here, $v_{\mathrm{F}}^{+} = v_{\mathrm{F}}^{-} = v_{\mathrm{F}}$. We have verified that the velocity v_{F} depends neither on the amplitude of the initial perturbation nor on the value of the threshold θ (varying the values of θ and/or δ_0 from 10^{-14} to 0.1 no appreciable variation of v_{F} have been observed).

Another velocity that is meaningful to define is the limit velocity v_{L} of propagation of disturbances within the *linear* approximation. Now we will consider the spreading of an infinitesimal perturbation associated no more with eqs. (1) and (2) but with their linearized expression

$$\delta x_i^{n+1} = f'(\tilde{x}_i^n)\left[(1-\varepsilon)\delta x_i^n + \frac{\varepsilon}{2}\left(\delta x_{i+1}^n + \delta x_{i-1}^n\right)\right] \tag{7}$$

where f' represents the derivative of the considered map. In the linear approximation, the time evolution of an initially localized disturbance $|\delta x_{i=0}^{n=0}| = \delta_0$ in a reference frame moving with velocity $v = i/n$ is given asymptotically by

$$\delta x_i^n = \delta_0 \mathrm{e}^{\Lambda(v)n} \tag{8}$$

where $\Lambda(v)$ represent the so–called maximal comoving Lyapunov exponent [9]. Within this scheme v_{L} corresponds to the limit velocity for which a disturbance can still propagate, i.e.

$$\Lambda(v_{\mathrm{L}}) = 0 \tag{9}$$

It is possible to show that $\Lambda(v)$ is concave (see Ref. [10]), hence eq. (9) has only two solutions corresponding to the forward and backward fronts. For coupled logistic, tent or cubic map we have found numerically that the fronts always propagate with velocity $v_{\mathrm{F}} = v_{\mathrm{L}}$ [11]. However, for other kinds of maps we have observed that the front speed can be higher than v_{L} [7, 11]. This phenomenon is found even in the simple piecewise linear map

$$f(x) = ax \quad \mathrm{mod} 1 \tag{10}$$

for suitable values of the coupling ε and of the map parameter a. For CML of the type (10), the measured front velocity v_{F} together with the linear velocity v_{L} are reported in Fig. 1 (for $\varepsilon = 1/3$ and $1 \leq a \leq 2$). It is evident from the figure that $v_{\mathrm{F}} \geq v_{\mathrm{L}}$ for any reported value of the parameter a.

This indicates that in some case, even if the associated maximal comoving Lyapunov exponent is negative, the disturbance can still propagate. Therefore, a linear stability analysis is not sufficient to indicate which front propagation speed will be selected. In the "usual" situation ($v_{\mathrm{F}} = v_{\mathrm{L}}$) the local degree of chaoticity pulls the leading–part of the front with a velocity higher then that associated to the non–linear mechanism of information production. In the "anomalous" case ($v_{\mathrm{F}} > v_{\mathrm{L}}$) the non–linear aspects prevail on the linear one (represented by the

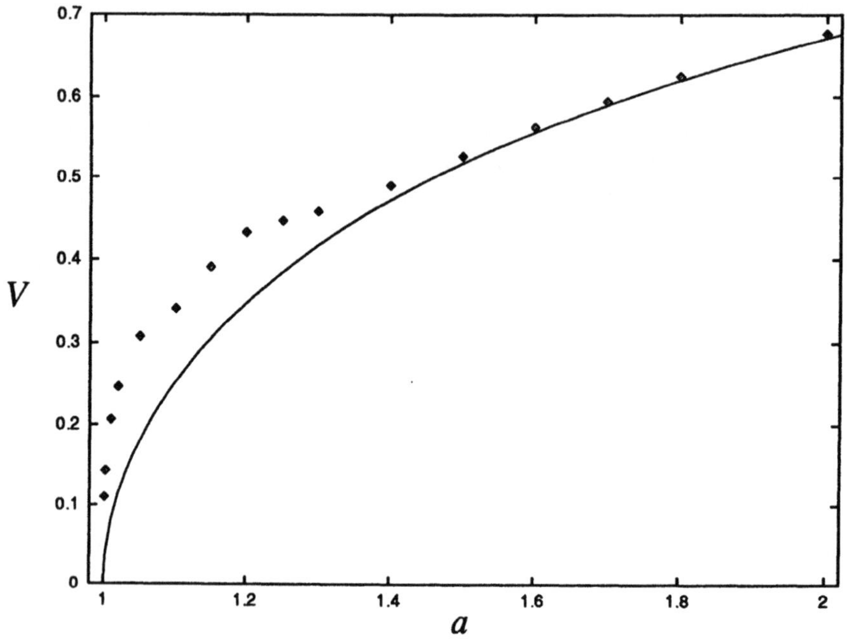

Fig. 1. Measured front velocities v_F (symbols) and linear velocities v_L (solid line) as a function of the parameter a for the piecewise linear map (10). The diffusive coupling is assumed to be $\varepsilon = 1/3$.

maximal comoving Lyapunov exponent) and are able to push the front with a velocity higher than v_L.

Analogous results have been found for front propagations into unstable (or metastable) steady states for the one dimensional nonlinear diffusion equation [4]

$$u_t = u_{xx} + g(u) \tag{11}$$

where $g(u) \in C^1[0,1]$, $g(0) = g(1) = 0$. If $g > 0$ in $(0,1)$, then $u = 0$ is an unstable fixed point, while $u = 1$ is the stable one. In this case, any localized initial condition $u(x, t = 0) = \delta(x)$ will generate a propagating front joining the unstable to the stable state. Depending on the function $g(u)$, the speed of the front can be selected by the behaviour of $g(u)$ at $u = 0$, i.e. by infinitesimal deviations from the unperturbed state, or by the global behaviour of $g(u)$. The former corresponds to our v_F, while the latter is analogous to v_L. Even if there are obvious differences between our system and those considered in [3, 4], it seems that the speed selection mechanisms are surprisingly similar. A more complete and comprehensive study will be reported elsewhere [11].

3 Power–Law Coupling

Long–range interactions arise naturally in epidemic models [2]. In particular for epidemic models in one dimension it has been observed that the condition to obtain a finite propagation speed for the epidemic is that the spatial coupling should decay exponentially or faster [2]. From recent results of Paladin and Vulpiani [6] it is possible to deduce that also for chaotic CML with non–local couplings a finite propagation velocity can been defined only if the interactions decay with an exponential law or faster. In particular, they observed that the time necessary for a localized disturbance to overcome a threshold value at a certain distance R from the initially perturbed site growths linearly with R if the coupling is local or exponentially decaying. But this time is independent of R for couplings decaying with power laws (with exponent not too large). This observation would indicate that the velocity is infinite in this case. In order to check it, we have considered CML models as described in eqs. (3) and (4).

In this case we observe an exponential increase with the number of iterations N for the quantity $R(N)$ defined in the previous section,

$$R(N) \propto \exp\left[cN\right] \qquad \text{for} \qquad N \geq N_t \tag{12}$$

where N_t is some transient time depending on the amplitude of the initial perturbation, on the value of the threshold and on the maximal Lyapunov exponent λ [11]. Therefore, a velocity definition can be recovered if in the expression (6) we substitute the distance $R(N)$ with its logarithm, in this case the "logarithmic velocity" will be represented by the rate c appearing in (12). In Fig. 2, the quantity $\log[R(N)]$ is reported as a function of N for some different power–law coupling. It is easily seen that the rate c is given by the expression:

$$c = \frac{\lambda}{\alpha} \ , \tag{13}$$

where α is the range exponent defined in eq. (3). This implies that the disturbance initially situated at $i = 0$ will have, at a certain time n, a spatial profile given by

$$\delta x_i^n \propto \frac{e^{\lambda n}}{i^{\alpha}} \tag{14}$$

For the chosen coupling ($c_1 = 0.3$, $c_2 = 0.01$), the maximal Lyapunov exponent λ seems not to depend on α and coincides with the exponent obtained in the nearest–neighbours case. This might results from the smallness of c_2, due to which the backreaction of the long–range coupling on the local chaotic dynamics can be neglected as a first approximation. In fact, the local degree of chaoticity is due essentially to the nearest–neighbours coupling, while the other interactions simply "model" the spatial profile of the disturbance. Further investigations are in order to clarify the effect on the dynamics of power–law interactions at very long times [11].

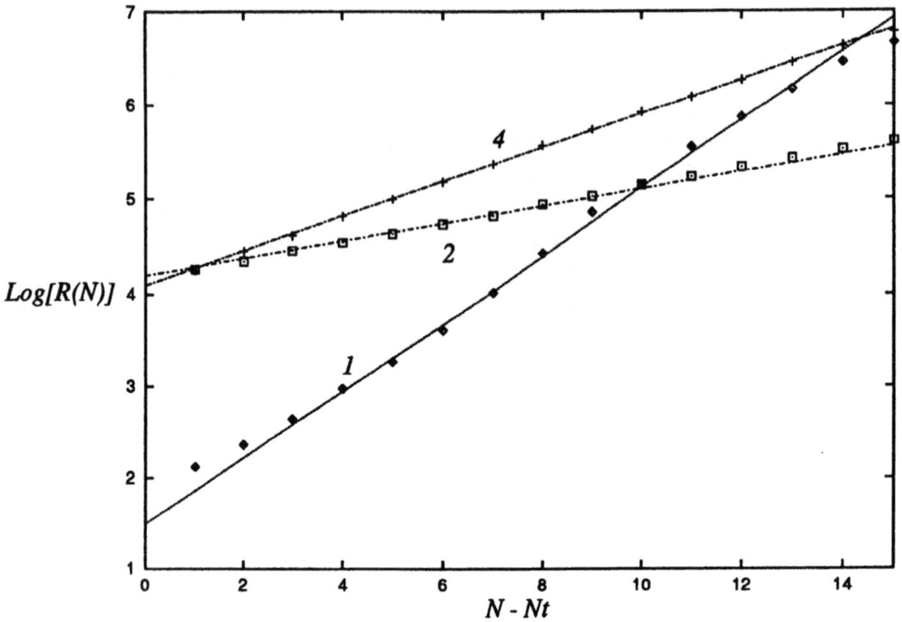

Fig. 2. Natural logarithm of the average distance $R(N)$ versus the number of iterates $N - N_t$ for the model (3). Symbols refer to our numerical data; the straight lines have slopes λ/α, where λ is the maximal Lyapunov exponent (in the present case $\lambda = 0.362$) and α is the exponent determining the power–law decay. The labels indicates the value of α for the corresponding symbols and line. The data refer to a chain of lenght $L = 16385$ with an initial disturbance $\delta_0 = 10^{-6}$ and a chosen threshold $\theta = 0.01$. It should be noticed that the transient time N_t depends on the α value.

Acknowledgments

We thank A. Politi, who first introduces us to these arguments, and P. Grassberger for extremely effective suggestions and continuous discussions. We are also indebted with M. Frese, who encourages us to carry on this research. We gratefully acknowledge the European Economic Community for the research fellowship No ERBCHBICT941569 "Multifractal Analysis of Spatio–Temporal Chaos".

References

1. M.C. Cross and P.H. Hohenberg, Rev. Mod. Phys. **65**, 851 (1993).
2. D. Mollison, J. Royal Statistical Society, **B39**, 283 (1977); in *Mathematical Aspects of Rabies Epizootic*, ed. P.J. Bacon (Academic Press, London, 1983).
3. A. Kolmogorv, I. Petrovsky and N. Piskunov, Bull. Univ. Moscow, Ser. Int. **A1**, 1 (1937);

D.G. Aronson and H.F. Weinberg, Adv. Math. **30**, 33 (1978).
4. W. van Saarlos, Phys. Rev. **A37**, 211 (1988); Phys. Rev. **A39**, 6367 (1989); R.D. Benguria and M.C. Depassier, Phys. Rev. Lett. **73**, 2272 (1994); W. van Saarlos, M. van Hecke and R. Holyst, preprint (1994).
5. K. Kaneko, Prog. Theor. Phys. **72**, 980 (1984); I. Waller and R. Kapral, Phys. Rev. **A30**, 2047 (1984).
6. G. Paladin and A. Vulpiani, J. Phys. **A27**, 4911 (1994).
7. A. Politi and A. Torcini, Europhys. Lett. **28**, 545 (1994).
8. K. Kaneko, Prog. Theor. Phys. **74**, 1033 (1985); A.S. Pikovsky, Phys. Lett. **A156**, 223 (1991).
9. R.J. Deissler and K. Kaneko, Phys. Lett. **A119**, 397 (1987).
10. A. Politi and A. Torcini, Chaos **2**, 293 (1992).
11. A. Torcini, P. Grassberger and A. Politi, to be published.

Lévy–Stable and Extreme Value Distributions in Modelling of Dynamical Phenomena in Complex Physical Systems

K. Weron[1], *K. Kosmulski*[1], *A. Jurlewicz*[2] *and S. Mercik*[2]

[1] Institute of Physics,
Technical University of Wrocław, PL–50–370 Wrocław, Poland
[2] Hugo Steinhaus Center for Stochastic Methods,
Technical University of Wrocław, PL–50–370 Wrocław, Poland

Abstract: Two examples of physical phenomena, dielectric relaxations and turbulent flows, where the Lévy–stable and extreme value distributions appear naturally, are considered. In the case of dielectric relaxations the common mathematical structure underlying four different definitions of relaxation function, responsible for generating the stretched exponential relaxation law, is found. In the case of turbulent flows, the distribution of velocities in a velocity field produced by randomly distributed sources of circulation is shown to be Lévy–stable with the characteristic exponent $\frac{D}{2}$, where D denotes the fractal dimension of the turbulence.

1 Introduction

In this paper, we present two mathematical models of physical phenomena in which the Lévy–stable distributions with the characteristic exponent α less than 2 are rigorously obtained. The considerations are motivated by the empirical observations which simply cannot be explained by means of the Gaussian (α=2) distribution.

Despite its very close relationship with the normal law, the family of Lévy–stable distributions for 60 years has not enjoyed the great interest of mathematicians in probability theory. The reserve met by the Lévy–stable laws was apparently due to the fact that they did not find any applications for a long time. However, there was one exception here. In 1919, several years before the appearance of Lévy's monograph introducing the concept of a stable law [1], the Danish astronomer Holtsmark published a paper where he found a probabilistic principle obeyed by random fluctuations of the gravitational field of stars in space [2]. As became clear much later, the distribution found by Holtsmark and now bearing his name, belongs to the collection of spherically symmetric Lévy–stable laws with parameter $\alpha = \frac{3}{2}$. In our times there has been a sharp increase of interest not only in laws stable with respect to summation (Lévy–stable) but also to other operations (max–stable). It is due to their appearance in certain socio–economic and physical models. Since the stable laws may not be familiar to a broad society of physicists, we review some basic facts about the Lévy–stable and the extreme value (max–stable) distributions in Section 2.

Section 3 is devoted to the theoretical description of the dynamical dielectric response in complex dipolar systems. As became clear from the studies on relaxation phenomena, the functions which describe the dynamical behaviour of such systems deviate considerably from Debye's predictions of the exponential relaxation law. A great part of these studies claim the "stretched exponential" form of the relaxation law [3]. However, the "double power" response law is also observed [4,5]. In order to explain the deviation of the experimental results from the classical Debye model several theoretical models involving different physical mechanisms have been proposed, see e.g. [6-8]. The recently introduced probabilistic representation [9-11] of the cluster model for dielectric relaxation [12] may be a promising theoretical approach since it uniquely leads to the experimentally observed double power form of the dielectric response as well as to the stretched exponential and the classical exponential forms as its special cases.

In Section 4 we propose a mathematically rigorous approach to the vortex models [13-15] which regard the turbulence as the velocity field produced by randomly distributed sources of circulation such as vortex filaments. The idea is inspired by Zolotarev's model of point sources of influence [16]. The model is of a fairly general character, and certain types of problems from areas of astronomy and physics turn out to be connected with it. For instance, the set of assumptions used by Holtsmark [2] corresponds completely to the conditions in that model. If we regard the stars as point vortices and the gravitational field as the velocity field, then we can expect that Zolotarev's method is also applicable to the turbulence. In order to consider the effect of the fractal structure of the turbulence [17] we generalize Zolotarev's model in the case that the point vortices distribute fractally.

2 Stable Laws

In this section we collect some basic facts from the theory of the laws stable with respect to an operation like summation, maximum, or minimum. Stability property means that under the linear normalization the operation over the set of independent, identically distributed random variables can only lead to the law of a single random variable from the set. It is known that the summation scheme produces Lévy–stable distributions [16, 18, 19] and the maximum and minimum schemes lead to extreme value laws [20]. The most important property of the stable distributions is that they have domains of attraction.

Lévy–Stable Distributions

Let $\beta_1, \beta_2, \ldots$ be a sequence of independent, identically distributed random variables and let $F(b)$ denote the distribution function of each β_i, i.e., $F(b) = \Pr(\beta_i \leq b)$. Let $S_n = \beta_1 + \beta_2 + \cdots + \beta_n$.

The distribution function F is said to be *Lévy-stable* if it is non–degenerate and if there exist some constants $A_n > 0$ and B_n such that

$$\Pr\left(\frac{S_n}{A_n} - B_n \le b\right) = F(b) \quad \text{for each } b > 0 \ .$$

In such case there exists constant α, called *the index of stability*, $0 < \alpha \le 2$ such that $A_n = n^{\frac{1}{\alpha}}$. The case $\alpha = 2$ can be recognized as a well-known normal law.

The Lévy–stable distributions are usually described by their characteristic functions (Fourier transforms) [16]. For example, the Fourier transform of the symmetric Lévy–stable law is of the form $\exp(-|At|^{\alpha})$ where $A > 0$ and the index of stability α is taken from the range (0,2].

The completely asymmetric Lévy–stable distributions are supported on the non–negative half–line and so can be described by their Laplace transforms which are shown [18] to be of the form $\exp(-(At)^{\alpha})$, $t \ge 0$, for some $A > 0$ and $0 < \alpha < 1$. Observe that in that case the index of stability α can only be taken from the smaller range (0,1).

Let us point that in the theory of the Lévy–stable laws any function of the form $\exp(-(At)^{\alpha})$, $t \ge 0$, for $0 < \alpha < 1$ can be interpreted in two different ways. If it is taken as the Fourier transform, then it is connected with a random variable which takes both positive and negative values. But if this function is the Laplace transform, then the random variable connected with it is non–negative.

The distribution function F belongs to *the domain of attraction of the Lévy-stable law* if for some constants $A_n > 0, B_n, A > 0, B$, we have

$$\Pr\left(\frac{S_n}{A_n} - B_n \le b\right) \xrightarrow{w} G(Ab + B) \tag{2.1}$$

for stable distribution function G with some index of stability α, $0 < \alpha \le 2$ (the notation "$\xrightarrow{w}$" denotes the weak convergence, i.e., the convergence at continuity points of the limiting function). It should be stressed that for any distribution function F if the non–degenerate limiting function G in (2.1) exists it is Lévy-stable. It is proved [18] that:

- The distribution function F belongs to the domain of attraction of the normal law if and only if

$$\lim_{x\to\infty} \frac{x^2 \int_{|b|>x} dF(b)}{\int_{|b|<x} b^2 dF(b)} = 0 \ .$$

- The distribution function F belongs to the domain of attraction of the Lévy -stable law with the index of stability α, $0 < \alpha < 2$, if and only if

$$\lim_{b\to\infty} \frac{1 - F(xb) + F(-xb)}{1 - F(b) + F(-b)} = x^{-\alpha} \qquad \text{for each } x > 0$$

  and if there exist limits

$$\lim_{b\to\infty} \frac{1 - F(b)}{1 - F(b) + F(-b)} \quad \text{and} \quad \lim_{b\to\infty} \frac{F(-b)}{1 - F(b) + F(-b)} \ .$$

Extreme Value Distributions

Let $\eta_1, \eta_2, \ldots$ be a sequence of independent, identically distributed random variables and let $F(t)$ denote the distribution function of each η_i. Let $M_n = \max(\eta_1, \ldots, \eta_n)$.

Extremal Types Theorem for Maxima, [20].

If for some constants $a_n > 0, b_n$, we have

$$\Pr(a_n(M_n - b_n) \leq t) \xrightarrow{w} G(t) \quad \text{as } n \to \infty \tag{2.2}$$

for some non–degenerate G, then for some constants $a > 0,\ b$, we have $G(t) = G_0(at + b)$, where G_0 is one of the three following extremal value types for maxima:

$$\text{Type I} \quad : G_0(t) = \exp(-e^{-t}) \ , \qquad -\infty < t < +\infty \ ;$$

$$\text{Type II} \quad : G_0(t) = \begin{cases} 0, & t < 0 \ , \\ \exp(-t^{-\gamma}), \quad \text{for some } \gamma > 0, & t \geq 0 \ , \end{cases}$$

$$\text{Type III} : G_0(t) = \begin{cases} \exp(-(-t)^{\gamma}), \quad \text{for some } \gamma > 0, & t < 0 \ , \\ 1, & t \leq 0 \ . \end{cases}$$

Conversely, each distribution function G_0 of extremal value type may appear as a limit in (2.2) and, in fact, appears when G_0 itself is the distribution function of each η_i.

If the non–degenerate limit in (2.2) exists we say that F, the distribution function of each η_i, belongs to *the domain of attraction of the max-stable law of adequate type.*

Now we will formulate the conditions under which a distribution function belongs to the domain of attraction of max-stable laws. Let $t_F = \sup\{t : F(t) < 1\}$. Necessary and sufficient conditions for the distribution function F to belong to the domain of attraction of max-stable law are :

Type I : there exists some strictly positive function $g(t)$ such that

$$\lim_{t \to t_F -} \frac{1 - F(t + xg(t))}{1 - F(t)} = e^{-x}, \qquad \text{for all real } x \ ;$$

Type II : $t_F = \infty$ and

$$\lim_{t \to \infty} \frac{1 - F(tx)}{1 - F(t)} = x^{-\gamma}, \qquad \gamma > 0, \text{ for each } x > 0 \ ;$$

Type III : $t_F < \infty$ and

$$\lim_{t \to 0+} \frac{1 - F(t_F - xt)}{1 - F(t_F - t)} = x^{\gamma}, \qquad \gamma > 0, \text{ for each } x > 0 \ .$$

To obtain $G = G_0$, the constants a_n, b_n in the convergence (2.2) should be taken in each case above to be :

$$\text{Type I} \;: a_n = [g(\delta_n)]^{-1} \quad , b_n = \delta_n \;;$$

$$\text{Type II} \;: a_n = (\delta_n)^{-1} \quad , b_n = 0 \;;$$

$$\text{Type III} : a_n = (t_F - \delta_n)^{-1} \quad , b_n = t_F \;,$$

where $\delta_n = \inf\{t : F(t) \geq 1 - \frac{1}{n}\}$.

Let us consider now the minimum $m_n = \min(\eta_1, \ldots, \eta_n)$. Since it is simply given as $m_n = -\max(-\eta_1, \ldots, -\eta_n)$, the limiting results for minima can clearly be derived from those for maxima. Moreover, the extremal value types H_0 for minima and G_0 for maxima fulfill the relation $H_0(t) = 1 - G_0(-t)$. This is useful in obtaining the possible limiting distributions for minima:

$$\text{Type I} \;: H_0(t) = 1 - \exp(-e^{-t}), \qquad -\infty < t < \infty$$

$$\text{Type II} \;: H_0(t) = \begin{cases} 1 - \exp(-(-t)^{-\gamma}), & \text{for some } \gamma > 0, \; t < 0 \;, \\ 1, & t \geq 0 \;, \end{cases}$$

$$\text{Type III} : H_0(t) = \begin{cases} 0, & t < 0 \;, \\ 1 - \exp(-t^{\gamma}), & \text{for some } \gamma > 0, \; t \geq 0 \;. \end{cases}$$

The criteria for domains of attraction for minima can be easily obtained from those for maxima by replacing the "tail" $1 - F(t)$ by $F(-t)$.

3 Dielectric Relaxation

The surprising empirical observation that the relaxation decay law for many diverse dipolar systems (including organic liquids and solutions, polymers, and many glass forming materials) follows the same stretched exponential pattern

$$\phi(t) = \exp[-(\frac{t}{\tau})^{\alpha}] \;, \quad \text{for } 0 < \alpha < 1, \;\; \tau > 0 \qquad (3.1)$$

has already focused much theoretical attention [6-8]. Several derivations of this ubiquitous decay law, involving various concepts such as percolation, fractals, hierarchical relaxation of constraints, multipolar interaction transitions or defect–diffusion processes, are known. Klafter and Shlesinger [7] analysed the underlying common factor in the case of three different physical models: the Főrster direct transfer model, the hierarchically constrained dynamics model, and the defect–diffusion model, and found that a common feature responsible for generating (3.1) is the existence of a scale–invariant distribution of relaxation times. The same explanation was found for the spin–flip dynamics on fractals and percolative structures [8].

In this paper, we try to provide some insight into the relaxation mechanism by searching for the equivalence of different, "evident" (repeated in the literature since few decades) definitions of relaxation function [21]. In the vast literature on relaxation phenomena one can find the following statements defining the relaxation function $\phi(t)$:

1°. The relaxation function expresses the average proportion of dipoles which did not change their aligned orientation up to certain time

$$\phi(t) = \left\langle \frac{N(t)}{N} \right\rangle , \tag{3.2}$$

where N is the number of dipoles aligned along the field lines at $t = 0$. This definition provides perhaps a closest description to what is experimentally measured.

2°. The relaxation function is a weighted average of exponential relaxation function

$$\phi(t) = \int_0^\infty \rho(\tau) \exp(-\frac{t}{\tau}) d\tau , \tag{3.3}$$

where $\rho(\tau)$ is the probability density of a distribution of relaxation time τ of an individual exponentially relaxing dipole. The choice of the weight function $\rho(\tau)$ is commonly determined by the empirical relaxation functions, but such an approach does not explain why (3.1) should be so universally applicable.

3°. The relaxation function fulfills a generalized master equation, which governs the evolution of polarisation $P(t)$,

$$\frac{d\phi(t)}{dt} = -w(t)\phi(t) , \tag{3.4}$$

where $w(t)$ represents the time–dependent transition probability per unit time for a relaxing system and its form is usually determined by the empirical results.

From the physical point of view relaxation function represents the evolution of polarisation $P(t) = P_0\phi(t)$ in a dipolar system after an electric constant field is turned off. Hence the above definitions should have an underlying common mathematical structure responsible for generating the universal relaxation law (3.1). In order to show the condition under which the equivalence of (3.2)-(3.4) holds, let us first introduce the notion of the probabilistic representation of the cluster model for dielectric relaxation.

According to the idea contained in the cluster model [12], the condensed phase systems which exhibit position or orientation relaxation are composed of spatially limited "defect" regions (clusters) constituted around the dipoles whose positions have been altered by an external field. Because the structural ordering within the cluster is incomplete, the equilibrium geometry cannot be maintained by displacement fluctuations and the spatial uniformity must be lost on relaxing as the imperfect equilibrium structure evolves. Moreover, it is expected that the strongly coupled intra–cluster motions followed by the weakly coupled inter–cluster motions will give their own characteristic contribution to the observed relaxation processes. Such a picture allows us to represent the relaxing system [9-11] by a set of N aligned dipoles which life–times in their imposed orientation are affected by the random intra– and inter–cluster influences.

Let us assume [9,10] that the probability that the i–th dipole has not changed its initial aligned position, imposed by an external field, during time t following removal of that field, is exponential

$$\Pr(\theta_i \geq t|\beta_i = b) = e^{-bt} \quad \text{for } t \geq 0, b > 0 \tag{3.5}$$

conditioned only, in the first approximation, by the value b taken by the relaxation rate $\beta = 1/\tau$. The latter reflects the random intra–cluster influence. Here θ_i and β_i represent the random life–time and the relaxation rate of an individual aligned dipole, respectively. Let the non–negative $\beta_1, \beta_2, \ldots$ and $\theta_1, \theta_2, \ldots$ form sequences of independent, identically distributed random variables (n–n. i.i.d. r.v's) with distribution functions $F_\beta(b)$ and $F(t)$, respectively. The total probability that the i–th dipole has not changed its aligned position up to time t equals

$$\Pr(\theta_i \geq t) = \int_0^\infty e^{-bt} dF_\beta(b) \ . \tag{3.6}$$

The right–hand side of equ.(3.6) is a weighted average of exponential relaxation with respect to the distribution of relaxation rates β_i. The crucial observation is that it is also the Laplace–Stieltjes transform $\mathcal{L}(F_\beta; t)$ of the distribution function F_β. It has to be stressed, that so far *there are no restrictions* on F_β, the only requirement is that it is a distribution function supported on the non–negative half–line. From the physical point of view, F_β is unknown!

For a finite ensemble of N aligned dipoles one can then define the survival probability, i.e., the probability that the system as a whole has not changed its initial state imposed by an external electric field up to time t

$$\Pr[\min(\theta_1, \ldots, \theta_N) \geq t] = [\Pr(\theta_i \geq t)]^N \ .$$

The equality is justified by the assumption that θ_i's are i.i.d. r.v's. However, the above quantity converges to zero as $N \to \infty$ to get a meaningful expression one has to further rescale time. This leads to the following definition of the "first passage" relaxation function [9] of a macroscopic dipolar system

$$\phi(t) = \lim_{N\to\infty} [\Pr(\theta_i \geq \frac{t}{A_N})]^N = \lim_{N\to\infty} \Pr\left(A_N \min(\theta_1, \ldots, \theta_N) \geq t\right) \tag{3.7}$$

where A_N is a sequence of suitable normalizing constants. Substituting equ.(3.6) into (3.7) we have

$$\phi(t) = \lim_{N\to\infty} [\mathcal{L}(F_\beta; \frac{t}{A_N})]^N \ . \tag{3.8}$$

The N-th power of the Laplace–Stieltjes transform of the non–degenerate distribution function F_β converges to the non–degenerate limiting transform $\mathcal{L}(F_{\tilde{\beta}}; t)$ if and only if F_β belongs to the domain of attraction of a completely asymmetric Lévy–stable law $F_{\tilde{\beta}}$ with $0 < \alpha < 1$. Then we get

$$\phi(t) = \int_0^\infty e^{-bt} dF_{\tilde{\beta}}(b) \ , \tag{3.9}$$

where the Lévy–stable random variable $\tilde{\beta}$ equals

$$\tilde{\beta} = \lim_{N\to\infty} \frac{\sum_{i=1}^{N} \beta_i}{A_N} \ .$$

Hence the first passage relaxation function is *uniquely determined* and equals

$$\phi(t) = \exp[-(At)^{\alpha}], \quad \text{where} \quad 0 < \alpha < 1 \ , \tag{3.10}$$

where A is a positive constant. The relaxation function obtained in (3.10) is of the form of the well known empirical stretched exponential relaxation function (3.1). In the case when $\alpha \to 1$ one obtains the relaxation function of Debye's exponential form and this corresponds to the case of degenerate limiting distribution of the relaxation rate $\tilde{\beta}$, [22].

The above results allow us to represent a system consisting of a large number of aligned dipoles, each relaxing exponentially in its own random medium with an unknown distribution F_β of the "microscopic" relaxation rates β_i, equ.(3.5), by an "average" dipole relaxing exponentially with the Lévy–stable distribution $F_{\tilde{\beta}}$ of the "macroscopic" relaxation rate $\tilde{\beta}$, equ.(3.9). This corresponds to the definition (3.3).

On the other hand, since

$$\frac{d}{dt}\left[\Pr\left(\theta_i \geq \frac{t}{A_N}\right)\right]^{N} = N\left[\Pr\left(\theta_i \geq \frac{t}{A_N}\right)\right]^{N-1} \frac{d}{dt}\left[\Pr\left(\theta_i \geq \frac{t}{A_N}\right)\right]$$

and analogical equation is fulfilled for $\frac{d}{dt}[\mathcal{L}(F_\beta; \frac{t}{A_N})]^N$, equ.(3.6) can be rewritten as follows

$$\frac{d}{dt}\left[\Pr\left(\theta_i \geq t\right)\right]^{N} = \left[\Pr\left(\theta_i \geq t\right)\right]^{N-1}\left[\mathcal{L}(F_\beta;t)\right]^{-N+1}\frac{d}{dt}\left[\mathcal{L}(F_\beta;t)\right]^{N} \ .$$

Now rescaling time $t \Rightarrow \dfrac{t}{A_N}$ and taking the limit as $N \to \infty$, we get

$$\frac{d\phi}{dt}(t) = -\alpha A(At)^{\alpha-1}\phi(t) \ , \quad \text{for } 0 < \alpha < 1 \ , \tag{3.11}$$

where we have used the results (3.7) - (3.10). Thus the assumption (3.5) leads to the generalized master equation of the form (3.4), with $w(t) = \alpha A(At)^{\alpha-1}$, being the intensity of transition probability for the relaxing system as a whole, and a solution of the form (3.10).

In order to find the relationship with the definition (3.2) consider a sample of i.i.d. r.v's

$$\theta_{1N}, \ldots, \theta_{NN}$$

representing life–times of N dipoles with a common distribution function $F(t)$ defined by (3.6):

$$F(t) = 1 - \int_0^{\infty} e^{-bt} dF_\beta(b) \ . \tag{3.12}$$

The dipoles switch in a certain order what can be reflected in the notion of the order statistics [23,24]

$$\theta_{1:N}, \ldots, \theta_{N:N} \ ,$$

which is simply a non–decreasing rearangement of the above life–times, i.e.,

$$\theta_{1:N} \leq \ldots \leq \theta_{k:N} \leq \ldots \leq \theta_{N:N} \ .$$

Traditionally, $\theta_{k:N}$ is called *the k-th order statistic* of the above sample. Note that

$$\theta_{1:N} = \min(\theta_{1N}, \ldots, \theta_{NN}) \ .$$

In terms of the above ordered life–times for a fixed size N of the system the ratio of dipoles which did not switch their aligned orientation up to time t is equal

$$\phi_N(t) = 1 - \frac{k_N(t)}{N} \ ,$$

where for any $t > 0$ the event

$$\{k_N(t) = k\} = \{\theta_{k:N} \leq t, \ \theta_{k+1:N} > t\}, \quad k = 0, 1, \ldots, N \ .$$

Thus we have

$$\begin{aligned} \Pr(k_N(t) = k) &= \binom{N}{k} [\Pr(\theta_i \leq t)]^k \, [\Pr(\theta_i > t)]^{N-k} \\ &= \binom{N}{k} [F(t)]^k \, [1 - F(t)]^{N-k} \ , \end{aligned} \tag{3.13}$$

where $F(t)$ is defined by (3.12). The random variable $k_N(t)$ has then the Bernoulli distribution with the parameters N and $F(t)$.

Hence

$$\langle \phi_N(t) \rangle = 1 - \frac{\langle k_N(t) \rangle}{N} = 1 - F(t) \ ,$$

and the relaxation function based on the definition (3.2) equals

$$\phi(t) = \Pr(\theta_i \geq t) \ , \tag{3.14}$$

what is rather a surprising result, since it recovers the original – unknown! – distribution $F(t)$ of life–times for individual dipoles.

When do the relaxation functions defined by the eqs.(3.2) - (3.4) coincide ?

The relaxation functions (3.2) - (3.4) coincide if and only if the distribution $F_\beta(b)$ of relaxation rates β_i on the microscopic level is equal to the distribution $F_{\tilde{\beta}}(b)$ of relaxation rate $\tilde{\beta}$ on the macroscopic level. Moreover, both of them have to be Lévy–stable with $0 < \alpha < 1$. Indeed, if we assume that the first passage relaxation function is well defined, then we have from formulas (3.6) that (3.14)

$$\phi(t) = \Pr(\theta_i \geq t) = \int_0^\infty e^{-bt} dF_\beta(b) = \mathcal{L}(F_\beta; t) \ .$$

On the other hand, by (3.9), $\phi(t) = \mathcal{L}(F_{\tilde{\beta}};t)$, and hence the equality of Laplace transforms holds $\mathcal{L}(F_\beta;t) = \mathcal{L}(F_{\tilde{\beta}};t)$. Since the Laplace transform uniquely determines the distribution function, the necessary and sufficient condition for the equality of (3.2) - (3.4) is that the distribution functions $F_\beta(b) = F_{\tilde{\beta}}(b)$, where by (3.9) the relaxation rate $\tilde{\beta}$ is completely asymmetric Lévy–stable random variable with $0 < \alpha < 1$. In the case of the definition (3.3) this means that weight function $\rho(\tau)$ has to be of the form

$$\rho(\tau) = \frac{1}{\tau^2}\rho_\alpha(\frac{1}{\tau}) \ ,$$

where $\rho_\alpha = p(\frac{1}{\tau},\alpha,1)$ is the completely asymmetric Lévy–stable probability density with $0 < \alpha < 1$.

Let us note that the first passage relaxation function, by its very existence implies that the unknown distribution function $F_\beta(b)$ of the relaxation rates β_i is in the domain of attraction of the completely asymmetric Lévy–stable law $F_{\tilde{\beta}}(b)$ [11]. Among commonly used one–sided distribution functions such as Burr, Pareto, generalized Pareto, gamma, loggamma, transformed gamma, lognormal and Weibull distribution functions only Burr, Pareto and Lévy–stable itself belong to the domain of attraction of the completely asymmetric Lévy–stable law by satisfying the necessary and sufficient condition

$$\lim_{b\to\infty} \frac{1-F(xb)}{1-F(b)} = x^{-\alpha} \qquad \text{for each } x > 0 \text{ and } 0 < \alpha < 1 \ .$$

In conclusion, the equivalence of (3.2) - (3.4), restricts the domain of attraction to the Lévy–stable distribution only. Thus the common factor responsible for generating the stretched exponential law (3.1) is the Lévy–stable distribution of microscopic relaxation rates β_i.

4 Turbulent Velocity Field

In this section, we consider the turbulent velocity field in $\mathbb{R}^3$ resulting from a random distribution of N point–vortices [13-15]. The velocity of a point $\mathbf{r}$ is given by the Biot–Savart law

$$\mathbf{v}_N(\mathbf{r}) = \sum_{j=1}^{N} \mathbf{v}_j(\mathbf{r}) \ , \tag{4.1}$$

$$\mathbf{v}_j(\mathbf{r}) = -\frac{\boldsymbol{\omega}_j \times (\mathbf{r}-\mathbf{r}_j)}{4\pi|\mathbf{r}-\mathbf{r}_j|^3} \ , \tag{4.2}$$

where $\mathbf{r}_j$ and $\boldsymbol{\omega}_j$ denote the position and the vorticity of the j–th vortex. Both, the position $\mathbf{r}_j$ and the corresponding vorticity $\boldsymbol{\omega}_j$ are assumed to be random.

Let us consider some region U of finite or infinite volume in $\mathbb{R}^3$ in which is dispersed a countable system C of "point objects" each capable of creating a "field of action" [16] described in the case of turbulence by means of eqs. (4.1) and (4.2). Equation (4.2) shows the velocity $\mathbf{v}_j(\mathbf{r})$ in the velocity field created at a point $\mathbf{r} \in U$ by a vortex located at $\mathbf{r}_j \in U$ and characterized by ω_j. The field generated by the whole system $\mathcal{C}$ will be understood as a certain limit. By definition, the velocity $\mathbf{V}(\mathbf{r})$ of the field generated at $\mathbf{r}$ by all the vortices in the system $\mathcal{C}$ is equal to the weak limit of $\mathbf{v}_R(\mathbf{r})$ as $R \to \infty$, where R is a radius of a sphere $U_R = U_R(\mathbf{r})$ with a fixed center at a point $\mathbf{r}$. To determine conditions for the existence of this limit and to find an analytic description of it, we can consider the question of the limit value as $R \to \infty$ of the characteristic function of the variable $\mathbf{v}_R(\mathbf{r})$

$$f_R(\mathbf{q}) = E\exp\{i(\mathbf{q}, \mathbf{v}_R(\mathbf{r})\}, \quad \text{where } \mathbf{q} \in I\!R^3 \ . \tag{4.3}$$

It is clear from (4.1) that $\mathbf{v}_R(\mathbf{r})$ is the sum of a random number N_R of identically distributed random variables $\mathbf{v}_j(\mathbf{r})$ – determined by a pair $(\mathbf{r}_j, \omega_j)$ – that are independent of N_R and of one another.

Let us consider U_1 and U_2 some disjoint regions in U of finite volumes u_1 and u_2. A countable system of randomly situated points in U is called a Poisson ensemble if it has the following properties:

1°. The numbers N_1 and N_2 of points falling in U_1 and U_2, respectively, are independent random variables.

2°. For any $k \geq 0$ the probability $\Pr(N_1 = k)$ depends on k and the volume u_1, but not on the shape of U_1.

3°. For small values of the volume u_1

$$\Pr(N_1 = 1) = \rho u_1 + o(u_1) \ ,$$

$$\Pr(N_1 \geq 2) = o(u_1) \ ,$$

where $\rho > 0$ is a constant signifying the average density of the concentration of points of the system $\mathcal{C}$ in U. By using these properties one can show [18] that N_1 has a Poisson distribution with parameter $\lambda = \rho u_1$, i.e.,

$$\Pr(N_1 = k) = \frac{\lambda^k}{k!} e^{-\lambda}, \quad k = 0, 1, \dots \ . \tag{4.4}$$

Subject to conditions used in vortex models, we assume that point vortices form a Poisson ensemble and hence the distribution of the random number N_R in the sum (4.1) of i.i.d. r.v's $\mathbf{v}_j(\mathbf{r})$ is of the type (4.4).

Therefore

$$\begin{aligned} f_R(\mathbf{q}) &= \sum_{k=0}^{\infty} \Pr(N_R = k) \ E\exp\{i(\mathbf{q}, \mathbf{v}_R(\mathbf{r})\} \\ &= \sum_{k=0}^{\infty} \exp[-\rho u_R] \frac{(\rho u_R)^k}{k!} \phi_R^k(\mathbf{q}) \\ &= \exp\{\rho u_R(\phi_R(\mathbf{q}) - 1)\} \ , \end{aligned} \tag{4.5}$$

where u_R is the volume of U_R and $\phi_R(\mathbf{q})$ the characteristic function of $\mathbf{v}_j(\mathbf{r})$.

In order to take into account the effect of fractal structure of turbulence [17], we postulate hereafter that the vortices are located in the D–dimensional fractal region $I\!F_D$. The number D may be considered as the fractal dimension of this turbulence. In the case that the vortices distribute uniformly over $I\!F_D$, the number N_R of vortices within a sphere of radius R is statistically proportional to R^D and may be expressed as [25]

$$N_R = \eta \frac{\pi^{\frac{D}{2}}}{\Gamma(\frac{D}{2}+1)} R^D \ , \tag{4.6}$$

where η is a constant density of vortices on $I\!F_D$ ($\eta \equiv \rho$) and $\Gamma(x)$ is the gamma function. The fractal dimension D of the real turbulence is estimated [26,27], both experimentally and theoretically, to be 2.6 ± 0.1. This type of fractal relationship can be easily produced numerically by means of the Lévy flight [28].

Considering only the vortices within a sphere U_R in $I\!F_D$ we may transform the expression in the exponent of equ.(4.5) to the following form

$$\log f_R(\mathbf{q}) = \rho \int_{I\!R^3} \int_{U_R} \{\exp[i(\mathbf{q}, \mathbf{v}(\mathbf{r}))] - 1\} \, d\mathbf{r} P_{\boldsymbol{\omega}}(d\boldsymbol{\omega}) \ , \tag{4.7}$$

where $P_{\boldsymbol{\omega}}$ is the distribution of $\boldsymbol{\omega}_j$ and $\mathbf{v}(\mathbf{r}) \equiv \mathbf{v}_j(\mathbf{r})$. Thus, the function $f_R(\mathbf{q})$ obtains the canonical Lévy representation for characteristic functions of infinitely divisible distributions on $I\!R^3$ [16]. The question of the convergence of the integral in (4.7), as $R \to \infty$, reduces to the question of the existence of the limit of the following integrals [16]:

$$\mathbf{M} = \rho \int_{I\!R^3} \int_{I\!F_D} \mathbf{v}(\mathbf{r}) d\mathbf{r} P_{\boldsymbol{\omega}}(d\boldsymbol{\omega}) \tag{4.8}$$

and

$$L = \rho \int_{I\!R^3} \int_{I\!F_D} \min\{1, |\mathbf{v}(\mathbf{r})|\} |\mathbf{v}(\mathbf{r})| d\mathbf{r} P_{\boldsymbol{\omega}}(d\boldsymbol{\omega}) \ . \tag{4.9}$$

After some computations, taking $\mathbf{r} = 0$ and considering the volume integral in the polar coordinates, i.e.,

$$\int_{I\!F_D} d\mathbf{r} \ldots = \int_0^\infty dr \int_0^{2\pi} d\phi \int_0^\pi d\theta \frac{\pi^{\frac{D}{2}} D}{4\pi \Gamma(\frac{D}{2}+1)} r^{D-1} \sin\theta \ldots$$

it can be shown that the conditions $|\mathbf{M}| < \infty$ and $L < \infty$ hold and the characteristic function $f(\rho) = \lim_{R\to\infty} f_R(\rho)$ of the limit distribution has the form

$$\log f(\mathbf{q}) = \frac{\rho \pi^{\frac{D}{2}} D}{4\pi \Gamma(\frac{D}{2}+1)} \int_{I\!R^3} P_{\boldsymbol{\omega}}(d\boldsymbol{\omega}) \int_0^\pi d\theta \int_0^{2\pi} d\phi \int_0^\infty dr \, r^{D-5} \times$$
$$\times \left\{ \exp[\frac{i}{4\pi}(\mathbf{q}, \boldsymbol{\omega} \times \mathbf{s}) \frac{1}{r^2}] - 1 - \frac{i}{4\pi}(\mathbf{q}, \boldsymbol{\omega} \times \mathbf{s}) \frac{1}{r^2} \right\} \ , \tag{4.10}$$

where $\mathbf{s}$ is the unit vector in the direction determined by θ and ϕ. Using the following relationship [16]

$$\int_0^\infty (e^{i\xi r} - 1 - i\xi r) r^{-\alpha-1} dr = \Gamma(-\alpha)|\xi|^\alpha \exp\{-i\frac{\pi}{2}\alpha \,\mathrm{sign}\xi\}, \quad \alpha \in (1,2] \ ,$$

and substituting $p = \frac{1}{r^2}$, equ.(4.10) can be rewritten as

$$\log f(\mathbf{q}) = \frac{\rho D \pi^{\frac{D}{2}} \Gamma(-\frac{D}{2})}{8\pi\Gamma(\frac{D}{2}+1)(4\pi)^{\frac{D}{2}}} \times$$

$$\times \int_{\mathbb{R}^3} P_\omega(d\omega) \int_0^\pi d\theta \int_0^{2\pi} d\phi \, |(\mathbf{q}, \boldsymbol{\omega} \times \mathbf{s})|^{\frac{D}{2}} \exp\{-i\tfrac{\pi}{4} D \mathrm{sign}(\mathbf{q}, \boldsymbol{\omega} \times \mathbf{s})\} =$$

$$= A \int_{\mathbb{R}^3} P_\omega(d\omega) \int_0^{2\pi} d\phi \int_0^\pi d\theta \, |(\mathbf{q}, \boldsymbol{\omega} \times \mathbf{s})|^{\frac{D}{2}} \ , \tag{4.11}$$

where

$$A = \frac{\rho \pi^{\frac{D}{2}} D \cos(\frac{\pi}{4}D) \Gamma(-\frac{D}{2})}{2(4\pi)^{\frac{D}{2}+1} \Gamma(\frac{D}{2}+1)} \ .$$

Changing the coordinates so that $(\mathbf{s}, \mathbf{k}) = \cos\theta$, where $\mathbf{k} = \frac{\mathbf{q} \times \boldsymbol{\omega}}{|\mathbf{q} \times \boldsymbol{\omega}|}$, the characteristic function (4.11) takes the form

$$\log f(\mathbf{q}) = -\frac{\rho\pi}{(D+2)D2^{D-1}\sin(\frac{\pi}{4}D)\Gamma^2(\frac{D}{2})} |\mathbf{q}|^{\frac{D}{2}} \left\langle \left| \frac{\boldsymbol{\omega} \times \mathbf{q}}{|\mathbf{q}|} \right|^{\frac{D}{2}} \right\rangle \ , \tag{4.12}$$

where

$$\left\langle \left| \frac{\boldsymbol{\omega} \times \mathbf{q}}{|\mathbf{q}|} \right|^{\frac{D}{2}} \right\rangle \equiv \int_{\mathbb{R}^3} P_\omega(d\omega) \left| \frac{\boldsymbol{\omega} \times \mathbf{q}}{|\mathbf{q}|} \right|^{\frac{D}{2}} \ .$$

When the distribution of the vorticity P_ω is isotropic equ.(4.12) reduces to the following form

$$\log f(\mathbf{q}) = -A_0 |\mathbf{q}|^{\frac{D}{2}} \ ,$$

where A_0 is a positive constant depending on the parameters ρ,D and $\langle|\omega|^{\frac{D}{2}}\rangle$ characterising the considered turbulent velocity field. Thus, only in the case of isotropic turbulent velocity fields, the limit distribution of $\mathbf{V}(\mathbf{r})$ is the stable distribution with the characteristic exponent $\frac{D}{2}$. The Holtsmark distribution is included in (4.12) as the special case $D = 3$, that is, the vortices distribute uniformly in $\mathbb{R}^3$. It should be stressed that, since the maximum value of D is 3, the characteristic exponent $\frac{D}{2}$ cannot be 2, i.e., the limit velocity distribution cannot be Gaussian. As the stable distribution [16,19] it has a long tail and the tail depends on the fractal dimension D of the turbulence.

This work was supported by KBN Grant No.2 P302 065 04.

References

1. P.Lévy, Calcul des Probabilitiés (Gauthier–Villars,Paris 1925).
2. J. Holtsmark, Ann. de Phys. **58** (1919).
3. T.V. Ramakrishnan and M. Raj Lakshmi, eds, *Non–Debye Relaxations in Condensed Matter* (World Scientific, Singapore, 1987).
4. A.K. Jonscher, *Dielectric Relaxation in Solids* (Chelsea Dielectric Press, London, 1983).
5. S. Havriliak and S.J. Havriliak, J. Non–Cryst. Solids **172–174**, 297 (1994).
6. H. Scher, M.F. Shlesinger, and J.T. Bendler, Physics Today **44**, 26 (1991).
7. J. Klafter and M.F. Shlesinger, Proc. Natl. Acad. Sci. USA **83**, 848 (1986).
8. R. Rammal, J. Physique **46**, 1857 (1985).
9. K. Weron, J. Phys.: Condens. Matter **3**, 9151 (1991).
10. K. Weron, J. Phys.: Condens. Matter **4**, 10507 (1992)
11. K. Weron and A. Jurlewicz, J. Phys. A: Math. Gen. **26**, 395 (1993).
12. L.A. Dissado and R.M. Hill, Proc. Roy. Soc. A **390**, 131 (1983).
13. V. Frisk, P.L. Sulem, and M.A. Nelkin, J, Fluid Mech. **87**, 719 (1978).
14. H. Fujisaka and H. Mori, Prog. Theor. Phys. **67**, 54 (1979).
15. H.G.E. Hentschel and J. Procaccia, Phys. Rev. Lett. **49**, 1158 (1982).
16. V.M. Zolotarev, *One dimensional Stable Distributions* (American Mathematical Society, Providence, R.J., 1986)
17. B.B. Mandelbrot, *The Fractal Geometry of Nature* (Freeman & Co., San Francisco, 1982).
18. W. Feller, *An Introduction to Probability and Its Applications*, vol. 2 (Wiley, New York, 1966).
19. A. Weron, *Stable Processes and Measures: A Survey*, Lecture Notes in Math. 1391 (Springer, Berlin, 1984).
20. M.R. Leadbetter, G. Lindgren, and H. Rootzen, *Extremes and Related Properties of Random Sequences and Processes* (Springer, New York, 1986)
21. C.J. Bőttcher and P. Bordewijk, *Theory of Electronic Polarisation*, vol 2 (Elsevier, Amsterdam, 1978).
22. A. Jurlewicz, A. Weron, and K. Weron, Appl.Math. to appear (1995)
23. J. Galambos, *The Asymptotic Theory of Extreme Order Statistics*, (Wiley, New York, 1978).
24. A. Weron, K.Weron, and A.W.Woyczynski, J. Stat. Phys. **78**, 1027 (1995).
25. H. Takayasu, Prog. Theor. Phys. **72**, 471 (1984).
26. A.S. Monin and A.M. Yaglom, *Statistical Fluid Mechanics: Mechanics of Turbulence*, (MIT Press, Cambridge, 1975).
27. S. Kida, Prog. Theor. Phys. **67**, 1630 (1982).
28. A. Janicki and A. Weron, *Simulation of Chaotic Behavior of α–Stable Stochastic Processes* (Marcel Dekker, New York, 1994).

Wigner or Non–Wigner: That Is the Question

J. Zakrzewski[1,2], *K. Dupret*[1] *and D. Delande*[1]

[1] Laboratoire Kastler–Brossel, Universite Pierre et Marie Curie, 75005 Paris, France
[2] Instytut Fizyki, Uniwersytet Jagielloński, PL–30–059 Kraków, Poland***.

Abstract: It is shown that, contrary to a common belief, statistical spectral properties of systems that exhibit classically chaotic behaviour may not obey the statistics typical for Random Matrix Theory. In particular, the nearest neighbor statistics for hydrogen atom in a strong static magnetic field does not obey the usual Wigner distribution. A simple model using a set of regular states coupled to a set of chaotic states modeled by a random matrix reproduces correctly the observed distributions both in the bound regime and for resonances above the ionization threshold.

It has been commonly believed and successfully tested on a variety of quantum systems whose classical counterpart is chaotic that the local statistical properties of their spectra follow the predictions of Random Matrix Theory [1, 2]. In particular, the statistical distribution of energy spacing between consecutive levels – also called Nearest Neighbor Spacing (NNS) distribution – has been shown to be in excellent agreement with the spacing distribution between consecutive eigenvalues of random matrices. For systems enjoying time–reversal invariance, the Hamiltonian is real in a suitable basis and the Gaussian Orthogonal Ensemble (GOE) of random matrices has to be used. The NNS distribution of this ensemble is very close to the Wigner surmise [1, 2]:

$$P(s) = \frac{\pi s}{2} \exp\left(-\frac{\pi s^2}{4}\right) , \tag{1}$$

where $P(s)\mathrm{d}s$ measures the probability of having a spacing in the interval $[s, s + \mathrm{d}s]$ (after renormalizing the spectra to unit mean level spacing).

In this contribution we show, that the spectra of a highly excited hydrogen atom in a static magnetic field *do not* obey predictions of the Random Matrix Theory. This is quite surprising since the system studied has served for many years as an exemplary "quantally chaotic" model for which a detailed theory has been confronted with careful experimental studies [3]. The Hamiltonian, for the

*** permanent address

magnetic field directed along the z axis reads (in atomic units):

$$H = \frac{\mathbf{p}^2}{2} - \frac{1}{r} + \frac{\gamma^2}{8}(x^2 + y^2) \tag{2}$$

where γ denotes the magnetic field in units of 2.35×10^5 T. We neglect all relativistic, spin, QED or finite mass of the nucleus effects as well as drop the standard paramagnetic term $\gamma L_z/2$ (L_z, the angular momentum along the z axis is a good quantum number, together with parity) which can exactly be taken into account by redefinition of the zero of the energy. The classical dynamics of this system depends only on the scaled energy, $\epsilon = E\gamma^{-2/3}$, and is almost fully chaotic for $\epsilon > -0.12$. Interestingly, previous investigations of NNS distribution [4, 5] – on spectra obtained either at constant magnetic field or at constant scaled energy – are in good agreement with the Wigner prediction above $\epsilon = -0.12$. Present contribution considers, however, the region very close to the classical ionization threshold, $E = 0$, which has not been accessible to computer analysis before [6]. We present here also preliminary results for statistics of resonances above the ionization threshold.

For reasons to be explained below, the eigenvalues of the system are found in regions of constant E/γ. We restrict ourselves to $L_z = 0$, odd parity with respect to z axis, spectra obtained at a constant magnetic field. Numerical diagonalization of the Hamiltonian in a huge Sturmian basis [4, 5] using the Lanczos algorithm [7] allows us to compute at low cost the few levels of interest in a narrow energy band. The typical matrix size was about 70000, from which circa 50 energy levels were computed (ranging between the 70th and the 380th excited state, depending on γ). The value of γ ranged from 9×10^{-5} to 6.35×10^{-4}. For each γ value, the energy spectrum was unfolded to determine the average density of states and the spacings were determined. To study levels in a narrow energy band, only 5 to 20 spacings were computed for each γ value. Different γ values were sufficiently separated one from the other to ensure that the spectra are not correlated.

The exemplary NNS distribution obtained is presented in Fig. 1 (inset) for energy levels such that $-0.2\gamma < E < -0.1\gamma$. The main body of the figure shows the corresponding cumulative NNS distribution, $\int_0^s P(x)\, dx$, together with the Wigner prediction. Clearly, there is a significant deviation; especially, there are no large spacings.

The relevant parameter E/γ is *not* the scaled energy, which indicates that the deviations observed here are of quantum origin. We have compared data obtained for two different stretches of γ values (differing in average by a factor 2.5) keeping the E/γ ration fixed. The two distributions coincide within the statistical errors, indicating that the observed deviation persists in the semiclassical limit $\gamma \to 0, E \to 0, E/\gamma$ fixed.

What is the physical origin of the observed phenomenon? The diamagnetic potential $\gamma^2\rho^2/8$ confines the motion transverse to the magnetic field, but not in the magnetic field direction. Hence, just below the ionization threshold, the electron can explore the region around the z axis very far from the nucleus. In

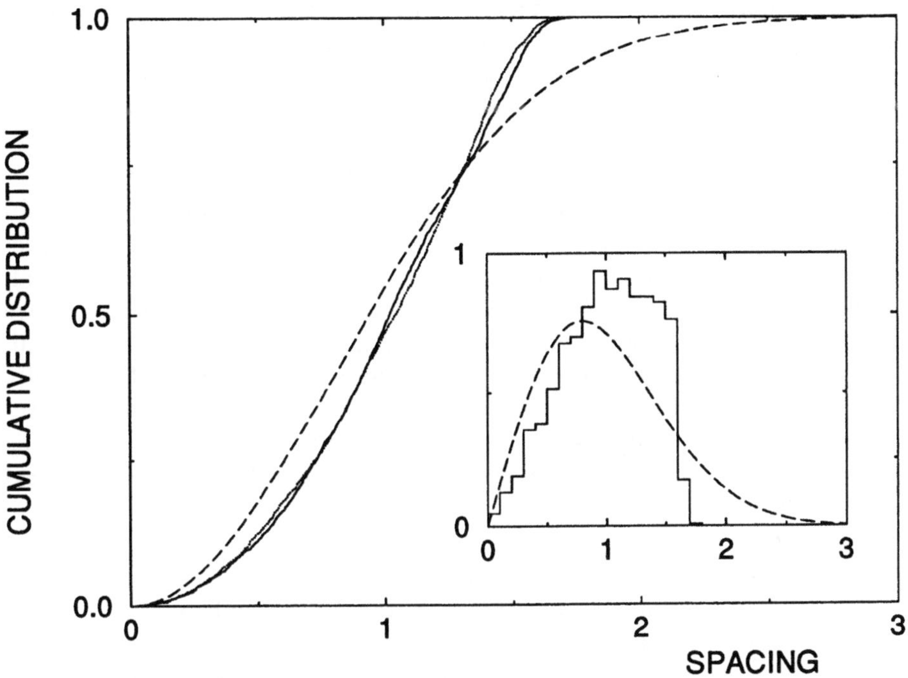

Fig. 1. Cumulative nearest neighbor spacing (NNS) distribution for the hydrogen atom in a magnetic field close to the ionization threshold. The NNS distribution itself in shown in the inset. The data are collected for 4293 spacings in the energy band $[-0.2\gamma; -0.1\gamma]$ for 478 values of γ. It deviates noticeably from the Wigner prediction, Eq. (1) shown as a dotted line. The dashed line represents the prediction of a simple model where a series of equally spaced levels is coupled to a GOE matrix (ratio of regular levels to chaotic levels = 1.72).

the region $|z| \gg \rho$, the Coulomb potential $1/\sqrt{\rho^2 + z^2}$ is almost equal to $1/|z|$, leading to the adiabatic-like separation [3]:

$$H \simeq H_{\text{sep}} = H_z + H_\rho = \frac{p_z^2}{2} - \frac{1}{|z|} + \frac{p_\rho^2}{2} + \frac{\gamma^2 \rho^2}{8} \quad . \tag{3}$$

H_{sep} is a sum of the 1-dimensional Coulomb, H_z, and of the harmonic oscillator, H_ρ, hamiltonians yielding the spectrum:

$$E(n_z, n_\rho) = (n_\rho + \frac{1}{2})\gamma - \frac{1}{2n_z^2} \tag{4}$$

with n_z and n_ρ integers ($n_z > 0$).

The various n_z Rydberg series associated with different n_ρ values are coupled by the nonadiabatic term $\frac{1}{|z|} - \frac{1}{\sqrt{\rho^2+z^2}}$. This coupling seems not to be strong enough to completely mix the various series. Thus modeling the full hamiltonian by a random matrix is not appropriate. A key observation is the fact that at a given energy partial densities of states (proportional to n_z^3) corresponding to different n_ρ series strongly differ due to the fact that the ionization thresholds for the various series are different. Thus the most abundant, $n_\rho = 0$ series, is perturbed by relatively few members of the other series and remnants of regularity are preserved even when the non–adiabatic coupling is strong.

To model this situation, we consider a Hilbert space composed of two subspaces which we baptize "regular" and "chaotic". A regular subspace is represented by a matrix diagonal in the regular subspace with equally spaced eigenvalues (this represents the $n_\rho = 0$ series discussed above). In the chaotic subspace (which represents the strongly mixed $n_\rho = 1, 2...$ series), we model the hamiltonian by a random matrix. The coupling of the regular states to the chaotic ones is taken constant throughout the regular series. Indeed, in the physical system considered here, the Rydberg states of the $n_\rho = 0$ series are coupled to the other series close to the nucleus, where all Rydberg states have almost identical wavefunctions.

In the limit of large matrices, this model has only two parameters: the relative weights of chaotic and regular states (i.e. the ratio of the dimensions of the chaotic and regular subspaces) and the strength of the coupling. We have not been able to deduce analytically the NNS distribution for our model but we have determined it numerically. As shown in Fig. 1 (in comparison with the NNS distribution for the hydrogen atom in a magnetic field) the model correctly reproduces observed deviations from fully random behaviour. The differences between the predictions of the model and the numerical results on the physical system are of the order of the statistical fluctuations. This makes us confident that the model catches the essential part of the physics involved.

Note that the observed deviations have an intrinsic *quantum* origin. It is because few series interact – one being dominant – that Wigner distribution is not observed. Thus E/γ has to be kept close to 0. This means that, at constant scaled energy $\epsilon = E\gamma^{-2/3}$ in the semiclassical limit $E \to 0, \gamma \to 0$, the ratio E/γ tends to $-\infty$ and the Wigner distribution is recovered. When speaking about "semiclassical limit", one has to be careful, indicating precisely how the various quantities tend to 0.

The situation discussed in this contribution has a tremendous practical importance, especially in atomic physics. Indeed, the Coulomb interaction has an infinite range and may easily produce such situations where an adiabatic separation is possible in some part of phase space. Let us give some simple examples.

The first example is quite spectacular and probably quite obvious for most atomic physicists. The helium atom is a mainly chaotic three–body system. Nevertheless, below every single ionization threshold, one observes regular Rydberg series whose NNS distribution is far from Wigner (it is actually a δ–peak at $s = 1$). This is just the extreme limit of our model when one series has a much larger density of states than all the other ones (associated with perturbers in the Rydberg series) and the chaotic subspace almost disappears.

For the hydrogen atom in a magnetic field close to the ionization threshold, either below or above, one is always close to a Landau threshold $(n_\rho + 1/2)\gamma$ where a Rydberg series converge to. Hence, strong deviations from random matrix predictions are expected everywhere in this region, either for bound states or resonances. This has in fact been observed experimentally [3]. These authors discovered windows where one Rydberg series appear to be accidentally decoupled from the other ones, giving rise to a locally regular series. Although this decoupling is exceptional throughout the spectra, our statistical observations rely on the same phenomenon.

We have performed an extensive calculation of energy positions and widths of resonances above the ionization threshold for the hydrogen atom in a magnetic field of $6T$ corresponding to the experiment [3] using a complex coordinate rotation approach [7]. The details of this work will be published elsewhere [8]. As in the bound region we collect the spectra for a small interval of E/γ values. This ratio determines the number of open ionization channels. Exemplary results are presented in Fig. 2 as an NNS statistics for resonance spacings and

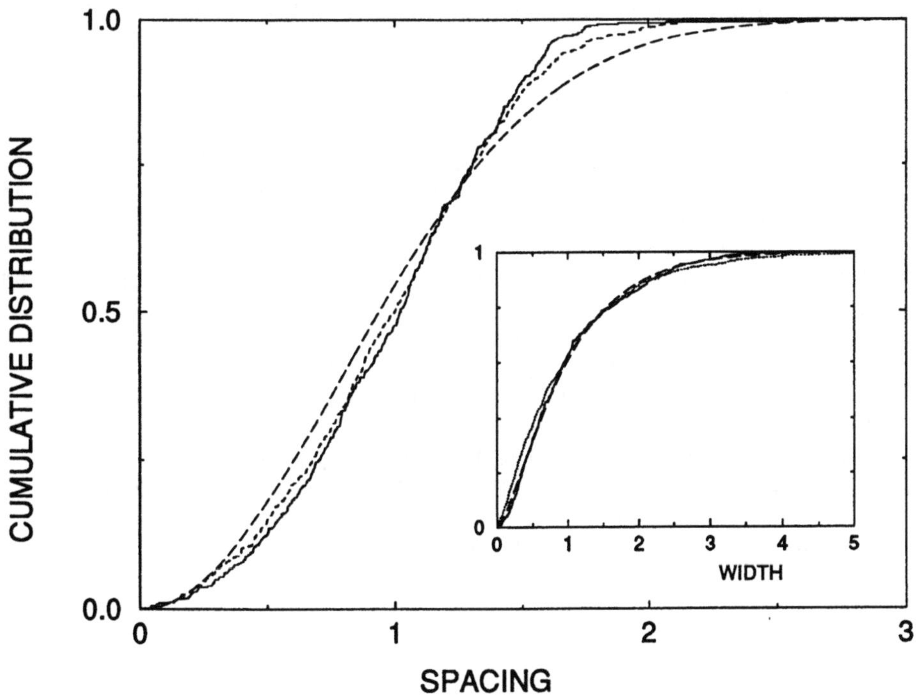

Fig. 2. Cumulative nearest neighbor spacing distribution for resonances in the case of 3 open channels at magnetic field around 6 Tesla. Spectra in the interval $[2.5\gamma, 2.9\gamma]$ taken for analysis. The inset shows the distribution of resonance widths.

the distribution of the widths. Results are compared with both the predictions of purely random matrix model in the limit of small width/spacing ratio (as is the case for our system) and a model based on mixing of the regular and chaotic series. In the latter we assume that only the regular states are coupled to the continuum (decay) since these states are extended far into the large z region – the ionization direction. The model has one additional parameter in comparison with the bound states model presented above – namely the strength of the decay. This parameter is fixed to reproduce the width/spacing ratio observed in physical system. As seen from the figure the model with regular series much better reproduces the observed distributions than the purely random model. The results presented in Fig.2 are obtained for 3 open channels but similar conclusions are reached analysing spectra obtained for one, two or four open channels. That confirms the claim given above that the deviations from purely random statistical behaviour occur everywhere above the ionization threshold.

Acknowledgments

J.Z. acknowledges partial support of KBN, grant P302 102 06. Laboratoire Kastler–Brossel, de l'Ecole Normale Supérieure et de l'Université Pierre et Marie Curie, is Unité Associée 18 du Centre National de la Recherche Scientifique.

References

1. O. Bohigas in *Chaos and quantum physics*, edited by M.–J. Giannoni, A. Voros and J. Zinn–Justin, Les Houches Summer School, Session LII (North–Holland, Amsterdam, 1991).
2. F. Haake, *Quantum Signatures of Chaos* (Springer, Berlin 1991).
3. C. Iu, G.R. Welch, M.M. Kash, K. Hsu, and D. Kleppner, Phys. Rev. Lett. **63**, 1133 (1989).
4. D. Delande and J.C. Gay, Phys. Rev. Lett., **57**, 2006 (1986);
   D. Wintgen and H. Friedrich, Phys. Rev. Lett., **57**, 571 (1986);
   G. Wunner, U. Woelk, I. Zech, G. Zeller, T. Ertl, P. Geyer, W. Schweizer and P. Ruder, Phys. Rev. Lett., **57**, 3261 (1986).
5. H. Friedrich and D. Wintgen, Phys. Rep. **183**, **37** (1989).
6. J. Zakrzewski, K. Dupret, and D. Delande, Phys. Rev. Lett. **74**, 522 (1995).
7. D. Delande, A. Bommier and J.C. Gay, Phys. Rev. Lett. **66**, 141 (1991).
8. K. Dupret, D. Delande and J. Zakrzewski, to be published.

Random Matrices of Circular Symplectic Ensemble

Karol Życzkowski

Instytut Fizyki im. M. Smoluchowskiego,
Uniwersytet Jagielloński, PL–30–059 Kraków, Poland

Abstract: Random unitary matrices of symplectic ensemble describe statistical properties of time-dependent, periodical quantum systems with a half-integer spin. We present a method of constructing random matrices typical to circular symplectic ensemble and show that the numerically generated unitary symplectic matrices display statistical properties of spectrum and eigenvectors according to the predictions of the random matrix theory.

1 Introduction

Random matrices, often used to describe statistical properties of complicated quantum systems with many degrees of freedom, are also applicable for simple quantum systems with few degrees of freedom, which exhibit chaos in the classical limit [1, 2]. A Hamiltonian of an autonomous quantum system may be represented by a Hermitian matrix of a Gaussian ensemble [3], whereas for a system periodically perturbed in time a more convenient characterization is provided by a unitary matrix representing the evolution operator propagating the wave function of the system over one period of the perturbation. Canonical ensembles of unitary matrices, invariant with respect to orthogonal, unitary or symplectic transformations where introduced by Dyson [4]. Such random matrices are also useful for investigating open scattering systems, described by a unitary S matrix [5].

Depending on the symmetry properties of the system one of the three canonical ensembles should be used in both cases. Systems possessing an antiunitary symmetry (mostly the time reversal invariance) display a linear repulsion of neighbouring energy levels (eigenphases) and are described by orthogonal ensembles. Unitary ensembles, appropriate for systems with the time reversal symmetry broken, exhibit quadratic level repulsion. Systems with a half-integer spin, a time-reversal invariance and no rotational symmetry pertain to the symplectic universality class, which is characterized by a quartic level repulsion [2]. Qualitatively speaking the presence of the Kramers degeneracy makes any additional accidental degeneracy very unlikely. This kind of spectral statistics was found for a periodically kicked top with a half integer spin [6] and an appropriately modified version of the kicked rotator [7, 8], Statistical properties of such time–dependent dynamical systems may be therefore described by random matrices of circular symplectic ensemble (CSE).

It is relatively easy to generate random Hermitian matrices pertaining to different universality classes - the matrix elements of such matrices are statistically independent random variables drawn according to a Gaussian distribution with zero mean. [3]. The only constrains are imposed by the algebraic conditions of symmetry (reality), hermiticity and symplecticity, involving pairs of elements.

Construction of unitary matrices typical of circular ensembles is more complicated, since unitarity imposes correlation between elements of the matrix. Recently we proposed a simple algorithm allowing one to construct random unitary matrices typical of circular unitary ensemble (CUE) and circular orthogonal ensemble (COE). In this work we present a method of generating random matrices characteristic of circular symplectic ensemble and show that obtained matrices conform to the predictions of random matrix theory (RMT).

2 Circular Unitary Ensemble

Circular unitary ensemble is defined by Haar measure in the space of $N \times N$ unitary matrices $U(N)$, invariant under the group of unitary transformations [4]. In order to construct numerically a unitary matrix typical of CUE we apply the parameterization of Hurwitz [9] and use the appropriate generalized Euler angles. An arbitrary unitary transformation U can be composed from elementary unitary transformations in two-dimensional subspaces. The matrix of such an elementary unitary transformation will be denoted by $E^{(i,j)}(\phi, \psi, \chi)$. The only nonzero elements of $E^{(i,j)}$ are

$$\begin{array}{llll} E_{kk}^{(i,j)} = 1 \ , & k = 1, \ldots, N; \, k \neq i, j & & \\ E_{ii}^{(i,j)} = \cos\phi e^{i\psi} \ , & & E_{ij}^{(i,j)} = \sin\phi e^{i\chi} \ , & \\ E_{ji}^{(i,j)} = -\sin\phi e^{-i\chi} \ , & & E_{jj}^{(i,j)} = \cos\phi e^{-i\psi} \ , & \end{array} \qquad (1)$$

From the above elementary unitary transformations one constructs the following $N-1$ composite rotations

$$\begin{aligned} E_1 &= E^{(1,2)}(\phi_{12}, \psi_{12}, \chi_{12}) \ , \\ E_2 &= E^{(2,3)}(\phi_{23}, \psi_{23}, 0) E^{(1,3)}(\phi_{13}, \psi_{13}, \chi_{13}) \ , \\ E_3 &= E^{(3,4)}(\phi_{34}, \psi_{34}, 0) E^{(2,4)}(\phi_{24}, \psi_{24}, 0) E^{(1,4)}(\phi_{14}, \psi_{14}, \chi_{14}) \qquad (2) \\ &\ldots \\ E_{N-1} &= E^{(N-1,N)}(\phi_{N-1,N}, \psi_{N-1,N}, 0) E^{(N-2,N)}(\phi_{N-2,N}, \psi_{N-2,N}, 0) \\ &\quad \ldots E^{(1,N)}(\phi_{1N}, \psi_{1N}, \chi_{1N}) \end{aligned}$$

and eventually forms the unitary transformation U as

$$U = e^{i\alpha} E_1 E_2 E_3 \ldots E_{N-1}. \qquad (3)$$

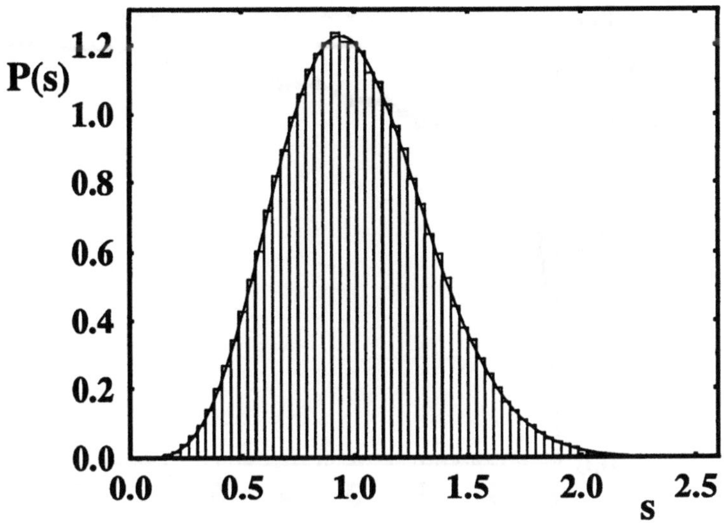

Fig. 1. Nearest neighbours distribution for 5000 self-dual unitary matrices of size $2N = 100$ typical to CSE.

If for a large matrix size N the angles $\alpha, \phi_{rs}, \psi_{rs}$, and χ_{1s} are taken uniformly from the intervals

$$0 \le \psi_{rs} < 2\pi \ , \quad 0 \le \chi_{1s} < 2\pi \ , \quad 0 \le \alpha < 2\pi \ , \tag{4}$$

whereas

$$\phi_{rs} = \arcsin(\xi_{rs}^{1/2r}), \quad r = 1, 2, \ldots, N-1 \ , \tag{5}$$

with ξ_{rs} uniformly distributed in

$$0 \le \xi_{rs} < 1 \ , \tag{6}$$

then the obtained matrix displays statistical properties characteristic to CUE [10].

3 Circular Symplectic Ensemble

A unitary matrix $S = (S^\dagger)^{-1}$ of the size $2N$ is called *self–dual* if

$$S^R = -ZS^TZ = S \ , \tag{7}$$

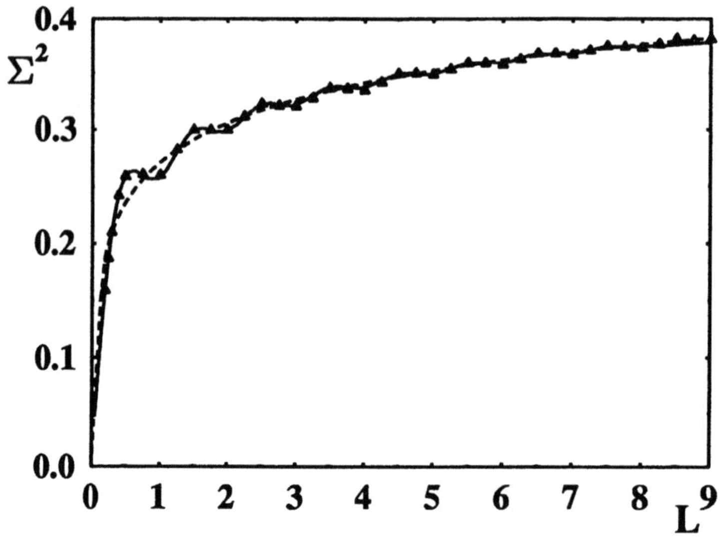

Fig. 2. Number variance $\Sigma^2(L)$ for unitary self-dual matrices ($\triangle$). Solid line stands for the exact result for CSE, while dotted line represents a large L approximation.

where the antisymmetric unitary matrix Z is

$$Z = \begin{pmatrix} 0 & 1 & 0 & 0 & \cdots & 0 & 0 \\ -1 & 0 & 0 & 0 & \cdots & 0 & 0 \\ 0 & 0 & 0 & 1 & \cdots & 0 & 0 \\ 0 & 0 & -1 & 0 & \cdots & 0 & 0 \\ \vdots & \vdots & \vdots & \vdots & \ddots & 0 & 0 \\ 0 & 0 & 0 & 0 & \cdots & 0 & 1 \\ 0 & 0 & 0 & 0 & \cdots & -1 & 0 \end{pmatrix} . \tag{8}$$

For each self dual unitary matrix S there exists a quaternion real (symplectic) matrix B of size N, such that $B^R B = 1$ and $S = B^{-1}EB$, and the diagonal matrix E consists of N complex numbers $[\exp(i\theta_j)]$ on the unit circle, each repeated twice.

The circular symplectic ensemble is uniquely defined in the space of self–dual quaternion matrices by the property of being invariant under every transformation $S \to B^R S B$, where B is any $N \times N$ unitary quaternion matrix [4]. Every self dual unitary matrix can be written as

$$S = U^R U \ , \tag{9}$$

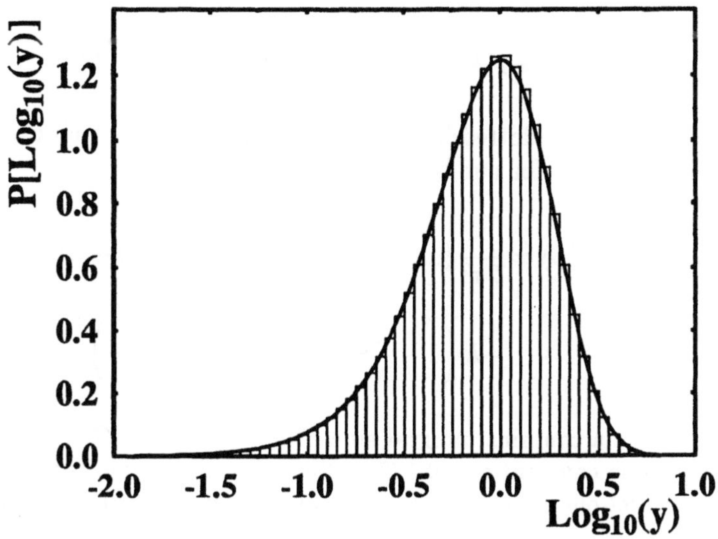

Fig. 3. Eigenvector statistics $P[\log_{10}(y)]$ for 5000 unitary matrices of symplectic ensemble and the corresponding $\chi^2_{\nu=4}$ distribution.

where U is unitary [3]. Note that the matrix U is determined to the extend of transformation $U \rightarrow UB$, it is the right multiplication by an arbitrary symplectic matrix B. This property defines the measure of the circular symplectic ensemble. Therefore, if matrices U are taken according to the Haar measure, the self dual matrix S are drawn uniformly with respect to the measure of CSE.

In order to obtain a CSE matrix it is sufficient to generate a $2N \times 2N$ unitary matrix according to equations (1 - 6) and to compute $S = U^R U$. To demonstrate practical importance of this algorithm we have numerically generated several CSE unitary matrices and analyzed the statistical properties of spectrum and eigenvectors. Fig. 1 presents the nearest neighbour distribution $P(s)$ obtained for 5000 matrices of size $2N = 100$. Due to Kramers degeneracy each matrix provides 50 data to the histogram. Note a good coincidence with the approximate Wigner formula [3]

$$P_S(s) = (\frac{64}{9\pi})^3 s^4 \exp\left[-\frac{64s^2}{9\pi}\right] \tag{10}$$

represented by a solid line.

To study the long range correlations of the spectrum we computed the average number of levels $\langle N_S(L)\rangle$ in an interval of the length L, and the number variance $\Sigma^2(L) = \langle N_S^2(L)\rangle - \langle N_S(L)\rangle^2$. Numerical results obtained from the same ensemble of 5000 matrices typical to CSE are represented by triangles in Fig. 2. Often used large L approximation

$$\Sigma^2_{app}(L) = \frac{1}{2\pi^2}\left(\ln(4\pi L) + 1 + \gamma + \pi^2/8\right), \tag{11}$$

is represented by a dotted line, while solid line stands for an exact, albeit more complicated result for number variance [11]

$$\begin{aligned}\Sigma^2_S(L) = \frac{1}{2\pi^2}\left[\ln(4\pi L) + 1 + \gamma - \cos(4\pi L) - \mathrm{Ci}(4\pi L)\right] \\ +L\left(1 - \frac{2}{\pi}\mathrm{Si}(4\pi L)\right) + \left(\frac{\mathrm{Si}(2\pi L)}{2\pi}\right)^2 .\end{aligned} \tag{12}$$

Above formula contains integral sine $Si(x)$ and cosine $Ci(x)$ functions and the Euler gamma constant $\gamma \approx 0.577....$ Note that the numerical results are faithful to the predictions of symplectic ensemble to a degree which requires usage of the exact formula (12) with small L corrections.

Diagonalizing self-dual unitary matrices we obtained not only theirs eigenvalues, but also the eigenvectors $|\varphi_l\rangle$, $l = 1, \ldots, N$; each represented in the initial basis by N real quaternions

$$\tau_{lk} = c^0_{lk}\tau_0 + c^1_{lk}\tau_1 + c^2_{lk}\tau_2 + c^3_{lk}\tau_3 , \tag{13}$$

where τ_0 is the 2×2 unit matrix and $\tau_i, i = 1, 2, 3$ are Pauli matrices. Localization properties of the eigenvectors depend on their components $y_{lk} = \sum_{j=0}^{3} |c^j_{lk}|^2$. Statistical properties of eigenvectors determine the distribution of eigenvector components $P(y)$. It is known [12, 13] that for Gaussian or circular ensembles eigenvector statistics tends in the limit of large N to the χ^2_ν distribution

$$P_\nu(y) = \frac{(\nu/2)^{(\nu/2)}}{\Gamma(\nu/2) < y >}\left(\frac{y}{< y >}\right)^{\nu/2-1} \exp\left[-\frac{\nu y}{2 < y >}\right] , \tag{14}$$

where the number of degrees of freedom ν equals to 1 for the orthogonal, 2 for the unitary and 4 for the symplectic ensemble. Since this distribution is peaked around zero it is convenient to use the logarithmic scale and study $P[\log(y)]$. Figure 3 shows the eigenvector distribution of CSE matrices with the mean value $< y >$ normalized to unity. Observe fine agreement with the distribution (14) (with $\nu = 4$) represented in the figure by a solid line.

We have presented a simple algorithm allowing one to generate a random unitary matrix typical to circular symplectic ensemble. Such matrices can be used to describe statistical properties of quantum dynamical systems with a half-integer spin or to test numerically various hypothesis concerning CSE, which may hardly be verified by means of analytical calculations.

References

1. O. Bohigas, in: *Chaos and Quantum Physics*, Les-Houches Session LII 1989, edited by M.J. Giannoni and A. Voros, North-Holland, Amsterdam, 1991.
2. F. Haake, *Quantum Signatures of Chaos*, Springer, Berlin, 1991.
3. M.L. Mehta, *Random Matrices* 2nd edn. (Academic Press, New York, 1990).
4. F.J. Dyson, J. Math. Phys. **3**, 140 (1962).
5. R. Blümel and U.Smilansky, Phys. Rev. Lett. **64**, 241 (1990).
6. R. Scharf, B. Dietz, M. Kuś, F. Haake and M.V. Berry, *Europhys. Lett.* **5**, 383 (1988).
7. R. Scharf, J.Phys. **A22**, 4223 (1989).
8. M. Thaha, R. Blümel and U. Smilansky, Phys. Rev. **E48**, 1764 (1993).
9. A. Hurwitz, Nachr. Ges. Wiss. Gött. Math.-phys. Kl. 71, (1897).
10. K. Życzkowski and M. Kuś, J. Phys. **A27**, 4235 (1994).
11. T.A. Brody, J. Flores, J.B. French, P.A. Mello, A. Pandey and S.S.M. Wong, Rev. Mod. Phys. **53**, 385 (1981).
12. C.E. Porter, *Statistical Theories of Spectra: Fluctuations*, edited by C.E. Porter (Academic Press, New York, 1965).
13. F. Haake and K. Życzkowski, Phys. Rev. **A42**, 1013 (1990).

References

[illegible]

LIST OF KARPACZ SCHOOLS
OF THEORETICAL PHYSICS
Bibliography: 1964 – 1994

I – 1964: R.S. Ingarden (unpubl.), *Particles, Fields and Superconductivity,*

II – 1965: W. Zi̧ętek (ed.), *Symmetries in Particle and Condensed Matter Physics,* Mimeographed Lecture Notes of the University of Wrocław

III – 1966: Z. Galasiewicz (ed.), *Statistical Physics of Condensed Matter,* Acta Universitatis Wratislaviensis, No. 80

IV – 1967: J. Rzewuski (ed.), *Functional Methods in Quantum Field Theory and Statistical Mechanics,* Acta Universitatis Wratislaviensis, No. 88, 89, 90

V – 1968: J. Łopuszański (ed.), *Axiomatic Approach to Quantum Field Theory and Many Body Problem,* Acta Universitatis Wratislaviensis, No. 98, 99, 113

VI – 1969: J. Przystawa (ed.), *Group Theory and Statistical Physics of Condensed Matter,* Mimeographed Lecture Notes of the University of Wrocław

VII – 1970: Z. Galasiewicz (ed.), *Liquid Helium and Many Body Problems,* Acta Universitatis Wratislaviensis, No. 141

VIII – 1971: J. Lukierski (ed.), *New Developments in Relativistic Quantum Field Theory and its Applications,* Acta Universitatis Wratislaviensis, No. 164, Vol. 1 and 2

IX – 1972: K. Wojciechowski (ed.), *Theory of Metals and the Many Body Problem,* Acta Universitatis Wratislaviensis, No. 181, Vol. 1 and 2

X – 1973: J. Lukierski (ed.), *New Developments in Relativistic Quantum Field Theory,* Acta Universitatis Wratislaviensis, No. 207

XI – 1974: J. Łopuszański, A. Pękalski, J. Przystawa (eds.), *Magnetism in Metals and Metallic Compounds,* Plenum Press, London

XII – 1975: W. Garczyński (ed.), *Functional and Probabilistic Methods in Quantum Field Theory,* Acta Universitatis Wratislaviensis, No. 368

XIII – 1976: J. Lukierski (ed.), *Recent Development in Relativistic Quantum Field Theory and Its Application,* Acta Universitatis Wratislaviensis, No. 389

XIV – 1977: Z. Galasiewicz (ed.), *Collective Effects in Condensed Media,* Acta Universitatis Wratislaviensis, No. 436

XV – 1978: W. Karwowski (ed.), *Mathematical Aspects of Quantum Field Theory,* Acta Universitatis Wratislaviensis, No. 519

XVI – 1979: A. Pękalski, J. Przystawa (eds.), *Modern Trends in the Theory of Condensed Matter,* Lecture Notes in Physics No. 115, Springer–Verlag, Berlin

XVII – 1980: L. Turko, A. Pękalski (eds.), *Developments in the Theory of Fundamental Interactions,* Studies in High Energy Physics, Vol. 3, Harwood Acad. Publ., Chur

XVIII – 1981: W. Garczyński (ed.), *Gauge Field Theories: Theoretical Studies and Computer Simulations,* Studies in High Energy Physics, Vol. 4, Harwood Acad. Publ., Chur

1982: *The School did not take place because of the Martial Law in Poland,*

XIX – 1983: B. Milewski (ed.), *Supersymmetry and Supergravity,* World Scientific, Singapore

XX – 1984: A. Pękalski, J. Sznajd (eds.), *Static Critical Phenomena in Inhomogeneous Systems,* Lecture Notes in Physics No. 206, Springer–Verlag, Berlin

XXI – 1985: L. Michel, J. Mozrzymas, A. Pękalski (eds.), *Spontaneous Symmetry Breakdown and Related Subjects,* World Scientific, Singapore

XXII – 1986: A. Jadczyk (ed.), *Fields and Geometry,* World Scientific, Singapore

XXIII – 1987: T. Paszkiewicz (ed.), *Physics of Phonons,* Lecture Notes in Physics No. 285, Springer–Verlag, Berlin

XXIV – 1988: R. Gielerak, W. Karwowski (eds.), *Stochastic Methods in Mathematical Physics,* World Scientific, Singapore

XXV – 1989: Z. Haba, J. Sobczyk (eds.), *Functional Integration, Geometry and Strings,* Birkhaeuser Verlag, Basel

XXVI – 1990: Z. Galasiewicz, A. Pękalski (eds.), *Ordering Phenomena in Condensed Matter Physics,* World Scientific, Singapore

XXVII – 1991: P. Garbaczewski, Z. Popowicz (eds.), *Nonlinear Fields: Classical, Random, Semicassical,* World Scientific, Singapore

XXVIII – 1992: R. Gielerak (ed.), *Infinite Dimensional Geometry in Physics,* Elsevier–North Holland, Amsterdam

XXIX – 1993: T. Paszkiewicz, K. Rapcewicz (eds.), *Die Kunst of Phonons,* Plenum Press, London

XXX – 1994: J. Lukierski, Z. Popowicz, J. Sobczyk (eds.), *Quantum Groups. Formalism and Applications,* Polish Scientific Publishers PWN, Warsaw

Springer-Verlag and the Environment

We at Springer-Verlag firmly believe that an international science publisher has a special obligation to the environment, and our corporate policies consistently reflect this conviction.

We also expect our business partners – paper mills, printers, packaging manufacturers, etc. – to commit themselves to using environmentally friendly materials and production processes.

The paper in this book is made from low- or no-chlorine pulp and is acid free, in conformance with international standards for paper permanency.

Lecture Notes in Physics

For information about Vols. 1–419
please contact your bookseller or Springer-Verlag

Vol. 420: F. Ehlotzky (Ed.), Fundamentals of Quantum Optics III. Proceedings, 1993. XII, 346 pages. 1993.

Vol. 421: H.-J. Röser, K. Meisenheimer (Eds.), Jets in Extragalactic Radio Sources. XX, 301 pages. 1993.

Vol. 422: L. Päivärinta, E. Somersalo (Eds.), Inverse Problems in Mathematical Physics. Proceedings, 1992. XVIII, 256 pages. 1993.

Vol. 423: F. J. Chinea, L. M. González-Romero (Eds.), Rotating Objects and Relativistic Physics. Proceedings, 1992. XII, 304 pages. 1993.

Vol. 424: G. F. Helminck (Ed.), Geometric and Quantum Aspects of Integrable Systems. Proceedings, 1992. IX, 224 pages. 1993.

Vol. 425: M. Dienes, M. Month, B. Strasser, S. Turner (Eds.), Frontiers of Particle Beams: Factories with $e^+ e^-$ Rings. Proceedings, 1992. IX, 414 pages. 1994.

Vol. 426: L. Mathelitsch, W. Plessas (Eds.), Substructures of Matter as Revealed with Electroweak Probes. Proceedings, 1993. XIV, 441 pages. 1994

Vol. 427: H. V. von Geramb (Ed.), Quantum Inversion Theory and Applications. Proceedings, 1993. VIII, 481 pages. 1994.

Vol. 428: U. G. Jørgensen (Ed.), Molecules in the Stellar Environment. Proceedings, 1993. VIII, 440 pages. 1994.

Vol. 429: J. L. Sanz, E. Martínez-González, L. Cayón (Eds.), Present and Future of the Cosmic Microwave Background. Proceedings, 1993. VIII, 233 pages. 1994.

Vol. 430: V. G. Gurzadyan, D. Pfenniger (Eds.), Ergodic Concepts in Stellar Dynamics. Proceedings, 1993. XVI, 302 pages. 1994.

Vol. 431: T. P. Ray, S. Beckwith (Eds.), Star Formation and Techniques in Infrared and mm-Wave Astronomy. Proceedings, 1992. XIV, 314 pages. 1994.

Vol. 432: G. Belvedere, M. Rodonò, G. M. Simnett (Eds.), Advances in Solar Physics. Proceedings, 1993. XVII, 335 pages. 1994.

Vol. 433: G. Contopoulos, N. Spyrou, L. Vlahos (Eds.), Galactic Dynamics and N-Body Simulations. Proceedings, 1993. XIV, 417 pages. 1994.

Vol. 434: J. Ehlers, H. Friedrich (Eds.), Canonical Gravity: From Classical to Quantum. Proceedings, 1993. X, 267 pages. 1994.

Vol. 435: E. Maruyama, H. Watanabe (Eds.), Physics and Industry. Proceedings, 1993. VII, 108 pages. 1994.

Vol. 436: A. Alekseev, A. Hietamäki, K. Huitu, A. Morozov, A. Niemi (Eds.), Integrable Models and Strings. Proceedings, 1993. VII, 280 pages. 1994.

Vol. 437: K. K. Bardhan, B. K. Chakrabarti, A. Hansen (Eds.), Non-Linearity and Breakdown in Soft Condensed Matter. Proceedings, 1993. XI, 340 pages. 1994.

Vol. 438: A. Pękalski (Ed.), Diffusion Processes: Experiment, Theory, Simulations. Proceedings, 1994. VIII, 312 pages. 1994.

Vol. 439: T. L. Wilson, K. J. Johnston (Eds.), The Structure and Content of Molecular Clouds. 25 Years of Molecular Radioastronomy. Proceedings, 1993. XIII, 308 pages. 1994.

Vol. 440: H. Latal, W. Schweiger (Eds.), Matter Under Extreme Conditions. Proceedings, 1994. IX, 243 pages. 1994.

Vol. 441: J. M. Arias, M. I. Gallardo, M. Lozano (Eds.), Response of the Nuclear System to External Forces. Proceedings, 1994, VIII. 293 pages. 1995.

Vol. 442: P. A. Bois, E. Dériat, R. Gatignol, A. Rigolot (Eds.), Asymptotic Modelling in Fluid Mechanics. Proceedings, 1994. XII, 307 pages. 1995.

Vol. 443: D. Koester, K. Werner (Eds.), White Dwarfs. Proceedings, 1994. XII, 348 pages. 1995.

Vol. 444: A. O. Benz, A. Krüger (Eds.), Coronal Magnetic Energy Releases. Proceedings, 1994. X, 293 pages. 1995.

Vol. 445: J. Brey, J. Marro, J. M. Rubí, M. San Miguel (Eds.), 25 Years of Non-Equilibrium Statistical Mechanics. Proceedings, 1994. XVII, 387 pages. 1995.

Vol. 446: V. Rivasseau (Ed.), Constructive Physics. Results in Field Theory, Statistical Mechanics and Condensed Matter Physics. Proceedings, 1994. X, 337 pages. 1995.

Vol. 447: G. Aktaş, C. Saçlıoğlu, M. Serdaroğlu (Eds.), Strings and Symmetries. Proceedings, 1994. XIV, 389 pages. 1995.

Vol. 448: P. L. Garrido, J. Marro (Eds.), Third Granada Lectures in Computational Physics. Proceedings, 1994. XIV, 346 pages. 1995.

Vol. 449: J. Buckmaster, T. Takeno (Eds.), Modeling in Combustion Science. Proceedings, 1994. X, 369 pages. 1995.

Vol. 450: M. F. Shlesinger, G. M. Zaslavsky, U. Frisch (Eds.), Lévy Flights and Related Topics in Physics. Proceedigs, 1994. XIV, 347 pages. 1995.

Vol. 452: A. M. Bernstein, B. R. Holstein (Eds.), Chiral Dynamics: Theory and Experiment. Proceedings, 1994. VIII, 351 pages. 1995.

Vol. 453: S. M. Deshpande, S. S. Desai, R. Narasimha (Eds.), Fourteenth International Conference on Numerical Methods in Fluid Dynamics. Proceedings, 1994. XIII, 589 pages. 1995.

Vol. 454: J. Greiner, H. W. Duerbeck, R. E. Gershberg (Eds.), Flares and Flashes, Germany 1994. XXII, 477 pages. 1995.

Vol. 455: F. Occhionero (Ed.), Birth of the Universe and Fundamental Physics. Proceedings, 1994. XV, 387 pages. 1995.

Vol. 456: H. B. Geyer (Ed.), Field Theory, Topology and Condensed Matter Physics. Proceedings, 1994. XII, 206 pages. 1995.

Vol. 457: P. Garbaczewski, M. Wolf, A. Weron (Eds.), Chaos – The Interplay Between Stochastic and Deterministic Behaviour. Proceedings, 1995. XII, 573 pages, 1995.

New Series m: Monographs

Vol. m 1: H. Hora, Plasmas at High Temperature and Density. VIII, 442 pages. 1991.

Vol. m 2: P. Busch, P. J. Lahti, P. Mittelstaedt, The Quantum Theory of Measurement. XIII, 165 pages. 1991.

Vol. m 3: A. Heck, J. M. Perdang (Eds.), Applying Fractals in Astronomy. IX, 210 pages. 1991.

Vol. m 4: R. K. Zeytounian, Mécanique des fluides fondamentale. XV, 615 pages, 1991.

Vol. m 5: R. K. Zeytounian, Meteorological Fluid Dynamics. XI, 346 pages. 1991.

Vol. m 6: N. M. J. Woodhouse, Special Relativity. VIII, 86 pages. 1992.

Vol. m 7: G. Morandi, The Role of Topology in Classical and Quantum Physics. XIII, 239 pages. 1992.

Vol. m 8: D. Funaro, Polynomial Approximation of Differential Equations. X, 305 pages. 1992.

Vol. m 9: M. Namiki, Stochastic Quantization. X, 217 pages. 1992.

Vol. m 10: J. Hoppe, Lectures on Integrable Systems. VII, 111 pages. 1992.

Vol. m 11: A. D. Yaghjian, Relativistic Dynamics of a Charged Sphere. XII, 115 pages. 1992.

Vol. m 12: G. Esposito, Quantum Gravity, Quantum Cosmology and Lorentzian Geometries. Second Corrected and Enlarged Edition. XVIII, 349 pages. 1994.

Vol. m 13: M. Klein, A. Knauf, Classical Planar Scattering by Coulombic Potentials. V, 142 pages. 1992.

Vol. m 14: A. Lerda, Anyons. XI, 138 pages. 1992.

Vol. m 15: N. Peters, B. Rogg (Eds.), Reduced Kinetic Mechanisms for Applications in Combustion Systems. X, 360 pages. 1993.

Vol. m 16: P. Christe, M. Henkel, Introduction to Conformal Invariance and Its Applications to Critical Phenomena. XV, 260 pages. 1993.

Vol. m 17: M. Schoen, Computer Simulation of Condensed Phases in Complex Geometries. X, 136 pages. 1993.

Vol. m 18: H. Carmichael, An Open Systems Approach to Quantum Optics. X, 179 pages. 1993.

Vol. m 19: S. D. Bogan, M. K. Hinders, Interface Effects in Elastic Wave Scattering. XII, 182 pages. 1994.

Vol. m 20: E. Abdalla, M. C. B. Abdalla, D. Dalmazi, A. Zadra, 2D-Gravity in Non-Critical Strings. IX, 319 pages. 1994.

Vol. m 21: G. P. Berman, E. N. Bulgakov, D. D. Holm, Crossover-Time in Quantum Boson and Spin Systems. XI, 268 pages. 1994.

Vol. m 22: M.-O. Hongler, Chaotic and Stochastic Behaviour in Automatic Production Lines. V, 85 pages. 1994.

Vol. m 23: V. S. Viswanath, G. Müller, The Recursion Method. X, 259 pages. 1994.

Vol. m 24: A. Ern, V. Giovangigli, Multicomponent Transport Algorithms. XIV, 427 pages. 1994.

Vol. m 25: A. V. Bogdanov, G. V. Dubrovskiy, M. P. Krutikov, D. V. Kulginov, V. M. Strelchenya, Interaction of Gases with Surfaces. XIV, 132 pages. 1995.

Vol. m 26: M. Dineykhan, G. V. Efimov, G. Ganbold, S. N. Nedelko, Oscillator Representation in Quantum Physics. IX, 279 pages. 1995.

Vol. m 27: J. T. Ottesen, Infinite Dimensional Groups and Algebras in Quantum Physics. IX, 218 pages. 1995.

Vol. m 28: O. Piguet, S. P. Sorella, Algebraic Renormalization. IX, 134 pages. 1995.

Vol. m 29: C. Bendjaballah, Introduction to Photon Communication. VII, 193 pages. 1995.

Vol. m 30: A. J. Greer, W. J. Kossler, Low Magnetic Fields in Anisotropic Superconductors. VII, 161 pages. 1995.

Vol. m 31: P. Busch, M. Grabowski, P. J. Lahti, Operational Quantum Physics. XI, 230 pages. 1995.

Vol. m 33: R. Alkofer, H. Reinhardt, Chiral Quark Dynamics. VIII, 115 pages. 1995.

FSC
www.fsc.org
MIX
Papier aus verantwortungsvollen Quellen
Paper from responsible sources
FSC® C105338

If you have any concerns about our products,
you can contact us on
ProductSafety@springernature.com

In case Publisher is established outside the EU,
the EU authorized representative is:
Springer Nature Customer Service Center GmbH
Europaplatz 3, 69115 Heidelberg, Germany

Printed by Libri Plureos GmbH
in Hamburg, Germany